Springer Proceedings in Physics

Volume 398

Indexed by Scopus

The series Springer Proceedings in Physics, founded in 1984, is devoted to timely reports of state-of-the-art developments in physics and related sciences. Typically based on material presented at conferences, workshops and similar scientific meetings, volumes published in this series will constitute a comprehensive up to date source of reference on a field or subfield of relevance in contemporary physics. Proposals must include the following:

- Name, place and date of the scientific meeting
- A link to the committees (local organization, international advisors etc.)
- Scientific description of the meeting
- List of invited/plenary speakers
- An estimate of the planned proceedings book parameters (number of pages/articles, requested number of bulk copies, submission deadline).

Please contact:

For Americas and Europe: Dr. Zachary Evenson; zachary.evenson@springer.com
For Asia, Australia and New Zealand: Dr. Loyola DSilva; loyola.dsilva@springer.com

Zhi Fang · Cheng Zhang · Danhua Mei ·
Shuai Zhang
Editors

Proceedings of the 5th International Symposium on Plasma and Energy Conversion

iSPEC2023, 27–29 October, Nanjing, China

Springer

Editors
Zhi Fang
Nanjing Tech University
Nanjing, China

Danhua Mei
Nanjing Tech University
Nanjing, China

Cheng Zhang
Institute of Electrical Engineering
Chinese Academy of Sciences
Beijing, China

Shuai Zhang
Institute of Electrical Engineering
Chinese Academy of Sciences
Beijing, China

ISSN 0930-8989 ISSN 1867-4941 (electronic)
Springer Proceedings in Physics
ISBN 978-981-97-2244-0 ISBN 978-981-97-2245-7 (eBook)
https://doi.org/10.1007/978-981-97-2245-7

This Springer imprint is published by the registered company Springer Nature Singapore Pte Ltd.
The registered company address is: 152 Beach Road, #21-01/04 Gateway East, Singapore 189721, Singapore

Preface

This proceedings book contains selected and expanded contributions presented at the 5th International Symposium on Plasma and Energy Conversion (ISPEC), held in Nanjing, China, on 27–29 October. 2023.

These proceedings highlight the latest research findings, technological advances, and innovative ideas in plasma and energy conversion which are critical to addressing the global energy and environmental challenges.

Oriented towards the international frontier of carbon emission reduction technologies and China's national strategic goals of carbon peak and carbon neutrality, plasma and related energy conversion has been gaining attention as a promising alternative to thermal-catalysis or electro-catalysis. The proceedings discuss and exchange cutting-edge scientific innovations and technological advances in fields such as the basic process of discharge plasmas, plasma-assisted synthesis of chemicals and fuels, plasma-controlled environmental pollution, plasma-assisted preparation and treatment of catalysts and function materials, plasma-interface interactions, plasma modelling and simulation technology, and high-voltage measurement and instrumentation, which show great industrial potentials in conversion and utilization of greenhouse gases (e.g. CO_2 and CH_4), nitrogen fixation, plasma deposition and chemical synthesis, environmental clean-up (e.g. gas cleaning and wastewater treatment), as well as the development of the corresponding plasma devices and driving power supplies.

This collection of papers presents the main contributions of plasma and energy conversion in the form of separate chapters, including cutting-edge studies on conversion technology, synthesis and treatment technology, complex mechanism simulation and modelling, as well as in-situ detection and diagnosis.

These proceedings are suitable for researchers engaged in fields like plasma chemistry, plasma-catalysis, discharge diagnosis and modelling, chemical modelling, and high-voltage applications.

Nanjing, China Zhi Fang
Beijing, China Cheng Zhang
Nanjing, China Danhua Mei
Beijing, China Shuai Zhang

Contents

Contributors

Xiaoye Bai ZTT Cable Accessories Co., Ltd., Nantong, China

Evgeny Baksht Institute of High Current Electronics, Siberian Branch of the Russian Academy of Sciences, Tomsk, Russia

Dmitry Beloplotov Institute of High Current Electronics, Siberian Branch of the Russian Academy of Sciences, Tomsk, Russia

Meiling Cai College of Electrical Engineering and Control Science, Nanjing Tech University, Nanjing, China

Kai Cao ZTT Cable Accessories Co., Ltd., Nantong, China

Xiangnan Cao Beijing International S&T Cooperation Base for Plasma Science and Energy Conversion, Institute of Electrical Engineering, Chinese Academy of Sciences, Beijing, China

Zhang Chaohai Department of Electrical Engineering, Nanjing University of Aeronautics and Astronautics, Nanjing, China

Bingyan Chen School of Mechanical and Electrical Engineering, Hohai University, Changzhou, Jiangsu, China;
School of Science, Hohai University, Changzhou, Jiangsu, China;
College of Science, Hohai University, Changzhou, China

Guangzhi Chen Hubei State Key Laboratory of Power Grid Environmental Protection, School of Electrical Engineering and Automation, Wuhan University, Wuhan, Hubei, China;
Hubei Key Laboratory of Power Equipment & System Security for Integrated Energy, School of Electrical Engineering and Automation, Wuhan University, Wuhan, Hubei, China

Hao Chen Fuzhou Power Supply Company, State Grid Fujian Electric Power Co., Ltd., Fuzhou, China

Hui Chen Fuzhou Power Supply Company, State Grid Fujian Electric Power Co., Ltd., Fuzhou, China

Pengju Chen Nanjing Tech University, Nanjing, China

Qing Chen College of Electrical Engineering and Control Science, Nanjing Tech University, Nanjing, China

Rirong Chen Hubei State Key Laboratory of Power Grid Environmental Protection, School of Electrical Engineering and Automation, Wuhan University, Wuhan, Hubei, China;
Hubei Key Laboratory of Power Equipment & System Security for Integrated Energy, School of Electrical Engineering and Automation, Wuhan University, Wuhan, Hubei, China

She Chen Hunan University, Changsha, China

Shuyu Chen Beijing International S&T Cooperation Base for Plasma Science and Energy Conversion, Institute of Electrical Engineering, Chinese Academy of Sciences, Beijing, China;
University of Chinese Academy of Sciences, Beijing, China

Tianheng Chen State Grid Zhejiang Electric Power Company, Ltd. Shaoxing Power Supply Company, Shaoxing, Zhejiang, China

Xiaoyue Chen State Key Laboratory of Power Grid Environment Protection, School of Electrical Engineering and Automation, Wuhan University, Wuhan, Hubei, China

Xin Chen State Key Laboratory of Advanced Power Transmission Technology, State Grid Smart Grid Research Institute Co. Ltd, Beijing, China;
State Grid Tangshan Power Supply Company, Tangshan, China

Xinyu Chen College of Electrical Engineering and Control Science, Nanjing Tech University, Nanjing, Jiangsu, China

Yuwei Chen Fuzhou Power Supply Company, State Grid Fujian Electric Power Co., Ltd., Fuzhou, China

Zehong Chen Hunan University, Changsha, China

Lu Cheng State Key Laboratory of Power Equipment and Electrical Insulation, Xi'an Jiaotong University, Xi'an, Shaanxi, China

Weiye Cheng The Key Laboratory of Engineering Dielectrics and Its Application, Ministry of Education, Harbin University of Science and Technology, Harbin, China;
The School of Electrical and Electronic Engineering, Harbin University of Science and Technology, Harbin, China

Jincheng Chu College of Electrical Engineering and Control Science, Nanjing Tech University, Nanjing, China

Qianle Cong State Grid Zhejiang Electric Power Company, Ltd. Shaoxing Power Supply Company, Shaoxing, Zhejiang, China

Yuntian Cong Beijing Institute of Control Engineering, Beijing, China

Xinglei Cui College of Electrical Engineering and Control Science, Nanjing Tech University, Nanjing, Jiangsu, China

Yunkun Deng Electric Power Research Institute of Yunnan Power Grid Co., Ltd., Kunming, Yunnan, China

Fenglin Ding Beijing Institute of Control Engineering, Beijing, China

Jiatao Ding ZTT Cable Accessories Co., Ltd, Nantong, Jiangsu, China

Yu Ding ZTT Cable Accessories Co., Ltd., Nantong, China

Yujian Ding China Electric Power Research Institute, Haidian District, Beijing, China

Yutong Ding School of Electrical Engineering, Xi'an Jiaotong University, Xi'an, China

Bing Duan China Academy of Space Technology (Xi'an), Xi'an, China

Yan Du Xi'an Key Laboratory of Interconnected Sensing and Intelligent Diagnosis for Electrical Equipment, Xi'an Polytechnic University, Xi'an, China

Yaping Du Department of Building Environment and Energy Engineering, The Hong Kong Polytechnic University, Hong Kong, China

Guiyou Fan EHV Branch Company of State Grid Fujian Electric Power Co. Ltd, Fuzhou, China

Yankun Fan EHV Branch Company of State Grid Fujian Electric Power Co. Ltd, Fuzhou, China

Yezeng Fan Key Laboratory for Power Machinery and Engineering of M. O. E, Shanghai Jiaotong University, Shanghai, China

Jinyong Fang Xi'an Branch of China Academy of Space Technology, Xi'an, China

Zhi Fang College of Electrical Engineering and Control Science, Nanjing Tech University, Nanjing, Jiangsu, China

Yu Feng The Key Laboratory of Engineering Dielectrics and Its Application, Ministry of Education, Harbin University of Science and Technology, Harbin, China;
The School of Electrical and Electronic Engineering, Harbin University of Science and Technology, Harbin, China;
Beijing International S&T Cooperation Base for Plasma Science and Energy Conversion, Institute of Electrical Engineering, Chinese Academy of Sciences, Beijing, China

Junhui Fu College of Electrical Engineering and Control Science, Nanjing Tech University, Nanjing, China

Lei Fu State Key Laboratory of Electrical Insulation and Power Equipment, Xi'an Jiaotong University, Xi'an, Shaanxi, China

Jingming Gao College of Advanced Interdisciplinary Studies, National University of Defense Technology, Changsha, China

Peng Gao Hubei Engineering Research Center for Safety Monitoring of New Energy and Power Grid Equipment, Hubei University of Technology, Wuhan, P. R. China

Wenhao Gao Shandong Provincial Key Laboratory of UHV Transmission Technology and Equipment, Shandong University, Jinan, China

Wenxin Gao School of Electronics and Information, Xi'an Polytechnic University, Xi'an, China;
Xi'an Key Laboratory of Interconnected Sensing and Intelligent Diagnosis for Electrical Equipment, Xi'an Polytechnic University, Xi'an, China

Yongqiang Gao State Grid Shandong Electric Power Company Dongying Power Supply Company, Dongyin, China

Zhan Gao Key Laboratory for Power Machinery and Engineering of M. O. E, Shanghai Jiaotong University, Shanghai, China

Jiahao Guo Electric Power Research Institute of State Grid Jiangsu Electric Power Co., Ltd, Nanjing, China

Ning Guo Electric Power Research Institute of State Grid Jiangsu Electric Power Co., Ltd, Nanjing, China

Shaowei Guo State Key Laboratory of Advanced Power Transmission Technology, State Grid Smart Grid Research Institute Co. Ltd, Beijing, China

Xiangbin Guo School of Mechanical and Electrical Engineering, Hohai University, Changzhou, China

Yunshun Guo School of Electrical Engineering, Xi'an Jiaotong University, Xi'an, China

Qi Han State Grid Shandong Electric Power Company Dongying Power Supply Company, Dongyin, China

Tianyu Han School of Electrical Engineering and Automation, Hefei University of Technology, Hefei, Anhui, China

Junjia He State Key Laboratory of Advanced Electromagnetic Technology, Huazhong University of Science and Technology, Wuhan, China

Junwen He School of Electrical Engineering, Dalian University of Technology, Dalian, People's Republic of China

Qiang He State Grid Zhejiang Electric Power Company, Ltd. Shaoxing Power Supply Company, Shaoxing, Zhejiang, China

Shunfan He South-Central Minzu University, Wuhan, People's Republic of China

Xiang He School of Science, Hohai University, Changzhou, Jiangsu, China

Biao Hu Nanjing University of Aeronautics and Astronautics, Nanjing, People's Republic of China

Di Hu State Grid Anhui Electric Power Research Institute, Hefei, Anhui Province, China

Taizhuang Hu College of Advanced Interdisciplinary Studies, National University of Defense Technology, Changsha, China

Xiaocheng Hu Xi'an Branch of China Academy of Space Technology, Xi'an, China

Zhengyong Hu State Grid Shanghai Municipal Electric Power Company, Shanghai, China

Bangdou Huang Beijing International S&T Cooperation Base for Plasma Science and Energy Conversion, Institute of Electrical Engineering, Chinese Academy of Sciences, Beijing, China

Huijun Huang China Academy of Space Technology (Xi'an), Xi'an, China

Zhen Huang Key Laboratory for Power Machinery and Engineering of M. O. E, Shanghai Jiaotong University, Shanghai, China

Lei Jia Electric Power Research Institute, United Laboratory of Advanced Electrical Materials and Equipment Support Technology, China Southern Power Grid Company Limited Guangzhou, Guangdong, China

Danyue Jiang School of Mechanical and Electrical Engineering, Hohai University, Changzhou, Jiangsu, China

Hua Jiang Hubei State Key Laboratory of Power Grid Environmental Protection, School of Electrical Engineering and Automation, Wuhan University, Wuhan, Hubei, China;
Hubei Key Laboratory of Power Equipment & System Security for Integrated Energy, School of Electrical Engineering and Automation, Wuhan University, Wuhan, Hubei, China

Nan Jiang School of Electrical Engineering, Dalian University of Technology, Dalian, People's Republic of China

Tao Jiang School of Metallurgy, Northeastern University, Shenyang, Liaoning, China

Wangjie Jiang Electric Power Research Institute of State Grid Jiangsu Electric Power Co., Ltd., Nanjing, China

Yongfeng Jiang School of Mechanical and Electrical Engineering, Hohai University, Changzhou, Jiangsu, China

Yuan Jiang Key Laboratory of Knowledge Automation for Industrial Processes of Ministry of Education, School of Automation and Electrical Engineering, University of Science and Technology Beijing, Beijing, China

Axin Jin China Academy of Space Technology (Xi'an), Xi'an, China

Menglei Jin School of Electrical Engineering and Automation, Wuhan University, Hubei, China

YanGuo Ke State Grid Anhui Electric Power Research Institute, Hefei, Anhui Province, China

Chuncai Kong Ministry of Education Key Laboratory for Non-Equilibrium Synthesis and Modulation of Condensed Matter, Shaanxi Province Key Laboratory of Advanced Functional Materials and Mesoscopic Physics, School of Physics, Xi'an Jiaotong University, Xi'an, China

Lei Lan State Key Laboratory of Power Grid Environment Protection, School of Electrical Engineering and Automation, Wuhan University, Wuhan, Hubei, China

Chengzhen Li Shandong Provincial Key Laboratory of UHV Transmission Technology and Equipment, Shandong University, Jinan, China

Deyan Li ZTT Cable Accessories Co., Ltd., Nantong, China

Feng Li State Grid Zhejiang Electric Power Company, Ltd. Shaoxing Power Supply Company, Shaoxing, Zhejiang, China

Guoliang Li South-Central Minzu University, Wuhan, People's Republic of China

Jie Li School of Electrical Engineering, Dalian University of Technology, Dalian, People's Republic of China

Ju Li School of Electrical Engineering, Dalian University of Technology, Dalian, People's Republic of China

Kunyi Li Southwest Jiaotong University, Chengdu, Sichuan, China

Li Li Electric Power Research Institute of Guangdong Power Grid Co., Ltd., Guangdong, China

Ni Li State Key Laboratory of Power Grid Environmental Protection, China Electric Power Research Institute, Wuhan, P. R. China

Qianwen Li School of Mechanical and Electrical Engineering, Hohai University, Changzhou, Jiangsu, China

Qinghao Li State Key Laboratory of Electrical Insulation and Power Equipment, Xi'an Jiaotong University, Xi'an, Shaanxi, China

Qingquan Li Shandong Provincial Key Laboratory of UHV Transmission Technology and Equipment, Shandong University, Jinan, China

Runhua Li College of Electrical Engineering and Control Science, Nanjing Tech University, Nanjing, Jiangsu, China

Xin Li Northwest Branch of State Grid Corporation of China, Xi'an, China

Xu Li School of Mechanical and Electrical Engineering, Hohai University, Changzhou, Jiangsu, China

Yalong Li State Key Laboratory of Power Grid Environmental Protection, China Electric Power Research Institute, Wuhan, P. R. China;
Hubei Engineering Research Center for Safety Monitoring of New Energy and Power Grid Equipment, Hubei University of Technology, Wuhan, P. R. China

Yang Li School of Mechanical and Automotive Engineering, South China University of Technology, Guangzhou, China

Yingbin Li State Key Laboratory of Electrical Insulation and Power Equipment, Xi'an Jiaotong University, Xi'an, Shaanxi, China

Yinglong Li Department of Stomatology, Beijing Chaoyang Hospital, Capital Medical University, Beijing, China

Yong Li Beijing Institute of Control Engineering, Beijing, China

Ze Li Shandong Provincial Key Laboratory of UHV Transmission Technology and Equipment, Shandong University, Jinan, China

Zhitao Li School of Mechanical and Automotive Engineering, South China University of Technology, Guangzhou, China

Zhiyuan Li State Key Laboratory of Power Equipment and Electrical Insulation, Xi'an Jiaotong University, Xi'an, Shaanxi, China

Zhuohang Li Key Laboratory for Power Machinery and Engineering of M. O. E, Shanghai Jiaotong University, Shanghai, China

Zhuoxiao Li Shanghai Jiao Tong University, Shanghai, China

Hongsheng Liang State Grid Corporation of China, Xicheng District, Beijing, China

Boyang Lin School of Mechanical and Electrical Engineering, Hohai University, Changzhou, Jiangsu, China

Chuanjie Lin Fuzhou Power Supply Company, State Grid Fujian Electric Power Co., Ltd., Fuzhou, China

Lin Lin Electric Power Research Institute of State Grid Jiangsu Electric Power Co., Ltd., Nanjing, China

Xiaosi Lin State Grid Fujian Technical Training Centre, State Grid Fujian Electric Power Co., Ltd., Quanzhou, China

Jianben Liu State Key Laboratory of Power Grid Environmental Protection, China Electric Power Research Institute, Wuhan, P. R. China

Jie Liu State Grid Corporation of China, Xicheng District, Beijing, China

Junjie Liu State Key Laboratory of Electrical Insulation and Power Equipment, Xi'an Jiaotong University, Xi'an, Shaanxi, China

Juyang Liu State Key Laboratory of Advanced Power Transmission Technology, State Grid Smart Grid Research Institute Co. Ltd, Beijing, China

Liguo Liu State Grid Jiangsu Electric Power Co., Ltd, Nanjing, China

Min Liu Northwest Branch of State Grid Corporation of China, Xi'an, China

Peiqiong Liu State Key Laboratory of Electrical Insulation and Power Equipment, Xi'an Jiaotong University, Xi'an, China

Qi Liu School of Mechanical and Electrical Engineering, Hohai University, Changzhou, Jiangsu, China

Qingxiang Liu Southwest Jiaotong University, Chengdu, Sichuan, China

Shiyun Liu College of Electrical Engineering and Control Science, Nanjing Tech University, Nanjing, China;
School of Electrical Engineering, Southeast University, Nanjing, China

Wenfeng Liu State Key Laboratory of Power Equipment and Electrical Insulation, Xi'an Jiaotong University, Xi'an, Shaanxi, China

Yan Liu State Key Laboratory of Power Grid Environmental Protection, China Electric Power Research Institute, Wuhan, P. R. China

Yang Liu State Key Laboratory of Advanced Electromagnetic Technology, Huazhong University of Science and Technology, Wuhan, China

Yongpeng Liu Shandong Provincial Key Laboratory of UHV Transmission Technology and Equipment, Shandong University, Jinan, China

Yongxin Liu State Grid Zhejiang Electric Power Company, Ltd. Shaoxing Power Supply Company, Shaoxing, Zhejiang, China

Mikhail Lomaev Institute of High Current Electronics, Siberian Branch of the Russian Academy of Sciences, Tomsk, Russia

Xinpei Lu Huazhong University of Science and Technology, Wuhan, People's Republic of China

Yunyang Lu College of Electrical Engineering and Control Science, Nanjing Tech University, Nanjing, China

Yan Luo CSG Electric Power Research Institute Co., Ltd., Guangdong, China

Zhaorui Luo College of Electrical Engineering and Control Science, Nanjing Tech University, Nanjing, China

Langlang Lv Hubei State Key Laboratory of Power Grid Environmental Protection, School of Electrical Engineering and Automation, Wuhan University, Wuhan, Hubei, China;
Hubei Key Laboratory of Power Equipment & System Security for Integrated Energy, School of Electrical Engineering and Automation, Wuhan University, Wuhan, Hubei, China

Bo Ma School of Electrical Engineering and Automation, Hefei University of Technology, Hefei, Anhui, China

Jingyi Ma College of Electrical Engineering and Control Science, Nanjing Tech University, Nanjing, China

Ruonan Ma Henan Key Laboratory of Ion-Beam Bioengineering, School of Agricultural Sciences, Zhengzhou University, Zhengzhou, China

Suliang Ma Energy Storage Technology Engineering Research Center, North China University of Technology, Beijing, China

Danhua Mei College of Electrical Engineering and Control Science, Nanjing Tech University, Nanjing, China

Fanbo Meng Shandong Provincial Key Laboratory of UHV Transmission Technology and Equipment, Shandong University, Jinan, China

Qianhang Ni State Grid Zhejiang Electric Power Company, Ltd. Shaoxing Power Supply Company, Shaoxing, Zhejiang, China

Kai Nie Shandong Provincial Key Laboratory of UHV Transmission Technology and Equipment, Shandong University, Jinan, China

Lanlan Nie Huazhong University of Science and Technology, Wuhan, People's Republic of China

Chao Pan State Grid Anhui Electric Power Research Institute, Hefei, Anhui Province, China

Alexei Panchenko Institute of High Current Electronics, Siberian Branch of the Russian Academy of Sciences, Tomsk, Russia

Yanze Pang State Key Laboratory of Electrical Insulation and Power Equipment, Xi'an Jiaotong University, Xi'an, China

Yuchen Ping Department of Electrical Engineering, Nanjing University of Aeronautics and Astronautics, Nanjing, China

Chengyan Ren Beijing International S&T Cooperation Base for Plasma Science and Energy Conversion, Institute of Electrical Engineering, Chinese Academy of Sciences, Beijing, China;
University of Chinese Academy of Sciences, Beijing, China

He Ren School of Electric Engineering, Southeast University, Nanjing, China

Jia Ren State Grid Zhejiang Electric Power Company, Ltd. Shaoxing Power Supply Company, Shaoxing, Zhejiang, China

Rongchen Shang School of Mechanical and Electrical Engineering, Hohai University, Changzhou, Jiangsu, China

Tao Shao Beijing International S&T Cooperation Base for Plasma Science and Energy Conversion, Institute of Electrical Engineering, Chinese Academy of Sciences, Beijing, China

Yongchun Shao Xi'an Branch of China Academy of Space Technology, Xi'an, China

Zhaohui Song Northwest Branch of State Grid Corporation of China, Xi'an, China

Dmitry Sorokin Institute of High Current Electronics, Siberian Branch of the Russian Academy of Sciences, Tomsk, Russia

Lei Su State Grid Hubei Electric Power Research Institute, Wuhan, China

Weiwei Su Electric Power Research Institute of State Grid Jiangsu Electric Power Co., Ltd., Nanjing, China

Yaotian Su State Key Laboratory of Advanced Power Transmission Technology, State Grid Smart Grid Research Institute Co. Ltd, Beijing, China

Yu Su China Electric Power Research Institute, Haidian District, Beijing, China

Dongwei Sun Electric Power Research Institute of Guangdong Power Grid Co., Ltd, Guangzhou, China;
Sulfur Hexafluoride Key Lab of China Southern Power Grid, Guangzhou, China

Hao Sun School of Electrical Engineering, Xi'an Jiaotong University, Xi'an, China

Jiantao Sun China Electrical Power Research Institute, Beijing, China

Qiming Sun State Key Laboratory of Electrical Insulation and Power Equipment, Xi'an Jiaotong University, Xi'an, Shaanxi, China

Wenhui Sun School of Electrical Engineering, Southeast University, Nanjing, China

Yongjian Sun State Grid Shandong Electric Power Company Dongying Power Supply Company, Dongyin, China

Jiaxin Tan Key Laboratory of Electrical Insulation and Power Equipment, Xi'an Jiaotong University, Xi'an, China

Nian Tang Electric Power Research Institute of Guangdong Power Grid Co., Ltd, Guangzhou, China;
Sulfur Hexafluoride Key Lab of China Southern Power Grid, Guangzhou, China;
Electric Power Research Institute of Guangdong Power Grid Co., Ltd., Guangdong, China

Zhicheng Tang School of Mechanical and Electrical Engineering, Hohai University, Changzhou, China

Victor Tarasenko Institute of High Current Electronics, Siberian Branch of the Russian Academy of Sciences, Tomsk, Russia

Wei Tian South-Central Minzu University, Wuhan, People's Republic of China

Zhengbing Tian ZTT Cable Accessories Co., Ltd, Nantong, Jiangsu, China

Weihao Tie Xi'an Branch of China Academy of Space Technology, Xi'an, China

Nikita Vinogradov Institute of High Current Electronics, Siberian Branch of the Russian Academy of Sciences, Tomsk, Russia

Kun Wan Hubei Engineering Research Center for Safety Monitoring of New Energy and Power Grid Equipment, Hubei University of Technology, Wuhan, P. R. China

Li Wan College of Electrical Engineering and Control Science, Nanjing Tech University, Nanjing, China

Baojun Wang Beijing Institute of Control Engineering, Beijing, China

Chao Wang Key Laboratory for Power Machinery and Engineering of M. O. E, Shanghai Jiaotong University, Shanghai, China

Dibo Wang CSG Electric Power Research Institute Co., Ltd., Guangdong, China

Feng Wang Hunan University, Changsha, China

Ge Wang Beijing Institute of Control Engineering, Beijing, China

Haozhou Wang Beijing International S&T Cooperation Base for Plasma Science and Energy Conversion, Institute of Electrical Engineering, Chinese Academy of Sciences, Beijing, China;
University of Chinese Academy of Sciences, Beijing, China

Jian Wang State Grid Corporation of China, Beijing, China

Jianan Wang Electric Power Research Institute of State Grid Jiangsu Electric Power Co., Ltd., Nanjing, China

Jiangping Wang Xi'an Branch of China Academy of Space Technology, Xi'an, China

Jiayu Wang School of Mechanical and Electrical Engineering, Hohai University, Changzhou, Jiangsu, China

Jiyang Wang College of Electrical Engineering and Control Science, Nanjing Tech University, Nanjing, China

Ke Wang China Electrical Power Research Institute, Beijing, China;
Electric Power Research Institute of Yunnan Power Grid Co., Ltd., Kunming, Yunnan, China

Ruixue Wang College of Mechanical and Electrical Engineering, Beijing University of Chemical Technology, Beijing, China

Sen Wang Nanjing Tech University, Nanjing, China

Xinkui Wang ZTT Cable Accessories Co., Ltd., Nantong, China

Yang Wang School of Electronics and Information, Xi'an Polytechnic University, Xi'an, China;
Xi'an Key Laboratory of Interconnected Sensing and Intelligent Diagnosis for Electrical Equipment, Xi'an Polytechnic University, Xi'an, China

Yukai Wang Southwest Jiaotong University, Chengdu, Sichuan, China

Zhiliang Wang State Grid Zhejiang Electric Power Company, Ltd. Shaoxing Power Supply Company, Shaoxing, Zhejiang, China

Zhuofei Wang Beijing International S&T Cooperation Base for Plasma Science and Energy Conversion, Institute of Electrical Engineering, Chinese Academy of Sciences, Beijing, China

Menggen Wei School of Electronics and Information, Xi'an Polytechnic University, Xi'an, China;
Xi'an Key Laboratory of Interconnected Sensing and Intelligent Diagnosis for Electrical Equipment, Xi'an Polytechnic University, Xi'an, China

Pengfei Wei College of Electrical Engineering and Control Science, Nanjing Tech University, Nanjing, China

Tao Wen School of Electrical Engineering and Automation, Hefei University of Technology, Hefei, Anhui, China

Xishan Wen State Key Laboratory of Power Grid Environment Protection, School of Electrical Engineering and Automation, Wuhan University, Wuhan, Hubei, China

Fan Wu Electric Power Research Institute of State Grid Jiangsu Electric Power Co., Ltd, Nanjing, China

Jianwen Wu School of Automation Science and Electrical Engineering, Beihang University, Beijing, China

Kai Wu State Key Laboratory of Electrical Insulation and Power Equipment, Xi'an Jiaotong University, Xi'an, Shaanxi, China

Peng Wu Beijing Institute of Control Engineering, Beijing, China

Shuqun Wu Nanjing University of Aeronautics and Astronautics, Nanjing, People's Republic of China

Yan Wu School of Electrical Engineering, Dalian University of Technology, Dalian, People's Republic of China

Yi Wu School of Electrical Engineering, Xi'an Jiaotong University, Xi'an, China

Shangwen Xia School of Automation Science and Electrical Engineering, Beihang University, Beijing, China

Yubin Xian Huazhong University of Science and Technology, Wuhan, People's Republic of China

Xiaolong Xiao Electric Power Research Institute of State Grid Jiangsu Electric Power Co., Ltd, Nanjing, China

Zhaoliang Xing State Key Laboratory of Advanced Power Transmission Technology, State Grid Smart Grid Research Institute Co. Ltd, Beijing, China

GuoWen Xiong State Key Laboratory of Power Grid Environment Protection, School of Electrical Engineering and Automation, Wuhan University, Wuhan, Hubei, China

Yi Xiong South-Central Minzu University, Wuhan, People's Republic of China

Lei Xu Southwest Jiaotong University, Chengdu, Sichuan, China

Long Xu State Grid Zhejiang Electric Power Company, Ltd. Shaoxing Power Supply Company, Shaoxing, Zhejiang, China

Peng Xu State Grid Shanghai Municipal Electric Power Company, Shanghai, China

Pengfei Xu ZTT Cable Accessories Co., Ltd., Nantong, China

Tongxin Xu School of Electrical Engineering, Southeast University, Nanjing, China

Xiaohui Xu Department of Astronomy, University of Science and Technology of China, Hefei, China

Chen Yang Nanjing University of Aeronautics and Astronautics, Nanjing, People's Republic of China

Fan Yang State Grid Hubei Electric Power Research Institute, Wuhan, China

Guangdong Yang School of Metallurgy, Northeastern University, Shenyang, Liaoning, China

Hao Yang The Key Laboratory of Engineering Dielectrics and Its Application, Ministry of Education, Harbin University of Science and Technology, Harbin, China;
The School of Electrical and Electronic Engineering, Harbin University of Science and Technology, Harbin, China

Jiaming Yang State Key Laboratory of Electrical Insulation and Power Equipment, Xi'an Jiaotong University, Xi'an, Shaanxi, China

Jingyu Yang State Grid Tangshan Power Supply Company, Tangshan, China

Xiaoxia Yang School of Mechanical and Electrical Engineering, Hohai University, Changzhou, China

Zhaodi Yang Hubei Engineering Research Center for Safety Monitoring of New Energy and Power Grid Equipment, Hubei University of Technology, Wuhan, P. R. China

Zhichun Yang State Grid Hubei Electric Power Research Institute, Wuhan, China

Zhimao Yang Ministry of Education Key Laboratory for Non-Equilibrium Synthesis and Modulation of Condensed Matter, Shaanxi Province Key Laboratory of Advanced Functional Materials and Mesoscopic Physics, School of Physics, Xi'an Jiaotong University, Xi'an, China

Cheng Yao Electric Power Research Institute, United Laboratory of Advanced Electrical Materials and Equipment Support Technology, China Southern Power Grid Company Limited Guangzhou, Guangdong, China;
State Key Laboratory of Power Equipment and Electrical Insulation, Xi'an Jiaotong University, Xi'an, Shaanxi, China

Xiuyuan Yao China Electric Power Research Institute, Haidian District, Beijing, China

Yuankun Ye Nanjing Tech University, Nanjing, China

Guanglin Yu School of Electrical Engineering, Dalian University of Technology, Dalian, People's Republic of China

Yong Yu State Grid Tangshan Power Supply Company, Tangshan, China

Dong Yue The Key Laboratory of Engineering Dielectrics and Its Application, Ministry of Education, Harbin University of Science and Technology, Harbin, China;

The School of Electrical and Electronic Engineering, Harbin University of Science and Technology, Harbin, China

Yiming Zang Shanghai Jiao Tong University, Shanghai, China

Fuping Zeng Hubei State Key Laboratory of Power Grid Environmental Protection, School of Electrical Engineering and Automation, Wuhan University, Wuhan, Hubei, China;
Hubei Key Laboratory of Power Equipment & System Security for Integrated Energy, School of Electrical Engineering and Automation, Wuhan University, Wuhan, Hubei, China

Min Zeng School of Mechanical and Automotive Engineering, South China University of Technology, Guangzhou, China

Haoyu Zhan State Key Laboratory of Power Grid Environment Protection, School of Electrical Engineering and Automation, Wuhan University, Wuhan, Hubei, China

Baolei Zhang ZTT Cable Accessories Co., Ltd., Nantong, China

Bo Zhang School of Electronics and Information, Xi'an Polytechnic University, Xi'an, China;
Xi'an Key Laboratory of Interconnected Sensing and Intelligent Diagnosis for Electrical Equipment, Xi'an Polytechnic University, Xi'an, China

Boya Zhang State Key Laboratory of Electrical Insulation and Power Equipment, Xi'an Jiaotong University, Xi'an, China

Chenchen Zhang State Grid Anhui Electric Power Research Institute, Hefei, Anhui Province, China

Cheng Zhang Beijing International S&T Cooperation Base for Plasma Science and Energy Conversion, Institute of Electrical Engineering, Chinese Academy of Sciences, Beijing, China

Chenxu Zhang School of Electrical Engineering, Southeast University, Nanjing, China

Chong Zhang State Key Laboratory of Advanced Power Transmission Technology, State Grid Smart Grid Research Institute Co. Ltd, Beijing, China

Chunyang Zhang School of Mechanical and Electrical Engineering, Hohai University, Changzhou, Jiangsu, China

Guangjin Zhang State Key Laboratory of Electric Insulation and Power Equipment, Xi'an Jiaotong University, Xi'an, Shaanxi, China

Huibo Zhang College of Advanced Interdisciplinary Studies, National University of Defense Technology, Changsha, China

Jianjun Zhang State Grid Tangshan Power Supply Company, Tangshan, China

Jingyun Zhang Department of Electrical Engineering, Nanjing University of Aeronautics and Astronautics, Nanjing, China

Junmin Zhang South-Central Minzu University, Wuhan, People's Republic of China

Qian Zhang International Student Center, University of Science and Technology Beijing, Beijing, China

Shuwen Zhang School of Software, Southeast University (Suzhou), Suzhou, China

Song Zhang State Grid Shanghai Municipal Electric Power Company, Shanghai, China

Wei Zhang School of Mechanical and Automotive Engineering, South China University of Technology, Guangzhou, China

Wenjing Zhang EHV Branch Company of State Grid Fujian Electric Power Co. Ltd, Fuzhou, China

Xiaojiao Zhang Department of Stomatology, Beijing Chaoyang Hospital, Capital Medical University, Beijing, China

Xiaoxing Zhang Hubei Engineering Research Center for Safety Monitoring of New Energy and Power Grid Equipment, Hubei University of Technology, Wuhan, P. R. China

Xinghui Zhang State Grid Corporation of China, Beijing, China

Xinyi Zhang Nanjing University of Aeronautics and Astronautics, Nanjing, People's Republic of China

Yanze Zhang State Key Laboratory of Power Grid Environment Protection, School of Electrical Engineering and Automation, Wuhan University, Wuhan, Hubei, China

Yiming Zhang State Grid Tangshan Power Supply Company, Tangshan, China

Yingzhe Zhang Huazhong University of Science and Technology, Wuhan, People's Republic of China

Zhengquan Zhang Southwest Jiaotong University, Chengdu, Sichuan, China

Zicheng Zhang College of Advanced Interdisciplinary Studies, National University of Defense Technology, Changsha, China

Zixiao Zhang State Key Laboratory of Advanced Electromagnetic Technology, Huazhong University of Science and Technology, Wuhan, China

Baoxi Zhao South-Central Minzu University, Wuhan, People's Republic of China

Dong Zhao School of Electronics and Information, Xi'an Polytechnic University, Xi'an, China;

Xi'an Key Laboratory of Interconnected Sensing and Intelligent Diagnosis for Electrical Equipment, Xi'an Polytechnic University, Xi'an, China

Tianjian Zhao State Grid Zhejiang Electric Power Company, Ltd. Shaoxing Power Supply Company, Shaoxing, Zhejiang, China

Xiangen Zhao Department of Building Environment and Energy Engineering, The Hong Kong Polytechnic University, Hong Kong, China

Yi Zhao School of Electrical Engineering and Automation, Hefei University of Technology, Hefei, Anhui, China

Yuesheng Zheng College of Electrical Engineering and Automation, Fuzhou University, Fuzhou, China

Linlin Zhong School of Electric Engineering, Southeast University, Nanjing, China

Cheng Zhou Beijing Institute of Control Engineering, Beijing, China

Daiming Zhou State Grid Zhejiang Electric Power Company, Ltd. Shaoxing Power Supply Company, Shaoxing, Zhejiang, China

Dongdong Zhou State Grid Zhejiang Electric Power Company, Ltd. Shaoxing Power Supply Company, Shaoxing, Zhejiang, China

Guliang Zhou State Grid Shanghai Municipal Electric Power Company, Shanghai, China

Jun Zhou State Key Laboratory of Electrical Insulation and Power Equipment, Xi'an Jiaotong University, Xi'an, Shaanxi, China

Zijian Zhou College of Electrical Engineering and Control Science, Nanjing Tech University, Nanjing, China

Kexin Zhu Hubei State Key Laboratory of Power Grid Environmental Protection, School of Electrical Engineering and Automation, Wuhan University, Wuhan, Hubei, China;
Hubei Key Laboratory of Power Equipment & System Security for Integrated Energy, School of Electrical Engineering and Automation, Wuhan University, Wuhan, Hubei, China

Lei Zhu Key Laboratory for Power Machinery and Engineering of M. O. E, Shanghai Jiaotong University, Shanghai, China

Libo Zhu School of Mechanical and Electrical Engineering, Hohai University, Changzhou, Jiangsu, China

Lingyu Zhu State Key Laboratory of Electrical Insulation and Power Equipment, Xi'an Jiaotong University, Xi'an, Shaanxi, China

Longyang Zhu Fuzhou Power Supply Company, State Grid Fujian Electric Power Co., Ltd., Fuzhou, China

Min Zhu Department of Electrical Engineering, Nanjing University of Aeronautics and Astronautics, Nanjing, China

Weiping Zhu State Grid Jiangsu Electric Power Co., Ltd, Nanjing, China

Xi Zhu College of Electrical Engineering and Control Science, Nanjing Tech University, Nanjing, China

Zihe Zhu State Key Laboratory of Electrical Insulation and Power Equipment, Xi'an Jiaotong University, Xi'an, Shaanxi, China

Ran Zhuo CSG Electric Power Research Institute Co., Ltd., Guangdong, China

Part I
Basic Process, Diagnosis and Simulation in Plasmas

Chapter 1
Research on a Self-balancing Pulsed AC Corona Ionizer

Wei Zhang, Yang Li, Zhitao Li, and Min Zeng

Abstract Corona ionizers are widely used in semiconductor industry to eliminate static electricity and reduce static damage. In order to improve the speed and ability of Corona ionizers, a pulsed AC corona ionizer which can automatically balance positive and negative ions is studied. In this paper, the composition and working principle of the ionizer are introduced, the main circuit of the voltage doubling rectifier circuit technology and the control circuit using STM32F103CBT6 as the control core are established. The main program, PWM generator an PI control algorithm software were designed. On this basis, the experimental platform of Corona ionizer is built and verified by experiments. The results shows that the peak value of the high-voltage power supply reaches 9.44 kV. When the ionizer works at the frequency of 50 Hz, the distance of 30 cm, and the airflow pressure of 0.1 MPa, the decay time is less than 1.4 s and the offset voltage is less than ±5 V. The amplitude and duty cycle are automatically adjusted to rebalance the positive and negative ions while the ionizer is in a non-equilibrium state. The designed ionizer has short electrostatic elimination time, good ion balance, and can automatically calibrate the ionizer balance state to reduce electrostatic damage.

1.1 Introduction

Eliminating static charges on devices to prevent electrostatic discharge problems in manufacturing industries such as semiconductors and electronic devices is important to improve safety and productivity [1–3]. The ions produced by the corona discharge are transported by the blown air to the insulator, where the charge on the insulator is neutralized by the ions. In view of the damage caused by static electricity, corona ionizers are widely used in semiconductor manufacturing and other industries [4, 5].

W. Zhang · Y. Li · Z. Li · M. Zeng (✉)
School of Mechanical and Automotive Engineering, South China University of Technology, Guangzhou 510641, China
e-mail: memzeng@scut.edu.cn

Z. Fang et al. (eds.), *Proceedings of the 5th International Symposium on Plasma and Energy Conversion*, Springer Proceedings in Physics 398,
https://doi.org/10.1007/978-981-97-2245-7_1

For the performance of the ion generator, three key conditions are required: short electrostatic elimination time, good ion balance (a low offset-voltage), and long-term performance stability [6]. The charge decay time, the time required to eliminate electrostatic charge on a charged object, is significantly affected by the amplitude of applied voltage to the needles, the voltage waveform, and its frequency. Therefore, the parameters of the high voltage power supply inside the ionizer must be improved to ensure the quality of electrostatic elimination. The performance of the electrostatic eliminator declines with the deterioration of the high-voltage electrode because of dust attraction and abrasion from the corona discharge. Therefore, the high-voltage power supply needs to be automatically adjusted according to the environment to ensure the ion balance of the electrostatic eliminator.

High-voltage power supply is divided into DC type and AC type high-voltage power supply, and AC high-voltage voltage also includes sine wave AC and pulse AC. Many scholars have studied different high-voltage power supply to improve the performance of ionizers [7–9]. Wang [10] studied the dual DC ion generator with feedback regulation system, the elimination time was < 4 s, and the offset voltage was controlled within ± 3 V, which met the requirements of electrostatic control in electrostatic protection area. Katsuyuki [11] developed a completely balanced fan-type electrostatic eliminator using an intermittent pulse AC voltage power supply. In the absence of sensor feedback, the offset voltage (ion balance) has a short-term fluctuation range of less than ± 2 V.

In order to monitor the normal operation of the ionizer, it is generally calibrated using a charging flat instrument. However, a simple desktop function with a charging plate monitor is not enough to guarantee safety performance. Byoungjin [12] proposed a new alternative method to directly measure the discharge current of the ionizer. Seidl [13] directly measured the discharge current of the ionizer, and judged the working state of the ionizer by measuring the asymmetry of positive and negative charges.

In this paper, a pulsed AC high voltage ionizer is proposed, which can automatically balance positive and negative ions. The peak value of the high-voltage power supply reaches 9.44 kV, and the power supply parameters can be controlled more accurately. The static dissipation time is short, the ion balance degree is high, and the balance state of the ionizer can be automatically calibrated to effectively reduce static damage.

1.2 The Structure of the Pulse High-Voltage Ionizer

The corona ionizer is mainly composed of a high-voltage power supply, an ion detection system and a human–computer interaction system, and the overall framework is shown in Fig. 1.1. The high-voltage power supply is composed of main circuit and control system, which is the core of the whole system. The main circuit of the power supply is the key to the ionization occurring in the ionizer, which is responsible for converting the low voltage into a high voltage of more than 3 kV. The control system

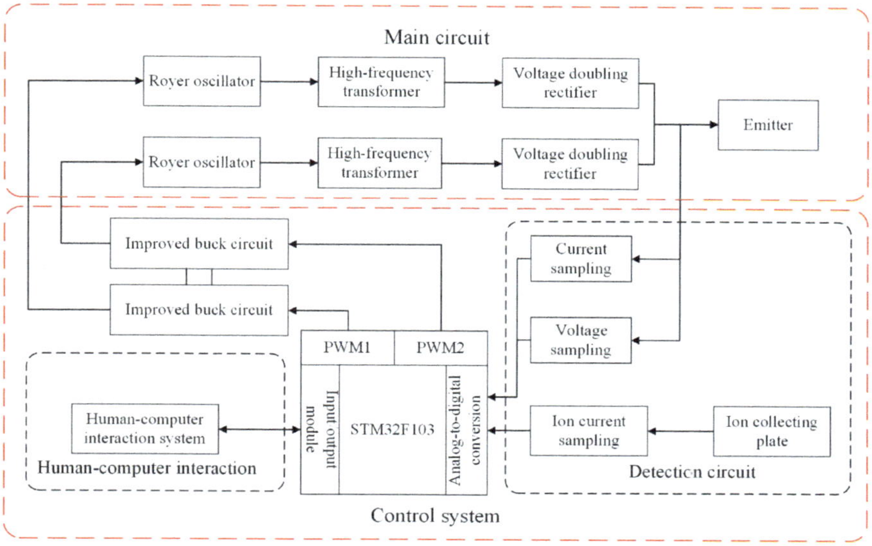

Fig. 1.1 Schematic diagram of corona ionizer structure

is used to control the whole ionization process, which not only needs to deal with the communication with the human–computer interaction system, real-time fault detection and protection of the main circuit and other functions, but also needs to adjust the duty cycle according to the ion current feedback and other signals for PID processing to realize closed-loop precise control and improve the dynamic response performance of the ionizer system. The ion detection system is mainly responsible for collecting positive and negative ions in the environment, monitoring and adjusting the working state of the ionizer, so that the ionizer always works in a balanced state. The human–computer interaction system is mainly responsible for setting the power supply parameters and displaying the frequency, duty cycle and working status of the ionizer.

1.3 Main Circuit Design

1.3.1 The Principle of the Main Circuit

The main circuit of the designed high voltage power supply positive high voltage is shown in Fig. 1.2. It consists of Royer oscillation circuit, high-frequency transformer, fourth-order voltage doubling rectifier circuit and other components. The front stage of the power supply is a DC input of 0–9 V, which is resonated by Royer to obtain a 20 kHz sine wave with amplitude of 0–37 V. After high-frequency transformer

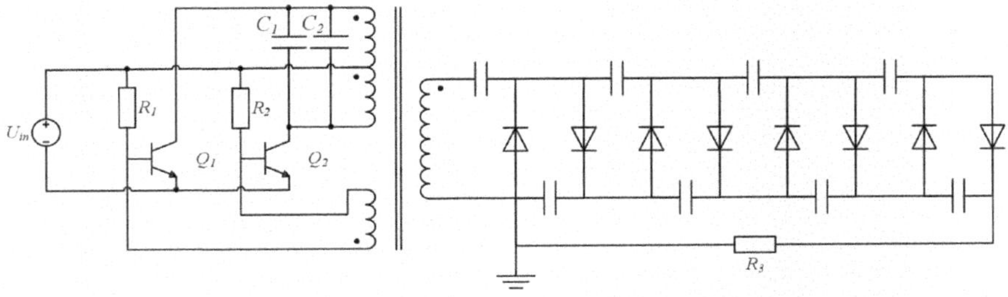

Fig. 1.2 Positive high voltage main circuit for high voltage power supplies

and fourth-order voltage doubling rectifier circuit to obtain the output 0–6 kV high-voltage power supply to meet the requirements of ionized air. By controlling the frequency, duty cycle and amplitude of the input signal of the power supply, different high-voltage outputs were obtained to meet the balancing requirements of the ionizer.

1.3.2 Royer Oscillator Circuit

A Royer oscillator circuit usually consists of two triodes, two base resistors, a resonant capacitor, and a transformer with three windings. Considering that the resonant capacitor is charged and discharged repeatedly and the temperature is high, two capacitors are used in parallel.

At the moment of switching on the power supply, the U_{in} injected current into the triode base, and the circuit can control the current of the triode through the change of the magnetic flux of the primary coil to generate sinusoidal signals, inverting the direct current to alternating current, and then achieving voltage boost through the transformer. The existence of C_1, C_2 in the circuit makes the oscillator circuit simply vibrate according to a specific frequency. The relationship between the frequency f of oscillation and the primary inductance L of the transformer coil and the resonant capacitance C_1, C_2 is shown in Eq. (1.1):

$$f = \frac{1}{2\pi}\sqrt{\frac{1}{4L(C_1 + C_2)}} \tag{1.1}$$

Considering the structure of the Royer oscillator circuit, the influence of the leakage inductance of the transformer and the selection of the capacitor in the subsequent voltage doubling circuit, the efficiency of the oscillation frequency is higher in the 20 kHz circuit, and the resonant capacitors were selected with a capacitance of 47 nF. Therefore, according to Eq. (1.1), a transformer with a resonant inductance of about 330 μH was designed.

1.3.3 Voltage Doubling Rectifier Circuit

The half-wave Coccroft-Walton (C-W) doubler circuit is also known as the basic CW circuit, and its structure is shown in Fig. 1.3. Its operating principle is: set the peak value of the power supply AC voltage is U, in the first negative half-cycle input voltage down positive up negative, C_1 through the D_1 for charging, in this cycle C_1 charging to U, the polarity of the left negative right positive, as shown in Fig. 1.3a. Positive half-cycle input voltage up positive down negative, C_1 and the power supply is connected in series, the power supply and C_1 through the D_2 at the same time to the charging of D_2, the voltage at the ends of C_2 finally $2U$, as shown in Fig. 1.3b. At the beginning of the next cycle, the voltage at both ends of the power supply again becomes $2U$, as shown in Fig. 1.3. The terminal voltage is finally $2U$, as shown in Fig. 1.3b. At the beginning of the next cycle, the voltage at both ends of the power supply becomes positive down and negative up again, and the power supply, C_1, and C_2 charge the C_3 through the D_3, and the voltage at the two ends of C_3 is charged to $2U$. After a few charging cycles, the voltage at the two ends of C_1 is U, and the voltages at the ends of both C_2–C_{2n} become $2U$, and the voltages at the two ends of C_2, C_4, and C_6... C_{2n} is connected in series, the left end of C_2 is grounded, and the output of the right end of C2n is $2nU$, realizing nth-order voltage doubling.

Assume that the load is R, the peak AC supply is U, the frequency is f, and each capacitor has the same capacitance size, both C. q is the charge transferred from C_{2n} to the load in each cycle, and the ripple voltage of C_{2n} is δV. The doubler-rectifier circuit has a conduction period of t_1, a non-conduction period of t_2 (and $t_1 \ll t_2$), and the current from C_{2n} to the load in the time of the non-conduction period is I.

Which has the relation:

$$I = \frac{q}{t_2} \tag{1.2}$$

Since the oscillation period $T = t_1 + t_2 = 1/f$, from Eq. (1.2):

$$q = \frac{I}{f} \tag{1.3}$$

Also, q satisfies the relation:

$$q = C \times \delta V \tag{1.4}$$

From Eq. (1.3) and (1.4), the ripple voltage across C_{2n} is

$$\delta V = \frac{I}{f \times C_2} \tag{1.5}$$

From the previous analysis of the voltage doubling principle, we know that the final output voltage is the superposition of the voltages on C_2, C_4...C_{2n-2}, C_{2n}, so

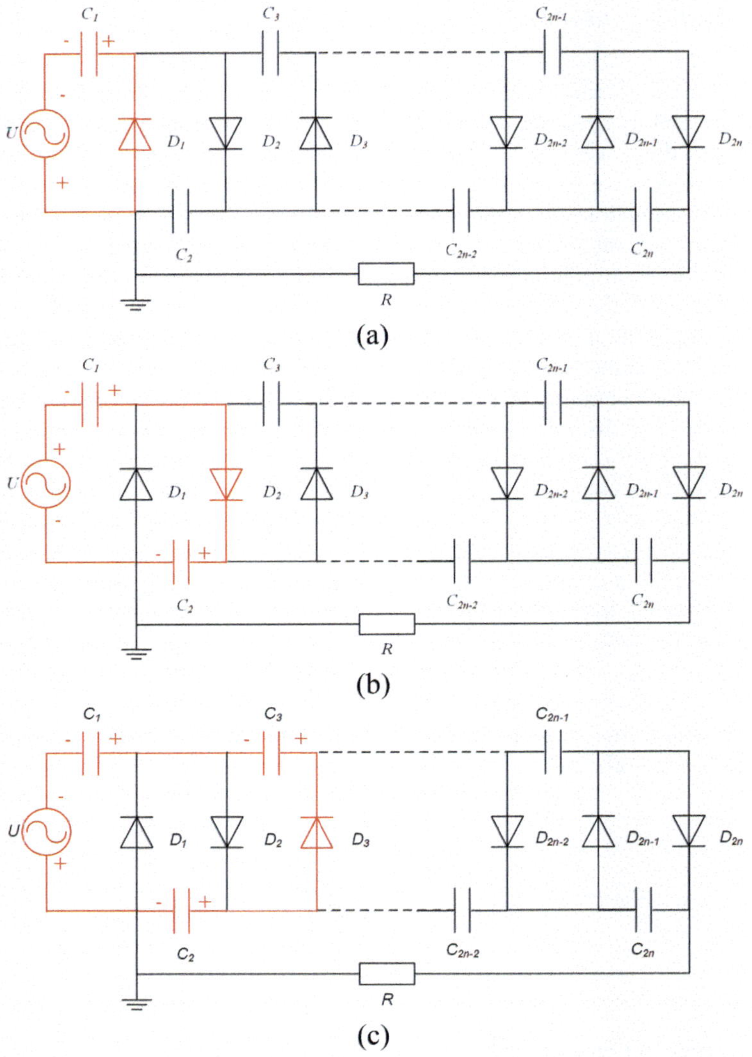

Fig. 1.3 Multi-stage voltage doubling rectifier circuit topology: **a** first negative half-cycle state, **b** first positive half-cycle state, **c** second cycle state

the total ripple voltage of the output circuit is:

$$\delta V_{total} = \frac{n(n+1)I}{2fC} \tag{1.6}$$

From Eq. (1.5), the pressure drop across C_{2n} is:

$$\Delta U_{2n} = \frac{nI}{fC} \tag{1.7}$$

The total voltage drop is:

$$\Delta U = \frac{I}{6fC}\left(4n^3 + 3n^2 - n\right) \tag{1.8}$$

When the high-frequency transformer wants to make a very high ratio, the process is very high, so the subsequent booster part uses a voltage doubling circuit. The transformer secondary output voltage is about 2 kV, taking into account the voltage drop, the voltage boost to ±5 kV requires at least 4 levels of voltage doubling circuit. The circuit adopted in this experiment is shown in Fig. 1.2.

If there is no load, the maximum voltage that the capacitor and diode are subjected to is about 2.8 times the secondary output voltage, so the capacitor and diode should choose a device with a higher voltage value. The capacitor voltage is selected by 50% of the input AC voltage, because the peak value of AC is about 1.4 times the effective value, and the maximum voltage of the general capacitor is about 70% of the withstand voltage value. For the size of the capacitor value depends on the size of the load current of the voltage doubling circuit, the larger the output current, the larger the capacity needs to be, otherwise it can not provide enough output current, but the larger the capacitor charging time is longer, the slower the voltage rise. The design of the actual selection of capacitors are 150 pF/3 kV, 100 pF/3 kV, 82 pF/3 kV three kinds of high-voltage capacitors, diode selection of 4 kV high-voltage diode 2CL2FD, which can fully meet the needs of the circuit.

1.4 Adjustment Control Method

1.4.1 Improved Buck Circuit Operating Principle

The Buck circuit converts the DC voltage provided by the input power supply into an adjustable low voltage output to meet the power supply requirements of different circuits. The improved Buck circuit topology is shown in Fig. 1.4. Conventional Buck circuits need to control the gate and source voltages of MOSFET to control on and off, while the improved Buck circuits ground the source directly, so that the drive signal from the gate input is more effective in controlling the switching of the MOSFET.

The model of the improved Buck circuit is shown in Fig. 1.5. When the MOSFET is on, the load is supplied by the power supply and charged by C and L. When the MOSFET is off, the load is powered by L and C. According to the principle of volt-second balancing there is:

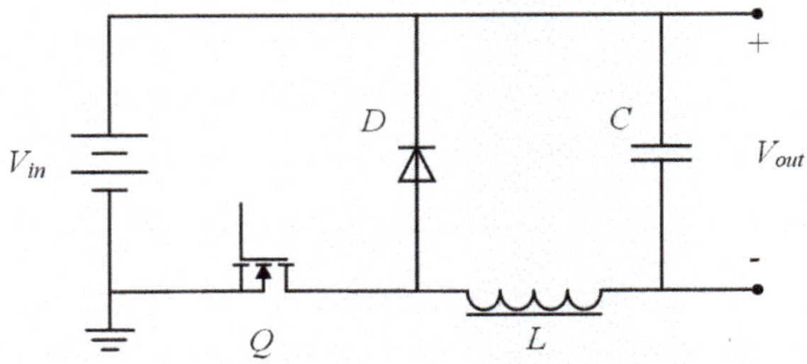

Fig. 1.4 Improved buck circuit topology

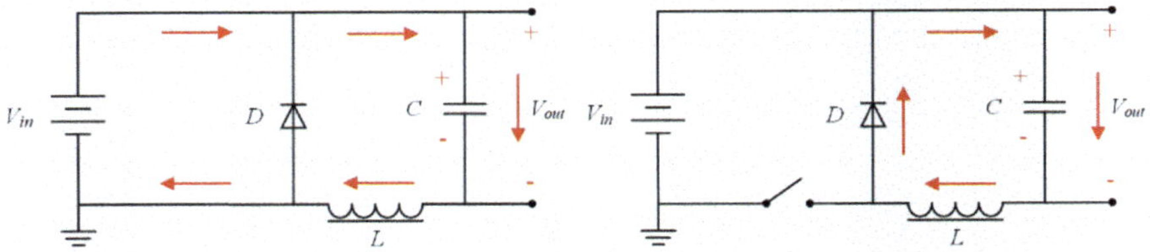

Fig. 1.5 Equivalent circuits of each operation stages: **a** Mode I, MOSFET on, **b** Mode II, MOSFET off

$$V_{in} - V_{out} \cdot DT = V_{out} \cdot (1 - D)T \tag{1.9}$$

where D is the MOSFET conduction duty cycle and T is the period. The final result is:

$$V_{out} = D \cdot V_{in} \tag{1.10}$$

1.4.2 Digital PWM Generation

PWM is the core of driving pulsed high-voltage power supply, and its main role is to compare the input feedback signal with the given value, modulate the output of a certain duty cycle of the pulse signal, and adjust the output of 0–9 V voltage signal to supply the Royer oscillation circuit to control and adjust the output of the high-voltage power supply.

The STM32F103CBT6 comes with an internal PWM generation module, and the operating mode is selected by setting the control registers in the PWM time base. Set PWM1 and PWM2 to work in continuous incremental counting mode, which are used as positive and negative high voltage drive signals. When working in auto-balance mode, the duty cycle register is constantly updated and its value is calculated by the feedback electricity through the PI control module, the PWM program flow chart is shown in Fig. 1.6.

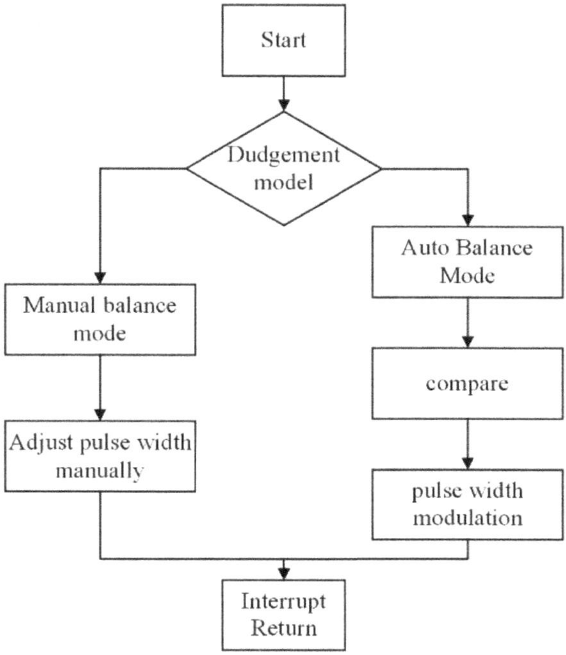

Fig. 1.6 PWM program flow chart

Positive and negative high voltages need to be turned on alternately to set a dead time for the PWM1 and PWM2 signals of the complementary outputs. In this design, the first dead time is inserted when PWM1 is off and PWM2 is on. The second dead time is inserted when PWM2 is off and PWM1 is on. Figure 1.7 shows the schematic diagram of 2-way PWM generation.

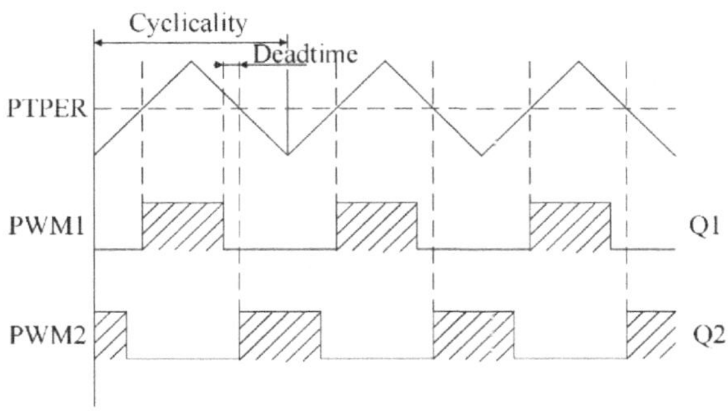

Fig. 1.7 PWM generation schematic

1.4.3 Incremental PI Adjustment

In order to ensure that the corona ionizer is in a balanced state for a long time, the amount of positive and negative ions in the air is monitored in real time to adjust the high-voltage amplitude, so that the amount of positive and negative ions output is as equal as possible. As shown in the Fig. 1.8, I_{in} is the reference value, I_f is the measured ion current, e is the deviation between I_{in} and I_f, the actuator is the PWM driver module, and the control object is the MOSFET in the Buck circuit, and the working principle is as follows: the detected I_f is compared with I_{in} to obtain e, and then e is used as the input, and then the PI controller outputs the PWM pulse signal to adjust the opening and closing of the MOSFET. Adjust the turn-on and turn-off time of the MOSFET to control the ion current.

The mathematical model of the PI controller is:

$$u = K_P\left(e + \frac{1}{T_I}\int_0^t edt\right) \tag{1.11}$$

where: K_P is the proportionality coefficient and T_I is the integration time parameter.

In the digital PI controller, the formula is discretized to obtain:

$$\Delta u_k = u_k - u_{k-1} = K_P(e_k - e_{k-1}) + T_I e_k \tag{1.12}$$

Equation (1.12) is the recursive relational equation for writing the digital PI controller program, and the digital PI subroutine flowchart is shown in Fig. 1.9.

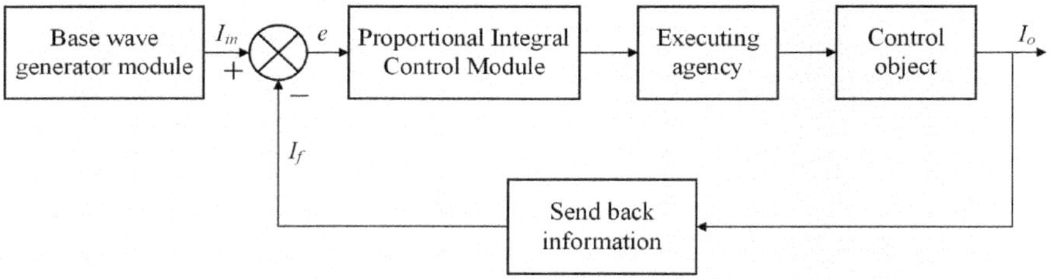

Fig. 1.8 PI control flowchart

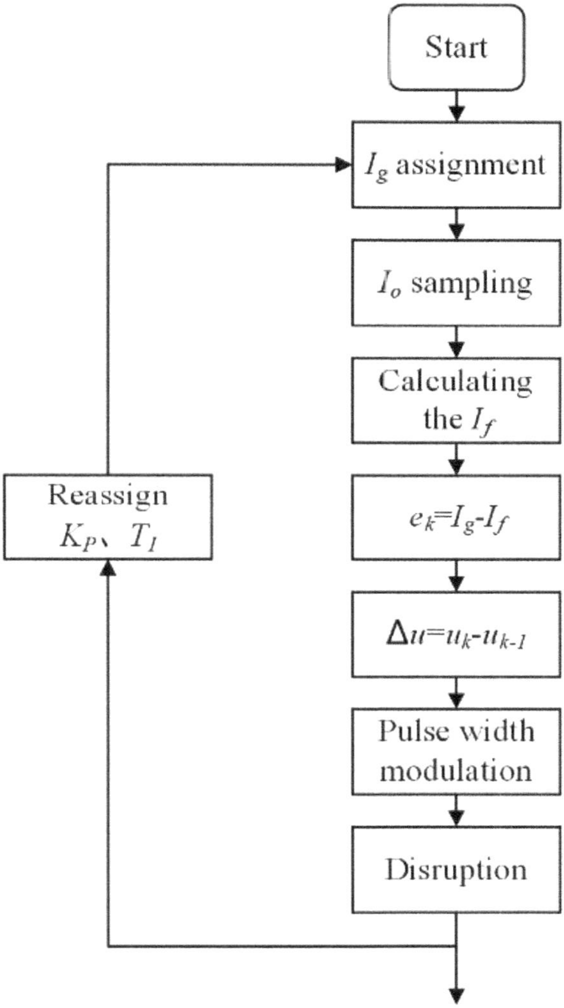

Fig. 1.9 Digital PI subroutine flowchart

1.5 Experiment

1.5.1 Pulsed AC High Voltage Power Supply Test

The pulsed high voltage power supply platform was built and installed into the ionizer housing, and the built test platform is shown in Fig. 1.10. Measurements were made on the output of the doubling and rectifying stages of positive high voltage and the final output of pulsed high voltage using a high voltage attenuation bar and an oscilloscope. When the power supply was operated at a frequency of 50 Hz and a duty cycle of

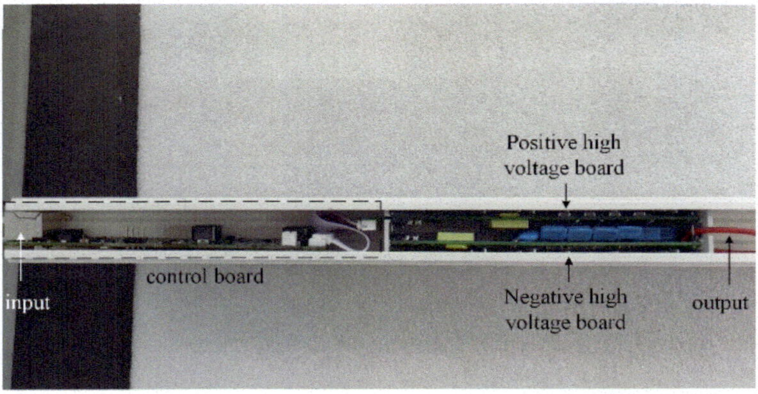

Fig. 1.10 Physical drawing of ionizer assembly

45%, the measured waveforms are shown in Fig. 1.11. Figure 1.11a for the voltage doubling rectifier circuit input voltage, amplitude of 2 kV; Fig. 1.11b for the first-order doubling rectifier circuit output, its amplitude of 3.76 kV; Fig. 1.11c for the second-order doubling rectifier circuit output, its amplitude of 4.72 kV; Fig. 1.11d for the third-order doubling rectifier circuit output, its amplitude of 5.52 kV; Fig. 1.11e for the fourth-order doubling rectifier circuit output, its amplitude of 6.64 kV; Fig. 1.11f shows the positive and negative high-voltage outputs, whose peak-to-peak value is 9.44 kV.

1.5.2 Performance Tests of Ionizer

The experimental platform mainly includes the prepared corona ionizer, the voltage-regulating power supply, the gas supply device and the measuring device. Figure 1.12 shows the CPM test configuration with the proposed ionizer, and the CPM plate stands vertically on the floor.

In the CPM test, the decay times of both polarities and the offset voltage were measured ten times for identical test conditions. The average value of the ten data samples for the identical test conditions was chosen as the final test result. Since the firmware of the used commercial CPM provides the value of the voltage measured after 30 s as the offset voltage, the average value of ten offset voltages is equal to the average value of the voltages measured for 5 min.

When the ionizer works at the frequency of 50 Hz, and the airflow pressure of 0.1 MPa, the CPM test result curve for the ionizer is shown in Fig. 1.13. As can be seen from the diagram, When the ionizer works at distance of 30 cm, the decay time is less than 1.4 s and the offset voltage is less than ± 5 V. The developed corona ionizer eliminates static electricity quickly and has a low offset voltage.

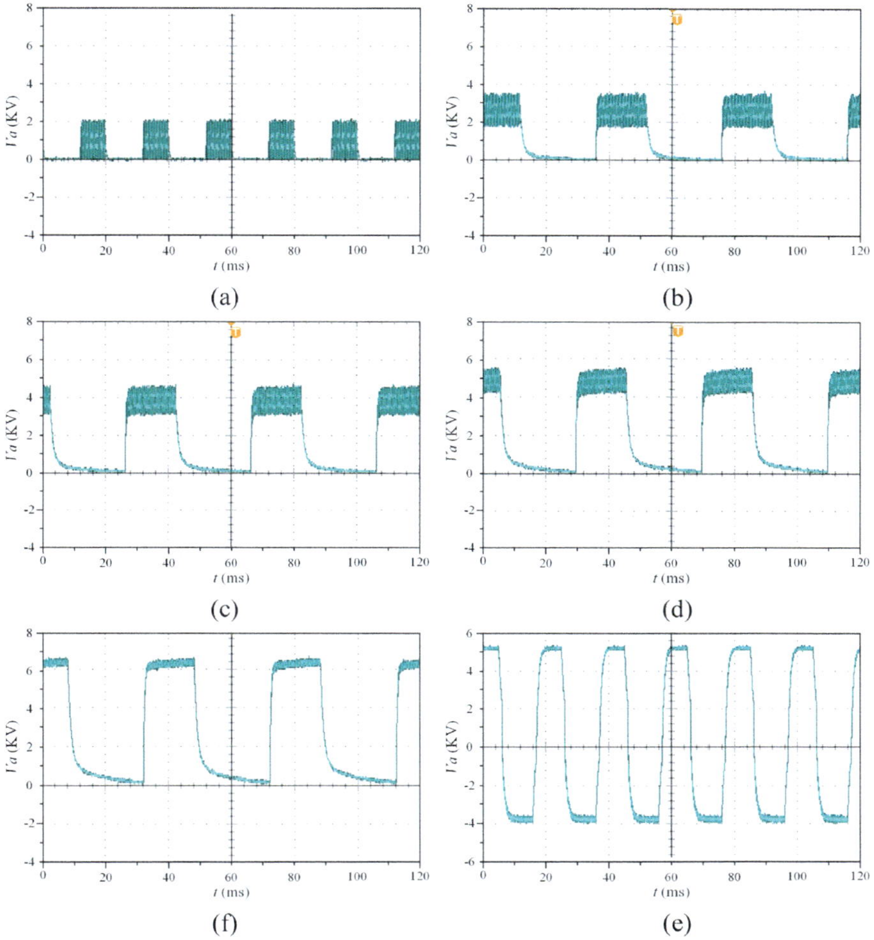

Fig. 1.11 Pulse high voltage power supply waveform: **a** multiplier rectifier circuit input for positive high voltage, **b** output of first-order doubling rectifier circuit, **c** output of second-order doubling rectifier circuit, **d** output of third-order voltage doubling rectifier circuit, **e** output of fourth-order voltage-doubling rectifier circuit, **f** high voltage output waveform

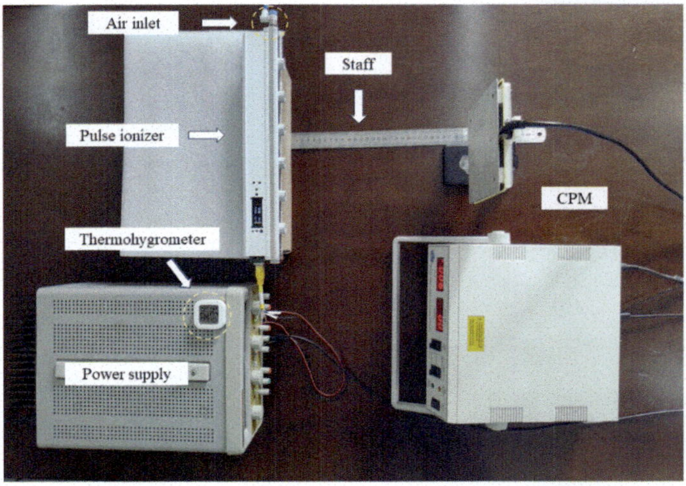

Fig. 1.12 CPM test setup for the proposed plasma ionizer with a vertical orientation (top view)

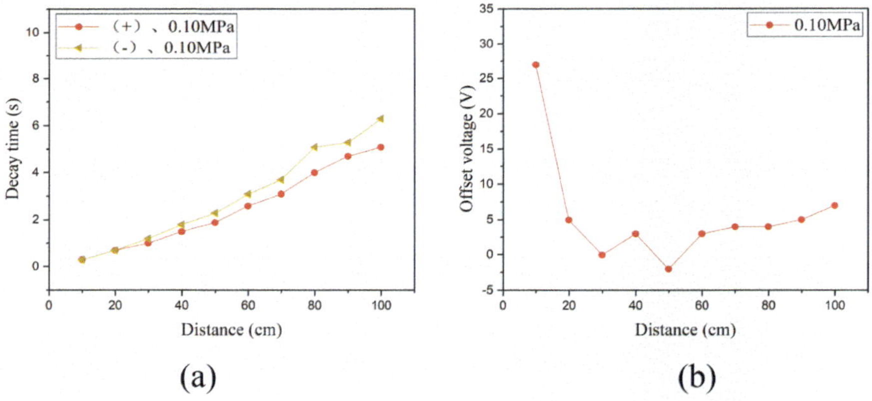

Fig. 1.13 Graphs plotting the CPM test results with the ionizer operating at 50 Hz and the airflow pressure of 0.1 MPa: **a** decay time, **b** offset voltage

1.6 Conclusion

In order to improve the speed and ability of static elimination of corona ionizer, this paper designs a pulse AC high-voltage ionizer that can automatically balance positive and negative ions, and the hardware and software structure design of the internal high-voltage power supply of the ionizer is described in detail. The PI control method is adopted to ensure that the ionizer works in a balanced state. The experimental platform of the corona ionizer was built for testing, and the main research results and conclusions are as follows:

(1) Adopting quadruple-voltage rectifier circuit for boosting, it realizes that the peak value of positive high-voltage output reaches 6.64 kV and the peak value of high-voltage output reaches 9.94 kV.
(2) The incremental PI adjustment has simple structure, good stability and easy adjustment, which can better eliminate the steady state error and make the ionizer work in a balanced state.
(3) When the ionizer works at a frequency of 50 Hz, a distance of 30 cm, and an air pressure of 0.1 MPa, the electrostatic elimination time is < 1.4 s, and the residual voltage is < ±5 V. The developed ionizer has a short dissipation time and a high degree of ionic balance.

Funding The Natural Science Foundation of Guangdong Province (2021A1515011773); The Science and Technology Program Project of Guangzhou (2023B03J1329).

References

1. T. Sugimoto, K. Furutachi, Y. Higashiyama, in *An Electrostatic Field Driven Ionizer for a Moving Charged Target*. 2014 IEEE Industry Application Society Annual Meeting (2014)
2. S. Plong-ngooluam, N. Jindapetch, P. Wounchoum, et al., A finite element analysis of multiple ion receiving plates for ionizer balance monitoring. J. Electrostat. **86**, 50–58 (2017)
3. S. Plong-ngooluam, N. Jindapetch, P. Wounchoum, et al., A proof-of-concept study demonstrating a multi-plate ion balance analyzer. Sens. Actuat. A: Phys. **257**, 118–124 (2017)
4. S. Cooray, T. Sato, T. Terashige, et al., in *Basic Characteristics of the Field Assisted Air Ionizer*. 2018 40th electrical overstress/electrostatic discharge symposium (EOS/ESD) (2018)
5. D. Nemoto, W. He, T. Matsuo et al., Static elimination performance of a corona ionizer in nitrogen-based electronegative gases. IEEE Trans. Ind. Appl. **52**(5), 4345–4350 (2016)
6. K. Takahashi, K. Takaki, I. Hiyoshi, et al., Development of a corona discharge ionizer utilizing high-voltage AC power supply driven by PWM inverter for highly efficient electrostatic elimination. Modern Appl. Electrostat. Dielect. (2019)
7. T.I.M. Okano, T. Terashige, An equivalent circuit simulation of an AC corona discharge air ionizer-frequency dependence of neutralization performance. Int. J. Plas. Environ. Sci. Tech. **13** (2019)
8. S. Toshiyuki, F. Kazutoshi, H. Yoshio, An electrostatic field-driven ionizer for a moving charged target. IEEE Trans. Ind. Appl. **51**(5), 4139–4144 (2015)
9. Z. Qun, L. Liangliang, B. Xiaotao, et al., Electrostatic elimination of charged particles by DC-type bipolar electrostatic eliminator. Powder Tech. **408** (2022)
10. W. Ronggang, H. Liuliang, W. Xiaoe, in *Experimental Study on Humidity Sensitivity of New DC Ionization*. 2022 Asia-Pacific International Symposium on Electromagnetic Compatibility (APEMC) (2022)
11. K. Takahashi, A. Goto, S. Yamaguchi, et al., in *Development of a Perfectly Balanced Electrostatic Eliminator Utilizing an Intermittent Pulse AC Voltage Power Supply*. 2015 37th Electrical Overstress/Electrostatic Discharge Symposium (EOS/ESD) (2015)

12. B. Byungjin, K. Jingook, A low-voltage microwave plasma ionizer with reduced electric overstress due to displacement current coupling from a high-ac voltage source. IEEE Trans. Electromagn. Compat. **63**(5), 1501–1511 (2021)
13. S. Seidl, F. Nieden, R. Gaertner, in *Study of the Discharge Current Created by an Ionizer*. 2018 40th Electrical Overstress/Electrostatic Discharge Symposium (EOS/ESD) (2018)

Chapter 2
Characteristic Analysis of Intermediate Frequency Vacuum Arc Based on Stereo Vision Technology

Yuan Jiang, Qian Zhang, Suliang Ma, Jianwen Wu, and Shangwen Xia

Abstract With the increasing demand of electric power for more-electric aircraft, current breaking and protection in intermediate frequency (360–800 Hz) AC power system becomes more and more difficult. Different from the case of power frequency at 50 Hz, the arc time is shortened due to the increase of current frequency, so that the anode spot is not significant when the intermediate frequency vacuum arc is generated. Therefore, a large number of metal droplets produced by contacts are important sources of metal vapor at post-arc stage, and the metal vapor will have a great influence on the recovery strength of dielectric. During the splashing process, the metal droplet continues to evaporate, and the Cu mass fraction on the surface of the droplet decreases from 65 to 10%. Stereo vision technology is used to calculate the velocity and acceleration of droplets sprayed by contacts in three-dimensional space, so as to obtain the pressure distribution in the intermediate frequency vacuum arc chamber during the process of post-arc breakdown. According to the parallax

Y. Jiang (✉)
Key Laboratory of Knowledge Automation for Industrial Processes of Ministry of Education, School of Automation and Electrical Engineering, University of Science and Technology Beijing, Beijing 100083, China
e-mail: jiangyuan@ustb.edu.cn

Q. Zhang
International Student Center, University of Science and Technology Beijing, Beijing 100083, China
e-mail: jess121@163.com

S. Ma
Energy Storage Technology Engineering Research Center, North China University of Technology, Beijing 100144, China
e-mail: msl13811581885@ncut.edu.cn

J. Wu · S. Xia
School of Automation Science and Electrical Engineering, Beihang University, Beijing 100191, China
e-mail: wujianwen@buaa.edu.cn

S. Xia
e-mail: hellowho@126.com

Z. Fang et al. (eds.), *Proceedings of the 5th International Symposium on Plasma and Energy Conversion*, Springer Proceedings in Physics 398,
https://doi.org/10.1007/978-981-97-2245-7_2

principle, the velocity of the droplets is about 10 m/s, and the acceleration is 105 m/s^2, and the pressure gradient is 1.2 MPa/mm.

2.1 Introduction

Applying vacuum switches to the variable frequency (intermediate frequency 360–800 Hz) power system of more electric aircraft can solve the problems of arc extinguishing and limited breaking ability caused by the increase in current frequency. An in-depth analysis of the complex discharge process of intermediate frequency vacuum arcs is not only a cutting-edge research topic in electrical engineering, but also a key point in improving the performance of arc chambers [1–3].

The activity level of the vacuum arc anode determines the breaking ability of the switching device, especially when the current is high, the anode can actively emit metal vapor, plasma, and metal droplets. The main mechanism of anode erosion theories are the thermal explosion mechanics, magnetohydrodynamic theory, and thermal stress theory [4]. At present, the main methods for studying intermediate frequency vacuum arcs include experimental methods and simulation calculations, where simulation calculations can quantitatively describe the internal changes of the plasma. Compared with experimental methods, research costs are lower and the microscopic characteristics of the arc are relatively easy to obtain. The simulation technology for anode erosion has evolved from an early one-dimensional heat transfer model to a two-dimensional magnetohydrodynamic model, and is moving towards a three-dimensional transient model [4]. The key parameters required for arc simulation still need to be attained through experimental methods [5, 6], and accurate simulation parameters are the basis for acquiring high-precision simulation calculation results. Ref. [7] used non-contact spectroscopy to obtain electron temperatures ranging from 0.5 to 3 eV and electron density levels ranging from 10^{20} to 10^{21} m^{-3}, providing an important basis for the simulation of intermediate frequency vacuum arcs. The experiment in Ref. [8] captured the dynamic evolution process of the arc in the arc chamber, and verified the accuracy of the 3D simulation model based on the arc extinguishing time. Ref. [9] observed in the experiment that the breakdown area of the intermediate frequency vacuum arc after arc coincides with the high energy region of the anode edge ion acquired in the simulation, confirming the rationality of the simulation model. When simulating the DC fault arc resistance in Ref. [10], when the medium is only air, the calculation error is larger than the experimental measurement value. After adding copper vapor composition, the maximum deviation is about 13%.

Arc pressure is an important parameter in the discharge process, which is a combined macroscopic state manifested by the action of electric field, magnetic field, and thermal field on the plasma. It not only directly affects the formation of plasma jet [11], the diffusion and agglomeration of vacuum arc [12, 13], but also acts on the metal droplets generated by the contact, affecting the post arc breakdown process through jet motion [14]. Regarding the fault arc pressure inside the medium voltage switchgear, Ref. [15] proposed a simplified model method for calculating the internal pressure rise of the switchgear, which shortened the calculation time of arc pressure rise by nearly 55%. Ref. [16] studied the effect of SF_6 substitute gas on fault arc pressure, and found out the rate of arc pressure rise in the switchgear under the same energy was closely related to gas molecules and initial inflation pressure.

Because of the special environment of vacuum arc chambers, it is difficult to us sensors to measure arc pressure directly, therefore non-contact measurement methods are generally used. Under the assumption that the metal droplets sprayed after the arc are driven by arc pressure, Ref. [17] proposed an indirect method for calculating arc pressure. Arc images were used to obtain the coordinates of metal droplets at each sampling time and calculate arc pressure based on acceleration. However, this method can only obtain the pressure in a two-dimensional image level and lacks the third dimensional spatial information. To solve the above issues, a method for solving the three-dimensional arc pressure gradient based on droplet motion using stereoscopic arc image is established in this paper, and the influence of metal vapor on the post arc breakdown process is analyzed also.

2.2 Experimental System

Imitate the three-phase variable frequency power system of the Boeing B787 aircraft, with a power supply capacity of 1450 kVA and a phase voltage of 230 V. It is estimated that the maximum short-circuit current instantaneous value is 5 times the rated current [1]. In the experiment, the highest current level reaches 10 kA, which is difficult to directly obtain using power in the laboratory. Therefore, an intermediate frequency vacuum switch arc experimental platform is built in this paper, as shown in Fig. 2.1a. By adjusting the C_1 and L_1, the intermediate frequency current of 360–800 Hz required for the experiment is generated. A Rogowski coil is used to measure the current with a range of 30 kA and a bandwidth range of 0.2 Hz–1 MHz.

In order to facilitate the acquisition of stable arc images, the "arc striking" strategy is adopted. The arcing branch is composed of an electrolytic capacitor C_2, a current limiting resistor R_1, a thyristor T_2, and a power diode D_1. It can draw a DC arc between the vacuum switch contacts and automatically turn off after introducing intermediate frequency current. In addition, R_0 and C_0 form a frequency modulation circuit to regulate the transient recovery voltage (TRV) after current zero.

An axial magnetic vacuum arc chamber is used in this paper with a contact diameter of 41 mm, and the contact material is CuCr50 with an opening distance of 3 mm. For the convenience of capturing images, the metal shielding cover was removed and only the glass cover with an outer diameter of 100 mm was kept. Two high-speed cameras are used to capture arc images, with a shooting interval of 1/40000 s and an exposure time of 5 μs, which could simultaneously capture and observe the intermediate frequency vacuum arc through stereo vision, as shown in Fig. 2.1b.

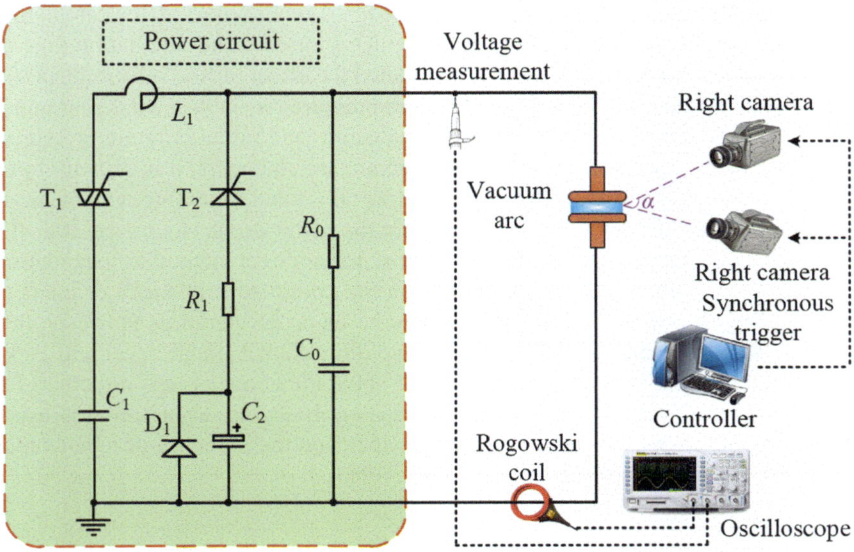

(a) Experimental system of intermediate frequency vacuum arc

(b) Stereoscopic shooting for intermediate vacuum arc

Fig. 2.1 Intermediate frequency vacuum arc experiment system with stereoscopic view

2.3 Experimental Phenomena

2.3.1 *Extinguishing of Intermediate Frequency Vacuum Arc*

Figure 2.2 shows the voltage, current, and arc image of the intermediate frequency vacuum arc, with a peak current of approximately 12 kA.

The changes in the images captured by the left and right cameras are basically consistent. As the current approaches zero, the arc column area of the intermediate frequency vacuum arc gradually decreases, and the cathode spots are always uniformly distributed on the electrode surface. After the current passes zero, the arc extinguishes, and then the arc voltage undergoes an oscillation with a frequency of 10 kHz. The oscillation time is about 0.5 ms, and the oscillation peak is close to −400 V. The final voltage stabilizes at about −250 V, and no post-arc breakdown occurs.

2.3.2 *Post-arc Breakdown of Intermediate Frequency Vacuum Arc*

Figure 2.3 shows the voltage, current, and arc image of post-arc breakdown of intermediate frequency vacuum arc, with a peak current of approximately 15 kA.

Different from previous experiment (extinction after current zero), in this experiment, when the current passes zero, post arc breakdown happens, and the arc voltage undergoes high-frequency oscillation with a frequency of about 50 kHz, and the voltage recovers to steady state for about 2 ms. What's more, a large amount of metal droplets is observed splashing outward in the images captured by both the left and right cameras. The spray of metal droplets is driven by arc pressure, which affects the post arc breakdown process. Some of the metal droplets have very similar features in two images, and it can be concluded that they are projections of the same droplet in the left and right image planes, as marked in Fig. 2.3b.

2.4 Process Analysis of Post Arc of Intermediate Frequency Vacuum Arc

2.4.1 *Calculation of Arc Pressure Gradient*

According to references [11] and [17], when post arc breakdown occurs in an intermediate frequency vacuum interrupter, the relation between the movement of metal droplets and the pressure is as follows:

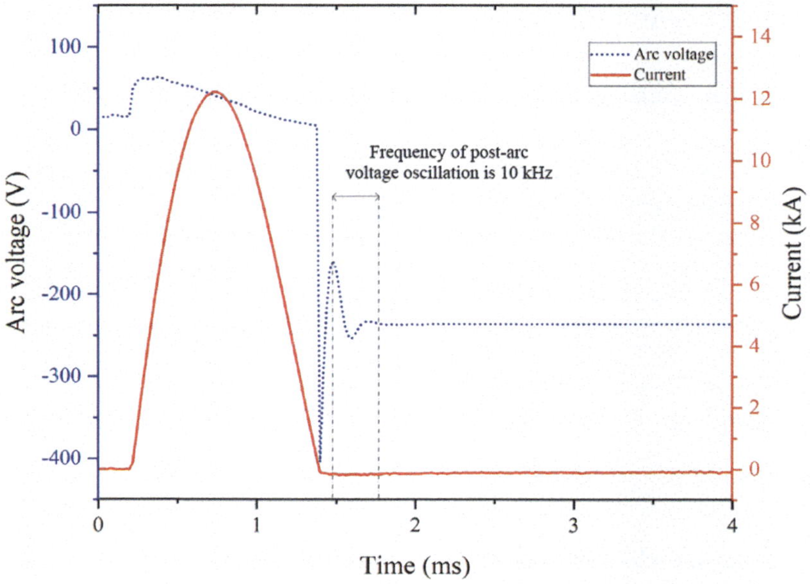

(a) Arc voltage and current

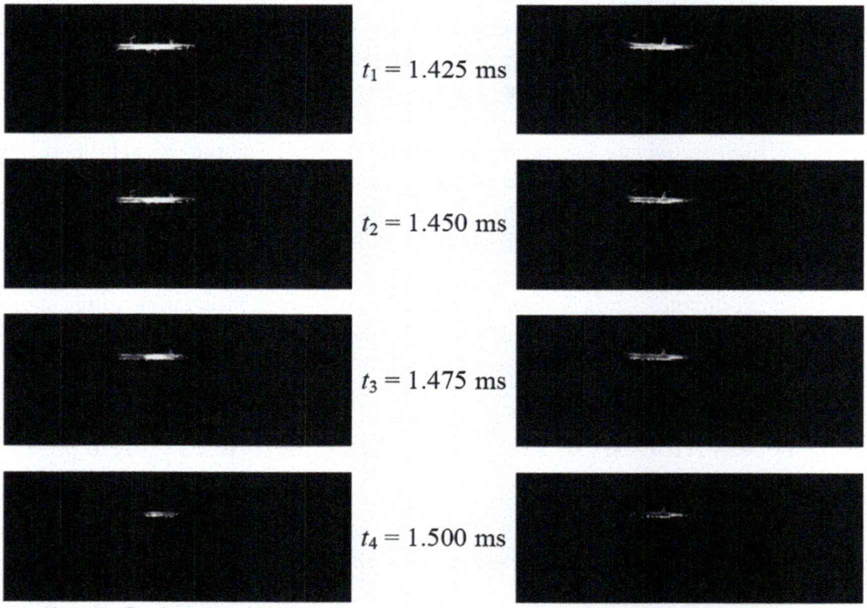

(b1) Left view arc image (b2) Right view arc image

(b) Process of arc extinguishing

Fig. 2.2 Arc voltage, current and arc image when intermediate frequency vacuum arc extinguishes after current zero

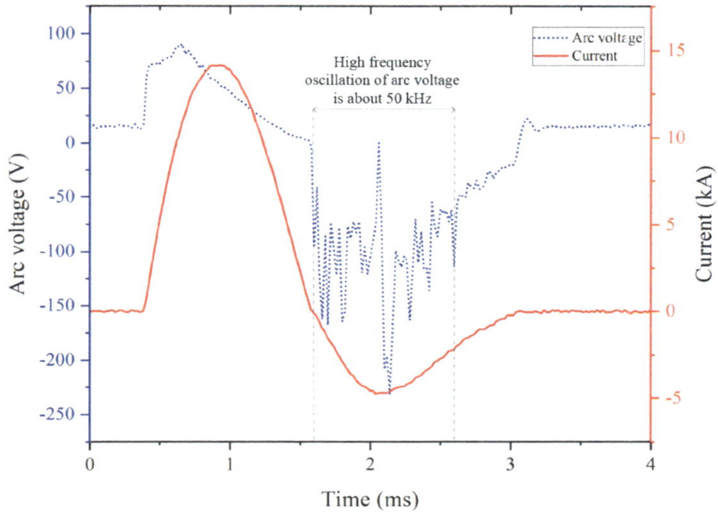

(a) Arc voltage and current

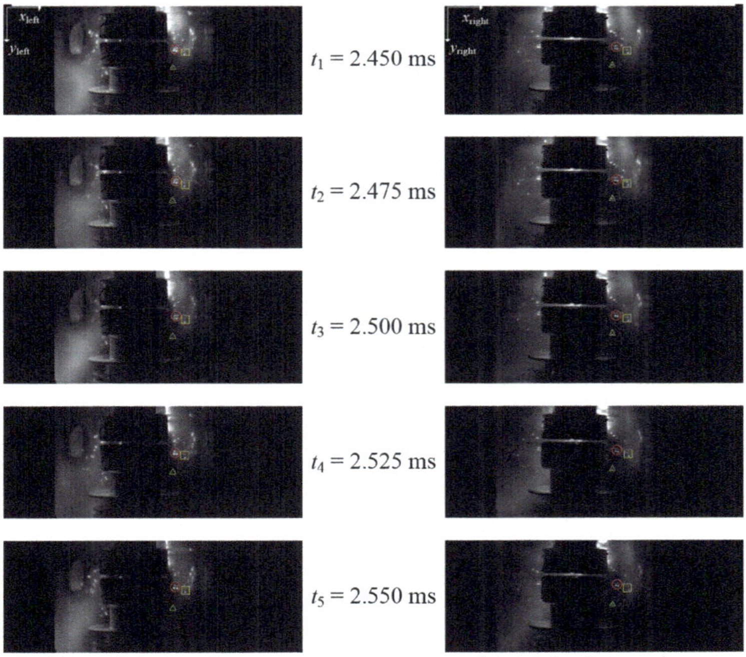

(b1) Left view arc image (b2) Right view arc image

(b) Process of arc voltage high frequency oscillation

Fig. 2.3 Arc voltage, current and arc image when post arc breakdown happens

$$ma = \boldsymbol{F} = \int_0^{2\pi} \mathrm{d}\theta \int_0^{\pi} r^2 \sin\varphi \cos\varphi(-\nabla P) r \cos\varphi \mathrm{d}\varphi = -\frac{4}{3}\pi r^3 \nabla P \quad (2.1)$$

Among them, m means the mass of the metal droplet; $\boldsymbol{a}$ stands for the acceleration of the metal droplet, which is a three-dimensional vector; $\boldsymbol{F}$ is the force acting on the metal droplet, which is also a three-dimensional vector; r means the radius of the metal droplet; θ means the zenith angle in a spherical coordinate system; φ represents the azimuth angle in the spherical coordinate system; ∇P stands for the pressure gradient along the direction of metal droplet movement, which is a three-dimensional vector. The arc pressure decreases outward from the center, while the droplets migrate from the inside of the arc to the outside, that is why a negative sign in the equation.

If the spherical coordinate system is converted to a Cartesian coordinate system, the acceleration $\boldsymbol{a}$ can be decomposed into the acceleration component a_x in the x-axis direction, the acceleration component a_y in the y-axis direction, as well as the acceleration component a_z in the z-axis direction. Similarly, the pressure gradient ∇P can be disassembled into pressure gradient ∇P_x in the x-axis direction, ∇P_y in the y-axis direction, and ∇P_y in the z-axis direction respectively.

Therefore, the following correlation can be obtained:

$$\begin{cases} \nabla P_x = -\dfrac{3ma_x}{4\pi r^3} = -\rho a_x \\ \nabla P_y = -\dfrac{3ma_y}{4\pi r^3} = -\rho a_y \\ \nabla P_z = -\dfrac{3ma_z}{4\pi r^3} = -\rho a_z \end{cases} \quad (2.2)$$

In the equation, ρ refers to the density of metal droplet. According to Eq. (2.2), when the droplet acceleration in a certain axis direction is measured, the pressure gradient of the intermediate frequency vacuum arc in that direction can be calculated.

2.4.2 Mapping of Arc Plane Coordinate System into Spatial Coordinate System

Figure 2.4 showcases an intermediate frequency vacuum arc by stereoscopic camera system. The axis of the dual camera forms an observation angle α. The observation is best when the angle α is between 0 and 90°, and in this experiment α is around 27°. The camera axis is naturally perpendicular to the respective imaging planes, and the arc image is formed on the left and right imaging planes respectively. In the left imaging plane, the left-perspective plane Cartesian coordinate system x_{left}-y_{left} is established, where the positive direction of x_{left} is horizontally to the right and the positive direction of the y_{left} axis is vertically downwards. Likewise, in the right imaging plane, the right-perspective plane Cartesian coordinate system x_{right}-y_{right} is established, where the positive direction of x_{right} is horizontally to the right and the positive direction of the y_{right} axis is vertically downwards. The acceleration of metal droplets in both the left and right planes is the projection of spatial acceleration $\boldsymbol{a}$ onto the two imaging planes.

For the convenience of calculating spatial acceleration, this paper makes the following provisions for the spatial coordinate system xyz: the x-axis is parallel to the left imaging plane and collinear with the x_{left} axis of the left imaging plane; the y-axis is perpendicular to the top view plane and the positive direction is perpendicular to the paper surface, facing inward, and in the same direction as the y_{left} axis and y_{right} axis; the positive direction of the z-axis is determined by the right-hand rule. According to the spatial geometric position relation, the acceleration components of the target droplet in the spatial coordinate system can be determined based on the accelerations of the metal droplet on the left and right imaging planes, as follows:

$$\begin{cases} a_{\text{x}} = a_{\text{x}_{\text{left}}} + a_{\text{x}_{\text{right}}} \cdot \cos\alpha \\ a_{\text{y}} = a_{\text{y}_{\text{left}}} = a_{\text{y}_{\text{right}}} \\ a_{\text{z}} = a_{\text{x}_{\text{right}}} \cdot \sin\alpha \end{cases} \tag{2.3}$$

$a_{\text{x}_{\text{left}}}$ represents the acceleration component of the metal droplet on the x_{left} axis in the left imaging plane; $a_{\text{y}_{\text{left}}}$ stands for the acceleration component of the metal droplet on the y_{left} axis in the left imaging plane; $a_{\text{x}_{\text{right}}}$ means the acceleration component of the metal droplet on the x_{right} axis in the right imaging plane; and $a_{\text{y}_{\text{right}}}$ is the acceleration component of the metal droplet on the right axis in the right imaging plane.

Select a target droplet, such as the droplet marked in Fig. 2.3. For the left angle arc imaging plane, at time t_1, t_2, and t_3, $x_{\text{left}}^{t_1}$, $x_{\text{left}}^{t_2}$, $x_{\text{left}}^{t_3}$ are obtained as abscissa and $y_{\text{left}}^{t_1}$, $y_{\text{left}}^{t_2}$, $y_{\text{left}}^{t_3}$ as ordinate respectively. According to the physical definition, the acceleration of the metal droplet on the left imaging plane can be calculated as:

$$\begin{cases} a_{\text{x}_{\text{left}}} = \dfrac{\frac{x_{\text{left}}^{t_3} - x_{\text{left}}^{t_2}}{t_3 - t_2} - \frac{x_{\text{left}}^{t_2} - x_{\text{left}}^{t_1}}{t_2 - t_1}}{\frac{t_3 - t_1}{2}} = \dfrac{x_{\text{left}}^{t_1} + x_{\text{left}}^{t_3} - 2x_{\text{left}}^{t_2}}{\Delta t^2} \\ a_{\text{y}_{\text{left}}} = \dfrac{\frac{y_{\text{left}}^{t_3} - y_{\text{left}}^{t_2}}{t_3 - t_2} - \frac{y_{\text{left}}^{t_2} - y_{\text{left}}^{t_1}}{t_2 - t_1}}{\frac{t_3 - t_1}{2}} = \dfrac{y_{\text{left}}^{t_1} + y_{\text{left}}^{t_3} - 2y_{\text{left}}^{t_2}}{\Delta t^2} \end{cases} \tag{2.4}$$

Since Δt stands for the image sampling interval, the acceleration of metal droplets on the right imaging plane can be attained as:

$$\begin{cases} a_{\text{x}_{\text{right}}} = \dfrac{x_{\text{right}}^{t_1} + x_{\text{right}}^{t_3} - 2x_{\text{right}}^{t_2}}{\Delta t^2} \\ a_{\text{y}_{\text{right}}} = \dfrac{y_{\text{right}}^{t_1} + y_{\text{right}}^{t_3} - 2y_{\text{right}}^{t_2}}{\Delta t^2} \end{cases} \tag{2.5}$$

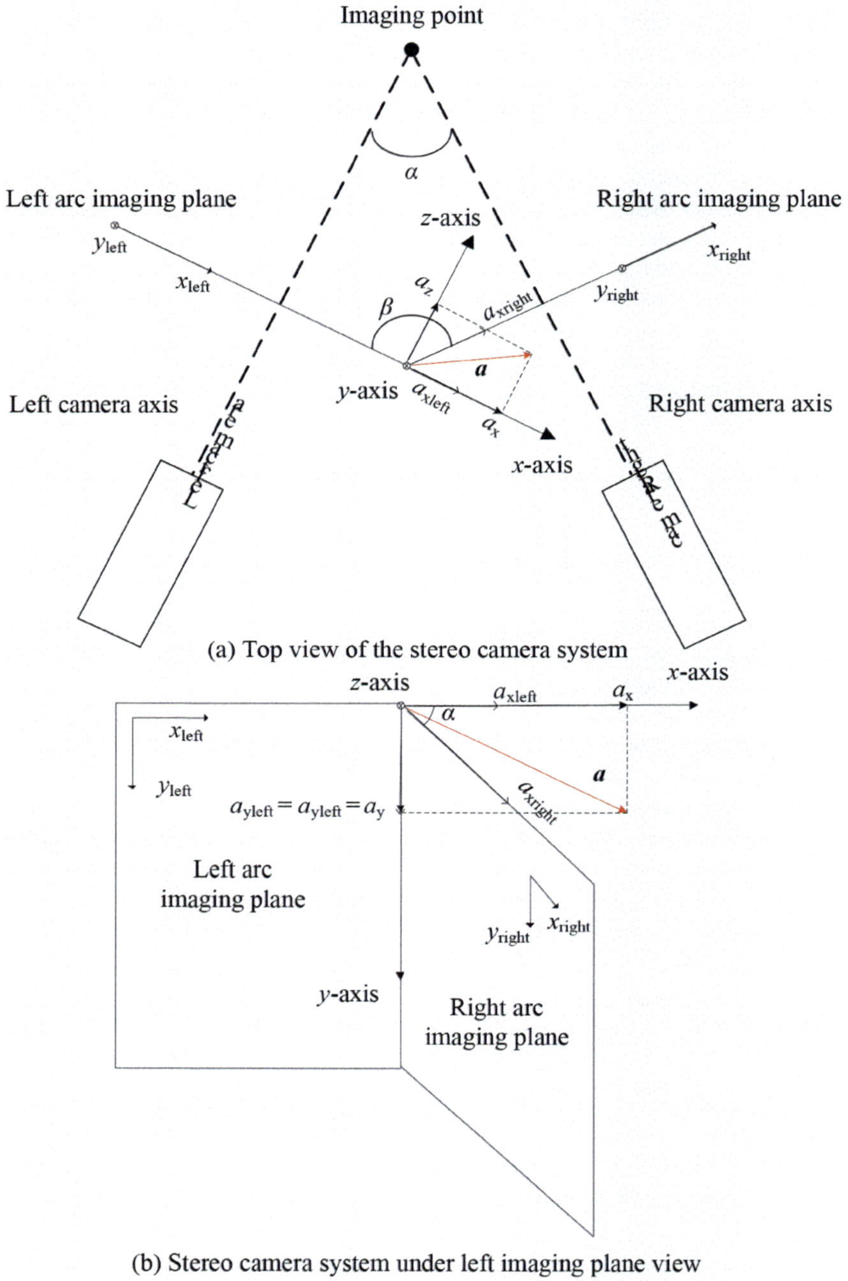

Fig. 2.4 Arc stereo camera system

2.4.3 Post Arc Pressure and Metal Evaporation of Intermediate Frequency Vacuum Arc

According to the spatial coordinate system *xyz* established in Figs. 2.4 and 2.5 shows the spatial motion path of the droplets, and it can be clearly observed that the droplets have displacement in all three-dimensional direction.

By using Eqs. (2.3) and (2.4), the acceleration components of each droplet on each axis in the spatial *xyz* coordinate system can be obtained, as displayed in Table 2.1. It can be noted that the amount of velocity and acceleration of each droplet is basically the same, indicating the credibility of the calculation and results. The occurrence of a droplet acceleration of 0 in the results is caused by the short image sampling interval and errors in reading coordinates.

The CuCr50 contact was used in this experiment, and due to the significant difference in melting points between Cu and Cr, the composition of the splashed metal droplets is unknown. The metal droplets eventually deposit on the inner glass wall of the arc chamber. After sampling the glass, scanning electron microscopy analysis was carried out, and the composition analysis results are illustrated in Fig. 2.6.

As per the analysis of speed, acceleration, and droplet density, taking the "metal droplet □" calibrated in Fig. 2.3b as an example, it can be calculated that the time for metal droplets to splash from the contact edge to the glass wall is about 2.8 ms, and the pressure gradient value in the arc extinguishing chamber during post arc breakdown is as below:

$$\nabla P = \sqrt{\nabla P_{\mathrm{x}}^2 + \nabla P_{\mathrm{y}}^2 + \nabla P_{\mathrm{z}}^2} = 1.2\ \mathrm{MPa/mm} \tag{2.6}$$

However, as a matter of fact, the above calculation is an ideal assumption for the actual situation. On one hand, the charge of the metal droplet is ignored, and the arc pressure is not the only driving force for the movement of the metal droplet; On the other hand, the pressure gradient in the arc chamber is not evenly-distributed. Despite these, the results still have significant reference value: when the intermediate frequency vacuum arc undergoes post arc breakdown, there is a large pressure gradient from the arc area to the outside, driving the metal droplets to spray outward, forming three-dimensional spatial motion, and accompanied by the motion, the droplets continue to release metal vapor. Assuming that the metal droplet is a symmetrical sphere with uniform mass distribution, according to the principles of diffusion and convection, the flow velocity q of vapor at any radius r outside the droplet is related to the mass fraction w_{Cu} of Cu vapor in the following way [14]:

$$\int \frac{4\pi D\rho}{w_{\mathrm{Cu}} - 1} \mathrm{d}w_{\mathrm{Cu}} = \int \frac{q}{r^2} \mathrm{d}r \tag{2.7}$$

Among them, D means the diffusion coefficient; ρ means the density of gaseous Cu. According to the above formula, the density of Cu vapor around metal droplets can be calculated. Supposing the radius of the metal droplet r_0 is 0.5 mm, the results

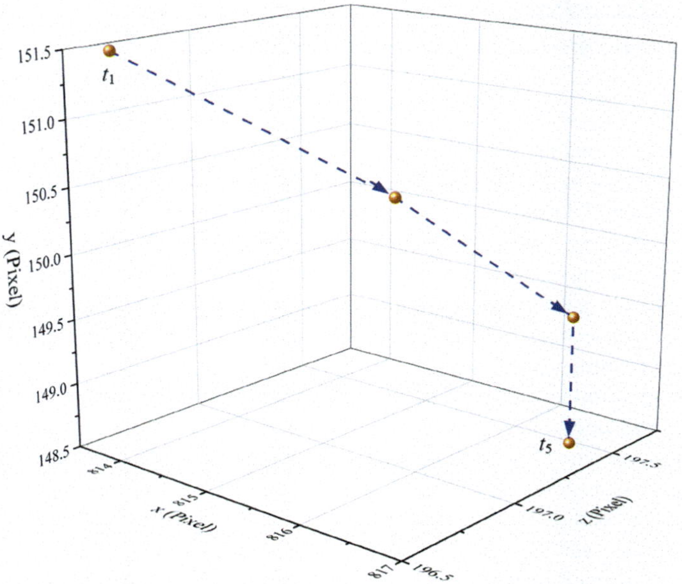

(a) Space motion trajectory of droplet 1

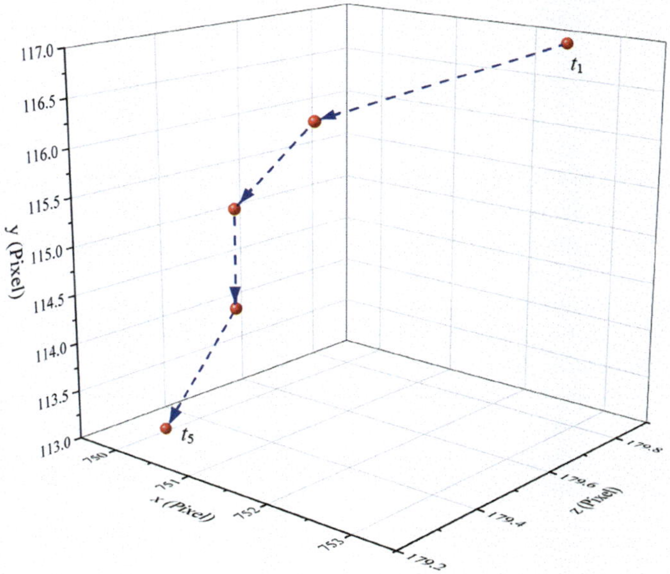

(b) Space motion trajectory of droplet 2

Fig. 2.5 The spatial motion trajectory of the droplet

Table 2.1 The acceleration component of metal droplets moving in space

Variable	Droplet 1	Droplet 2	Droplet 3
v_x/(m/s)	4.7	7.0	−7.2
v_y/(m/s)	−5.0	−7.5	−10.0
v_z/(m/s)	1.1	2.3	−1.1
v/(m/s)	7.0	10.5	12.4
a_x/(10^5 m/s^2)	1.9	−1.0	0.9
a_y/(10^5 m/s^2)	0	−1.0	−1.0
a_z/(10^5 m/s^2)	0.4	−0.5	0
a/(10^5 m/s^2)	1.9	1.5	1.3

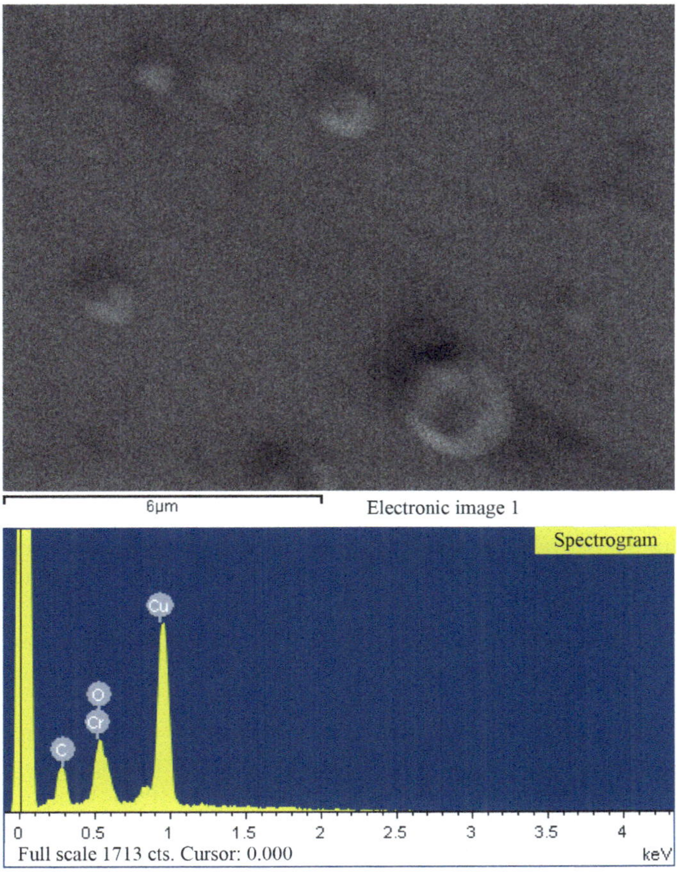

Fig. 2.6 SEM results of metal droplet composition

are revealed in Fig. 2.7a (Given the symmetry, only 1/4 of the plane distribution of the sphere is drawn). The maximum density of Cu vapor on the surface of the droplet can reach 2.2×10^{19} m^{-3}.

Figure 2.7b shows the mass change of Cu vapor around the metal droplet during the evaporation. When the droplet begins to evaporate, the radius r_0 is set to be 0.5 mm. For the initial state, the mass fraction of Cu vapor on the surface of the droplet (the distance from the center is the droplet radius r_0) is the highest, close to 65%; At a distance of 5 times r_0 from the center, the mass fraction of Cu vapor decreases to below 20%. As the evaporation goes on, the radius of the metal droplet continues to drop, and the Cu mass fraction on the droplet surface also declines. When the radius goes down to 0.25 mm, the Cu mass fraction on the droplet surface also drops to 40%. When the radius further descends to 0.05 mm, the Cu mass fraction on the droplet surface becomes only about 10%. During the 2.8 ms movement period of the metal droplet, a large amount of Cu vapor enters the arc chamber through diffusion and convection, weakening the recovery strength of the post arc medium. During this period, post arc breakdown and high-frequency voltage oscillation occur.

2.5 Conclusion

The post arc breakdown process of intermediate frequency vacuum arc is investigated in this paper and the following conclusions are drawn.

It can be seen from the experimental results, when the intermediate frequency current passes zero and post arc breakdown happens, the arc voltage undergoes high-frequency oscillation with a frequency of about 50 kHz. The time for the voltage to return to steady state is about 2 ms, and a large number of metal droplets are observed in the stereo view arc images during the post arc period.

According to the stereo visual analysis of the arc image, the splashing of metal droplets during post arc breakdown is a three-dimensional movement, driven by the arc pressure gradient. The pressure gradient value in the arc extinguishing chamber can reach 1.2 MPa/mm. The acceleration in three directions reaches the order of magnitude of 10^5 m/s^2, and the splashing speed is 10 m/s. The time for the metal droplets to reach the inner wall of the arc chamber glass from the contact edge is about 2.8 ms.

The Cu vapor density on the surface of metal droplets is 2.2×10^{19} m^{-3}. During a splashing period of 2.8 ms, the metal droplets continuously evaporate, and the Cu mass fraction on the droplet surface declines from 65 to 10%. A large amount of Cu vapor enters the arc chamber in the form of diffusion and convection, weakening the dielectric recovery strength. During this period, post arc breakdown and high-frequency voltage oscillation occur.

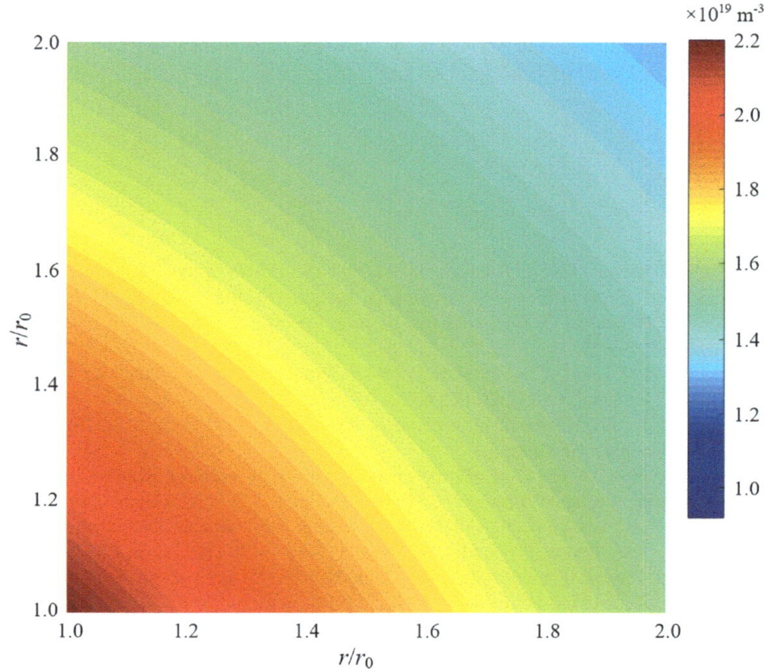

(a) Cu vapor density around metal droplets

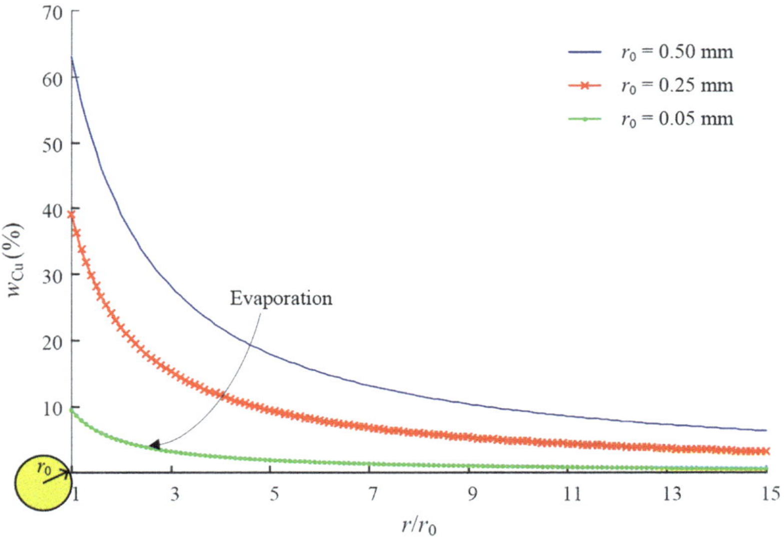

(b) Mass change of Cu vapor around metal droplet during evaporation

Fig. 2.7 The process by which metal droplets evaporate to release Cu vapor

Acknowledgements This work is supported by National Natural Science Foundation of China (52177127), Guangdong Basic and Applied Basic Research Foundation (No. 2020A1515110725), and Aviation Science Fund (2020Z025074001).

References

1. Y. Jiang, Q. Li, *Vacuum Circuit Breaker for Aviation Variable Frequency Power System: Theory and Application of Arc in Electrical Apparatus*, 1st edn. (Springer, Singapore, 2021)
2. G. Buticchi, P. Wheeler, D. Boroyevich, The more-electric aircraft and beyond. Proc. IEEE **111**(4), 356–370 (2023)
3. Y. Jiang, J. Wu, Q. Li et al., Characteristics of intermediate frequency vacuum arc in butt contact with short gap. Proceed. CSEE **43**(16), 6517–6525 (2023). (in Chinese)
4. M. He, C. He, L. Jin, Research on erosion of gas spark switch electrodes. High Volt. Eng. **30**(10), 54–55 (2004). (in Chinese)
5. X. Cheng, H. Chen, G. Ge et al., Probe array diagnosis of vacuum circuit breaker post-arc sheath development process. Proceed. CSEE **43**(8), 3227–3237 (2023). (in Chinese)
6. C. Huang, X. Liu, Y. Cao et al., Analysis of vacuum arc characteristics with double-break DC-VCB on asynchronous. Proceed. CSEE **42**(9), 3480–3490 (2022). (in Chinese)
7. J. Wang, J. Wu, Plasma characteristics of intermediate-frequency vacuum arc. Proceed. CSEE **31**(36), 145–152 (2011). (in Chinese)
8. W. Jia, W. Sima, T. Yuan et al., 3D simulation and experiment research on arc motion characteristics in the semi-enclosed arc-extinguishing chamber. Trans. China Electrotech. Soc. **36**(S1), 321–329 (2021). (in Chinese)
9. Y. Jiang, S. Ma, J. Wu, et al., Experiment and simulation for edge breakdown in intermediate frequency vacuum arc. Trans. China Electrotech. Soc., 1–9 (2023) (in Chinese)
10. Y. Zhong, L. Xiong, Z. Yang et al., Numerical simulation and experiment of small current DC arc considering copper vapor medium. Trans. China Electrotech. Soc. **35**(13), 2913–2921 (2020). (in Chinese)
11. Y. Jiang, J. Wu, S. Ma et al., Appearance of vacuum arcs in axial magnetic field and butt contacts at intermediate frequencies. IEEE Trans. Plasma Sci. **47**(2), 1405–1412 (2019)
12. H.C. Miller, Anode modes in vacuum arcs: Update. IEEE Trans. Plasma Sci. **45**(8), 2366–2374 (2017)
13. L. Wang, X. Huang, X. Zhang et al., Modeling and simulation of high-current vacuum arc considering the micro process of anode vapor. J. Phys. D Appl. Phys. **50**(9), 265201 (2017)
14. Y. Jiang, J. Wu, Q. Li et al., Influence of metal vapor on post-arc breakdown for intermediate frequency vacuum arc. Vacuum **193**(11), 110551 (2021)
15. T. Wu, D. Yang, P. Li, et al., Study on the model simplification method for the calculation of arc pressure rise in mv switchgear. High Volt. Apparat. **56**(3), 39–45+53 (2020) (in Chinese)
16. J. Zhang, D. Yuan, M. Li et al., Influence of insulating gas on pressure rise in switch cabinet due to internal arc fault. High Volt. Apparat. **53**(8), 100–104 (2017). (in Chinese)
17. Y. Jiang, Y. Liu, Q. Li et al., Interruption and droplets emission in intermediate-frequency vacuum arc of aviation power supply system. Proceed. CSEE **40**(2), 684–692 (2020). (in Chinese)

Chapter 3
The Effect of Various Coil Parameters on Inductively Coupled Plasma Reactor: A Numerical Simulation

Kunyi Li, Zhengquan Zhang, Lei Xu, Yukai Wang, and Qingxiang Liu

Abstract The study of the profiles of temperature and flow field in inductively coupled plasma under varying coil parameters can serve as a valid reference for the design of inductively coupled plasma devices used for the subject of carbon dioxide. First, a two-dimensional axisymmetric model of inductively coupled plasma (ICP) is established. Then, based on the local thermodynamic equilibrium conditions, a multi-field coupling approach including the flow field, the electromagnetic field, and the temperature field is used for the numerical simulation of a pure argon plasma with the COMSOL fine element simulation software and the effects of coil parameter variations on the spatial distribution of temperature and plasma density in the cavity were studied. The research results show that a suitable coil radius and spacing can significantly improve the uniformity of the plasma, while also affecting the core temperature of the plasma. Thus, the temperature and flow field distribution in the torch can be controlled by adjusting the parameters of each coil to meet the requirements of efficient treatment of carbon dioxide.

3.1 Introduction

In the plasma treatment of greenhouse gases, plasma parameter distribution has an important influence on the efficiency of greenhouse gas treatment, and it is very important to control the electron density distribution of plasma to improve the conversion efficiency [1, 2]. In recent decades, mathematical models for simulating the temperature and flow field distribution in ICP torches have evolved considerably. Initially, McKelliget et al. [3] used two-dimensional magnetic vector equations to solve the distribution of electromagnetic fields in space, and on this basis investigated in detail the electromagnetic, thermal, mass, and fluid flow phenomena from

K. Li (✉) · Z. Zhang · L. Xu · Y. Wang · Q. Liu
Southwest Jiaotong University, Chengdu 610031, Sichuan, China
e-mail: likunyi0406@foxmail.com

Z. Fang et al. (eds.), *Proceedings of the 5th International Symposium on Plasma and Energy Conversion*, Springer Proceedings in Physics 398,
https://doi.org/10.1007/978-981-97-2245-7_3

the perspective of coil geometry and coil current. Joshi et al. [4] performed a numerical simulation analysis of an inductively coupled plasma with a frequency of 3 MHz at atmospheric pressure using ANSYS software, and based on this they investigated the influence of different working gas flow rates on the properties of argon thermal plasma. Breznes et al. [5] simulated an argon plasma at a frequency of 6.78 MHz using COMSOL software and investigated the effects of different operating parameters on the distribution of the flow and temperature field within the ICP torch. The results were in good agreement with the experimental curves. Zhu et al. [6] established a strongly coupled mathematical and physical model of a three-dimensional high-frequency thermal plasma with thermal electromagnetic flux and used the software FLUENT to calculate and analyze the temperature and flux field distribution of the high-frequency thermal plasma.

Since the energy supply of the plasma discharge is achieved by effectively coupling the energy of electromagnetic waves with the plasma through a coil [7], changing the geometric size of the coil may have some influence on the characteristics of the plasma discharge. Therefore, it is particularly important to investigate the high-frequency plasma discharge laws of ICP discharge devices at different coil sizes. The reason for choosing the simulation method of finite element computation is that, on the one hand, if an experimental approach is chosen, changes in coil turns and coil radius require a large number of coils to be remanufactured, and problems in matching between the coils and the matching box may occur during the experimental process, which may affect normal discharge. On the other hand, it is difficult and costly to accurately determine the flow characteristics and temperature distribution of ICP by experimental means, and the finite element computational simulation can also eliminate the inevitable external disturbances in the experiment, so the simulation method is comprehensively considered.

Therefore, in this paper, a high-frequency inductively coupled plasma device for greenhouse gas processing is used as a research object and the finite element simulation software COMSOL is employed to perform calculations for coupling multiple physical fields. The influence of changes in coil parameters on the discharge effect of the inductively coupled plasma is investigated, which provides a guide for the redesign of plasma torches.

3.2 Model Description

In this article, a fluid model is used to study the discharge process of argon gas. To simplify the calculation, we assume that the inductively coupled plasma is a two-dimensional axisymmetric model, with an optically thin ionomer in the torch under local thermodynamic equilibrium conditions, and that the plasma is in a steady laminar flow state. The multiphysics field coupling model of an inductively coupled plasma is composed of electromagnetic field equations, Navier–Stokes equations for quasi-electrically neutral gases (including conservation of mass, conservation of momentum, and conservation of total energy), and models of chemical reaction kinetics of gases. Electron number density and average electron energy were calculated by solving the drift–diffusion equation and the charged particle electron energy conservation equation. The equations used in the model of coupling of several physical fields are as follows [7].

3.2.1 Electromagnetic Field Equation

The magnetic field interface is used to calculate the magnetic field and electric field distribution around the coil and plasma. The calculation process uses the magnetic vector potential as the dependent variable to solve the system of Maxwell's equations:

$$\nabla \times \boldsymbol{H} = \boldsymbol{J} \tag{3.1}$$

$$\boldsymbol{B} = \nabla \times \boldsymbol{A} \tag{3.2}$$

$$\boldsymbol{J} = \boldsymbol{\sigma E} + \boldsymbol{\sigma v} \times \boldsymbol{B} + \boldsymbol{J}_e + \boldsymbol{j\omega D} \tag{3.3}$$

$$\boldsymbol{E} = -\boldsymbol{j\omega A} \tag{3.4}$$

In the formula, E and H are the electric field strength and magnetic field strength respectively; B is the magnetic induction strength; D is the potential displacement vector; A is the magnetic vector potential; σ is the conductivity of plasma; and J is the total current density.

3.2.2 Flow Field and Temperature Field Equations

According to the general laws of fluid mechanics, all flow phenomena satisfy the fundamental equations of mass continuity and momentum. At the same time, the equations of conservation of energy must be considered, since there are heat transfer processes.

Mass continuity equation:

$$\nabla \cdot (\boldsymbol{\rho u}) = 0 \tag{3.5}$$

where, ρ is the gas density; u is the speed.

Momentum conservation equation:

$$\rho\left(\frac{\partial \boldsymbol{u}}{\partial t} + \boldsymbol{u} \cdot \nabla \boldsymbol{u}\right) = \nabla \cdot (-p\boldsymbol{I} + \boldsymbol{K}) + \boldsymbol{F} \tag{3.6}$$

where, $\boldsymbol{K} = (\mu + \mu_T)\left[\nabla u + (\nabla u)^T\right]$; μ is the dynamic viscosity of flow; μ_T is turbulent viscosity; p is pressure; F is the Lorentz force; I is the identity matrix;

Energy conservation equation:

$$\rho C_p \frac{\partial T}{\partial t} + \rho C_p \boldsymbol{u} \cdot \nabla T = \nabla \cdot (k \nabla T) + Q_J - Q_r \tag{3.7}$$

$$Q_J = \frac{1}{2}\sigma\omega^2 AA^* \tag{3.8}$$

where, C_p is the specific heat capacity under constant pressure; T is the temperature; k is thermal conductivity; Q_J and Q_r are Joule heat and volume radiation loss respectively.

3.2.3 Establishment of Theoretical Model

In the simulation of RF inductively coupled plasma discharge, the plasma is mainly used as a heat source, and heat exchange occurs between the plasma and the quartz wall and between the quartz wall and the expanded air region. Figure 3.1 is a schematic diagram of the longitudinal section of the two-dimensional axisymmetric model device. The simulated ICP discharge device is a cylindrical structure with a chamber height of 200 mm, an outer diameter of 108 mm, and an inner diameter of 100 mm. The sidewall is made of a dielectric material with a thickness of 4 mm, and the composition is quartz. The cylindrical discharge chamber is filled with argon gas, and a cylindrical coil wound from hollow copper tubes is located in the lower part of the side wall of the dielectric quartz window. The coil is 6 mm in diameter and is 43 mm from the aluminum alloy base of the substrate. The coil is connected to a high frequency power source with a frequency of 13.56 MHz. When the high frequency source is turned on, the argon gas inside the cylinder is ionized, resulting in plasma generation.

3.3 Results and Discussion

3.3.1 The Effect of Coil Numbers of ICP

First of all, the first coil is fixed at a distance of 43 mm from the entrance of the ICP torch, and the distance between the two adjacent coils to 20 mm, the coil radius to 3 mm. When the number of coils turns changes from 1 to 4, the spatial distribution of

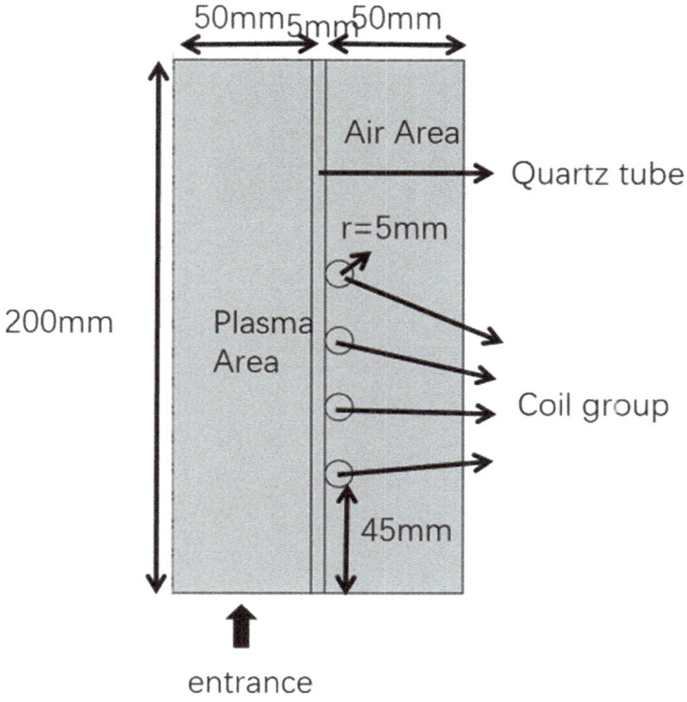

Fig. 3.1 Schematic diagram of a two-dimensional axisymmetric model of an inductively coupled plasma device

electron temperature and electron density in the plasma cavity is shown in Figs. 3.2, 3.3, 3.4, and 3.5.

From the calculation results of the electron temperature and conductivity distribution in the argon plasma, it can be seen that the electron temperature distribution coincides with the conductivity distribution. As the number of coils turns increases, the core area of electron density gradually increases, the peak electron temperature

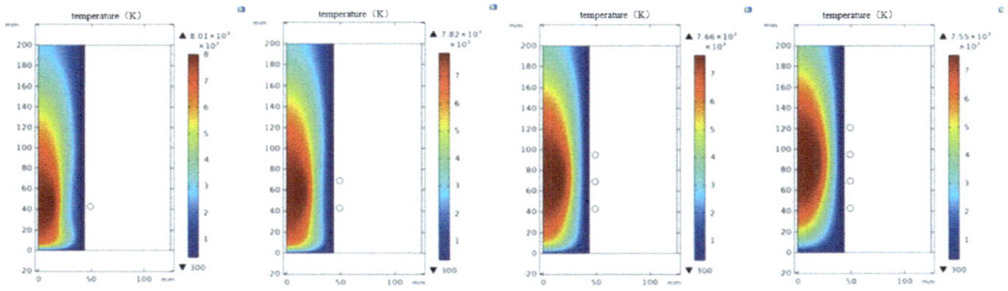

Fig. 3.2 Electronic temperature under different coil turns

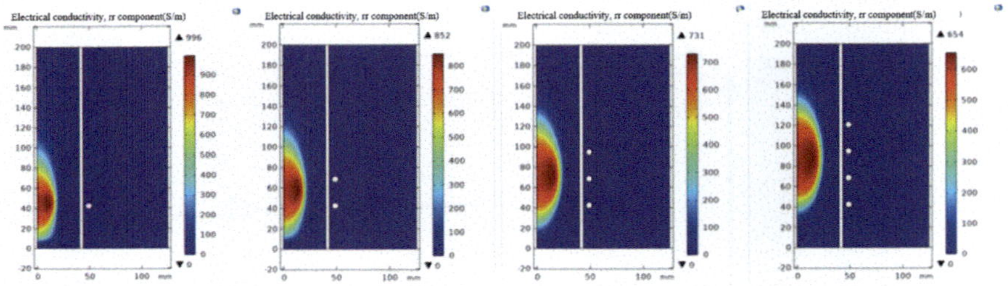

Fig. 3.3 Conductivity under different coil turns

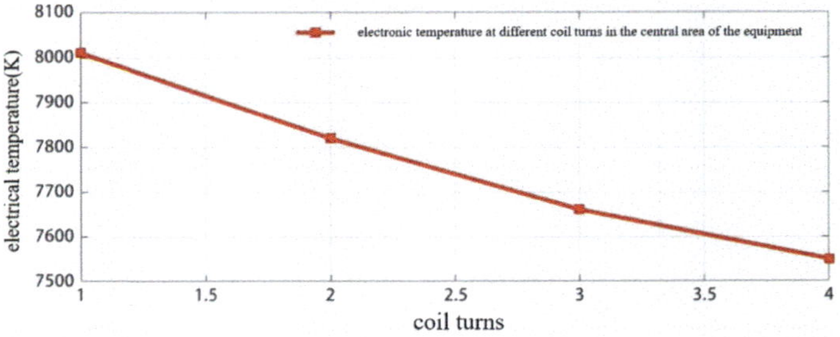

Fig. 3.4 Electronic temperature at different coil turns in the central area of the equipment

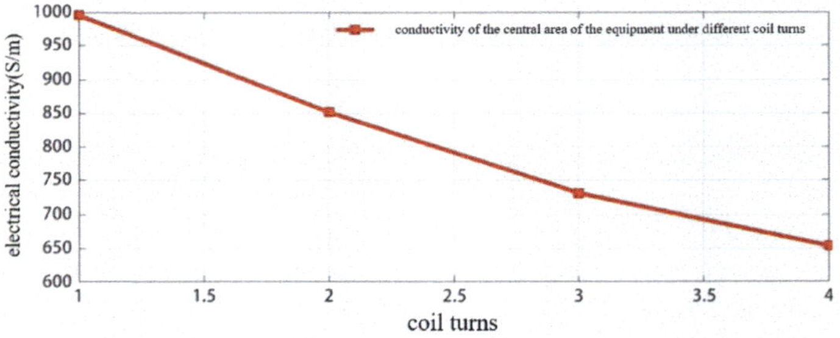

Fig. 3.5 Conductivity of the central area of the equipment under different coil turns

and conductivity both decreases, and the electron density distribution shifts toward the outlet.

At the same time, increasing the number of coil turns increases the potential well area of the plasma, increases the drift and diffusion area of the high-energy electrons, and thus improves the uniformity of the density distribution, and lower the core temperature zone of the plasma. Perhaps it is because as the number of turns in the coil increases, the concentration of electrons is higher and the collision between electrons and neutral particles in the cavity becomes more intense, so the electrons must expend more energy for excitation, ionization, and other processes.

The continuous loss of energy during the collision process leads to a decrease in electron temperature.

3.3.2 The Effect of Coil Distance of ICP

Then, the first coil is fixed at a distance of 43 mm from the entrance of the ICP torch, and the number of coil turns is fixed at 4, and the coil radius to 3 mm. When the distance between the two adjacent coils changes from 10 to 30 mm, the spatial distribution of electron temperature and electron density in the plasma cavity is shown in Figs. 3.6, 3.7, 3.8, and 3.9.

As can be seen from the calculation results of the temperature distribution and the flow field distribution of the argon plasma, the electron temperature distribution agrees with the conductivity distribution. As the coil distance increases, the peak conductivity gradually decreases, the potential well area of the plasma increases,

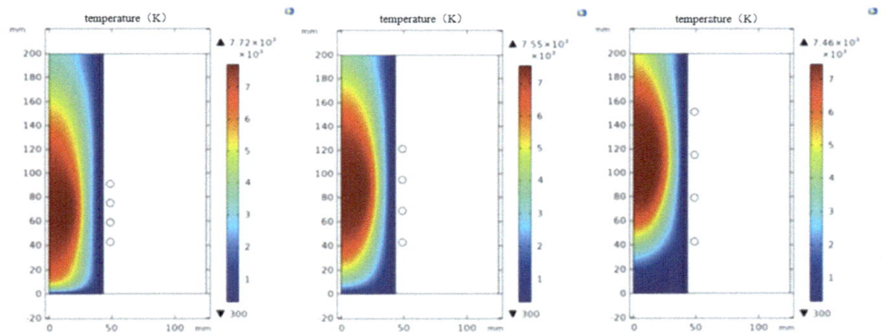

Fig. 3.6 Electronic temperature under different coil spacing

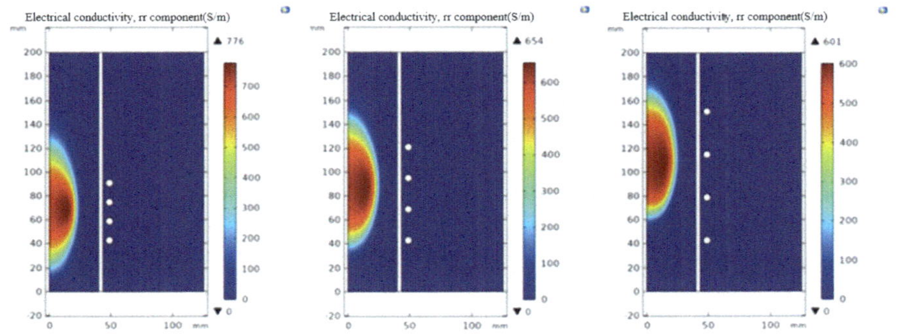

Fig. 3.7 Conductivity under different coil spacing

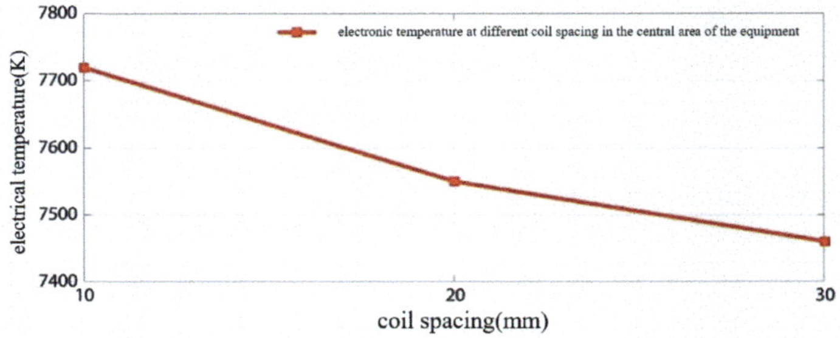

Fig. 3.8 Electronic temperature at different coil spacing in the central area of the device

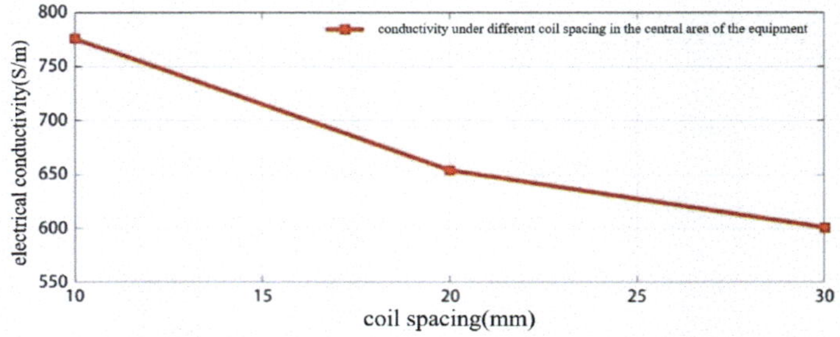

Fig. 3.9 Conductivity under different coil spacing in the central area of the equipment

and the core area of the plasma gradually shifts toward the outlet. The core area of the plasma changes significantly with distance. Moreover, the peak temperature of the plasma also shows a gradually decreasing trend with the increase of the coil distance. The possible reason is that increasing the coil spacing increases the discharge area near the coil, but the input power of the coil is constant, so the Joule thermal power per unit volume decreases and the plasma temperature decreases.

3.3.3 The Effect of Coil Radius of ICP

Then, the first coil is fixed at a distance of 41 mm from the entrance of the ICP torche, and the number of coil turns is fixed at 4, and the distance between the two adjacent coils to 30 mm. When the coil radius changes from 1 to 4 mm. The spatial distribution of electron temperature and electron density in the plasma cavity is calculated, and the results are shown in Figs. 3.10, 3.11, 3.12, and 3.13.

From the calculation results of the electron temperature and the conductivity distribution in the argon plasma, it can be seen that with increasing coil radius, the peak electron temperature and plasma conductivity decrease, but change only slightly.

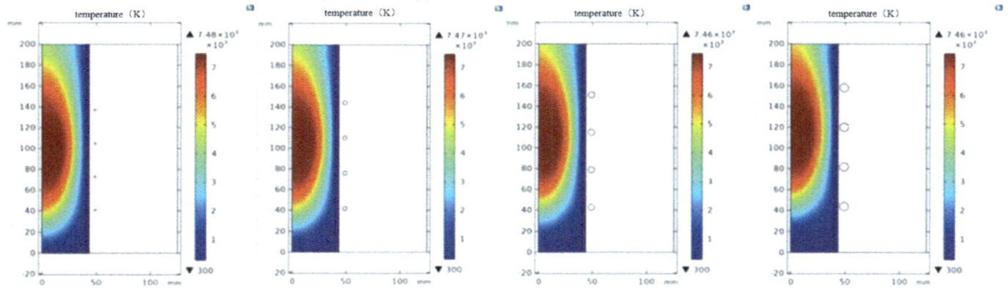

Fig. 3.10 Electron temperature at different coil radius

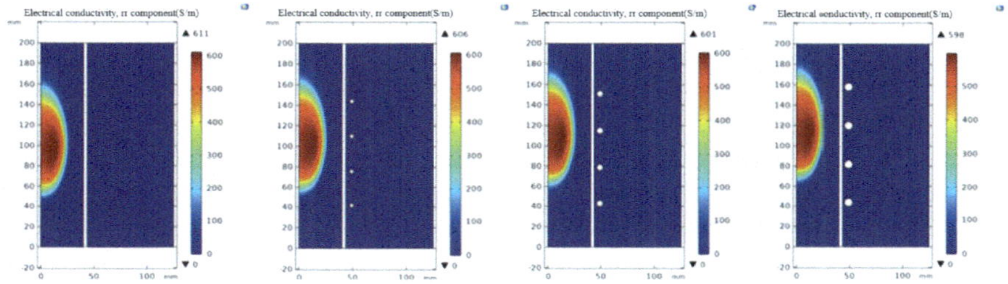

Fig. 3.11 Conductivity under different coil radius

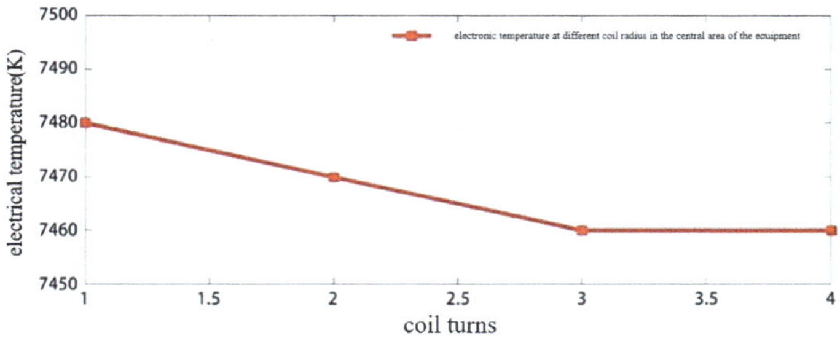

Fig. 3.12 Electron temperature at different coil radius

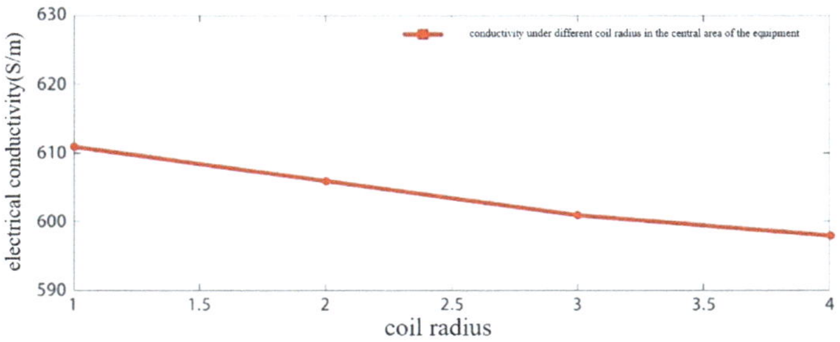

Fig. 3.13 Conductivity under different coil radius

3.4 Conclusion

This article investigates the effect of changes in coil parameters on the spatial distribution of temperature and turbulent flow fields in an inductively coupled argon plasma torch. The main conclusions obtained are as follows:

1. Increasing the number of coils turns has a significant effect on the plasma properties: as the number of coils turns increases, the electron density core gradually shifts toward the outlet, the electron density core area increases significantly, and both the peak electron temperature and conductivity decrease, improving the uniformity of the plasma density distribution.
2. As the coil spacing increases, the potential well area of the plasma increases, and the core area of the plasma changes significantly with spacing. The maximum electron temperature and the conductivity inside the core decrease. When the distance is 30 mm, the plasma has a uniform distribution.
3. As the coil radius increases, the peak electron temperature and plasma conductivity decrease, but the amplitude is small, and the plasma core area also changes slightly. Overall, the change in coil radius has little effect on the plasma.

References

1. J.W. Mckelliget, N. El-Kaddah, Theoretical prediction of the effect of coil configuration on gas mixing in an inductively coupled plasma torch. MRS Online Proceed. Lib. (OPL) **98**, 21 (1987)
2. S. Xue, P. Proulx, M.I. Boulos, Effect of the coil angle in an inductively coupled plasma torch: a novel two-dimensional model. Plasma Chem. Plasma Process. **23**, 245–263 (2003)
3. R. Ye, A.B. Murphy, T. Ishigaki, Numerical modeling of an Ar–H_2 radio-frequency plasma reactor under thermal and chemical nonequilibrium conditions. Plasma Chem. Plasma Process. **27**, 189–204 (2007)
4. S.B. Punjabi, S.N. Sahasrabudhe, N.K. Joshi, et al., Comparative study of laminar and turbulent flow model with different operating parameters for radio frequency-inductively coupled plasma torch working at 3 MHz frequency at atmospheric pressure. Phys. Plasmas. **21**(1) (2014)
5. A.O. Brezmes, C. Breitkopf, Fast and reliable simulations of argon inductively coupled plasma using COMSOL. Vacuum **116**, 65–72 (2015)
6. Z. Hai-Long, L. Xue-Ping, T. Hong-Hui, Three-dimensional numerical simulation of physical field distribution of radio frequency thermal plasma. Acta Phys. Sin. **70**(15) (2021)
7. Y. Ming-Hao, Numerical investigation on interaction mechanisms between flow field and electromagnetic field for nonequilibrium inductively coupled plasma. Acta Phys. Sin. **68**(18) (2019)

Chapter 4
A Simulation Model for Dielectric Barrier Discharge Considering Finite Trap Density

Yingzhe Zhang, Lanlan Nie, Yubin Xian, and Xinpei Lu

Abstract A simulation model for dielectric barrier discharge considering the finite trap density is established to investigate the discharge characteristics and the variation of trap electrons in He and N_2 discharges. Instead of a constant, secondary electrons emission coefficient γ varying with trap electron filling is obtained by Auger electron theory. In addition to the gas ionization degree, the secondary electron emission coefficient might also be the reason for the different discharge characteristics of He and N_2. Compared to He^+ and $He_2{}^+$ ions, the γ of $N_2{}^+$ and $N_4{}^+$ are very small and cannot emit intrinsic electrons in the valence band of the dielectric. However, due to the existence of localized energy levels, $N_2{}^+$ and $N_4{}^+$ can emit the electrons in the traps, causing $\gamma \neq 0$. It is found that the cathode trap density gradually decreases during the discharge process in N_2 discharge, while keeps 100% filling in He discharge. Changing the band gap width of the dielectric does not effectively regulate the discharge characteristics of N_2 discharge. However, reducing the band gap width of the dielectric can effectively increase the secondary electron emission coefficient in He discharge, thus increasing the seed electron density and allowing the discharge to occur at lower gap electric fields, which is beneficial for uniform discharge. Therefore, for glow and Townsend type dielectric barrier discharge, targeted optimization of dielectric material properties should be combined with discharge mechanisms to achieve uniform discharge.

4.1 Introduction

Homogeneous Dielectric Barrier Discharge at Atmospheric Pressure has a wide range of applications in materials surface treatment, agriculture, energy, environmental protection, and bio-medicine [1]. As there are various spatial non-uniformities and time instabilities during the discharge process, which reduce energy efficiency and

Y. Zhang · L. Nie (✉) · Y. Xian · X. Lu
Huazhong University of Science and Technology, Wuhan 430074, People's Republic of China
e-mail: nielanlan2017@163.com

Z. Fang et al. (eds.), *Proceedings of the 5th International Symposium on Plasma and Energy Conversion*, Springer Proceedings in Physics 398,
https://doi.org/10.1007/978-981-97-2245-7_4

cause fluctuations in active particles, it is of significant importance to study the mechanism of dielectric barrier discharge at atmospheric pressure and the factors affecting its uniformity.

For N_2 homogeneous dielectric barrier discharge, a prevalent understanding posits that seed electrons originate from the shallow trap electrons emitted by the metastable particles generated in the previous discharge [2]. Previous researches have indicated that the shallow traps and surface microstructure of the dielectric material greatly affect the uniformity of discharge [3–5]. For physical quantities that are difficult to measure in experiments, simulation models are often combined for research. Massines et al. explained the discharge mechanism by establishing a one-dimensional model to calculate particle density and spatiotemporal distribution of the electric field [6, 7]. With the introduction of the trap electron desorption mechanism, Golubovskii et al. researched on the influence of various electronic emission mechanisms on the discharge characteristic [8, 9]. Current simulation models commonly assume that the secondary electron emission coefficient γ is a constant. However, it should be noted that γ is not only related to the energy level structure of the dielectric material but also associated with the trap electron density on the dielectric surface. Literature [10] established a relationship between the secondary electron emission coefficient and the surface charge density based on Auger electron theory, finding that the narrower the band gap of the dielectric, the larger the secondary electron emission coefficient, making the DBD works more uniformly in weaker average gap electric fields. However, none of the above simulation works considered the finiteness of trap density, which is inconsistent with the actual physical process. Plasma electrons typically have energies in the range of a few electron volts and exhibit a penetration depth on the nanometer scale, so the trap electrons capture demonstrates saturation characteristics [11].

In this work, a simulation model for dielectric barrier discharge considering the finite trap density is established and verified by comparing with the experimental results. Firstly, utilizing the Auger electron theory, we calculate the variation of the secondary electron emission coefficient with the trap electrons filling degree. Additionally, we explore the impact of the dielectric material's band gap width on the secondary electron emission coefficient. Subsequently, we incorporate the secondary electron emission coefficient into the simulation model and investigate the characteristics of dielectric barrier discharge and the variations in trap electron behavior for different gas compositions.

4.2 Numerical Simulation Model

The simulation model is based on the parallel flat electrode structure, with a plate area of 12.56 cm^2. The two electrodes are covered with insulating dielectric, with one side connected to an external AC power source and the other grounded. The relative dielectric constant of dielectric is 9, and the thickness is 1 mm. The discharge gap width for N_2 is 1 mm, while for He it is 5 mm. The particles considered for N_2

discharge include e, N_2^+, N_4^+, $N_2(A^3\Sigma_u^+)$, $N_2(a'^1\Sigma_u^-)$, $N_2(B^3\Pi_g)$, and $N_2(C^3\Sigma_u)$, and He discharge species include e, He^+, He_2^+, N_2^+, He^*, He_2^*. The chemical reactions and rates used in the simulation refer to the references [12–14]. By discretizing equations using the finite difference method and programming the model in Python, we obtained the discharge characteristics of dielectric barrier discharge considering the finite trap density.

4.2.1 Model Equation

Assuming that the size of the electrode plate is much larger than the discharge gap, a one-dimensional fluid model of dielectric barrier discharge plasma is established. The governing equations of this model include the continuity equation of particles, and the Poisson's equation.

$$\frac{\partial n_{e,p,m}(x,t)}{\partial t} + \nabla \cdot \Gamma_{e,p,m}(x,t) = S_{e,p,m}(x,t) \tag{4.1}$$

$n_{e,p,m}$ represent the densities of electrons, ions, and neutral particles, respectively, $S_{e,p,m}$ are the source terms of particles, and $\Gamma_{e,p,m}$ are their flux densities, in the following form:

$$\Gamma_e(x,t) = -\mu_e n_e(x,t)E(x,t) - D_e \nabla n_e(x,t) \tag{4.2}$$

$$\Gamma_p(x,t) = \mu_p n_p(x,t)E(x,t) - D_p \nabla n_p(x,t) \tag{4.3}$$

$$\Gamma_m(x,t) = -D_m \nabla n_m(x,t) \tag{4.4}$$

E is the electric field, μ_e, μ_p are the mobility of particles, and D_e, D_p, D_m are the diffusion coefficients. The model solves the potential by coupling Poisson's equation:

$$\frac{\partial^2 V(x,t)}{\partial x^2} = -\frac{q}{\varepsilon_0}\left(n_p(x,t) - n_e(x,t)\right) \tag{4.5}$$

The total discharge current density obtained from Sato's formula is the sum of conduction current and displacement current [15, 16]:

$$I_d(x,t) = S \times \left[e \int_0^d (\Gamma_p(x,t) - \Gamma_e(x,t))dx + \varepsilon_0 \frac{dV_{app}}{dt} \right] \times \left(\frac{d_1}{\varepsilon_1} + d_g + \frac{d_2}{\varepsilon_2} \right)^{-1} \tag{4.6}$$

4.2.2 Boundary Condition

The boundary conditions in the model mainly include Poisson's equation boundary conditions, particle density boundary conditions and Surface charge accumulation boundary conditions. The left boundary voltage is the external power supply on the electrode, and the right boundary voltage is 0. At the interface between plasma and dielectric, the charge accumulates on the surface and is calculated through Gauss's law:

$$\vec{n} \cdot (\varepsilon_0 \varepsilon_1 E_d - \varepsilon_0 E_g) = \sigma \tag{4.7}$$

The boundary condition for electron flux on the surface of the dielectric is [17]:

$$\begin{aligned} n \cdot \Gamma_e = & -(2a-1) n_e \mu_e E \cdot \vec{n} + \frac{1}{2} v_{th,e} n_e \\ & - \frac{1}{2} v_{th,e} n_\gamma - 2(1-a) \sum_p \gamma_p \Gamma_p \cdot \vec{n} \end{aligned} \tag{4.8}$$

$$a = \begin{cases} 1 & \mathrm{sgn}(q) \mu E \cdot \vec{n} > 0 \\ 0 & \mathrm{sgn}(q) \mu E \cdot \vec{n} \leq 0 \end{cases} \tag{4.9}$$

Here, a is the switching function and vth,e is the electron thermal motion velocity. Assuming that ions and metastable particles are immediately adsorbed at the boundary [18], we can infer that they satisfy Neumann boundary conditions. The dielectric surface charge accumulation of the grounding electrode and power electrode side is given by Eqs. (4.10) and (4.11).

$$\sigma(t) = \int_0^t i_T(0, t)\, dt \tag{4.10}$$

$$\sigma(t) = -\int_0^t i_T(0, t) dt \tag{4.11}$$

Taking into account the finite trap density, assuming E_t is the trap energy level, σ_N is the trap density, σ_{Et} is the electron density captured by the trap, then the change of trap electron density satisfies Eq. (4.12). The first term on the right side of the equation represents the trap electron capture, the second term represents the trap electron desorption process, the third term represents the loss caused by recombination with ions, and the last two terms represent the secondary electrons emission of trap electrons, which will be described in detail in the next section.

$$\frac{d\sigma_{et}}{dt} = n_{ce} B_t (\sigma_N - \sigma_{et}) - \sigma_{et} v_e^{des}$$

$$- \alpha_{rw}\sigma_{et}\sigma_{+} - P_{vd}\gamma_{vd}\Gamma_p - 2P_{dd}\gamma_{dd}\Gamma_p \tag{4.12}$$

Here, B_t is the capture coefficient, α_{rw} is the surface recombination coefficient, v_e^{des} is the trapped electrons desorption frequency, including thermal desorption and particle collision desorption. σ_e, σ_+ is the density of positive and negative charges accumulated on the dielectric surface [9].

4.2.3 *Calculation of Secondary Electrons Emission Coefficient*

According to Hagstrum's research [19], the secondary electrons emission of low-speed ions mainly depends on the potential energy of ions, and the emission process mainly includes the one-step process of Auger neutralization and the two-step process of resonance neutralization and Auger deexcitation [20]. Because resonance neutralization is a tunneling process that requires particles to approach the surface, while Auger neutralization can occur further away from the surface of the dielectric [21], it is believed that Auger neutralization is the dominant process.

Traps refer to localized energy levels formed in the band gap between the conduction and valence bands due to impurities and structural defects in the dielectric. As is shown in Fig. 4.1, ε_0 is the vacuum energy level, ε_V is the top of the valence band, ε_C is the bottom of the conduction band, with the band gap width of $E_g = \varepsilon_C - \varepsilon_v$, and the electron affinity $\chi = \varepsilon_0 - \varepsilon_C$. Traps include traps occupied by electrons and empty traps, $\varepsilon_{tbottom}$ and ε_{ttop} represent the bottom and top of the empty trap energy level, and $\varepsilon_{ebottom}$ and ε_{etop} represent the bottom and top of the energy level occupied by electrons [22]. After considering trap energy level, in addition to the secondary electron emission in the valence band γ_{VV}, trapped electrons may also participate in the secondary electrons emission process as is shown in Fig. 4.1b, c, which are expressed as γ_{Vd} and γ_{dd}. Then the total secondary electrons emission coefficient is:

$$\gamma_i = \gamma_{vv}(1 - P_{vd} - P_{dd}) + \gamma_{vd}P_{vd} + \gamma_{dd}P_{dd} \tag{4.13}$$

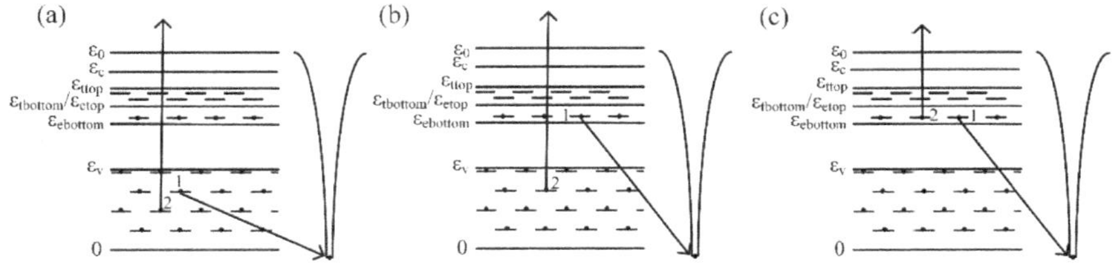

Fig. 4.1 Schematic diagram of secondary electrons emission process considering trap energy level

P_{vd} and P_{dd} are the probabilities of corresponding processes occurring, and are proportional to the trap electron density [23]. When electrons accumulate on the dielectric surface, P_{vd} and P_{dd} will change and affect the secondary electrons emission coefficient. Assuming the trap energy level occupied by electrons is $d\varepsilon = \varepsilon_{etop} - \varepsilon_{ebottom}$, the surface trap electron density is $\sigma = ed\varepsilon N_e l$, where l is the penetration depth, taking 10^{-10} m. Then, P_{vd} and P_{dd} can be derived from the surface trap electron density and the secondary electrons emission coefficient varying with trap electron filling degree could be obtained.

4.3 Results and Discussion

By comparing with experimental results, the reliability of the simulation model can be verified and the simulated current waveforms of He and N_2 discharges obtained are shown in Fig. 4.2. For He discharge, the voltage amplitude is 1300 V, the frequency is 10 kHz, and the N_2 impurities is 100 ppm. The simulation predicts a discharge current peak of 26.6 mA, with a pulse width of approximately 4.4 μs, which is in alignment with the voltage and current waveform featured in reference [7, 24]. For N_2 discharge, the voltage amplitude is set at 7000 V with a frequency of 10 kHz. The simulation yields a discharge current peak of 16 mA and a pulse width of approximately 29.6 μs, which corresponds with the voltage and current waveform featured in reference [25]. When the polarity of the spatial electric field is reversed, there is a small increase in current with the moving charged particles [7]. Although both experiments use alumina as the dielectric, we must consider the alumina's microstructure differences in different experiments. Fitting through simulation and experimental results, we obtain that the total trap electron density of He discharge is 10^9 cm^{-2} and the total trap density of N_2 discharge is 10^{10} cm^{-2}, all of which are within a reasonable range. Consistently, the alumina dielectric used in the simulation has a trap energy level of 1.4 eV, a band gap width of $E_g = 8$ eV, which is based on the values in reference [26]. Although there are still certain discrepancies between the trap parameters and the actual physical situation, it is worthwhile to investigate the effect of trap parameters on discharge characteristics to understand deeply the discharge mechanism.

The band gap width of alumina is between 4–9 eV, and can be changed by changing temperature or doping. To investigate the impact of band gap width on discharge characteristics, we calculate the variation of the secondary electron emission coefficient with the trapped electron filling degree for band gap widths of 7, 8, and 9 eV. As is shown in Fig. 4.3, the secondary electrons emission coefficient of N_2^+ and N_4^+ is very small and cannot emit the intrinsic electrons in the valence band compared with He^+ and He_2^+. However, with the introduction of local energy levels, N_2^+ and N_4^+ can emit trapped electrons so that the secondary electrons emission coefficient is not zero. For ions with large ionization energy, such as He^+ and He_2^+, the secondary electrons emission coefficient increases significantly with the decrease of the band gap width, but for ions with small ionization energy, such as N_2^+ and N_4^+, the secondary electrons

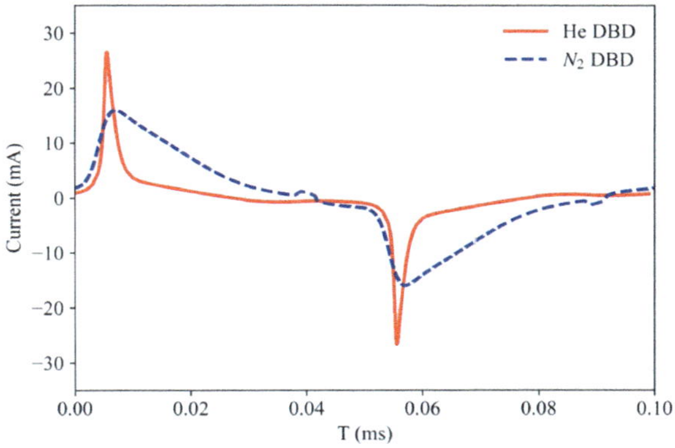

Fig. 4.2 The simulated current waveforms of He and N_2 discharge

emission coefficient does not change significantly with the decrease of the band gap width.

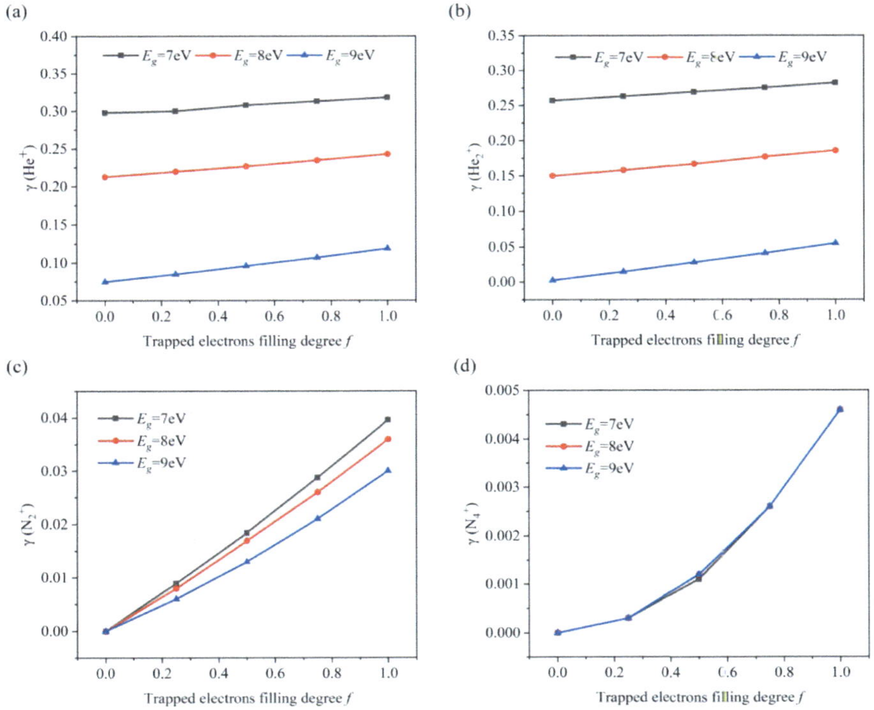

Fig. 4.3 Variation of secondary electrons emission coefficient with the trapped electrons filling degree under different band gap widths. **a** He^+ **b** $He_2{}^+$ **c** $N_2{}^+$ **d** $N_4{}^+$

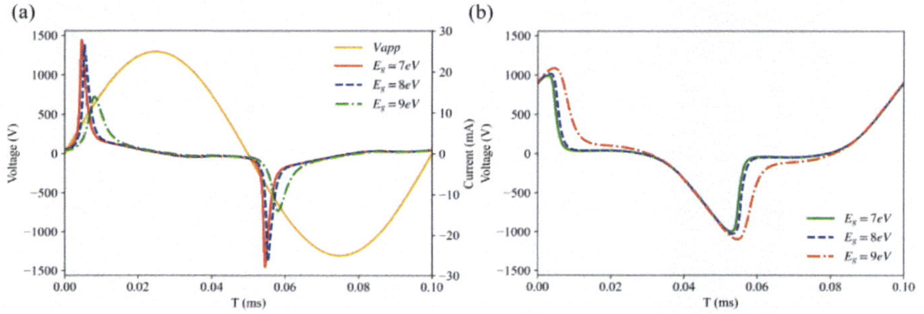

Fig. 4.4 Waveforms of He dielectric barrier discharge current (**a**) and air gap voltage (**b**) at different band gap widths

Introduce the relationship between the secondary electrons emission coefficient and the trap electron filling degree into the He discharge simulation model, we obtained the discharge current and air gap voltage waveforms for different band gap widths as it is shown in Fig. 4.4. As the band gap width of the dielectric decreases, the secondary electron emission coefficient increases, resulting in the release of more secondary electrons. Consequently, the air gap breakdown voltage decreases from 1099.4 to 998.8 V, and the discharge current peak slightly advances and increases from 14.1 to 27.95 mA. During the discharge process, due to the capture rate of trapped electrons being much higher than the desorption rate, the trapped electrons always maintain a filling degree of 100%. When the band gap widths of the dielectric are 7 eV, 8 eV, and 9 eV, the corresponding seed electron densities at the cathode are 7.686×10^5 cm^{-3}, 4.525×10^5 cm^{-3}, 2.017×10^5 cm^{-3}, respectively. As the decrease of band gap width, the seed electron density increases, which is beneficial for discharge uniformity.

Consistent with the trap density of 10^9 cm^{-2} used for He discharge, we obtained the N_2 discharge current waveform and the variation of cathode trap electron density under different band gap widths, as shown in Fig. 4.5. As the band gap width increases, the discharge current peak and the variation of cathode trap electron density remain basically unchanged, indicating that the band gap width does not affect the N_2 discharge characteristics. During the discharge process, the cathode trap electron density gradually decreases. When the band gap widths are 7 eV, 8 eV, and 9 eV, the cathode seed electron densities are 2.58×10^6 cm^{-3}, 2.496×10^6 cm^{-3}, 2.434×10^6 cm^{-3}, with little change in seed electron density compared to He discharge. Therefore, changing the band gap width of the dielectric material cannot effectively adjust the discharge N_2 discharge characteristics, because the band gap width only affects the secondary electrons emission coefficient, while the secondary electrons emission process has little effect on the seed electron density.

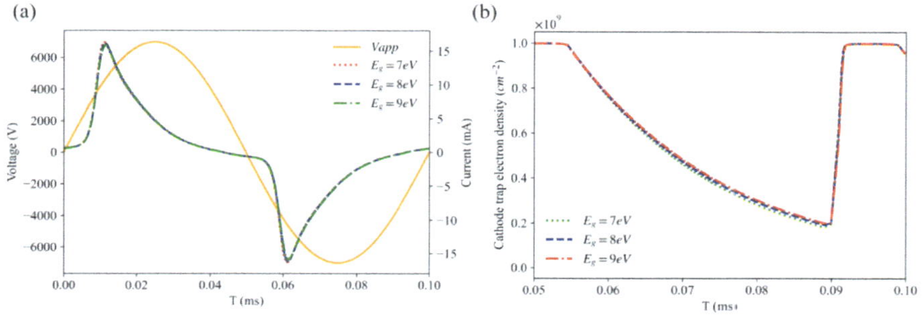

Fig. 4.5 Waveforms of N_2 dielectric barrier discharge current (**a**) and the variation of cathode trap electron density (**b**) at different band gap widths

4.4 Conclusion

In this work, a simulation model of dielectric barrier discharge considering the finite trap density is established. The simulation model reveals that altering the band gap width of the dielectric material does not effectively regulate the characteristics of the N_2 discharge. However, reducing the band gap width effectively increases the secondary electron emission coefficient of the He discharge, which makes it discharge at a lower electric field, and finally enhances the discharge uniformity. The difference of He and N_2 discharge characteristics is not only related to the gas ionization degree, but also related to the secondary electrons emission coefficient. Compared with He^+ and He_2^+, N_2^+ and N_4^+ cannot emit the intrinsic electrons in the valence band, However, with the introduction of trap electrons, their secondary electrons emission coefficient are no longer zero. For N_2 discharge, the cathode trapped electrons density gradually decreases during the discharge process, while for He discharge, due to the capture rate of trapped electrons being much faster than the desorption rate, the cathode trapped electrons always maintain a filling rate of 100%. In the future, we will continue to explore the impact of trap characteristics of dielectric materials on discharge characteristics and trapped electrons.

References

1. Q. Liu, Z. Cui, W. Hu, Z. Cai, in *Brief Review: Characteristics and Applications of Homogeneous Atmospheric Pressure Dielectric Barrier Discharge*. 2021 6th Asia Conference on Power and Electrical Engineering (ACPEE). pp. 1149–1153 (2021)
2. H. Luo, Z. Liang, X. Wang, Z. Guan, L. Wang, Homogeneous dielectric barrier discharge in nitrogen at atmospheric pressure. J. Phys. D: Appl. Phys. **43** (2010)
3. M. Li, C. Li, H. Zhan, J. Xu, X. Wang, Effect of surface charge trapping on dielectric barrier discharge. Appl. Phys. Lett. **92** (2008)

4. S. Suzuki, H. Itoh, Gradual increase in secondary ionization coefficient γ and charge accumulation on a dielectric electrode during DBD with repeated breakdown. Plasma Sources Sci. Technol. **24**, 055016 (2015)
5. J. Ran, C. Li, D. Ma, H. Luo, X. Li, Homogeneous dielectric barrier discharges in atmospheric air and its influencing factor. Phys. Plasmas **25** (2018)
6. F. Massines, P. Ségur, N. Gherardi, C. Khamphan, A. Ricard, Physics and chemistry in a glow dielectric barrier discharge at atmospheric pressure: diagnostics and modelling. Surf. Coat. Technol. **174–175**, 8–14 (2003)
7. F. Massines, A. Rabehi, P. Decomps, R.B. Gadri, P. Ségur, C. Mayoux, Experimental and theoretical study of a glow discharge at atmospheric pressure controlled by dielectric barrier. J. Appl. Phys. **83**, 2950–2957 (1998)
8. Y.B. Golubovskii, V.A. Maiorov, J. Behnke, J.F. Behnke, Modelling of the homogeneous barrier discharge in helium at atmospheric pressure. J. Phys. D Appl. Phys. **36**, 39 (2002)
9. Yu.B. Golubovskii, V.A. Maiorov, J. Behnke, Influence of interaction between charged particles and dielectric surface over a homogeneous barrier discharge in nitrogen. J. Phys. D Appl. Phys. **35**, 751–761 (2002)
10. C. Yao, S. Chen, Z. Chang, H.B. Mu, G.J. Zhang, Atmospheric pressure dielectric barrier discharge involving ion-induced secondary electron emission controlled by dielectric surface charges. J. Phys. D Appl. Phys. **52**, 455202 (2019)
11. P.F. Ambrico, M. Ambrico, A. Colaianni, L. Schiavulli, G. Dilecce, S.D. Benedictis, Thermoluminescence study of the trapped charge at an alumina surface electrode in different dielectric barrier discharge regimes. J. Phys. D Appl. Phys. **43**, 325201 (2010)
12. Y.H. Choi, J.H. Kim, Y.S. Hwang, One-dimensional discharge simulation of nitrogen DBD atmospheric pressure plasma. Thin Solid Film. **506–507**, 389–395 (2006)
13. M. Li, C. Han, W. Liu, Numerical simulation of the multi-pulse phenomena in atmospheric-pressure dielectric barrier glow discharges in He/N_2 mixture. Eur. Phys. J. Appl. Phys. **82**, 30801 (2018)
14. Y. Zhang, W. Ning, D. Dai, Influence of nitrogen impurities on the performance of multiple-current-pulse behavior in a homogeneous helium dielectric-barrier discharge at atmospheric pressure. J. Phys. D Appl. Phys. **52**, 045203 (2018)
15. R. Morrow, N. Sato, The discharge current induced by the motion of charged particles in time-dependent electric fields; Sato's equation extended. J. Phys. D Appl. Phys. **32**, L20 (1999)
16. N. Sato, Discharge current induced by the motion of charged particles. J. Phys. D Appl. Phys. **13**, L3 (1980)
17. G.J.M. Hagelaar, F.J. De Hoog, G.M.W. Kroesen, Boundary conditions in fluid models of gas discharges. Phys. Rev. E Stat. Phys. Plasmas Fluids Relat Interdiscip. Topics **62**, 1452–1454 (2000)
18. S. Nemschokmichal, R. Tschiersch, J. Meichsner, The influence of negative ions in helium–oxygen barrier discharges: II. 1D fluid simulation and adaption to the experiment. Plas. Sources Sci. Tech. **25**, 055024 (2016)
19. H.D. Hagstrum, *Theory of Auger Ejection of Electrons from Metals by Ions* (1954)
20. Y. Motoyama, H. Matsuzaki, H. Murakami, A study of the secondary electron yield γ of insulator cathodes for plasma display panels. IEEE Trans. Electron Devices **48**, 1568–1574 (2001)
21. M. Daksha, A. Derzsi, Z. Mujahid, D. Schulenberg, B. Berger, Z. Donkó, J. Schulze, Material dependent modeling of secondary electron emission coefficients and its effects on PIC/MCC simulation results of capacitive RF plasmas. Plasma Sources Sci. Tech. **28**, 034002 (2019)
22. S. Zhang, G.-Y. Sun, A. Volčokas, G.-J. Zhang, A.-B. Sun, Unveiling the role of dielectric trap states on capacitively coupled radio-frequency plasma discharge: dynamic charging behaviors. Plasma Sources Sci. Technol. **30**, 055007 (2021)
23. S.J. Yoon, I.L.I. Lee, J.-W.L. Lee, B.O.B. Oh, Intrinsic and effective secondary electron emission coefficients in AC plasma display panel. Jpn. J. Appl. Phys. **40**, 809 (2001)
24. A. Saridj, A.W. Belarbi, Numerical modeling of a DBD in glow mode at atmospheric pressure. J. Theo. Appl. Phys. **13**, 179–190 (2019)

25. M.C. Bouzidi, X. Bonnin, N. Naudé, H. Piquet, A. Belinger, N. Gherardi, Maximization of the working domain of an Atmospheric Pressure Townsend Discharge (APTD) using a current-source static converter. J. Phys.: Conf. Ser. **550**, 012044 (2014)
26. P.S. Pickard, M.V. Davis, Analysis of electron trapping in alumina using thermally stimulated electrical currents. J. Appl. Phys. **41**, 2636–2643 (2003)

Chapter 5
Investigation on the Discharge Characteristics of Multi-Channel Spark Plasma Jet and Its Triggered Gas Switch

Weihao Tie, Xiaocheng Hu, Yongchun Shao, Jinyong Fang, Axin Jin, Huijun Huang, and Bing Duan

Abstract We designed a six-channel annular micro-plasma-gun to generate a stable multiple spark plasma jet. The multi-channel plasma jet characteristics were investigated at different pulse parameters. The PIC-MCC particle model of the micro-plasma-gun discharge was established, which can well describe the physical process of the initial development of spark plasma jet and the competitiveness between multiple jet channels. Finally, the micro-plasma-gun was embedded into the triggered electrode of a three-electrode gas switch, and the discharge characteristics of the gas switch under multi-channel spark plasma jet triggering, UV pre-ionization triggering and filed distortion triggering were investigated. The results show that, at the initial electron avalanche, the electron temperature starts to increase gradually, and electron number also increases exponentially. When it transitions to the streamer discharge, the electron temperature drops to several eV, and the growth rate of electron number tends to saturation, approaching to 10^{12}–10^{13}. Under the fast and high energy voltage pulse, it is easier to form the multi-channel spark plasma jet. Comparing to field distortion, UV pre-ionization and single spark plasma jet, the lower delay time and jitter can be obtained at the multi-channel spark plasma jet triggering method. The triggered delay time and jitter is 43.5 and 1.5 ns at the working coefficient of 50%.

5.1 Introduction

In recent years, high-density plasma jet with fast propulsion velocity was widely adopt in a variety of applications, such as high power laser, electronthermal-chemical launcher, plasma thruster etc. Therefore, more and more attention was focused on the generation, development and physical essence of the plasma jet.

W. Tie (✉) · X. Hu · Y. Shao · J. Fang · A. Jin · H. Huang · B. Duan
China Academy of Space Technology (Xi'an), Xi'an 710100, China
e-mail: twh.110.666@163.com

Z. Fang et al. (eds.), *Proceedings of the 5th International Symposium on Plasma and Energy Conversion*, Springer Proceedings in Physics 398,
https://doi.org/10.1007/978-981-97-2245-7_5

Weihao Tie *ect* investigated the effect of pulse energy on the dynamic development of spark plasma jet, and proposed a novel of spark plasma jet triggered discharge mode for gas switch [1, 2]. Peifei Zhang etc. studied the electron temperature and density of spark plasma jet with time [3]. Chen Li and Xiaoang Li etc. designed the spark plasma jet triggered gas switch respectively, and the discharge characteristics of the gas switch under different working condition were studied [4, 5]. The previous researches mainly concentrate on the discharge characteristics of the gas switch triggered by single spark plasma jet. There are few reports on the physical mechanism of the development of spark plasma jet, especially the comparative analysis between plasma jet triggering mode and other triggering modes, which is not only very important to improve triggered discharge characteristics of gas switch, but also deepen the understanding of the physical mechanism of the spark plasma jet.

In this paper, firstly, a six-channel annular micro-plasma-gun to generate a stable multiple spark plasma jet was designed. Then, the PIC-MCC particle model of the micro-plasma-gun discharge was established, which can well describe the physical process of the initial development of spark plasma jet and the competitiveness between multiple jet channels. Finally, the triggered discharge characteristics of gas switch at multichannel plasma jet, single plasma jet, UV pre-ionization and filed distortion triggering were investigated respectively.

5.2 Experimental Setup and Theoretical Model

5.2.1 Experimental Setup

Figure 5.1 shows the schematic of the experimental setup employed for testing the discharge characteristics of the spark plasma jet triggered gas switch. The structure of the spark plasma jet triggered gas switch was similar to that of a typical three-electrode electric field distortion gas switch. A six-channel annular micro-plasma-gun composed by needle and plate electrodes was embedded into the middle triggered electrode. Two 40-nF pulse capacitor (C_1 and C_2) were charged to $\pm$ 25 kV. They were connected to the positive and negative high-voltage electrodes. The load (R_L) was a 3Ω water resistor. In order to increase the coupling energy, the needle electrode in the micro-plasma-gun was grounded by a 1.2 μH inductor (L_g). A trigger pulse with the amplitude of + 40kV and the rise time of 25 ns was imposed on the triggered electrode by a 1 μH isolation inductor (L_i). The working coefficient of the switch was changed by adjusting the gas pressure.

The trigger voltage was measured by using a resistance voltage divider. Pearson 5046 Coils were used to monitor the load current and needle current, respectively. And the voltage and current waveforms were recorded in a Tektronix DPO4104 digital oscilloscope.

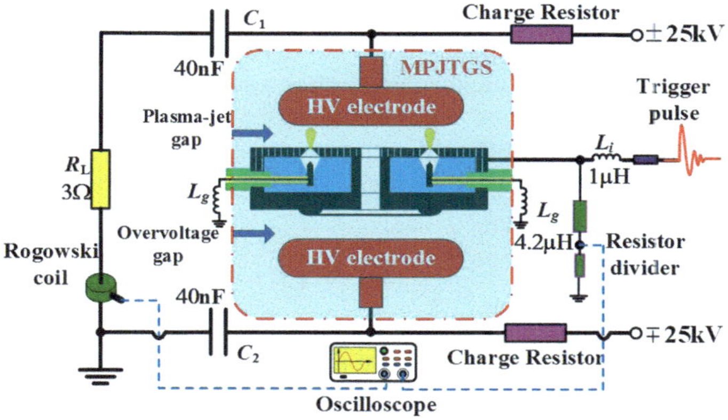

Fig. 5.1 Schematic diagram of experimental setup of spark plasma jet triggered gas switch

5.2.2 *Theoretical Model*

In the development of spark plasma jet, the electric field E acting on the charged particle can be calculated using Poisson's equation.

$$\begin{aligned} \nabla^2\phi &= -\frac{\rho}{\varepsilon_0} \\ E &= -\nabla\phi \end{aligned} \tag{5.1}$$

where ϕ is electric potential; ρ is space charge density; ε_0 is permittivity of vacuum. The magnetic flux density B can be calculated by the following equation.

$$\frac{\partial B}{\partial t} = -\nabla \times E \tag{5.2}$$

The motion of charged particles satisfies Newton's equation.

$$\frac{\mathrm{d}(m\gamma v)}{\mathrm{d}t} = qE + qv \times B \tag{5.3}$$

where γ is relativistic factor. And the velocity of the charged particle can be expressed as the following equation.

$$\frac{\mathrm{d}x}{\mathrm{d}t} = v \tag{5.4}$$

where x is displacement distance. In this simulation, 47 collision processes between electrons and nitrogen molecules are considered, including elastic collision, rotation excitation, vibration excitation, electronic state excitation and collision ionization. The specific reaction kinetics equations are shown in the Table 5.1 [6].

In the process of solving time-domain electromagnetic fields, in order to ensure the stability and convergence of the calculation, the CFL condition must be met between the time step size of the MCC simulation and the unit grid size.

$$c\mathrm{d}t < \frac{1}{\sqrt{\left(\frac{1}{\mathrm{d}x}\right)^2 + \left(\frac{1}{\mathrm{d}y}\right)^2}} \tag{5.5}$$

where c is velocity of light; x and y represent the moving distance in x and y directions in unit time respectively.

5.3 Results and Discussion

Figure 5.2 shows the geometric model of the two-channel micro-plasma-gun. Two plasma-gun are symmetrically established to facilitate the numerical study of the development process of multichannel spark plasma jet. The electric field distribution in the micro-plasma-gun is approximately V-shaped, as shown in Fig. 5.3.

In order to increase the deposited energy, the needle electrode is set to ground potential and a positive voltage pulse is applied to the plate electrode. The waveform of the Gaussian voltage pulse is shown in Fig. 5.4.

Figure 5.5 shows the space–time evolution of electrons and ions in the spark plasma jet. The initial electrons are continuously emitted from the head of the cathode (needle electrode). Because the applied voltage is in the form of pulse, the background electric field of the micro-plasma-gun is very weak at the initial stage, so the discharge develops slowly before 2 ns. With the increase of time, the background electric field gradually increases, therefore the discharge development speed is significantly accelerated. Before 2.8 ns, the electron number density at the head of the avalanche is the highest. After 3 ns, the highest electron number density is located inside the electron avalanche. As the distribution of electric field is V-shaped, at 3.6 ns, the distribution of electron avalanches is branched and the electrons have reached the anode (plate electrode). With the further development of the discharge, the number density of electron in avalanche increases rapidly and runs through to the anode. At the moment of 4.8 ns, the electron has been ejected from the micro-plasma-gun under the action of electric field force, and the weight of the particle reaches 10^{13}. From space–time distribution of ions, it can be seen that the number of electrons at the initial stage is very small, significantly lower than the number of electrons, and there is a small distance between the ion formation position and the needle electrode (on the order of 0.01 mm). This space electrostatic charge region is the cathode sheath of pulse discharge. At the initial stage, the electron energy is low, and it needs to be accelerated in the electric field for a period of time to accumulate kinetic energy, in order to produce effective ionization reaction and form positive

Table 5.1 The reaction kinetics equations of spark plasma jet

No	Reaction kinetics equations	Excitation or ionization energy
1	$e + N_2 \rightarrow e + N_2$	–
2	$e + N_2 \rightarrow e + N_2{*}$ (Rot)	$E = 0.02$eV
3	$e + N_2 \rightarrow e + N_2{*}$ (VIB V1)	$E = 0.2889$eV
4	$e + N_2 \rightarrow e + N_2{*}$ (VIB 2V1)	$E = 0.5742$eV
5	$e + N_2 \rightarrow e + N_2{*}$ (VIB 3V1)	$E = 0.8559$eV
6	$e + N_2 \rightarrow e + N_2{*}$ (VIB 4V1)	$E = 1.1342$eV
7	$e + N_2 \rightarrow e + N_2{*}$ (VIB 5V1)	$E = 1.4088$eV
8	$e + N_2 \rightarrow e + N_2{*}$ (VIB 6V1)	$E = 1.6801$eV
9	$e + N_2 \rightarrow e + N_2{*}$ (VIB 7V1)	$E = 1.9475$eV
10	$e + N_2 \rightarrow e + N_2{*}$ (VIB 8V1)	$E = 2.2115$eV
11	$e + N_2 \rightarrow e + N_2{*}$ (VIB 9V1)	$E = 2.4718$eV
12	$e + N_2 \rightarrow e + N_2{*}$ (VIB 10V1)	$E = 2.7284$eV
13	$e + N_2 \rightarrow e + N_2{*}$ (VIB 11V1)	$E = 2.9815$eV
14	$e + N_2 \rightarrow e + N_2{*}$ (VIB 12V1)	$E = 3.231$eV
15	$e + N_2 \rightarrow e + N_2{*}$ (VIB 13V1)	$E = 3.4769$eV
16	$e + N_2 \rightarrow e + N_2{*}$ (VIB 14V1)	$E = 3.7191$eV
17	$e + N_2 \rightarrow e + N_2{*}$ (VIB 15V1)	$E = 3.9576$eV
18	$e + N_2 \rightarrow e + N_2{*}$ (A3SIG V = 0–4)	$E = 6.725$eV
19	$e + N_2 \rightarrow e + N_2{*}$ (A3SIG V = 5–9)	$E = 7.36$eV
20	$e + N_2 \rightarrow e + N_2{*}$ (B3PI V = 0–3)	$E = 7.744$eV
21	$e + N_2 \rightarrow e + N_2{*}$ (W3DEL V = 0–5)	$E = 8.05$eV
22	$e + N_2 \rightarrow e + N_2{*}$ (A3SIG V = 10–21)	$E = 8.217$eV
23	$e + N_2 \rightarrow e + N_2{*}$ (B3PI V = 4–16)	$E = 8.451$eV
24	$e + N_2 \rightarrow e + N_2{*}$ (W3DEL V = 6–10)	$E = 8.729$eV
25	$e + N_2 \rightarrow e + N_2{*}$ (A1PI V = 0–3)	$E = 8.95$eV
26	$e + N_2 \rightarrow e + N_2{*}$ (B!3SIG V = 0–6)	$E = 8.974$eV
27	$e + N_2 \rightarrow e + N_2{*}$ (A1SIG V = 0–6)	$E = 9.191$eV
28	$e + N_2 \rightarrow e + N_2{*}$ (W3DEL V = 11–19)	$E = 9.562$eV
29	$e + N_2 \rightarrow e + N_2{*}$ (W1DEL V = 0–5)	$E = 9.59$eV
30	$e + N_2 \rightarrow e + N_2{*}$ (A1PI V = 4–15)	$E = 9.665$eV
31	$e + N_2 \rightarrow e + N_2{*}$ (B3SIG V = 7–18)	$E = 9.933$eV
32	$e + N_2 \rightarrow e + N_2{*}$ (B3SIG V = 7–19)	$E = 10.174$eV
33	$e + N_2 \rightarrow e + N_2{*}$ (W1DEL V = 6–18)	$E = 10.536$eV
34	$e + N_2 \rightarrow e + N_2{*}$ (C3PI V = 0–4)	$E = 11.188$eV
35	$e + N_2 \rightarrow e + N_2{*}$ (E3SIG)	$E = 11.875$eV
36	$e + N_2 \rightarrow e + N_2{*}$ (A1SIG V = 0–1)	$E = 12.289$eV
37	$e + N_2 \rightarrow e + N_2{*}$ (B1PI V = 0–6)	$E = 12.771$eV

(continued)

Table 5.1 (continued)

No	Reaction kinetics equations	Excitation or ionization energy
38	$e + N_2 \rightarrow e + N_2{}^*$ (C1SIG V = 0–3)	$E = 12.95$eV
39	$e + N_2 \rightarrow e + N_2{}^*$ (G3PI V = 0–3)	$E = 13.001$eV
40	$e + N_2 \rightarrow e + N_2{}^*$ (C3 1PI V = 0–3)	$E = 13.093$eV
41	$e + N_2 \rightarrow e + N_2{}^*$ (F3PI V = 0–3)	$E = 13.174$eV
42	$e + N_2 \rightarrow e + N_2{}^*$ (B1SIG V = 0–10)	$E = 13.371$eV
43	$e + N_2 \rightarrow e + N_2{}^*$ (B1PI V = 7–14)	$E = 13.564$eV
44	$e + N_2 \rightarrow e + N_2{}^*$ (O3 1PI V = 0–3)	$E = 13.5644$eV
45	$e + N_2 \rightarrow e + N_2{}^*$ (B!1SIG V = 6–18)	$E = 14$eV
46	$e + N_2 \rightarrow e + N_2{}^*$ (SUM SINGLETS)	$E = 14.2$eV
47	$e + N_2 \rightarrow e + e + N_2{}^+$	$E = 15.581$eV

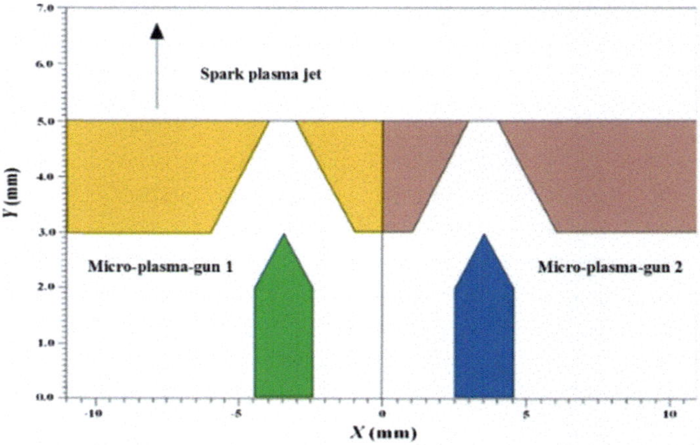

Fig. 5.2 Geometric model of the micro-plasma-gun

Fig. 5.3 Electric field distribution of the micro-plasma-gun

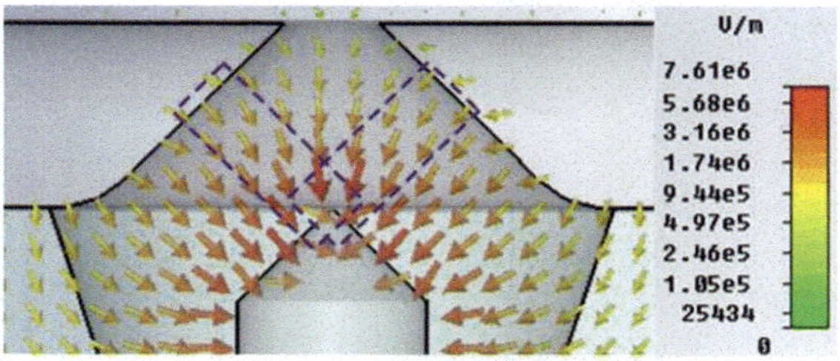

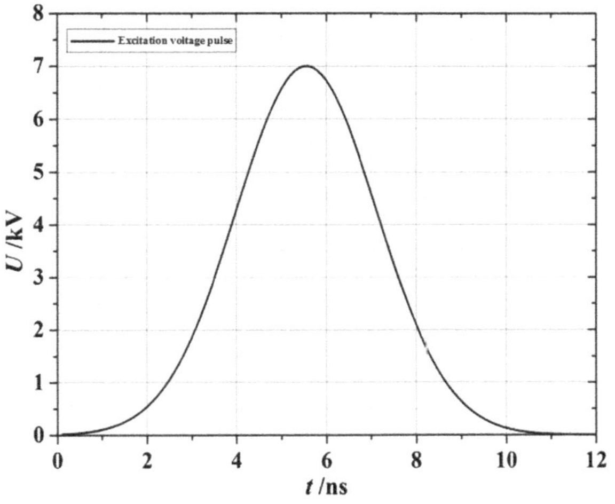

Fig. 5.4 Applied voltage pulse waveform

ions, which is shown as a positive ion blank area in space. With the development of discharge, the number of ions continues to increase, the spatial distribution of ions is similar to that of electrons, and the thickness of the cathode sheath is also gradually thinning. After about (3.8–4) ns, the cathode sheath basically disappears. In addition, the migration speed of positive ions is more than two orders of magnitude lower than that of electrons, so once ions are ionized and formed, their spatial position changes very little. As can be seen from Fig. 5.5, ions can be observed outside the micro-plasma-gun at 4.8 ns. Because the electric field outside is very weak and ionization efficiency is relatively low, the number of ions is significantly lower than that of electrons, and electrons are mainly concentrated in the jet head.

Figure 5.6 shows the electric field at the axis of the micro-plasma-gun with time. At 2 ns, the maximum field strength is located at the tip of the needle electrode, and then gradually decreases. From 2 to 3.6 ns, the amplitude of the electric field increases geometrically, but the envelope of the electric field is basically the same. When the discharge develops to about 4 ns, the electric field envelope has changed significantly. A "step" with high field strength is formed at (0.38–0.4) cm, and the field strength is 30 kV/cm, accounting for 37.5% of the peak background field strength. When the discharge develops to 4.8 ns, the space charge field in the discharge channel is very obvious because the number of electrons is significantly higher than that of ions, accounting for about 50% of the peak background field strength.

Figure 5.7 shows the curve of average electron energy and number with time. It can be seen from the figure that the electron energy gradually increases from 1 ns, accompanied by weak oscillation, and the maximum electron energy is close to 18 eV. At 3.7 ns, the electron energy drops rapidly, and the electron energy at critical streamer stage is about 4.3 eV. The number of electrons increases exponentially. At the time of (0–2) ns, due to the low initial electron energy, the increase of the number of electrons mainly depends on the cathode field emission, and the number is in order

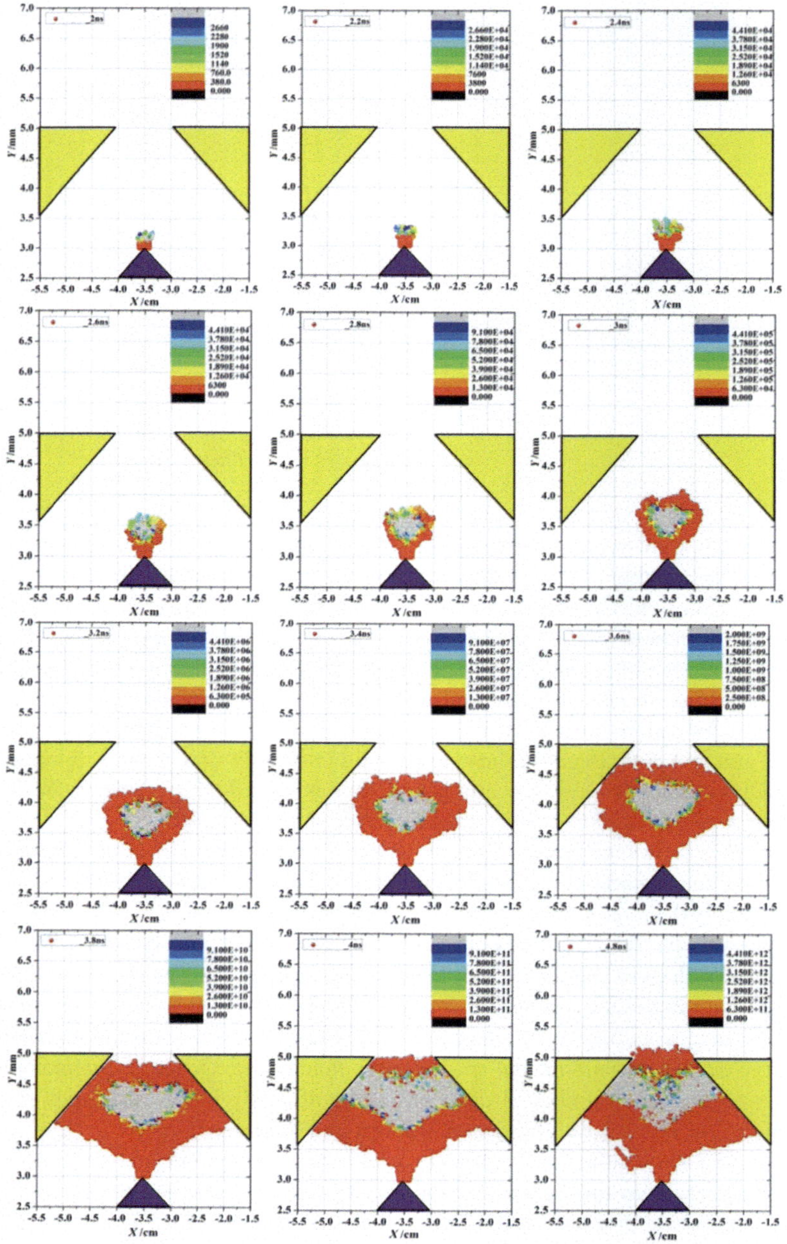

(a) Space-time distribution of electron in spark plasma jet

Fig. 5.5 Space–time distribution of charged particle in spark plasma jet

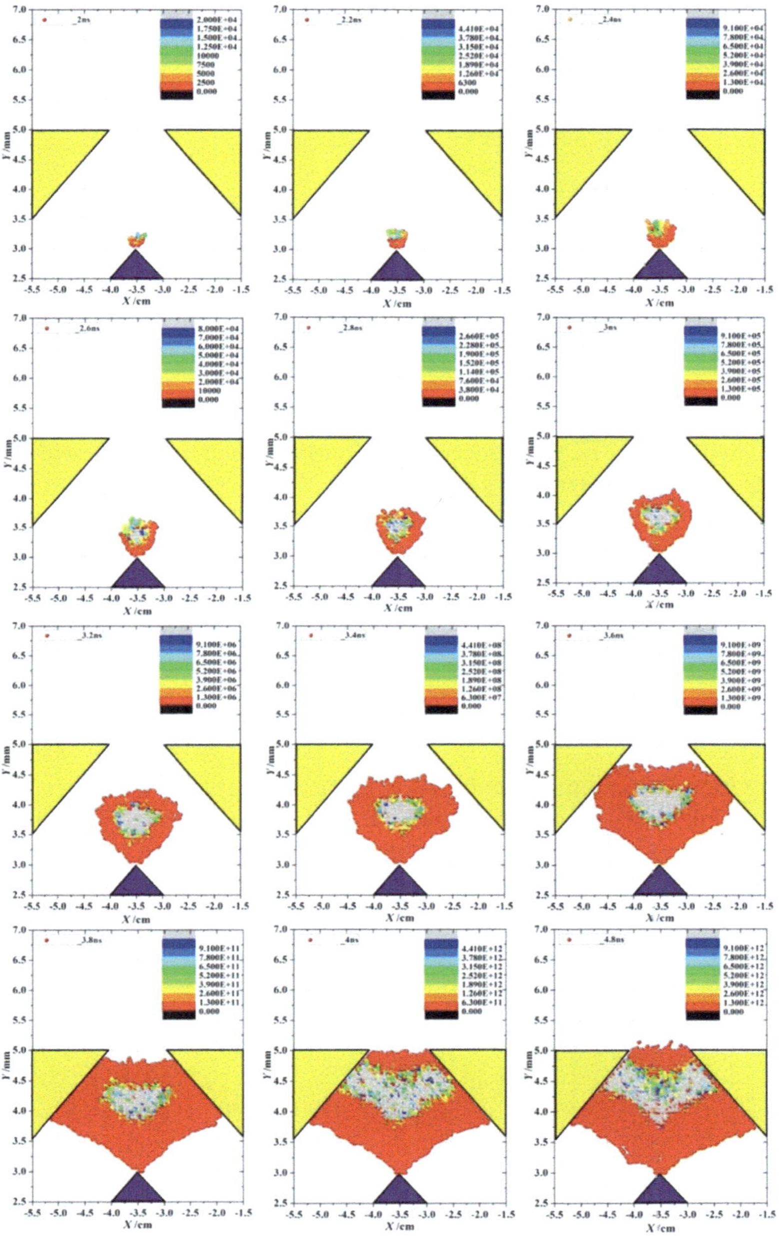

(b) Space-time distribution of ion in spark plasma jet

Fig. 5.5 (continued)

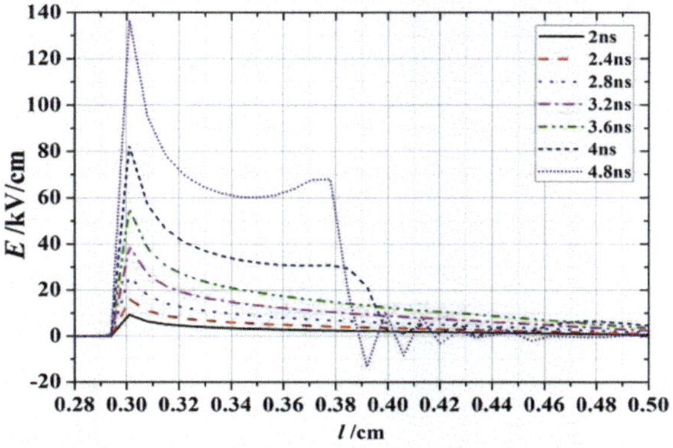

Fig. 5.6 Distribution of electric field in the micro-plasma-gun

of 10^2. From 2 to 4.2 ns, the number of electrons entered a period of rapid growth, showing an order of magnitude growth. After 4.2 ns, the number of electrons tends to be stable, about 10^{12} order of magnitude. Then, it gradually increased to 10^{13}. By comparing the electron energy and number curves, it can be inferred that the discharge has been transformed into streamer at about 4.5 ns.

J. C. Martin's multi-channel discharge judgment formula points out that for the gas discharge occurring at the pulse front stage, the voltage rise rate is the key factor affecting the multi-channel discharge, especially determining the statistical delay of the gas gap. This means, effective initial electrons can be generated earlier under that fast leading edge pulse [7]. Figure 5.8 shows the fast and slow voltage pulse waveforms.

Figure 5.9 shows the time-varying curve of the average electron energy of the two channel spark plasma jet under fast pulse. It can be seen from the figure that when the two discharge channels have good synchronization characteristics in the time domain, the electron energy variation trends of the two channels are very similar. In the critical streamer stage, the electron energy of the successive discharge channels

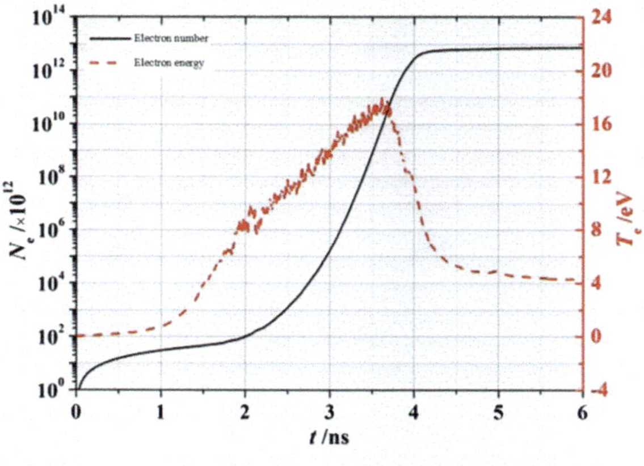

Fig. 5.7 Electron number and energy in spark plasma jet with time

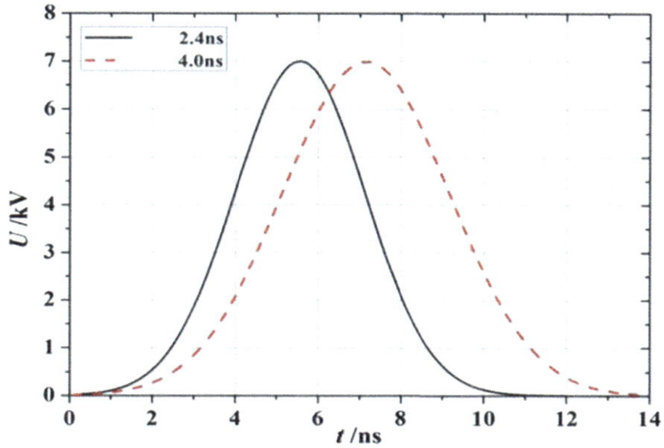

Fig. 5.8 Fast and slow two voltage pulse waveforms

are approximately the same. The curve of the number of electrons with time is shown in Fig. 5.10. The number of electrons of the two channels is also very close, and the time difference of converting into streamer discharge is no more than 0.3 ns. For nanosecond pulse discharge, the discharge voltage of gas gap is generally several times of DC static discharge voltage, and the voltage drop time is usually several ns. If the distance between two adjacent discharge channels is in the range of cm, the propagation delay of electromagnetic wave is about 10^{-1} ns. The discharge jitter of micro-plasma-gun is 0.2 ns at one atmospheric pressure. According to Martin's multi-channel discharge formula, the probability that the post discharge channel will eventually form streamer discharge under fast pulse is great.

Figure 5.11 shows the space–time evolution of electrons and ions in the development process of two-channel spark plasma jet. When the time difference between the formations of effective initial electrons in the two micro-plasma-guns is small,

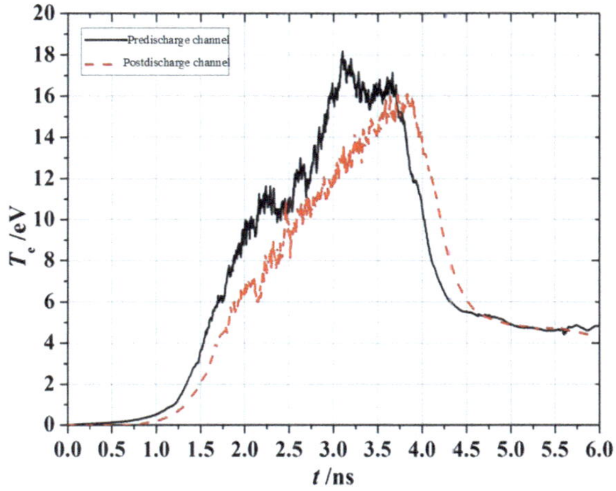

Fig. 5.9 Electron energy in fast and slow voltage pulse with time

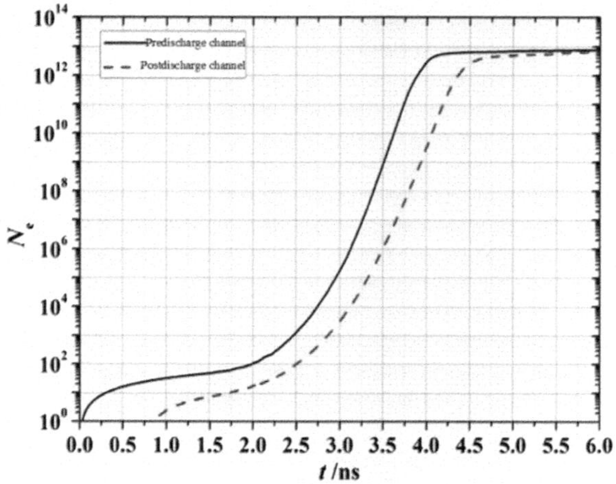

Fig. 5.10 Electron number in fast and slow voltage pulse with time

the development speed of the two plasma-guns at the same time is very similar, and the space–time distribution characteristics of charged particles are also similar. It is noted that at the moment of 4.5 ns, both the first and second discharge channels penetrate the gas gap of the plasma-gun. At this time, the number of electrons in the plasma jet reaches 10^{12}, and the radius of the discharge channel is in the order of 0.1 mm. It can be further estimated that the electron density reaches $(10^{15}–10^{16})/cm^3$. It can be seen that a higher density of discharge plasma is formed in both discharge channels.

Figure 5.12 shows the discharge delay time and jitter of gas switch triggered by different modes. It can be seen from the figure that under the same working coefficient, the spark plasma jet triggered gas switch has the lowest discharge delay and jitter. While under the field distortion trigger mode, the discharge characteristics of the gas switch is the worst. Further study found that compared with the single channel spark plasma jet trigger mode, the gas switch has the lowest delay time and jitter at the multichannel spark plasma jet trigger mode with the working coefficient higher than 50%. However, under the action of single channel spark plasma jet trigger mode, the reliable operation range of the switch is relatively wider.

5.4 Conclusions

A multichannel spark plasma jet particle model based PIC-MCC principle is built to study physical process. During the electron avalanche discharge stage, the electron energy gradually increases to over 10 eV, while the number of electrons increases exponentially. After being converted into streamer discharge, the space charge field in the plasma channel is significantly enhanced, and the electron energy rapidly drops to several eV, and the growth rate of the electron number tends to saturation

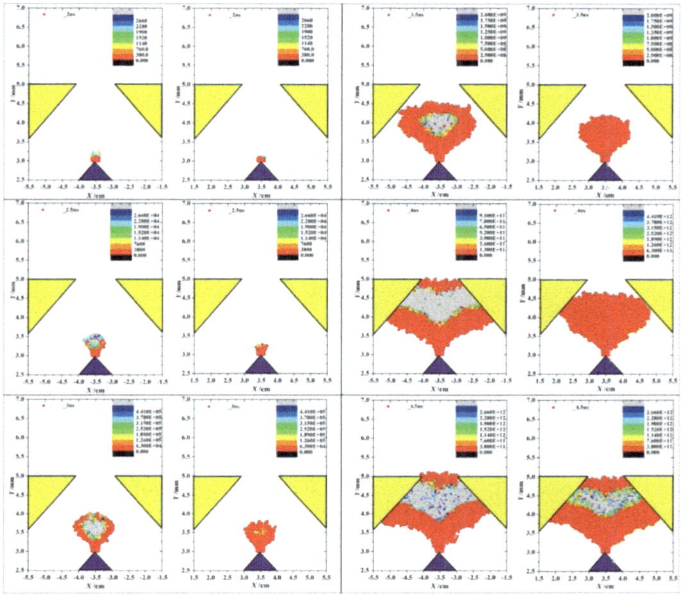

(a) Space-time distribution of electron in two-channel spark plasma jet

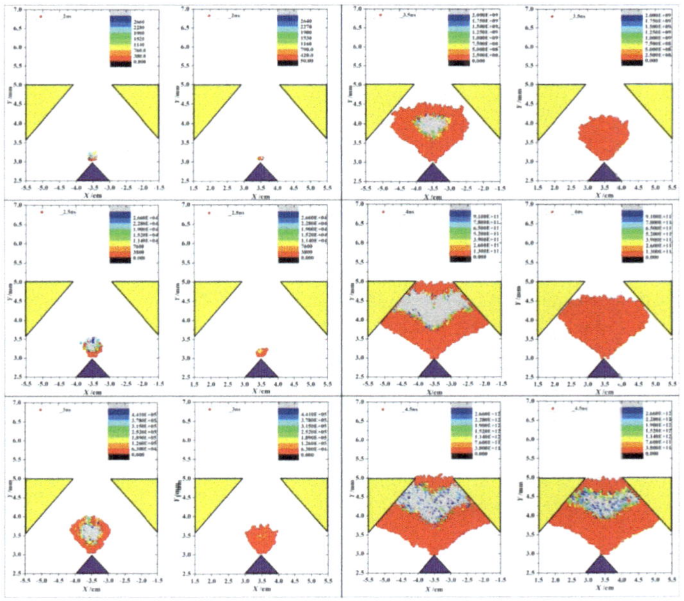

(b) Space-time distribution of ion in two-channel spark plasma jet

Fig. 5.11 Space–time distribution of charged particle in two-channel spark plasma jet

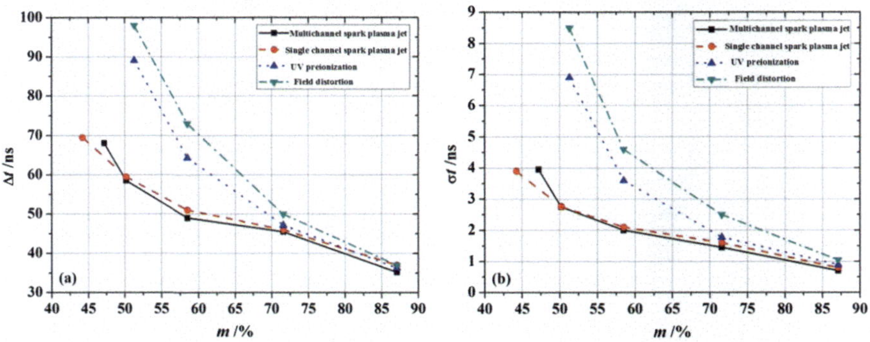

Fig. 5.12 Discharge delay time and jitter of gas switch triggered by different modes

about 10^{12}–10^{13}. Accelerating the voltage pulse at the forefront helps to increase electron energy and number, and makes it easier to form multichannel plasma jet. Compared to UV preionization, field distortion and single channel spark plasma jet trigger modes, gas switch has lower discharge delay and jitter under multichannel spark plasma jet trigger mode. However, the gas switch triggered by single channel spark plasma jet can be operated at lower working coefficient.

References

1. W. Tie, Y. Zhang, C. Meng, Q. Zhang, Z. Yan, P. He, Effect of pulse energy on the dynamic development of spark-plasma-jet driven by nanosecond high-voltage pulse. Plasma Sources Sci. Technol. **27**(1), 015005 (2019)
2. W. Tie, S. Liu, X. Liu, Q. Zhang, L. Pang, L. Liu, A novel low-jitter plasma-jet triggered gas switch operated at a low working coefficient. Rev. Sci. Instrum.Instrum. **85**(2), 023504 (2014)
3. P. Zhang, X. Liu, M. Lin, J. Zhang, Research on the spatio-temporal characteristics of high energy pulsed plasma jets. Phys. Plasmas **29**(9), 093505 (2022)
4. X. Li, G. Gao, P. Jiang, N. Zhang, Z. Li, Q. Zhang, An enhanced plasma injection method for triggering a 10 cm-magnitude high-pressure SF 6 gas gap for a megavolt ultrafast bypass switch. Rev. Sci. Instrum.Instrum. **92**(10), 104701 (2021)
5. C. Li, H. Fan, R. Cao, W. Yang, Y. Li, Study on the breakdown characteristics of the trigatron spark gap triggered by plasma jet. AIP Adv.Adv **10**(1), 015002 (2020)
6. J. Chang, Study on Multichannel Discharge Characteristics of a Three Electrode Coaxial Field Distortion Gas Spark Switch. Xian Jiaotong University (2016)
7. http://www.lxcat.laplace.univ-tlse.fr

Chapter 6
A PIC Simulation Method for Ultra-Long-Range Transport of High-Energy Charged Particle Beams

Yongchun Shao, Weihao Tie, and Xiaocheng Hu

Abstract In some situation where precise simulation of phase space or radius of beam are necessary during the propagation of charged particle beams, we must include the effects of space charge forces on the beam. However, for the commonly used particle-in-cell(PIC) algorithm in those situation, directly solving the static electromagnetic field based on particle distribution and current distribution will lead to a poor accuracy of simulation results because retarded potential was not considered. While calculating retarded potential directly in the laboratory frame will result in a very high computing cost. Here we propose a simulation method based on relativistic coordinate system transformation for high-energy particle beam propagation. In this method, the particles and external electromagnetic fields are transformed to the beam rest frame where the PIC method are carried out, and the results are transformed back to the laboratory frame finally. Because of coordinate transformation, the influence of retarded potential can be reduced, and the simulation will require shorter propagation time and smaller modeling domain with a ratio of nearly one in relativity factor γ. Therefore, using this simulation method will greatly reduce the computational cost of beam transmission and achieve higher computational accuracy for simulation of propagation.

6.1 Introduction

The simulation and analysis of beam dynamics of high-energy and strong-current charged particle beams is an important part of beam optics research and accelerator design. One of the challenging problems is to calculate the nonlinear space charge effect of the particle beam, that is, the electromagnetic field produced by the particle beam itself. There are some models used to calculate or simulate space charge effects, including envelope equation, particle–particle (PP), PIC, and so on.

Y. Shao (✉) · W. Tie · X. Hu
Xi'an Branch of China Academy of Space Technology, Beijing, China
e-mail: shao_yongchun@163.com

Z. Fang et al. (eds.), *Proceedings of the 5th International Symposium on Plasma and Energy Conversion*, Springer Proceedings in Physics 398,
https://doi.org/10.1007/978-981-97-2245-7_6

The envelope equation is a set of ordinary differential equations about particle radius and axial position based on a series of assumptions, which can achieve fast calculation [1]. The assumptions of this method are far from reality, so the accuracy is low, and it cannot provide detailed data on specific particles in the beam. The PP method directly calculates the vector sum of the electromagnetic fields of all other particles at a certain particle position, which can achieve a high calculation accuracy. While this method requires a huge calculational cost. The computational complexity of the PP method is proportional to the $O(N_p^2)$, where N_P are total number of computational macroparticles used in the simulation. The most widely used method is PIC, which uses a computational grid to replace the charge density distribution and current distribution of macroscopic particles, and solves the Poisson equation on the grid at each time step. The computational complexity of the PIC method is proportional to the $O(N_p)$, which makes simulation faster for many applications [2, 3].

However, the above methods are all calculations performed in the laboratory coordinate system. In this paper, we propose a PIC beam simulation calculation method based on special relativity. In some specific scenarios, such as the transmission of particle beams in free space, the particles and external electromagnetic fields are transferred to a high-speed coordinate system by using Lorentz transformation. The PIC calculation is then performed in this coordinate system, and the results are converted back to the laboratory coordinate system finally. This method can avoid the influence of the retarded potential of high-energy charged particles, reduce the transport time that needs to be calculated, simplify the calculation of self-space charge fields. Thus, this method can achieve higher computational accuracy and require lower computational costs.

6.2 Method

6.2.1 Initial Conditions for Beam Dynamics Simulation

In PIC simulations, macroparticles are usually used to replace real particles for simulation, that is, the charge and mass of multiple real particles are replaced by a simulated particle with the same charge-to-mass ratio of a real charged particle, so as to greatly reduce the number of particles that need to be calculated in simulation. These charged macroparticles are affected by external fields as well as self-generated space charge fields from Coulomb interactions between charged particles. The trajectory and phase space of particle beams are nalysed through the calculation of macroparticles. The initial state of the macroparticles can be obtained by sampling based on the initial phase space of the particle beam, or based on the PIC simulation results of the macroparticles in the previous process.

6.2.2 Coordinate System Conversion of Particle Beam and External Electromagnetic Fields

Establish a laboratory coordinate system s, in which the z-axis direction is the transport direction of the particle beam, x-axis and y-axis are the cross-sectional directions of the particle beam, time 0 is the initial moment of the particle, and z = 0 is the coordinate of the initial particle beam center of mass. The s' system is an inertial coordinate system determined based on the initial position and velocity of the center of mass of the particle beam. The time when the origins of the two coordinate axes coincide is the 0 time of the s' system. The initial position and velocity of the macroparticle in the s' system are obtained by the Lorentz transformation of the special theory of relativity.

However, due to the different initial z position in s system of the particle beam, the converted time for different particles in s' system is different. While, since the time difference is quite small, the space charge effect within the time difference can be ignored, and a charged particle tracking calculation that only considers the external field is performed in the original coordinate system. Particles at different z-axis positions in the s system choose different moments, so that the particles in the s' system are at the same moment.

Since the external electromagnetic field is almost unaffected by the particle beam, the external field is usually solved first and then used as a constraint for beam dynamics calculations. In addition, since most external fields are generated by beam optical components that are stationary in the laboratory coordinate system, it can be easily calculated in the laboratory coordinate system. After the calculation of the external electromagnetic field is completed in s system, the external field is then converted to the s' coordinate system by the Lorentz transformation.

6.2.3 Calculation of Self-Space Charge Field in S' System

For beams with a low energy dispersion, the particle speed will be much smaller than the speed of light after the Lorentz transformation, so the influence of the retarded potential can be ignored. So the real electric field is very close to the electrostatic field under this particle distribution, and there is almost no self-generated magnetic field. Therefore, the space charge field can be obtained by solving Poisson's equation, and the boundary conditions of the potential for beam transporting in the pipeline can be set as

$$\begin{cases} \varphi(r = R) = 0 \\ \varphi(z = a) = 0 \\ \varphi(z = b) = 0 \end{cases} \tag{6.1}$$

For beam transporting in the free space, the boundary conditions of the potential can also be set as above, or be set to a spherical boundary as

$$\varphi(r = R) = \frac{Q}{4\pi\varepsilon_0 R} \tag{6.2}$$

The charge distribution of the grid can be obtained directly by summing the number of macroparticles in the corresponding grid, or for some certain regular mesh grid, a macroparticle can distribute the charge amount to the grid where it is located and surrounding grids according to a certain proportion [4].

After solving the Poisson equation to obtain the electric potential through the finite element method or integrated Green function method [5], the corresponding electric field distribution can be obtained. Smaller mesh accuracy will result in more accurate solution and higher computational costs. The size of the grid can be adaptively increased as the size of the particle beam increases.

Compared with the previous beam dynamics PIC solution algorithm, the boundary conditions of this method are still, the computational domain in the z direction is smaller, and only the self-generated electrostatic field with no self-generated magnetic field need to be solved. If the self-space charge field data in the s system is needed, you can convert the corresponding space charge electric field in the s' system back to the s system using Lorentz transformation.

6.2.4 Charged Particle Tracking

Kinetic equation of particles is

$$\frac{\mathrm{d}\vec{p}}{\mathrm{d}t} = q(\vec{E}_{sc} + \vec{E}_{ext} + \vec{v} \times \vec{B}_{ext}) \tag{6.3}$$

where $\vec{E}_{sc}$ is the vector of the space charge electrostatic field, $\vec{E}_{ext}$ and $\vec{B}_{ext}$ are the vectors of external electromagnetic fields in s' system.

In some cases where the energy dispersion is small, since the particle speeds in the s' coordinate system are low, Newton's kinematic equations can be directly used to calculate the motion of the particles. Otherwise, calculations need to be performed using relativistic kinematics.

Shorter calculation steps will result in higher calculation accuracy and calculation cost. The time step can be appropriately increased as the particle beam disperses and the space charge effect decreases. After the particle's position at the next moment is determined, the space charge electromagnetic field is recalculated based on the updated position.

6.2.5 Results Expressed in Laboratory Coordinate System

Due to the time dilation, the transport time required for simulation in s' system is only about

$$\frac{1}{\gamma} = \sqrt{1 - \frac{v^2}{c^2}} \tag{6.4}$$

times of that in the laboratory coordinate system, a large amount of calculations can be saved. When the simulation reaches the s' system's final state, the position and speed of particles is converted back to the s system by the Lorentz transformation. The particles are not at the same moment in s system after transformation, but since most of the particle beams are in a relatively dispersed state at this time, the space charge effect is small. Charged particle tracking can be directly calculated in the s system without considering the space charge force. For more accurate, different moments in s' system are extracted respectively for particles of different energies, so that they are all at the same moment in the s system when converted back.

6.2.6 Scope of Application of the Method

This method is suitable for simulation calculations of high-energy particle beams with relatively low energy dispersion and small emission angles under relatively small acceleration conditions. Otherwise, the particles still have a large velocity in the s' system, and the self-generated electromagnetic field cannot be solved directly based on Poisson's equation.

When the beam velocity changes significantly, a threshold can also be set. When the velocity of beam in the s' system exceeds the threshold, the s' system can be changed to another high-speed inertial coordinate system.

6.2.7 Flow Chart

Based on the above process, the flow chart of this method is drawn as follows (Fig. 6.1).

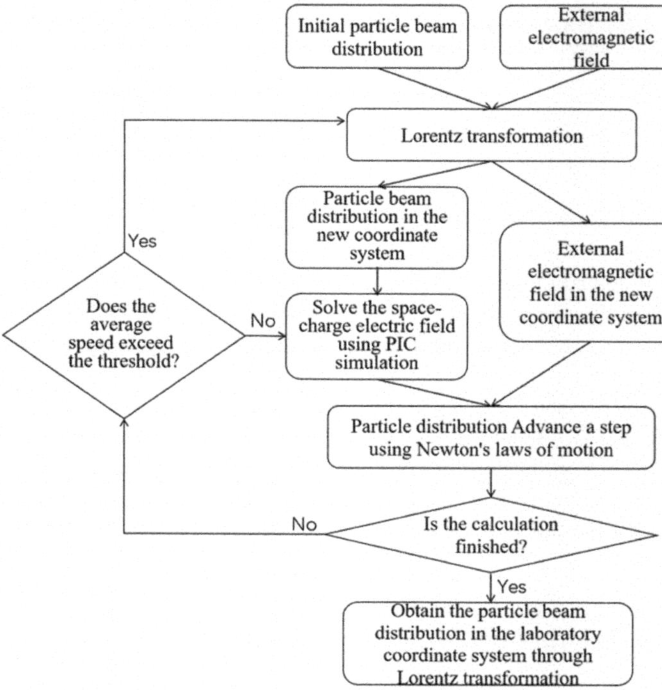

Fig. 6.1 The flow chart of the beam dynamics PIC simulation based on special relativity

6.3 Examples

We use the PIC method in high-speed coordinate system to simulate a long-range transport of a high-energy electron beam in free space. The initial shape of electron beam is columnar and the electron distribution is uniform. The transverse radius is 5 mm, the axial length is 20 mm and the current is 1 A. The energy of electron is 10 meV with no energy dispersion, and the velocity direction is along the axial direction. The simulated transport distance is 1 km, the transmission process time in the laboratory coordinate system is 3.34 ps, and the time in the particle beam center of mass coordinate system is 0.16 ps. The variation of transverse rms radius along the transport distance and the beam transverse shape at 1 km are as follows (Figs. 6.2 and 6.3).

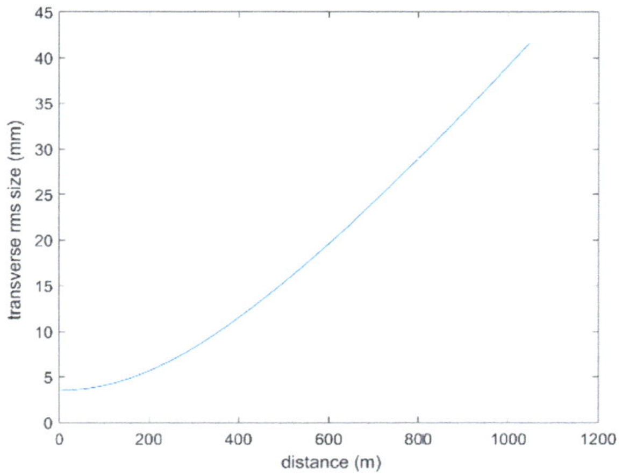

Fig. 6.2 The variation of transverse rms radius along the transport distance

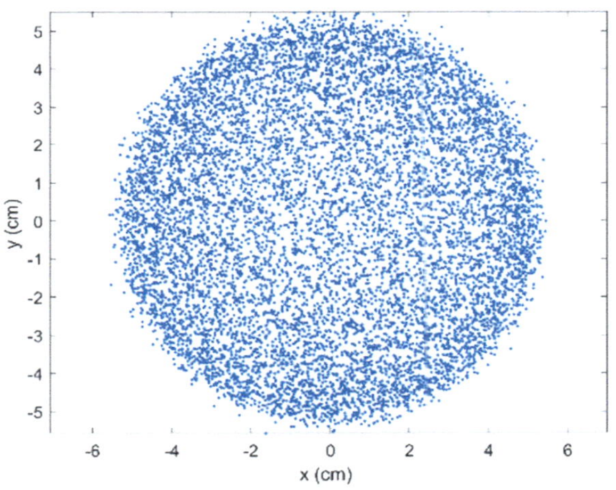

Fig. 6.3 The beam transverse shape at 1km

6.4 Conclusion

In this paper, we propose a high-energy particle beam dynamics simulation method based on coordinate system transformation, which can effectively solve the simulation of ultra-long-distance beam transmission. In addition, this method is also suitable for simulation calculations of high-energy particle beams with relatively low energy dispersion and small emission angles under relatively small acceleration conditions. Compared with the PIC simulation of beam dynamics in the laboratory coordinate system, this method can effectively eliminate the influence of the retardation potential, omit the calculation of the endogenous magnetic field, the boundary conditions of the Poisson equation are simpler, and use Newtonian mechanics rather than the theory of relativity. Mechanics simplifies calculations, and the physical time required for calculation is shortened to $1/\gamma$. Therefore, using this method can better reduce the computational cost of beam dynamics.

References

1. G.A. Nagy, M. Szilágyi, Introduction to the Theory of Space-charge Optics. NASA STI/Recon Technical Report A (1974)
2. J. Qiang, Symplectic particle-in-cell model for space-charge beam dynamics simulation. Phys. Rev. Accel. Beams (2018)
3. Y.K. Batygin, Particle-in-cell code BEAMPATH for beam dynamics simulations in linear accelerators and beamlines. Nucl. Instrum. Methods Phys. Res., Sect. A **539**(3), 455–489 (2005)
4. D. Phadte, C.B. Patidar, M.K. Pal, Development of a relativistic particle in cell code PARTDYN for linear accelerator beam transport. Nucl. Instrum. Methods Phys. Res., Sect. A **851**, 20–28 (2017)
5. J. Qiang, S. Lidia, R.D. Ryne et al., Three-dimensional quasistatic model for high brightness beam dynamics simulation. Phys. Rev. Special Topics-Accel. Beams (2006)

Chapter 7
Influence of O_2 on Thermophysical Properties of C_4F_7N–CO_2 Arc Plasmas

Chenxu Zhang, Tongxin Xu, Wenhui Sun, and Linlin Zhong

Abstract C_4F_7N is a potential SF_6 replacement as insulating and arc quenching medium in power system. It is usually mixed with CO_2 and O_2 to further reduce the global warming potential (GWP) and liquefaction temperature of gas mixtures and also to prevent the condensation of graphite after the arc discharge. In this work, we perform a numerical study on thermophysical properties of C_4F_7N–CO_2–O_2 arc plasmas at different proportions of O_2. We first obtain the compositions of C_4F_7N–CO_2–O_2 arc plasmas using the method of Gibbs free energy minimization. Based on the plasma compositions, we further calculate the thermodynamic properties, namely mass density, specific heat, and specific enthalpy, at temperatures ranging from 300 to 30,000 K. Also, the transport coefficients including electrical conductivity, viscosity, and thermal conductivity of C_4F_7N–CO_2–O_2 arc plasmas are calculated based on the Chapman-Enskog theory. Lastly, we investigate the influence of O_2 on the compositions, thermodynamic properties, and transport coefficients of C_4F_7N–CO_2–O_2 arc plasma, which could provide basic knowledge and data for the further study of C_4F_7N gas as arc quenching medium.

7.1 Introduction

Sulfur hexafluoride (SF_6) is widely used in electrical equipment due to its exceptional insulation and arc quenching properties. However, despite numerous advantages, SF_6 possesses a significant drawback: its greenhouse warming potential (GWP) is estimated to be 23,900 times higher than that of CO_2. Consequently, SF_6 was designated as a restricted-use greenhouse gas, as mentioned in The Kyoto Protocol [1]. Reducing the reliance on SF_6 gas and exploring environmentally friendly alternative gases is becoming an important research focus in the power industry.

Numerous research studies have been conducted to identify more environmentally friendly alternatives to SF_6 gas for insulation purposes. The ideal SF_6 alternative

C. Zhang · T. Xu · W. Sun · L. Zhong (✉)
School of Electrical Engineering, Southeast University, Nanjing 210096, China
e-mail: 213213526@seu.edu.cn

Z. Fang et al. (eds.), *Proceedings of the 5th International Symposium on Plasma and Energy Conversion*, Springer Proceedings in Physics 398,
https://doi.org/10.1007/978-981-97-2245-7_7

gases should possess low GWP, ensure safety, and be environmentally sustainable. The decades-long search for an alternative identified a new class of compounds that incorporates this type of functionality: C_4F_7N, which exhibits dielectric breakdown strength that is nearly twice that of SF_6, and also atmospherically much shorter lives than SF_6 [2]. Ahmed et al. have conducted a feasibility analysis of C_4F_7N as a substitute to SF_6. Their research shows that the dielectric strength of 20% C_4F_7N and 80% CO_2 reaches the same level as SF_6 [3]. However, the potential uses of pure C_4F_7N are impacted by its high liquefaction temperature (4.7 °C, compared to SF_6's −63 °C). To address this limitation, C_4F_7N gas should be mixed with buffer gases like CO_2 and N_2 [4]. Ye et al. have investigated the impact of O_2 content on the partial discharge (PD) characteristic of C_4F_7N–CO_2–O_2 gas mixtures [5]. In order to prevent the condensation of graphite after the arc discharge, C_4F_7N is typically mixed with CO_2 and O_2. The experimental results [6] show that the mixture of C_4F_7N–CO_2 decomposes into C_3F_8, CF_3CN, C_3F_6, C_4F_{10}, C_2F_5CN and C_4F_8 at an overheated temperature of 500 °C, and the mixture of C_4F_7N–CO_2 decomposes into C_3F_8, CF_3CN, C_3F_6 after adding O_2. However, previous studies have primarily focused on the insulating properties of C_4F_7N without thoroughly investigating its characteristics as an arc quenching medium.

This paper presents a numerical study on thermophysical properties of C_4F_7N–CO_2–O_2 arc plasma at different proportions of O_2. The methods used to calculate the composition, thermodynamic properties, and transport coefficients are described in Sect. 7.2. In Sect. 7.3, we focus on the influence of O_2 on the composition, thermodynamic properties and transport coefficients of C_4F_7N–CO_2 arc plasmas. Finally, the conclusion is presented in Sect. 7.4, summarizing the findings of our research.

7.2 Calculation Method

7.2.1 Composition and Thermodynamic Properties

The calculation of plasma composition lays the basis for the research of thermodynamic and transport properties. As can be seen from previous studies [7], it is often assumed that the system is in a local thermodynamic equilibrium (LTE) state to simplify the calculations, which means the electrons, ions and neutral particles have the same kinetic temperature. Considering that we are required to solve for the composition of particle components at different temperatures, we combine chemical equilibrium, Dalton's partial pressure law, and electrical neutrality with a system of non-linear equations to solve the problem [8]. In this paper, the Gibbs free energy method is applied to calculate the composition. The thermodynamic function is the Gibbs free energy that has to be minimal if the thermodynamic system reached the thermodynamic equilibrium [9]. The Gibbs free energy of the plasma system is written as G:

$$G = \sum_{i=1}^{N} n_i \left(\mu_i^0 + RT \ln \left(\frac{n_i}{\sum_{j=1}^{N} n_j} \right) + RT \ln \left(\frac{P}{P^0} \right) \right) \tag{7.1}$$

where n_i is the mole number, N the number of different species present in the plasma, and μ_i^0 is the chemical potential of i species at reference pressure P^0.

The thermodynamic properties, typically including mass density ρ, specific enthalpy h and specific heat at constant pressure C_p, can be derived from standard thermodynamic relationships. The relevant formulas are given as follows:

$$\rho = \frac{\sum_{i=1}^{N} m_i c_i}{V} \tag{7.2}$$

$$h = \frac{\sum_{i=1}^{N} x_i H_i^0}{M} \tag{7.3}$$

$$C_p = \left(\frac{\partial h}{\partial T} \right)_p \tag{7.4}$$

where m_i and c_i are the mass and amount of particle i respectively in the system and V is the total volume of the gaseous phase. The value of V can be determined using ideal gas law [10].

7.2.2 Transport Coefficients

Transport coefficients, such as electrical conductivity, viscosity, and thermal conductivity, are essential in understanding the behavior of plasmas. These coefficients are calculated using the Chapman-Enskog method, which is based on the assumption that the distribution function of the substance changes due to particle collisions and the macroscopic flow variables or parameters (such as temperature, pressure, fluid velocity, and composition) are the first moments of the distribution function. In the calculation, the Maxwell distribution function of first order perturbation is usually used. The Chapman-Enskog solution to the Boltzmann equation is usually represented by a Sonning polynomial expansion, resulting in a system of linear equations. By solving these linear equations to a selected order, the transport coefficients can be determined [11].

Conductivity refers to the transport of electrons and ions in a plasma under concentration, pressure, and temperature gradients. When calculating the electrical conductivity, the contribution of ions can be ignored, and only the contribution of electrons can be calculated using the three-order approximation. Viscosity represents the momentum transfer caused by velocity gradients. The viscosity coefficient is calculated using a first-order approximation, where the contribution of electrons is omitted and only the influence of heavy particles is considered [12]. Thermal conductivity is the heat transfer caused by thermal gradients, chemical reactions, and internal degrees of freedom. It can be divided into three components: translation thermal conductivity caused by motion energy transfer, internal thermal conductivity caused by internal energy transfer, and reaction thermal conductivity caused by energy transfer due to molecular decomposition and recombination, ionization, and other chemical reactions [13]. In the calculation process, the translation thermal conductivity treats electrons and heavy particles differently, using second and third-order approximations respectively. The internal thermal conductivity is calculated using a second-order approximation. The thermal conductivity can be expressed as the sum of the three components: translation thermal conductivity, internal thermal conductivity, and reaction thermal conductivity [11].

7.3 Results and Discussion

7.3.1 *Influence of O_2 on Compositions of C_4F_7N–CO_2 Arc Plasmas*

The minimization of the Gibbs free energy is used to obtain the equilibrium compositions. Figure 7.1a shows the compositions of C_4F_7N–CO_2 (mixing ratio 1:9) arc plasma mixed with 10% O_2 (volume proportion) at ambient pressure, compared with the compositions in which O_2 fraction is 90% as shown in Fig. 7.1b. At temperatures below 5000 K, oxygen mainly exists in the forms of O_2, CO_2, CO, NO, and F_2CO. However, at temperatures above 5000 K, oxygen mainly exists in the form of O, O^+ and O^{2+}. As the percentage of O_2 increases, the molar fractions of CO_2, CO and NO decrease.

7.3.2 *Influence of O_2 on Thermodynamic Properties of C_4F_7N–CO_2 Arc Plasmas*

The impact of O_2 concentration on the mass density of C_4F_7N–CO_2 arc plasmas at ambient pressure is illustrated in Fig. 7.2. From the graph, it is evident that the mass density of the arc mixtures remains relatively stable as the percentage of O_2 increases.

Figure 7.3 shows the specific heat of C_4F_7N–CO_2 arc plasmas with different O_2 ratios at the same pressure, within the temperature range of 300–30,000 K. In mixtures containing low or medium O_2 concentrations ($\leq$ 50%), four distinct peaks in specific heat are observed at 2400, 3300, 7200, and 15,600 K. As more O_2 is added, these four peaks gradually merge into three peaks. This discrepancy can be attributed

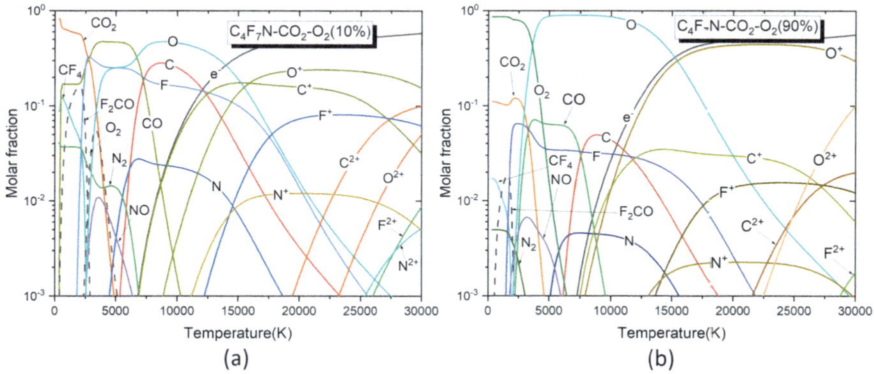

Fig. 7.1 Composition of $C_4F_7N–CO_2$ (mixing ratio 1:9) arc plasma at temperature of 300–30,000 K with O_2 volume fractions of **a** 10% and **b** 90%

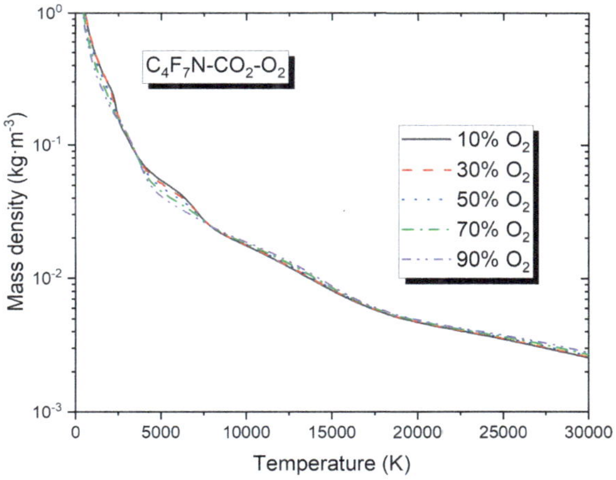

Fig. 7.2 Mass density of $C_4F_7N–CO_2$ (mixing ratio 1:9) arc plasma mixed with different proportions of O_2

to the different species and chemical reactions considered in the calculation. Varying ratios of reactants also lead to different dissociation temperatures and thus different peaks.

Figure 7.4 presents the specific enthalpy of $C_4F_7N–CO_2$ arc plasmas mixed with different proportions of O_2. The specific enthalpy decreases with increasing O_2 content. Additionally, roughly at temperatures below 7500 K, the specific enthalpy slightly increases with the increase of O_2, while above 7500 K, the specific enthalpy slightly decreases with the increase of O_2.

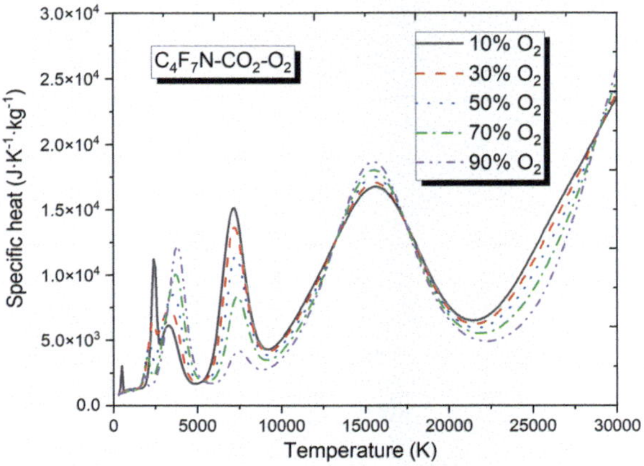

Fig. 7.3 Specific heat of C_4F_7N–CO_2 (mixing ratio 1:9) arc plasma mixed with different proportions of O_2

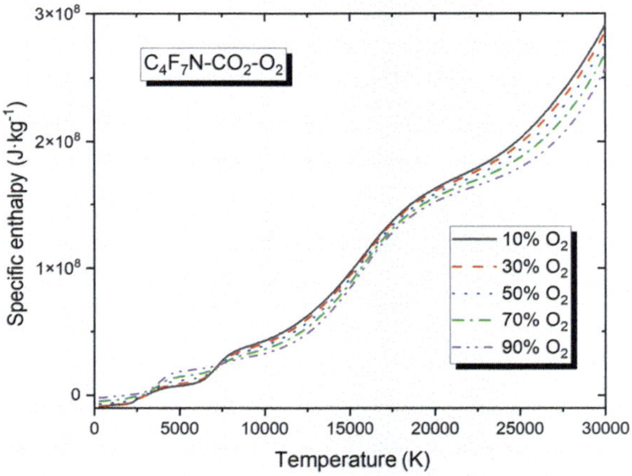

Fig. 7.4 Specific enthalpy of C_4F_7N–CO_2 (mixing ratio 1:9) arc plasma mixed with different proportions of O_2

7.3.3 Influence of O_2 on Transport Coefficients of C_4F_7N–CO_2 Arc Plasmas

Figure 7.5 displays the influence of O_2 on the thermal conductivity of C_4F_7N–CO_2 mixtures. As can be seen, the arc plasma with different O_2 content present four peaks in specific heat at around 2400, 3300, 7000 and 15,500 K. In addition, the peak changes in conductivity flatten out with increasing temperature.

As observed in Fig. 7.6, with the increase of temperature, the viscosity coefficient has a peak at about 10,000 K. Additionally, when the temperature is lower than 12,500 K, the viscosity coefficient increases with the increase of O_2; when the temperature is 12,500–22,500 K, the viscosity coefficient decreases with the

Fig. 7.5 Thermal conductivity of C_4F_7N–CO_2 (mixing ratio 1:9) arc plasma mixed with different proportions of O_2

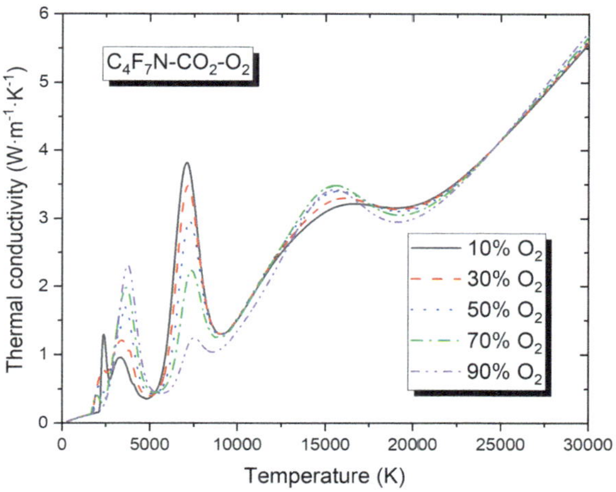

increase of O_2; when the temperature is higher than 22,500 K, the viscosity coefficient increases with the increase of O_2. More O_2 leads to less ions and also the molar mass of O_2 is smaller than C_4F_7N and CO_2, which leading to lower viscosity.

Figure 7.7 describes the electrical conductivity of the C_4F_7N–CO_2 mixtures. With the increase of temperature, the conductivity increases obviously. As seen in Fig. 7.7, the O_2 content does not have a significant effect on the electrical conductivity of the C_4F_7N–CO_2 mixtures below 21,500 K. Above 21,500 K, the electrical conductivity increases with increasing O_2 content.

Fig. 7.6 Viscosity of C_4F_7N–CO_2 (mixing ratio 1:9) arc plasma mixed with different proportions of O_2

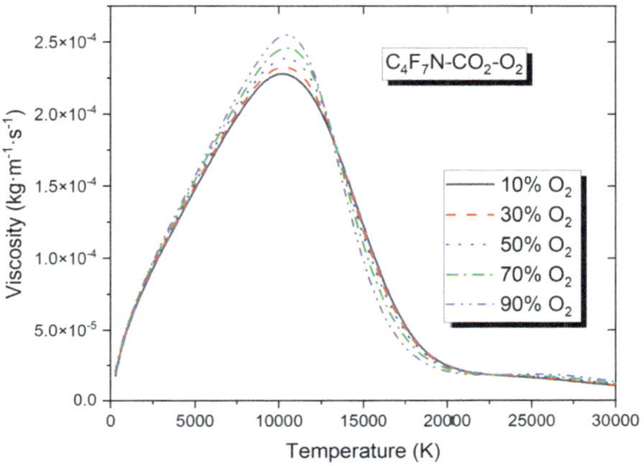

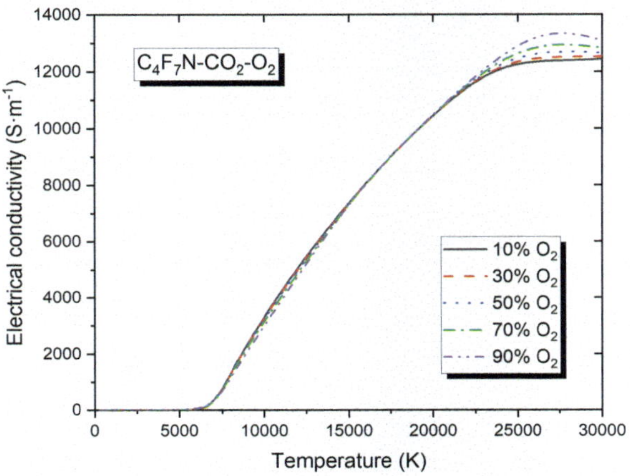

Fig. 7.7 Electrical conductivity of C_4F_7N–CO_2 (mixing ratio 1:9) arc plasma mixed with different proportions of O_2

7.4 Conclusion

The presence of O_2 in the C_4F_7N–CO_2–O_2 arc plasma has a significant impact on the equilibrium compositions, thermodynamic properties and transport coefficients. The important conclusions for the influence of O_2 can be drawn as follows:

1. The increase in oxygen content leads to a more thorough oxidation reaction in the arc plasma, which results in a decrease in the content of CO, NO and C, while increasing the content of oxygen-containing ions such as O^+ and O^{2+}.
2. In general, the addition of O_2 does not alter the overall trend of thermodynamic properties and transport coefficients with the change of temperature. However, it does slightly modify the values at specific temperatures. For example, the change of O_2 concentration changes the peak value of viscosity.
3. The findings of this paper can further provide a database for simulating the arc quenching performance of C_4F_7N–CO_2 mixtures, which can help to understand the differences in the arc quenching performance of different gases as well as facilitate the selection of good performance arc quenching medium.

Acknowledgements This work was supported in part by the Zhishan Young Scholar Project of Southeast University (2242022R40022), and the Student Research Training Program of Southeast University (202316038).

References

1. L. Zhong, J. Wang, J. Xu, X. Wang, M. Rong, Effects of buffer gases on plasma properties and arc decaying characteristics of C_4F_7N–N_2 and C_4F_7N–CO_2 arc plasmas. Plasma Chem. Plasma Process. **39**(6), 1379–1396 (2019)

2. N. Tang, L. Chen, B. Zhang, X. Li, Experimental and theoretical exploration of C_4F_7N gas decomposition under partial discharge, in *2020 IEEE International Conference on High Voltage Engineering and Application*, Beijing, China (2020), pp. 1–4
3. R. Ahmed, R.A. Rahman, M.S. Kamarudin, M.F. Yousof, H. Ahmad, A. Salem, Feasibility of fluoronitrile (C_4F_7N) as a substitute to sulphur hexafluoride (SF_6) in gas insulated application: a review, in *2022 IEEE International Conference on Power and Energy*, Langkawi, Kedah, Malaysia (2022), pp. 391–396
4. H.S. Song, H.S. Kim, B.W. Lee, Comparison of DC PDIV and BDV of GIS defects between SF_6 gas and C_4F_7N/CO_2 gas mixture, in *2022 9th International Conference on Condition Monitoring and Diagnosis*, Kitakyushu, Japan (2022), pp. 529–532
5. F. Ye, X. Zhang, Y. Li, Y. Yao, S. Xiao, X. Zhang, C. Xie, X. Shao, X. Sun, Effect of O_2 on AC partial discharge and decomposition behavior of $C_4F_7N/CO_2/O_2$ Gas Mixture. IEEE Trans. Dielectr. Electr. Insul. **28**(4), 1440–1448 (2021)
6. R. Zhuo, J. Liu, M. Fu, Y. Jing, Y. Luo, H. Song, Effect of oxygen on the thermal decomposition of eco-friendly gas insulating medium C_4F_7N–CO_2, in *2020 IEEE International Conference on High Voltage Engineering and Application*, Beijing, China (2020), pp. 1–4
7. F. Yang, M. Rong, Y. Wu, A. Murphy, S. Chen, Z. Liu, Q. Shi, Numerical analysis of arc characteristics of splitting process considering ferromagnetic plate in low-voltage arc chamber. IEEE Trans. Plasma Sci. **38**(11), 3219–3225 (2010)
8. M.I. Boulos, P. Fauchais, E. Pfender, *Thermal Plasmas: Fundamentals and Applications*, 1st edn. (Springer, New York, USA, 1994)
9. P. André, Z. Koalaga, Composition of a thermal plasma formed from PTFE with copper in non-oxidant atmosphere. High Temp. Mater. Processes (London) **14**(3), 285–294 (2010)
10. M. Rong, L. Zhong, Y. Cressault, A. Gleizes, X. Wang, F. Chen, H. Zheng, Thermophysical properties of SF_6–Cu mixtures at temperatures of 300–30,000 K and pressures of 0.01–1.0 MPa: Part 1. Equilibrium compositions and thermodynamic properties considering condensed phases. J. Phys. D: Appl. Phys. **47** (2014)
11. X. Wang, L. Zhong, Y. Cressault, A. Gleizes, M. Rong, Thermophysical properties of SF_6–Cu mixtures at temperatures of 300–30,000K and pressures of 0.01–1.0 MPa Part 2. Collision integrals and transport coefficients. J. Phys. D: Appl. Phys. **47** (2014)
12. W. Wang, M. Rong, A. Murphy, W. Yi, H. Su, F. Yang, Computation analysis on properties of high temperature nitrogen arc plasmas, in *High Voltage Engineering* (2010)
13. A.B. Murphy, C.J. Arundell, Transport coefficients of argon, nitrogen, oxygen, argon-nitrogen, and argon-oxygen plasmas. Plasma Chem. Plasma Process. **14**, 451–490 (1994)

Chapter 8
The Physicochemical Properties of Pure Water Treated by DBD Reactor Packed with Porous Foam Ceramic

Yuchen Ping, Jingyun Zhang, Min Zhu, and Chaohai Zhang

Abstract Plasma activated water (PAW) has excellent biochemical activity and high bactericidal performance, which is an important application of atmospheric discharge plasma in biomedical field. In this work, a method for preparing DBD plasma activated water based on porous foam ceramic packed was proposed. The effects of pore size of porous foam ceramic, applied voltage and water flow rate on discharge characteristics were investigated. Physical and chemical parameters such as pH value, electrical conductivity value and concentration of H_2O_2, NO_2^-, NO_3^- were measured in PAW. The results show that packing porous foam ceramic in DBD can effectively reduce the discharge threshold and increase the content of reactive nitrogen oxides in PAW. At the same time, it was found that the content of H_2O_2 and other substances in PAW obtained from foam ceramics with a pore of 10ppi was the highest. This study provides a basis for the design of plasma activated water device and the method of increasing the concentration of reactive nitrogen oxides.

8.1 Introduction

In recent years, atmospheric pressure discharge plasma has been widely researched in many fields such as energy conversion [1], material modification [2], wastewater treatment [3], biomedical [4] and so on. As a green, clean, safe and effective new biochemical technology [5], plasma activated water treatment technology produces a variety of active nitrogen oxides (·OH, O_3, H_2O_2, NO_2^-, NO_3^-, etc.) in the water through air discharge, which can have strong biochemical activity and excellent bactericidal performance in a short period of time without the addition of any chemical reagents. The synergistic effect between multiple active substances has been well applied in the sterilization and preservation of fruits and vegetables [6], promotion of seed germination [7], wound surface inhibition [8] and so on.

Y. Ping · J. Zhang · M. Zhu (✉) · C. Zhang
Department of Electrical Engineering, Nanjing University of Aeronautics and Astronautics, Nanjing 211106, China
e-mail: zhumin@nuaa.edu.cn

Z. Fang et al. (eds.), *Proceedings of the 5th International Symposium on Plasma and Energy Conversion*, Springer Proceedings in Physics 398,
https://doi.org/10.1007/978-981-97-2245-7_8

The content of reactive nitrogen oxides in water greatly affects the effect of PAW. Compared with gas-phase discharge alone, gas–liquid two-phase discharge increases the contact area between plasma and water, which can effectively improve the mass transfer efficiency and the content of active substances [9]. Currently, researchers use more jet discharge devices. Konstantin F. et al. used atmospheric pressure induction microwave-excited needle-water discharge to generate high-purity argon plasma for the study of the physicochemical properties of PAW [7]. Under near-optimal irradiation conditions with impurities excluded the production rates were 200 μM/min for H_2O_2, ·OH for 800 μM/min, and 2 mM/min for NOx. Liu et al. used a high-frequency high-voltage power supply to stimulate the multi-needle array-water discharge to generate plasma, and investigated the change rule of active particle concentration under different gas flow rate conditions [10]. The experiment shows that when the air flow rate is 0 L/h, the concentration of NO_2^- can reach 2.44 mmol/L at the highest.

However, jet treatment of PAW requires a large gas flow rate, and the contact area with water is small. Through the multi-channel array form of jet discharge can expand the action area to a certain extent, but the complexity of the device will also increase. DBD has the advantages of simple structure, stable generation of large area uniform discharge, easy to realize a large area of water treatment. MAN et al. designed a micro-bubble technology based on the activation of the water generating device [11], so that the discharge occurs in the attached to the surface of the high-voltage electrodes of 0.5–2 μm bubbles. Under nanosecond-scale pulse excitation, the concentration of NO_3^- can reach 36×10^{-2} mmol/L, the concentration of NO_2^- can reach 18×10^{-2} mmol/L, and the amount of activated nitrogen oxides produced per unit of energy can reach up to 7.94×10^{-5} mmol/J. Jong et al. proposed the structure of water surface plasma source (WSPS) for large-area water treatment [12]. One electrode was immersed in water and the other electrode was suspended in air to realize the direct interaction between plasma and water over a large area. After the device treated distilled water for 3 min, the pH of 2 L of activated water decreased from 6.25 to 4.24, and the EC increased from 2.00 to 22.33 μS·cm. Niels [13], Jelena [14] et al. have fabricated their own homemade coaxial liquid-membrane type reactor that can form a flow-through thin liquid film in the discharge area and investigated the degradation of organic pollutants such as atrazine, nicotine and others.

In this paper, a flat-plate DBD reactor based on the filling of porous foam ceramics is proposed. During the discharge process, water flows between the pores inside the porous foam ceramics to increase the gas–liquid contact area and stagnation time, which in turn increases the dissolution of active substances in the PAW. The effects of discharge voltage, pore size of the foam ceramic, water flow rate and other parameters on the characteristics of gas–liquid two-phase pulsed discharge inside the porous foam ceramic were investigated, and the effects of different pore sizes of the foam ceramic on the generation characteristics of the active substance components were explored.

Fig. 8.1 Structure of DBD reactor: **a** Schematic of the reactor; **b** porous foam ceramics (left to right:10 ppi, 30 ppi, 60 ppi); **c** Physical drawing of the reactor

8.2 Experimental Methods

8.2.1 Experimental Setup and Platform

The schematic diagram of the flat plate DBD reactor used for the experiment is shown in Fig. 8.1a. The reactor shell as a whole was customized from quartz glass, and the high and low electrodes were made of 0.5 mm thick copper sheets (80 * 45 mm) laminated to the quartz outer wall; the discharge area was filled with porous foam ceramics (100 * 50 * 10 mm) with different pore sizes (10 ppi, 30 ppi, and 60 ppi) as shown in Fig. 8.1b; and the top of the reactor was equipped with a micro-porous nozzle, with the outlet pore size less than 1 mm, as shown in Fig. 8.1c.

The platform used for the experiment consists of a microsecond pulse power supply, a reactor, a peristaltic pump and a parameter measurement system, as shown in Fig. 8.2. Among them, the parameter measurement system includes an oscilloscope, current probe, high-voltage probe, UV–Vis spectrophotometer, pulse power supply frequency 0–2 kHz adjustable, peristaltic pump water flow rate accuracy of up to 0.1 mL/min. The water used to be treated are deionized water.

8.2.2 Active Substance Detection Methods

According to the Beer-Lambert law, the concentration of the partial reactive nitrogen oxides in PAW can be determined directly or indirectly by absorbance [15]. The Beer-Lambert law is shown as follows:

$$C_{Abs}(\lambda) = \varepsilon lc \tag{8.1}$$

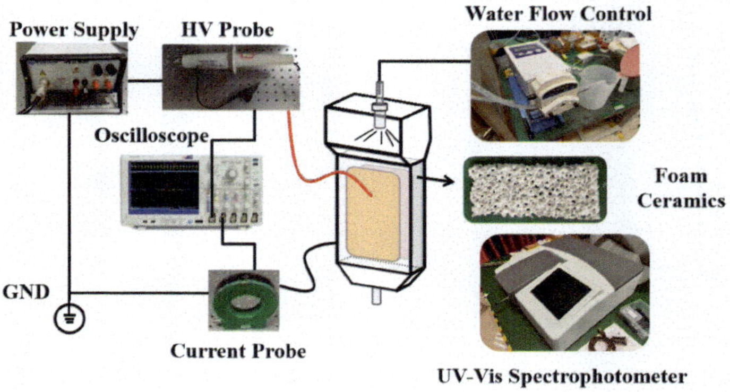

Fig. 8.2 Schematic diagram of the experimental platform

where C_{Abs} is the absorbance of the test substance, λ is the specific wavelength corresponding to the substance, ε is the molar absorption of the substance at the specific wavelength, *l* is the length of the optical path, and *c* is the concentration of the substance.

The experiments were carried out by UV–Vis spectrophotometer supplemented with relevant detection reagents to detect the active substances in PAW. The standard optical path of the instrument is 10 mm, the precision is 0.001, and the wavelength display range is 190–1100 mm. The concentration of H_2O_2 was measured by titanium sulfate [16], and the corresponding absorbance at the wavelength of 410 nm was taken for the measurement; the concentration of NO_2^- was measured by sulfanilic acid and N-ethylenediamine hydrochloride [17], and the absorbance at the wavelength of 540 nm was taken for the measurement; and the concentration of NO_3^- was measured by the direct detection of the difference in absorbance between 210 and 275 nm [11]. In addition, in order to quantify the concentration of the active substance in this experiment, a standard curve was linearly fitted by measuring the absorbance of different concentrations of standard solutions, and the fitting equations are shown in Eqs. (8.2) and (8.3).

$$C_1 = 1.4348A_1 \tag{8.2}$$

$$C_2 = 1.8826A_2 \tag{8.3}$$

$$C_3 = 0.9175A_3 \tag{8.4}$$

where C_1, C_2, C_3 are the concentrations of H_2O_2, NO_3^-, NO_2^- in the PAW, and A_1, A_2, A_3 are the corresponding absorbance of H_2O_2, NO_3^-, NO_2^- in the PAW respectively.

8.3 Results and Discussion

8.3.1 Discharge Characteristics

Figure 8.3 shows the waveforms of the discharge voltage and current collected at a pore size of 10 ppi of the foam ceramic. The applied voltage was 15 kV, the pulse frequency was 1 kHz, and the pulse width was 10 us. After filling the foam ceramics in the reactor, the discharges are in the form of filamentary micro discharges between the pores of the foam ceramics and between the inner wall of the quartz and the foam ceramics. As can be seen in Fig. 8.3, the filamentary micro discharge occurs several times during the rising edge of the pulse, but the current peak is not high. During the falling edge of the pulse, the number of filamentary micro discharge becomes significantly less, but the current peak is higher.

Figure 8.4 shows the variation of discharge power with voltage. It can be seen that the discharge power gradually increases with the increase of voltage. The discharge power obtained by filling 10 ppi pore foam ceramics is close to saturation at 14 kV, and the power reaches 6 W.

Figure 8.5 demonstrates the effect of water flow rate on the discharge power. Compared with the pure gas-phase discharge, the discharge power increases significantly after adding a certain amount of water flow. The water molecules dissociate under higher intensity discharges, resulting in an increase in local conductivity, which in turn has an effect on the discharge power [18]. It was found that the 13 kV voltage is close to the initial discharge voltage, so the filamentary discharge is still unstable. The increase of water flow rate will lead to the continuous enhancement of the local electric field, which promotes the increase of the discharge power. It can be seen from Fig. 8.5a that under the condition of 13 kV voltage, with the increase of water flow rate, the discharge power increases gradually. While in Fig. 8.5b, c, after the voltage is increased, with the increase of water flow rate, the change of the discharge power is smaller, and even has a tendency to decrease. This is due to the fact that the

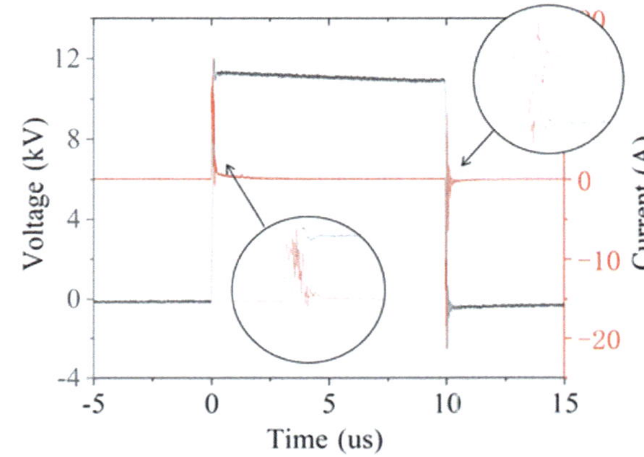

Fig. 8.3 Discharge voltage-current waveform

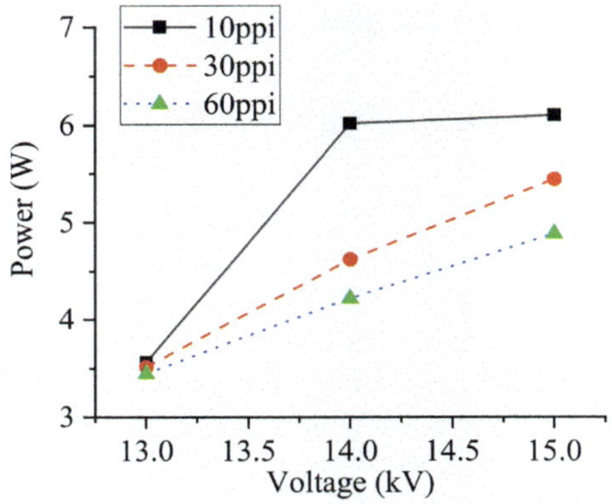

Fig. 8.4 Variation of discharge power with voltage

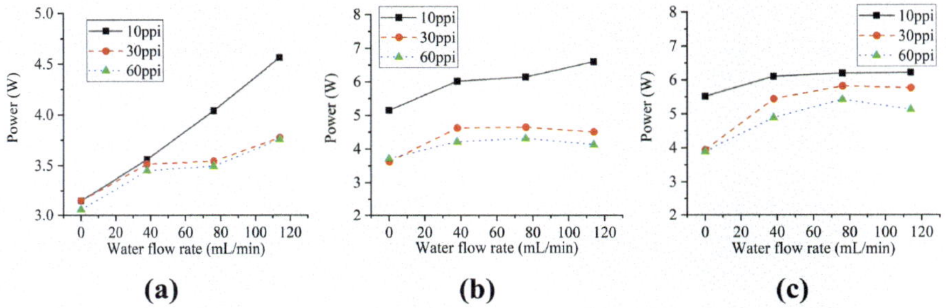

Fig. 8.5 Variation of discharge power with water flow rate: **a** 13 kV; **b** 14 kV; **c** 15 kV

filamentary discharge tends to be stabilized after the voltage is increased, the effect of water flow is greatly reduced, so the discharge power tends to be stabilized. And too large water flow will produce a large disturbance to the discharge electric field, reducing the number of filamentary discharges. Therefore, when the water flow rate reaches 117 mL/min the discharge power will decrease to a certain extent.

Meanwhile, comparing the magnitude of discharge power under the condition of filling different pore space foam ceramics, it can be seen that larger pore space is favorable to the occurrence of filamentary discharge. In the case of discharge stabilization, with the increase of foam ceramic pores, the higher the discharge power, which is shown as the foam ceramic reactor filled with 10 ppi pores has the highest discharge power, followed by the foam ceramic reactor filled with 30 ppi pores.

8.3.2 Characterization of Active Substance Production

The water flow rate was varied to detect the reactive nitrogen oxides contained in the PAW after treatment in the reactor filled with foam ceramics of different porosity, and the results are shown in Fig. 8.6. The applied voltage was 15 kV, the frequency was 2 kHz, the pulse width was 10 μs, and the treated water volume was 50 mL. It can be seen that, with the increase of water flow rate, the stay time of the treated PAW in the discharge area becomes shorter, and the concentration of reactive nitrogen oxides in the PAW basically shows a decreasing tendency. Especially, during the period that the water flow rate is increased from 9.5 mL/min to 19 mL/min, the decreases of H_2O_2 and NO_2^- are larger. The content of reactive nitrogen oxides in the PAW after treatment in the reactor with 10 ppi pores is higher than that in the reactors with 30 ppi and 60 ppi pores. The concentration of H_2O_2 can reach up to 90.0 μmol/L, the concentration of NO_2^- can reach up to 5.6 μmol/L, and the concentration of NO_3^- can reach up to 84.2 μmol/L. This is in line with the previous studies on the discharge characteristics of the reactors. Under the same conditions, the reactor with 10 ppi pores has the highest discharge intensity and produces a relatively high concentration of reactive substances. In particular, for the reactor with 30 ppi pores, the concentration of NO_3^- increases instead with the increase of water flow rate. This may be caused by the oxidation reaction between H_2O_2 and NO_2^- generated during the discharge process. The reaction equation is as follows:

$$NO_2^- + H_2O_2 \rightarrow NO_3^- + H_2O \tag{8.5}$$

At the same time, the experiment also tested the physicochemical properties of the PAW, as shown in Fig. 8.7. With the increase of water flow rate, the pH value showed a slow increasing trend, while the electrical conductivity gradually decreased. As the plasma generated by the gas discharge is in direct contact with the surrounding water, a part of the energized electrons react with the water molecules to generate H^+, so the solution shows acidity [19]. In this case, the high pH of the 30 ppi pore reactor may be caused by the decrease in the concentration of NO_3^- and NO_2^- in the PAW. The electrical conductivity depends on the ion concentration in the water. The trend of the conductivity in the experiment is also consistent with the trend of the concentration of the active substance, which decreases faster in the early stage and then tends to level off.

In addition, the concentration of the reactive nitrogen oxides in the PAW was detected by cycling 50 mL of deionized water for a certain period of time, and the relationship between the changes in the concentration of the reactive nitrogen oxides was obtained as shown in Fig. 8.8. The applied voltage was 15 kV, frequency was 2 kHz, and the water flow rate was 38 mL/min. Similar to the previous experimental results, the PAW after treatment in the reactor with 10 ppi pores has a relatively high concentration of reactive nitrogen oxides. Zhou et al. pointed out in their study that, at the low pH value, NO_2^- and H_2O_2 in the PAW had a certain effect on the sterilizing effect, and the effect of H_2O_2 was more obvious [5]. From Fig. 8.8a, it can be seen that the concentration of H_2O_2 generated during the discharge process of the reactor was high and the rising trend was obvious. Under the condition of 10 ppi pore, the concentration of H_2O_2 could reach up to 282.7 μmol/L, which may be helpful for the study of the sterilizing effect of the plasma-activated water. As can be seen from Fig. 8.8a, c, after a period of treatment, the concentration of H_2O_2 in the

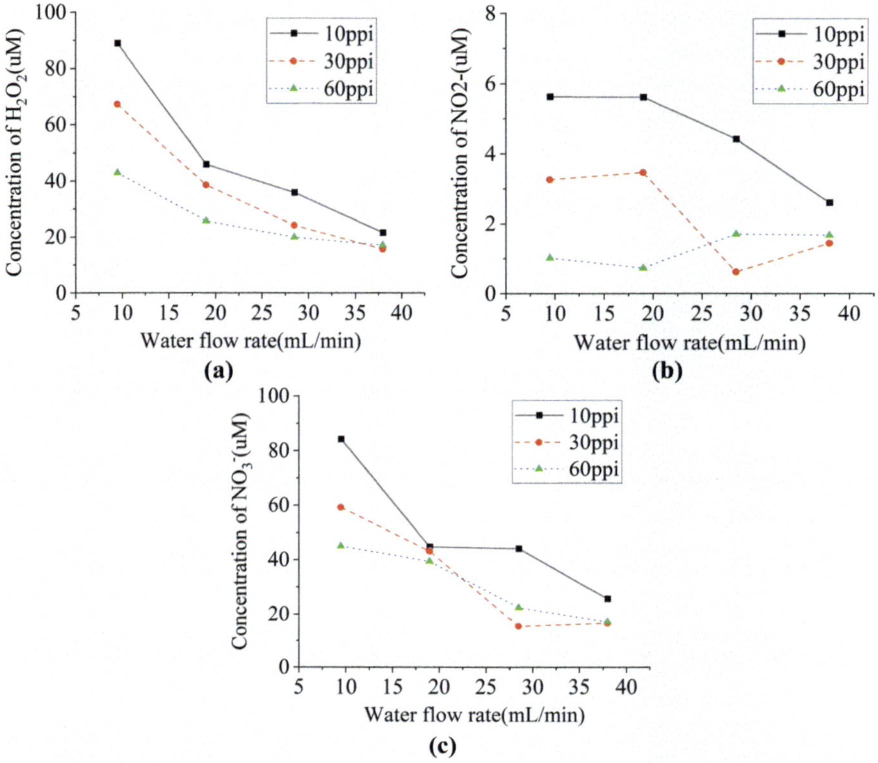

Fig. 8.6 Changes in the concentration of reactive nitrogen oxides in PAW treated under different water flow conditions: **a** the concentration of H_2O_2; **b** the concentration of NO_2^-; **c** the concentration of NO_3^-

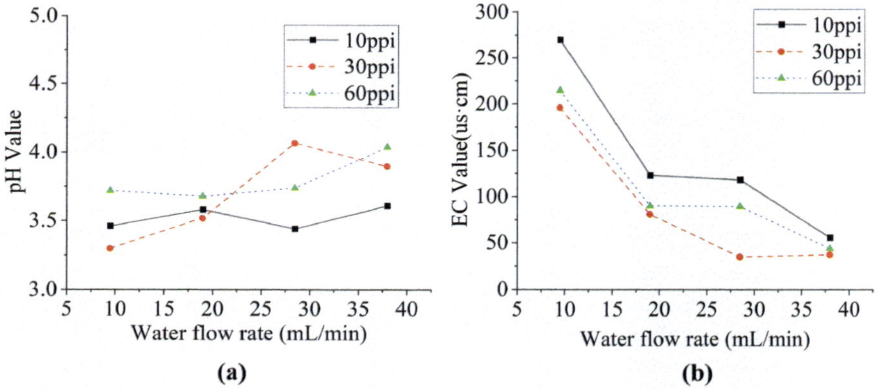

Fig. 8.7 Physicochemical properties of PAW treated by different pore size foam ceramic reactors: **a** Changes in pH; **b** changes in EC

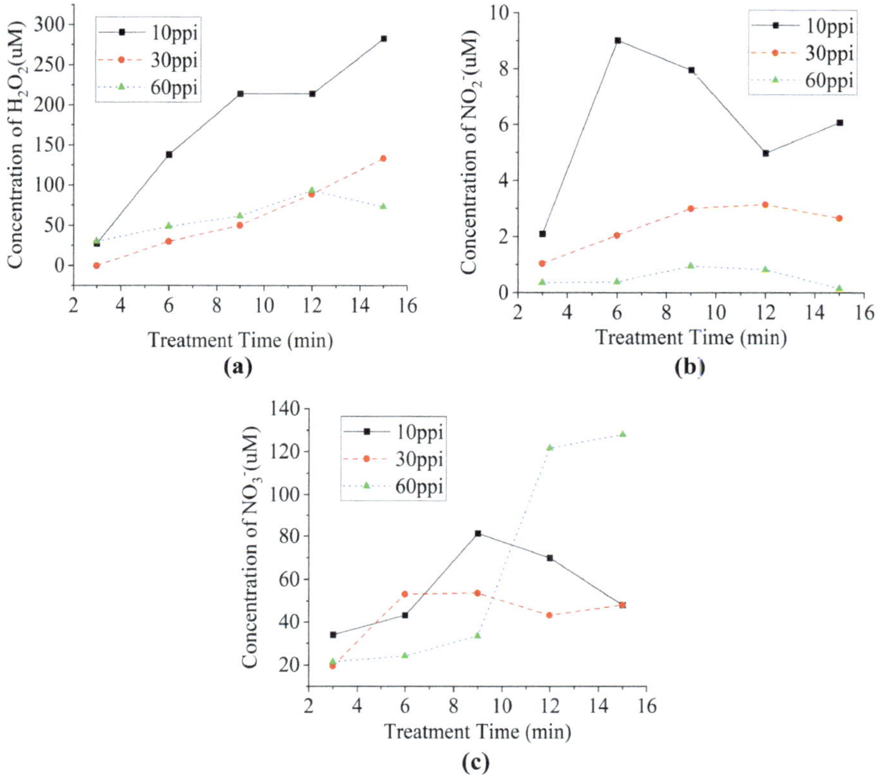

Fig. 8.8 Changes in the concentration of reactive nitrogen oxides in PAW at different discharge times: **a** The concentration of H_2O_2; **b** the concentration of NO_2^-; **c** the concentration of NO_3^-

PAW treated by the 60 ppi pore reactor decreased, while the concentration of NO_3^- increased dramatically, which may also be due to the oxidative reaction between H_2O_2 and NO_2^-.

8.4 Summary

- A flat-plate DBD reactor based on porous foam ceramic filling was proposed, where water flowed between the pores within the porous foam ceramic during the discharge process, increasing the gas–liquid contact area and stagnation time, which in turn increased the dissolution of active substances in the activated water.
- A study of the discharge characteristics of the reactor found that when the discharge voltage was low, the increase of water flow rate would lead to a continuous enhancement of the local electric field and promote the increase of the

discharge intensity. But when the discharge voltage was increased, the excessive water flow rate would produce a disturbance to the discharge electric field, reducing the number of filamentary discharges and decreasing the discharge intensity. In addition, experiments showed that larger pore space is favorable to the occurrence of filamentary discharge.

- The use of porous foam ceramic filling is effective for the production of the reactive nitrogen oxides in PAW. Choose a smaller water flow, larger pores of the foam ceramic is conducive to the production of the reactive nitrogen oxides, especially H_2O_2. The results showed that in the foam ceramic pore 10 ppi conditions, The concentration of H_2O_2 in the PAW after a single treatment could reach up to 90.0 μmol/L, the concentration of NO_2^- could reach up to 5.6 μmol/L, the concentration of NO_3^- could reach up to 84.2 μmol/L; The concentration of H_2O_2 in the PAW after the cycle of treatment could reach up to 282.7 μmol/L.

Acknowledgements The authors would like to express their sincere gratitude for the financial supports through 173 Program of China.

References

1. A. George, B. Shen et al., A review of non-thermal plasma technology: a novel solution for CO_2 conversion and utilization. Renew. Sustain. Energy Rev. **135**, 109702 (2021)
2. J.F. Wu, H.M. Yuan et al., Research progress in the preparation of functional materials by low temperature plasma modification technology. Mater. Rep. **36**(21):224–232 (2022)
3. G. Iervolino et al., Enhanced removal of water pollutants by dielectric barrier discharge non-thermal plasma reactor. Sep. Purif. Technol. **215**, 155–162 (2019)
4. D. Liu, T. He, H. Zhang, Penetration effect of gas plasmas on human tissues: state-of-the-art and current issues. High Voltage Eng. **45**(7), 2329–2342 (2019)
5. R.W. Zhou, R.S. Zhou, P.Y. Wang et al., Plasma-activated water: generation, origin of reactive species and biological applications. J. Phys. D Appl. Phys. **53**(30), 303001 (2020)
6. J.F. Chen et al., *The Effects of DBD Plasma Treatment on the Main Postharvest Pathogenic Fungi of Goji Berries and Its Preservation Properties* (Nanjing Forestry University, 2023)
7. K.F. Sergeichev, N.A. Lukina et al., Physicochemical properties of pure water treated by pure argon plasma jet generated by microwave discharge in opened atmosphere. Front. Phys. **8**, 614684 (2021)
8. F. Tan, Y. Wang et al., Plasma dermatology: skin therapy using cold atmospheric plasma. Front. Oncol. **12**, 918484 (2022)
9. X.R. Li, Z.B. Li et al., Research progress in application of coaxial DBD technology for organic wastewater treatment. Appl. Chem. Ind. **49**(10), 2662–2669 (2020)
10. K. Liu, W. Ren, C.F. Ran et al., Long-lived species in plasma-activated water generated by an AC multi-needle-to-water discharge: effects of gas flow on chemical reactions. J. Phys. D Appl. Phys. **54**(6), 065201 (2021)
11. C. Man, C. Zhang et al., Nanosecond-pulsed microbubble plasma reactor for plasma activated water generation and bacterial inactivation. Plasma Process. Polym.Polym. **19**, e2200004 (2022)
12. J.K. Yang, C.H. Cho et al., Water surface plasma source for large area water treatment by using volume dielectric barrier discharge. IEEE Trans. Plasma Sci. **50**, 3209312 (2022)

13. N. Wardenier et al., Removal of micropollutants from water in a continuous-flow electrical discharge reactor. J. Hazard. Mater. **362**, 238–245 (2019)
14. J. Krupež, V.V. Kovačević, et al., Degradation of nicotine in water solutions using a water falling film DBD plasma reactor: direct and indirect treatment. J. Phys. D: Appl. Phys. **51**, 174003 (2018)
15. J.S. Oh et al., How to assess the plasma delivery of RONS into tissue fluid and tissue. J. Phys. D Appl. Phys. **49**(30), 304005 (2016)
16. X.W. Liu, Z.L. Chen et al., Spectrophotometric determination of low concentration of hydrogen peroxide in O_3/H_2O_2 system using titanium sulfate. China Water Waste Water **26**(16), 126–129 (2010)
17. C. Li, S. Fu. Brief analysis on the research of nitrite test method in "Green Food–Edible Salt". China Well Rock Salt **43**(04), 31–33 (2012)
18. A.W. Bright, B. Makin, Polar liquids: a survey of purification, conduction mechanisms, and interfacial effects. J. Mater. Sci. **2**(2), 184–193 (1967)
19. Z.J. Liu et al., Quantifying the concentration and penetration depth of long-lived RONS in plasma-activated water by UV absorption spectroscopy. AIP Adv. **9**(1), 015014 (2019)

Chapter 9
Determining a Collision Cross-Section Set from Electron Swarm Parameters Using Machine Learning Method

Ke Wang, Peiqiong Liu, Yunkun Deng, and Boya Zhang

Abstract A complete collision cross section set of eco-friendly gases is very important for the study of the micro-discharge mechanism of these gases. According to the self-consistent physical connection between the electron swarm parameters and the collision cross sections, the prediction of the collision cross sections are made through the analysis of electron swarm parameters by pulsed Townsend experiments. To automate this process, we use the machine learning method to establish the mapping relationship from electron swarm parameters to collision cross sections. Firstly, we present a suitable neural network with electron swarm parameters as input and collision cross sections as output. Then, We train the neural network using collision cross sections from the LXCat project, paired with electron swarm parameters calculated by Boltzmann equation. Finally, We successfully apply this machine learning approach to obtain a set of collision cross sections of C_4F_7N gas, that refines the set published in previous study (XJTUAETLab database in http //www.lxcat.net [1]).

9.1 Introduction

In recent years, the new eco-friendly gases represented by C_4F_7N and $C_5F_{10}O$ have attracted much attention due to their excellent insulating properties. These eco-friendly gases have large molecular weights, special and asymmetric molecular structures, and they need to be mixed with buffer gases. Therefore, it is necessary to study their microscopic discharge processes and mechanisms in order to analyze the core issues such as their discharge mechanisms and synergistic effect mechanisms.

K. Wang · Y. Deng
Electric Power Research Institute of Yunnan Power Grid Co., Ltd., Kunming 650217, Yunnan, China

P. Liu · B. Zhang (✉)
State Key Laboratory of Electrical Insulation and Power Equipment, Xi'an Jiaotong University, Xi'an 710049, China
e-mail: zhangby@xjtu.edu.cn

Z. Fang et al. (eds.), *Proceedings of the 5th International Symposium on Plasma and Energy Conversion*, Springer Proceedings in Physics 398,
https://doi.org/10.1007/978-981-97-2245-7_9

To study the microscopic discharge processes of new eco-friendly gases, it is important to get hold of important microscopic parameters like electron swarm parameters and collision cross sections.

We can obtain the electron swarm parameters through electron swarm experiment [2]. However, the main methods for directly obtaining collision cross sections are beam experiments and quantum chemical calculations, they are difficult to carry out and have limited results. And the obtained results need to be continuously corrected according to the self-consistent physical connection between the electron swarm parameters and the collision cross section. Therefore, the formation of the cross section set for each gas studied previously has undergone a long process. The electron swarm parameters can be obtained directly from the collision cross section by solving the Boltzmann equation, but there is no explicit mapping from the electron swarm parameters to the collision cross section. However, with the development of machine learning and the improvement of collision cross-section databases, it is promising to directly establish a nonlinear mapping relationship between two sets of data [3–5]. On the basis of obtaining the accurate electron swarm parameters and utilizing the physical self-consistency between the collision cross sections and the electron swarm parameters, we adopt the machine learning method to build an accurate inversion model to accelerate the acquisition of the complete collision cross section set for the new eco-friendly gases.

In our previous work, we published a complete collision cross section set of C_4F_7N in the LXCat-XJTUAETLab database as shown in Fig. 9.1 [1]. In this paper, we apply a machine learning method to optimize the manually corrected collision cross section set rationally, predict a new set of C_4F_7N gas, and evaluate it's the accuracy, completeness and self-consistency.

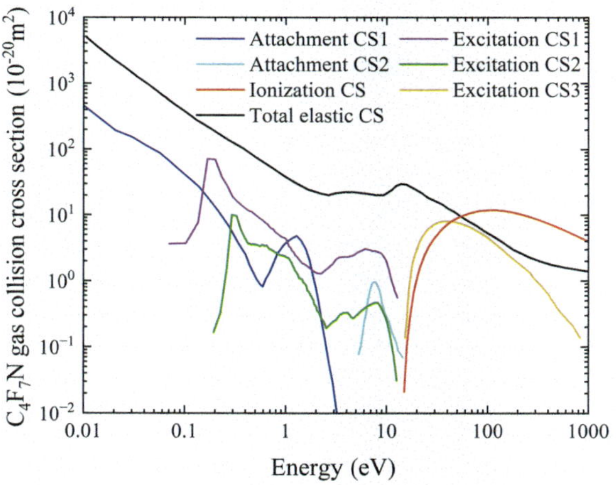

Fig. 9.1 Manual correction of C_4F_7N gas collision cross section

9.2 Machine Learning Mothed

In this paper, we use a machine learning mothed to predict the collision cross section from the electron swarm parameters. This mothed is divided into 3 main processes, which are building the neural network model, preparing the input and output training data and training the model. To obtain higher prediction accuracy, neural network models with different types of cross sections were trained separately in this paper. Collision cross sections are mainly divided into elastic collision cross sections, attachment cross sections, vibrational excitation cross sections, electronical excitation cross sections and ionization cross sections. And the prediction process for each type of cross section is roughly the same. We taking elastic impact cross section and low threshold vibration excitation cross section as examples, the above three processes are briefly summarized in turn.

9.2.1 Neural Network Model

In this paper, a fully connected neural network (FCNN) is used to predict the collision cross section, as shown in Fig. 9.2. The model consists of an input layer, three hidden layers, and an output layer. The hidden layer consists of a fully connected layer and an activation function.

The input of the neural network is a feature vector x containing electron swarm parameters, and the output is a feature vector y containing collision cross section. The nonlinear mapping relationship between the two is described through training, as shown in Formula (9.1):

$$y = \sigma(x) \tag{9.1}$$

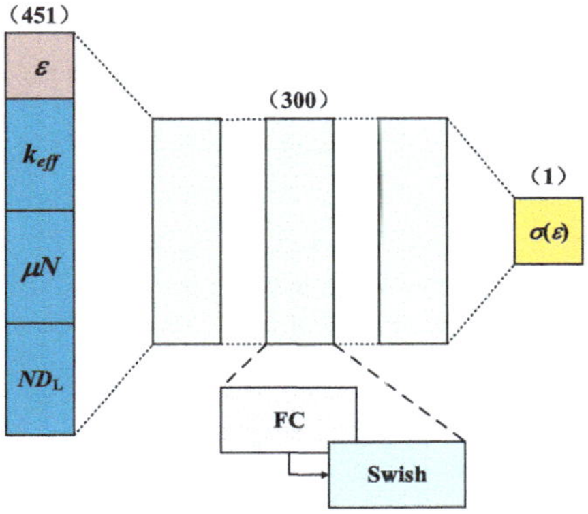

Fig. 9.2 Neural network model

At the same time, this paper considers the prediction of cross sections under discrete energy points, so the value of each output y is a function of energy ε, and the input x contains the swarm parameters and ε, as shown in Formula (9.2):

$$x = \begin{bmatrix} \varepsilon \\ (k_{eff})_1 \\ (k_{eff})_2 \\ \vdots \\ (\mu N)_1 \\ (\mu N)_2 \\ \vdots \\ (ND_L)_1 \\ (ND_L)_2 \\ \vdots \end{bmatrix} \tag{9.2}$$

where, k_{eff} is the effective ionization rate coefficient, μN is the mobility, and ND_{L} is the density-normalized longitudinal electron diffusion coefficient. The subscript represents the electron swarm parameters measured under different reduced electric fields.

The mapping relationship between x–y described by this neural network is mathematically represented by the following function:

$$y(x) = (A_4 \circ swish \circ A_3 \circ swish \circ A_2 \circ swish \circ A_1)(x) \tag{9.3}$$

$$swish(x) = x/\left(1 + e^{-x}\right) \tag{9.4}$$

$$A_n(x) \equiv W_n x + b_n \tag{9.5}$$

where *swish* is a nonlinear activation function. $A_{\text{n}}(\text{x})$ is the relationship between two neighbouring layers of neurons defined by the weight matrix W_{n} and the vector b_{n} matrix, which are continuously optimized during the training of the neural network. The number of input neurons is 451, each hidden layer contains 300 neurons, and the number of output is 1. The size of the W_{n} matrix and b_{n} vector corresponds to the number of neurons.

9.2.2 Preparing Training Data

In order to achieve effective training of neural networks, it is very important to select appropriate input and output data. That is a large number of physically significant collision cross sections and their calculated electron swarm parameters. We can obtain the actual collision cross sections of various gases from the LXCat database, but most of the data cannot fully cover the energy domain we consider, which is (10^{-2} eV, 10^{3} eV). Therefore, we need to choose the appropriate analytic expression to fit the known data and derive the value at the unknown energy point. Taking elastic collision cross section as an example, we choose the power-law relation described by Morgan [3] to fit the relation between electron energy and collision cross section:

$$\sigma(\varepsilon) = k\varepsilon^{p} \tag{9.6}$$

where ε is the electron energy, $\sigma(\varepsilon)$ is the cross-section size at ε, and k and p are the best fitting parameters obtained by fitting data. For Low threshold vibration excitation cross sections, we need a special expression that can take into account its special data characteristics, including threshold and peak value, such as the function proposed by Rost and Pattard in literature [6]:

$$\sigma(\varepsilon) = \frac{k\varepsilon^{\alpha}}{(\varepsilon+\varepsilon_M/\alpha)^{\alpha+1}} \tag{9.7}$$

where, ε_{M} is the electron energy corresponding to the maximum cross-section, k and a are the best fitting parameters obtained through fitting data, and $\sigma(\varepsilon)$ is the cross-section size at ε. For other types of collision cross sections, other more suitable analytical expressions are considered.

The effective prediction of cross sections by deep neural networks necessitates a substantial quantity of data, and the data available in the LXCat database is relatively restricted. Therefore, we choose to use the interpolation method to synthesize more training examples that meet the actual physical characteristics through the existing cross-section data. A new composite cross-section is calculated by taking a weighted geometric mean of the actual collision cross-section using Eq. (9.8):

$$\sigma_{new}(\varepsilon) = \sigma_1^{1-r}\left(\varepsilon+\varepsilon_1 - \varepsilon_1^{1-r}\varepsilon_2^{r}\right)\sigma_2^{r}\left(\varepsilon+\varepsilon_2 - \varepsilon_1^{1-r}\varepsilon_2^{r}\right) \tag{9.8}$$

where, $\sigma_1(\varepsilon)$ and $\sigma_2(\varepsilon)$ are cross sections of different gases, ε_1, ε_2 and $\varepsilon 1 - \mathrm{r}\ 1 \varepsilon \mathrm{r}\ 2$ are the threshold energies of $\sigma_1(\varepsilon)$, $\sigma_2(\varepsilon)$ and the new collision cross section $\sigma_{new}(\varepsilon)$, respectively, and $0 \le r \le 1$ are uniformly distributed random variables. The collision cross section generated by this formula retains the correlation between the cross section size and its threshold energy.

For C_4F_7N gas, we choose to place explicit constraints on the collision cross section of the training set to make the neural network output in a direction closer to the actual measurement results. Based on a complete set of collision cross sections

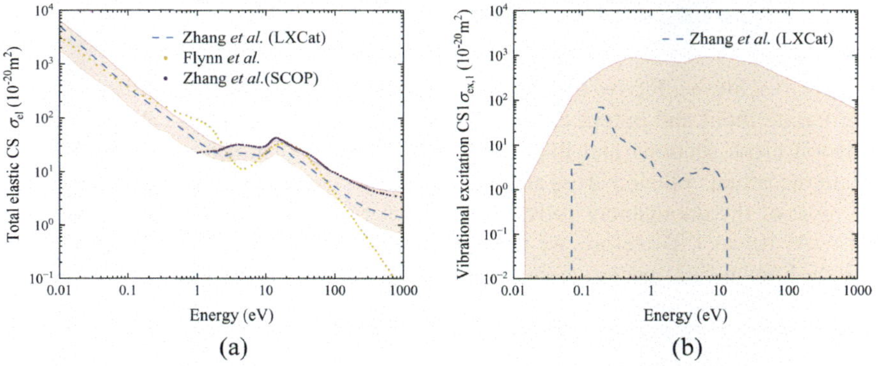

Fig. 9.3 Training set cross section confidence band

manually corrected by our group, as shown in Fig. 9.1, 7 types of collision cross sections are set in the training set. They are elastic cross section $\sigma_{el}(\varepsilon)$, low threshold vibration excitation cross section $\sigma_{ex,1}(\varepsilon)$, high threshold vibration excitation cross section $\sigma_{ex,2}(\varepsilon)$, electron excitation cross section $\sigma_{ex,3}(\varepsilon)$, ionization cross section $\sigma_{io}(\varepsilon)$ and attachment cross section $\sigma_{at,1}(\varepsilon)$, $\sigma_{at,2}(\varepsilon)$. Assuming that $\sigma_{el}(\varepsilon)$, $\sigma_{ex,2}(\varepsilon)$, $\sigma_{ex,3}(\varepsilon)$, $\sigma_{io}(\varepsilon)$ are known cross sections, these data are geometrically weighted averaged with the generated training set corresponding to the section type by Formula (9.7). By setting the weight of known C_4F_7N cross sections in the training set to $r = 0.9$, the neural network can realize the correction of these cross sections. Assuming that $\sigma_{ex,1}(\varepsilon)$, $\sigma_{at,1}(\varepsilon)$, $\sigma_{at,2}(\varepsilon)$ are unknown cross sections, these types of cross section training sets are not constrained and are freely predicted by the neural network. Figure 9.3a shows the confidence band of the training set of elastic cross sections generated by the above process, and the manually corrected cross sections. And Fig. 9.3b shows the same content of low threshold vibration excitation cross sections. The results of other papers are also presented in Fig. 9.3a, b [7, 8].

As described in the above process, 5×10^4 samples are generated for each type of collision cross section, forming 5×10^4 complete collision cross sections sets. Finally, for each set of collision cross sections, the two term approximate Boltzmann equation is used to calculate the electron swarm parameters. For C_4F_7N gas, the calculation of the electron swarm parameters of the training set should be consistent with the data obtained from the experimental measurement of C_4F_7N gas [2], so as to facilitate verification. Therefore, we respectively calculated that k_{eff}, μN and ND_L of the training sample accounted for 4.93, 9.74, 20.08, 48.32 and 100% in Ar, in the reduced electric fields *(E/N)* range of 110–1020Td. Finally, 5×10^4 input–output training pairs are obtained.

Finally, the input and output data are normalized after logarithm. The eigenvectors of electron swarm parameters and collision cross sections are fed into the neural network for effective learning. For special data k_{eff}, which is positive and negative and spans multiple orders of magnitude, logarithmic processing cannot be carried out directly. We apply the following formula to process it and then normalize it:

$$y = \begin{cases} \log(x), x \geq 0 \\ -\log(-x), x < 0 \end{cases} \quad (9.9)$$

9.2.3 Training Neural Network Model

Next, the neural network is trained. Through the above process, we generate 5×10^4 electron swarm parameters-collision cross section pairs. The collision cross sections are sampled in the range of [10^{-2} eV, 10^3 eV] with 200 random energies according to Eq. (9.9), and a total of 1×10^7 training examples are generated:

$$\varepsilon = \varepsilon_{\min}^{1-s} \varepsilon_{\max}^{s} \quad (9.10)$$

where, s is a uniformly distributed random number.

The training samples were randomly divided into 100 batches of repeated training neural networks. The average absolute error was selected as the loss function, and the Adam optimizer [9] was used to train the neural network by minimizing the loss function (13). The step size α was set to 10^{-4}, and the exponential decay rate β_1 and β_2 were set to 0.9 and 0.999, respectively. The parameters in b_n are initialized to zero and those in W_n are initialized to uniform random numbers, and 7 types of collision cross sections are predicted respectively.

In order to prevent overfitting, we constantly predict the cross section of C_4F_7N gas during the training process, calculate the electron swarm parameters by the Boltzmann equation. And compare the calculated valued with the experimental values of electron swarm parameters used for fitting. The instruction to stop training is issued when the mean square error is no longer decreasing.

Finally, the seven types of cross sections mentioned above are trained separately. When training each type of collision cross section, the predicted cross section and other cross sections together form the test set, assuming that other cross sections are known. Every time a type of collision cross section is successfully trained, the cross section is used to replace the corresponding cross section in the manually corrected cross section set. After 7 training sessions, a new complete collision cross section set of C_4F_7N is obtained.

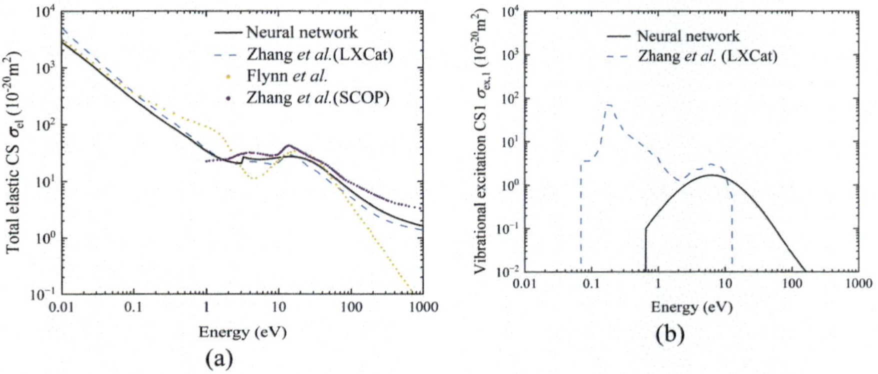

Fig. 9.4 Prediction of C_4F_7N gas collision cross section

9.3 Result Analysis

9.3.1 Prediction of Collision Cross Sections

For the total elastic cross section $\sigma_{el}(\varepsilon)$, due to constraints placed on it in the training data set, the prediction of this cross section is quantitatively limited to the vicinity of the known manually corrected elastic cross section, as shown in Fig. 9.4a. The average relative error between prediction and manual correction is 21%. At lower electron energies, it is closer to the corrected result of FLynn et al. For the low threshold vibration excitation cross section $\sigma_{ex,\,1}(\varepsilon)$, it is freely predicted by the neural network, so it differs significantly from the manual correction results. The variation of peak size and threshold energy indicates the non-uniqueness of this inverse problem, as shown in Fig. 9.4b. In addition, Fig. 9.4a, b also shows results from other papers.

9.3.2 Self-consistency of the Predicted Cross Section

Figure 9.5 shows the electron swarm parameters calculated by the machine prediction cross section and manually corrected cross section, and compares them with the experimental values. As shown in Fig. 9.5, The results show that in pure C_4F_7N gas, the electron swarm parameters calculated by the predicted cross-section are more consistent with the experimental values. In other proportions, the effect is not obvious compared with the manual correction cross section.

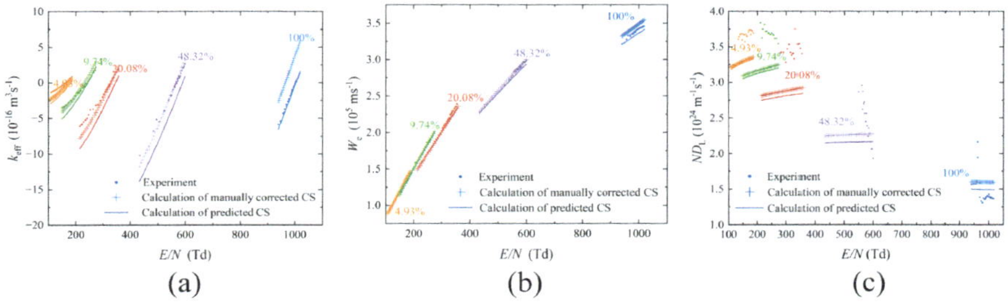

Fig. 9.5 Calculated electron swarm parameters compared with experimental values

9.4 Discussion

As described in the above procedure, a complete collision cross section set of C_4F_7N gas obtained by the machine is further advanced than the set of manually corrected results. From the cross-section characteristics, the neural network gives a reasonable improvement constrained by the manually corrected cross-section. Furthermore, the neural network prediction demonstrates a significant difference for the prediction of unconstrained collision cross sections, highlighting the non-uniqueness of this inverse problem. This problem can be improved by upgrading the neural network architecture and training data generation. For example, the mixed density network is used to reduce the uncertainty of the prediction process, and the generative adversarial network is used to generate more characteristic training data. From the point of view of self-consistency, the calculation of electron swarm parameters of the predicted collision cross section for pure gas is more consistent with the experimental values, and it slightly deteriorates or remains unchanged under other proportions, which is a trade-off made by the neural network to improve the accuracy of pure C_4F_7N. Thus, very reliable input data is expected to maximize the accuracy of the neural network prediction outcomes.

9.5 Conclusions

Through effective training, the neural network model generates a reliable set of C_4F_7N electron-molecular collision cross sections. Importantly, it gets rid of the subjectivity inherent in the traditional analysis of electron swarm parameters by manual experience, and improves the efficiency of obtaining collision cross sections. It can be proved that the nonlinear mapping relationship between the two sets of data from electron swarm parameters to electron-molecule collision cross sections can be well established by using machine learning method, and the problem of local optimization can be avoided, which puts forward a new idea for obtaining the complete collision cross section set of new eco-friendly gases.

Acknowledgements This manuscript is sponsored by the Science and Technology Project of China Southern Power Grid (YNKJXM20220051).

References

1. XJTUAETLab database, http://www.lxcat.net, last accessed 2023/5/23
2. J. Xiong, B. Zhang, X. Li et al., Study on measurements of swarm parameters in SF_6 alternative gases by pulsed Townsend method. Proc. CSEE **41**(2), 759–769 (2021)
3. W.L. Morgan, The feasibility of using neural networks to obtain cross sections from electron swarm data. IEEE Trans. Plasma Sci. **19**(2), 250–255 (1991)
4. P.W. Stokes, D.G. Cocks, M.J. Brunger et al., Determining cross sections from transport coefficients using deep neural networks. Plasma Sources Sci. Technol. **29**(5), 055009 (2020)
5. V. Jetly, B. Chaudhury, Extracting electron scattering cross sections from swarm data using deep neural networks. Mach. Learn.: Sci. Technol. **2**(3), 035025 (2021)
6. J. M. Rost, T. Pattard, Analytical parametrization for the shape of atomic ionization cross sections. Phys. Rev. A, **55**(1), R5 (1997)
7. M. Flynn, J. Agan, A. Neuber et al., Generation and optimization of cross-sections for electron-C4F7N collisions. J. Phys. D: Appl. Phys. **56**(48), 485207 (2023)
8. J. Zhang, N. Sinha, M. Jiang et al., DC breakdown characteristics of C_4F_7N/CO_2 mixtures with particle-in-cell simulation. IEEE Trans. Dielectr. Electr. Insul.Insul. **29**(3), 1005–1010 (2022)
9. D.P. Kingma, J. Ba, Adam: a method for stochastic optimization. arXiv preprint arXiv: 1412. 6980 (2014)

Chapter 10
Numerical Simulation of 50 Hz/5 kHz Dual-Frequency Dielectric Barrier Discharge in Atmospheric-Pressure Air

Junwen He, Guanglin Yu, Ju Li, Nan Jiang, Jie Li, and Yan Wu

Abstract A dual-frequency (DF) excitation, where the low-frequency voltage (U_{LF}) is introduced into a stable intermediate-frequency (IF) discharge, is proposed to improve discharge. A one-dimensional (1D) plasma fluid model has been developed to investigate the effect of the U_{LF} and IF voltage (U_{IF}) on the current pulse amplitude, electron density, and dielectric surface charge. The simulated results show that the current pulse amplitude, the maximal electron density of the breakdown, and the dielectric surface charge are modulated by the U_{LF} and vary periodically with the U_{LF} magnitude. When the U_{LF} transits from positive to negative polarity, the maximum current pulse amplitude and electron density of the breakdown process is observed within a DF cycle. The maximum electron density of the breakdown within a DF cycle is enhanced by increasing the U_{LF}. In the DF discharge with a U_{LF} of 12 kV, the maximum electron density of the breakdown within a DF cycle is 17.91% higher than that at the same time in the single IF discharge. This is due to the residual voltage between the gap enhanced by the increase of U_{LF}, which causes more charged particles to accumulate on the dielectric surface, thereby increasing the discharge intensity. The distribution range of the current pulse amplitude expands as the U_{IF} amplitude increases. When the U_{IF} is 12 kV, the distribution range of the current pulse amplitude is expanded by 37.5% compared with that when the U_{IF} is 9 kV. This is caused by the generating more transferred charges at the higher U_{IF}.

J. He · G. Yu · J. Li · N. Jiang (✉) · J. Li · Y. Wu
School of Electrical Engineering, Dalian University of Technology, Dalian 116024, People's Republic of China
e-mail: jiangnan@dlut.edu.cn

J. Li
e-mail: lijie@dlut.edu.cn

Z. Fang et al. (eds.), *Proceedings of the 5th International Symposium on Plasma and Energy Conversion*, Springer Proceedings in Physics 398,
https://doi.org/10.1007/978-981-97-2245-7_10

10.1 Introduction

Atmospheric pressure air dielectric barrier discharge (DBD) is a promising method utilized in various applications, such as gas conversion [1–3], plasma-assisted combustion [4–6], active flow control [7–9], and food preservation [10–12]. The DBD reactor is typically sustained by alternating current (AC) power with a frequency of a few kilohertz (intermediate frequency, IF), due to the high frequency of gap breakdown [13–15]. In practical applications, there is a need to enhance the discharge energy coupled to the plasma. Increasing the applied voltage and driving frequency are the traditional approaches for increasing discharge energy, but they lead to a reduction in energy efficiency [16–18]. Therefore, as the requirements for electrical energy efficiency continue to increase, the demand for efficient DBD excitation methods is also engaged.

The dielectric surface charge has two significant effects on the electric field and energy coupling in plasma. Firstly, the dielectric surface charge suppresses the continuous increase of the electric field, thereby restricting the energy coupled to plasma [19, 20]. Secondly, the dielectric surface charge provides the initial electric field before the breakdown [21, 22]. Various methods have been explored to modulate the dielectric surface charge, such as the utilization of the dielectric material with special properties [23, 24]. Another approach involves utilizing the multi-source in a DBD reactor, such as combining AC voltage with nanosecond pulse voltage (NS + AC) [25, 26] or coupling AC discharge with a direct current (DC) bias voltage (DC + AC) [27]. However, these methods lead to a monotonic increase or decrease in the dielectric surface charge, which inevitably suppresses one of two effects caused by the dielectric surface charge. To overcome this limitation, the dual-frequency (DF) excitation consisting of two different frequencies is proposed to flexibly modulate the dielectric surface charge and improve the discharge performance. In the DF DBD, the gap breakdown is sustained by the IF voltage (U_{IF}). The charged particles produced from the IF discharge are regulated by the low-frequency voltage (U_{LF}) to achieve flexible control over the dielectric surface charge [28]. Benard et al. [29] have shown that a larger frequency difference between the low frequency and the intermediate frequency of the DF waveform leads to a more significant enhancement of the electric wind. Our previous study has indicated that the residual voltage between the gap is controlled by the U_{LF}, which leads to modulating the accumulation of the residual charged species [28]. However, the effect of the U_{IF} is not yet clear. Thus, further investigation is required to gain a deeper understanding of the DF discharge.

Nowadays, it is still a challenge to obtain the spatial–temporal evolution of the microscopic parameters in the experiment. Nevertheless, these microscopic parameters, including the dielectric surface charge and electron density, serve as crucial evidence for the plasma modulation in the DF discharge. The numerical simulation provides an alternative tool to research the DF discharge. Therefore, a one-dimensional (1D) plasma model is developed to explore the DF discharge in this study. The DF waveform is composed of 5000 and 50 Hz. The effect of the U_{LF} and

U_{IF} on the discharge parameters is investigated in this study. These results may pave the way for optimizing the energy efficiency of the DF-DBD systems.

10.2 Numerical Modeling

The plasma fluid model (PFM) is commonly utilized for simulating atmospheric pressure air plasma due to its ability to provide accurate solutions with less resource consumption compared to the particle-in-cell (PIC) model [30–33]. The PFM is based on the drift–diffusion-reaction framework, where the densities of species and electron energy are determined by coupling the drift–diffusion equations with the Poisson equation. A one-dimensional (1D) PFM is carried out by the COMSOL Multiphysics software built-in plasma module and the Helmholtz equation solver. A detailed description of the PFM has been provided in the previous studies [34]. Hence, a brief description is given in this study.

The drift–diffusion equation of the species and energy are listed below,

$$\frac{\partial}{\partial t}\left(n_{\mathrm{e,i,uc}}\right) + \nabla \cdot \Gamma_{\mathrm{e,i,uc}} = S_{\mathrm{e,i,uc}} \tag{10.1}$$

$$\frac{\partial}{\partial t}(n_{\varepsilon}) + \nabla \cdot \Gamma_{\varepsilon} = S_{\varepsilon} \tag{10.2}$$

Here, the subscripts e, i, uc, and ε denote electrons, ions, uncharged neutral species, and electron energy, respectively. The variables n, Γ, and S represent the density, flux, and source term of the subscript. The species flux $\Gamma_{\mathrm{e,i,uc}}$ and electron energy flux Γ_{ε} are written as,

$$\Gamma_{\mathrm{e,i,uc}} = \mathrm{sign}\left(q_{\mathrm{e,i,uc}}\right)\mu_{\mathrm{e,i,uc}} n_{\mathrm{e,i,uc}}\mathbf{E} - D_{\mathrm{e,i,uc}}\nabla n_{\mathrm{e,i,uc}} \tag{10.3}$$

$$\Gamma_{\varepsilon} = -\mu_{\varepsilon} n_{\varepsilon}\mathbf{E} - D_{\varepsilon}\nabla n_{\varepsilon} \tag{10.4}$$

where μ and D denote the mobility and diffusivity coefficient of the species, respectively. The reduced electron mobility ($\mu_{\mathrm{e}}N$) is determined by solving the Boltzmann equation with the LoKI-B code [35]. The electron diffusivity coefficient, electron energy mobility, and electron energy diffusivity coefficient are obtained by the Einstein relationship. **E** represents the electric field vector. The electric field is determined by the space charge distribution and the applied voltage, which is expressed as,

$$\nabla \cdot (\varepsilon_0 \varepsilon_r \mathbf{E}) = \rho_q \tag{10.5}$$

where ε_0 and ε_r represent the vacuum permittivity and the relative permittivity of each region, respectively. ρ_q denotes the space charge density.

The dielectric is set to charge conservation, and its surface is treated as the charge accumulation. The boundary condition of charge accumulation is defined as,

$$d\rho_s / dt = \mathbf{n} \cdot \mathbf{J}_i + \mathbf{n} \cdot \mathbf{J}_e \tag{10.6}$$

where ρ_s denotes the surface charge density. $\mathbf{n} \cdot \mathbf{J}_i$ and $\mathbf{n} \cdot \mathbf{J}_e$ represent denote the normal components of the total ion current density and the total electron current density, respectively. The secondary electron emission induced by the positive ions is considered on the gas–solid interfaces in the model. The secondary electron flux and secondary electron energy flux of the gas–solid interfaces are written as,

$$\mathbf{n} \cdot \Gamma_e = \left(\frac{1}{2} v_{e,th} n_e\right) - \sum_p \gamma_p (\Gamma_p \cdot \mathbf{n}) \tag{10.7}$$

$$\mathbf{n} \cdot \Gamma_\varepsilon = \left(\frac{5}{6} v_{e,th} n_\varepsilon\right) - \sum_p \varepsilon_p \gamma_p (\Gamma_p \cdot \mathbf{n}) \tag{10.8}$$

Here, $v_{e,th}$, ε_p, and γ_p are the electron thermal velocity, secondary electron energy, and secondary electron emission coefficient, respectively. The secondary electron emission and the average secondary electron energy are set up to 0.1 and 0.5 eV on the metal electrode. However, these parameters are 0.006 and 2.5 eV on the dielectric surface. The dual-frequency voltage is applied on the high-voltage electrode and its expression is given below,

$$U_{DF} = U_{LF} \sin(2\pi f_{LF} t) + U_{IF} \sin(2\pi f_{IF} t) \tag{10.9}$$

In this work, the gas pressure and temperature are kept constant at 760 Torr and 300 K, respectively. The increment of gas heat is ignored. The axis of the 1D model is presented in Fig. 10.1. At the starting point of the axis, the HV electrode is positioned. Within the range of 0–1.5 mm of the axis, a dielectric material with a dielectric coefficient of 7.8 is situated. An air gap with a distance of 3.5 mm is set up to the remaining part of the axis. The position of the grounding electrode is at the endpoint of the axis. The air gap region is discrete into a grid of 450 elements. However, a fine grid is employed on the dielectric material. The air chemistry process with 99 chemical reactions and 19 species is considered. Moreover, the photoionization of oxygen molecules is taken into account in the model. The photoionization rate is determined by solving three Helmholtz differential equations [36–38].

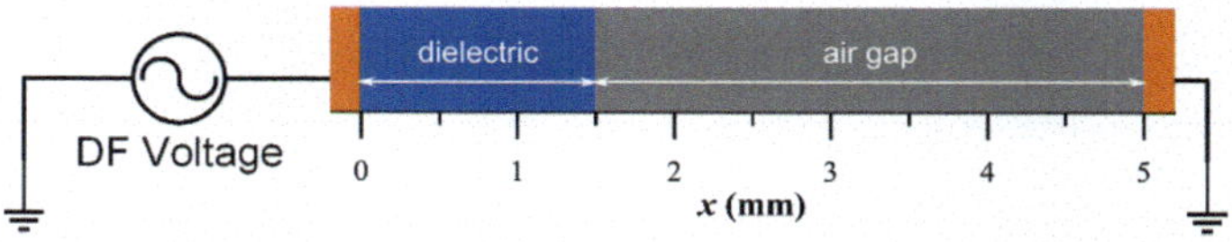

Fig. 10.1 Schematic of the model

The backward differentiation formula (BDF) is employed for the time integration. A parallel sparse direct solver (PARDISO) is chosen to solve the direct linear system. The time steps are automatically determined by the BDF solver, adjusting for both relative and absolute tolerances to manage the solution error. During the breakdown phase, the maximum time step interval is set to 20 ps, while in other phases, it is increased to 25 ns to minimize time consumption.

10.3 Results

10.3.1 Characteristic of Voltage and Ampere Under the DF Excitation

The simulated current is commonly utilized for validating the accuracy and applicability of the numerical model [39, 40]. In this study, the simulated results are taken from a time range of 100–120 ms to minimize the influence of the initial conditions. Figure 10.2 illustrates the representative waveforms of the voltage and current. The current waveform exhibits multiple current peaks within 20 ms (a DF cycle). Notably, only one current peak is observed during each half-period (HP) of IF. This observation is consistent with the previous research on air breakdown in a fixed position [19, 41]. Additionally, the amplitude of these current peaks ranges from 8.01 to 10.51 mA in the positive HP of IF and from −17.08 to −12.45 mA in the negative HP of IF. These pulse amplitudes are in the same order of magnitude as the previous reports [34]. Therefore, the computed results in the study qualitatively agree with those experimental observations.

Figure 10.2 illustrates that the current peak value is influenced by both the applied voltage and the time. In the IF discharge, the current peak value gradually diminishes as the breakdown occurrence increases (shown in the left column of Fig. 10.2). This decline can be attributed to the residual charged particles generated by the prolonged discharge [42–44], which lasts for 120 ms. In the DF discharge (the right column of Fig. 10.2), the amplitude of negative current pulses initially increases. When the U_{LF} switches from positive to negative polarity, the negative current pulse reaches its maximum. Afterward, the amplitude of the negative current pulse gradually decreases over time. However, the positive current pulses exhibit the opposite trend compared to the negative ones. These observations suggest that the current pulse amplitude is regulated by the U_{LF} in the DF discharge. Since the amplitude of negative current pulses is higher than that of positive current pulses, further research focuses on the negative current pulses.

Figures 10.3 and 10.4 show the statistical results of the negative current pulse amplitude. Figure 10.3 demonstrates that the current pulse amplitude of the IF discharge is concentrated between −14.55 and −16.05 mA. The introduction of the U_{LF} widens the range of the current pulse amplitude. A wider range of the current amplitude implies an increased possibility of high current pulses. Therefore,

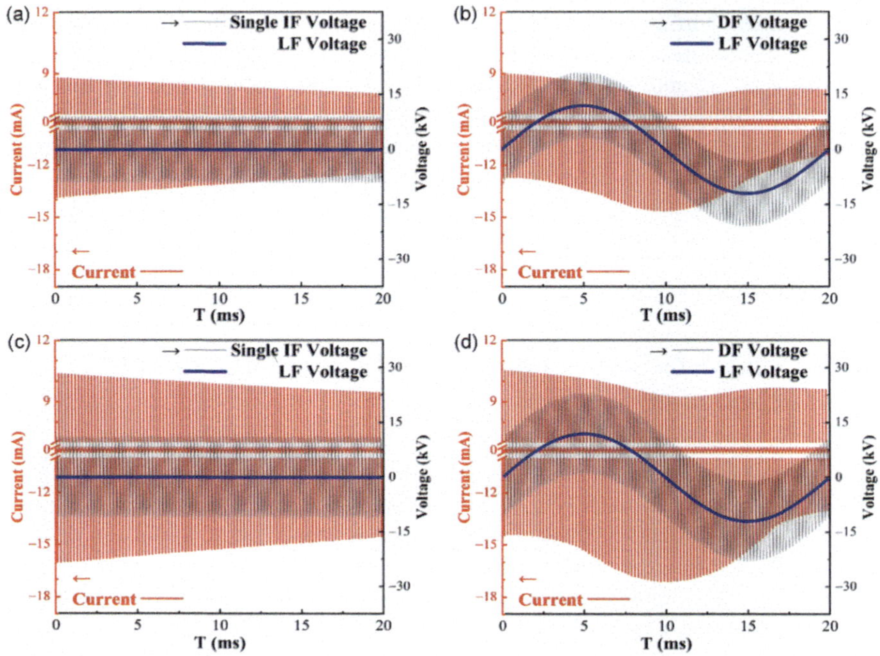

Fig. 10.2 Characteristic of volt and current obtained by the simulation. **a** Single IF excitation with $U_{\mathrm{IF}} = 9$ kV. **b** DF excitation with $U_{\mathrm{IF}} = 9$ kV and $U_{\mathrm{LF}} = 12$ kV. **c** Single IF excitation with $U_{\mathrm{IF}} = 11$ kV. **c** DF excitation with $U_{\mathrm{IF}} = 11$ kV and $U_{\mathrm{LF}} = 12$ kV

the addition of LF voltage is beneficial to improve the current pulse amplitude for more discharge energy injected into the plasma. Furthermore, the range of the current pulse amplitude further broadens with the increase of the U_{LF}. The average value of the current pulse amplitude drifts towards higher values as the U_{IF} increases (shown in Fig. 10.4). For instance, the current pulse amplitude ranges from −11.6 to −14.8 mA in the DF discharge with a U_{IF} of 9 kV and a U_{LF} of 12 kV. As the U_{IF} grows, the amplitude range of the current pulse gradually broadens under the DF excitation. Specifically, the range of the current pulse amplitude expands by 37.5% as the U_{IF} increases from 9 to 11 kV. These results mean that increasing either the U_{IF} or the U_{LF} improves the regulated effect of the current pulse amplitude.

10.3.2 Discharge Dynamic of the DF Excitation

Figure 10.5a, b exhibit the spatial–temporal evolution of electron density distribution during the breakdown process under the DF excitation. Initially, the electron density gradually increases due to the Townsend discharge. Once the Meek's number reaches

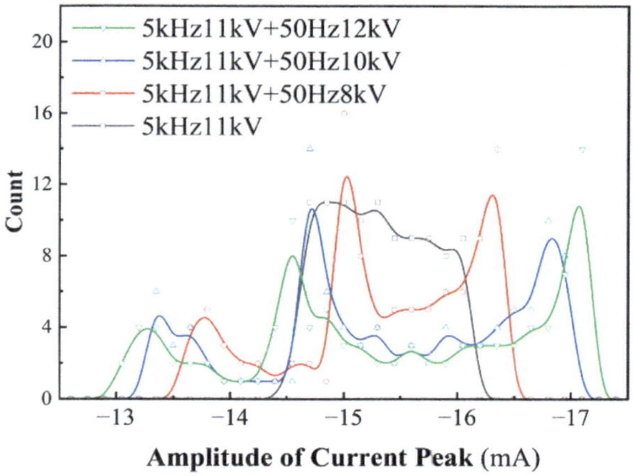

Fig. 10.3 Amplitude distribution of the negative current pulse under the DF excitation with a fixed U_{IF} of 11 kV and the U_{LF} of 0, 8, 10, 12 kV, respectively

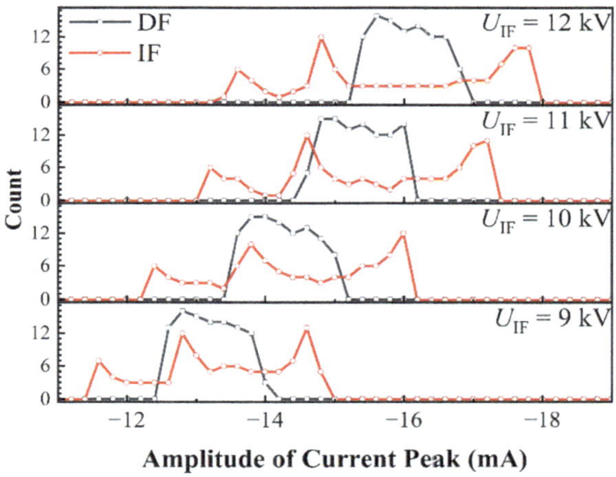

Fig. 10.4 Amplitude distribution of the negative current pulse under the DF excitation with a fixed U_{LF} of 12 kV and the U_{IF} of 9, 10, 11, 12 kV, respectively

18 to 20 [45, 46], the Townsend phase transitions to the streamer propagating phase. The Meek's number is determined by $\alpha|\mathbf{E}|d$, where α is the Townsend ionization coefficient, and d denotes the length of the electron avalanche. The electron avalanche occurs in the entire plasma region during the Townsend period, therefore d is set to 3.5 mm. The purple line represents a contour line with a Meek's number of 18. During the streamer discharge period, an intensive electric field is observed at the front of the space charge, referred to as the 'virtual anode'. The virtual anode propagates toward the cathode driven by the electric field, and its electron density rapidly increases with the propagation of the virtual anode [40]. When the virtual anode arrives at the cathode, the electron density reaches its maximum during the breakdown process. The discharge quenches when the virtual anode arrives at the cathode. Afterwards, the electron density sharply decreases. Additionally, the transient glow discharge is also observed at the anode during the virtual anode propagation. This developing

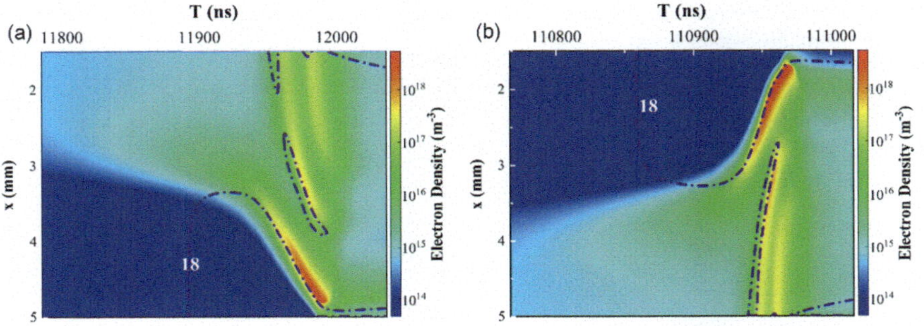

Fig. 10.5 Spatial–temporal evolution of electron density during the air breakdown taken from the 50th IF cycle of the DF excitation with the $U_{IF} = 11$ kV and the $U_{LF} = 12$ kV. **a**, **b** in the positive and negative HP, respectively

process of the discharge exhibits the characteristic feature of a streamer discharge [47].

Figure 10.6 exhibits the spatial–temporal evolution of the electric density distribution around the cathode in the DF discharge with the $U_{IF} = 11$ kV and $U_{LF} = 12$ kV at the following periods: labeled as the 7th, 20th, 32nd, 45th, 57th, 70th, 82nd, and 95th IF periods. It can be seen that the maximum electron density of the breakdown process varies with the number of IF periods. Although the discharge model remains the same in the DF excitation, the microscopic parameters of the air breakdown vary across different periods.

Figure 10.7 plots the relationship between the count of the IF cycle and the maximum electron density of the breakdown process in the DF discharge with a fixed U_{IF} of 11 kV. During the negative HP of IF, the maximum electron density of the breakdown in the DF discharge is lower compared to that in the IF discharge at first. Subsequently, the maximum electron density of the breakdown process increases with the increase in the IF periods. When the U_{LF} switches from the positive polarity to the negative polarity, the maximum electron density of the breakdown process reaches its peak value. At this moment, the maximum electron density of the breakdown in the DF discharge is 10.30–17.91% higher than that in the IF discharge. Conversely, the trend of the maximum electron density during the breakdown in the positive HP of IF is opposite to that in the negative HP. Overall, the maximum electron density of each breakdown in the DF discharge is higher than that in the IF discharge during half a DF cycle. These outcomes suggest a temporal regulation of the maximum electron density of the breakdown in the DF discharge. The maximum electron density of the breakdown presents a periodic variation within a DF cycle. Furthermore, the regulated effect of the maximum electron density during the breakdown becomes more pronounced with a higher amplitude of the U_{LF} in the DF discharge.

Figure 10.8 shows the variation in the maximum electron density difference of each breakdown between the DF discharge and the IF discharge as a function of the number of the IF cycle, with the U_{LF} amplitude of the DF discharge maintained at a

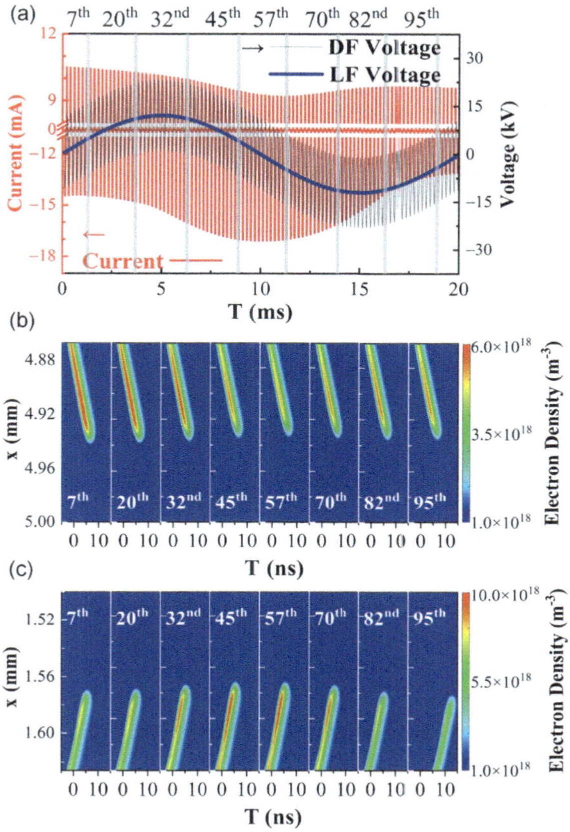

Fig. 10.6 Spatial–temporal evolution of electron density around the cathode and its corresponding time period under the DF excitation with the U_{IF} of 11 kV and the U_{LF} of 12 kV. **a** the volt-ampere waveforms and the labeled periods. **b**, **c** the spatial–temporal evolution of electron density at the labeled periods in the positive and negative HP, respectively

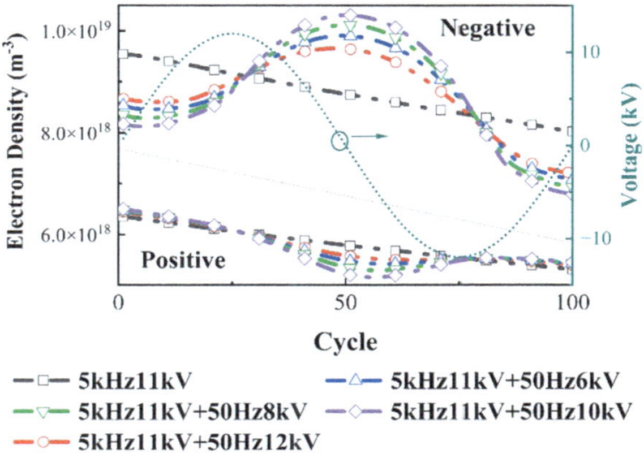

Fig. 10.7 The relationship between the number of the breakdown occurring and the maximal electron density of the air breakdown under the DF excitation with a fixed U_{IF} of 11 kV and the U_{LF} of 0, 6, 8, 10, and 12 kV, respectively

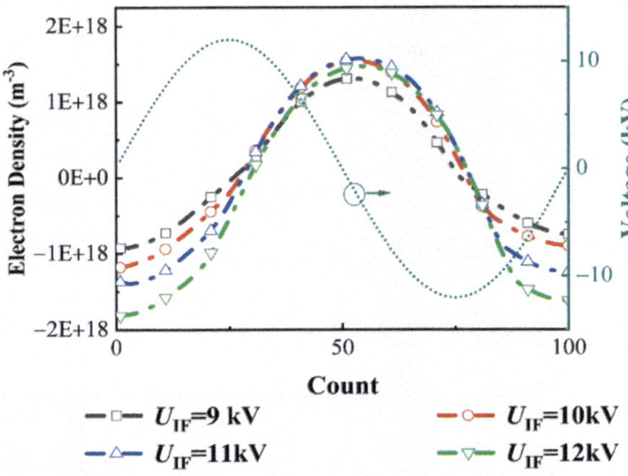

Fig. 10.8 The curve of the difference value in the maximal electron density of each breakdown between the DF discharge and the single IF discharge with the number of the IF cycle when the IF component of the DF discharge is set up to a fixed amplitude of 12 kV

constant of 12 kV. All results are taken from the negative HP of IF due to its higher maximum electron density of the breakdown than that in the positive HP. Generally, the difference value between the DF discharge and the IF discharge is a periodic variation. The maximum electron density difference of the breakdown between the DF discharge and the IF discharge varies between -7.49×10^{17} and 1.31×10^{17} m^{-3} at the DF discharge with the U_{IF} of 9 kV and the U_{LF} of 12 kV. When the U_{IF} increases to 12 kV, the difference value falls within the range of -1.61×10^{18} to 1.48×10^{18} m^{-3}. Notably, the regulated effect of the difference value becomes more obvious with the U_{IF} increase. These findings imply that the regulated effect of the maximal electron density can be improved by increasing either the U_{IF} or the U_{LF} in the DF discharge.

10.3.3 Dielectric Surface Charge of the DF Excitation

In Fig. 10.9, the right column plots the curve of the dielectric surface charge density within 2.5 IF cycles. The surface charge density changes within a certain range over an IF cycle in both the single IF discharge and the DF discharge. The transferred charge is around 60 nC/cm^2 during each air breakdown. This value is the same order magnitude as previous findings [40]. Moreover, the maximum and minimum of the dielectric surface charge density within an IF cycle depend on the U_{LF} and period in the DF discharge. The maximum and minimum values of the dielectric surface charge density at an IF cycle vary inversely with the time trend of the U_{LF}. It is attributed to more charged particles accumulated on the dielectric for the neutralization of the extra voltage from the U_{LF}. These results reproduce the modulated effect of the dielectric surface in the experiment [28]. Thus, the simulated results of the dielectric surface charge are credible in this study.

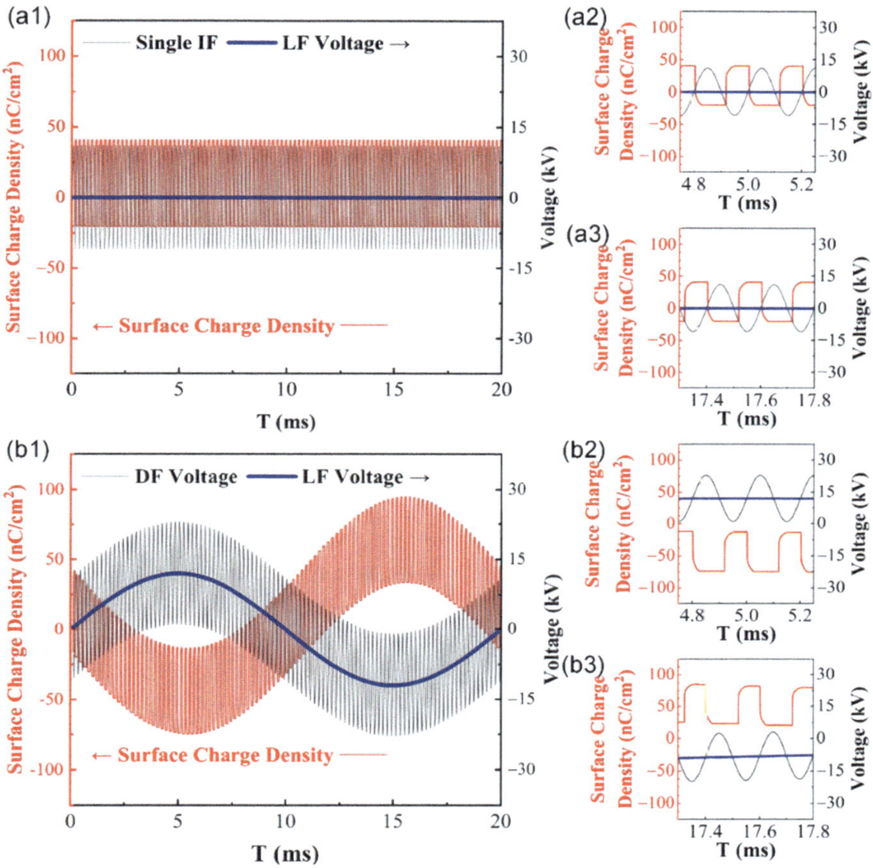

Fig. 10.9 The behaviors of the dielectric surface charge density at **a** the single IF discharge with the U_{IF} of 11 kV and **b** the DF discharge with the U_{IF} of 11 kV and the U_{LF} of 12 kV

Figure 10.10 demonstrates a periodic variation in the difference value of the dielectric surface charge density between the single IF discharge and the DF discharge at a constant U_{IF} of 11 kV. The difference value of the dielectric surface charge between the single IF discharge and the DF discharge increases with an increase of the U_{LF} amplitude in the DF discharge while maintaining a fixed U_{IF}. The residual voltage between the gap is modulated by the U_{LF} [28]. A higher amplitude of the U_{LF} leads to a higher residual voltage across the gap (as shown in Fig. 10.11), leading to a greater accumulation of charged particles on the dielectric surface, thereby generating a stronger initial electric field. Consequently, enhancement of ULF in the DF-DBD intensifies discharge intensity.

When the ULF is fixed at 12 kV, there is no significant difference in the distribution range of dielectric charge density in the DF discharge with UIF of

Fig. 10.10 The difference value of surface charge density between the DF discharge and the single IF discharge at a fixed U_{IF} of 11 kV

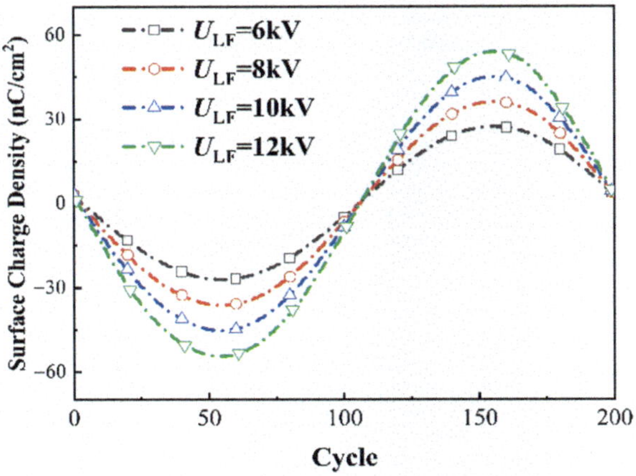

Fig. 10.11 The residual voltage between the gap taken from the negative HP of 50th cycle. The time points of the breakdown are aligned

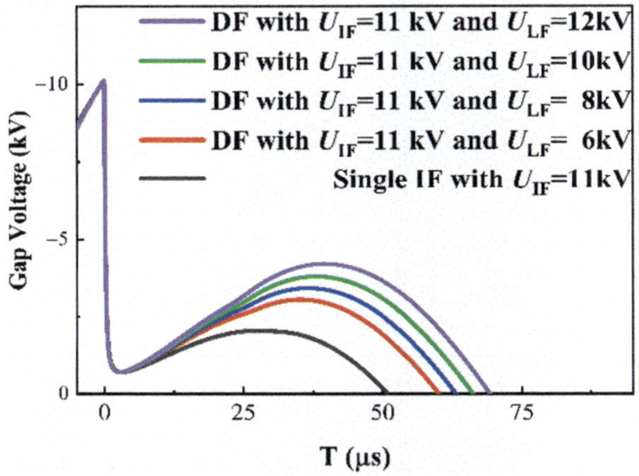

9 and 11 kV (shown in Fig. 12a). However, it should be noted that the transferred charge of each breakdown increases proportionally with an increase in the U_{IF} amplitude (in Fig. 12b). Figure 10.13 shows the relationship between the difference in the extreme value of the dielectric surface charge density within each HF between the IF discharge and the DF discharge with a fixed U_{LF} of 12 kV under the different U_{IF}. It can be observed that the difference in the extreme value of the dielectric surface charge density between the IF discharge and DF discharge almost overlap at the different amplitudes of the U_{IF}. These results mean that the U_{LF} magnitude plays a key role in the modulation of the dielectric surface charge. The modulated effect on the dielectric surface becomes more pronounced with higher values of the transferred charge.

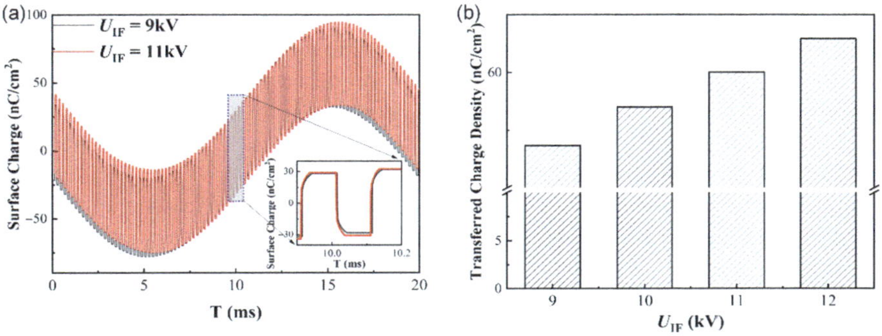

Fig. 10.12 **a** The curves of the dielectric surface charge at the DF discharge with a fixed U_{LF} of 12 kV and the U_{IF} of 9 and 11 kV. **b** The transferred charge density within 50th cycle at a fixed U_{LF} of 12 kV and the U_{IF} of 9, 10, 11, and 12 kV

Fig. 10.13 The difference value of surface charge density between the DF discharge and the single IF discharge at a fixed U_{LF} of 12 kV

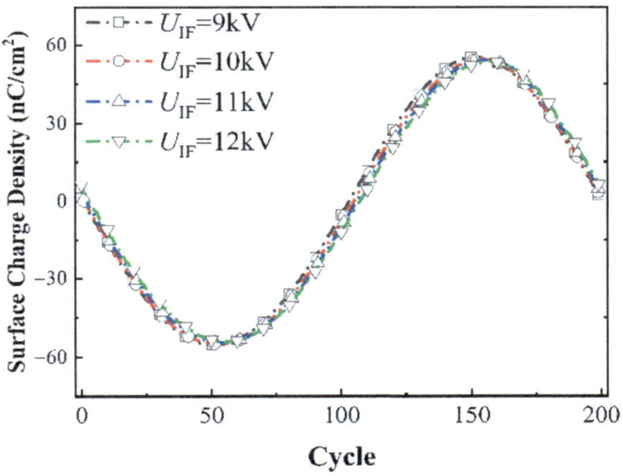

10.4 Conclusion

In this work, a DF-DBD plasma fluid model is implemented. The effect of the U_{IF} and U_{LF} on plasma parameters has been studied. The main conclusions are summarized as follows:

(1) The plasma parameters, including the current pulse amplitude, maximum electron density, and dielectric surface charge of each breakdown, exhibit a periodic variation within a DF cycle. The plasma parameters of the breakdown in the DF discharge exhibit comparatively higher values than those in the IF discharge within half a DF cycle, under the same U_{IF} amplitude. The maximum value of the current pulse and electron density is achieved in the DF discharge when the

U_{LF} switches from positive to negative polarity. These findings suggest that the plasma parameters of each breakdown are modulated in the DF discharge.

(2) The maximum electron density of the breakdown within the DF discharge with a fixed U_{IF} of 11 kV exhibits an increase corresponding to the U_{LF} amplitude. Specifically, when the DF discharge is composed of a U_{IF} of 11 kV and a U_{LF} of 12 kV, the maximum electron density of the breakdown is 17.91% higher than that at the same time within the IF discharge with the same amplitude of the U_{IF}. These results can be attributed to the increase in residual voltage across the air gap caused by the increment of the U_{LF} amplitude, which causes more charged particles to accumulate on the dielectric surface, leading to an increase in the initial electric field before the breakdown, thereby enhancing the discharge.

(3) The distribution range of the current pulse amplitude gradually widens as the U_{IF} amplitude increases during the DF discharge with a fixed U_{LF}. When the DF discharge has a U_{LF} of 12 kV, the distribution range of the current pulse amplitude expands by 37.5% as the U_{IF} increases from 9 to 11 kV. This result indicates that a more pronounced modulated effect of the plasma parameters is also observed at increasing the U_{IF} of the DF discharge with a fixed U_{LF}. More charge is transferred during the breakdown process with the higher amplitude of the U_{IF}, which may play a critical role in the plasma parameter modulation of the DF discharge.

Acknowledgements The authors acknowledge the support of National Natural Science Foundation of China (Grant NO: 52177130). This research was also financially supported by the Industrial Prospects and Key Core Technology Projects of Suzhou city (No. SYC2022029).

References

1. A. Zhou, J. Liu, B. Zhu, X. Li, A. Zhu, Plasma catalytic removal of VOCs using cycled storage-discharge (CSD) mode: an assessment methodology based on toluene for reaction kinetics and intermediates. Chem. Eng. J. **433**, 134338 (2022)
2. G. Ge, H. Lei, X. Yao, Y. Fang, X. Cheng, Investigation of hybrid plasma-catalytic degradation of toluene over FeOOH/γ-Al_2O_3 catalysts. J. Environ. Chem. Eng. **11**(3), 109756 (2023)
3. S.M. Allabakshi, P.S.N.S. Srikar, S. Gomosta, R.K. Gangwar, S.M. Maliyekkal, UV-C photon integrated surface dielectric barrier discharge hybrid reactor: a novel and energy-efficient route for rapid mineralisation of aqueous azo dyes. J. Hazard. Mater. **446**, 130639 (2023)
4. R. Patel, J. van Oijen, N. Dam, S. Nijdam, Low-temperature filamentary plasma for ignition-stabilized combustion. Combust. Flame **247**, 112501 (2023)
5. R. Patel, J. van Oijen, S. Nijdam, N. Dam, On pulse-to-pulse coupling in low-temperature filamentary plasma-assisted ignition in methane-air flows. Plasma Sources Sci. Technol. (2023)
6. R. Paulauskas, D. Martuzevičius, R.B. Patel, J.E.H. Pelders, S. Nijdam, N.J. Dam, M. Tichonovas, N. Striūgas, K. Zakarauskas, Biogas combustion with various oxidizers in a nanosecond DBD microplasma burner. Exp. Therm. Fluid Sci. **118**, 110166 (2020)
7. J.G. Zheng, Y.D. Cui, B.C. Khoo, A comparative study of alternating current and nanosecond plasma actuators in flow separation control. Int. J. Heat Mass Transf. **135**, 1097–1117 (2019)
8. D. Chen, K. Asada, S. Sekimoto, K. Fujii, H. Nishida, A high-fidelity body-force modeling approach for plasma-based flow control simulations. Phys. Fluids **33**(3) (2021)

9. A. Hatamoto, K. Emori, H. Nishida, Surface temperature characterization of dielectric barrier discharge plasma actuator in quiescent air. J. Phys. D-Appl. Phys. **56**(15), 155203 (2023)
10. Q. Li, F. Shen, X. He, C. Xing, W. Yan, Y. Fang, Q. Hu, Modification of soy protein isolate using dielectric barrier discharge cold plasma assisted by modified atmosphere packaging. Food Chem. **401**, 134158 (2023)
11. J. Wang, C. Xing, J. Xia, H. Chen, J. Zhang, W. Yan, Degradation of carbendazim in aqueous solution by dielectric barrier discharge cold plasma: Identification and toxicity of degradation products. Food Chem. **403**, 134329 (2023)
12. P.J. Asl, V. Rajulapati, M. Gavahian, I. Kapusta, P. Putnik, A. Mousavi Khaneghah, K. Marszałek, Non-thermal plasma technique for preservation of fresh foods: a review. Food Control **134**, 108560 (2022)
13. A.A. Abdelaziz, T. Ishijima, T. Seto, N. Osawa, H. Wedaa, Y. Otani, Characterization of surface dielectric barrier discharge influenced by intermediate frequency for ozone production. Plasma Sources Sci. Technol. **25**(3), 35012 (2016)
14. X. Zhang, Y.D. Cui, C.J. Tay, B.C. Khoo, Ultrasound generated by alternating current dielectric barrier discharge plasma in quiescent air. Plasma Sources Sci. Technol. **29**(1), 15017 (2020)
15. K. Kourtzanidis, G. Dufour, F. Rogier, Self-consistent modeling of a surface AC dielectric barrier discharge actuator: in-depth analysis of positive and negative phases. J. Phys. D-Appl. Phys. **54**(4), 45203 (2020)
16. Y. Zhang, L. Wei, X. Liang, M. Aimek, Ozone production in coaxial DBD using an amplitude-modulated AC power supply in air. Ozone Sci. Eng. **41**(5), 437–47 (2019)
17. L. Wei, W. Lin, W. Liao, Y. Zhang, Y. Chen, P. Yuan, Current pulses and their influences on ozone production in oxygen-fed dielectric barrier discharge. J. Electrost. **119**, 103748 (2022)
18. G. Ge, H. Lei, X. Yao, Y. Fang, X. Cheng, Investigation of high-concentration toluene degradation by DBD plasma under different operating parameters. J. Phys. D-Appl. Phys. **56**(31), 315201 (2023)
19. Y. Akishev, G. Aponin, A. Balakirev, M. Grushin, V. Karalnik, A. Petryakov, N. Trushkin, Memory' and sustention of microdischarges in a steady-state DBD: volume plasma or surface charge? Plasma Sources Sci. Technol. **20**(2), 24005 (2011)
20. Y. Akishev, G. Aponin, A. Balakirev, M. Grushin, V. Karalnik, A. Petryakov, N. Trushkin, DBD surface streamer expansion described using nonlinear diffusion of the electric potential over the barrier. J. Phys. D-Appl. Phys. **46**(46), 464014 (2013)
21. T. Li, H. Yan, J. Li, J. Schulze, S. Yu, J. Song, Q. Zhang, The role of surface charge and its decay in surface dielectric barrier discharges. Plasma Sources Sci. Technol. **31**(5), 55016 (2022)
22. T. Li, H. Yan, S. Yu, Y. Wang, J. Li, J. Song, Q. Zhang, Measurements of surface charge dynamics and surface-breakdown characteristics of surface dielectric barrier discharges. J. Appl. Phys. **132**(15), 153301 (2022)
23. A. Starikovskiy, R. Miles, in *Conference on 51st AIAA Aerospace Sciences Meeting including the New Horizons Forum and Aerospace Exposition*, Grapevine, Texas, USA. pp. 754–70 (2013)
24. S.B. Leonov, V. Petrishchev, I.V. Adamovich, Dynamics of energy coupling and thermalization in barrier discharges over dielectric and weakly conducting surfaces on μs to ms time scales. J. Phys. D-Appl. Phys. **47**(46), 465201 (2014)
25. B.M. Goldberg, I. Shkurenkov, I.V. Adamovich, W.R. Lempert, Electric field in an AC dielectric barrier discharge overlapped with a nanosecond pulse discharge. Plasma Sources Sci. Technol. **25**(4), 45008 (2016)
26. K. Takashima, T. Kaneko, Ozone and dinitrogen monoxide production in atmospheric pressure air dielectric barrier discharge plasma effluent generated by nanosecond pulse superimposed alternating current voltage. Plasma Sources Sci. Technol. **26**(6), 65018 (2017)
27. D. Opaits, G. Neretti, S. Zaidi, M. Shneider, R. Miles, A. Likhanskii, S. Macheret, in *Conference on 46th AIAA Aerospace Sciences Meeting and Exhibit*, pp. 1372. Reno Nevada (2008)
28. J. He, B. Peng, G. Yu, R. Wang, N. Jiang, J. Li, Y. Wu, Experimental and numerical study on atmospheric-pressure air dielectric barrier discharge via 50 Hz/5000 Hz dual-frequency excitation. High Volt. **9**(2), 391 (2023)

29. N. Benard, E. Moreau, Capabilities of the dielectric barrier discharge plasma actuator for multi-frequency excitations. J. Phys. D-Appl. Phys. **43**(14), 145201 (2010)
30. B. Bagheri et al., Comparison of six simulation codes for positive streamers in air. Plasma Sources Sci. Technol. **27**(9), 95002 (2018)
31. T. Wong, I. Timoshkin, S. Macgregor, M. Wilson, M. Given, The design of a python library for the automatic definition and simulation of transient ionization fronts. IEEE Access **11**, 26577–26592 (2023)
32. A.P. Jovanović, D. Loffhagen, M.M. Becker, Streamer–surface interaction in an atmospheric pressure dielectric barrier discharge in argon. Plasma Sources Sci. Technol. **31**(4), 4 (2022)
33. K. Lin, S. Chuang, W. Guo, C. Cheng, C. Ou, Thermal characterization of a single microdischarge in atmospheric pressure air dielectric barrier discharges. Plasma Sources Sci. Technol. **29**(7), 75002 (2020)
34. J. He, B. Peng, N. Jiang, K. Shang, N. Lu, J. Li, Y. Wu, Experimental and simulated investigation of microdischarge characteristics in a pin-to-pin dielectric barrier discharge (DBD) reactor. Plasma Sci. Technol. **24**(10), 105402 (2022)
35. A. Tejero-Del-Caz, V. Guerra, D. Gonçalves, M.L. Da Silva, L. Marques, N. Pinhão, C.D. Pintassilgo, L.L. Alves, The LisbOn KInetics Boltzmann solver. Plasma Sources Sci. Technol. **28**(4), 43001 (2019)
36. G.V. Naidis, On photoionization produced by discharges in air. Plasma Sources Sci. Technol. **15**(2), 253–255 (2006)
37. A. Bourdon, V.P. Pasko, N.Y. Liu, S. Célestin, P. Ségur, E. Marode, Efficient models for photoionization produced by non-thermal gas discharges in air based on radiative transfer and the Helmholtz equations. Plasma Sources Sci. Technol. **16**(3), 656–678 (2007)
38. A. Luque, U. Ebert, C. Montijn, W. Hundsdorfer, Photoionization in negative streamers: Fast computations and two propagation modes. Appl. Phys. Lett. **90**(8), 81501 (2007)
39. Y. Zhang, D. Dai, W. Ning, L. Li, Influence of electron backflow on discharge asymmetry in atmospheric helium dielectric barrier discharges. AIP Adv. **8**(9), 95327 (2018)
40. K. Lin, Y. Huang, W. Guo, Y. Chang, Development of a semi-empirical 1.5D plasma fluid model for a single microdischarge in atmospheric pressure dielectric barrier discharges. Plasma Sources Sci. Technol. **28**(11), 115014 (2019)
41. Y.S. Akishev, G. Aponin, A. Balakirev, M. Grushin, V. Karalnik, A. Petryakov, N. Trushkin, Role of the volume and surface breakdown in a formation of microdischarges in a steady-state DBD. Eur. Phys. J. D **61**(2), 421–429 (2011)
42. Z. Zhao, Z. Dai, A. Sun, J. Li, Streamer-to-precursor transition in N2–SF6 mixtures under positive repetitive submicrosecond pulses. High Volt. **7**(2), 382–389 (2022)
43. Z. Zhao, Z. Huang, X. Zheng, C. Li, A. Sun, J. Li, Evolutions of repetitively pulsed positive streamer discharge in electronegative gas mixtures at high pressure. Plasma Sources Sci. Technol. **31**(7), 75006 (2022)
44. B. Peng, N. Jiang, K. Shang, N. Lu, J. Li, Y. Wu, Streamer dynamics and charge self-erasing of two counter-propagating plasmas in repetitively pulsed surface dielectric barrier discharge. High Volt. **7**(4), 730–743 (2022)
45. R.V. Hodges, R.N. Varney, J.F. Riley, Probability of electrical breakdown: evidence for a transition between the Townsend and streamer breakdown mechanisms. Phys. Rev. A **31**(4), 2610–2620 (1985)
46. C. Montijn, U. Ebert, Diffusion correction to the Raether-Meek criterion for the avalanche-to-streamer transition. J. Phys. D-Appl. Phys. **39**(14), 2979–2992 (2006)
47. S. Nijdam, J. Teunissen, U. Ebert, The physics of streamer discharge phenomena. Plasma Sources Sci. Technol. **29**(10), 103001 (2020)

Chapter 11
A Design of Flexible Dielectric Barrier Discharge Device

Biao Hu, Shuqun Wu, Chen Yang, and Xinyi Zhang

Abstract Plasma is rich in active and charged particles and has been used in medical fields such as sterilization. The plasma generator based on the flexible volume dielectric barrier discharge (VDBD) has the advantages of being soft and thin and can be applied to wounds of different shapes. Currently, most research focuses on experiments and applications of flexible VDBD devices, and there are few quantitative studies on device structural parameters. To achieve quantification of structural parameters, this article first proposes a VDBD structure based on silicone, then obtains the relationship between the initial discharge voltage and structural parameters through the equivalent circuit model and electrostatic field simulation, and then discusses the range of the obtained parameters. Finally, a verification test was conducted, and the experimental results agreed with expectations. The results show that: (1) discharge first occurs from the hole's edge in the structure with holes punched in silicone. (2) The electric field strength calculated using the equivalent circuit model exhibits a minor deviation from the electric field strength obtained through simulation at the hole's edge.

11.1 Introduction

Plasma, renowned for its abundance of active and charged particles, is significant in the medical field, particularly in sterilization, disinfection, and wound healing [1]. Nevertheless, conventional plasma treatment presents challenges of rigidity, unwieldiness, and limited adaptability. This paper discusses the transformative impact of flexible plasma generators, utilizing materials like polyimide (PI) to create pliable, slim, and versatile devices suitable for diverse wound geometries. These plasma

B. Hu · S. Wu (✉) · C. Yang · X. Zhang
Nanjing University of Aeronautics and Astronautics, Nanjing, People's Republic of China
e-mail: wushuqun@nuaa.edu.cn

B. Hu
e-mail: hubiao1@nuaa.edu.cn

Z. Fang et al. (eds.), *Proceedings of the 5th International Symposium on Plasma and Energy Conversion*, Springer Proceedings in Physics 398,
https://doi.org/10.1007/978-981-97-2245-7_11

generators predominantly employ the dielectric barrier discharge (DBD) structure, including SDBD, CDBD, and VDBD [2]. Among these, VDBD enables the delivery of plasma deep into wound surfaces, thus providing improved therapeutic outcomes. The adoption of this advanced technology by medical companies such as PlasmaDerm underscores its promising potential in the medical field.

In recent years, domestic and international scholars have researched flexible VDBD plasma generators. Weiwu [3] conducted safety verification for a flexible VDBD device, using Polydimethylsiloxane as the insulating medium and a conductive film as the high-voltage electrode. Through simulations and experiments, they studied the current flowing through the human body under different insulating mediums. Jun-hyun [4] investigated the light intensity, spectrum, and electron temperature of plasma under various materials and different power supply voltages. Meanwhile, in the study of Boekema [5], human skin was employed as a suspended ground electrode for antibacterial and safety experiments. It is evident that, on the one hand, most of the structures currently utilized by scholars employ the human body or materials to be processed as suspended ground electrodes. This approach needs to adequately ensure the uniformity of the air gap in a bent state. Furthermore, the current research primarily focuses on the practicality and safety of the device, with limited attention given to its inherent structural parameters, particularly their quantification. It is worth noting that structural parameters often exert a significant influence on the process of discharge.

To enhance the design of the plasma generator based on flexible VDBD, we introduced a structure. This structure ensures air gap uniformity by incorporating holes in the silicone. We then established the proportional relationship between the applied voltage and the air gap voltage by analyzing the equivalent circuit model under non-discharge conditions. This analysis also provided us with a range of parameters. Lastly, experimental tests were performed to validate whether the designed device meets the specified performance criteria.

11.2 Models and Methods

11.2.1 Device Structure

Figure 11.1 depicts the hierarchical structure of the flexible VDBD device detailed in this article. The device operates in ambient air at atmospheric pressure, with high voltage applied to the electrodes and the grounding of a metal mesh film. The high-voltage electrode is constructed from high-purity, flexible copper foil, with a square area of 65 cm*65 cm and a protruding part measuring 15 cm*15 cm. The dielectric layer comprises PI (Polyimide), known for its flexibility, thermal stability, and high breakdown strength. To ensure insulation and safety, an additional layer of PI is applied above the high-voltage electrode, providing insulation and protection. Below the dielectric layer lies a silicone layer with double-sided adhesive, facilitating

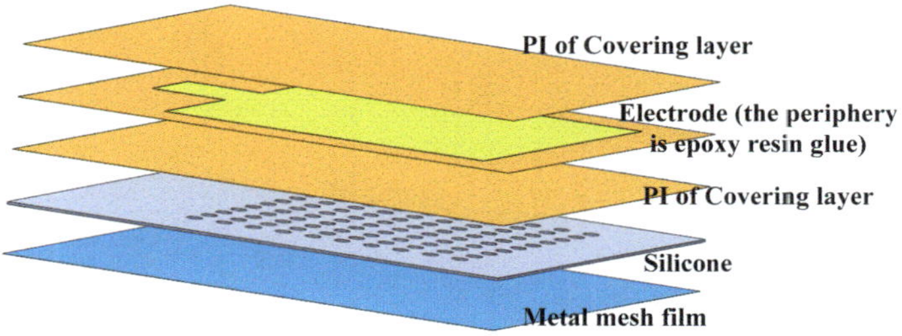

Fig. 11.1 The Layered diagram of device structure

a secure bond between the upper and lower components. The silicone features an array of 10*10 circular holes with a center-to-center spacing of 6 mm. Beneath the silicone layer is a transparent conductive material in a metal mesh film. This film is constructed on a PET-based material with a thickness of 0.1 mm. The dimensions of the cover film, dielectric film, silicone, and metal mesh films are 85 cm in width and 100 cm in height. Structural parameters of the device, including dielectric thickness, silicone thickness, and silicone hole radius, are determined through simulation and calculation.

11.2.2 Simulation Model

This article uses COMSOL multi-physics to simulate the electrostatic field. Figure 11.2 displays a three-dimensional model constructed within COMSOL, precisely mirroring the designed device structure. The model's origin is at the center of the device's bottom surface. An air domain, measuring 150 cm in length, 100 cm in width, and 25 cm in height, surrounds the device. This model is employed for electrostatic field simulation. For ease of observation, two parts are designated: the XZ section and the perpendicular line intersecting the YZ plane at $y = 1.5$ cm. The pertinent simulation parameters for the device's constituent materials are summarized in Table 11.1.

11.3 Results and Discussion

11.3.1 Equivalent Circuit Model

Structural parameters, such as dielectric thickness, exert a significant influence on discharge voltage. Thus, it is vital to define the initial discharge voltage for the device. Drawing from prior experimental findings, this article sets the initial discharge voltage to be below 1500 V. This threshold ensures that the voltage stays within a

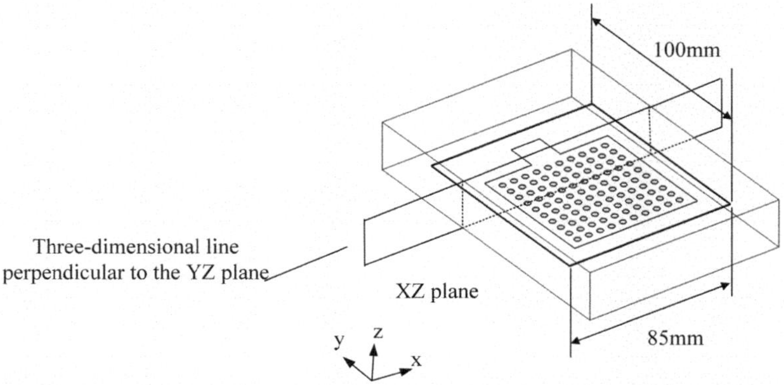

Fig. 11.2 COMSOL simulation 3D model

Table 11.1 Material parameters of each part of the device

Material	Relative permittivity
PI	3.4
PET	4
Copper	
Silicone	3.2

safe range and mitigates the risk of excessive temperature rise during discharge. According to the literature [6], an equivalent circuit model for DBD, as illustrated in Fig. 11.3, has been proposed. In this model, C_d represents the capacitance of the dielectric layer, while C_g signifies the capacitance of the air gap layer. As the air gap layer experiences breakdown during discharge, R_g is introduced to account for the parallel resistance at both ends of the air gap layer. In the context of the device structure described in this article, the dielectric layer's capacitance includes the capacitance achieved by the series connection of the dielectric layer PI and the metal mesh PET, denoted as C_d. The air gap layer's capacitance comprises the capacitance obtained by the parallel connection of silicone and the air gap, labeled as C_g. It's important to note that this article exclusively focuses on the initial discharge scenario, and therefore, the equivalent circuit model does not consider the parallel resistance R_g.

The calculation formula for capacitance is

$$C = \frac{\varepsilon_0 \varepsilon_r k S}{d} \tag{11.1}$$

where $\varepsilon_0 = 8.854 \times 10^{-12}$ F/m, ε_r is the relative dielectric constant of the material, and S is the electrode area, which is 44.5 cm^2. k is the coefficient of the wire area of the metal mesh membrane in the total area, which is 9.75%; d is the thickness.

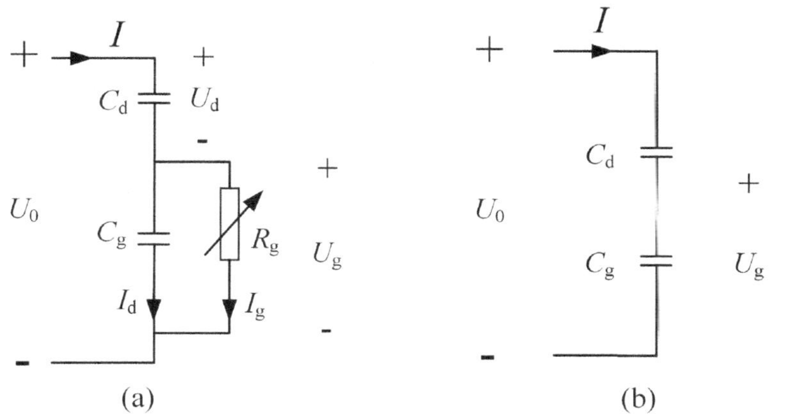

Fig. 11.3 Equivalent circuit model. **a** Classical equivalent circuit model; **b** The equivalent circuit model without discharge in this article

The device structure proposed in this article deviates from the conventional VDBD structure. The breakdown field strength of free space air at a temperature of 20 °C and atmospheric pressure is approximately 30 kV/cm. In contrast, the field strength of flashover that occurs on the surface of silicone ranges from 7 to 25 kV/cm [7]. Assuming uniform electric field strength inside the hole, discharge should initiate from the hole's edge. To ascertain the distribution and numerical magnitude of the electric field within the hole under a specified applied voltage, an electrostatic field simulation is conducted using COMSOL.

11.3.2 COMSOL Electrostatic Field Simulation

As shown in Fig. 11.2, the simulation model uses a dielectric thickness of 50 μm, a silicone hole radius of 1.5 mm, and a silicone thickness of 0.3 mm. The device and the surrounding air domain are configured as electrostatic regions, with the copper electrodes designated terminals and the terminal voltage set to 1 kV. The lower surface of the metal mesh membrane is grounded, and the results are presented in Fig. 11.4.

Figure 11.4a indicate that the electric field distribution within each hole is relatively uniform. In Fig. 11.4b, it can be observed that the electric field strength decreases when the distance to the center is more significant than 1.32 mm, while it remains relatively constant when the distance is less than 1.32 mm.

To visualize the electric field strength within the hole, we created three three-dimensional lines perpendicular to the XY plane inside the hole. These lines were generated by selecting a hole intersected by the earlier cross-section, with center coordinates at y = 1.5 mm and x = 3 mm. These lines were positioned as follows:

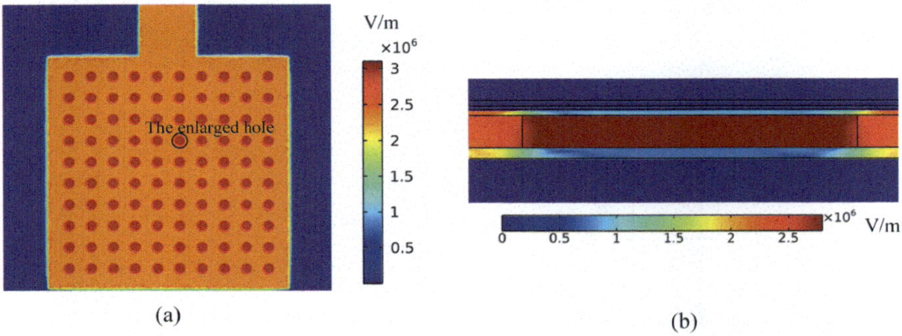

Fig. 11.4 Electrostatic field distribution. **a** electric field distribution in XZ section where y = 1.5 mm; **b** electric field distribution of the enlarged hole in XZ section where y = 1.5 mm; electric field distribution in XY section where z = -0.2 mm;

one at the hole center, another at 0.5 mm from the hole center, and the third at 1.49 mm from the hole center. The horizontal axis represents the distance along the positive z-axis, covering a range from 0.1 mm to 0.4 mm (corresponding to the range of the hole), as depicted in Fig. 11.5.

The findings in Fig. 11.5 indicate that as the distance along the positive z-axis increases, the electric field strength at the hole's center and the line 0.5 mm from the center remains steady at 29.3 kV/cm. The electric field strength at the line 1.49 mm from the center stabilizes at 26 kV/cm. This outcome aligns with the two-dimensional electric field distribution. The stable electric field strength value represents the electric field strength of the cross-section in the hole. With a dielectric thickness of 50 μm, silicone thickness of 0.3 mm, and a hole radius of 1.5 mm, the calculated capacitance for the dielectric layer is 242.6 pF, the PET capacitance is 142.7 pF,

Fig. 11.5 The electric field strength curve of three-dimensional lines at different positions in the hole changes with distance from the z-axis

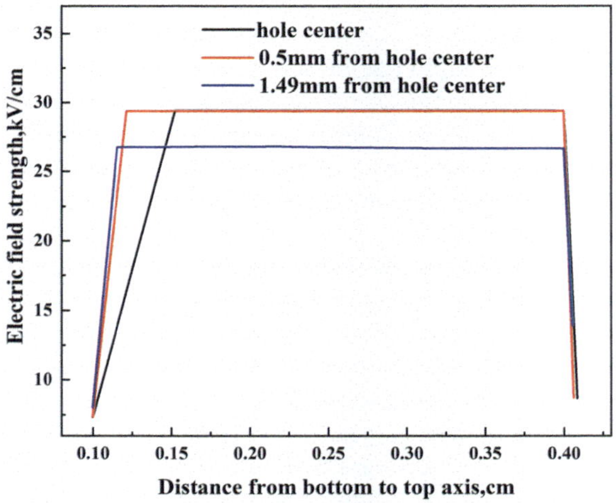

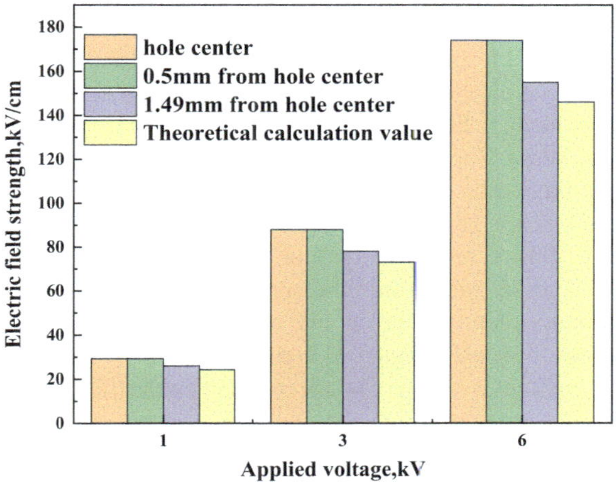

Fig. 11.6 Comparison chart of the electric field strength of the line at different positions and the calculated electric field strength

and the capacitance of the silicone parallel air gap is 33.6 pF. Thus, we have C_d, which is equal to 89.8 pF, and C_g, which is equal to 33.6 pF. Taking an external 1 kV as an example, Consider the voltage distribution relationship U_g equals 0.73 kV. According to $E = U_g/d$, the calculated value of electric field strength at this point is 24.33 kV/cm. Subsequent external voltages of 3 kV and 6 kV are explored, as demonstrated in Fig. 11.6.

Taking into account the earlier analysis, it is likely that discharge commences from the edge of the hole. Consequently, our primary concern lies in the deviation between the electric field strength at the hole's edge and the calculated electric field strength. As shown in Fig. 11.6, under three voltages of 1 kV, 3 kV, and 6 kV, the theoretically calculated electric field strength is used to replace the actual electric field strength at the hole's edge (1.49 mm from the hole's center). The errors are 6.5%, 6.8%, and 5.8%, respectively, below 10%. This suggests that the equivalent circuit model proposed in this article accurately calculates the electric field strength. Furthermore, under the three voltages of 1 kV, 3 kV, and 6 kV, the difference ratios between the electric field strength at the hole's center and the electric field strength at the hole's edge are 12.7%, 12.8%, and 10.3%, respectively. Therefore, it is implausible that air breakdown in the middle of the hole would occur before the surface flashover along the silicone at the hole's edge.

11.3.3 Parameters Calculation

Subsequent calculations involve replacing the electric field strength at the hole's edge with the voltage of the air gap divided by the silicone thickness. The experimental device is fabricated using Flexible Printed Circuit Boards (FPCB), providing options for dielectric thickness of 25 μm, 50 μm, and 75 μm. Initially, with the silicone thickness fixed at 0.3 mm and the hole radius set at 3 mm, values for C_1 and C_2 are computed for different dielectric layer thicknesses, as presented in Table 11.2.

If we aim for discharge in silicone with a 0.3 mm thickness, as per the earlier conclusion, the field strength of surface flashover needs to be in the range of 7–25 kV/cm. Considering the maximum value to ensure a certain discharge, this implies a voltage drop of 0.75 kV in the air gap. Calculations indicate that when the dielectric thickness is 25 μm, 50 μm, and 75 μm, the initial discharge voltage is 979 V, 1031 V, and 1082 V, respectively. This demonstrates that the influence of the three dielectric thicknesses on the initial discharge voltage is within a range of a few dozen volts. Thinner dielectric layers enhance device flexibility but may be more susceptible to breakdown and damage at higher voltages. Following a comprehensive evaluation, a PI thickness of 50 μm is selected. When the dielectric thickness is 50 μm, C_d is 89.8 pF. Assume the silicone thickness is d_g (unit: cm), the hole radius is r (unit: cm), and calculate C_g. The expression for C_g is:

$$C_g = \frac{114.2 - 596r^2}{100d_g} \tag{11.2}$$

Assuming that the applied voltage is U_0, and we want to ensure that U_0 can just produce flashover along the silicone surface, then:

$$U_0 \times \frac{C_d}{C_d + C_g} = Ed \tag{11.3}$$

Simplify to get:

$$U_0 = Ed + \frac{E}{8980}(114.2 - 596r^2) \tag{11.4}$$

Assuming the distance from the hole edge is X mm, and the center-to-center distance has been established as 6 mm, then $20\ r + X = 6$. If the hole radius is too small, the area where plasma is generated is minimal, resulting in a shallow effective utilization rate of the device. Conversely, if the hole is too large, the area of supporting silicone becomes too small, making it harder to ensure the overall stability of the device. On this basis, it is determined that the range of r should be between 0.1 and 0.2 cm.

If d is 0 cm, and r takes the maximum value, the theoretical minimum value of U_0 is 150.9 V. However, it is not possible to achieve discharge without silicone. Therefore, a certain safety margin is included for the minimum voltage value, which

Table 11.2 C_d and C_g under different media thicknesses

Item	C_d with dielectric thickness of 25 μm	C_d with dielectric thickness of 50 μm	C_d with dielectric thickness of 75 μm	C_g
Capacitance value/pF	110.3	89.8	75.8	33.6

is set to 300 V, d_g is 0.01 cm. According to the previously designed U_0, which is no more than 1500 V, let r assume the minimum value at this point, d_g is 0.088 cm. Therefore, the final parameters determine the dielectric thickness as 50 μm, the silicone thickness ranging from 0.1 to 0.88 mm, and the hole radius varying from 1 to 2 mm.

11.4 Experimental Verification

The experimental device is shown in Fig. 11.7. The silicone thickness is 0.45 mm; the dielectric thickness is 50 μm, and the hole radius is 1.5 mm. The remaining parameters are the same as the simulation part. Turn the side of the metal mesh toward the camera during the experiment to capture the discharge phenomenon. The high-voltage power supplies used in the experiment include a pulse power supply (Lingfengyuan HV-2015) and an AC power supply (Suman CTP-2000K). Photos of the plasma were taken by a digital camera (Nikon d7100); the sensitivity and exposure time of the digital camera were fixed at 6400 and 2 s, respectively. A current probe (Pearson 2877) and a voltage probe (Tektronix P6015A) detected current and voltage signals, respectively. Current and voltage waveforms were recorded by a Tektronix oscilloscope (MDO3034).

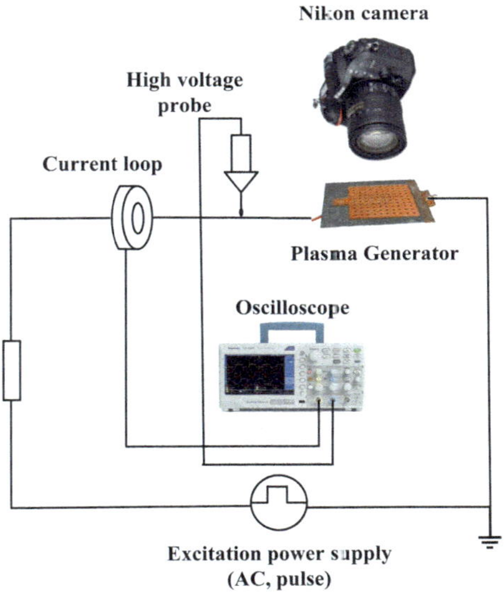

Fig. 11.7 Experimental set-up

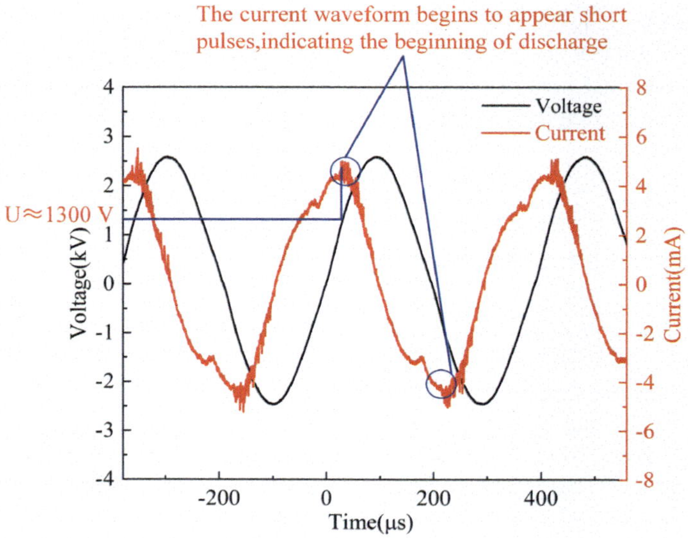

Fig. 11.8 The voltage and current waveform

11.4.1 Initial Discharge Voltage Verification

AC power serves as the excitation source to validate the initial discharge voltage. Discharge takes the form of multiple breakdowns and extinguishments in each half-cycle of the applied voltage. These occurrences are evident in the voltage-current waveform as numerous short-duration pulses appear in each half-cycle of the applied voltage [8]. In the experiment, the output voltage of the power was incrementally increased until the peak load voltage reached 2.6 kV, and the device exhibited distinct discharges. The waveform at this stage is depicted in Fig. 11.8. During the positive half-cycle of the voltage, discharges initiate at around 1.3 kV and persist until the voltage reaches its peak at 2.6 kV. The duration of the discharge event is approximately 75 μs, with a maximum current pulse amplitude of 5 mA. A similar phenomenon occurs during the negative half-cycle of the voltage. Consistent with the calculation Eq. (11.4) for determining the initial discharge voltage, as previously summarized, the calculated initial discharge voltage is 1.405 kV. The experimental results are very close to those calculated by Eq. (11.4).

11.4.2 Discharge Mode Verification

Because the frequency of the AC power supply used is relatively high, the discharge temperature can become quite elevated. Therefore, experiments employ a pulse power supply. The parameters of this pulse power supply include a frequency of

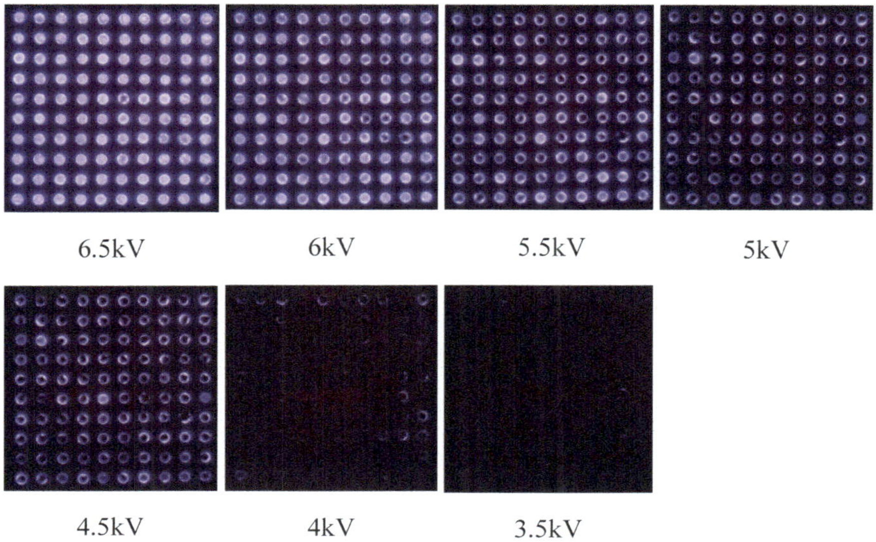

Fig. 11.9 Discharge photos under different pulse voltages

400 Hz and a pulse width of 10 μs. Variations in voltage are made to investigate the discharge patterns.

As depicted in Fig. 11.9, when the power supply voltage is increased to 3.5 kV, discharges become noticeable. Raising the power supply voltage to 4.5 kV reveals plasma formation within most holes, with the plasma predominantly concentrated at the periphery of each hole, creating a ring-like pattern; further increasing the voltage to 6.5 kV results in the plasma filling all the holes, with higher brightness observed at the hole edges, indicating that the discharge intensity at the hole's edge surpasses that at the center of the hole. These discharge phenomena validate the previous simulation conclusion that discharge initially initiates at the edge of the hole and, with increasing voltage, gradually extends toward the hole's center.

11.5 Conclusion

This article initiates by selecting initial discharge voltages below 1.5 kV as reference points. It subsequently verifies the feasibility of deriving the air gap voltage through the equivalent circuit model and calculates the electric field strength via electrostatic field simulation. The various structural parameters of the device are obtained from the initial discharge voltage range. The main conclusions are as follows:

(1) In the VDBD structure with holes punched in silicone, discharge first occurs from the hole's edge. As the voltage increases, the plasma gradually propagates toward the center of the hole.
(2) Electrostatic simulation reveals that the electric field distribution within each hole is relatively uniform. However, the electric field strength at the edge of an individual hole is lower than at the center of the hole. The electric field strength calculated using the equivalent circuit model exhibits a minor deviation from the electric field strength at the hole's edge obtained through simulation.

References

1. G. Cho, Y. Kim, Flexible plasma sheets. Appl. Sci. Conv. Technol. **27**(2), 23–25 (2018)
2. Y. Sun, B. Zhang, C. Wang, Polyimide-based flexible plasma sheet and surface ionization waves propagation. Adv. Electron. Mater. **7**(11) (2021)
3. D. Weiwu, *Research on Air Discharge Flexible Plasma Generation Device* (North China University of Science and Technology, China, 2022)
4. K. Jun-hyun, P. Chang Jin, J.P. Chang, Atmospheric pressure floating electrode-dielectric barrier discharges (FE-DBDs) having flexible electrodes. Korean J. Chem. Eng. **57**(3), 432–437 (2019)
5. B. Boekema, M. Stoop, M. Vlig, Antibacterial and safety tests of a flexible cold atmospheric plasma device for the stimulation of wound healing. Appl. Microbiol. Biotechnol. **105**(5), 2057–2070 (2021)
6. M. Laroussi, X. Lu, V. Kolobov, Power consideration in the pulsed dielectric barrier discharge at atmospheric pressure. J. Appl. Phys. **96**(5), 3028–3030 (2004)
7. G. Gradinaru, V.P. Madangarli, T.S. Sudarshan, Surface flashover in silicon-vacuum systems. IEEE Trans. Electr. Insul. **28**(4), 555–565 (1993)
8. Z. Cheng, F. Zhi, Hu, J.: Measurement of dielectric barrier discharge electrical parameters and reactor parameters. Insul. Mater. **4**, 52–55 (2007)

Chapter 12
Generations of Reactive Species Under High-Voltage Nanosecond Pulse Discharge in Gas–Liquid–Solid Three-Phase

Zhicheng Tang, Qi Liu, Xiaohui Xu, Xiaoxia Yang, Xu Li, Xiangbin Guo, and Bingyan Chen

Abstract In this paper, a high-voltage nanosecond pulsed power supply was used to excite a wire-tube reactor in a gas–liquid-solid three-phase to generate reactive species in treated water. Especially, the concentrations of O_3, H_2O_2 and HNO_x in treated water were examined. The results show that the concentrations and energy efficiency ratios of the H_2O_2 and HNO_x in the treated water showed a decreasing trend, while those of the O_3 did not change much by fixing the liquid conductivity as 100.0 μS/cm and increasing the water content (5–85%), under the experimental conditions of the applied voltage as 18.5 kV, the total gas–liquid flow rate as 2.00 LPM, and using SiO_2 spheres as the solid-phase filler. Moreover, the concentrations and energy efficiency ratios of H_2O_2 and HNO_x in the treated water showed a tendency to increase and then decrease by fixing the water content of gas–liquid at 45% and increasing the liquid conductivity from 2.00 to 800.0 μS/cm. The maximum concentrations of both H_2O_2 and HNO_x were occurred at the conductivity as 100.0 μS/cm. The concentrations of O_3 showed a slow increase, and its energy efficiency ratios were first constant and then increased slowly. This work can provide an experimental reference for the application of plasma in the environmental field.

Z. Tang · Q. Liu · X. Yang · X. Li · X. Guo · B. Chen (✉)
School of Mechanical and Electrical Engineering, Hohai University, Changzhou 213200, China
e-mail: cbyan@foxmail.com

X. Xu
Department of Astronomy, University of Science and Technology of China, Hefei 230026, China

B. Chen
College of Science, Hohai University, Changzhou 213200, China

Z. Fang et al. (eds.), *Proceedings of the 5th International Symposium on Plasma and Energy Conversion*, Springer Proceedings in Physics 398,
https://doi.org/10.1007/978-981-97-2245-7_12

12.1 Introduction

Plasma is an electrically neutral substance that exists in an ionized gas state, composed of a large number of electrons, atoms, positive and negative ions, and molecules. It is considered the fourth state of matter in addition to the three states of gas, liquid, and solid in nature and the universe. Depending on the temperature conditions that generate plasma, it can be classified as high-temperature and low-temperature plasma [1].

Pulsed gas discharge is a form of generating low temperature plasma [2]. Compared to AC and DC discharges, pulsed discharges consume less energy. Meanwhile high-voltage pulsed discharges are characterized by steep fronts and narrow pulses, which can improve the processing capacity of the reactor at wide electrode spacings increase the stability of the discharge, enhance the mass transfer efficiency of the discharge plasma reactor [3], and consistently produce a variety of active components such as nitrogen oxides (NO_x or $NO_x{}^{-y}$), ozone (O_3), hydrogen peroxide (H_2O_2), etc. [4]. In recent years, pulsed discharge plasma has been widely used in microbial inactivation [5, 6], environmental protection [7, 8], biomedicine [9, 10], and nanomaterials fabrication [11, 12].

Pulsed discharge plasma is often used in aqueous solution in environmental protection, biomedicine and other fields. The ratio of liquid volume in aqueous solution to the total volume of gas–liquid two phases will directly affect the pulse discharge process as well as the content of active components in the plasma. In addition, the composition and content of particles in the solution in different environments can lead to large differences in the conductivity of the solution. There are complex interaction processes between the plasma and the water molecules and microscopic particles in the solution [13, 14]. Therefore, the study of the interaction between pulsed discharge plasma and the liquid phase has become a popular problem in this field.

In this study, a line-tube configuration was employed as the pulsed reactor, with quartz beads as the solid-phase filler for high-voltage nanosecond pulsed-discharge excitation in a gas–liquid-solid three-phase environment. By measuring the concentration and energy efficiency ratio of active components in the liquid phase after discharge treatment, the effects of fluid water content and solution conductivity on the electrical parameters and the formation of active components in the high-voltage pulse discharge were studied.

12.2 Devices and Methods

12.2.1 Experimental Device

In order to investigate the generation pattern of active constituents in a gas–liquid-solid three-phase discharge under high voltage nanosecond pulsed excitation, an experimental research platform, as shown in Fig. 12.1, was constructed.

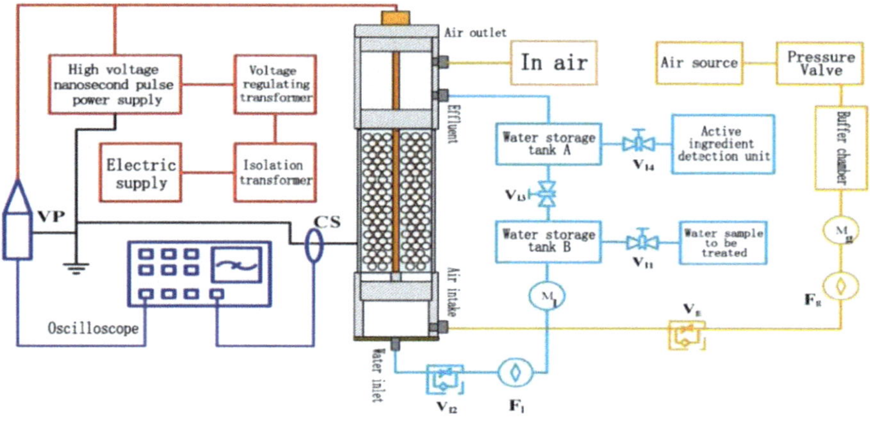

Fig. 12.1 Experimental platform

The experimental platform consists of a gas–liquid-solid three-phase discharge reactor, a high voltage nanosecond pulsed power supply, a high voltage voltage probe, a high-frequency current probe, a mixed-domain digital oscilloscope, a voltage regulator transformer, an isolation transformer, an active constituent detection unit, a gas pump, a gas flowmeter, a buffer chamber, a liquid pump, a liquid flowmeter, water storage tank A, water storage tank B, and a pressure regulating valve assembly.

The gas is introduced into the buffer chamber through a pressure valve, pumped into the discharge reactor by a gas pump (M_g), and the instantaneous gas flow is monitored by a gas flowmeter (F_g).

The water sample to be treated flows into water storage tank B via a check valve (V_{11}), pumped into the discharge reactor by a liquid pump (M_l), and its flow is monitored by a liquid flowmeter (F_l). Check valves (V_g and V_{12}) are used to control the unidirectional flow of gas and liquid, preventing backflow. After the liquid undergoes the discharge treatment, it is sent to water storage tank A, while the gas is discharged through the gas outlet. The unidirectional throttle valve (V_{14}) controls the flow direction of the water sample from water storage tank A to the active constituent detection unit.

During the experiment, if it is necessary to cyclically treat the water sample, the unidirectional throttle valve (V_{13}) is opened to allow the water sample from water storage tank A to flow into water storage tank B for further discharge treatment.

The high-voltage nanosecond pulse power supply provides high-voltage pulse excitation to the discharge reactor, and the output voltage of the high-voltage nanosecond pulse power supply can be varied by adjusting the rotary knob on the voltage regulator transformer. Voltage and current parameters at the discharge reactor end were acquired by a high voltage probe VP (North Star: PVM-5), a high frequency current probe CS (Pearson: 6595), and a mixed domain digital oscilloscope (Tektronix: MDO3054).

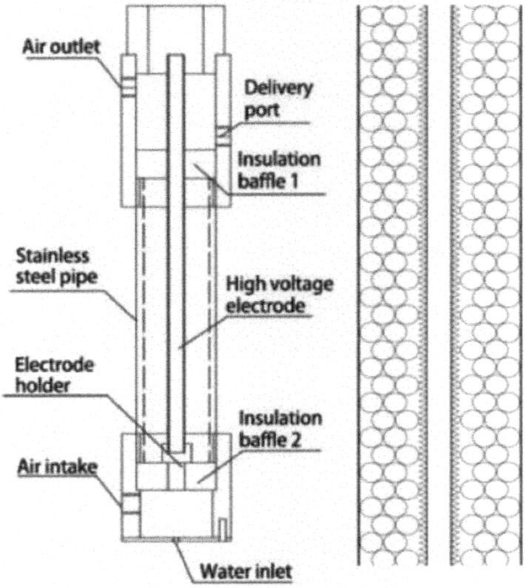

Fig. 12.2 Overall and internal schematic diagram of the reactor

The active ingredients such as O_3, NO_x, and H_2O_2 are measured respectively using an O_3 detector (GDYS-101SC2, measurement range: 0–2.50 mg/L, resolution: 0.01 mg/L), nitrite nitrogen detector (GDYS-101SX, measurement range: 0–0.45 mg/L, resolution: 0.01 mg/L), and H_2O_2 detector (Kyoritsu chemical-check Lab. Corp, DPM-H_2O_2, measurement range: 0–2.00 mg/L, resolution: 0.01 mg/L).

Figure 12.2 is a schematic diagram of the reactor. The gas–liquid-solid three-phase discharge reactor consists of a top, middle, and bottom. The middle part is a stainless steel cylinder, and the top and bottom are made of PTEF material. The top has a water outlet, a gas outlet, and an upper insulating baffle, while the bottom has an air inlet, a water inlet, a lower insulating baffle, and an electrode bracket. The top and bottom are connected to the stainless steel cylinder in the middle via threads.

A high-voltage electrode is installed inside the stainless steel cylinder, and solid-phase small balls are filled. The high-voltage electrode passes through the upper insulating baffle from the top and the lower end is fixed above the lower insulating baffle by the electrode bracket.

The high-voltage electrode is threaded, the radius of curvature of the protrusion on the threaded surface is small, making the local electric field stronger, and it is easier to breakdown the medium between the electrodes than the smooth surface. The high-voltage electrode can be made of materials such as tungsten, molybdenum, titanium, stainless steel, tungsten-molybdenum alloy, or titanium alloy. This study chose to use stainless steel material.

Figure 12.3a is a schematic diagram of the insulating baffle. The insulating baffle material can be PTFE or organic glass, and this study chose to use PTFE. The insulating baffle is equipped with arrayed small holes.

The arrayed small holes in the insulating baffle 1 allow the water and gas in the discharge reactor to flow out, while the arrayed small holes in insulating baffle

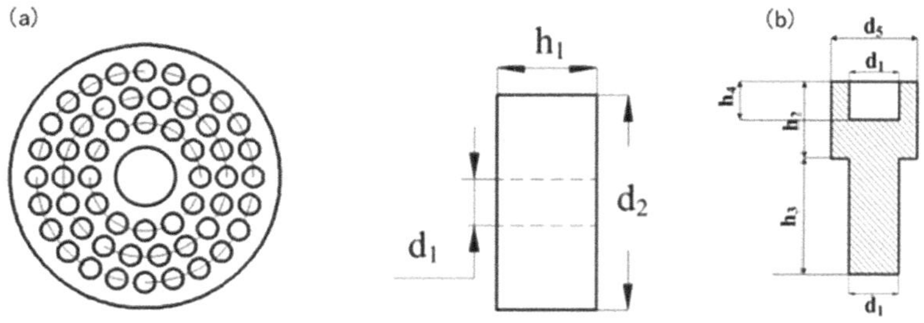

Fig. 12.3 Schematic diagram of insulation baffle and electrode support **a** Insulation baffle; **b** Electrode support

2 serve two purposes. The first purpose is to disperse the air entering the reactor to produce more uniform bubbles, thereby reducing the difficulty of the reaction. The second function is to fix the electrode to prevent tilting. The insulating baffle and the threaded metal electrode are embedded in the electrode bracket, which serves to fix the electrode and prevent it from tilting.

The outer diameter of the insulating baffle is d_2, and its thickness is h_1, with a light hole in the center with a diameter of d_1 for placing the electrode bracket. The surface is equipped with an array of small holes, which are evenly distributed in 3 layers. The diameter of the small holes is d_3.

To prevent quartz balls with a diameter of d_4 from blocking the small holes when placed inside the reactor, affecting the ingress of water and gas, the diameter of the small holes and the diameter of the quartz balls should satisfy the following relationship: the center distance between any two adjacent small holes should be less than the center distance between any two adjacent quartz balls. This relationship is governed by the following equation:

$$\Delta d_i < d_4 \tag{12.1}$$

where Δd_i represents the center distance between any two small holes.

Figure 12.3b shows a schematic diagram of the electrode bracket. The material for the electrode bracket can be made from one of the following: quartz, glass, or ceramic

As high-voltage discharge generates local high temperature, the electrode bracket helps to prevent local high temperature damage at the contact point between the metal electrode and the insulating baffle. The small-diameter cylindrical end of the electrode bracket is inserted into the light hole with a diameter of d_1 in the insulating baffle at the bottom of the device. The materials, such as glass, quartz, or ceramic, have high insulation properties, which prevent high-voltage electrodes from discharging at the reactor's bottom.

Moreover, glass, quartz, or ceramic materials have a relatively high melting point, so they will not melt due to the high temperature of the reaction. The electrode sleeve features a double-step structure with an upper cylindrical diameter of d_5, a height of h_2, a lower cylindrical diameter of d_1, a height of h_3, and a cylindrical groove at the upper end with a diameter of d_1 and a depth of h4 for securing the high-voltage electrode.

The solid-phase spheres are made of glass or quartz material. The gap between the two electrodes is filled with these spheres, providing conditions for easier discharge during the feeding of water and gaseous substances, as the gas bubbles trapped in the gaps of the spheres can decrease the difficulty of discharge. To study the impact of the catalyst's characteristics on the application of discharge plasma, the catalyst can be loaded on the surface of the solid-phase spheres.

The diameter of the spheres (2R) is generally designed within the range of 0.50–5.00 mm. If the spheres are too small, it is not conducive for the gas entering the reactor to become bubbly, which can easily lead to blockages inside the reactor. If the spheres are too large, they have a smaller surface area, therefore carrying a less amount of surface-loaded catalyst, which can decrease the effect of synergistic catalysis.

12.2.2 Experimental Methods

To study the generation characteristics of active components of high-voltage nanosecond pulse discharge plasma in a gas–liquid-solid tri-phase environment, this article investigates the impacts of different fluid water contents antrid solution conductivities on active component generation and energy efficiency ratio by controlling variables.

12.2.3 Method of Regulating Solution Conductivity

The procedure was initiated by setting a target conductivity and required solution volume. Subsequently, the mass of the necessary electrolyte was calculated. The solution was then prepared, and its conductivity was measured using a DDS-11A device. Finally, the electrolyte's mass and solution volume were fine-tuned.

The calculation process for the required electrolyte mass is as follows:

$$\Lambda_m = \Lambda_m^{\infty}(1 - \beta\sqrt{c}) \tag{12.2}$$

The Λ_m is the molar conductivity, in units of $S.m^2.mol^{-1}$. $\Lambda_m{}^{\infty}$ is the molar conductivity at infinite dilution, with units $S.m^2.mol^{-1}$. For a specific substance, its molar conductivity at infinite dilution is a constant, where for instance, $\Lambda_m{}^{\infty}(NaCl) = 0.012645\ S.m^2.mol^{-1}$. β is a constant, corresponding to the temperature, electrolyte, and solvent; c is the mole concentration of the solute, in units of $mol.m^{-3}$.

The conductivity k is measured in $S.m^{-1}$. The relationship between molar conductivity Λ_m, conductivity k, and mole concentration c of the solute is as follows:

$$\Lambda_m = \frac{k}{c} \tag{12.3}$$

By combining Formulas (12.2) and (12.3), the mole concentration c of the solute can be deduced from the conductivity. The required electrolyte mass m can then be calculated from the equation below:

$$m = c \cdot V \cdot \mathrm{M} \tag{12.4}$$

In Eq. (12.4), m is the required mass of the electrolyte in grams (g); V is the volume of deionized water (L); M is the molar mass of the solute, in g.mol^{-1}. For NaCl, M (NaCl) = 58.5 g.mol^{-1}.

12.2.4 Energy Efficiency Ratio Evaluation

Figure 12.4a shows a typical voltage and current waveform. By multiplying the voltage and current, you can obtain the instantaneous power spectrum shown in Fig. 12.4b. The supplied energy can be calculated using the full-width at half-maximum (FWHM) method. The energy Esp within a single pulse period can be represented by the following expression:

$$E_{sp} = \sum P_i \cdot t_i \tag{12.5}$$

In Eq. (12.5), P_i is the ith power peak value, and t_i is the time occupied by the ith power half-peak value.

The expression for the total supply energy Et within the continuous discharge time t is:

$$E_t = \mathrm{t} \cdot \mathrm{f} \cdot E_{sp} \tag{12.6}$$

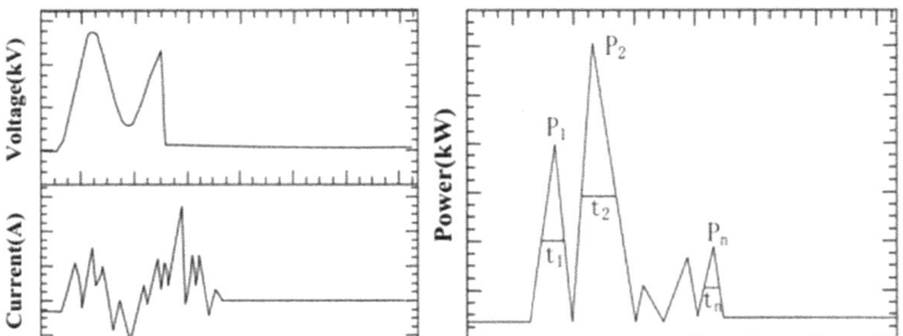

Fig. 12.4 Electrical parameter waveforms **a** Typical supply voltage and current waveform; **b** Instantaneous power spectrum

In this equation, f is the pulse repetition frequency of the high-voltage nanosecond pulse power supply. The pulse repetition frequency used in this study is fixed at 300 Hz.

The energy efficiency ratio (η) of the entire experimental platform is the ratio of the generation amount of active components to the consumed electrical energy (E_t):

$$\eta = \frac{(c_1 - c_0)V_l}{E_t} \tag{12.7}$$

In this equation, c_0 and c_1 are the initial and post-treatment concentrations of the active components in the solution, and V_l is the volume of the solution to be treated, in liters.

12.2.5 Mechanism of Active Ingredient Generation

Figure 12.5 shows the discharge process and the generation mechanism of active components under high voltage pulse excitation in the experiment setup described in this article. When a high-voltage pulse power supply discharges in a gas–liquid dual phase, electrons are instantaneously accelerated into high-energy electrons by a strong electric field. These high-energy electrons collide with gas molecules (mainly N_2 and O_2 in the air) and water molecules, resulting in ionization, excitation, decomposition, and other reactions. Influenced by ultraviolet light, shock waves and other physical effects, various free radicals and O_3, H_2O_2 strong oxidative active components are generated. The addition of a solid phase aids in retaining gas bubbles within the reactor, further reducing the difficulty of discharge.

Electrons within the energy range of 0–5 eV can dissociate oxygen molecules int ·O, and ·O can further react with water molecules to form hydroxyl radicals, ·OH [13, 15, 16]:

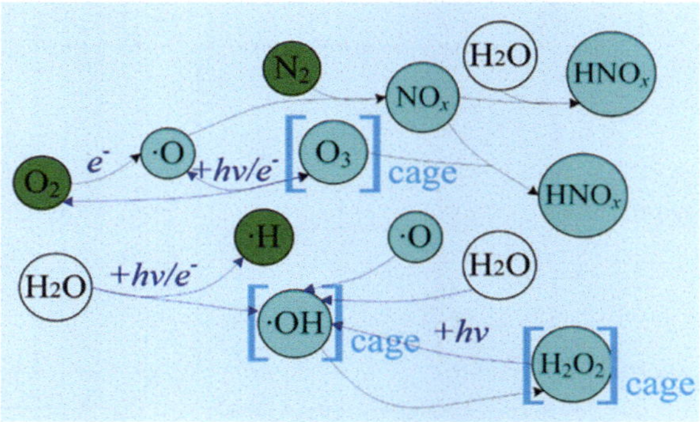

Fig. 12.5 Pulse discharge process and active ingredient generation mechanism

$$e^- + O_2 \rightarrow 2 \cdot O(^1D) + e^- \tag{12.8}$$

$$\cdot O(^1D) + H_2O \rightarrow 2 \cdot OH \tag{12.9}$$

1–2 eV electrons can dissociate a water molecule into hydroxyl-radical, ·OH, and hydrogen radical, ·H [17, 18]:

$$e + H_2O \rightarrow e + \cdot OH + \cdot H \tag{12.10}$$

$$\cdot OH + \cdot OH \rightarrow H_2O_2 \tag{12.11}$$

The oxygen radical ·O, with the involvement of third molecules M (N_2 or H_2O), which act as thermal energy carriers, can react with O_2 to form ozone, O_3 [19, 20]:

$$\cdot O + O_2 + M \rightarrow O_3 + M \tag{12.12}$$

Electrons can excite nitrogen molecules (N_2) in the air, turning them into excited state nitrogen molecules ($N_2{}^*$), which can then react further with O_2 [13, 16, 18]:

$$e + N_2 \rightarrow N_2^* + e \tag{12.13}$$

$$N_2^* + O_2 \rightarrow N_2 + 2 \cdot O(^1D) \tag{12.14}$$

Excited state oxygen atoms $\cdot O(^1D)$ can react with N_2 to form nitric oxide (NO) and nitrogen radicals (N), whereas N atoms can react with ·OH to produce NO and ·H [13, 16, 17]:

$$O(^1D) + N_2 \rightarrow NO + N \tag{12.15}$$

$$N + \cdot OH \rightarrow NO + N \tag{12.16}$$

During the discharge process, NO can be oxidized by ozone to form nitrogen dioxide (NO_2) and O_2 [13, 21, 22]:

$$NO + O_3 \rightarrow NO_2 + O_2 \tag{12.17}$$

NO_2 can react with H_2O to form an acidic substance [13, 21, 22], and the main reaction is as follows:

$$3NO_2 + H_2O \rightarrow 2NO_3^- + 2H^+ + NO \tag{12.18}$$

Also, the UV light produced by the discharge process can directly dissociate H_2O molecules, producing ·OH and ·H [13, 23].

$$H_2O + hv \rightarrow \cdot OH + \cdot H \ (\lambda = 145 - 246\,\text{nm}) \tag{12.19}$$

12.3 Experimental Results and Analysis

12.3.1 Active Ingredients and Water Content

Different gas–liquid volume ratios will directly affect the difficulty of discharge, and then have an impact on the electrical parameters and the generation of active components. In order to explore the effect of different water content on the concentration of active components generated by gas–liquid-solid tri-phase pulse discharge, the total gas–liquid flow Q_t in this experiment was set to 2.00 LPM, and the water content was adjusted to 15%, 25%, 45%, 65%, 85% in turn. The corresponding gas–liquid flow rates under different water content are shown in Table 12.1. During the experiment, the preset water content was maintained by fine-tuning the valve.

Figure 12.6 shows the changes in the concentration of active ingredients after discharge at different water contents. In the experiment, the solution conductivity is fixed at 100.0 uS/cm, the volume of the solution is 2.00 L, and a high-voltage nanosecond pulse power supply is used for discharge processing for 5 min. During the experiment, the supply voltage at the high-voltage end of the reactor is kept within the range of 18.0–19.0 kV.

From Fig. 12.6, it can be seen that when the solution conductivity is fixed at 100.0 uS/cm, as the water content increases, the concentration of H_2O_2 and NO_x (nitric oxide) shows a decreasing trend, while the concentration of O_3 basically does not change. As the initial volume of the solution is 1.00 L and the solution is recycled during the experiment, the content and concentration of the active ingredients are numerically the same, that is, the increase in water content is not conducive to the production of H_2O_2 and NO_x, and has no effect on the production of O_3.

In the experiment, the supply voltage is kept constant, the energy given to the electrons by the strong electric field does not change, and the reaction process of high-energy electrons colliding with gas molecules does not change. However, when the water content increases, the gas volume in the reactor decreases, the number of

Table 12.1 Corresponding gas–liquid two-phase flow rates under different water contents

Water content α (%)	Gas phase flow (L/min)	Liquid phase flow (L/min)
5	1.90	0.10
25	1.50	0.50
45	1.10	0.90
65	0.70	1.30
85	0.30	1.70

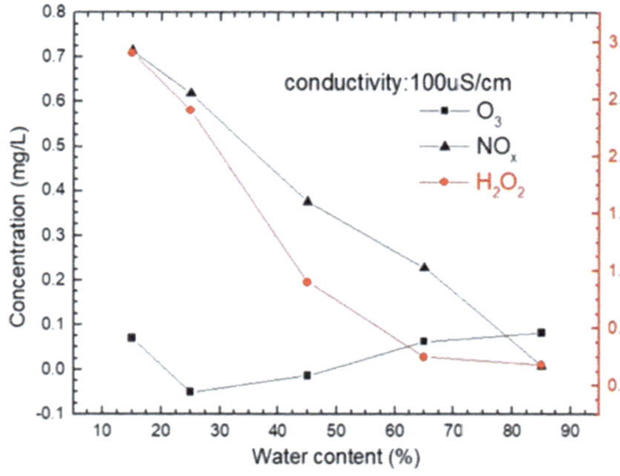

Fig. 12.6 Relationship between water content and active ingredient concentration

discharges per unit volume of gas decreases, leading to a decrease in the content of N_2 and O_2 and the number of discharges.

According to formulas (12.8) and (12.9), the production of ·O and ·OH decreases; According to formula (12.11), the generation of H_2O_2 decreases; According to formula (12.10), in the presence of gas, although high-energy electrons colliding with water molecules can also produce ·OH, the pulse voltage will preferentially excite the gas, so the ·OH produced by water molecules is less. Overall, with the increase in water content, the H_2O_2 content decreases.

Theoretically, according to formula (12.8), the content of ·O in the reactor decreases with the increase in water content, and then according to formula (12.12), the content of O_3 will decrease with the increase in water content. However, in the experiment, the gas is not recycled, a small part of the generated O_3 is dissolved in the liquid, and the rest is dissipated into the air. Therefore, with the increase in water content, the amount of O_3 dissolved in the liquid is always very small. Because of the small base, the decrease in O_3 is not very obvious, so it appears to be roughly constant.

From formula (12.18), it can be known that the fundamental reason for the production of NOx is that NO_2 reacts with H_2O to produce acidic substances. According to formulas (12.13) to (12.17), the generation of NO_2 requires the cooperative reaction between N_2, ·O, NO and O_3. According to the previous explanation, as the water content increases, the content of N_2 and ·O decreases, which leads to a decrease in the content of NO, and ultimately leads to a decrease in the content of NO_x.

12.3.2 Energy Efficiency Ratio and Water Content

The relationship between water content and power supply energy is shown in Fig. 12.7. It can be observed from the diagram that, when the electrolyte conductivity

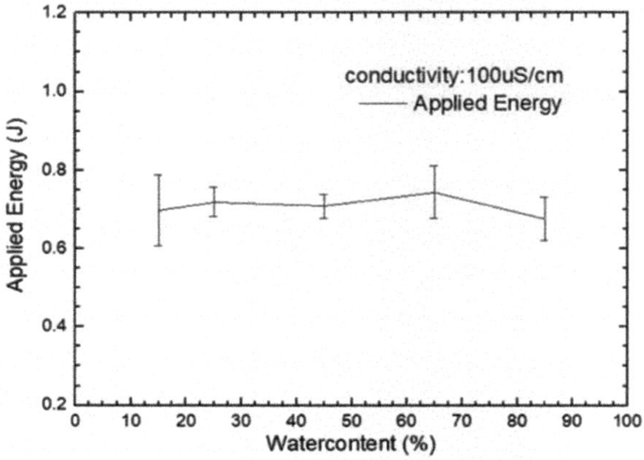

Fig. 12.7 Relationship between water content and power supply energy

is fixed at 100.0 uS/cm, the power supply energy remains essentially unchanged with the increase in water content.

As the water content increases, the gas flow decreases, the flow rate slows down, the volume of gas in the reactor decreases, and under the same experimental conditions, the difficulty of discharge increases, showing a trend of declining breakdown frequency.

However, this only indicates that the number of times the gas is broken down under high-voltage pulse power supply within 5 min is decreasing. Once the gas breaks down, the average power supply voltage and average power supply current displayed on the oscilloscope are similar, so the energy supplied by a single pulse breakdown of the gas is essentially the same. This suggests to a certain extent that the water content has little impact on the power supply energy.

Figure 12.8 shows the relationship between the water content and the energy efficiency ratio of the three active components. From the figure, it can be seen that when the solution conductivity is fixed at 100.0 uS/cm, with the increase in water content, the energy efficiency ratio of the generated active components H_2O_2 and NO_x shows a decreasing trend, while the energy efficiency ratio of the generated active component O_3 remains basically unchanged, which is consistent with the change in the concentration of active components with the water content.

A possible explanation for the above experimental results is as follows: With the increase in moisture content, the gas flow rate decreases, the gas volume in the reactor decreases, and the content of O_2 and N_2 decreases, resulting in a reduction in the concentration of generated H_2O_2 and O_3. Within the 5-min discharge time, although the number of gas breakdowns decreases with the increase in moisture content, the number of breakdowns remains on the same order of magnitude, and the energy consumed is also basically the same. Therefore, the change in energy efficiency ratio with the moisture content follows the same pattern as the concentration of active components.

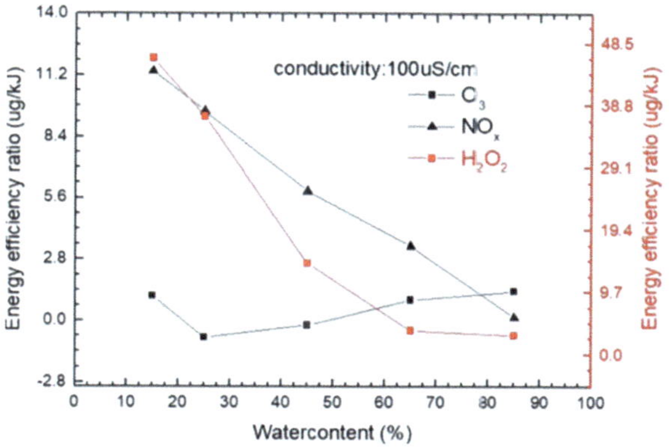

Fig. 12.8 Relationship between water content and energy efficiency ratio

12.3.3 Active Ingredients and Conductivity

The different conductivity indicates different ion concentrations in the liquid phase. The ions in the liquid phase participate in the conduction to form an ion stream, which makes it difficult to establish a plasma in the liquid phase. Therefore, it is crucial to explore the effects of conductivity on the pattern of pulse discharge electrical parameters and the generation rules of active components.

As the conductivity increases, the critical breakdown voltage decreases. The higher the front-end input voltage of the source, the more necessary it is to maintain the supply voltage under different conductivities between 18.0 and 19.0 kV, which requires the front-end input voltage of the power source to reach a high value. Due to the maximum input voltage of the front end of the power supply that can be provided in this study is 210.0 V, the range of adjustable conductivity is relatively narrow. The adjusted conductivities are 2.00 uS/cm, 50.0 uS/cm, 100.0 uS/cm, 150.0 uS/cm, and 200.0 uS/cm respectively, each for a discharge duration of 5.0 min.

Figure 12.9 shows the relationship between conductivity and the concentration of active components. From the figure, it can be seen that fixed at a 45% moisture content, with the increase in conductivity, the concentrations of H_2O_2 and NO_x show a trend of first increasing and then decreasing, reaching a peak at 100.0 uS/cm, while the concentration of O_3 shows a slow increasing trend.

The possible reasons for the above experimental phenomena are as follows: With the increase in conductivity, the ion concentration in the liquid phase increases, and the ions participate in the conduction to form an ion stream, making it difficult to establish a plasma in the liquid phase. To maintain the supply voltage under different conductivities between 18.0 and 19.0 kV, the front-end input voltage of the power supply increases, causing the electric field strength between the two electrodes to increase and the energy possessed by the electric field to boost.

When the conductivity is in the range of 2.0–100.0 uS/cm, the strong electric field breaks through the gas medium, and electrons gain energy to collide with gas

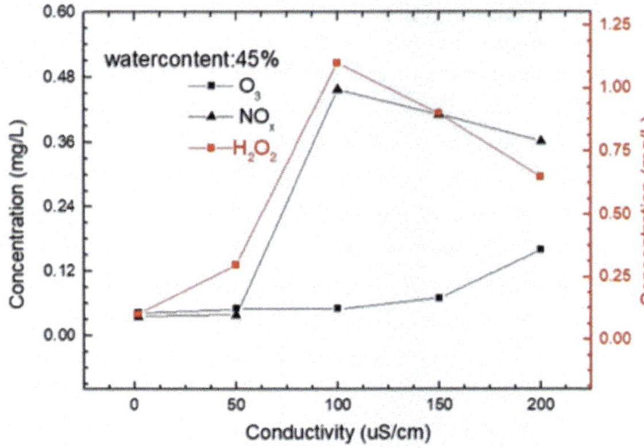

Fig. 12.9 Relationship between conductivity and active ingredient concentration

molecules and water molecules. According to formulas (12.8) to (12.18), the concentrations of ·OH, ·O, ·H, and N increase, leading to a rise in the concentrations of active components H_2O_2, O_3, and NO_x. When the conductivity is in the range of 100.0–200.0 uS/cm, the ion concentration significantly increases, and the energy gained by the ions from the electric field increases. The energy obtained by electrons reduces compared to the conductivity in the range of 2.0–100.0 uS/cm, making the probability of collision with gas molecules and water molecules lower. This results in a decrease in the ability to excite, decompose, and ionize them, which leads to reductions in the concentrations of ·OH, ·O, ·H, and N and consequently, decreases in H_2O_2 and NO_x.

The reason for the increase in O_3 concentration instead of a decrease may be due to the reduction in collisions between electrons and O_2, causing O_2 to mostly react with ·O. At the same time, collisions between electrons and N_2 are also reduced, resulting in no reactions between NO and O_3. Therefore, the concentration of generated O_3 increases.

12.3.4 Energy Efficiency Ratio and Conductivity

The relationship between conductivity and power supply energy is shown in Fig. 12.10. It can be seen from the figure that when the water content is fixed at 45%, with the increase in conductivity, the power supply energy shows an upward trend.

A plausible explanation for this experimental phenomena might be that when holding the power supply voltage within the range of 18.0–19.0 kV, with the increase of conductivity, the ion concentration in the solution elevates. The ions in the liquid take part in conduction to form an ionic flow, which adds to the difficulty of discharge. As such, the energy consumed in the breakdown of the gas is likely to increase with the rise in conductivity.

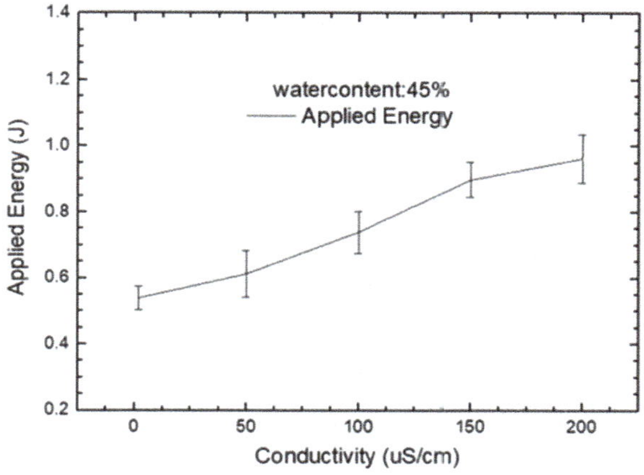

Fig. 12.10 The relationship between conductivity and power supply energy

Figure 12.11 shows the relationship between conductivity and the energy efficiency ratio of the three active components. From the graph, it can be observed that when the water content of the fluid is fixed at 45%, with the increase in conductivity, the energy efficiency ratio of active components H_2O_2 and NO_x exhibits a trend of first increasing and then decreasing. It reaches its maximum when the conductivity is at 100.0 uS/cm. The energy efficiency ratio of active component O_3 remains relatively stable when the conductivity ranges from 2.0 to 100.0 uS/cm. When the conductivity is between 100.0 and 200.0 uS/cm, the energy efficiency ratio of O_3 exhibits a slow upward trend.

The possible reasons for these experimental phenomena are as follows:

When the conductivity is between 2.0 and 100.0 uS/cm, as the conductivity increases, the ion concentration does not show a significant increase. Electrons can gain enough energy to collide with a large number of gas and water molecules, and the consumed power energy is not too great, leading to an increase in the content

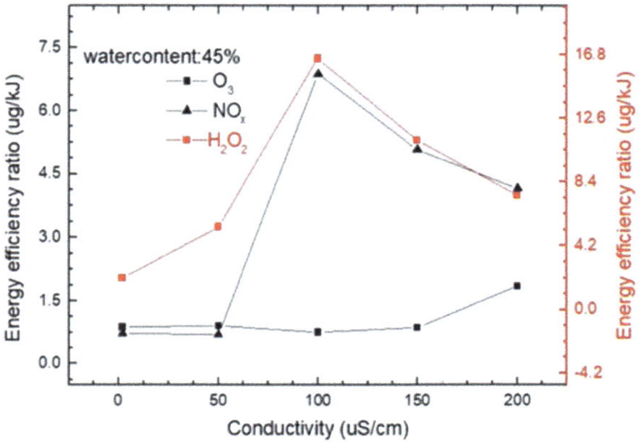

Fig. 12.11 Relationship between conductivity and energy efficiency ratio

of H_2O_2 and NO_x. Hence, the energy efficiency ratio of H_2O_2 and NO_x increases. Although O reacts with O_2 to develop O_3, the process of generating NO_x consumes O_3, so the O_3 content does not change significantly, and accordingly, the energy efficiency ratio of O_3 remains basically unchanged.

When the conductivity is between 100.0 and 200.0 uS/cm, as the conductivity increases, the ion concentration significantly increases. Ion conduction leads to an ion flow, making it more difficult for gas break down and more power energy is required. As electrons cannot collide with a large number of gas and water molecules, the content of H_2O_2 and NO_x decreases, leading to a drop in energy efficiency ratio.

The slow increase in the energy efficiency ratio of O_3 after 100.0 uS/cm might be due to a decrease in the collision between electrons and O_2, resulting in a larger number of O_2 reacting with O. At the same time, the collisions between electrons and N_2 are reduced, leading to less NO reacting with O_3. Consequently, the concentration of generated O_3 increases, leading to an increased energy efficiency ratio.

12.4 Conclusion

(1) The water content in the gas–liquid two phases directly affects the difficulty of gas–liquid mixed discharge. The conductivity of the liquid will have a direct effect on the discharge mechanism and discharge mode during the pulse discharge process. Expanding the gas phase two-phase into gas–liquid-solid three-phase is conducive to reducing the difficulty of underwater pulse discharge.

(2) With a fixed solution conductivity of 100.0 uS/cm and a total gas–liquid flow rate Q_t of 2.00 LPM, the increase of water content is unfavorable to the generation of H_2O_2 and NO_x, and has no effect on the generation of O_3.

(3) Fixed the water content of gas–liquid at 45%, with the increase of conductivity, the concentrations of H_2O_2 and NO_x showed a tendency of increasing and then decreasing, and the concentration of O_3 showed a tendency of slowly increasing; when the conductivity was 100.0 uS/cm, it was favorable to the generation of H_2O_2 and NO_x.

References

1. N. Shaba, B.K. Tiwari, A.R. Jambrak, et al., Extraction yield and biological activity of phycobiliproteins from porphyridium purpureum using atmospheric cold plasma discharge and jet systems. LWT **187** (2023)
2. A.A. Ivanova, I.V. Mursenkova, Experimental study of the motion of a shock wave in the plasma of a pulsed volume discharge in air. Mosc. Univ. Phys. Bull. **78**(2), 204–213 (2023)
3. R.B. Zhang, Y. Wu, G.F. Li. Enhanced plasma chemical processes in gas-liquid-solid three-phase discharge reactor. J. Dalian Univ. Technol. (05), 625–629 (2005)

4. H. Wang. Study on synergistic degradation of xanthate in flotation wastewater by high-voltage pulse discharge with catalysts. Jiangxi Univ. Sci. Technol. (2019)
5. L. Guo, P. Zhao, Y. Jia, et al., Inactivation of airborne pathogenic microorganisms by plasma-activated nebulized mist. J. Hazard. Mater. **459**, 132072–132072 (2023)
6. R. Sun, W. Xu, L. Xiong, et al., The combined effects of ultrasound and plasma-activated water on microbial inactivation and quality attributes of crayfish during refrigerated storage. Ultrasonics Sonochem. **98**, 106517–106517 (2023)
7. B. Chen, X. Gao, K. Chen, et al. Regulation characteristics of oxide generation and formaldehyde removal by using volume DBD reactor. Plasma Sci. Technol. **20**, 024009 (11pp) (2018)
8. D.D. Gao, M.W. Meng, J.L. Xie, et al. Removal of formaldehyde by modified activated carbon fiber synergized with plasma. Environ. Eng. 36(12) (2018)
9. P. Seri, S. Nici, M. Cappelletti, et al., Validation of an indirect nonthermal plasma sterilization process for disposable medical devices packed in blisters and cartons. Plasma Process. Polym. **20**(9) (2023)
10. A.Z. Zain, M. Restiwijaya, A.R. Hendrini, Development of ozone reactor for medicine base on Dielectric Barrier Discharge (DBD) plasma. J. Phys. Conf. Ser. **1153**(1), 012089 (8pp) (2019)
11. Liu, F., Li, S., Zhao, Y., et al., in The effect of pulse voltage rise rate on the polypropylene surface hydrophilic modification by ns pulsed nitrogen DBD. Plasma Sci. Technol. **25**(10) (2023)
12. S. Schiavi, M. Parmigiani, P. Galinetto, et al., Plasmonic nanomaterials for micro- and nanoplastics detection. Appl. Sci. **13**(16), 5 (2023)
13. K. Chen. Design of positive and negative corona discharge reactors and regulation of their liquid-phase reaction products. Hohai Univ. (2019)
14. B. He, Y. Ma et al., Simultaneous quantification of aqueous peroxide, nitrate, and nitrite during the plasma-liquid interactions by derivative absorption spectrophotometry. J. Phys. D Appl. Phys. **50**(44), 5207 (2017)
15. S. Yukinori, B. David et al., Plasma chemistry model of surface microdischarge in humid air and dynamics of reactive neutral species. J. Phys. D Appl. Phys. **45**, 425201 (2012)
16. J. Yan, S. Dai, X. Li et al., Emission spectroscopy diagnosis of the radicals generated in gas-liquid phases gliding arc discharge. Spectrosc. Spect. Anal. **28**, 1851–1855 (2008)
17. B. Penetrante, M. Hsiao, J. Bardsley et al., Identification of mechanisms for decomposition of air pollutants by non-thermal plasma processing. Plasma Sources Sci. Technol. **6**, 251–259 (1997)
18. W. Wang, F. Liu, J. Zhang et al., Study of radicals in pulsed corona discharge by emission spectroscopy. Spectrosc. Spect. Anal. **24**, 1288–1292 (2004)
19. L. Chuan Ho, P. Babu, R. Kumar, et al., HBGS (hydrate based gas separation) process for carbon dioxide capture employing an unstirred reactor with cyclopentane. Energy **63**, 252–259 (2013)
20. A. Tas Marnix, R. van Hardeveld, E. van Veldhuizen, Reactions of NO in a positive streamer corona plasma. Plasma Chem. Plasma P. **17**, 371–391 (1997)
21. A. Haddouche, M. Lemerini, Chemical analysis of NO_x removal under different reduced electric. Plasma Sci. Technol. **17**(7), 589–594 (2015)
22. K. Takita, N. Abe, G. Masuya et al., Ignition enhancement by addition of NO and NO_2 from a N_2/O_2 plasma torch in a supersonic flow. P. Combust. Inst. **31**, 2489–2496 (2007)
23. G. Tezcanli-Güyer, N. Ince, Individual and combined effects of ultrasound, ozone and UV irradiation: a case study with textile dyes. Ultrasonics **42**, 603–609 (2004)

Part II
Plasma Assisted Processing

Chapter 13
An Experimental Study on the Characteristics of Gliding Arc Discharge in CO_2 Dissociation with an Inlet Laval Nozzle

Zixiao Zhang, Yang Liu, Xiangen Zhao, Yaping Du, and Junjia He

Abstract The Gliding Arc Discharge (GAD), a kind of non-equilibrium plasma, has gathered substantial interest in the domain of CO_2 decomposition due to its high energy efficiency. Current research has demonstrated that the flow rate significantly impacts the CO_2 dissociating, but the discharge characteristic is not well understood. In this paper, experiments were carried out to compare the GAD characteristics in CO_2 dissociating between the traditional reactor and the reactor featuring an inlet Laval nozzle, in which current and voltage waveforms, high-speed optical images, and Schlieren images were recorded simultaneously. The experimental results show that in the reactor with an inlet Laval nozzle, the re-ignition voltage, discharge period, and power decreased by 75%, 39%, and 87%, respectively, and a higher probability of back-breakdown occurring at the same flow rate (4 L/min), compared to the typical GAD reactor. In addition, although the luminosity of the GAD plasma diminished, the intensified turbulence caused by the high flow rate inside the reactor through the Laval nozzle expanded the diffusion region of the hot gas. These findings highlight the multi-perspective discharge characteristics of high-flow-rate GAD, thereby enhancing comprehension of the diverse mechanisms through which high flow rate affects CO_2 conversion.

Z. Zhang (✉) · Y. Liu · J. He
State Key Laboratory of Advanced Electromagnetic Technology, Huazhong University of Science and Technology, Wuhan 430074, China
e-mail: m202272050@hust.edu.cn

X. Zhao · Y. Du
Department of Building Environment and Energy Engineering, The Hong Kong Polytechnic University, Hong Kong 999077, China

Z. Fang et al. (eds.), *Proceedings of the 5th International Symposium on Plasma and Energy Conversion*, Springer Proceedings in Physics 398,
https://doi.org/10.1007/978-981-97-2245-7_13

13.1 Introduction

With the increase in fossil fuel consumption, the concentration of carbon dioxide has increased sharply all over the world, resulting in the global environmental challenge. Therefore, the reduction and utilization of CO_2 emissions have become a common concern for all of mankind. Particularly, the most effective solution appears to be the decomposition and transformation of CO_2 into high-value added chemicals [1].

The strong thermodynamic stability of the CO_2 molecule necessitates high energy consumption to break the C=O chemical bonds [2, 3]. The traditional method for CO_2 dissociation typically relies on thermal pyrolysis [4–6]. However, this method requires the gas temperature to be within the range of 2000–3000 K. The energy consumption for CO_2 dissociation, solely induced by gas heating, is extremely high and not viable for industrial applications. In recent years, non-equilibrium plasmas have gained considerable attention due to their high chemical activity and operation at ambient temperature and pressure. A kind of non-equilibrium plasma, the gliding arc discharge (GAD), has become particularly appealing for CO_2 decomposition due to its high energy efficiency [7, 8].

Flow rate has been widely recognized as a key operational factor for CO_2 dissociation as it determines the interactions between the flow and the plasma. At the low flow rate, the interaction time of the plasma with the feed gas is sufficient, thus CO_2 conversion performance can be enhanced. However, it was found that by implementing high flow rates in rotating gliding arc discharges, the CO_2 conversion performance can be enhanced [9]. In addition, gliding arc reactors equipped with Laval Nozzle have exhibited superior heat transfer performance for CO_2 convection, comparable to the vortex form [10, 11]. In summary, existing studies mainly focuses on the CO_2 conversion performance including conversion and energy efficiency, at varying flow rates. However, a study of the discharge characteristics at flow rates with large variations is still lacking.

With this in mind, this paper will conduct experiments to compare the GAD characteristics in CO_2 dissociating a traditional reactor and a reactor featuring an inlet Laval nozzle. In these experiments, the current and voltage waveforms, high-speed optical images, and Schlieren images are recorded simultaneously. The effect of high flow rate by a Laval nozzle on the discharge characteristic is obtained by comparing the electrical parameters, luminosity intensity and thermal characteristics.

13.2 Experimental Setup

Figure 13.1 shows the experimental setup for investigating the discharge characteristics for CO_2 dissociation in gliding arc discharge with various flow rate. The gliding arc discharge is powered by a high-voltage DC supply, with a 100 kΩ resistor to limit the current. The discharge voltage and current waveforms were recorded with a digital oscilloscope through a high-voltage probe (PINTECH P6028A) and a current

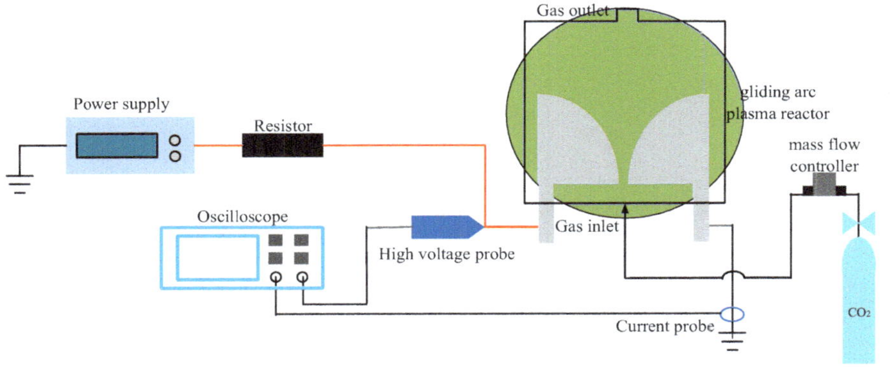

Fig. 13.1 Schematic of the experimental setup

probe (CYBERTEK HCP8030D), respectively. The reaction gas is supplied by pure CO_2 (99.99%), the flow rate of which is regulated by a mass flow controller.

Due to spatial constraints, the direct imaging system and Schlieren imaging system are not depicted in Fig. 13.1. The direct imaging system was used to photograph the luminosity of the gliding arc and was achieved with a high-speed camera (CHRONOS 1.4-color). The camera was facing the front of the reactor and shot at 9000 frames per second (fps). The Schlieren imaging system was used to visualize gas density variations within the reactor [12–14], employing a Z-type optical path with an observation range indicated by the green circle in Fig. 13.1. An additional high-speed camera (iX-Cameras, i-SPEED 716) was used for the Schlieren imaging system, which shoots at 9000 fps.

To investigate the characteristics of gliding arc discharge at high flow rates, a Laval nozzle is installed at the gas inlet to boost the flow rate. Its specific structure is depicted in Fig. 13.2. Based on the detailed dimensions of the Laval nozzle presented in Fig. 13.2, it can be inferred that the flow rate increases approximately 400 times when the reactor is equipped with the Laval nozzle.

13.3 Results and Discussion

13.3.1 Electrical Characteristics

Figure 13.3 shows typical voltage and current waveforms during gliding arc discharge at a flow rate of 1.5 L/min. Results in Fig. 13.3a demonstrate a discharge period of approximately 180 ms, consistently repeating over time. During each discharge period, there is a noticeable trend of the discharge voltage rapidly decreasing, and then slowly increasing as the gliding arc undergoes a cycle of breakdown, elongation and dissipation. Although the tendency of the voltage with the Lava nozzle is consistent

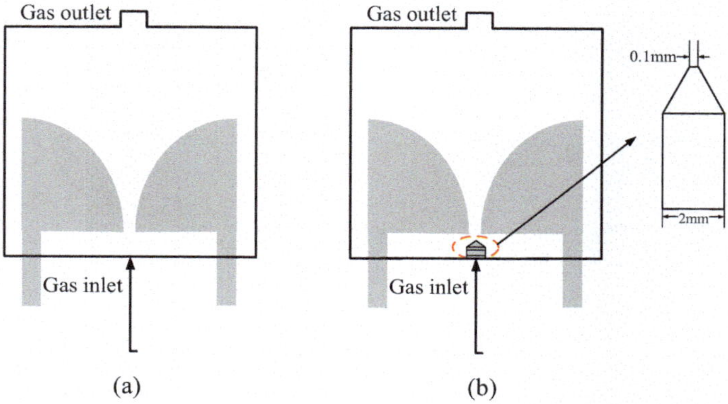

Fig. 13.2 Front views of **a** the traditional reactor and **b** the reactor with the Laval nozzle

with that in Fig. 13.3a, the periodicity deteriorates when using the Laval nozzle, and a back-breakdown event is more frequently observed.

For a quantitative comparison of gliding arc discharges in the traditional reactor and the reactor with the Laval nozzle, Tables 13.1 and 13.2 present statistics on the re-ignition voltage and discharge period for a flow rate of 1.5 L/min and 4 L/min, respectively. As can be inferred from Tables 13.1 and 13.2, as the flow rate of inlet gas increases, both the re-ignition voltage and discharge period progressively decrease. At a flow rate of 4 L/min, the re-ignition voltage and discharge period are reduced by 75% and 39% respectively.

To gain deeper insights into the input energy of gliding arc discharges, the discharge power is estimated from the current and the voltage, as demonstrated in Fig. 13.4. At a high flow rate, the discharge power significantly declines due to a reduction in discharge voltage. The power for the reactor with the Laval nozzle drops by up to 87% compared to the traditional reactor at a flow rate of 4 L/min.

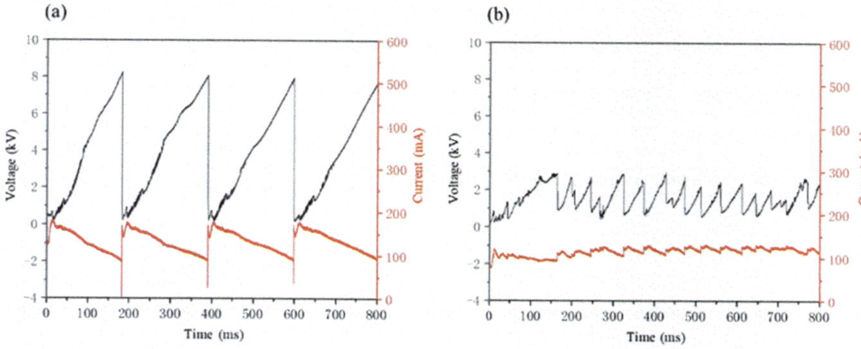

Fig. 13.3 Typical voltage-current waveform during gliding arc discharge with a gas flow rate of 1.5 L/min in **a** the traditional reactor and **b** the reactor with the Laval nozzle

Table 13.1 Statistical results of electrical characteristics of gliding arc discharges at flow rate of 1.5 L/min

		Re-ignition voltage (V)	Discharge period (ms)
Typical	Average value	9048.3	179.0
	Standard deviation	324.0	35.6
With Laval nozzle	Average value	3175.8	48.4
	Standard deviation	637.7	21.0

Table 13.2 Statistical results of electrical characteristics of gliding arc discharges at a flow rate of 4 L/min

		Re-ignition voltage (V)	Discharge period (ms)
Typical	Average value	8652.7	36.2
	Standard deviation	400.3	6.8
With Laval nozzle	Average value	2185.9	22.5
	Standard deviation	1182.5	9.7

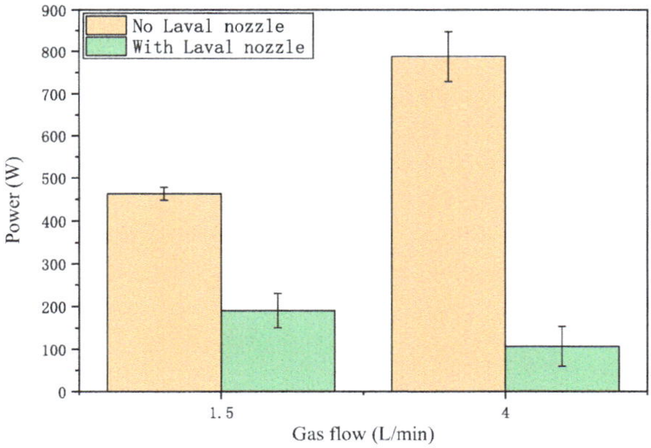

Fig. 13.4 Gliding arc discharge power in different reactors

This finding may be advantageous for enhancing the energy efficiency of gliding arc plasma in CO_2 dissociation.

13.3.2 Optical Characteristics

Figure 13.5 illustrates the typical optical morphology of the gliding arcs in the reactors without and with the Laval nozzle. It can be seen from Fig. 13.5a that in the traditional reactor, the gliding arc is characterized by intense blue luminosity, and the arc roots at the anode and cathode electrodes glide nearly simultaneously. However, according

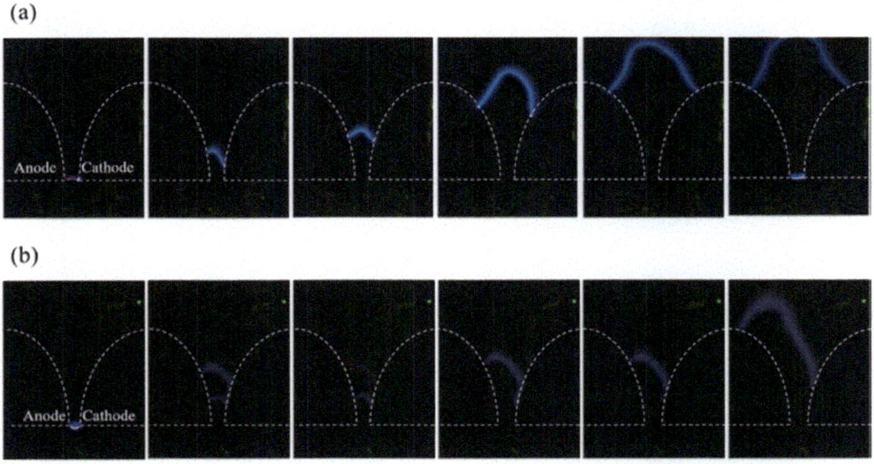

Fig. 13.5 The optical morphology of the gliding arc in **a** traditional reactor and **b** the reactor with the Laval nozzle

to Fig. 13.5b, the luminosity is less intense in the reactor with the Laval nozzle, and the anode arc root slides more rapidly than the cathode root. Moreover, the back-breakdown phenomenon is more liable to occur with a high flow rate.

Comparing these two cases in Fig. 13.5, the discharge power in the reactor with the Laval nozzle drastically reduced compared to the traditional one. Furthermore, due to the inconsistent spatial uniformity of the flow field caused by the strong turbulence of the high flow rate, the anode arc root glides upward faster than the cathode root. The temperature in the near cathode region sharply increases due to the fixed cathode position.

13.3.3 Thermal Characteristics

Figure 13.6 presents combined observations of thermal and optical morphology of gliding arc discharges in the traditional reactor and the reactor with the Laval nozzle. For Fig. 13.6a or Fig. 13.6b, the above are direct optical images and the below are Schlieren images. In the Schlieren images, the luminosity area is represented by the red circles, while the orange circles illustrates the gas heating area. The joint observations demonstrates that the luminosity area correlates closely with the gas heating area in the traditional reactor. However, Fig. 13.6b indicates that the gas heating area in the reactor with the Laval nozzle is larger due to the intense turbulence at high flow rates. Furthermore, the augmentation of heat convection at high rates of flow increases the gliding arc resistance, leading to rapid dissipation of the gliding arc.

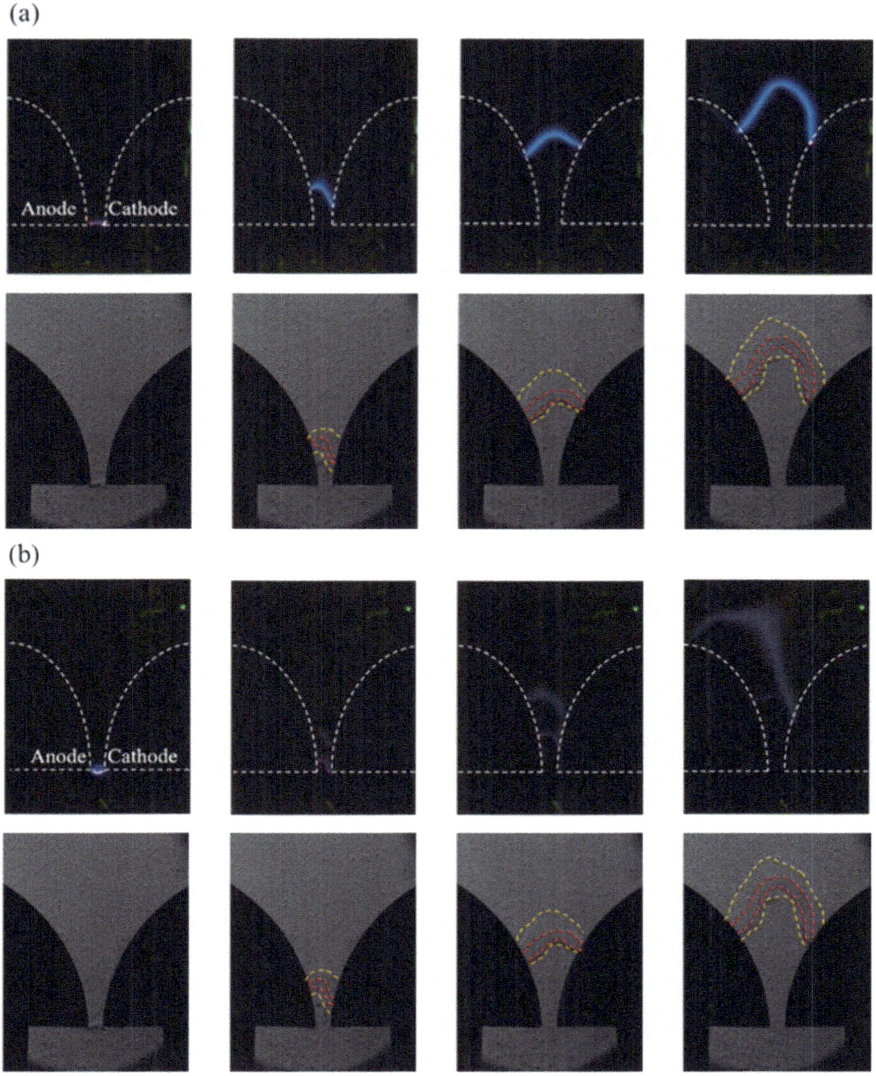

Fig. 13.6 Simultaneous observation of optical morphology (above) and thermal morphology (below) of gliding arc discharge in **a** a traditional reactor and **b** a reactor with a Laval nozzle

13.4 Conclusion

In this paper, an experimental platform is constructed to study the characteristics of gliding arc discharge at high flow rates. Electrical parameters (i.e., voltage and current) and optical and thermal morphologies of gliding arc discharges are recorded simultaneously. Experimental findings are compared between the traditional reactor and the reactor with the Laval nozzle. Main conclusions drawn are as follows:

(1) In the reactor equipped with the Laval nozzle, the re-ignition voltage, discharge period, and power decreased by 75%, 39%, and 87%, respectively, compared to the traditional reactor. Furthermore, the probability of back-breakdown occurrence increases when the reactor features a Laval nozzle.
(2) Compared to the traditional reactor, in the reactor equipped with the Laval nozzle, the anode arc root slides more rapidly than the cathode root, and the gas heating area is larger due to the intense turbulence at high flow rates. The augmentation of heat convection at high rates of flow increases the gliding arc resistance, leading to rapid dissipation of the gliding arc.

Acknowledgements The work of Xiangen Zhao and Yaping Du is supported by the Environment and Conservation Fund (Environmental Research, Technology Demonstration and Conference Projects; Project No. 26/2022).

References

1. R.N.E. Huaman, T.X. Jun, Energy related CO_2 emissions and the progress on CCS projects: a review. Renew. Sustain. Energy Rev. **31**, 368–385 (2014)
2. D.S. Laitar, P. Müller, J.P. Sadighi, Efficient homogeneous catalysis in the reduction of CO_2 to CO. J. Am. Chem. Soc. **127**(49), 17196–17197 (2005)
3. J.L. Liu, H.W. Park, W.J. Chung, D.W. Park, High-efficient conversion of CO_2 in AC-pulsed tornado gliding arc plasma. Plasma Chem. Plasma Process. **36**, 437–449 (2016)
4. A. Megaritis, R.C. Messenböck, I.N. Chatzakis, D.R. Dugwell, R. Kandiyoti, High-pressure pyrolysis and CO_2-gasification of coal maceral concentrates: conversions and char combustion reactivities. Fuel **78**(8), 871–882 (1999)
5. M. Xu, S. Li, Y. Wu, L. Jia, Q. Lu, Effects of CO_2 on the fuel nitrogen conversion during coal rapid pyrolysis. Fuel **184**, 430–439 (2016)
6. S.R. Lee, J. Lee, T. Lee, Y.F. Tsang, K.H. Jeong, J.I. Oh, E.E. Kwon, Strategic use of CO_2 for co-pyrolysis of swine manure and coal for energy recovery and waste disposal. J. CO2 Utiliz. **22**, 110–116 (2017)
7. A. Indarto, D.R. Yang, J.W. Choi, H. Lee, H.K. Song, Gliding arc plasma processing of CO_2 conversion. J. Hazard. Mater. **146**(1–2), 309–315 (2007)
8. M. Ramakers, G. Trenchev, S. Heijkers, W. Wang, A. Bogaerts, Gliding arc plasmatron: providing an alternative method for carbon dioxide conversion. Chemsuschem **10**(12), 2642–2652 (2017)
9. H. Zhang, L. Li, X. Li, W. Wang, J. Yan, X. Tu, Warm plasma activation of CO_2 in a rotating gliding arc discharge reactor. J. CO_2 Utiliz. **27**, 472–479 (2018)
10. S.Y. Lu, X.M. Sun, X.D. Li, J.H. Yan, C.M. Du, Physical characteristics of gliding arc discharge plasma generated in a Laval nozzle. Phys. Plasmas **19**(7), 072122 (2012)
11. H. Sekiguchi, S. Kodama, N. Yahagi, Behavior of gliding arc discharge in a supersonic flow. IEEE Trans. Plasma Sci. **49**(1), 98–103 (2020)
12. Y. Liu, P. Xiao, X. Zhao, X. Liu, X. Wang, Y. Du, J. He, Numerical simulation of the positive stem during dark period in long air gap discharge, in *2022 IEEE International Conference on High Voltage Engineering and Applications (ICHVE)* (IEEE, Chongqing, 2022), pp. 1–5

13. J. He, X. Wang, X. Zhao, J. Lee, Y. Du, X. Liu, Y. Liao, Schlieren techniques for observations of long positive sparks: review and application. High Volt. **7**(5), 825–839 (2022)
14. Y. Liu, X. Wang, X. Zhao, S. Xie, X. Liu, Y. Yao, J. He, Observation of a narrow thermal channel connected to the anode during dark periods in long positive sparks. Appl. Phys. Express **16**(4), 046002 (2023)

Chapter 14
Novel Plasma Discharge Structure Driven Hydrogenation of CO_2 at Room Temperatures and Atmospheric Pressure

Guangdong Yang and Tao Jiang

Abstract A novel coaxial dielectric barrier discharge (DBD) reactor has been developed for plasma-driven CO_2 hydrogenation at low temperatures and atmospheric pressure. Reverse water–gas shift reaction have been found dominant in the plasma CO_2 hydrogenation process. The results demonstrated a significant influence of the auxiliary electrode on both the CO_2 conversion rate and CO yield. This study aimed to investigate the impact of supported metal catalyst ($Cu/\gamma\text{-}Al_2O_3$) and auxiliary electrode on the hydrogenation performance of CO_2. Compared to plasma-only CO_2 hydrogenation without a catalyst, the combination of plasma with these catalysts resulted in a significant increase in the CO_2 conversion rate ranging from 50.1% to 90.7%. Furthermore, when plasma was combined with an auxiliary electrode, an approximately eight-fold increase in the conversion rate was observed, attributed to a transition from filamentary discharge to arc discharge mode. Additionally, it was found that the CO_2/H_2 molar ratio significantly influenced both the conversion rate of CO_2 and production of CO and CH_4. Notably, integrating an auxiliary electrode with plasma enabled processing higher flow rates of CO_2, achieving a yield of 63.92 ml/min at a flow rate of 250 ml/min for CO_2.

14.1 Introduction

CO_2 is an inherent constituent of the atmosphere and arises from diverse processes, encompassing respiration by animals and plants, gradual oxidation decomposition of organic matter, as well as combustion of fossil fuels. Historically, the concentration of CO_2 in the atmosphere has exhibited relative stability. However, with the advent of industrialization and widespread utilization of fossil fuels since the Industrial Revolution, global emissions have experienced a substantial surge. Over the past two

G. Yang · T. Jiang (✉)
School of Metallurgy, Northeastern University, Shenyang 110819, Liaoning, China
e-mail: jiangt@smm.neu.edu.cn

Z. Fang et al. (eds.), *Proceedings of the 5th International Symposium on Plasma and Energy Conversion*, Springer Proceedings in Physics 398,
https://doi.org/10.1007/978-981-97-2245-7_14

centuries, there has been an increase exceeding 30% in atmospheric CO_2 concentration [1]. As the predominant greenhouse gas, CO_2 accounts for approximately 55% of the greenhouse effect [2]. The continuous accumulation of this phenomenon disrupts the delicate equilibrium between energy absorption and radiation within Earth's atmosphere, leading to a progressive buildup of heat. Consequently, global warming ensues, triggering an array of environmental concerns that pose significant threats to human health as well as biodiversity. The issue surrounding CO_2 emissions is garnering increasing attention from diverse sectors of society due to its paramount importance in addressing these challenges.

CO_2 serves as both a greenhouse gas and the most abundant global carbon resource. The collection and utilization of CO_2 can be achieved through geological storage, fixation, and chemical conversion pathways. However, due to the high costs and potential environmental hazards associated with geological storage and fixation methods, scholars worldwide have increasingly focused on the chemical conversion of CO_2. This approach has garnered attention for its convenience, effectiveness, and environmentally friendly nature in addressing this issue. The hydrogenation of carbon dioxide to generate CO and/or CH_4 is regarded as a promising approach for the conversion and utilization of CO_2 in achieving a carbon–neutral cycle, given that both CO and CH_4 serve as crucial chemical feedstocks employed in the production of higher-energy products such as hydrocarbons and liquid fuels [3].

However, CO_2 is a thermodynamically stable and nonpolar molecule, necessitating high energy input for its conversion process. Theoretically, temperatures exceeding 1600 K are required to disrupt the chemical bond (C=O bond energy being 783 kJ/mol) [4]. Presently, there is a burgeoning interest in high-temperature thermochemical, photochemical, electrochemical methods as well as their synergistic combinations. Nevertheless, the selective activation and directed transformation of C=O chemical bonds under mild conditions continue to present formidable challenges.

Plasma, an electrically charged state of matter composed of atoms and atomic clusters that have undergone electron loss, as well as positive and negative ions formed by ionization, can facilitate catalytic reactions by providing a plethora of neutral species, high-energy electrons, photons, and positive/negative ions at low temperatures [5]. Moreover, the integration of plasma with catalysts enables precise control over micro- and [6] nano-scale particle behavior to enhance selectivity for specific products. This approach holds great promise in surface activation and coupling applications [7, 8].

Currently, extensive research has been conducted on the impact of various discharge types on reforming activity and efficiency, encompassing warm plasmas such as sliding arc discharge [9], spark discharge [10], microwave discharge [11], and radio frequency discharge [12], as well as cold plasmas like glow discharge [13] and dielectric barrier discharge [14]. In warm plasmas specifically, the energy transfer for CO_2 dissociation predominantly relies on the vibrational energy possessed by CO_2 molecules. However, research has demonstrated a positive correlation between electron number density and the conversion rate as well as energy efficiency of reforming reactions. Consequently, in high-density warm plasma systems (approximately 10^{23} m^{-3}), these reactions exhibit significantly enhanced energy efficiency compared to low-density dielectric barrier discharge systems (approximately 10^{19} m^{-3}). Nevertheless, when considering practical applications, it is crucial to evaluate reaction performance alongside system complexity, operational convenience, channel temperature, and discharge stability. Of particular importance is the examination of plasma-catalyst interactions; in this regard, dielectric barrier discharges remain indispensable.

Ozkan et al. [15, 16] conducted an investigation on the impact of discharge frequency, discharge power, insulation thickness, and material type on the efficiency of CO_2 conversion in a tubular flow DBD reactor. Their findings revealed that with an increase in discharge frequency, there was a decrease in effective voltage resulting in reduced rates of CO_2 conversion and energy efficiency. Moreover, when alumina was employed as the insulating material, it led to the generation of higher quantities of charge and micro-discharge filaments due to its elevated relative dielectric constant. Conversely, using quartz as the medium resulted in stronger electric field intensity and higher electron energy production. It is noteworthy that augmenting insulation thickness from 2.0 mm to 2.8 mm yielded approximately a 50% enhancement in CO_2 conversion rate. This improvement can be attributed to increased capacitance reduction and overall system voltage amplification while also leading to an upsurge in micro-discharge quantity. Mei [17] employed a coaxial cylindrical DBD reactor and utilized response surface methodology with multi-objective optimization to investigate the impact of various process parameters on CO_2 conversion rate. The findings revealed that enhancing discharge power, extending discharge length, and reducing inlet flow rate significantly enhanced the CO_2 conversion rate. In contrast, Xu [18] examined the influence of mixed Ar gas on CO_2 conversion in a packed bed DBD reactor. The experimental results demonstrated that when maintaining an inlet flow ratio of CO_2 to Ar at 1:4, it achieved the highest CO_2 conversion rate of 46%. The utilization of a distinct electrode structure holds significant potential for enhancing the rate of CO_2 conversion into other substances. In their study, Wang [19] demonstrated remarkable reactivity in terms of both CO_2 conversion and methanol output by employing a meticulously designed DBD reactor equipped with specialized water electrodes. By integrating plasma technology with either Cu/γ-Al_2O_3 or Pt/γ-Al_2O_3, there was a substantial increase observed in both the conversion rate and overall production of methanol, resulting in an impressive peak yield of 11.3% along with an improved selectivity reaching approximately 53.7%. Liu [20] synthesized a series of $LaNiO_3$ perovskite catalysts by calcining them at various temperatures and subsequently evaluated their performance in a DBD plasma reactor for the reverse water–gas shift (RWGS) reaction. The results revealed that the $LaNiO_3$ perovskite catalyst, calcined at 600 °C, exhibited an impressive CO selectivity of 76.6%, while the CO_2 conversion rates remained consistent across different calcination temperatures. In another study, Sun [21] successfully prepared a Pd/ZnO catalyst using the co-precipitation method and assessed its efficacy in plasma-catalyzed RWGS reaction. Experimental findings demonstrated that the "Pd/ZnO + Plasma" system achieved a remarkable CO_2 conversion rate of 32.5%, which was twice as high as that observed in the "ZnO + Plasma" system, with an accompanying CO selectivity exceeding 90%.

Jiang [22] investigation has effectively addressed the issue of limited gas residence time and inadequate gas excitation in the discharge zone by modifying the reactor/electrode structure and catalyst arrangement. Moreover, significant enhancement of catalytic performance in the synergistic system can be achieved through alteration of plasma discharge characteristics. Although plasma-coupled catalysts have demonstrated the potential to enhance the conversion rate and selectivity of CO, their incorporation also results in a reduction in gas residence time, thereby leading to a diminished overall yield of CO. In this investigation, our objective is to explore the influence of H_2/CO_2 molar ratio and catalyst type ($Cu/\gamma\text{-}Al_2O_3$) on the performance of plasma-catalyzed CO_2 hydrogenation within a dielectric barrier discharge system by introducing an auxiliary electrode. We comprehensively evaluate various parameters including CO_2 conversion rate, selectivity, and yield of both CO and CH_4, as well as energy efficiency. Expanded the range of novel application forms for dielectric barrier discharge plasma.

14.2 Experiment

14.2.1 Introduction of Experimental Equipment

The experiment was conducted in a coaxial packed-bed dielectric barrier discharge (DBD) reactor, as illustrated in Fig. 14.1. Circulating water served as the grounding electrode on the outer surface of a quartz tube with an outer diameter of 20 mm and wall thickness of 2 mm. A stainless steel rod with a diameter of 6 mm was centrally positioned within the quartz tube and employed as the high-voltage electrode. The DBD impedance device was connected to a high-voltage AC power supply with adjustable voltage and a frequency of 9 kHz. The applied voltage was measured using a high-voltage probe (Tektronix P6015A), while the current was recorded by a current monitor (Pearson 6585). Additionally, the voltage across the external capacitors was also measured. All electrical signals were sampled using a four-channel digital oscilloscope (Tektronix, DPO 3012). To calculate the discharge power (P) of the DBD reactor, we employed the Q-U Lissajous method. In this study, we maintained a fixed discharge power of 35 W for plasma processing. Temperature in the catalyst bed below 150 °C was monitored using fiber optic thermometers (Omega FOB102). Reactants and products were analyzed utilizing an online gas analyzer (Wuhan Sifang Optoelectronic Technology, Gasboard-3000). A spiral auxiliary electrode, constructed from a 1 mm diameter aluminum wire, is utilized. The placement of both the auxiliary electrode and catalyst follows the arrangement illustrated in Fig. 14.1.

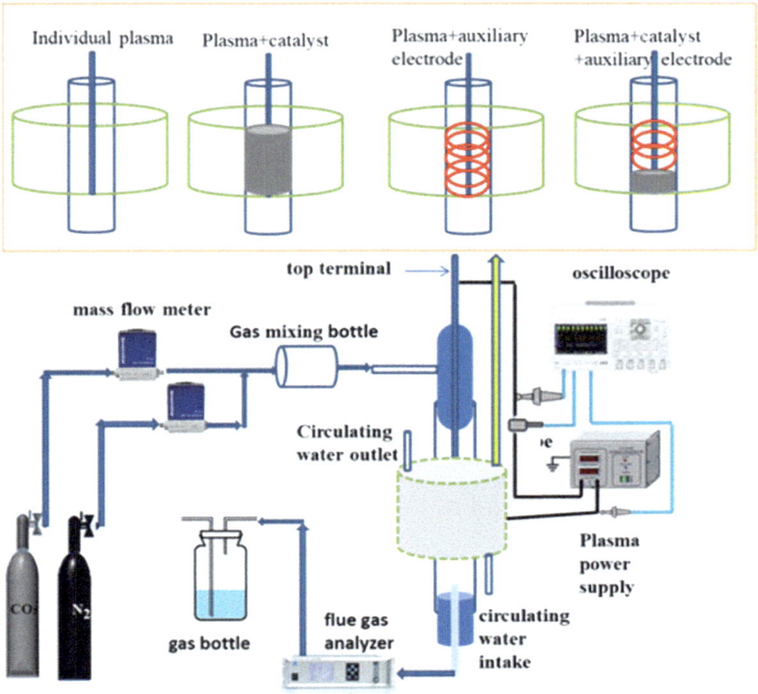

Fig. 14.1 Schematic of the experimental setup

14.2.2 Catalyst Preparation

The 10 wt.% Cu/γ-Al_2O_3 catalysts were prepared by incipient wetness impregnation using nitrate salts (Alfa Aesar, 99.5%) as the metal precursors. An appropriate weight of support (γ-Al_2O_3 beads, 1.5 mm in diameter) was added to the solution of nitrite salts. The mixture slurry was continuously stirred at 80 °C for 4 h and then dried at 110 °C overnight, followed by calcination at 600 °C for 6 h. In the packed-bed DBD reactor, plasma can be formed in both the catalyst bed and the packing material area. Filaments can be generated in the small gap between the bead– bead wall and the bead–quartz wall, while surface discharges can be formed on the surface of beads near the contact points between the beads. Prior to the plasma-catalytic CO_2 hydrogenation, the catalysts were reduced in an argon– hydrogen discharge at a discharge power of 7.5 W (50 ml/min and 20 vol.% H_2) for 30 min in the same DBD reactor.

14.2.3 Definition of Parameters

For the plasma-catalytic CO_2 hydrogenation, the conversion (C) of CO_2 is defined as

$$C_{CO2}\,(\%) = \frac{CO_2\,\text{converted (mol/s)}}{CO_2\,\text{input (mol/s)}} \times 100\% \tag{14.1}$$

The selectivities (S) and yields (Y) of the main products are calculated as

$$S_{CO}\,(\%) = \frac{\text{CO produced (mol/s)}}{CO_2\,\text{converted (mol/s)}} \times 100\% \tag{14.2}$$

$$S_{CH4}(\%) = \frac{CH_4\,\text{produced (mol/s)}}{CO_2\,\text{converted (mol/s)}} \times 100\% \tag{14.3}$$

14.3 Results and Discussion

14.3.1 Effect of Voltage

Figure 14.2 illustrates the impact of different voltages on the catalytic performance of plasma for CO_2 hydrogenation, employing separate plasma, Cu/γ-Al_2O_3 catalyst, auxiliary electrode, and auxiliary electrode + catalyst. Under a total flow rate of 200 ml/min and a CO_2:H_2 ratio of 1:1, it is observed that the conversion rate of CO_2 exhibits an almost linear increase with rising voltage (Fig. 14.2a). This finding suggests that augmenting discharge power can significantly enhance the conversion rate of CO_2. Furthermore, comparing the growth in CO_2 conversion rates between utilizing plasma alone and combining plasma with a catalyst reveals a relatively gradual progression. On the contrary, the inclusion of an auxiliary electrode leads to a significant acceleration in the rate of CO_2 conversion. For instance, at a voltage of 200 V, compared to solely using the catalyst, there was an increase from 10.33% to 46.76% in the conversion rate of CO_2; However, when the catalyst is combined with the auxiliary electrode, the conversion rate of CO_2 decreases to 26.45%. This phenomenon can be elucidated from two perspectives: firstly, the inclusion of a catalyst diminishes gas residence time; secondly, there is a modification in discharge mode. In the absence of an auxiliary electrode, under discharge mode, as voltage fluctuates, current exhibits an initial decline followed by an increase (as depicted in Fig. 14.2b). This transition progresses from glow discharge to filamentary discharge. Moreover, in the presence of catalysts, the discharge mode shifts towards surface discharge. Nevertheless, upon introducing an auxiliary electrode and increasing voltage levels observed are characterized by a sharp surge followed by steady growth trend and gradual evolution of filamentary discharges into arc discharges. The heightened energy associated with

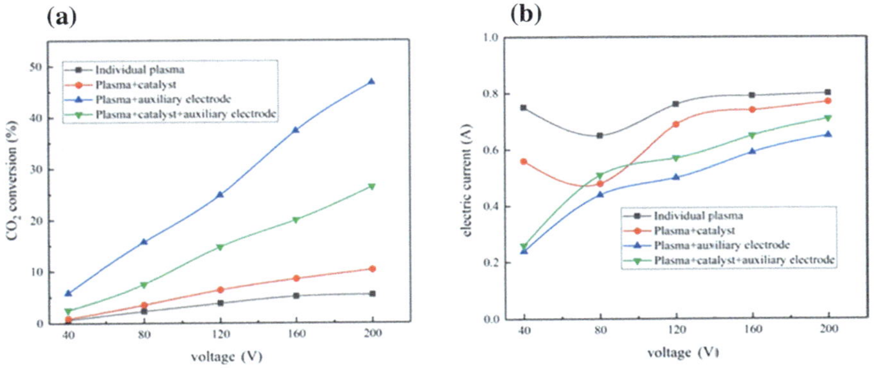

Fig. 14.2 Effect of voltages on the plasma-catalytic CO_2 hydrogenation. **a** Conversion of CO_2. **b** Change of current

arc excites ion reactions between hydrogen gas and CO_2 significantly enhancing CO_2 conversion rate.

14.3.2 Effect of CO_2/H_2 Molar Ratio

The introduction of auxiliary electrodes significantly enhances the CO_2 conversion rate, resulting in a more selective product, primarily CO. Furthermore, the ratio of CO_2 to hydrogen also plays a crucial role in determining the efficiency of plasma-catalyzed hydrogenation of CO_2 on the auxiliary electrode, as depicted in Fig. 14.3a at a discharge voltage of 160 V. The observed main products are CO and H_2O, with minor traces of CH_4 detected in the effluent. Under constant flow rate conditions, an increase in H_2 content initially leads to an improvement in the CO_2 conversion rate followed by a subsequent decline. Notably, when maintaining a stoichiometric proportion with a CO_2:H_2 molar ratio of 1:1, the conversion rate reaches its maximum value (37.37%). This finding highlights that achieving optimal raw gas conversion can be attained when reactant gases participate according to stoichiometric proportions based on chemical reactions.

Figure 14.3b illustrates the impact of CO_2/H_2 molar ratio on CO selectivity. As the molar ratio ranges from 3:1 to 1:1, a marginal increase in CO selectivity is observed, reaching an impressive value of 98.65%, accompanied by a substantial rise in CO yield by 79.13%. However, when transitioning from a molar ratio of 1:1 to 1:3, not only does the CO selectivity decline but it also leads to a reduction in overall CO yield. Moreover, augmenting H_2 content further enhances CH_4 selectivity. These findings hold significant potential for optimizing plasma-catalytic CO_2 hydrogenation processes and effectively controlling both CO conversion and CH_4 generation through precise adjustments of the H_2/CO_2 molar ratio.

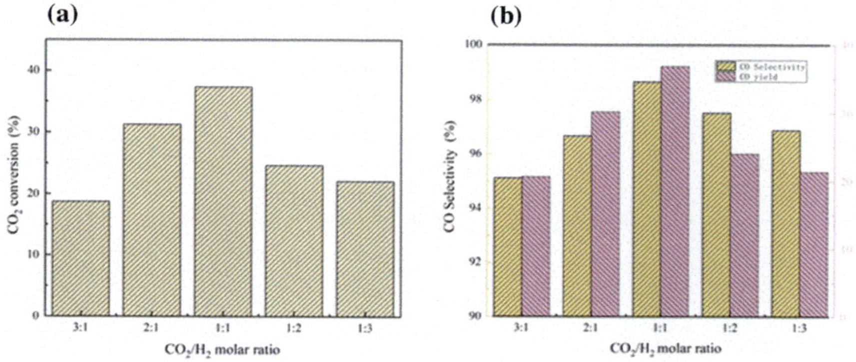

Fig. 14.3 Effect of CO_2/H_2 molar ratio on the plasma-catalytic CO_2 hydrogenation. **a** Conversion of CO_2. **b** Selectivity and yield of CO

14.3.3 Effect of Flow Rate

Figure 14.4a illustrates the impact of auxiliary electrodes on the hydrogenation performance of plasma under varying CO_2 flow rates, with a CO_2:H_2 ratio of 1:1. Despite the inherent limitations in scale for DBD systems, particularly when combined with catalysts, this investigation reveals that auxiliary electrodes exert only a marginal influence on gas flow rate. Simultaneously, as the CO_2 flow rate decreases, there is an increase in CO_2 conversion rate. Notably, at a CO_2 flow rate of 50 ml/min, the highest conversion rate (67.73%) is achieved due to the significant effect of residence time for raw gas within plasma. Additionally, due to the limitations of the experimental instrument detection, it was not possible to obtain data at lower flow rates. However, it can be reasonably inferred that a higher conversion rate would be achieved under lower feed gas flow rates. Furthermore, Figure (b) illustrates a continuous increase in CO production as the CO_2 flow rate steadily increases. This phenomenon occurs because the decrease in CO_2 conversion rate is slower than the increase in flow rate, resulting in an overall rise in CO production. Nevertheless, there appears to be a certain limit to increasing the overall flow rate on CO production based on the graph analysis. To further explore this direction and optimize plasma-catalyzed CO_2 hydrogenation performance, more efficient plasma excitation methods should be selected and implemented for improvement purposes within this study.

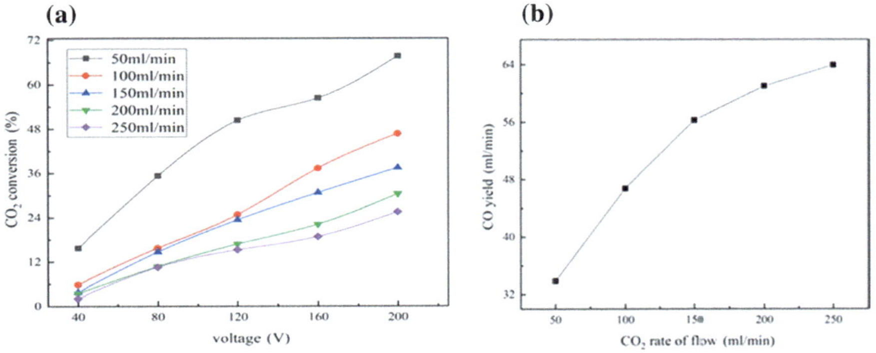

Fig. 14.4 Effect of flow rate on the plasma-catalytic CO_2 hydrogenation. **a** Conversion of CO_2. **b** Selectivity and yield of CO

14.4 Conclusion

The combination of plasma and auxiliary electrodes facilitates the low-temperature hydrogenation of CO_2. The voltage exerts a significant influence on the conversion rate of CO_2, exhibiting a maximum increase of 90.7% when plasma is combined with a catalyst compared to no catalyst conditions. However, when combined with auxiliary electrodes, the conversion rate of CO_2 experiences an impressive enhancement by 300%. The molar ratio plays a crucial role in both the conversion rate of CO_2 and the production of CO and CH_4 within the range from 3:1 to 1:3. Increasing the H_2 content in the reactant mixture substantially augments the conversion rate of CO_2. While higher flow rates result in increased production yield for CO, they also lead to a decrease in the overall conversion rate of CO_2. Moreover, under instrument detection limits (100 ml/min), it is possible to achieve an outstandingly high CO_2 conversion rate reaching up to 67.73%, accompanied by a corresponding production yield for CO amounting to 63.93 ml/min.

References

1. V.D.P. Vuuren, J. Edmonds, M. Kainuma, et al., The representative concentration pathways: an overview. Clim Change **109**(1–2) (2011)
2. A.A. Lacis, G.A. Schmindt, D. Rind et al., Atmospheric CO_2: principal control knob governing Earth's temperature. Science **330**(6002), 356–359 (2010)
3. S. Yang, M.Z. Li, M.N. Asif, et al., High selectivity to aromatics by a Mg and Na Co-modified catalyst in direct conversion of syngas. ACS Omega **5**(20) (2020)
4. T. Tomai, K. Katahira, H. Kubo et al., Carbon materials syntheses using dielectric barrier discharge microplasma in supercritical carbon dioxide environments. J. Supercrit. Fluids **41**(3), 404–411 (2007)
5. T. Shao, R.X. Wang, C. Zhang et al., Atmospheric-pressure pulsed discharges and plasmas: mechanism, characteristics and applications. High Volt. **3**(1), 14–20 (2018)

6. A. Bogaerts, X. Tu et al., The 2020 plasma catalysis roadmap. J. Phys. D Appl. Phys. **53**(44), 443001 (2020)
7. T. Shao, C. Zhang, et al., Atmospheric-pressure pulsed gas discharge and pulsed plasma application. High Volt. Eng. **42**(3), 685–705 (2016)
8. B. Ashford, X. Tu, Non-thermal plasma technology for the conversion of CO_2. Curr. Opin. Green Sustain. Chem. **3**, 45–49 (2017)
9. N. Lu, M.G. Liu, Y.D. Liu, et al., Research progress of hydrogen production by rotating gliding arc plasma reforming. High Volt. Eng. **47**(8), 3001–3011 (2021)
10. B. Zhu, X.S. Li, J.L. Liu et al., Optimized mixed reforming of biogas with O_2 addition in spark-discharge plasma. Int. J. Hydrogen Energy **37**(22), 16916–16924 (2012)
11. H. Sun, J. Lee, M.S. Bak, Experiments and modeling of atmospheric pressure microwave plasma reforming of a methane-carbon dioxide mixture. J. CO_2 Utiliz. 46, 101464 (2021)
12. Z.H. Liu, B.D. Huang, W.C. Zhu, et al., Phase-resolved measurement of atmospheric-pressure radio-frequency pulsed discharges in Ar/CH_4/CO_2 mixture. Plasma Chem. Plasma Process. **40**(4), 937–953 (2020)
13. G. Sribala, D. Michiels, C. Leys et al., Methane reforming to valuable products by an atmospheric pressure direct current discharge. J. Clean. Prod. **209**, 655–664 (2019)
14. Rao, M.U., Bhargavi, K.V.S.S., Chawdhury, P., et al., Non-thermal plasma assisted CO_2 conversion to CO: influence of non-catalytic glass packing materials. Chem. Eng. Sci. **267** (2023)
15. A. Ozkan, T. Dufour et al., The influence of power and frequency on the filamentary behavior of a flowing DBD-application to the splitting of CO_2. Plasma Sources Sci. Technol. **25**(2), 025013 (2016)
16. A. Ozkan, T. Dufour et al., How do the barrier thickness and dielectric material influence the filamentary mode and CO_2 conversion in a flowing DBD? Plasma Sources Sci. Technol. **25**(4), 045016 (2016)
17. D.H. Mei, Y.L. He et al., Optimization of CO_2 conversion in a cylindrical dielectric barrier discharge reactor using design of experiments. Plasma Processes Polym. **13**(5), 544–556 (2016)
18. S.J. Xu et al., CO_2 dissociation in a packed-bed plasma reactor: effects of operating conditions. Plasma Sources Sci. Technol. **27**(7), 075009 (2018)
19. L. Wang., Y.H. Yi., H.C. Guo, et al., Atmospheric pressure and room temperature synthesis of methanol through plasma-catalytic hydrogenation of CO_2. Acs Cataly. **8**(1), 90–100 (2017)
20. L. Liu, Z.K. Zhang, S. Das, et al., $LaNiO_3$ as a precursor of Ni/La_2O_3 for reverse water-gas shift in DBD plasma: effect of calcination temperature. Energy Conv. Manage. **206**,112475 (2020)
21. Y.H. Sun, J.J. Li, P.R. Chen, et al., Reverse water-gas shift in a packed bed DBD reactor: investigation of metal-support interface towards a better understanding of plasma catalysis. Appl. Cataly. A General **591**,117407 (2020)
22. N. Jiang, H. Jian, L. Jie et al., Plasma-catalytic degradation of benzene over Ag-Ce bimetallic oxide catalysts using hybrid surface/packed-bed discharge plasmas. Appl. Cataly. B Environ. **184**, 355–363 (2016)

Chapter 15
Non-thermal Plasma Approach for Methane Partial Oxidation to Syngas and Value-Added Oxygenates Under Ambient Conditions

Yunyang Lu, Li Wan, Danhua Mei, Shiyun Liu, and Zhi Fang

Abstract In recent years, partial oxidation of methane has attracted attention for its ability to convert cheap C1 source into fuels and high-value-added chemicals. In this study, a coaxial dielectric barrier discharge (DBD) reactor was developed to efficiently convert methane to syngas and liquid compounds under mild conditions through partial oxidation. On this basis, the effects of different input power and O_2/CH_4 ratio on the discharge characteristics and reaction performances were systematically evaluated. It was found that O_2/CH_4 ratio was the main factor affecting the distribution of products. The proper O_2/CH_4 ratio would significantly promote the H_2 yield (15.9%) and CH_3OH selectivity (12.2%). The possible mechanism in the plasma process for methane partial oxidation was deduced based on the analysis of discharge characteristics and reaction performance.

15.1 Introduction

Methane represents a prolific greenhouse gas implicated in global warming and climate alterations. Consequently, the transformation of methane into chemically valuable compounds stands as a novel energy strategy within the framework of addressing energy demand and environmental preservation [1]. Partial oxidation of methane for the synthesis of syngas and oxygenated compounds including methanol and formaldehyde has received significant attention owing to its sustainable properties, elevated energy density and high cost-effectiveness [2, 3]. The traditional thermal processes for methane conversion were still facing the challenges of high

Y. Lu · L. Wan · D. Mei · S. Liu (✉) · Z. Fang
College of Electrical Engineering and Control Science, Nanjing Tech University, Nanjing 211816, China
e-mail: shiyunliulisza@163.com

S. Liu
School of Electrical Engineering, Southeast University, Nanjing 210096, China

Z. Fang et al. (eds.), *Proceedings of the 5th International Symposium on Plasma and Energy Conversion*, Springer Proceedings in Physics 398,
https://doi.org/10.1007/978-981-97-2245-7_15

energy cost due to the high stability of the C–H bonds [4]. Therefore, it is highly desirable to develop environmentally friendly approach for converting methane into valuable chemicals under mild conditions.

Non-thermal plasmas (NTPs) have been demonstrated as a promising process to address the challenges inherent in traditional thermal process for methane partial oxidation [5]. In NTP process, the high energetic electrons typically within the energy range of 1–10 eV would be generated, while the global temperature still maintained in a low level [6]. This non-equilibrium character of NTP allows reacting substances to be activated and forming a large number of active radicals to efficiently trigger the chemical reactions at mild conditions [7]. In recent years, NTPs have attracted great interest in the field for the formation of fuels and chemicals via methane oxidation and reforming [8–10]. Among the various NTP sources, Dielectric barrier discharge (DBD) is widely used for chemical reactions since it offers a wider electron energy distribution, uniform distribution of discharge and easy to be prepared on a large scale.

Up to till now, the effectiveness of DBD plasma for methane conversion to syngas have been successfully demonstrated, while very limit work focused on the direct synthesis of oxygenates. Zhou et al. developed two DBD reactors with single and double dielectric layers for methane oxidation into valuable chemicals and reported that changing the structure of DBD reactor could shift the products form H_2O and CO_x ($x = 1, 2$) towards H_2O_2 and organic oxygenates [11]. Gao et al. used a parametrized DBD to direct convert CH_4 with air and argon to range of liquid chemicals including alcohol, acetic acid and aldehydes at ambient conditions [12]. Fathollahi et al. performed a study on the methane partial oxidation using a DBD reactor and found that increasing the oxygen amount in a constant methane flow rate favors the formation of methanol [13]. Despite some results have been performed previously, further enhancement in the conversion of methane and the selectivity of liquid chemicals remains crucial in the plasma methane partial oxidation process.

In the present study, a coaxial DBD reactor was developed to direct conversion of methane and oxygen into syngas and various liquid oxygenates. On this basis, the performance of methane conversion, products distribution and energy efficiency of the plasma process were studied at various input power and O_2/CH_4 ratio. The primary reaction pathway involved in the plasma methane partial oxidation process was also discussed.

15.2 Experimental Setup

Figure 15.1 depicts the schematic diagram of the experimental setup. A DBD reactor was developed for methane partial oxidation. In this study, the inner diameter of 10 mm, the outer diameter of 12 mm quartz tube was used as the dielectric material. A copper mesh with length of 90 mm wrapped around the tube acted as the high voltage electrode. A stainless steel rod with a diameter of 4 mm in the center of tube was grounded via an external capacitor and formed a discharge gap of

3 mm. A 10 kHz AC high-voltage power supply (DTP2000K) was connected to the reactor. The applied voltage and current were measured using a Tektronix P6015a high-voltage probe and a Pearson 6585 current monitor. The power deposited in the DBD reactor was calculated using the Lissajous curve. The voltage across on the capacitor was determined by using a Pintech N1070A differential probe to generate the Lissajous curve. These electrical signals were collected by a Tektronix TDS2014 digital oscilloscope. The amount and proportion of feed gas CH_4 and O_2 were controlled by two individual mass flow controllers. Prior to entering the DBD reactor, the gas mixture was homogenized with a mixer and the overall flow rate of 50 ml/min was used for the whole experiment. The outlet gas was firstly through a homemade ice trap to collect the liquid products. Thermo Scientific Nicole Fourier Transform Infrared Spectroscopy and Agilent GC–MS (GC 7820A and MSD 5975C) were used for qualitative and quantitative analysis of the liquid products. An online Pinna A0 gas chromatograph fitted with a flame ionization detector (FID) and a thermal conductivity detectors (TCD) was utilized to examine the reactants and gaseous products of the plasma reaction. Based on analysis from GC and GC–MS, the determination of CH_4 conversion, yield and selectivity of primary gaseous and liquid products, along with the energy efficiency of the plasma process, could be achieved through the following equations. CxHyOz represents oxygenates and when z equals 0 it stands for gaseous hydrocarbons.

$$C_{CH_4}(\%) = \frac{\text{Moles of } CH_4 \text{ converted}}{\text{Moles of } CH_4 \text{ input}} \times 100 \tag{15.1}$$

$$S_{H_2}(\%) = \frac{\text{Moles of } H_2 \text{ produced}}{2 \times \text{Moles of } CH_4 \text{ converted}} \times 100 \tag{15.2}$$

$$S_{CO}(\%) = \frac{\text{Moles of CO produced}}{\text{Moles of } CH_4 \text{ converted}} \times 100 \tag{15.3}$$

$$S_{C_xH_yO_z}(\%) = \frac{x \times \text{Moles of } C_xH_yO_z \text{ produced}}{\text{Moles of } CH_4 \text{ converted}} \times 100 \tag{15.4}$$

$$Yield(\%) = \text{selectivity} \times CH_4 \text{ conversion} \tag{15.5}$$

$$Energy\ efficiency\ (\text{mmol/kJ}) = \frac{CH_4 \text{ converted (mmol/ min)}}{\text{discharge power (W)}} \times \frac{1000}{60} \tag{15.6}$$

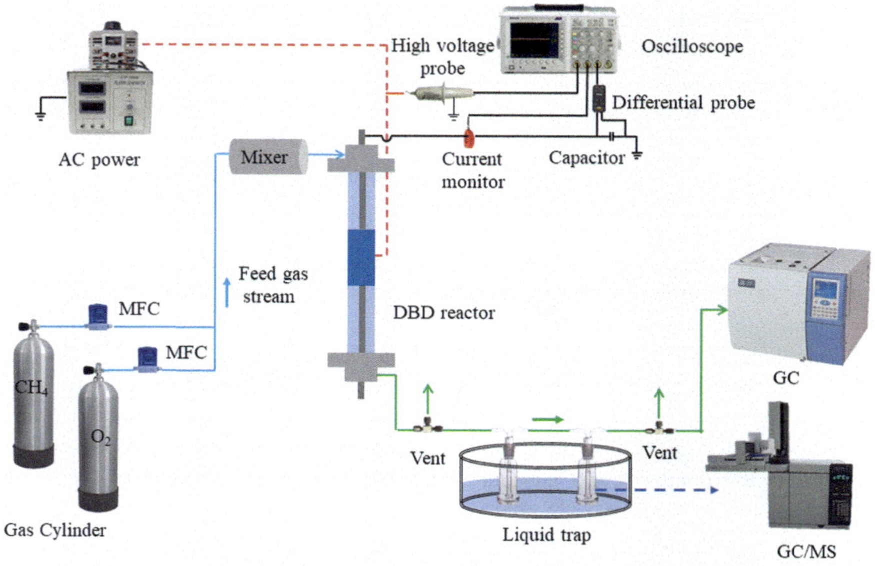

Fig. 15.1 Experimental setup schematic diagram of plasma methane partial oxidation

15.3 Results and Discussion

15.3.1 Discharge Characteristics

The typical Lissajous curves of the DBD reactor at constant input power of 90W was shown in Fig. 15.2. The power disposed on the DBD reactor could be reflected by the area of the curves. Obviously, the Lissjous curves appear as parallelograms, indicated the discharge pattern with the mixture of CH_4 and O_2 in the DBD reactor belongs to the filamentary micro-discharge mode [14]. The area first decreased as O_2/CH_4 ratio increases to 0.7, and further increasing the amount of oxygen extending the area of the curve. The discharge power of the DBD plasma under various input power and O_2/CH_4 ratio calculated from the Lissajous curves were summarized in Fig. 15.3. Clearly, the amount of oxygen significantly affects the discharge power and the effect was more pronounced at higher levels of input power. As O_2/CH_4 ratio increased from 0.5 to 0.9 with input power fixed at 50 W and 90 W, the discharge power varied in the range of 33.6 W to 35.8 W and 56.1W to 63.5 W, respectively. The minimum value of the discharge power was reached when O_2/CH_4 ratio set to 0.7 regardless the input power.

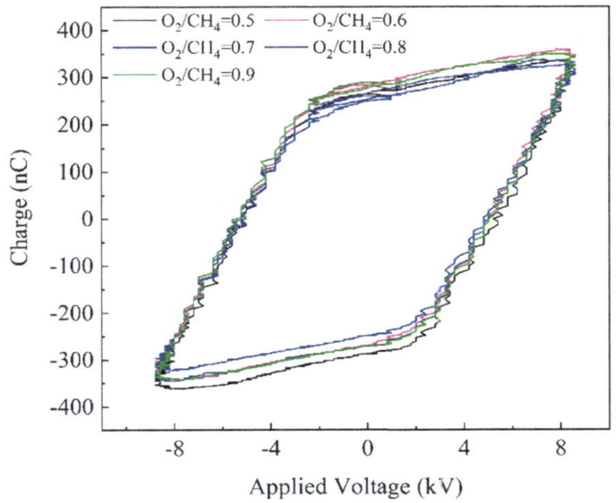

Fig. 15.2 Typical V-Q Lissajous curves with various O_2/CH_4 ratio

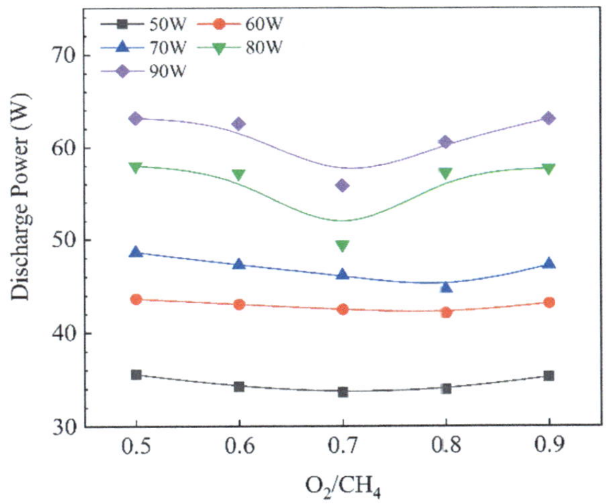

Fig. 15.3 The variation of discharge power as the function of input power and O_2/CH_4 ratio

15.3.2 Plasma Driven Partial Oxidation of Methane

The discharge power and O_2/CH_4 ratio are two important factors for plasma process of methane partial oxidation and their effects on methane conversion and products distribution were investigated. Figure 15.4a shows the variation of methane conversion as the function of input power and O_2/CH_4 ratio. As illustrated, the CH_4 conversion gradually increased with increasing the input power and the O_2/CH_4 ratio, and it increased up to 72% at the input power of 90 W and O_2/CH_4 ratio of 0.9. The result can be explained by the fact that raising the input power enhanced the electric

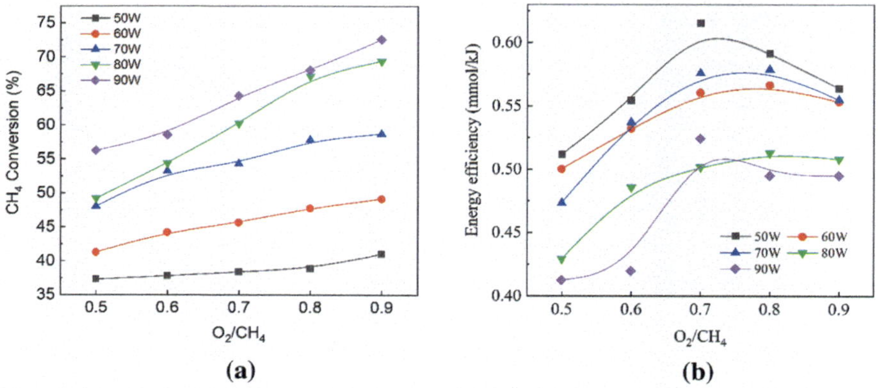

Fig. 15.4 The effect of input power and O_2/CH_4 ratio on **a** methane conversion and **b** energy efficiency of plasma-assisted methane partial oxidation process

field and the number of micro discharges, resulting in the formation of more energetic electrons in the discharge gap and causing the CH_4 dissociation via collisions of energetic electrons with methane molecules [13]. Additionally, increasing O_2/CH_4 ratio more oxygen atoms in the discharge region were generated through the collision between oxygen and energetic electrons. The formed oxygen atoms would further react with methane via oxidation and consequently improving the methane conversion. Energy efficiency is another important performance indicator for plasma methane partial oxidation. As depicted in Fig. 15.4b, the energy efficiency increased by a factor of 1.2 with the raising of O_2/CH_4 ratio from 0.5 to 0.7 at the input power of 90 W. While, the energy efficiency was proved to decrease with the input power loaded on the DBD reactor, the maximum energy efficiency of the plasma process was achieved 0.62 mmol/kJ at input power of 50 W.

Figures 15.5 and 15.6 show the yield and selectivity of main gaseous products with different input power and O_2/CH_4 ratio. H_2 and CO were detected as the major gaseous products, while no CO_2 were detected in this work, suggesting the oxygen amount was limited to trigger the full oxidation. The high input power has an obvious promotion effect on the yield of H_2, CO and hydrocarbons. The yield of H_2, CO and C_2H_2 was about a factor of 1.8, 1.6 and 13 higher for the input power of 90 W compared to 50 W, respectively at the same O_2/CH_4 ratio of 0.7. The primary hydrocarbon were found as C_2H_2 and C_2H_6 with the maximum yield of 1.1% and 2.1%, respectively. This is consistent with previous results that C_2H_6 is the major hydrocarbons in either methane dry reforming or direct decomposition in DBD plasma [15]. The amount of oxygen in the reactants was essential for the methane oxidation process so that the distribution of gaseous products were sensitive to O_2/CH_4 ratio. The yield of CO was increased by increasing O_2/CH_4 ratio from 0.5 to 0.9, and reaching the maximum value of 21.8% at input power of 90 W. In contrast, increasing O_2/CH_4 ratio shows a negative effect on the yield of C_2H_2 and C_2H_6. The maximum H2 yield of 15.9% was achieved at an optimal O_2/CH_4 ratio of 0.7

and input power of 90 W. This result could be ascribed to the fact that the formed H_2 be further oxidized to H_2O when higher O_2/CH_4 ratio was used in the process. Similar observation was reported by Moshrefi et. al. who confirmed that the formation of H_2O in a DC discharge plasma process for methane partial oxidation [16]. The selectivity of H_2 and CO showed the same tendencies as the yield of H_2 and CO, respectively. As a result of raising the O_2/CH_4 ratio, the number of active oxygen become more dominate in the discharge region, which react with methane molecules and leaded to the formation of H_2 and CO. Further increasing the O_2/CH_4 ratio the generated active hydrogen would be consumed by the oxygen radicals to generate water, resulting in the decrease of H_2 selectivity.

Plasma methane partial oxidation was initiated by the collisions between high energetic electrons and reactants [12]. The mean energy of the generated energetic electrons (1–10 eV) in DBD reactor is high enough to breakdown the chemical bonds in methane molecules (8.8 eV) and oxygen molecules (5.1 eV) [10]. The mixtures

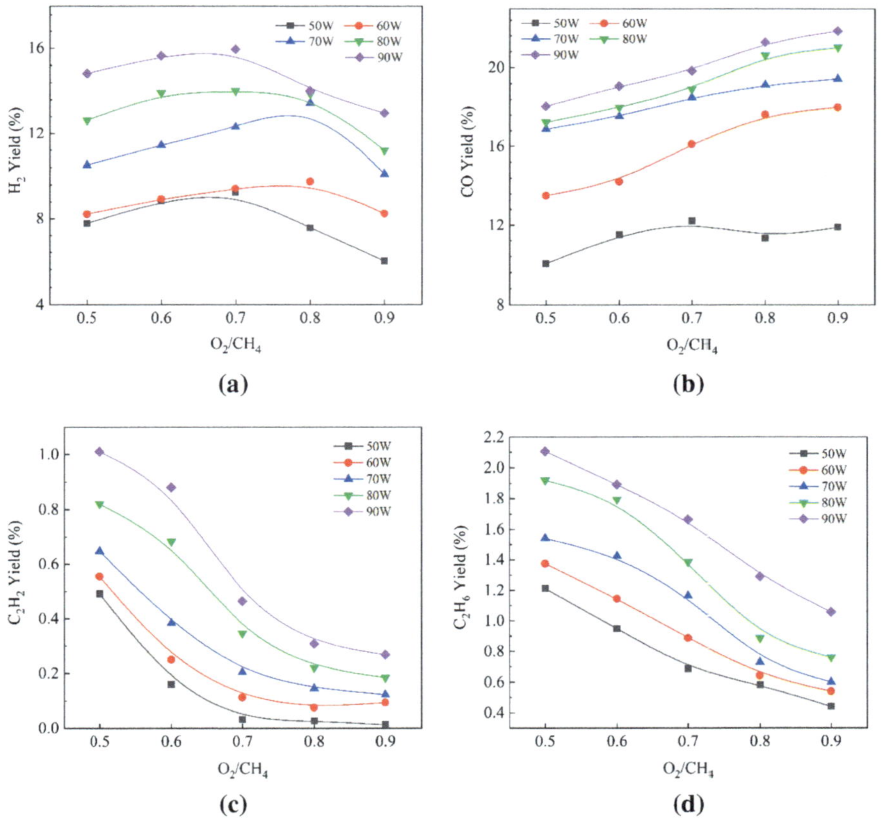

Fig. 15.5 Yield of the gaseous products **a** H_2; **b** CO; **c** C_2H_2 and **d** C_2H_6 as a function of input power and O_2/CH_4 ratio

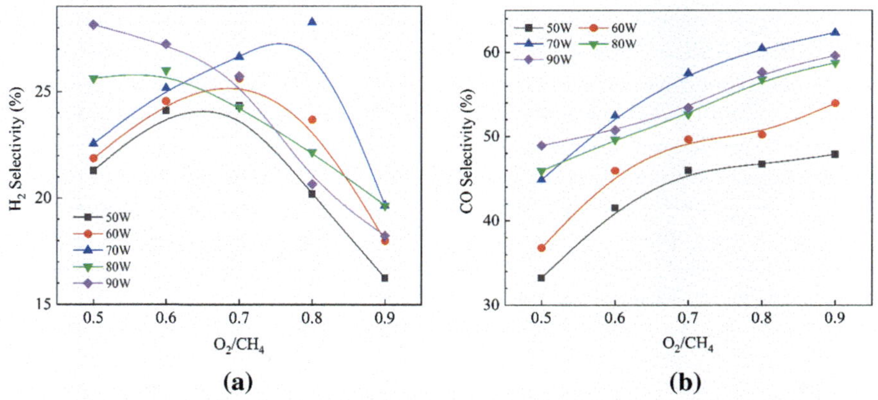

Fig. 15.6 Selectivity of the gaseous products **a** H_2 and **b** CO as a function of input power and O_2/CH_4 ratio

of methane and oxygen could be decomposed to various intermediate species such as $CH_x(x = 1\text{–}3)$ radicals, excited H and oxygen atoms in the discharge region, which could be described by the following reactions. In addition, the methane can be directly oxidized to CO and H_2 in the presence of proper amount oxygen, as represented by R8.

$$CH_4 + e^- \rightarrow CH_3 + H + e^- \tag{R1}$$

$$CH_4 + e^- \rightarrow CH_2 + H_2 + e^- \tag{R2}$$

$$CH_4 + e^- \rightarrow CH + 3H + e^- \tag{R3}$$

$$CH_4 + e^- \rightarrow C + 2H_2 \tag{R4}$$

$$O_2 + e^- \rightarrow O + O + e^- \tag{R5}$$

$$O + H \rightarrow OH \tag{R6}$$

$$CH_4 + O \rightarrow CH_3 + OH \tag{R7}$$

$$CH_4 + \frac{1}{2}O_2 \rightarrow 2H_2 + CO \tag{R8}$$

These active species could further react with each other such as H and CH_x radicals to produce H_2 and CO (R9–R11). Based on the analysis of plasma chemical kinetic modeling that CH_3 was the main radical from the electron impact dissociation of methane and directly influenced the formation of higher hydrocarbons [17]. The recombination of CH_3 with itself (R12) and methane with CH_2 (R13) were the most possible pathway for C_2H_6 formation. The main formation path of C_2H_2 could be described through reaction R14.

$$H + H \rightarrow H_2 \tag{R9}$$

$$CH_3 + O \rightarrow CO + H_2 + H \tag{R10}$$

$$CH_2 + O \rightarrow CO + 2H \rightarrow CO + H_2 \tag{R11}$$

$$CH_3 + CH_3 \rightarrow C_2H_6 \tag{R12}$$

$$CH_4 + CH_2 \rightarrow C_2H_6 \tag{R13}$$

$$2CH_4 \rightarrow C_2H_2 + 3H_2 \tag{R14}$$

15.3.3 Distribution of Liquid Productions

The chemical properties of the liquid compounds were analyzed by using FTIR and the corresponding infrared spectra was shown in Fig. 15.7. Several sharp peaks were clearly detected which represented the specific functional groups. The strong peak in the range of 3200–3500 cm^{-1} and 1630–1680 cm^{-1} was assigned to the –OH and $-C = O$ stretching vibrations, respectively, which indicated that a large amount of alcohols, ketones, carbonyl acids and/or esters were formed in the methane partial oxidation process [18]. The shape peak at 1021 cm^{-1} was attributed to the C–O stretching vibration, suggesting the existence of the ethers and/or ester in the liquid products. The peak at around 2920 cm^{-1} and 1429 cm^{-1} with weaker intensity were assigned to the symmetrical and asymmetrical $-CH_3$, which reveals that the aliphatic alkanes were present in the liquid products.

The liquid products form the plasma process of methane partial oxidation were further quantitative analyzed by GC–MS. Methanol (CH_3OH), acetic acid (CH_3COOH), acetone (CH_3COCH_3) and propionic acid (CH_3CH_2COOH) were the major liquid products. Figure 15.8 shows the selectivity of main liquid product as the function of O_2/CO_4 ratio when the input discharge fixed at 70 W. Clearly, when the O_2/CO_4 ratio increased to 0.9, the selectivity of CH_3OH, CH_3COOH and CH_3CH_2COOH were gradually increased up to 12.2%, 10.1% and 10.3% respectively, while the CH_3COCH_3 selectivity decreased to near zero. The CH_3 and OH

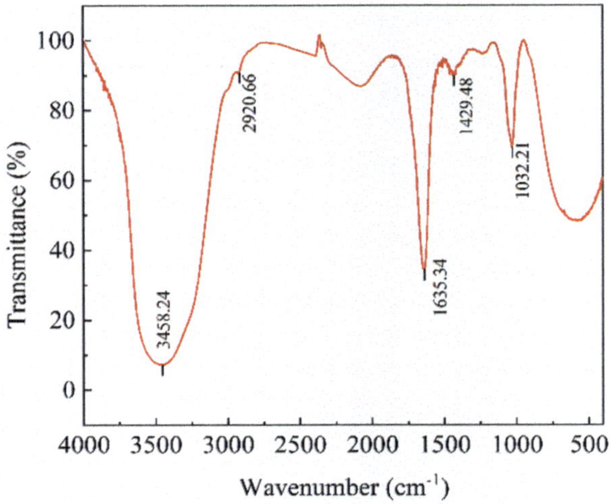

Fig. 15.7 Infrared spectra of liquid products (input power: 70 W, O_2/CO_4 ratio: 0.7)

were identified as the critical species for the methanol formation in the plasma process for methane oxidation based on the analysis of plasma chemical kinetic modelling [17]. The increased amount of oxygen in the feed gas contributed to the formation of CH_3 and OH species via the CH_4 dissociation with O atoms (R7), and then the generated CH_3 and OH species reacted with each other formed CH_3OH (R15), causing the increase selectivity of CH_3OH. In addition, the generated CO in the methane partial oxidation process may also be converted to methanol (R16–R19) [13]. Further oxidation of alcohols would formed ketones and acids. Meanwhile, the C_2 and C_3 oxygenates would be formed through the C–C coupling reactions between methanol and other intermediates.

$$CH_3 + OH + M \rightarrow CH_3OH + M \tag{R15}$$

$$CO + H \rightarrow CHO \tag{R16}$$

$$CHO + CHO \rightarrow CH_2O + CO \tag{R17}$$

$$CH_2O + H \rightarrow CH_3O \tag{R18}$$

$$CH_3O + H \rightarrow CH_3OH \tag{R19}$$

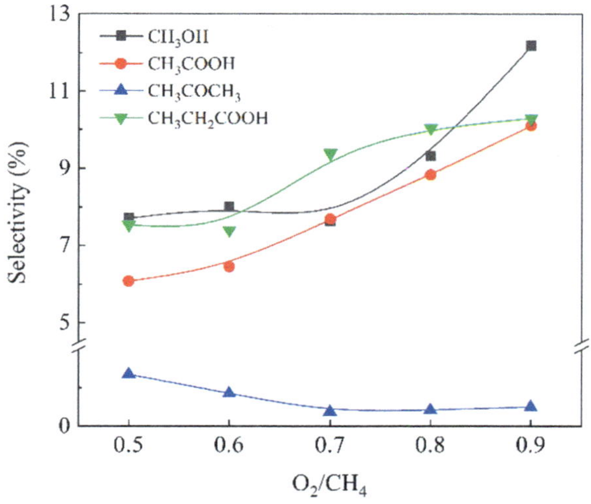

Fig. 15.8 Selectivity of main liquid products as the function of O_2/CO_4 ratio

15.4 Conclusion

In the present study, a coaxial DBD reactor has been developed to direct conversion methane with oxygen into syngas and liquid chemicals. Various parameters such as input power and O_2/CH_4 ratio were investigated with the purpose of obtaining high methane conversion, and selectivity of syngas and liquid chemicals. It was found that increasing the input power has a positive effect on the conversion of methane, the yield and selectivity of H_2 and CO. The increased input power results in a rise of electric field in the DBD reactor, the energetic electrons acquired enough energy to breakdown the chemical bonds of methane molecules, causing the increase of methane conversion and hydrogen and CO yield. The distribution of gaseous and liquid products were sensitive to the O_2/CH_4 ratio. As observed, by increasing O_2/CH_4 ratio from 0.5 to 0.9 at input power of 90 W, the CO yield and selectivity gradually increased up to 21.9% and 59.7%, respectively, while yield of C_2H_2 and C_2H_6 decreased to about 0.27% and 1.1%, respectively. The highest selectivity of methanol, acetic acid and propionic acid was reported to be 12.2%, 10.1% and 10.3% respectively in O_2/CH_4 ratio of 0.9 and input power of 70 W. Increasing the amount of oxygen in the feed gas contributed the formation of O atoms by the collision of energetic electrons with oxygen in the plasma region, which further react with the intermediates species from the CH_4 dissociation to produce the OH radicals and consequently increased the selectivity of methanol.

Acknowledgements We gratefully acknowledge the financial support from the National Natural Science Foundation of China (Grant No. 52377147, 51907087) and the Postdoctoral Science Foundation of China (No. 2020M671289).

References

1. X.Y. Li, C.L. Pei, J.L. Gong, Shale gas revolution: Catalytic conversion of C1–C3 light alkanes to value-added chemicals. Chem **7**(7), 1755–1801 (2021)
2. M.M. Rao, Z.P. Chen, Y.H. Ling, H. Huang, Z.P. Peng, M.F. Li, J.B. Dong, S.J. Li, Partial oxidation of methane enabling complete coking resistance of Ni-YSZ anode supported solid oxide fuel cells. Mater. Lett. **328**, 133130 (2022)
3. P. Chawdhury, Y.L. Wang, D. Ray, S. Mathieu, N. Wang, J. Harding, F. Bin, X. Tu, C. Subrahmanyam, A promising plasma-catalytic approach towards single-step methane conversion to oxygenates at room temperature. Appl. Catal. B **284**, 119735 (2021)
4. Y.H. Yi, S.K. Li, Z.L. Cui, Y.Z. Hao, Y. Zhang, L. Wang, P. Liu, X. Tu, X.M. Xu, H.C. Guo, A. Bogaerts, Selective oxidation of CH_4 to CH_3OH through plasma catalysis: insights from catalyst characterization and chemical kinetics modelling. Appl. Catal. B **296**, 120384 (2021)
5. D.H. Mei, Y.L. He, S.Y. Liu, J. Yan, X. Tu, Optimization of CO_2 conversion in a cylindrical dielectric barrier discharge reactor using design of experiments. Plasma Processes Polym. **13**(5), 544–556 (2016)
6. S.Y. Liu, D.H. Mei, M.A. Nahil, S. Gadkari, S. Gu, P.T. Williams, X. Tu, Hybrid plasma-catalytic steam reforming of toluene as a biomass tar model compound over Ni/Al_2O_3 catalysts. Fuel Process. Technol. **166**, 269–275 (2017)
7. L.J. Song, Y.L. Kong, X.H. Li, Hydrogen production from partial oxidation of methane over dielectric barrier discharge plasma and $NiO/\gamma\text{-}Al_2O_3$ catalyst. Int. J. Hydrogen Energy **42**(31), 19869–19876 (2017)
8. D.H. Mei, X.Q. Shen, S.Y. Liu, R.S. Zhou, X.C. Yuan, Z.Q. Rao, Y.F. Sun, Z. Fang, X.S. Du, Y. Zhou, X. Tu, Plasma-catalytic reforming of biogas into syngas over Ni-based bimetallic catalysts. Chem. Eng. J. **462**, 142044 (2023)
9. T. Abbas, H.S.M. Yahya, N.A.S. Amin, Non-thermal plasma catalytic dry reforming of methane over $Ni\text{-}Co_3O_4$ supported modified-titania catalysts: effect of process conditions on syngas production. Fuel Process. Technol. **248**, 107836 (2023)
10. H. Puliyalil, D.L. Jurkovic, V.D.B.C. Dasireddy, B. Likozar, A review of plasma-assisted catalytic conversion of gaseous carbon dioxide and methane into value-added platform chemicals and fuels. RSC Adv. **8**(48), 27481–27508 (2018)
11. J.S. Zhou, J.C. Zhou, Y. Xu, B.C. Yan, X.Y. Yu, Control of methane plasma oxidative pathways by altering the contribution of oxygen species. Fuel **284**, 118944 (2021)
12. Y. Gao, L.G. Dou, B.W. Feng, C. Zhang, T. Shao, Catalyst-free activation of CH_4 and air into platform chemicals and H_2 using parametrized nanosecond pulsed plasma. Energy Convers. Manage. **276**, 116570 (2023)
13. P. Fathollahi, M. Farahani, R.H. Rad, M.R. Khani, A. Asadi, M. Shafiei, B. Shokri, Selective oxidation of methane to methanol by NTP plasma: the effect of power and oxygen on conversion and selectivity. J. Electrostat. **112**, 103594 (2021)
14. X.M. Zhang, M.S. Cha, Partial oxidation of methane in a temperature-controlled dielectric barrier discharge reactor. Proc. Combust. Inst. **35**(3), 3447–3454 (2015)
15. S.Y. Liu, D.H. Mei, Z. Shen, X. Tu, Nonoxidative conversion of methane in a dielectric barrier discharge reactor: prediction of reaction performance based on neural network model. J. Phys. Chem. C **118**(20), 10686–10693 (2014)
16. M.M. Moshrefi, F. Rashidi, H.R. Bozorgzadeh, Use of a DC discharge in a plasma reactor with a rotating ground electrode for production of synthesis gas by partial oxidation of methane. Res. Chem. Intermed. **41**(9), 5941–5959 (2014)
17. C. De Bie, J. van Dijk, A. Bogaerts, The dominant pathways for the conversion of methane into oxygenates and syngas in an atmospheric pressure dielectric barrier discharge. J. Phys. Chem. C **119**(39), 22331–22350 (2015)
18. D.H. Mei, S.Y. Liu, S. Wang, R.W. Zhou, R.S. Zhou, Z. Fang, X.H. Zhang, P.J. Cullen, K. Ostrikov, Plasma-enabled liquefaction of lignocellulosic biomass: balancing feedstock content for maximum energy yield. Renew. Energy **157**, 1061–1071 (2020)

Chapter 16
CH_4 Reforming with CO_2 in a Segmented Packed-Bed Dielectric Barrier Discharge Reactor

Jiyang Wang, Jingyi Ma, Shiyun Liu, Junhui Fu, Danhua Mei, and Zhi Fang

Abstract CO_2 and CH_4 are the top two greenhouse gases for global warming and climate change, which has brought disastrous effects on human health, social economy and environment. The greenhouse effect can be lessened by converting greenhouse gases into useful compounds. In this work, plasma assisted CH_4 reforming with CO_2 has been carried out in a coaxial DBD reactor with segmented electrodes. γ-Al_2O_3 and Zeolite particles were packed in the reactor with different configurations. The influences of packing configurations on the electrical characteristics and reforming reaction performance have been investigated. The voltage-current waveforms, Lissajous figures, and charge generation and transfer characteristics have been compared under different packing configurations to understand the variations in electrical characteristics. The corresponding reaction performance has also been evaluated in terms of reactant conversion, product selectivity, carbon balance and the energy efficiency for reactant conversions.

16.1 Introduction

The escalating progress of industrial technologies on a global scale has brought forth a myriad of environmental challenges [1]. Human activities including transportation, energy production, industry and agriculture represent the primary sources of CO_2 emissions [2]. Excessive CO_2 emissions have resulted in various environmental issues, notably the greenhouse effect and ocean acidification [3]. Addressing the imperative goal of mitigating global warming requires a concerted effort to reduce both atmospheric CO_2 concentrations and anthropogenic CO_2 emissions. There exist various strategies to diminish CO_2 emissions, such as restricting vehicular travel and embracing clean secondary energy sources like wind and solar power [4]. However, these approaches fall short of addressing the fundamental cause of excessive CO_2

J. Wang · J. Ma · S. Liu · J. Fu · D. Mei (✉) · Z. Fang
College of Electrical Engineering and Control Science, Nanjing Tech University, Nanjing 211816, China
e-mail: danhuam@126.com

Z. Fang et al. (eds.), *Proceedings of the 5th International Symposium on Plasma and Energy Conversion*, Springer Proceedings in Physics 398,
https://doi.org/10.1007/978-981-97-2245-7_16

emissions due to the extensive global vehicle base and the current underdevelopment of clean secondary energy technologies [5]. CO_2 conversion stands out as an effective approach to tackle excessive CO_2 emissions. Rather than just a greenhouse gas, CO_2 emerges as a potential carbon resource that can be transformed into a valuable chemicals through direct conversion and utilization methods [6]. This transformative process not only mitigates the concentration of CO_2 in the atmosphere but also converts it into novel energy substances, providing a solution to both the energy and environmental crises [7].

Currently, CO_2 conversion mainly relies on thermal catalysis, photocatalysis and plasma catalysis [8]. Thermal catalysis imposes higher demands on reactor equipment, escalating costs, and enhancing the likelihood of catalyst deactivation, especially for non-precious metal catalysts, although it shows relatively high efficacy. Moreover, it necessitates the maintenance of high-temperature and high-pressure conditions, contributing to carbon emissions. For photocatalysis, it can be cost-effective and safe, but encounters challenges such as a sluggish reaction rate, a limited spectral response range, and non-compact assembly of photocatalytic devices [9]. Dry reforming of methane (Eq. 16.1) presently stands as a popular method for CO_2 conversion. Two major greenhouse gases CO_2 and CH_4 can be amalgamated to produce CO and H_2, along with other value-added chemical products [10]. Extensive research has been performed on dry reforming of methane over the past decades, there still exist persistent issues like carbon deposition and subsequent catalyst activity reduction, which hinder the commercialization of this process [11].

Non-thermal plasma (NTP) technology has been proved to be an effective strategy in facilitating the CO_2 reforming of CH_4, owing to its high activity and significantly higher electron temperature compared to heavy particles. Various discharge methods exist for generating NTP, including dielectric barrier discharge (DBD), glow discharge, gliding arc discharge and corona discharge [12]. DBD exhibits promising potential due to its capacity to generate NTP at room temperature and pressure. This method boasts advantages such as providing high electron energy density and a simple, uniform discharge structure in the reactor. It is also characterized by low cost and ease of implementation [13]. In addition, combining DBD with catalysts might generate the plasma catalysis synergy. Introducing catalytic particles into the discharge region will enhance the electric field strength and achieve a more effective discharge, and also change the reaction process through the physical structure of catalytic particles, including their effective pore structure and specific surface area. These factors are beneficial for improving the reactant conversion and desired product selectivity by applying suitable catalysts [14]. Gallon et al. [15] studied the impact of packing materials on the CO_2 reforming of CH_4 in DBD plasma. They compared the reaction performance when packing quartz wool, γ-Al_2O_3, and 3A molecular sieve with that without any packing. The findings revealed that packing quartz wool resulted in a significant increase in methane conversion and hydrogen yield. In contrast, packing 3A molecular sieve improved the selectivity of hydrogen and some light hydrocarbons like C_2H_2/C_2H_4 [16]. Wang et al. [17] have reported that using the DBD reactor with segmented electrodes could improve the reactant conversion and energy efficiency for plasma CO_2 conversion processes. However, there is limited researches related to the plasma-catalytic reforming process in the DBD reactor with segmented electrodes. It is urgent to study the discharge characteristics and reaction performance in the DBD reactor with segmented electrodes using different packing configurations.

In this work, plasma CO_2 reforming of CH_4 was performed in a DBD reactor with segmented electrodes. Two typical catalytic materials γ-Al_2O_3 and Zeolite were packed in the DBD reactor with different configurations. The discharge properties and the reaction performance were thoroughly discussed under these conditions. These studies would provide valuable insights for the design of DBD reactors and the optimization of CO_2 reforming CH_4 reactions for further advancements.

16.2 Experimental

16.2.1 Experimental Setup

The experimental setup contains a power supply, DBD reactor, gas supplying system, discharge diagnostic system, as well as product collection and analysis system. Figure 16.1 shows the connection diagram of the experimental system. Figure 16.2 displays the structure of the DBD reactor with two segmented electrodes. The reactor contains an inner electrode with a stainless steel rod (diameter: 4 mm) and two outer electrodes with stainless steel mesh (mesh size: 120 meshes; wire diameter: 0.07 mm; length: 3 cm). The distance between these two outer electrodes is 3 cm. The barrier tube is made up of a quartz glass tube with the inner and outer diameters of 8 and 10 mm, respectively. The DBD reactor is powered by a custom high-frequency AC power supply (13 kV, 10 kHz). Electrical diagnosis are conducted by using a Tektronix high voltage probe (P6015A, 75 MHz, 1000:1 voltage division) to measure the applied voltage, a Pearson current coil (6585, 250 MHz) to measure the total current, and a Pintech differential probe (N1000A, DC-50 MHz, 1000:1 voltage division) to measure the voltage on the external capacitor. All measurements are recorded using four-channel Tektronix digital oscilloscopes (TDS-2014B, 100 MHz). Discharge power and transferred charge are calculated based on these electrical signals.

The reactant gases are supplied by the gas cylinders. Mass flow controllers are employed to regulate the gas flow of CO_2 and CH_4. The total flow rate is fixed at 50 ml/min with a CO_2/CH_4 molar ratio of 2:3. These two gases are thoroughly mixed to form a homogeneous gas flow before being introduced into the reactor. γ-Al_2O_3 and Zeolite particles with a diameter around 1 mm are used as the catalytic packing materials. They are packed in the DBD reactor with different configuration as shown in Table 16.1, which is designed to explore their individual and synergistic effects of γ-Al_2O_3 and Zeolite on the discharge reaction process.

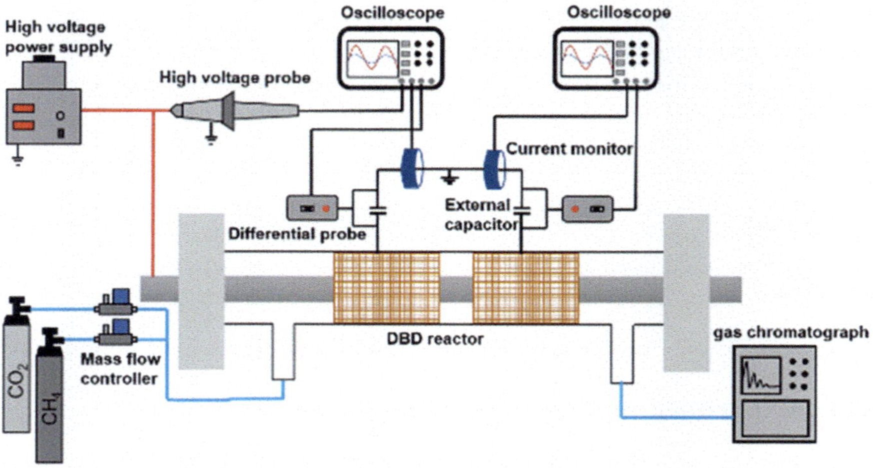

Fig. 16.1 Experimental setup and measurement system

Fig. 16.2 Schematic diagram of reactor structure

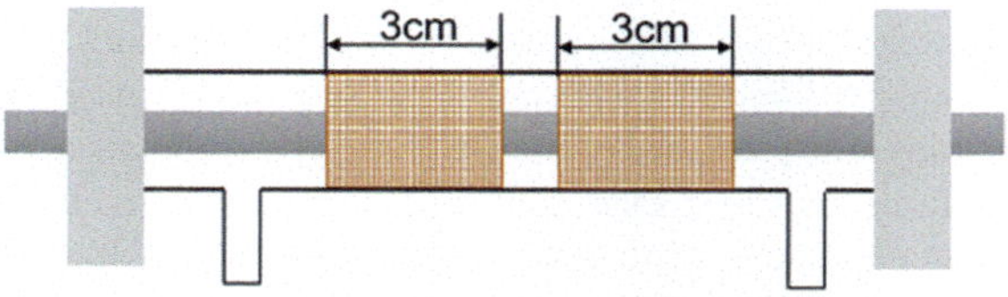

Table 16.1 Packing configuration of the DBD reactor with segmented electrodes

No	First discharge region	Second discharge region	Denoted as
1	γ-Al_2O_3	γ-Al_2O_3	A–A
2	γ-Al_2O_3	Zeolite	A–Z
3	Zeolite	γ-Al_2O_3	Z–A
4	Zeolite	Zeolite	Z–Z

16.2.2 Analysis Methods

In terms of electrical characteristics, the primary parameters under consideration are the discharge power and transferred charge, which can be determined by the Lissajous figure based on the signals of applied voltage and voltage on the external capacitor.

The gas components before and after reaction are analyzed via gas chromatography (GC) using a PANNAA60 model, which is featured a flame ionization detector

(FID) and a thermal conductivity detector (TCD). This setup facilitates the detection of CO_2, H_2, O_2, and C_2–C_4 hydrocarbon compounds. The gas flow rate before and after the discharge reaction was quantified using a soap film flow meter. The performance evaluation indices for the reaction includes reactant conversion, product yield, product selectivity and energy efficiency, which are calculated by the following equations.

CO_2 conversion:

$$C_{CO_2}(\%) = \frac{CO_2 \text{ converted (mol/s)}}{CO_2 \text{ input (mol/s)}} \times 100 \tag{16.1}$$

CH_4 conversion:

$$C_{CH_4}(\%) = \frac{CH_4 \text{ converted (mol/s)}}{CH_4 \text{ input (mol/s)}} \times 100 \tag{16.2}$$

H_2 yield:

$$Y_{H_2}(\%) = \frac{H_2 \text{ produced (mol/s)}}{2 \times CH_4 \text{ input (mol/s)}} \times 100 \tag{16.3}$$

H_2 selectivity:

$$S_{H_2}(\%) = \frac{H_2 \text{ produced (mol/s)}}{2 \times CH_4 \text{ converted (mol/s)}} \times 100 \tag{16.4}$$

CO yield:

$$Y_{CO}(\%) = \frac{CO \text{ produced (mol/s)}}{CH_4 \text{ input (mol/s)} + CO_2 \text{ input (mol/s)}} \times 100 \tag{16.5}$$

CO selectivity:

$$S_{CO}(\%) = \frac{CO \text{ produced (mol/s)}}{CH_4 \text{ converted (mol/s)} + CO_2 \text{ converted (mol/s)}} \times 100 \tag{16.6}$$

Hydrocarbon yield:

$$Y_{C_xH_y}(\text{mmol/kJ}) = \frac{C_xH_y \text{ produced (mol/s)}}{CH_4 \text{ input (mol/s)} + CO_2 \text{ input (mol/s)}} \times 100 \tag{16.7}$$

Hydrocarbon selectivity:

$$S_{C_xH_y}(\text{mmol/kJ}) = \frac{x \times C_xH_y \text{ produced (mol/s)}}{CH_4 \text{ converted (mol/s)} + CO_2 \text{ converted (mol/s)}} \times 100 \tag{16.8}$$

Carbon balance:

$$CB(\%) = \frac{\begin{array}{c}[CH_4 + CO_2]\,\text{unconverted (mol/s)}+\\ [CO + \sum x \times C_xH_y]\,\text{produced (mol/s)}\end{array}}{CO_2\,\text{input (mol/s)} + CH_4\,\text{input (mol/s)}} \tag{16.9}$$

Energy efficiency:

$$EE_{CO_2}(\text{mmol/kJ}) = \frac{CO_2\,\text{converted (mol/s)}}{\text{Discharge power (kW)}} \tag{16.10}$$

$$EE_{CH_4}(\text{mmol/kJ}) = \frac{CH_4\,\text{converted (mol/s)}}{\text{Discharge power (kW)}} \tag{16.11}$$

16.3 Results and Discussion

16.3.1 Effect of Packing Configurations on the Electrical Characteristics

Figure 16.3 presents the voltage and current waveforms under different packing configurations, including the applied voltage and current in two regions. Current1 and current2 denote the current in the first and second discharge region, respectively. Compared Fig. 16.3a and b, much more current pulsed are observed in the current signals of both regions when packing γ-Al_2O_3 than that packing Zeolite under the same applied voltage of 13 kV. This indicates that packing γ-Al_2O_3 is beneficial for the formation of discharge channels in the gap, which also can be seen in Fig. 16.3c and d. In addition, the magnitude of current in the γ-Al_2O_3 region is always higher than that in the Zeolite region in the case of packing two different materials, no matter it is in the first or second region. Moreover, the current signals when packing Zeolite shows higher displacement current than that when packing γ-Al_2O_3. All these phenomena suggest that packing γ-Al_2O_3 and Zeolite with different configurations results in the significant difference in the electrical characteristics, which will consequently lead to the different reaction performance.

Figure 16.4 shows the Lissajous figures for each discharge region in different packing configurations. Clearly, the shape of the Lissajous figure looks like parallelogram in the first discharge region, regardless of the packing configuration. While the Lissajous figure shape in the second discharge region turns to flat ellipse and the area of the Lissajous shape in the second region is significantly lower than that in the first region, except the condition when packing Zeolite in the first and γ-Al_2O_3 in the second region. It is clear from Fig. 16.4d, the Lissajous figures in both region exhibit the normal parallelogram, and the figure area in the second region is larger than that in the first area, which indicates that this packing configuration might enhance

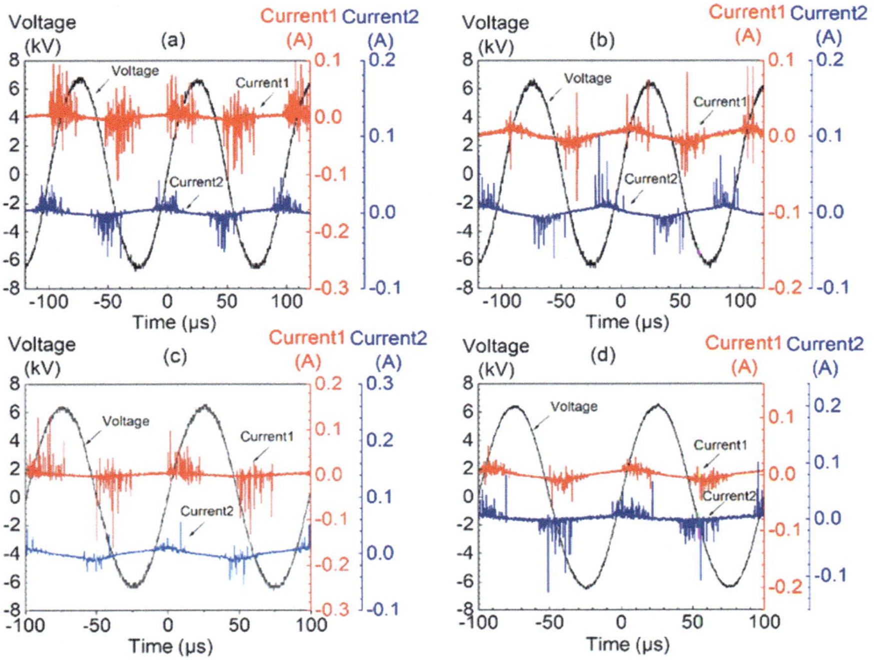

Fig. 16.3 Voltage-current waveforms in different packing configurations: **a** A–A; **b** Z–Z; **c** A–Z; **d** Z–A

the discharge power as well as the charge generation and transfer properties of the plasma reaction system.

Figure 16.5 displays the charge generation and transfer properties in different packing configuration. It is noteworthy that the peak-to-peak charge (Q_{pk-pk}) in the condition when packing Zeolite in the first region is higher than that in the other packing configurations. The highest Q_{pk-pk} is obtained when packing Zeolite in the first and γ-Al_2O_3 in the second region. The generated charge in each half-cycle (Q_d) exhibits no obvious difference in all packing configurations. For the transferred charge (Q_{trans}), large values are observed when packing γ-Al_2O_3 in the first discharge region. Figure 16.6 shows the discharge power in different packing configurations. The discharge power varies around 20 W when the applied voltage is 13 kV. The lowest discharge power of 19.3 W was obtained when packing γ-Al_2O_3 in the first and Zeolite in the second region.

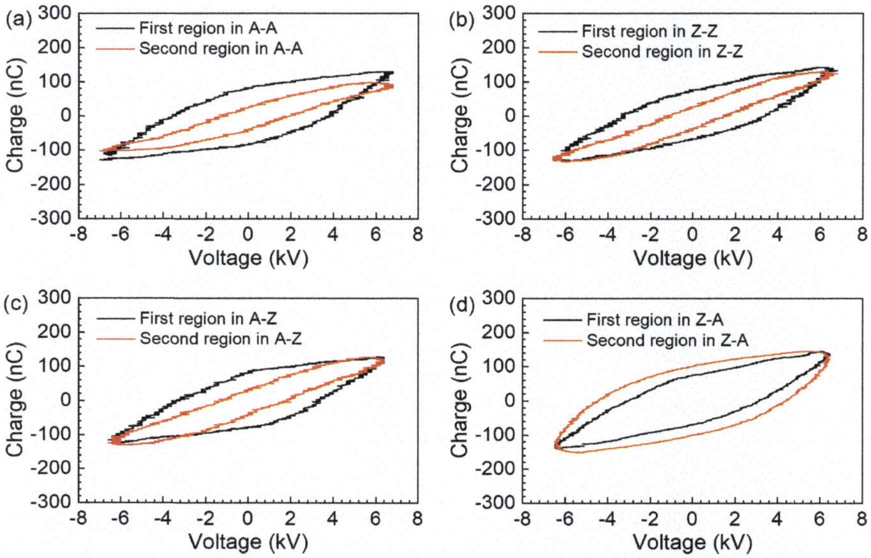

Fig. 16.4 Lissajous figures in different packing configurations: **a** each region in A–A; **b** each region in Z–Z; **c** total in A–A and Z–Z; **d** each region in A–Z; **e** each region in Z–A; **f** total in A–Z and Z–A

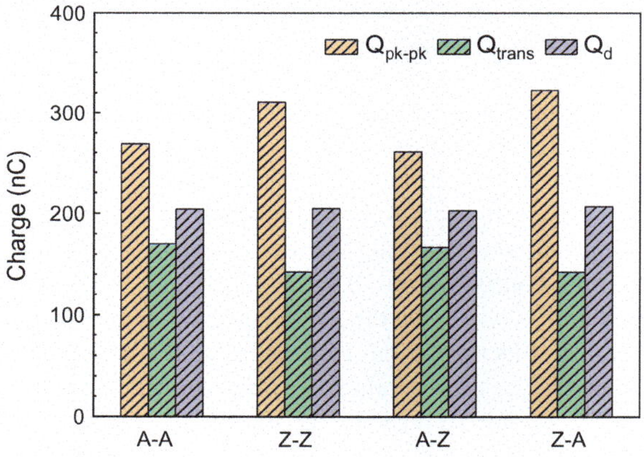

Fig. 16.5 Charge generation and transfer properties in different packing configurations

16.3.2 Effect of Packing Configurations on Reaction Performance

In this work, we use γ-Al_2O_3 and Zeolite with different packing configurations and evaluate their performance in the plasma reforming process, as both materials exhibit commendable adsorption capabilities, which renders them viable candidates as catalysts. γ-Al_2O_3 has been proved versatile in a variety of applications, as it

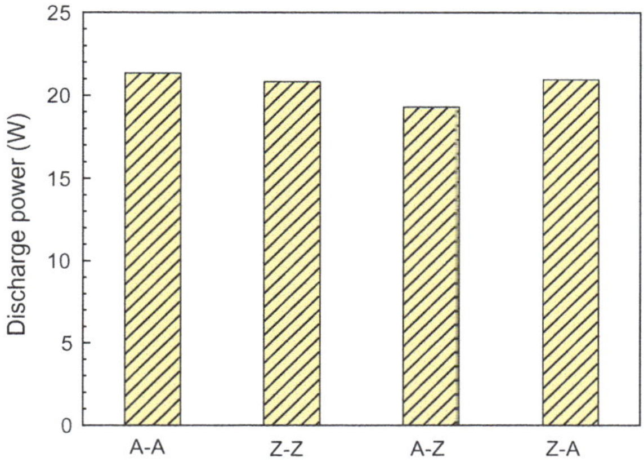

Fig. 16.6 Discharge power in different packing configurations

is characterized by its porous structure, expansive specific surface area, and robust adsorption capabilities. It serves not only as a support for loading noble metals or other substances for catalytic purposes but also exhibits the capability to activate chemical bonds (e.g., H–H bonds and C–H bonds). Consequently, it can be utilized directly as a catalyst in chemical reactions. For Zeolite, it is a 5A-type molecular sieve, featuring a pore size of 0.5 nm and a uniform distribution. This molecular architecture imparts a heightened selective adsorption capacity compared to γ-Al_2O_3. In this case, both γ-Al_2O_3 and Zeolite offer promising avenues for catalytic applications.

The packing configuration brings significant influence not only on discharge characteristics but also on the plasma chemical reactions. Figure 16.7 meticulously portrays the CO_2 and CH_4 conversions under diverse packing configurations. Strikingly, packing γ-Al_2O_3 in first and second discharge regions yields the highest conversions for both CO_2 and CH_4. In addition, packing the same materials in these two discharge regions results in the higher reactant conversion than that obtained when packing different materials. This disparity may be attributed to the current pulses with higher magnitude and a greater number of current pulses in the discharge region packing γ-Al_2O_3, compared to that packing Zeolite. This heightened pulse activity signifies a higher current density and consequently a greater electron density, which will generate more chemical reaction species for plasma reactions. Intriguingly, the conversion of CH_4 is higher than that of CO_2 in all packing conditions. The higher CH_4 conversion can be ascribed to its multifaceted conversion channels. Electron impact reactions of CH_4 initiate its dissociation, generating CH_3, CH_2, and CH radicals, followed by recombination reactions leading to higher hydrocarbons or further electron impact dissociation reactions. In addition, electron and surface reactions also play a role directly in driving plasma decomposition of CH_4. Conversely, the lower CO_2 conversion may find its roots in the presence of the water gas shift reaction (WGSR). Furthermore, the lower CO_2 concentration compared to CH_4 contributes to this discrepancy. The intricate interplay of these factors underscores the complex landscape of chemical reactions within the discharge area, unveiling the multifaceted impact of packing materials on both the discharge pattern and ensuing chemical transformations.

Fig. 16.7 Reactant conversion in different packing configurations

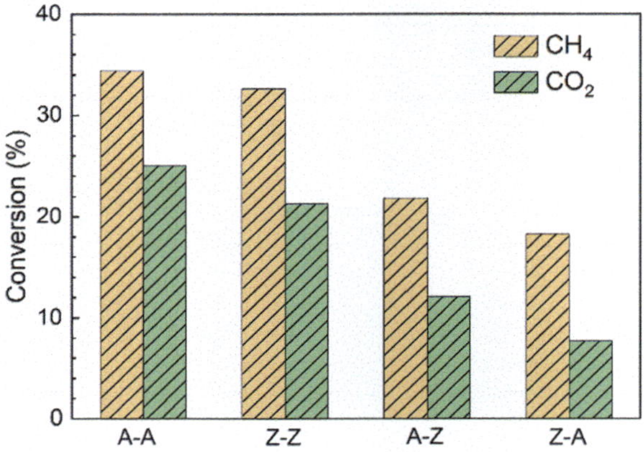

Figure 16.8 unveils the product selectivity from plasma CO_2 reforming of CH_4 in different packing configurations. Clearly, packing the same materials in the first and second discharge regions leads to the relatively lower selectivity for all gas products. The highest product selectivities are achieved in the condition when packing γ-Al_2O_3 and Zeolite in sequence. Besides the main gas product H_2 and CO, the selectivities of C_2H_6 are significantly higher than that of other hydrocarbon compounds in all packing configurations. Insights from Zhang et al. [18] emphasized the role of pore size in influencing product selectivity. According to their findings, smaller pore sizes inhibit the synthesis of long-chain species and fortify the formation of shorter-chain species, and consequently enhancing their selectivity. The higher selectivity of C_2H_6 in this work suggests that the pore size of γ-Al_2O_3 and Zeolite is suitable for generating C_2 hydrocarbons.

Figure 16.9 elucidates the carbon balance in different packing configurations. Packing different materials in the first and second discharge regions results in the higher carbon balance. The highest carbon balance of 90.3% is attained when the

Fig. 16.8 Product selectivity in different packing configurations

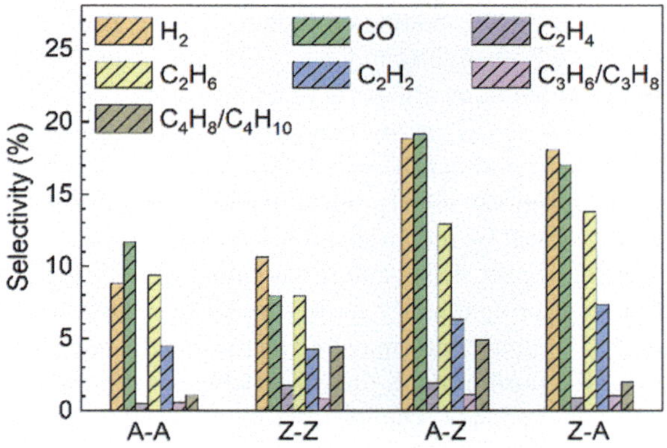

first region is packed with γ-Al_2O_3 and the second region with Zeolite. The carbon balances in all packing configurations are below 100%, especially in the condition when packing the same materials in both regions (around 80%). The variation in carbon balance primarily stems from the deposition of solid carbon on the reactor and catalyst, alongside the potential generation of some liquid hydrocarbons.

Figure 16.10 displays the energy efficiency in different packing configurations. Packing the same materials in both regions results in the higher energy efficiency, compared with other conditions. This is in accordance with the variations of reactant conversion in different packing configurations. The highest energy efficiency for the conversion of CH_4 and CO_2 is 0.36 and 0.18 mmol/kJ, respectively, which is achieved by packing γ-Al_2O_3 in both the first and second discharge regions. Packing the different materials regardless of the order leads to the lower energy efficiency. However, these configurations can give birth to higher selectivities of gas products, as shown in Fig. 16.8. Therefore, it is beneficial to regulate the product distribution of plasma CO_2 reforming of CH_4 by optimizing the packing configuration with different materials.

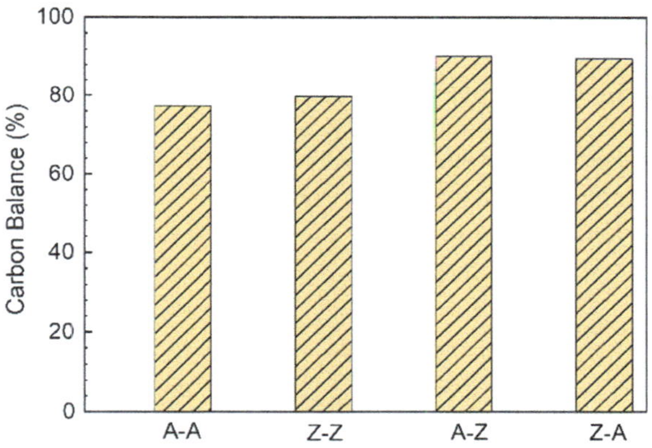

Fig. 16.9 Carbon balance in different packing configurations

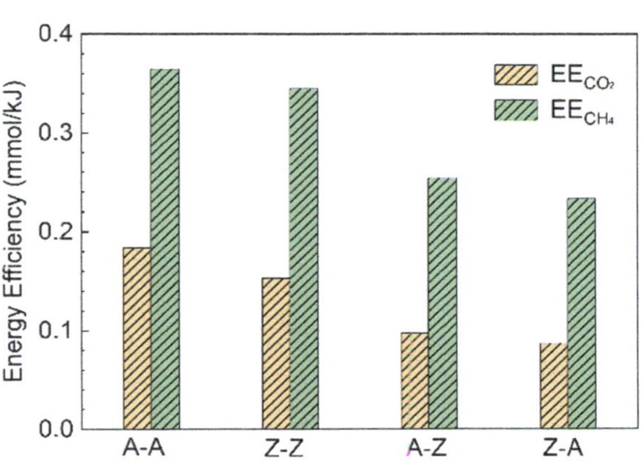

Fig. 16.10 Energy efficiency in different packing configurations

16.4 Conclusion

In this work, we conduct a deep exploration of the discharge characteristics and reaction performance of CH_4 reforming with CO_2 in a DBD reactor with segmented electrodes. γ-Al_2O_3 and Zeolite are used as the packing materials in the first and second discharge regions with different packing configurations. The introduction of different packing materials into the segmented packed-bed DBD reactor exerts an influence on the discharge characteristics. Notably, the amplitude and number of current pulse in the discharge region packed with γ-Al_2O_3 are higher than that in the region packed with Zeolite. This indicates that more discharge channels are formed and the chances of collisions between gas molecules and active particles are improved in the discharge region, ultimately enhancing the reactant conversion in the plasma reforming reaction process.

The highest conversions for both CH_4 and CO_2 as well as the corresponding energy efficiency are achieved when packing γ-Al_2O_3 in both the first and second discharge region. For the selectivity of gas products, the highest values are achieved when packing different materials in these two discharge regions. It is suggested that the high reactant conversion (energy efficiency) and the high selectivity of specified products can be achieved simultaneously by optimizing the packing configurations in the segmented packed-bed DBD reactor using different catalytic materials.

Acknowledgements This work is financially supported by the National Natural Science Foundation of China (No. 51807087 and No. 52177149).

References

1. F. Zhang, Z. Song, J. Zhu et al., Process of CH_4–CO_2 reforming over Fe/SiC catalyst under microwave irradiation. Sci. Total. Environ. **639**, 1148–1155 (2018)
2. D. Pakhare, J. Spivey, A review of dry (CO_2) reforming of methane over noble metal catalysts. Chem. Soc. Rev. **43**(22), 7813–7837 (2014)
3. Y. Gao, J. Jiang, Y. Meng et al., A review of recent developments in hydrogen production via biogas dry reforming. Energy Conver. Manag. **171**, 133–155 (2018)
4. M.M. Zain, A.R. Mohamed, An overview on conversion technologies to produce value added products from CH_4 and CO_2 as major biogas constituents. Renew. Sustain. Energy Rev. **98**, 56–63 (2018)
5. X. Tu, H.J. Gallon, M.V. Twigg, et al., Dry reforming of methane over a Ni/Al_2O_3 catalyst in a coaxial dielectric barrier discharge reactor. J. Phys. D. Appl. Phys. **44**(27) (2011)
6. C. Xu, X. Tu, Plasma-assisted methane conversion in an atmospheric pressure dielectric barrier discharge reactor. J. Energy Chem. **22**(3), 420–425 (2013)
7. Y.X. Zeng, X.B. Zhu, D.H. Mei et al., Plasma-catalytic dry reforming of methane over γ-Al_2O_3 supported metal catalysts. Catal. Today **256**, 80–87 (2015)
8. Y. Gao, S. Zhang, H. Sun et al., Highly efficient conversion of methane using microsecond and nanosecond pulsed spark discharges. Appl. Energy **226**, 534–545 (2018)
9. D.H. Mei, B. Ashford, Y.L. He et al., Plasma-catalytic reforming of biogas over supported Ni catalysts in a dielectric barrier discharge reactor: effect of catalyst supports. Plasma Process. Polym. **14**(6), 1600076 (2017)

10. D.H. Mei, S.Y. Liu, X. Tu, CO_2 reforming with methane for syngas production using a dielectric barrier discharge plasma coupled with Ni/γ-Al_2O_3 catalysts: process optimization through response surface methodology. J. CO_2 Utiliz. **21**, 314–326 (2017)
11. L. Wang, Y. Yi, C. Wu et al., One-step reforming of CO_2 and CH_4 into high-value liquid chemicals and fuels at room temperature by plasma-driven catalysis. Angew. Chem. Int. **56**(44), 13679–13683 (2017)
12. R. Vakili, R. Gholami, C.E. Stere et al., Plasma-assisted catalytic dry reforming of methane (DRM) over metal-organic frameworks (MOFs)-based catalysts. Appl. Catal. B Environ. **260**, 118195 (2020)
13. R. Snoeckx, R. Aerts, X. Tu et al., Plasma-based dry reforming: a computational study ranging from the nanoseconds to seconds time scale. J. Phys. Chem. C **117**(10), 4957–4970 (2013)
14. A. George, B. Shen, M. Craven et al., A review of non-thermal plasma technology: a novel solution for CO_2 conversion and utilization. Renew. Sustain. Energy Rev. **135**, 109702 (2021)
15. H.J. Gallon, X. Tu, J.C. Whitehead, Effects of reactor packing materials on H_2 production by CO_2 reforming of CH_4 in a dielectric barrier discharge. Plasma Processes Polym. **9**(1), 90–97 (2012)
16. W.C. Chung, K.L. Pan, H.M. Lee et al., Dry reforming of methane with dielectric barrier discharge and ferroelectric packed-bed reactors. Energy Fuels **28**(12), 7621–7631 (2014)
17. B.W. Wang, X. Wang, B. Zhang, Dielectric barrier micro-plasma reactor with segmented outer electrode for decomposition of pure CO_2. Front. Chem. Sci. Eng. **11**, 548 (2020)
18. Y. Zhang, Y. Qu, J. Wang, Effect of crystal size on the catalytic performance of HZSM-5 Zeolite in the methanol to aromatics reaction. Petrol. Sci. Technol. **36**(12), 898–903 (2018)

Chapter 17
Effect of Gliding Arc Plasma Reforming of Methane on Large-Bore Low-Speed Two-Stroke Marine Engine

Yezeng Fan, Zhan Gao, Zhuohang Li, Chao Wang, Lei Zhu, and Zhen Huang

Abstract As a clean fuel from a wide range of sources, natural gas can effectively improve the emission performance of large-bore low-speed marine engines. The high ignition energy and low flame propagation speed of natural gas can be effectively solved by using hydrogen or syngas from fuel reforming. In this study, a gliding arc plasma reforming reactor was utilized to reform methane, the main component of natural gas. Reforming was studied for different O/C molar ratios and input power. Based on the obtained ratios of reforming product components (H_2 and CO, etc.), numerical simulation was used to study the combustion and emission characteristics of the engine under different fuel reforming ratios. The reforming gas enters the engine through the natural gas intake valve. The results show that the products of gliding arc plasma reforming can effectively enhance the flame propagation in the cylinder, forward the center of exothermic heat release, and shorten the combustion duration. Meanwhile, it can reduce emissions of unburned natural gas.

17.1 Introduction

The shipping industry consumes large amounts of conventional fossil fuels every year. It is one of the major sources of greenhouse gas emissions [1–3]. Electrification of power units for the shipping industry, especially for ocean shipping, is currently difficult to achieve. Therefore, the shipping industry still has a large market for internal combustion engines (ICE). The large-bore, low-speed, two-stroke diesel engines are widely used in current marine engines due to their high output power and reliability in continuous operation [4, 5]. In 2018, the International Marine Organization (IMO) adopted a target to reduce greenhouse gas emissions from ships. It aims to reduce greenhouse gas emissions by at least half of the 2008 levels by the year 2050 [6]. At the same time, IMO has introduced the strict Tier III emission standard,

Y. Fan · Z. Gao · Z. Li · C. Wang · L. Zhu (✉) · Z. Huang
Key Laboratory for Power Machinery and Engineering of M. O. E, Shanghai Jiaotong University, Shanghai, China
e-mail: tonyzhulei@sjtu.edu.cn

Z. Fang et al. (eds.), *Proceedings of the 5th International Symposium on Plasma and Energy Conversion*, Springer Proceedings in Physics 398,
https://doi.org/10.1007/978-981-97-2245-7_17

which limits NO_x emissions to 3.4 g/kW*h [7–9]. Using clean alternative fuels is a necessary step towards reducing emissions from the shipping industry [10].

As an alternative fuel, natural gas has the advantage of abundant reserves and clean combustion and emissions [11, 12]. In addition, methane, the primary fuel for natural gas, has a high octane number and good resistance to knock, allowing engines to run at higher compression ratios and enhancing downsizing capabilities [13]. Compared to conventional diesel engines, using natural gas as fuel can effectively reduce Soot and NO_x emissions [14, 15]. However, the narrow, lean combustion limits and low laminar flame speeds of natural gas can easily lead to incomplete combustion or misfire when used in large-bore engines that require long flame propagation distance [16–18]. H_2, as a highly reactive zero-carbon fuel, can effectively solve these problems by blending it with natural gas. Arslan et al. [19] found that hydrogen-rich natural gas was effective in reducing HC emissions from engines. Ma et al. [20] investigated the effect on combustion by adding hydrogen and found that the lean combustion limit of the engine was significantly extended, and the thermal efficiency was higher.

However, the cost of storage and transportation of hydrogen is high, and there are some safety problems with carrying it directly on ships [21–24]. Fuel reforming can effectively solve these problems by obtaining hydrogen on the basis of carrying only one kind of fuel [25, 26]. Common methane fuel reforming methods include steam reforming, dry reforming, autothermal reforming, and partial oxidation reforming. However, conventional reforming methods usually require catalysts combined with heating and pressurization [27–30].

Plasma contains a significant number of high-energy electrons and radicals. The use of plasma for methane reforming allows for rapid response while effectively avoiding some issues associated with traditional catalytic reforming methods, such as catalyst deactivation due to carbon accumulation [31, 32]. Gliding arc plasma is a non-thermal equilibrium plasma with both hot and cold plasma characteristics, which is a research hotspot in the field of hydrogen production from plasma reforming fuel. Xia et al. [33] used gliding arc plasma to reform CO_2 and CH_4 and optimized the H_2/CO ratio in syngas by adding steam. Wang et al. [34] investigated methane steam reforming using gliding arc plasma and obtained the relevant reaction mechanism. Zhang et al. [35] investigated methane cracking using a rotating gliding arc plasma reactor and used optical emission spectroscopy (OES) to observe the radicals of the process. Although there have been many studies on gliding arc plasma reforming of methane, there have been fewer studies on its application to engines, especially large-bore low-speed two-stroke marine engines.

Considering the practical application in engines, this study uses gliding arc plasma for partial oxidation reforming of methane. H_2 produced by the reforming process can improve the flame propagation speed and contribute to the stabilization of combustion in the engine. The effects of different input power and O/C molar ratios on methane conversion, H_2, and CO yield were investigated. A three-dimensional computational fluid dynamics (CFD) model of a large-bore low-speed two-stroke engine was constructed. The effects of different fuel reforming ratios on the combustion and emission characteristics of the engine were investigated based on the ratio of each component of the reforming product under suitable operating conditions by numerical simulation. This study can provide reasonable guidance for the development of low-speed marine natural gas engines.

17.2 Experimental and Numerical Methods

17.2.1 Plasma Reforming System

The diagram of the gliding arc plasma reforming experimental system is shown in Fig. 17.1. The plasma reformer is cylindrical in shape. One is the high-voltage electrode, and the other is the low-voltage electrode. The high-voltage AC power supply provides energy to the reformer. It is capable of frequency variations from 5 to 20 kHz. The flow rate of the gas in the experiment is controlled by a mass flow meter.

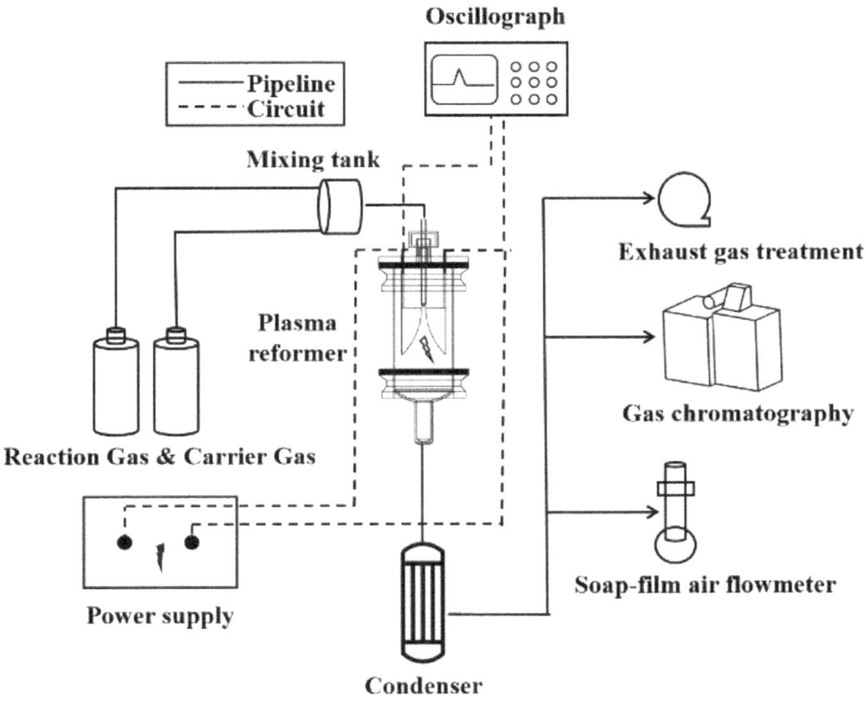

Fig. 17.1 Diagram of plasma reforming system

17.2.2 Product Analysis Methods

The reforming gas of the reaction was cooled down to room temperature by the condenser. Measure the total flow of product gas using a soap-film flow meter. The concentration of components of the product (H_2, CO and CH_4, etc.) was measured using gas chromatography equipped with thermal conductivity detector (TCD) and hydrogen flame ionization detector (FID). The discharge parameters of the reformer were obtained from an oscilloscope. To better evaluate the reaction performance, we define the following parameters. The oxygen/carbon molar ratio (O/C) is calculated as shown in Eq. (17.1).

$$O/C = \frac{0.21 * 2 * Q_{in}(\text{Air})}{Q_{in}(\text{CH}_4)} \tag{17.1}$$

The conversion of methane (C_{CH_4}) is calculated as shown in Eq. (17.2).

$$C_{CH_4} = \frac{Q_{in}(\text{CH}_4) - Q_{out}(\text{CH}_4)}{Q_{in}(\text{CH}_4)} * 100\% \tag{17.2}$$

The yields of H_2 (Y_{H_2}) and CO (Y_{CO}) are calculated as shown in Eq. (17.3) and Eq. (17.4), respectively.

$$Y_{H_2} = \frac{Q_{out}(\text{H}_2)}{Q_{in}(\text{CH}_4) * 2} * 100\% \tag{17.3}$$

$$Y_{CO} = \frac{Q_{out}(\text{CO})}{Q_{in}(\text{CH}_4)} * 100\% \tag{17.4}$$

$Q_{in}(\text{Air})$ and $Q_{in}(\text{CH}_4)$ represent the flow rate of air and methane being delivered into the system, respectively. $Q_{out}(\text{H}_2)$, $Q_{out}(\text{CO})$ and $Q_{out}(\text{CH}_4)$ represent the flow rate of each substance at the outlet of the system.

17.2.3 Numerical Model of the Engine

In this study, a numerical model of a large bore low-speed two-stroke natural gas/diesel dual-fuel engine, RT-flex 50DF, was constructed based on the computational fluid dynamics (CFD) platform CONVERGE. The engine's geometry is shown in Fig. 17.2. It includes structures such as injection valves for natural gas and reforming gas, pre-chambers, and cylinder. The key parameters of the engine are shown in Table 17.1.

We set the 360° Crank Angle (CA) as the crank angle corresponding to the top dead center (TDC). One cycle of a dual-fuel low-speed engine was simulated from 110° CA to 470° CA. As the piston moves upward, CH_4 and reforming product

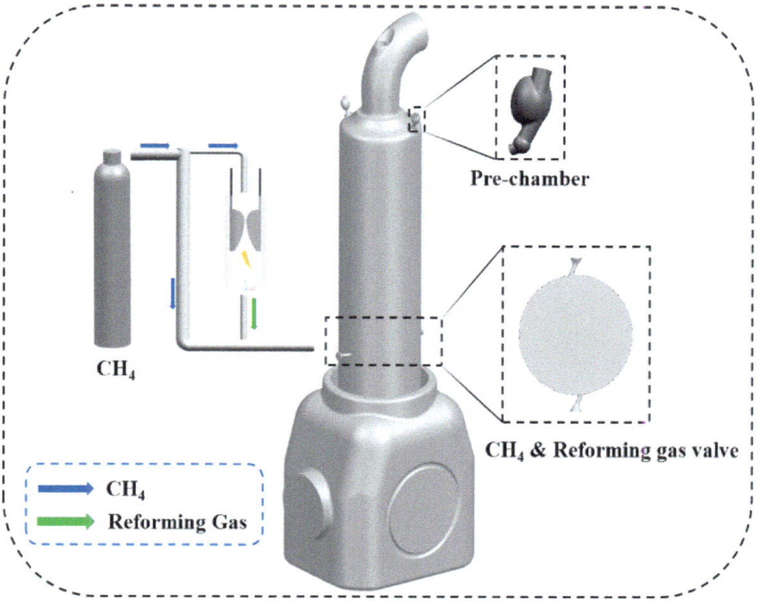

Fig. 17.2 The structure of the engine

Table 17.1 The key parameters of the engine

Parameters (unit)	Value
Bore (mm)	500
Stroke (mm)	2050
Connection rod length (mm)	2050
Number of injectors	2
Number of nozzles on injectors	3
Speed (r/min)	102

gas enter the cylinder through the intake valve. When the piston moved up to 8.5° CA before the top dead center (BTDC), a small amount of pilot fuel (diesel) was injected into the pre-chamber to form a jet flame to ignite the premixed gas. Since diesel and n-heptane have similar cetane number, n-heptane was used instead of diesel in the simulation. The mass of the pilot fuel (n-heptane) is about 0.2148 g/cycle. The physical models used in the simulation are shown in Table 17.2. The combustion mechanism for methane and n-heptane used in the Sage model contains 79 components and 262 reaction steps. The NO_x emissions were calculated by using the extended Zeldovich mechanism.

The model adopted Adaptive Mesh Refinement (AMR), which was solved using a variable time-step algorithm. Validation of the model with experiments has been shown in previous work [7]. To study the ignition process in the pre-chamber, the

Table 17.2 Physical model

Model	Name
Droplet collision model	NTC collision [36]
Droplet interaction model	O'Rourke and Amsden [37]
Turbulence model	RNG k–ε [38]
Combustion model	SAGE [39]
Spray breakup model	KH-RT [40]

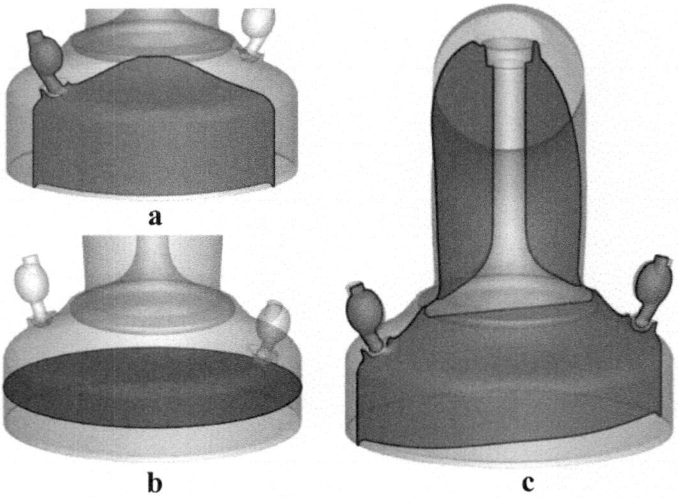

Fig. 17.3 Schematic diagram of section location

development of the jet flame, and the combustion process of the fuel in the main chamber in a better way, cross sections, as shown in Fig. 17.3, were chosen to reflect the combustion process.

17.3 Results and Discussion

17.3.1 Gliding Arc Plasma Reforming

Effect of Input Power

Under conditions of partial oxidation reforming, the mixture of methane and air was passed into the reformer at a fixed flow rate of 500 ml/min (methane and air). The O/C molar ratio was fixed at 1. Figure 17.4. illustrates how the conversion of methane (C_{CH_4}) and the yields of H_2 (Y_{H_2}) and CO (Y_{CO}) change as the input power increases. With increasing input power, the methane conversion is increasing. This is due to the increase in input power, which leads to better development of gliding arc in the reformer. This contributes to the formation of a sufficiently non-equilibrium plasma region. The energy and electron densities in the reaction region rise, and the interactions between the substances are enhanced. The partial oxidation reforming reaction of methane is promoted. When the input power is 45.9 W, the C_{CH_4} reaches 54.3%. The Y_{H_2} and Y_{CO} increased gradually with increasing input power. The transformation of methane to the final reforming product is promoted with the increase of energy density. By observing the experimental phenomena, it is found that a stable gliding arc cannot be formed when the input power is low. Oscillations of the arc occur only in parts of the area between the electrodes. As the input power increases, the discharge area expands. And the contact time between the arc and the reaction gas increases, it also contributes to the increase of C_{CH_4}, Y_{H_2} and Y_{CO}.

Effect of O/C

The O/C molar ratio is an important factor influencing the partial oxidation reforming process. According to the results in previous section, the discharge region is already well developed when the input power is at 27.9 W. Meanwhile, the rate of increase of H_2 yield and methane conversion slows down as the input power rises further. Therefore, 27.9 W was selected for the analysis of the O/C molar ratio. The variation of O_2 is realized by adjusting the flow rate of air. It was studied for O/C molar ratios of 0.7, 1, and 1.3. As shown in Fig. 17.5, the conversion of methane gradually increases with the increase of the O/C molar ratio. It increased from 38.3 to 45.9%. The Y_{CO} also increases gradually with the increase of O/C molar ratio. The rate of the methane oxidation reaction is promoted with the increase of O_2 in the reaction

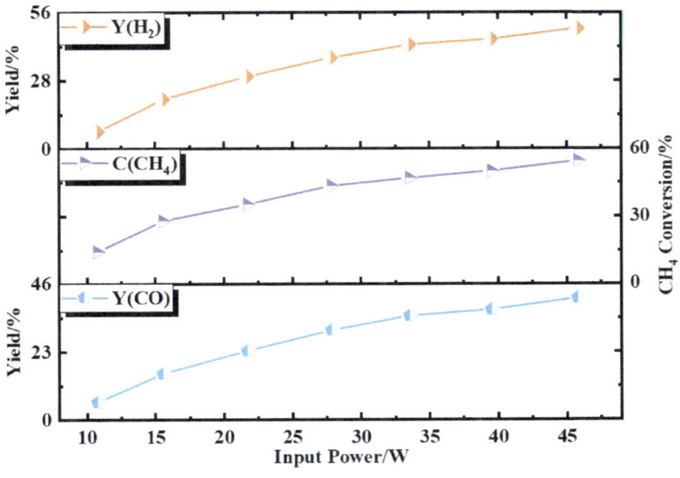

Fig. 17.4 Variation of yields and conversion with input power

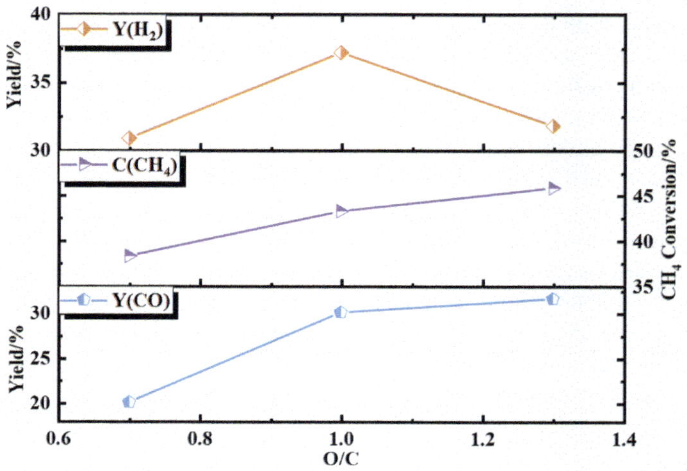

Fig. 17.5 Variation of yields and conversion with O/C

system. The Y_{H_2} shows an increasing and then decreasing trend with increasing O/C molar ratio. The conversion of methane to H_2 is promoted when O/C increases from 0.7 to 1. With a further increase of O, H_2 is gradually oxidized to H_2O, which reduces the hydrogen yield.

17.3.2 Effect of Reforming Products on Engine

Based on the results of Sect. 3.1, concentration of each component of the reforming products in the condition with an O/C molar ratio of 1 and an input power of 27.9 W was selected as the ratio of the reforming product gas supplied into the engine. H_2, CO, N_2, O_2, and unreformed CH_4 were considered to be the main components of the reforming products. The reforming ratio (R_{re}) is calculated according to Eq. (17.5).

$$R_{re} = \frac{m(reformed\ \mathrm{CH_4})}{m(\mathrm{CH_4}\ of\ base\ condition)} * 100\% \quad (17.5)$$

As the R_{re} increases, the temperature, pressure, and peak of heat release rate in the cylinder increase, as shown in Fig. 17.6. At the same time, the combustion process is accelerated and the exothermic interval is shifted forward as a whole. When the R_{re} is 15%, the peak of pressure in the cylinder rises by 12.7% compared to the condition without the addition of reforming gas. The peak of in-cylinder temperature increased by 8%. With the gradual increase of reforming gas, the percentage of H_2 increases, enhancing the flame propagation speed of the fuel in the cylinder. Besides, the addition of H_2 enhances the reactivity of the fuel in the cylinder. It contributes to

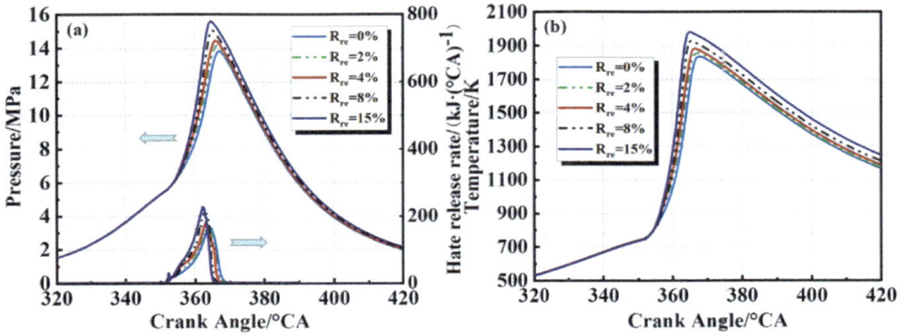

Fig. 17.6 Variation of **a** pressure and heat release rate, **b** temperature at different R_{re}

the production of more reactive radicals, such as H and OH, during the initial stage of combustion. This accelerates combustion and results in more complete combustion.

CA50 indicates the crank angle corresponding to the complete combustion of 50% mass-fraction fuel. It can represent the center of the combustion process in the cylinder. According to Fig. 17.7, CA50 gradually moves forward as the R_{re} increases. This also indicates that the addition of reforming gas promotes the combustion process and brings forward the combustion phase. When the R_{re} is 15%, CA50 advances to 1° CA after the top dead center (ATDC). Meanwhile, the indicated mean effective pressure (IMEP) of the engine gradually rises, and the power output capacity increases.

According to the results of Fig. 17.8, the ignition of n-heptane in the pre-chamber is advanced with the increase of the R_{re}. This is mainly due to the increase of H_2 concentration in the pre-chamber with the increase of R_{re}, which promotes the generation of OH and H radicals. This can shorten the ignition delay time of n-heptane. Since the H_2 concentration in the main chamber also increases, the propagation speed of the jet flame formed by n-heptane into the main chamber is also accelerated.

Fig. 17.7 Variation of IMEP and CA50

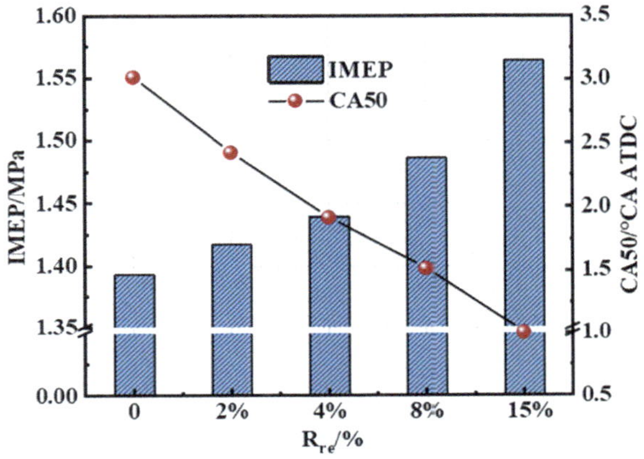

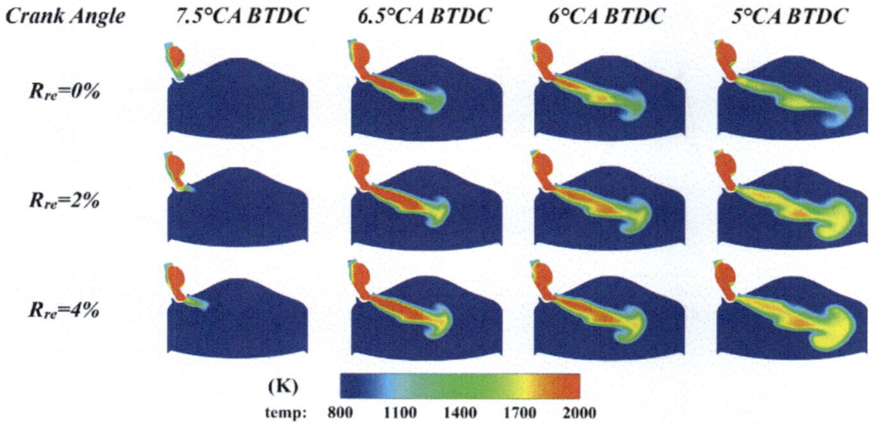

Fig. 17.8 Contours of temperature distribution at section A

As the concentrations of H_2 in the main combustion chamber increase, the combustion process is accelerated, as shown in Fig. 17.9a. The combustion phase of the engine is advanced and the combustion duration is shortened. OH is generally found in the high temperature region and their location can be a good reflection of the formation of the flame surface. According to the results in Fig. 17.9b, the flame propagation is accelerated as the R_{re} increases.

For natural gas engines, soot emissions are relatively low, so our primary focus is on the emissions of NO_x and unburned HC emissions. NO and NO_2 are the major NO_x emissions. According to Fig. 17.10a, the production of both NO and NO_2 increases with crank angle. And with the increase of R_{re}, its final generated mass also increases gradually. The final NO_x emissions increase with the increasing R_{re} but remain below the limit of Tier III standard according to Fig. 17.9c. This is mainly due to the fact that with the increase of H_2, the in-cylinder combustion rate is accelerated and the temperature rises leading to an increase in thermal-NO_x production. As the R_{re} increases, HC emissions decrease. At R_{re} of 4%, HC emissions are reduced by 10% compared to the condition without reforming gas addition as shown in Fig. 17.10b. This is mainly because as the R_{re} increases, the combustion in the cylinder is further improved. CH_4 is burned more entirely, and its final emissions are reduced. The primary component in unburned HC emissions is unburned CH_4. Its reduction contributes to lowering the greenhouse gas emissions from the engine. According to the NO distribution shown in Fig. 17.11, combined with Fig. 17.9a, it can be seen that the NO generation region is coupled with the high temperature region in the cylinder.

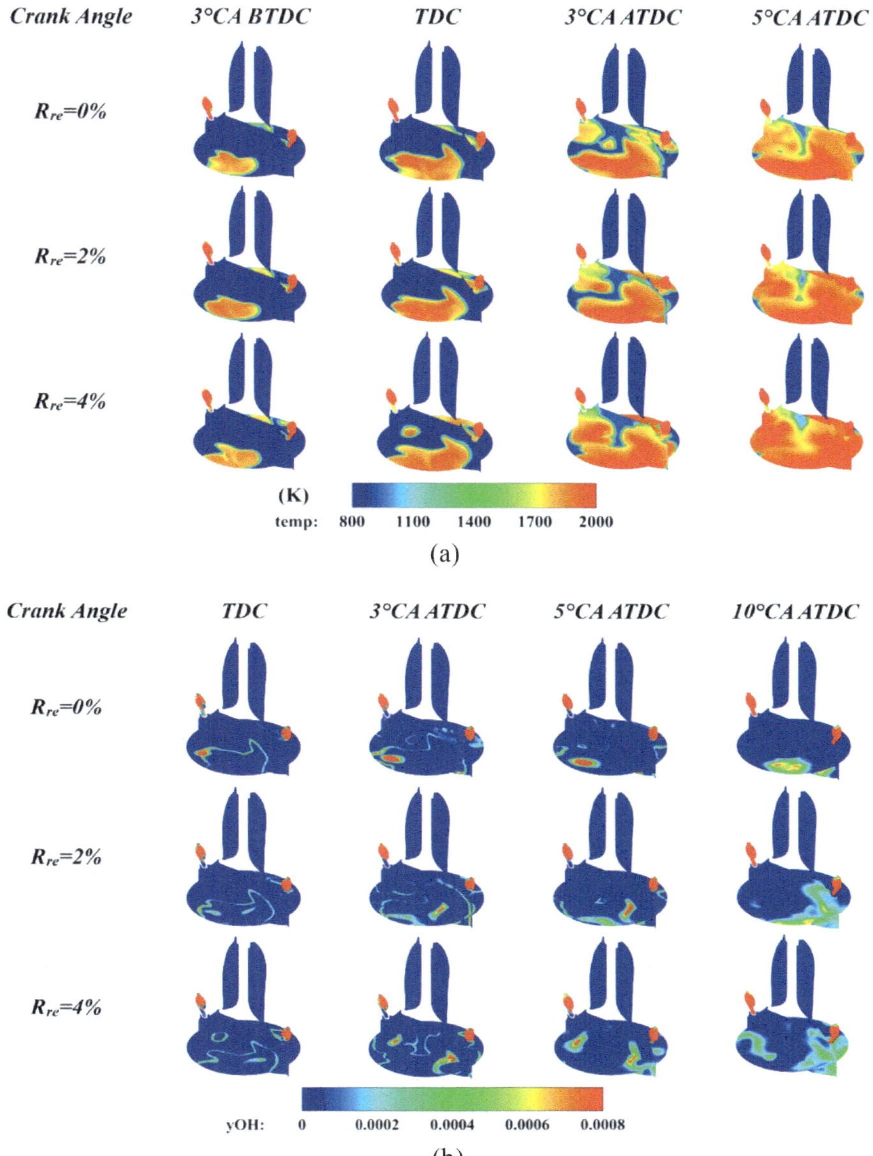

Fig. 17.9 Contours of **a** temperature and **b** OH mass fraction distribution at section B&C

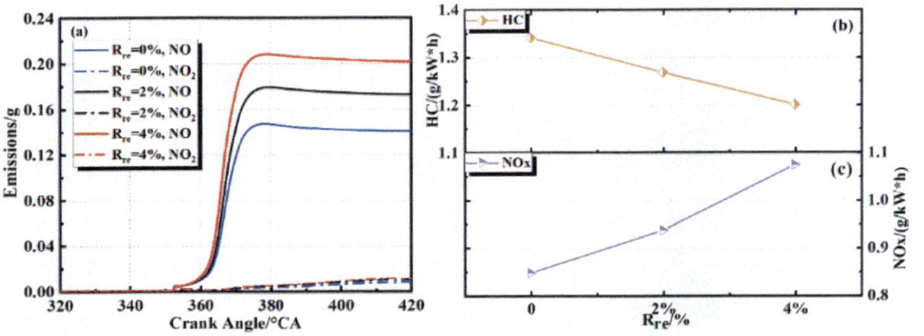

Fig. 17.10 Variation of **a** the mass of NO&NO_2, **b** HC emissions and **c** NO_x emissions

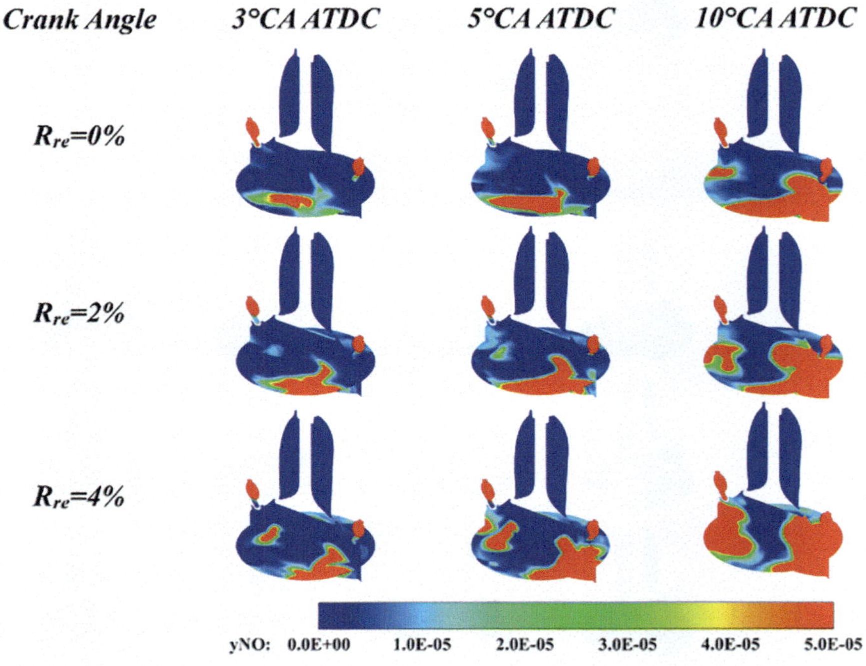

Fig. 17.11 Contours of NO mass fraction distribution at section B&C

17.4 Conclusion

This study investigated the effect of input power and O/C molar ratio on the reforming of methane by gliding arc plasma. Based on the components of the reforming products, their effects on the combustion and emissions of a large-bore low-speed marine engine were investigated using numerical simulations. Here are the main findings.

The products of plasma methane reforming in the partially oxidized condition are mainly H_2 and CO. As the input power increases, the methane conversion, the yields of hydrogen and CO yields increase. When the input power is 45.9 W, the ratio of methane conversion reaches 54.3%.

As the O/C molar ratio increases, the conversion of methane rises. However, an excess of O_2 will lead to a decrease in the yield of H_2, prompting the formation of H_2O from H.

The addition of methane plasma reforming products can effectively increase the in-cylinder temperature, pressure and peak of heat release rate. This can achieve an increase in the generation of highly reactive radicals, such as H and OH, which can enhance the intensity of jet flame and shorten the combustion duration.

The addition of reforming gas can reduce the engine's unburned HC emissions, especially methane. However, NO_x emissions rise due to the increase in-cylinder temperature. With the addition of 4% reforming product, NO_x emissions are 1.07 g/kW*h, which also meets the limit of Tier III.

Various strategies need to be explored in the future to realize the efficient operation of the coupled system of gliding arc plasma reforming and marine engines.

References

1. H. Liu, Z.-H. Meng, Z.-F. Lv, X.-T. Wang, F.-Y. Deng, Y. Liu et al., Emissions and health impacts from global shipping embodied in US–China bilateral trade. Nat. Sustain. **2**, 1027–1033 (2019). https://doi.org/10.1038/s41893-019-0414-z
2. N. Olmer, B. Comer, B. Roy, X. Mao, D. Rutherford, Greenhouse gas emissions from global shipping, 2013–2015 Detailed methodology (2017)
3. M.D. Kass, Z. Abdullah, M.J. Biddy, C. Drennan, Z. Haq, T. Hawkins, et al., Understanding the opportunities of biofuels for marine shipping. Oak Ridge National Lab. (ORNL), Oak Ridge, TN (United States (2018). https://doi.org/10.2172/1490575
4. Y. Lu, Z. Pan, F. He, G. Zhou, Research on the characteristics of the exhaust energy splitting method for the two-stroke low speed marine diesel engine. Appl. Therm. Eng. **237**, 121769 (2024). https://doi.org/10.1016/j.applthermaleng.2023.121769
5. Z. Lu, M. Liu, L. Shi, T. Wang, T. Lu, H. Wang, Numerical research of the injected exhaust gas recirculation strategy on a two-stroke low-speed marine diesel engine. Energy **244**, 122731 (2022). https://doi.org/10.1016/j.energy.2021.122731
6. Initial IMO GHG Strategy (n.d.). https://www.imo.org/en/MediaCentre/HotTopics/Pages/Reducing-greenhouse-gas-emissions-from-ships.aspx
7. L. Zhu, B. Li, A. Li, W. Ji, Y. Qian, X. Lu et al., Effects of fuel reforming on large-bore low-speed two-stroke dual fuel marine engine combined with EGR and injection strategy. Int. J. Hydr. Energy **45**, 29505–29517 (2020). https://doi.org/10.1016/j.ijhydene.2020.07.266
8. DieselNet: Engine Emission Standards (n.d.). https://dieselnet.com/standards/
9. W. Ji, A. Li, X. Lu, Z. Huang, L. Zhu, Numerical study on NO_x and ISFC co-optimization for a low-speed two-stroke engine via Miller cycle, EGR, intake air humidification, and injection strategy implementation. Appl. Therm. Eng. **153**, 398–408 (2019). https://doi.org/10.1016/j.applthermaleng.2019.03.035
10. Z. Huang, L. Zhu, A. Li, Z. Gao, Renewable synthetic fuel: turning carbon dioxide back into fuel. Front. Energy **16**, 145–149 (2022). https://doi.org/10.1007/s11708-022-0828-6

11. Z. Wang, X. Fu, D. Wang, Y. Xu, G. Du, J. You, A multilevel study on the influence of natural gas substitution rate on combustion mode and cyclic variation in a diesel/natural gas dual fuel engine. Fuel **294**, 120499 (2021). https://doi.org/10.1016/j.fuel.2021.120499
12. C. Hall, M. Kassa, Advances in combustion control for natural gas–diesel dual fuel compression ignition engines in automotive applications: a review. Renew. Sustain. Energy Rev. **148**, 111291 (2021). https://doi.org/10.1016/j.rser.2021.111291
13. T. Kar, Z. Zhou, M. Brear, Y. Yang, M. Khosravi, J. Lacey, A comparative study of directly injected, spark ignition engine performance and emissions with natural gas, gasoline and charge dilution. Fuel **304**, 121438 (2021). https://doi.org/10.1016/j.fuel.2021.121438
14. C.D. Huang, A.G. Riess, S.L. Hoffmann, C. Klein, J. Bloom, W. Yuan et al., A near-infrared period–luminosity relation for miras in NGC 4258, an anchor for a new distance ladder. APJ **857**, 67 (2018). https://doi.org/10.3847/1538-4357/aab6b3
15. X. Duan, B. Deng, Y. Liu, S. Zou, J. Liu, R. Feng, An experimental study the impact of the hydrogen enrichment on cycle-to-cycle variations of the large bore and lean burn natural gas spark-ignition engine. Fuel **282**, 118868 (2020). https://doi.org/10.1016/j.fuel.2020.118868
16. X. Wu, K. Liu, Q. Liu, J. Fu, J. Liu, Effects of direct water injection timings on knock suppression, combustion, and emission performance of high compression ratio hydrogen-enriched natural gas engine. Energy Conver. Manag. **250**, 114887 (2021). https://doi.org/10.1016/j.enconman.2021.114887
17. H. Yu, J. Chen, S. Duan, P. Sun, W. Wang, H. Tian, Effect of natural gas injection timing on performance and emission characteristics of marine low speed two-stroke natural gas/diesel dual-fuel engine at high load conditions. Fuel **314**, 123127 (2022). https://doi.org/10.1016/j.fuel.2021.123127
18. S. Di Iorio, P. Sementa, B.M. Vaglieco, Experimental investigation on the combustion process in a spark ignition optically accessible engine fueled with methane/hydrogen blends. Int. J. Hydr. Energy **39**, 9809–9823 (2014). https://doi.org/10.1016/j.ijhydene.2014.04.065
19. E. Arslan, N. Kahraman, The effects of hydrogen enriched natural gas under different engine loads in a diesel engine. Int. J. Hydr. Energy **47**, 12410–12420 (2022). https://doi.org/10.1016/j.ijhydene.2021.09.016
20. F. Ma, M. Wang, L. Jiang, R. Chen, J. Deng, N. Naeve et al., Performance and emission characteristics of a turbocharged CNG engine fueled by hydrogen-enriched compressed natural gas with high hydrogen ratio. Int. J. Hydr. Energy **35**, 6438–6447 (2010). https://doi.org/10.1016/j.ijhydene.2010.03.111
21. C. Zamfirescu, I. Dincer, Ammonia as a green fuel and hydrogen source for vehicular applications. Fuel Process. Technol. **90**, 729–737 (2009). https://doi.org/10.1016/j.fuproc.2009.02.004
22. A.M. Abdalla, S. Hossain, O.B. Nisfindy, A.T. Azad, M. Dawood, A.K. Azad, Hydrogen production, storage, transportation and key challenges with applications: a review. Energy Conver. Manag. **165**, 602–627 (2018). https://doi.org/10.1016/j.enconman.2018.03.088
23. M.G. Rasul, M.A. Hazrat, M.A. Sattar, M.I. Jahirul, M.J. Shearer, The future of hydrogen: challenges on production, storage and applications. Energy Conver. Manag. **272**, 116326 (2022). https://doi.org/10.1016/j.enconman.2022.116326
24. D. Wen, S. Liu, Z. Ning, M. Aziz, Techno-economic evaluation of hydrogen and ammonia as energy carriers in a multi-generation system. Energy Conver. Manag. **277**, 116670 (2023). https://doi.org/10.1016/j.enconman.2023.116670
25. W. Xu, Z. Zhang, Y. Huang, H. Qian, G. Li, Effect of catalyst shapes on hydrogen production from exhaust gas-fuel reforming for NG engine: a mesoscale 3D simulation. Fuel **348**, 128492 (2023). https://doi.org/10.1016/j.fuel.2023.128492
26. K. Toshinaga, N. Izumi, T. Shudo, A study on fuel reforming waste heat recovery engine system fueled with methane. Appl. Therm. Eng. **235**, 121325 (2023). https://doi.org/10.1016/j.applthermaleng.2023.121325
27. S. Sengodan, R. Lan, J. Humphreys, D. Du, W. Xu, H. Wang et al., Advances in reforming and partial oxidation of hydrocarbons for hydrogen production and fuel cell applications. Renew. Sustain. Energy Rev. **82**, 761–780 (2018). https://doi.org/10.1016/j.rser.2017.09.071

28. K.-W. Lin, H.-W. Wu, Thermodynamic analysis and experimental study of partial oxidation reforming of biodiesel and hydrotreated vegetable oil for hydrogen-rich syngas production. Fuel **236**, 1146–1155 (2019). https://doi.org/10.1016/j.fuel.2018.09.090
29. T. Blumberg, Y.D. Lee, T. Morosuk, G. Tsatsaronis, Exergoenvironmental analysis of methanol production by steam reforming and autothermal reforming of natural gas. Energy **181**, 1273–1284 (2019). https://doi.org/10.1016/j.energy.2019.05.171
30. R. Peters, J. Pasel, R.C. Samsun, F. Scharf, A. Tschauder, D. Stolten, Heat exchanger design for autothermal reforming of diesel. Int. J. Hydr. Energy **43**, 11830–11846 (2018). https://doi.org/10.1016/j.ijhydene.2018.03.085
31. H. Wang, Y. Yang, Z. Li, X. Kong, P. Martin, G. Cui et al., Plasma-assisted Ni catalysts: toward highly-efficient dry reforming of methane at low temperature. Int. J. Hydr. Energy **48**, 8921–8931 (2023). https://doi.org/10.1016/j.ijhydene.2022.11.287
32. P. Navascués, J. Cotrino, A.R. González-Elipe, A. Gómez-Ramírez, Plasma assisted dry reforming of methane: Syngas and hydrocarbons formation mechanisms. Fuel Process. Technol. **248**, 107827 (2023). https://doi.org/10.1016/j.fuproc.2023.107827
33. Y. Xia, N. Lu, J. Li, N. Jiang, K. Shang, Y. Wu, Combined steam and CO_2 reforming of CH_4 for syngas production in a gliding arc discharge plasma. J. CO2 Utiliz. **37**, 248–259 (2020). https://doi.org/10.1016/j.jcou.2019.12.016
34. B. Wang, Y. Cheng, C. Wang, J. Zou, Steam reforming of methane in a gliding arc discharge reactor to produce hydrogen and its chemical kinetics study. Chem. Eng. Sci. **253**, 117560 (2022). https://doi.org/10.1016/j.ces.2022.117560
35. H. Zhang, C. Du, A. Wu, Z. Bo, J. Yan, X. Li, Rotating gliding arc assisted methane decomposition in nitrogen for hydrogen production. Int. J. Hydr. Energy **39**, 12620–12635 (2014). https://doi.org/10.1016/j.ijhydene.2014.06.047
36. D.P. Schmidt, C.J. Rutland, A new droplet collision algorithm. J. Comput. Phys. **164**, 62–80 (2000). https://doi.org/10.1006/jcph.2000.6568
37. P.J. O'Rourke, A.A. Amsden, A spray/wall interaction submodel for the KIVA-3 wall film model, p. 2000-01–0271 (2000). https://doi.org/10.4271/2000-01-0271
38. Z. Han, R.D. Reitz, Turbulence modeling of internal combustion engines using RNG κ–ε models. Combust. Sci. Technol. **12**, 45–99 (2007). https://doi.org/10.1080/00102209508907782
39. P.K. Senecal, E. Pomraning, K.J. Richards, T.E. Briggs, C.Y. Choi, R.M. McDavid et al., Multi-dimensional modeling of direct-injection diesel spray liquid length and flame lift-off length using CFD and parallel detailed chemistry. SAE Trans. **112**, 1331–1351 (2003)
40. J.C. Beale, R.D. Reitz, Modeling spray atomization with the kelvin-helmholtz/rayleigh-taylor hybrid model. AAS **9**, 40 (1999). https://doi.org/10.1615/AtomizSpr.v9.i6.40

Chapter 18
Plasma-Enabled Liquefaction of Cellulose and Lignin as Lignocellulosic Biomass Model Compounds

Qing Chen, Jingyi Ma, Shiyun Liu, Meiling Cai, Danhua Mei, and Zhi Fang

Abstract Low-temperature plasma has shown the promising performance in biomass liquefaction with short liquefaction time and high liquefaction yield. In this study, we investigated the performance of plasma-enabled liquefaction of cellulose and lignin as lignocellulosic biomass model compounds. Different cellulose-to-lignin mass ratios were set to determine the optimum condition from the viewpoints of biomass raw materials. Elemental analysis, infrared spectroscopy, thermogravimetric analysis, and gas chromatography–mass spectrometry were performed to analyze the solid residues and liquid products obtained under optimum condition. Results showed that the highest liquefaction yield (98.8%) was achieved at a cellulose-to-lignin mass ratio of 2/3 and a liquefaction time of 3 min. Abundant hydroxyl groups, as well as functional groups such as carbon–hydrogen bonds and carboxyl groups, were detected in the liquid products. The liquid products contained a rich variety of alcohols, along with traces of organic acids, lipids, and aromatic compounds. These liquid products exhibit a promising potential to act as the cleaning fuels.

18.1 Introduction

Biomass refers to all organic substances derived from green plants by photosynthesis, which constitutes a renewable and clean energy source. Due to the unique characteristics of abundance and cleanness, biomass energy exhibits a promising potential for replacing fossil fuel energy sources in the future [1]. Moreover, carbon dioxide generated from the combustion of biofuels is recycled in the process of photosynthesis, which can achieve the goal of carbon neutrality. Biomass acts in an important position for the development of the circular economy, especially lignocellulosic biomass which is abundant, diverse and easy to collect.

Q. Chen · J. Ma · S. Liu · M. Cai · D. Mei (✉) · Z. Fang
College of Electrical Engineering and Control Science, Nanjing Tech University, Nanjing 211816, China
e-mail: danhuam@126.com

Z. Fang et al. (eds.), *Proceedings of the 5th International Symposium on Plasma and Energy Conversion*, Springer Proceedings in Physics 398,
https://doi.org/10.1007/978-981-97-2245-7_18

Lignocellulosic biomass is composed of three natural polymers (e.g., cellulose, hemicellulose and lignin). It has garnered significant attention as a raw material for chemical industry due to its abundant annual production and low emissions of harmful gases such as NO_x and SO_x [2]. However, due to its complicated components and the considerable variations in the composition of different lignocellulosic biomass sources, it poses substantial challenges to the conversion processes for the application of lignocellulosic biomass.

Thermochemical conversion has been regarded as a promising method for converting lignocellulosic biomass into fuels, chemicals and energy materials. The thermochemical conversion methods for lignocellulosic biomass include gasification, pyrolysis and liquefaction. Liquefaction is performed under relatively mild reaction conditions and is the only method capable of producing transportable liquid fuels. Direct liquefaction techniques, including hydrothermal liquefaction, atmospheric catalytic liquefaction and ultrasound/microwave-assisted liquefaction [3–5], have attracted great attention due to the mild reaction conditions, generating products with high added-value and promising overall efficiency [6, 7]. However, these methods usually suffer from the drawbacks of stringent reaction conditions, long liquefaction time and high energy consumption. Therefore, there is a need to explore energy-efficient liquefaction methods with high liquefaction rate.

Low-temperature plasma represents a promising option for the conversion and utilization of biomass. It is rich in high-energy particles such as energetic electrons and reactive free radicals, which enable it to occur at room temperature and have the properties of fast startup and quick response [8–10]. Liu et al. developed a plasma-catalyzed liquefaction system for rapid liquefaction of wood chips. The liquefaction process was performed using a co-solvent made up of polyethylene glycol 200 (PEG 200) and glycerol and the acidic catalyst of H_2SO_4. The impact of processing parameters on liquefaction yield and solution temperature were investigated, and the results confirmed that the enhanced liquefaction performance was associated with the advantages of electric field heating and the presence of abundant reactive species during the discharge plasma process [11]. Mei et al. performed plasma liquefaction of three typical lignocellulosic materials—sawdust, corn cobs and rice straw, which have different feedstock compositions. They found that the chemical composition of lignocellulosic biomass, particularly the content of lignin and ash, showed important impacts on the liquefaction yield, product distribution and quality. They also performed chemical analysis to deduce potential reaction pathways for the liquefaction of lignocellulosic biomass [12]. Previous investigations indicated that the plasma liquefaction process is significantly influenced by power supply parameters, reaction conditions and the composition of lignocellulosic biomass. However, there is still limited research on how to determine the optimum liquefaction condition from the viewpoints of raw materials, which restricts the further improvement in the liquefaction performance. The lignocellulosic biomass contains different contents of cellulose, hemicellulose and lignin, which significantly affects the reaction pathway and performance of plasma liquefaction. Therefore, it is benefit to perform the plasma liquefaction of each composition as model compounds of lignocellulosic biomass, which will help us obtain the optimum reaction condition for lignocellulosic biomass liquefaction using plasma technology.

Herein, we carried out the plasma-enabled liquefaction of cellulose and lignin as the model compounds of lignocellulosic biomass. These two compounds were mixed in different mass ratios. The liquefaction processes were performed under optimal reaction conditions determined in our previous work [11]. The elemental composition, functional groups, thermal properties, and compositions in the liquid products were investigated thoroughly to understand the influence of lignocellulosic biomass components on the liquefaction process.

18.2 Experimental

18.2.1 Materials

Cellulose powder and dealkalized lignin were used the raw materials for plasma-enabled liquefaction. Chemically pure grade Polyethylene glycol 200 (PEG 200) and analytical grade glycerol were mixed to form the liquefaction solvent. Analytical grade concentrated sulfuric acid was selected as the catalyst. Chemically pure grade dichloromethane was utilized to filter and wash the remaining liquid in the solid residues.

18.2.2 Experimental Setup and Processes

Figure 18.1 presents the experimental setup for plasma-enabled liquefaction. A DC power source (MAO-10) served as the excitation power source. The power conditions of the voltage, the duty cycle and the frequency were set at 700 V, 50% and 500 Hz, respectively. The plasma-enabled liquefaction reaction occurred in a 250 ml three-neck flask. Tungsten rods (99% purity, diameter: 1.5 mm, length: 200 mm) were selected as two electrodes, which were connected to both ends of the power source. These two electrodes were fixed using insulating rubber stoppers, and the gap distance between them was set at 10 mm. The vertical distance from the bottom of the reactor to the head of the tungsten rod was 10 mm. Table 18.1 shows the reaction conditions in detail for the plasma-enabled liquefaction process. In this work, the total mass of cellulose and lignin was 5 g. The cellulose-to-lignin mass ratio was set as 1/4, 2/3, 3/2 and 4/1 by altering the mass of component. The plasma-enabled liquefaction of pure cellulose and lignin was also performed for comparison.

After liquefaction, the liquefied product was diluted using dichloromethane (analytical grade) to reduce viscosity. The obtained solid–liquid mixture was separated with a vacuum filtration apparatus to separate unreacted raw materials and other solid residues. The residue was then washed with dichloromethane continuously to

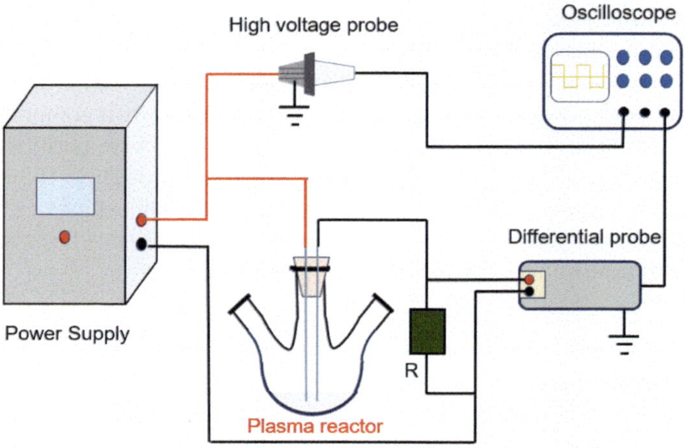

Fig. 18.1 Schematic diagram of the experimental setup for plasma-enabled liquefaction

Table 18.1 Reaction conditions for the plasma-enabled liquefaction process

Catalyst percentage content	Liquid–solid ratio	PEG-200/ glycerol molar ratio	Material mass	Cellulose-to-lignin mass ratio
1.0 wt.%	7	1.38	5	1/4, 2/3, 3/2, 4/1

form a transparent and colorless filtrate. After that, the residue was then transferred to an evaporation dish and dried in an oven at 105 °C for 12 h, followed by weighing to determine the solid residue mass. The liquid product was collected by using a rotary evaporator to remove dichloromethane, which was introduced during the washing process of the filter residue, at a temperature of 60 °C. For plasma-enabled liquefaction, the liquefaction yield is commonly used as an indicator for evaluating the performance of liquefaction reaction. It is calculated based on the percentage of solid residue mass to the raw material mass (Eq. (18.1)):

$$f(\%) = 100 - \frac{m_1}{m_2} \times 100 \tag{18.1}$$

where f is liquefaction yield, while m_1 and m_2 is the mass of solid residue and raw material, respectively.

18.2.3 Material Characterizations

Elemental analysis was conducted using the Vario EL Cube elemental analyzer to determine the content of C, H and N, while the O content was obtained by the difference method. Dulong equation was used calculated the high heating value (HHV) of the raw material, solid residue and liquid products [13].

Chemical qualitative analysis of the raw materials, solid residues, and liquid products was conducted using IRAffinity-1 Fourier Transform Infrared Spectroscopy. Thermal gravimetric analysis, a technique to assess the thermal stability of samples by measuring weight changes during heating, was performed using a Mettler Toledo TGA2 thermogravimetric analyzer with air as the protective gas to compare the thermal performance of the raw materials and solid residues. An Agilent gas chromatography-mass spectrometer (GCMS, 7890A-5975C series) was employed for qualitative analysis of the liquid products. The GCMS was equipped with an HP-5 capillary column and helium was used as the carrier gas. In the analysis process, temperatures for the injection port, ion source, and quadrupole were fixed at 250, 230 and 150 °C, respectively. The temperature program started at an initial temperature hold at 50 °C for 5 min, followed by a ramp to 90 °C at 5 °C/min with a 3-min hold, and a final ramp to 250 °C at 10 °C/min with a 10-min hold. The split ratio was set at 10:1, and the injection volume was 1 μL. The standard mass spectra from National Institute of Standards and Technology (NIST) library were used to determine the liquid compounds from the liquefaction process.

18.3 Results and Discussion

18.3.1 Liquefaction Yield

Figure 18.2 illustrates the liquefaction yield from the plasma-enabled liquefaction process for pure lignin. After the liquefaction reaction running for 1 min, the liquefaction yield was only 70.6%. When the liquefaction time increase to 3 min, the majority of lignin was liquefied, which results in a liquefaction yield of 97.5%. However, when the liquefaction time further increased to 4 min, the liquefaction yield decreased to 89.0%. In the separation process of liquefied products, longer filtering time was required for the products obtained from the liquefaction process with longer liquefaction time, and the products under the corresponding condition showed high viscosity. This phenomenon is attributed to the condensation reactions in the solution when prolonging the liquefaction reaction time, which generated more sugar-containing substances [14].

The variation in the liquefaction yield of pure cellulose by extending the liquefaction time was shown in Fig. 18.3. Clearly, the liquefaction yield enhanced quickly in the first two minutes. After that, the liquefaction yield was slightly increased to 90.5% by further extending the liquefaction time to 4 min. Compared with Fig. 18.2,

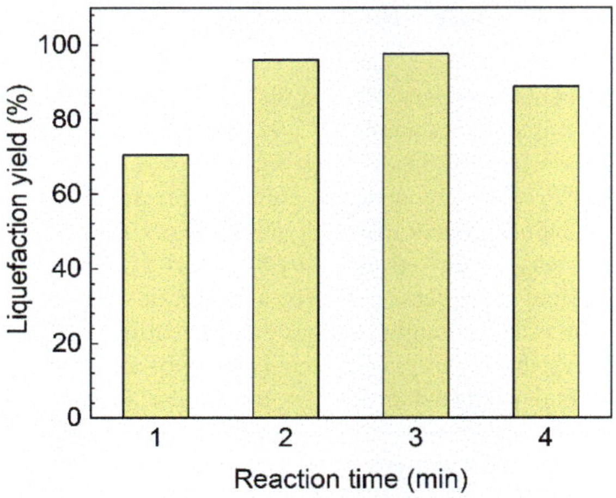

Fig. 18.2 Liquefaction yield of pure lignin with the variation of reaction time

the liquefaction yield of pure cellulose was generally lower than that of pure lignin under all reaction conditions in this work.

The lignocellulosic biomass have different contents of lignin and cellulose. In order to evaluate their influence on the liquefaction of actual lignocellulosic biomass, we performed the liquefaction of the mixture of lignin and cellulose with different cellulose-to-lignin mass ratios. The corresponding liquefaction yield is shown in Fig. 18.4, which exhibited a continuous decline trend by increasing the cellulose content when the reaction time was 1 and 2 min. When the reaction time was 3 min, it increased firstly and then declined when the cellulose-to-lignin mass ratio was increased from 1/4 to 4/1. While in the liquefaction reaction for 4 min, the liquefaction yield was slightly increased by raising the cellulose-to-lignin mass ratio. Overall, the

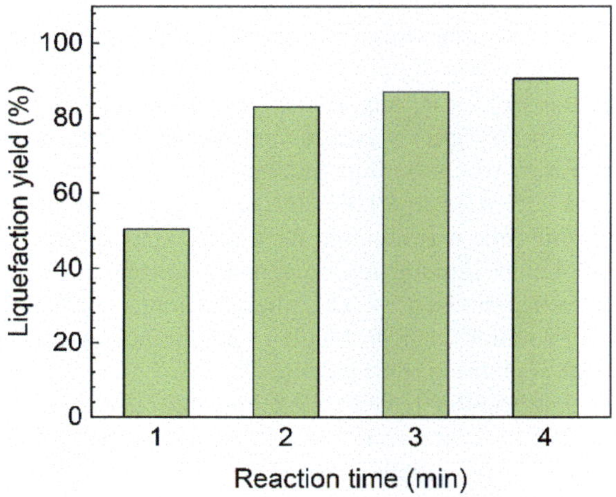

Fig. 18.3 Liquefaction yield of pure cellulose with the variation of reaction time

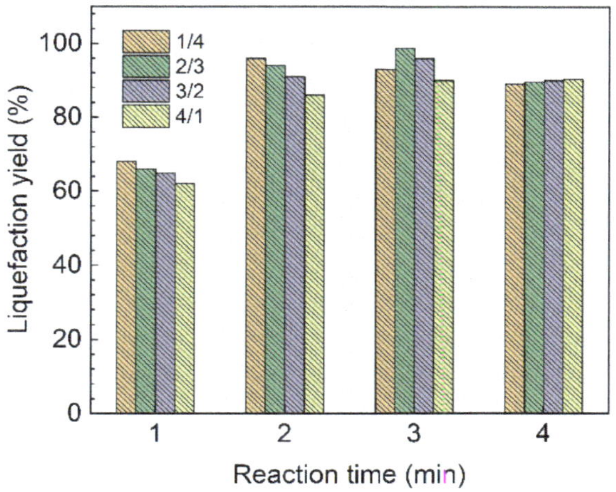

Fig. 18.4 Variations in the liquefaction yield of the mixture with different cellulose-to-lignin mass ratios

highest liquefaction yield of 98.8% was achieved at a liquefaction time of 3 min and a cellulose/lignin mass ratio of 2/3. In the early stage of liquefaction process, the solution temperature in the reaction system was relatively low, and lignin began to react, which is due to the broad reaction temperature range of lignin. However, higher temperatures are required for the reaction of cellulose. Therefore, as the proportion of cellulose content in the raw material solution increased, more residual residue remained after the reaction, which led to a poor liquefaction performance. When the reaction time exceeded 3 min, the solution temperature reached to a high level [14], and more cellulose was liquefied. The above phenomenon was precisely confirmed by the liquefaction performance of pure lignin and cellulose as shown in Figs. 18.2 and 18.3.

18.3.2 Elemental Analysis

Table 18.2 provides the elemental composition and corresponding HHV of the raw materials (cellulose and lignin), as well as the solid residues and liquid products under optimal conditions in Fig. 18.4. It is obvious that carbon (C) and oxygen (O) are the major composition in both cellulose and lignin. For the elements with small content, the hydrogen (H) content is significantly higher than that of nitrogen (N) and sulfur (S). These element compositions favoured the conversion of lignocellulosic biomass into hydrocarbons with high molecular weight.

Both cellulose and lignin exhibited relatively high oxygen content, indicating the presence of functional groups such as methoxy, hydroxy and ether bonds in their structures. After the liquefaction reaction, the liquid product showed a carbon content of 49.0%, significantly higher than that in the residue (40.9%), while the oxygen content in the liquid product (43.8%) was notably lower than that in the

Table 18.2 Elemental composition and HHV of raw material, solid residue and liquid products

	C (%)	H (%)	O (%)	N (%)	S (%)	HHV (MJ/kg)
Cellulose	41.04	6.54	52.4	0	0.02	13.85
Lignin	49.42	4.96	41.47	0.06	4.09	16.38
Solid residue	40.87	6.62	52.42	0	0.08	13.92
Liquid products	49.04	6.36	43.82	0.14	0.64	17.84

residue (52.4%). The hydrogen content in both solid residue and liquid products is similar around 6%. Compared to solid residues, the increase in carbon content of the liquid products suggested that more carbon is fixed in the liquid products. While the reduction in oxygen content indicated that deoxygenation reactions occurred. The above phenomenon contributed to the enhanced quality of the bio-oil, which was demonstrated by the HHV of liquefied products17.8 MJ/kg, higher than that of raw materials and solid residue.

18.3.3 FTIR Analysis

To investigate the chemical changes in cellulose/lignin as raw materials before and after plasma-enabled liquefaction, Fourier Transform Infrared Spectroscopy (FTIR) was employed for characterization and comparative analysis. Figure 18.5 displays the FTIR spectra of cellulose and lignin before liquefaction. The spectra show that cellulose and lignin share common functional groups but also possess unique functional groups individually. Clearly, the common absorption peaks were observed at the wavelengths of 3500, 2900, 1370 and 1060 cm^{-1}. The broadest absorption peak at 3500 cm^{-1} corresponds to the O–H stretching vibration band, indicating the presence of hydroxyl groups [15]. The absorption peak at 2900 cm^{-1} corresponds to the C–H stretching vibration band, with the corresponding characteristic groups of methyl and methylene [16]. The absorption peak at 1370 cm^{-1} is associated with the C–H bending vibration band, which is characteristic group of the aromatic ring structure [17]. The absorption peak at 1060 cm^{-1} corresponds to C–O stretching and O–H bending vibration bands, which represents the secondary alcohol and ether functional groups [18]. Lignin-specific the absorption peaks unique to lignin are located at 1610–1600 and 1520–1500 cm^{-1}, which represent the aromatic ring skeleton vibrations [19]. In this work, the absorption peaks at 1600 and 1510 cm^{-1} in the spectrum of lignin confirmed the presence of lignin. In contrast, cellulose has a unique characteristic peak at 897 cm^{-1}, corresponding to the β-glucosidic linkage [20]. If the absorption peak at 897 cm^{-1} weakens or disappears, it suggests the breakage of the cellulose molecular chain.

Figure 18.6 presents the FTIR spectra of the solid residues and liquid products. Clearly, both solid residues and liquid products exhibited characteristic peaks related to hydroxyl groups (3400 cm^{-1}), carbon-hydrogen bonds (2900, 1370 cm^{-1}), and

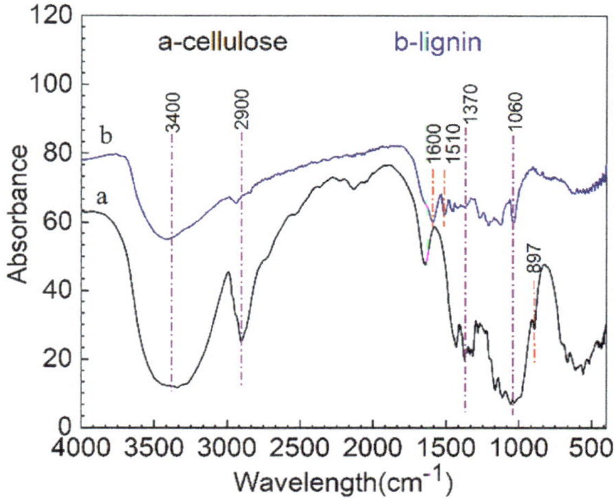

Fig. 18.5 Infrared spectra of cellulose and lignin before liquefaction reaction

carboxyl groups (1060 cm^{-1}). In the range of 3200 ~ 3700 cm^{-1}, a strong –O–H stretching vibration absorption peak indicates the presence of carboxylic acids, ketones, aldehydes, and esters in the liquid products. Comparison with the solid residues, the absorption peak at 1680 cm^{-1} in the spectrum of the liquid products corresponds to the C=O stretching vibration. This is attributed to the formation of carbonyl groups resulting from the degradation of lignin during the liquefaction process [21]. The absorption peak at 1300 cm^{-1} in the spectrum of the liquid products is attributed to the C–O–C stretching vibration of ether groups, which is generated not only from the raw materials but also from the products of dehydration and condensation reactions between raw materials and solvents [22]. From Figs. 18.5 and 18.6, it is evident that complex decarboxylation reactions occurred. The liquid products contain numerous oxygen-containing functional groups, which indicated that deoxygenation is required to improve the quality of the obtained bio-oil.

18.3.4 TG/DTG Analysis

To investigate the reaction behaviour of cellulose and lignin at high temperatures, weight loss curves of the respective raw material and solid residues under the optimal condition for the liquefaction of cellulose/lignin mixture were obtained by TG-DTG analysis. Figure 18.7 presents the TG-DTG curves for the individual thermal decomposition of cellulose and lignin. Significant difference was observed in the thermal stability of cellulose and lignin. In the first step when the temperature increased from 30 to 300 °C, cellulose underwent a dehydration stage, which resulted in a relatively gradual weight loss curve. In the temperature range of 300–370 °C, the TG curve for cellulose rapidly dropped from 96 to 10% with a weight loss rate of 1.23%/°C, while the DTG curve in this range exhibited a steep weight loss peak. This is attributed to

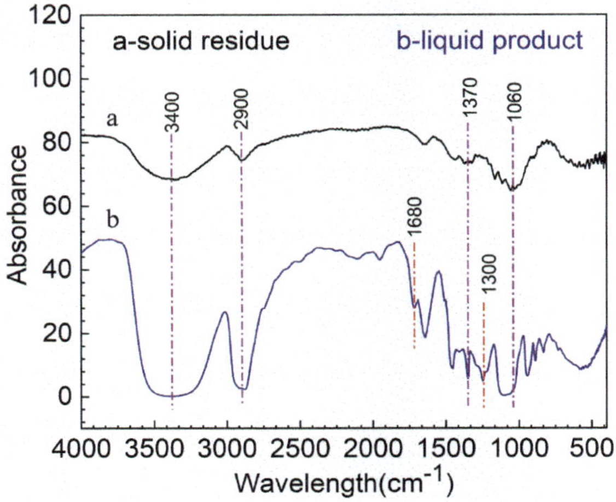

Fig. 18.6 Infrared spectra of solid residues and liquid products after liquefaction of cellulose/lignin mixture (mass ratio of 2/3)

the fact that cellulose is a linear and long-chain polymer compound. It is composed of D-glucose units linked by β-1-4 glycosidic bonds with a high degree of polymerization exceeding 7000, which leads the high molecular weight and strong thermal stability [23].

The temperature range for the weight loss of lignin was significantly wider than that of cellulose. The DTG curve for lignin reveals that there existed three main stages in the thermal decomposition of lignin, denoted as I, II, and III, respectively. In stage I, the DTG peak is relatively small, which is associated to the removal of moisture, resulting in a minor weight loss rate from 0 to 5%. In Stage II, the volatilization of lignin started, which resulted in a weight loss of 55%. This represented the main stage of lignin pyrolysis with the broadest temperature increase. In stage III, lignin was thermal cracked, and the weight loss peak was much sharper than that in the former two stages [24].

The weight loss of cellulose was around 99.5% when the temperature reached 800 °C, suggesting the complete degradation. In contrast, the weight loss of lignin was only 58.0%. This is mainly because the fact that lignin is a highly polymerized substance with a three-dimensional spatial structure. It exhibits higher stability than cellulose and is hard to be degraded completely. The unreacted portion of lignin would contribute to a higher content of char in the products.

Figure 18.8 presents the TG-DTG curves of the solid residues obtained under the optimal liquefaction condition. Typically, the entire pyrolysis process is divided into three stages, namely the heating stage, the rapid weight loss stage, and the slow weight loss stage. In the first stage of residue pyrolysis, the temperature increased from the initial value to 180 °C, and the mass loss was primarily due to the water loss in the residue. In the second stage, the temperature increased from 180 to 550 °C, which reached the critical temperature for thermal weight loss of the residue. The weight loss of the residue increased from 2 to 85% in this temperature range, which indicated a reduction of cellulose within the residue. This reduction is attributed

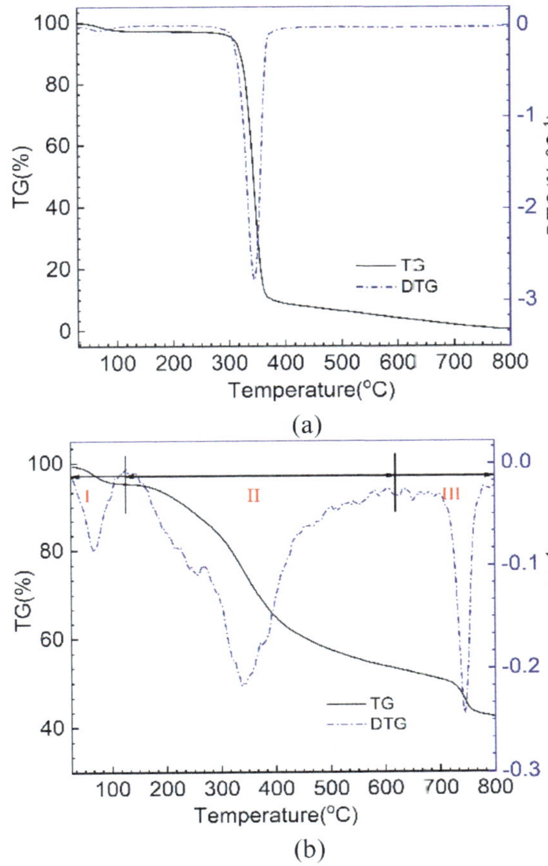

Fig. 18.7 TG and DTG curves of raw materials: **a** cellulose, **b** lignin

to incomplete liquefaction reactions resulting in residual cellulose. The DTG curve indicated that the thermal weight loss of the residue peaked in the temperature range of 180–300 °C, where volatile and organic small molecules were degraded. However, non-volatile large molecules were degraded in the temperature range of 300–550 °C. In the final stage within the temperature range of 550–800 °C, the negligible changes were observed in the TG-DTG curves, and the remaining substance mainly contained ash and some undecomposable carbon. The weight loss of the solid residue was 88% when the temperature was further increased to 800 °C. This value suggested that some undecomposed substances remained, which corresponded to the unreacted lignin in the raw material.

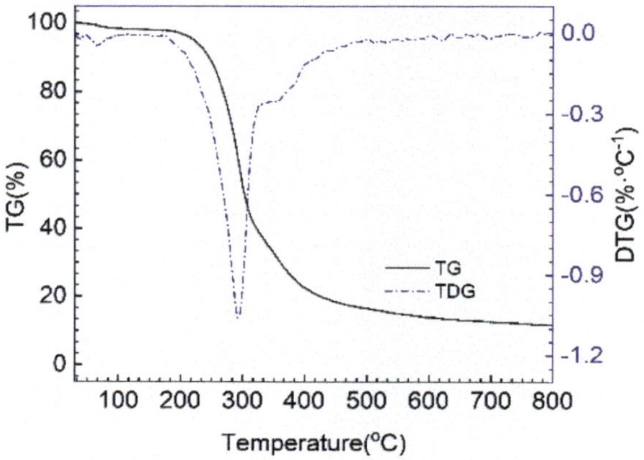

Fig. 18.8 TG and DTG curves of solid residue

18.3.5 GC–MS Analysis

The GC–MS analysis was performed to identify qualitatively the liquid products obtained from the plasma-enabled liquefaction of cellulose/lignin with a mass ratio of 2/3. Table 18.3 shows the typical compounds from the liquefaction process. These compounds mainly contained PEG 200, glycerol and their derivatives. Additionally, several new compounds were identified including organic acids, alcohols, ethers, esters, aldehydes, and phenolic aromatic compounds. The detection of aromatic hydrocarbons indicates the decomposition of lignin during the liquefaction process [25]. The results in Table 18.3 revealed that the plasma-enabled liquefaction led to the degradation of cellulose to form organic acids and aldehydes, which would further react to form esters.

Compared with the traditional liquefaction methods, the liquid products from plasma process exhibited fewer types but higher concentrations. The liquid products were rich in a variety of alcohol compounds, and all of them contain hydroxyl groups. The liquid products rich in hydroxyl groups can be directly used in agriculture and industrial applications, such as the production of polyurethane foam and polyurethane adhesives. Separation and purification methods, such as fractional extraction and thin-layer chromatography could be employed to obtain the desired liquid products [26].

Table 18.3 Typical compounds in the liquid products from the liquefaction of cellulose/lignin mixture

	Peak time	Compounds	Percentage of peak area
1	6.53	Glafenin	0.3655
2	8.03	1,3-Dioxolane-4-methanol,2-ethenyl-	2.378
3	9.12	1,3-Dioxolane-4-methanol, 2,2-dimethyl-	2.442
4	13.65	4-Pentenoic acid,3-hydroxy-,ethyl ester	4.981
5	16.08	Glycerin	17.25
6	16.84	Ethyl 2-butoxyacetate	2.937
7	16.95	Bis(2-chloro-1-methylethyl) ether	2.073
8	18.76	2-Propanol,1-propoxy-	1.892
9	19.98	12-Crown-4	1.822
10	22.01	Hydroxyethoxy)ethoxy]ethoxy]ethoxy]ethoxy]ethoxy]ethoxy]ethanol	4.37
11	22.08	15,15'-Bi-1,4,7,10,13-pentaoxacyclohexadecane	2.49
12	22.48	1,8-Naphthyridine, 2,4,7-trimethyl-	2.73
13	24.84	Heptaethylene glycol	3.62
14	26.54	9-Octadecenamide,(Z)-	3.71
15	31.92	Glycoursodeoxycholic acid	11.7

18.4 Conclusions

In this work, the efficient liquefaction of lignocellulosic biomass model compounds (lignin and cellulose) was conducted the plasma-enabled liquefaction process with different cellulose-to-lignin mass ratios, and the impact of liquefaction time on liquefaction performance was analyzed. With the increase in liquefaction time, the liquefaction yield of pure lignin exhibited an initial rise followed by a slight decline, while liquefaction yield of pure cellulose continuously enhanced. In the liquefaction of cellulose/lignin mixture, the liquefaction yield displayed a trend of initial decrease followed by an increase with the cellulose-to-lignin mass ratios. The turning point turned up at a liquefaction time of 3 min and a cellulose/lignin mass ratio of 2/3, where the highest liquefaction yield (98.8%) was achieved. After the plasma-enabled liquefaction process, the liquid product with a HHV of 17.34 MJ/kg was obtained, which is higher than that of the raw materials and solid residue. Abundant hydroxyl groups, carbon-hydrogen bonds, carboxyl groups, and various functional groups were detected in the liquid product. The liquid product was detected rich in alcohols, along with trace amounts of organic acids, lipids and aromatic compounds, which can be directly used as fuel, and also has the potential for further processing into chemical products.

References

1. L. Zhang, C. Xu, Overview of recent advances in thermo-chemical conversion of biomass. Energy Conver. Manag. **51**(5), 969–982 (2010)
2. J. Kim, H. Lee, Overview of the recent advances in lignocellulose liquefaction for producing biofuels, bio-based materials and chemicals. Bioresour. Technol. **279**, 373–384 (2019)
3. K. Tekin, S. Karagoz et al., A review of hydrothermal biomass processing. Renew. Sustain. Energy Rev. **40**, 673–687 (2014)
4. Z.X. Lu, L.W. Fan et al., Efficient liquefaction of woody biomass in polyhydric alcohol with acidic ionic liquid as a green catalyst. Biomass Bioenergy **8**, 154–161 (2015)
5. T. Mitani, Recent progress on microwave processing of biomass for bioenergy production. J. Jpn. Petrol. Instit. **61**(2), 113–120 (2018)
6. H.J. Huang, X.Z. Yuan, Recent progress in the direct liquefaction of typical biomass. Prog. Energy Combust. Sci. **49**, 59–80 (2015)
7. S.Q. Zhang, X. Yang et al., Liquefaction of biomass and upgrading of bio-oil: a review. Molecules **24**(12), 2250 (2019)
8. A. Inayat, R. Tariq, Z. Khan et al., A comprehensive review on advanced thermochemical processes for bio-hydrogen production via microwave and plasma technologies. Biomass Conver. Biorefinery **13**, 8593–8602 (2023)
9. J. Vanneste, T. Ennaert, A. Vanhulsel et al., Unconventional pretreatment of lignocellulose with low-temperature plasma. Chemsuschem **10**(1), 14–31 (2017)
10. A. Keudell, V. Gathen, Foundations of low-temperature plasma physics—an introduction. Plasma Sour. Sci. Technol. **26**(11), 113001 (2017)
11. F. Liu, M.L. Cai, S.Y. Liu et al., The optimization of plasma catalytic liquefaction technique for the conversion of sawdust into value-added chemicals. IEEE Access **8**, 2621–2630 (2019)
12. D.H. Mei, S.Y. Liu, S. Wang et al., Plasma-enabled liquefaction of lignocellulosic biomass: balancing feedstock content for maximum energy yield. Renew. Energy **157**, 1061–1071 (2020)
13. A. Demirbas, Calculation of higher heating values of biomass fuels. Fuel **76**(5), 431–434 (1997)
14. D.K. Xi, C.C. Jiang, R.W. Zhou et al., The universality of lignocellulosic biomass liquefaction by plasma electrolysis under acidic conditions. Bioresour. Technol. **268**, 531–538 (2018)
15. S. Oh, D. Yoo, Y. Shin et al., FTIR analysis of cellulose treated with sodium hydroxide and carbon dioxide. Carbohydr. Res. **340**(3), 417–428 (2005)
16. Y.J. Chi, FTIR analysis on function groups of David poplar wood and lignin degraded by 6 species of wood white-rot fungi. Sci. Silvae Sin. **41**(2), 136–140 (2005)
17. C. Cao, Z.L. Yang, L.J. Han et al., Study on in situ analysis of cellulose, hemicelluloses and lignin distribution linked to tissue structure of crop stalk internodal transverse section based on FTIR microspectroscopic imaging. Cellulose **22**(1), 139–149 (2015)
18. M. Sasaki, Z. Fang, Y. Fukushima et al., Dissolution and hydrolysis of cellulose in subcritical and supercritical water. Ind. Eng. Chem. Res. **39**(8), 2883–2890 (2000)
19. L.S. Yang, J.Y. Zhang, W.L. Zhu et al., The preparation of iron lignosulfonate with Na_2SO_3. Jiangsu Agri. Coll **13**(3), 83–85 (1992)
20. L. Xiong, W.D. Yu, Analysis of the cellulose macromolecule structure after acid treatment by FTIR microspectroscopy. J. Cell. Sci. Technol. **21**(2), 59–62 (2013)
21. T. Zhang, Y.J. Zhou, D.H. Liu et al., Qualitative analysis of products formed during the acid catalyzed liquefaction of bagasse in ethylene glycol. Bioresour. Technol. **98**(7), 1454–1459 (2007)
22. X.W. Zou, T.F. Qin, L.H. Huang et al., Mechanisms and main regularities of biomass liquefaction with alcoholic solvents. Energy Fuels **23**(10), 5213–5218 (2009)
23. Z.Q. Ma, J.Y. Xu, J.W. Ye et al., Determination of pyrolysis characteristics and kinetics of *Pinus sylvestris* var. mongolica using TGA-FTIR and distributed activation energy model. J. Southwest For. Univ. **35**(3), 90–96 (2015)
24. D.Y. Che, Y.P. Sun, Y.X. Sun, TG-FTIR experimental study on lignin pyrolysis. Sci. Technol. Eng. **16**(3), 225–238 (2016)

25. J.L. Liu, J.C. Jiang, H.T. Huang, Study on thermal transformations of lignin under curie-point pyrolysis-GC-MS conditions. Chem. Ind. For. Prod. **29**, 1–11 (2009)
26. M. Garcia-Perez, A. Chaala, H. Pakdel et al., Characterization of bio-oils in chemical families. Biomass Bioenergy **31**(4), 222–242 (2007)

Chapter 19
Research on the Electrical and Hydrophobicity Properties of BN/RTV Composite Materials by Plasma Treatment

Menggen Wei, Wenxin Gao, Dong Zhao, Yang Wang, Bo Zhang, and Yan Du

Abstract The room temperature vulcanized silicone rubber (RTV) is widely used as an anti-pollution flashover coating in power system due to its excellent insulating properties. To further improve the insulating properties of RTV materials, the BN/RTV composite insulating materials were prepared. The BN nanoparticles were modified by plasma. This study investigated the impact of different plasma modification time on the electrical properties of the BN/RTV composite insulating materials, including the material interface structure, hydrophobicity, charge-trap characteristic, and electrical strength were mainly studied. The results showed that after plasma modification, the BN nanoparticles disperse more uniformly within the RTV matrix. The dispersion of nanoparticles in RTV and the hydrophobicity of BN/RTV composite material were optimized when the plasma treatment time was 10 min. The introduction of BN nanoparticles leads to an increase in the number of deep traps in BN/RTV composites material and an increase in the pollution flashover voltage. The pollution flashover voltage of BN/RTV composites material was highest when the plasma treatment was 10 min.

M. Wei · W. Gao · D. Zhao · Y. Wang · B. Zhang · Y. Du (✉)
School of Electronics and Information, Xi'an Polytechnic University, Xi'an 710048, China
e-mail: duyan@xpu.edu.cn

Xi'an Key Laboratory of Interconnected Sensing and Intelligent Diagnosis for Electrical Equipment, Xi'an Polytechnic University, Xi'an 710048, China

M. Wei
e-mail: MneggenWei@stu.xpu.edu.cn

D. Zhao
e-mail: zhaodong2022@stu.xpu.edu.cn

Y. Wang
e-mail: wangyang@xpu.edu.cn

B. Zhang
e-mail: zhangbo@xpu.edu.cn

Z. Fang et al. (eds.), *Proceedings of the 5th International Symposium on Plasma and Energy Conversion*, Springer Proceedings in Physics 398,
https://doi.org/10.1007/978-981-97-2245-7_19

19.1 Introduction

The Room temperature vulcanized silicone rubber (RTV) is widely used as an anti-pollution flashover coating in transmission and distribution lines due to its excellent dielectric properties and hydrophobicity. RTV insulation performance has decreased dramatically, seriously affecting the safety and stable operation of the power system because its works in a complex environment for a long time [1]. In order to significantly improve the electrical properties of composites, it is common to add a certain percentage of nanoparticles to the RTV matrix [2]. However, the bonding strength and interfacial effect between the nanoparticles and the polymer matrix are not satisfactory, which may adversely affect the insulating properties of the nanocomposites [3].

Research has shown that plasma treatment technology is often used in material modification thanks to its low application cost and significant treatment effect. Chen et al. [4] found that the flashover voltage of the epoxy resin surface was significantly improved after plasma treatment with a mixture of helium and carbon tetrafluoride gas. Wang et al. [5] used a plasma jet to enhance the interfacial bond properties of the RTV recoating with sandwiched contaminant layer. Zhang et al. [6] used plasma jet to enhance the surface hydrophobicity of HTV.

In previous research, the insulating materials was manifested by direct treatment of their surfaces with plasma. In this paper, the BN/RTV composite insulating materials were prepared, the BN nanoparticles were treated by plasma and then modified with coupling agent KH570. The impact of different plasma modification times on the properties of electrical and hydrophobicity of the BN/RTV composite materials was investigated. This study explores the impact of plasma treatment on the insulation properties of BN/RTV composite materials.

19.2 Materials and Methods

19.2.1 Experimental Materials and Setup

The BN nanoparticles used in this paper were purchased from Shanghai Macklin Biochemical Co., Ltd, with an average particle diameter of 200 nm, a density of 2.27 g/cm^3, and a white powder appearance. The RTV rubber samples were purchased from Hebei ZhonglianHuayu Power Technology Co., Ltd., China. The HTV material was from Shanghai Aladdin Hengsheng Chemical Technology Co., Ltd. The Electrode of dielectric barrier discharge and plasma power supply were provided by Nanjing Suman Plasma Technology Co., Ltd, the experimental platform for plasma treatment in Fig. 19.1.

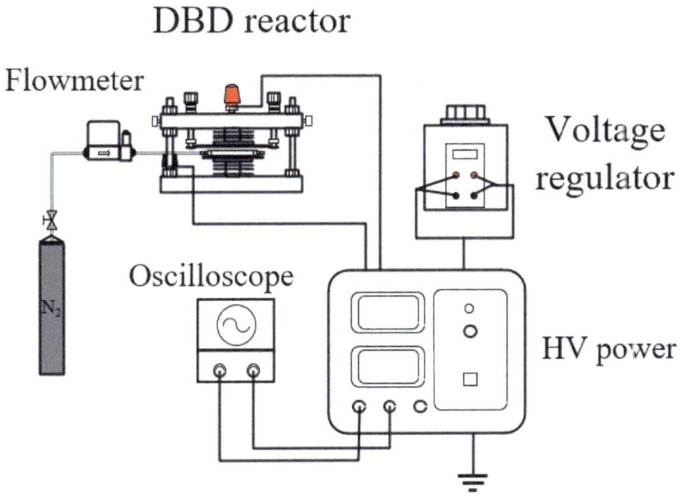

Fig. 19.1 The experimental platform of plasma

19.2.2 Sample Preparation Process

Treatment of Nanoparticles by Plasma

During the experiment, an appropriate amount of BN particles was placed in the plasma reaction vessel, an electrode spacing has 11 mm. The glass cover was closed, and N_2 was introduced into the reaction vessel. The plasma power supply was then activated with a working voltage set to 60 V and a working frequency of 19.00 kHz. The treatment time was set to 0, 3, 6, and 10 min respectively. After the treatment, the BN particles were placed in a vacuum oven at 80 °C for 1 h to eliminate the influence of air and moisture.

Preparation of BN/RTV Composites

Firstly, the processed BN particles, deionized water, and ethanol are mixed in a ratio of 1:5:3. Subsequently, 0.2 g of coupling agent KH570 is added to the mixture and stirred for 20 min with a magnetic stirrer, and a water bath oscillator for 20 min. Than the modified BN particles are added to the RTV coating material at 5 wt% and mixed well. Finally, the mixture was brushed on HTV substrate measuring 40 mm*40 mm*3 mm, after removing any air bubbles in a vacuum chamber. The names of the samples are shown in Table 19.1.

Table 19.1 Details of sample names

Sample	Plasma treatment time (minutes)	Content unit
RTV	0	0
P0	0	5 wt%
P1	3	5 wt%
P2	6	5 wt%
P3	10	5 wt%

19.3 Results and Discussion

19.3.1 Surface Potential Attenuation Test

Figure 19.2 shows the surface potential decay curves of RTV and BN/RTV composite material. It can be seen from Fig. 19.2 that the surface potential of all samples experienced a rapid decrease at first and then a slow decrease with the increase of measurement time. RTV exhibits a considerably faster decay rate compared to the other samples in the early stage, the surface potential approaching of RTV is 0 V after 5000 s. The experimental results show that the surface potential decay rate shows a trend of slow-fast-slow with the increase of plasma treatment time. Remarkably, the composite material treated for 3 min exhibits the slowest decay rate.

Figure 19.3 shows the trap parameter distribution of the samples. it is evident that the energy level of shallow trap of RTV is about 0.83 eV, and the deep trap center level is about 0.92 eV, with a significantly higher density of shallow traps compared to deep traps. However, the composite materials only have deep traps because the doping of nanoparticles, the center energy level of deep trap is between 0.89 and

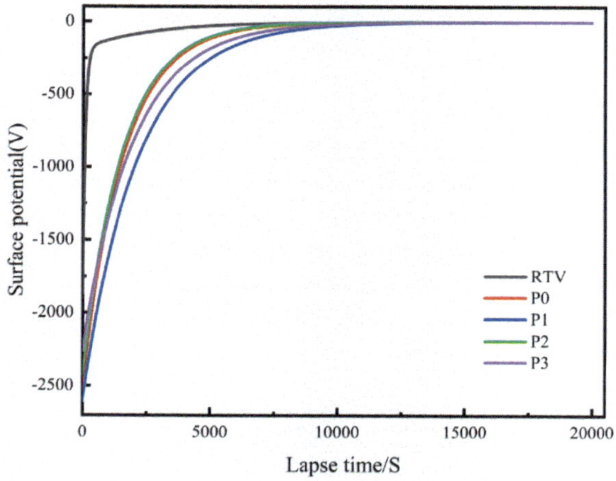

Fig. 19.2 Normalized surface potential decay curve of different samples

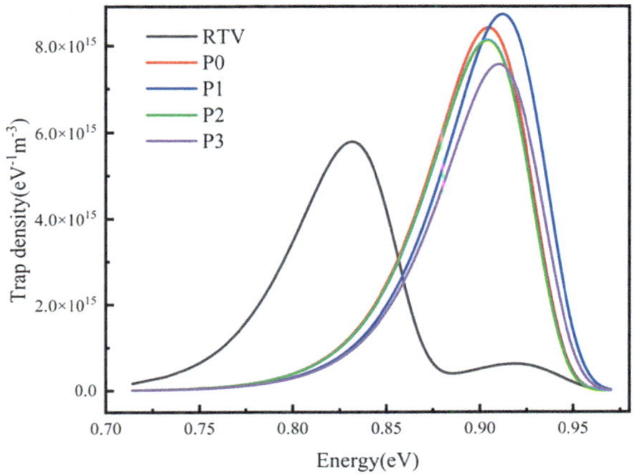

Fig. 19.3 Trap distribution of composite insulating materials

0.92 eV. Notably, the trap energy levels of the plasma-treated composite material are slightly higher than the untreated composite material.

The rapid decline in surface potential of RTV is mainly caused by the charge detrapping phenomenon occurring in shallow traps [7]. The BN nanoparticles of Plasma-treated leads to the introduction of a considerable number of deep traps into the composites, Charges in deep traps need to gain a large amount of energy to escape from the trap, the surface potential decay rate of BN/RTV composites becomes slower.

19.3.2 Flashover Voltage

Figure 19.4 shows the average values of the flashover voltage of pure RTV and BN/RTV composites. From Fig. 19.4, it can be seen that the flashover voltage of RTV is 15 kV, which is slightly higher than composite P0. The flashover voltage of the composite material increases to 12.7 kV after 3 min of plasma treatment, but it is still lower than the RTV. With the extension of the treatment time, the flashover voltage is 19.6 kV at 6 min of plasma treatment, which is 31% higher than the pure RTV. However, the flashover voltage decreased to 11.3 kV, with a treatment time of 10 min.

19.3.3 Pollution Flashover Voltage

Figure 19.5 shows the average value pollution flashover voltage [8] of RTV and BN/RTV composite materials. It can be seen from Fig. 19.5 that the pollution flashover voltage of pure RTV is the lowest than BN/RTV composite. However, the pollution flashover voltage of the composites is improved to 4.6 kV while nanoparticle is

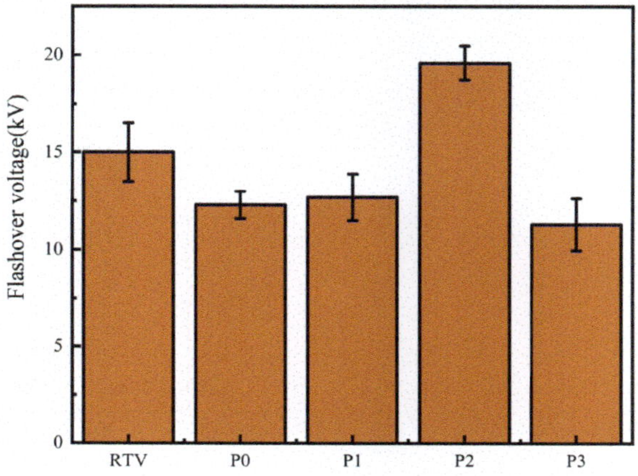

Fig. 19.4 Flashover voltages of composite insulating materials

added. The pollution flashover voltage of the composite material is 5.2 kV after 3 min of plasma treatment, which is about 37% higher than the RTV. Furthermore, the pollution flashover voltage continues to increase with increasing treatment time, and the maximum value of 6.4 kV is reached at 10 min of treatment.

Based on the analysis of the part 3.1, the interaction between the filler and the RTV matrix can be disrupted by plasma treatment. The pollution flashover voltage of the composite material increases because the transportation of charge to the filler is impeded, which leading to the formation of deep traps, inhibiting the accumulation of surface charges on the materials.

In order to evaluate the effect of plasma on pollution flashover voltage, the pollution flashover voltage of different materials is also analyzed by Weibull distribution, and the results are shown in Fig. 19.6. The data of each material are well fitted,

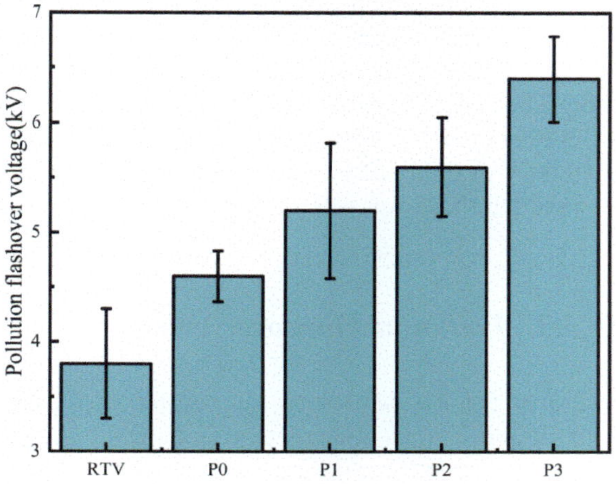

Fig. 19.5 Pollution flashover voltages of composite insulating materials

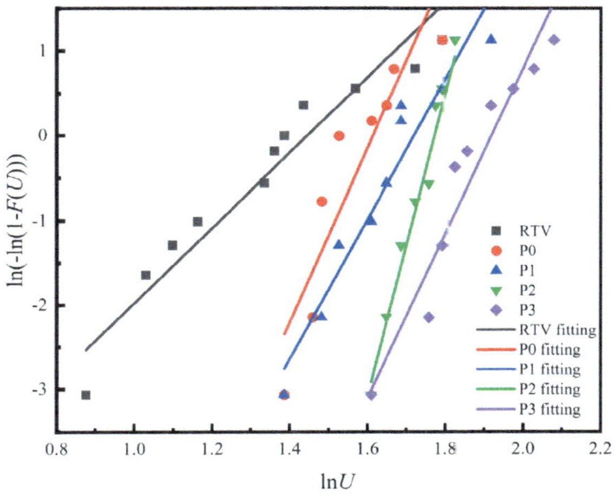

Fig. 19.6 Weibull fitting curves of pollution flashover voltages of composite insulating materials

Table 19.2 Weibull distribution parameters for pollution flashover voltages

Sample	α/kV	B
RTV	4.24	4.44
P0	5.02	10.36
P1	5.58	8.29
P2	5.89	17.91
P3	6.81	9.76

which confirms that the pollution flashover voltage of the composite material obeys the Weibull distribution.

The parameters of the Weibull distribution [9] of the pollution flashover voltage are reported in Table 19.2. Where α can be used to characterize the breakdown voltage of the insulating material and β can indicate the sensitivity of the insulating material to the applied voltage. As we can see from Table 19.2, material of P3 has the largest value of α, which means that the pollution flashover voltage of BN/RTV composites can improve by plasma. Compared to other materials, P0 and P2 have larger β, which explains that the probability of the occurrence of pollution flashover voltage of these two materials is more sensitive, because the change of the applied voltage.

19.3.4 Water Contact Angle Measurements

The Fig. 19.7 shows the average value of water contact angles of the RTV and plasma-treated BN/RTV composite materials. It can be observed that the pure RTV has the smallest contact angle, which is only 104.204°. The addition of BN nanoparticles slightly increases the hydrophobic contact angle of the composite material, but the

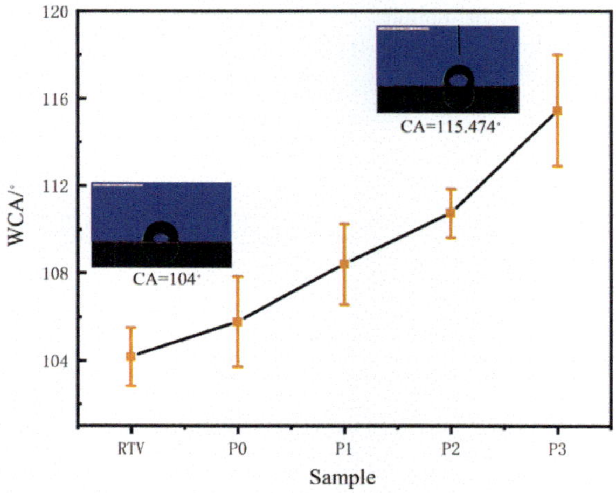

Fig. 19.7 The water contact angle of different sample

effect is not significant, measuring only 105.803°. The hydrophobic contact angle of the composite was 108.43° after 3 min of plasma treatment. With the extension of plasma treatment time, the hydrophobic angle of composite materials increases, the angle reaching 110.771° at 6 min. The hydrophobicity angle reaches a maximum value of 115.474° when the plasma treatment time is 10 min, which is about 11% higher than the pure RTV. The experiment results indicate that the hydrophobicity angle of BN/RTV composite material can be enhanced effectively with plasma.

19.4 Conclusion

In this paper, BN/RTV composites were prepared by treating BN nanoparticles with plasma technology. Growth of the number of deep traps in BN/RTV composites due to plasma treatment, which directly increases the pollution flashover voltage, the BN/RTV composites treated for 10 min showed 68% improvement compared to RTV. The water contact angle of the composites increased significantly with the extension of plasma treatment time, and reached a maximum value of 115.474° at 10 min of treatment. All the results pointed out that the electrical properties and hydrophobicity of BN/RTV composites are enhanced positively because of plasma.

References

1. J. Chen, B. Li, X. Zeng, Z. Li, Y. Wen, Q. Hu et al., Study on the influence of accelerated aging on the properties of an RTV anti-pollution flashover coating. Polymers **15**(3), 14453 (2023)
2. L. Wu, X. Wang, L. Ning, J. Han, Z. Wan, M. Lu, Improvement of silicone rubber properties by addition of nano-SiO_2 particles. J. Appl. Biomater. Funct. Mater. **14**(1), 11–14 (2016)

3. W. Zhou, Z. Wang, L. Dong, X. Sui, Q. Chen, Dielectric properties and thermal conductivity of PVDF reinforced with three types of Zn particles. Compos. A Appl. Sci. Manufact. **79**(1), 183–191 (2015)
4. S. Chen, S. Wang, Y. Wang, B. Guo, G. Li, Z. Chang et al., Surface modification of epoxy resin using He/CF_4 atmospheric pressure plasma jet for flashover withstanding characteristics improvement in vacuum. Appl. Surf. Sci. **414**, 107–113 (2017)
5. S. Wang, S. Li, J. Li, R. Zhang, H. Liu, Interfacial bonding enhancement of the RTV recoating with sandwiched contaminant by plasma jet. High Volt. **4**(4), 345–348 (2019)
6. R. Zhang, Q. Han, Y. Xia, S. Li, Plasma jet array treatment to improve the hydrophobicity of contaminated HTV silicone rubber. Plasma Sci. Technol. **19**, 105505–105513 (2017)
7. J. Yan, G. Liang, H. Lian, Y. Song, C. Peng, Y. Kang et al., Effect of plasma step gradient modification on surface electrical properties of epoxy resin. Plasma Sci. Technol. **23**(6), 76–99 (2021)
8. M.T. Gençoğlu, M. Cebeci, The pollution flashover on high voltage insulators. Electr. Power Syst. Res. **78**(11), 1914–1921 (2008)
9. R.W. Macpherson, M.P. Wilson, I.V. Timoshkin, S.J. Macgregor, M.J. Given, Impulsive flashover characteristics and weibull statistical analysis of gas-solid interfaces with varying relative humidity. IEEE Access **8**, 228454–228465 (2020)

Chapter 20
Preparation and Modification of Mesocarbon Microbeads Materials Using Magnetic Rotating Thermal Plasma Processing

Yuntian Cong, Cheng Zhou, Chuncai Kong, Yong Li, Zhimao Yang, Ge Wang, Baojun Wang, and Peng Wu

Abstract Mesocarbon microbeads (shortened by MCMB) are excellent precursors for preparing high-strength and high-density graphite materials, which can be treated by high temperature to achieve high quality carbon-based materials. At low pressure, about 10 Pa, by using Magnetic Rotating Thermal Plasma (shortened by MRTP) processing technology, MCMB can be rapidly heated to the high temperature at more than 3200 K in few minutes. The vacuum environment is beneficial for the precipitation of volatile substances in raw MCMB, meanwhile high temperature is conducive to the rearrangement of internal structures, which means significant improvement of graphitization. Thus, it is shown that the volume of MCMB sample treated by MRTP processing decreases, while its density and purity increase, especially the latter changed from 70 to 94%. Subsequently, the results of charge–discharge test of batteries made by the treated sample show that the discharge capacity is up to 360 mAh/g, while through the composition with Si powder can improve the performance to 1200 mAh/g. The MRTP processing made the surface of MCMB particles disordered, and gave rise to an improvement in the thermal stability and electrochemical properties of the powders. It also made the powders suitable for further use in energy-storage materials.

Y. Cong (✉) · C. Zhou · Y. Li · G. Wang · B. Wang · P. Wu
Beijing Institute of Control Engineering, Beijing 100190, China
e-mail: 15201169466@139.com

C. Kong · Z. Yang
Xi'an Jiaotong University, Xi'an 710049, China

Z. Fang et al. (eds.), *Proceedings of the 5th International Symposium on Plasma and Energy Conversion*, Springer Proceedings in Physics 398,
https://doi.org/10.1007/978-981-97-2245-7_20

20.1 Introduction

High density and purity graphite materials have excellent chemical stability, mechanical properties, conductivity and thermal conductivity, which widely used in fields such as electrical and electronic, mechanical, chemical, metallurgical, nuclear energy, military, aerospace, etc. They also play an important role in some cutting-edge fields. In the domestic field of new energy vehicles, the demand for the negative electrode quality of graphite batteries is gradually increasing; Driven by Industry 4.0, the market's demand for new graphite materials will continue to grow. In fields such as aerospace and equipment applications, the demand for high-end graphite materials is becoming increasingly urgent. With the improvement of human living standards, countries are increasingly paying attention to the research and development of new and sustainable energy. China urgently needs to carry out relevant technological breakthroughs and equipment development work, solving key technical problems that constrain the development of related technology fields, and have significant practical significance for improving the overall technical level of China's new material manufacturing equipment (Fig. 20.1).

Mesocarbon microbeads (MCMB) are excellent precursors for preparing high-strength and high-density carbon graphite materials. They are a disc-shaped nematic liquid crystal structure formed during the carbonization process of polycyclic aromatic compounds. In 1973, Honda H and Yamada Y first separated and obtained isotropic carbon materials from the intermediate phase of asphalt, and prepared them without adding any binder. They have good chemical stability, high packing density, good thermal stability, and have broad application and development prospects. Due to its outstanding performance similar to fullerene, MCMB can be used as diamond film materials, lubricating materials, catalysts, solar collectors, special rubber additives, stealth materials, negative electrode materials for lithium-ion batteries, etc. According to foreign research and development experience, MCMB needs to undergo ultra-high temperature heat treatment to remove low melting point impurities and optimize the internal atomic arrangement sequence of the material (as shown in Fig. 20.2) in order to prepare high-quality graphite materials with high density and purity [1–4].

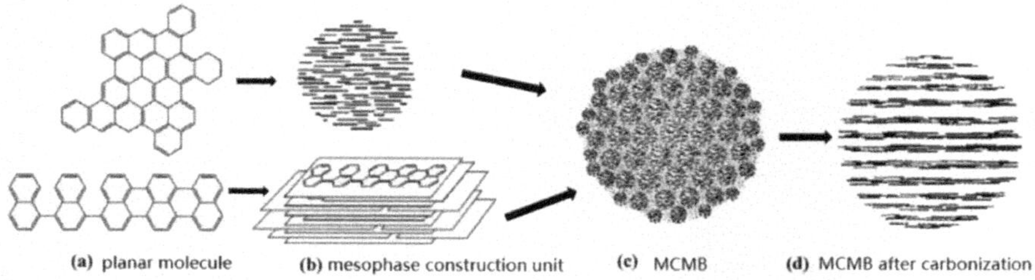

Fig. 20.1 Construction theory of mesocarbon microbeads

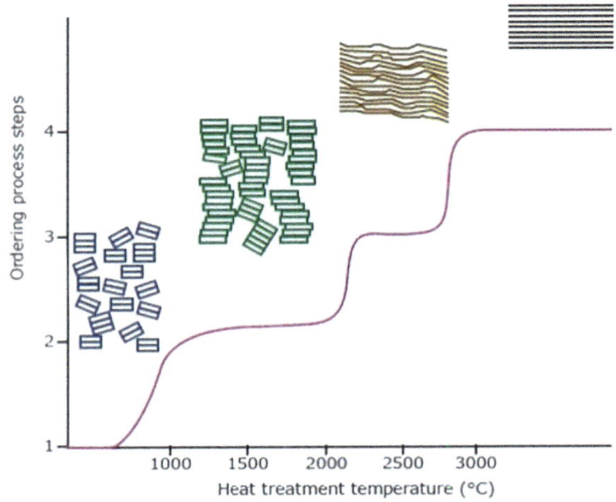

Fig. 20.2 Relationship between orderliness and temperature of carbon-based materials

After graphitization with ultra-high temperature equipment, mesophase carbon microspheres can obtain graphite materials with excellent chemical stability, mechanical properties and thermal conductivity. At present, the heat treatment of MCMB includes carbonization and graphitization, with the most chemical changes occurring during the carbonization process, which is also one of the main processes affecting material properties. The bulk density and graphitization degree of high-density and high-strength carbon/graphite materials prepared using MCMB as raw material increase with temperature, and graphite materials with excellent performance can be obtained when the temperature exceeds 3000 °C. Traditional preparation methods have complex processes, high energy consumption, and long production cycles, resulting in low efficiency throughout the entire production process. However, plasma can achieve rapid high temperatures, making this method possible for further improvement [5, 6].

This study uses the Magnetic Rotating Thermal Plasma (MRTP) developed by the Beijing Institute of Control Engineering as a plasma heat source to treat and modify MCMB materials. Its high-density and high-energy plasma jet is used to directionally bombard the material surface, producing ultra-high temperatures that alter the material's properties.

20.2 MRTP Principle, Development Status, and Technical Advantages

20.2.1 Development Status

The Beijing Institute of Control Engineering, affiliated to the Chinese Academy of Space Technology, was founded on October 11, 1956, formerly known as the Institute of Automation of the Chinese Academy of Sciences, and is one of the earliest units engaged in satellite development in China. The research institute is mainly engaged in the design and development of spacecraft attitude and orbit control systems, propulsion systems and their components, as well as the research and application of industrial control systems, and enjoys a high reputation both domestically and internationally. In the field of high-power plasma generator technology, early planning and layout were carried out during the 12th Five Year Plan period, and basic theoretical research on the electromagnetic acceleration mechanism of thrusters was conducted through independent research and development, laying a solid research and development foundation. In 2017, the BICE completed the development of a prototype and performance ignition tests, with a maximum ignition power of 114 kW, filling the gap in China.

In 2020, the Beijing Institute of Control Engineering officially carried out research on the application of high-power plasma generator technology. The research team spent two years in Huailai to conduct a validation study on the principle of ultra-high temperature treatment of carbon-based materials represented by mesophase-carbon microspheres.

In 2021, the technical team completed the preliminary verification of high-temperature plasma technology and conducted MCMB plasma graphitization treatment experiments, achieving preliminary experiments in stable high-temperature areas above 3000 °C.

20.2.2 MRTP Working Principle and Advantages

The working mechanism of MRTP is shown in the following figure. During operation, an arc is formed by discharging between the cathode and anode, and the current induces a magnetic field when passing through the high-temperature plasma between the cathode and anode. The interaction between the induced magnetic field and the current passing through the plasma generates high-speed, high-density, and high-energy plasma. Under high intensity heat sources, the activity energy of elementary particles is much greater than that of intermolecular chemical bonding. The microscopic motion of matter is mainly atomic thermal motion, and the original material is shattered by high-energy particles, thereby purifying the complex composition or structure of the material.

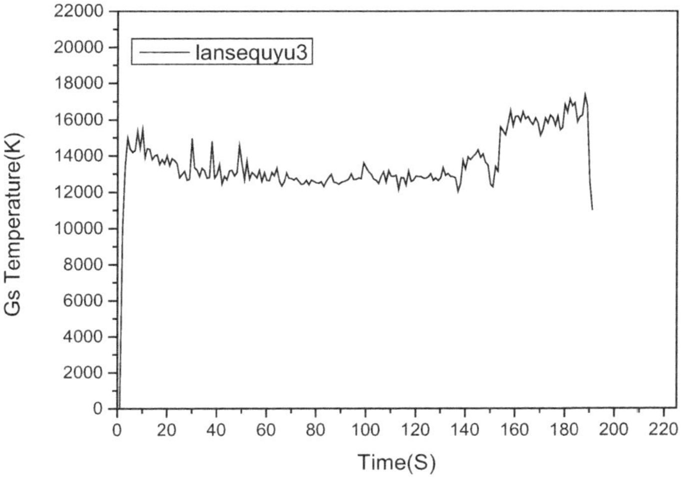

Fig. 20.3 Macro temperature test curve of 30 kW equilibrium plasma

Compared with traditional high-temperature heat source technologies such as resistance heating, radio frequency induction, chemical combustion, and arc heating, high-power Magnetic Rotating Thermal Plasma (MRTP) device technology has the following characteristics:

(1) High temperature. The high-power MRTP device has a highly ionized plasma, where electrons rotate at high speeds in a strong magnetic environment and collide with ions at high speeds and frequencies to transfer energy, resulting in extremely high macroscopic temperatures of the plasma as a whole. The equilibrium plasma macroscopic temperature test was conducted on a 30 kW magnetic rotary thermal plasma device (as shown in Fig. 20.3), and the temperature at a distance of 150 mm from the outlet can reach 15,000 K, far exceeding the temperature range that existing heat source technology can reach, providing the possibility for industrial high-temperature applications above 3300 K.
(2) Fast heating-up. The plasma density in the core plume area of the high-power MRTP device is about 10^{21} n/cm^3, which is more than two orders of magnitude higher than the traditional radio frequency ion isoelectric propulsion plasma density. This can then form ultra-high heat flux density on the material surface, achieving rapid heating of the treated material and greatly improving material preparation efficiency. The carbon material was heated using a 50 kW magnetic rotary plasma device, with temperatures ranging from 0 to 3000 °C. The actual measurement took only 15 s.
(3) Good uniformity. The high-power MRTP device adopts a specially designed strong magnet design scheme. The strong magnetic field causes the arc to rotate counter clockwise at high speed on the outer surface of the cathode and inner surface of the anode of the plasma device, greatly improving the uniformity

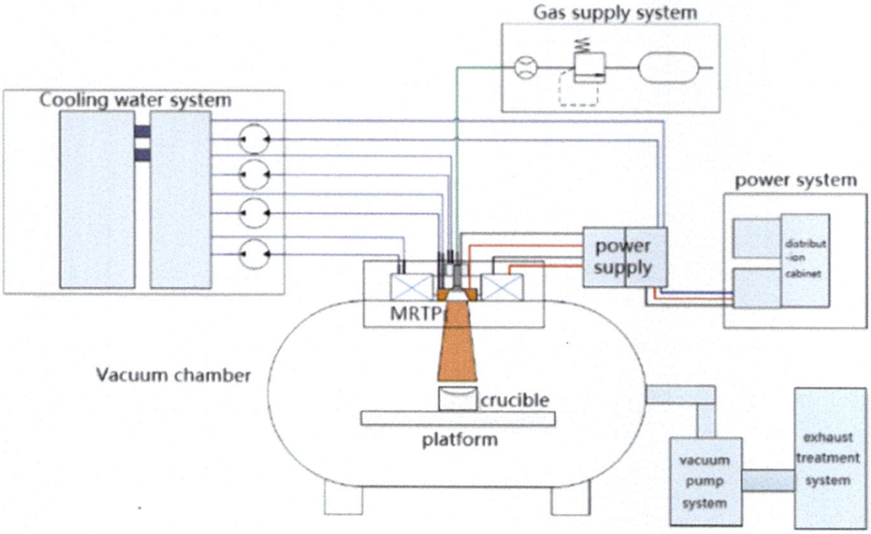

Fig. 20.4 System composition diagram

of energy distribution in the radial direction of the plasma, thereby achieving uniform heating of the surface of the treated material, which is conducive to improving product quality.

20.3 MRTP Plasma Treatment Test System and Its Composition

The MRTP plasma treatment test system is mainly composed of four parts: MRTP plasma generation device, high power processing unit, vacuum chamber, and process package (including comprehensive measurement and control software) (as shown in Fig. 20.4). MRTP can provide high-density and high-energy plasma, and the power processing unit provides matching high-voltage ignition, excitation, and anode power supply for the plasma device, The vacuum chamber provides a high vacuum environment for the stable operation of the plasma device, and the process package mainly solves core problems such as product ultra-high temperature control, continuous production, and quality control.

20.3.1 MRTP Plasma Source

The 50kW MRTP plasma engine (with magnetic coil) is mainly composed of anode components, cathode components, inter pole insulation seals, insulation mounting

Fig. 20.5 Schematic diagram of a 50 kW MRTP plasma engine with coil

parts, auxiliary components, and excitation magnetic coils. Its structure is shown in the following figure. The heat flux borne by the anode is relatively large, so in order to ensure the normal operation of the plasma device, it is necessary to focus on cooling the anode. The structural form and size of the anode directly determine the shape and size of the discharge chamber; The cathode is located at the center of the discharge zone and needs to directly withstand ion bombardment, strong thermal radiation, and Joule heat brought by the discharge current. The design of the cathode has a decisive impact on the performance and lifespan of the device. The additional magnetic field is designed based on engine configuration matching and plasma simulation analysis, and a convergence expansion type magnetic field is adopted to effectively improve the efficiency of the device. At the same time, an additional magnetic field with good magnetic field uniformity can improve ionization degree, with a larger core area range and higher temperature (Fig. 20.5).

20.3.2 Supporting Power Supply

The supporting startup power supply is designed according to the requirements of high-temperature plasma, and the development of additional coil power supply, high-frequency pulse power supply, and 50 kW power module is completed. The high-voltage high-frequency pulse power supply and high-power power supply work together and are connected to the plasma load. Its function is to generate high-voltage pulses after receiving the ignition command, ionize and breakdown the working fluid

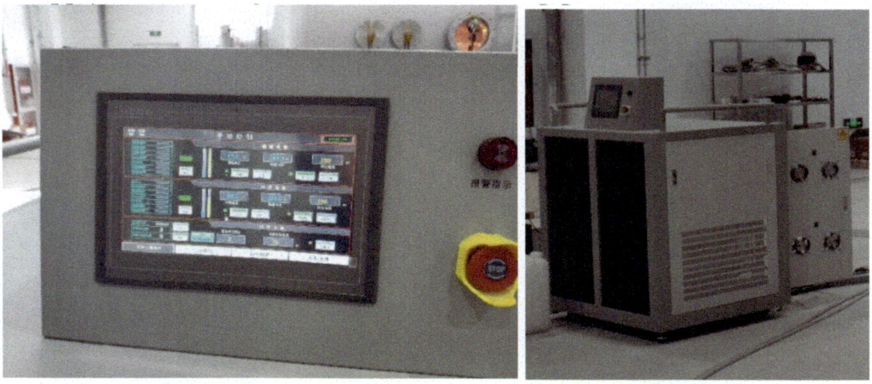

Fig. 20.6 50 kW supporting starting power module

in a short period of time, and generate arc ignition. Then, the high-power power supply supplies power to the plasma load. In the case of dynamic changes in plasma load, on the one hand, it meets the load requirements, and on the other hand, it maintains the conversion efficiency of high-power power supply in the optimal efficiency range. This requires the control system to be able to cooperate with plasma operation and respond quickly to changes in load. Therefore, it is necessary to add soft start control measures during the power supply time-sharing and hierarchical startup process (Fig. 20.6).

20.3.3 Vacuum Chamber and Maintenance System

The vacuum device includes the main structure of the vacuum chamber, a vent valve, a vacuum gauge, an observation window, a 50 kW MRTP interface, a cabin door, a driving mechanism interface, a configured vacuum pump, and a plug-in valve. The structure is shown in the following figure. The main structure of the vacuum chamber is a cylindrical vacuum container, and the design and manufacturing quality of the container will directly affect the equipment performance, which is also an important link in technical implementation. The vacuum chamber is a horizontal structure, consisting of a gate, a straight tube section, and flange interfaces. The main material is 0Cr18Ni9, and accessories such as the saddle and gate rotating mechanism are made of Q235 carbon steel. Two reinforcing ribs are arranged on the straight section of the container. Inner diameter of container Φ 3000 mm, maximum outer diameter Φ 3200 mm (gate flange), straight section length 6000 mm, supported by saddle. The cross-section of the gate sealing ring is ø 20 mm, and the material is B-3 fluororubber. Other flange sealing rings also use B-3 fluororubber. In order to improve the purification effect, product quality, and efficiency of graphite powder, while fully considering the energy efficiency ratio, the vacuum chamber adopts a combination of oil-free screw roots unit and molecular pump to pump the background vacuum degree below 0.01 Pa before proceeding with work (Fig. 20.7).

In order to meet the needs of high-temperature plasma technology in this project, it is necessary to develop a special heat sink structure designed vacuum chamber and a supporting high pumping speed pump group. At the same time, a vacuum pump combination filtration device is added to further filter the exhaust gas of the pump group (Fig. 20.8).

Fig. 20.7 Main structure of vacuum chamber

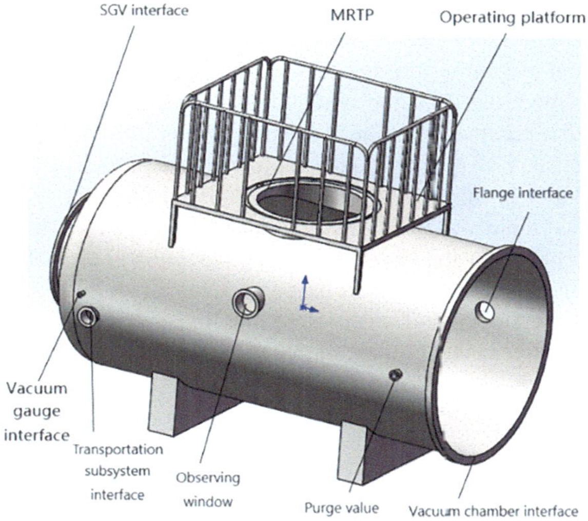

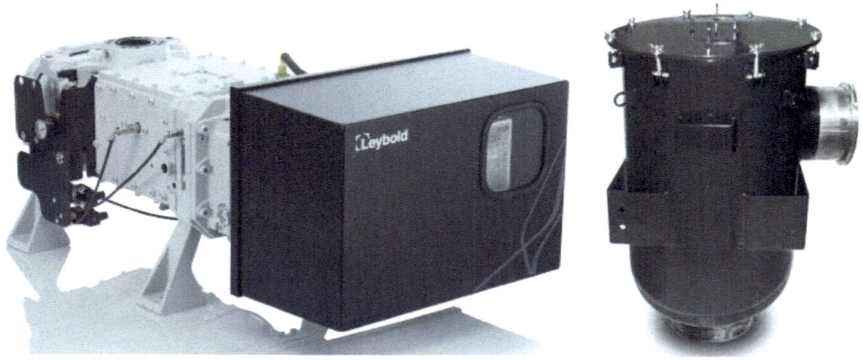

Fig. 20.8 Vacuum pump and filtration device

20.3.4 Water Supply and Gas Supply

The cooling water system provides cooling water sources for ultra-high temperature plasma generation devices, transportation subsystems, thermal subsystems, high-efficiency intelligent hybrid multi-stage starting power devices, etc., achieving cold and hot balance, ensuring normal operation of equipment, and improving production efficiency. The system mainly consists of a chiller, water supply unit, etc., and can provide 50KW of cooling capacity per unit time (Fig. 20.9).

The gas supply system provides working gas for the ultra-high temperature plasma generation device, usually argon or nitrogen. The system automatically adjusts the intake volume through an electric pressure reducing valve and a mass flow meter to meet the dynamic adjustment requirements of plasma plume density, size, and other parameters in the ultra-high temperature plasma generation device. The system consists of gas cylinder group, manual pressure reducing valve, electric pressure reducing valve, electromagnetic valve, mass flow meter, etc. (Fig. 20.10).

It mainly includes a chiller and a cooling water supply device. The chiller is used to maintain the cooling water temperature below 15 °C, and the chiller power is calculated and selected based on the system heat consumption power. The cooling water supply device is used for cooling water supply and recovery, mainly consisting of water tanks, pumps, pipelines, valves, etc. (Fig. 20.11).

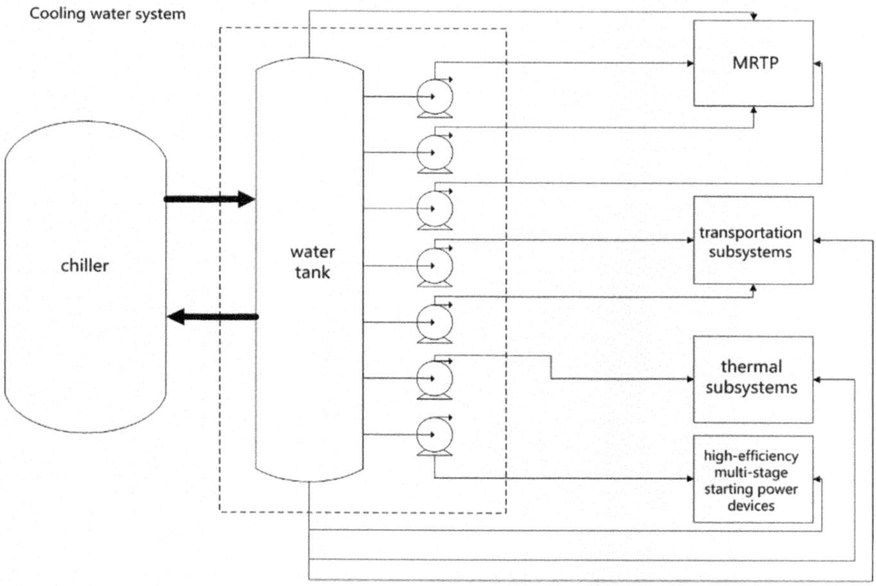

Fig. 20.9 Schematic diagram of cooling water system

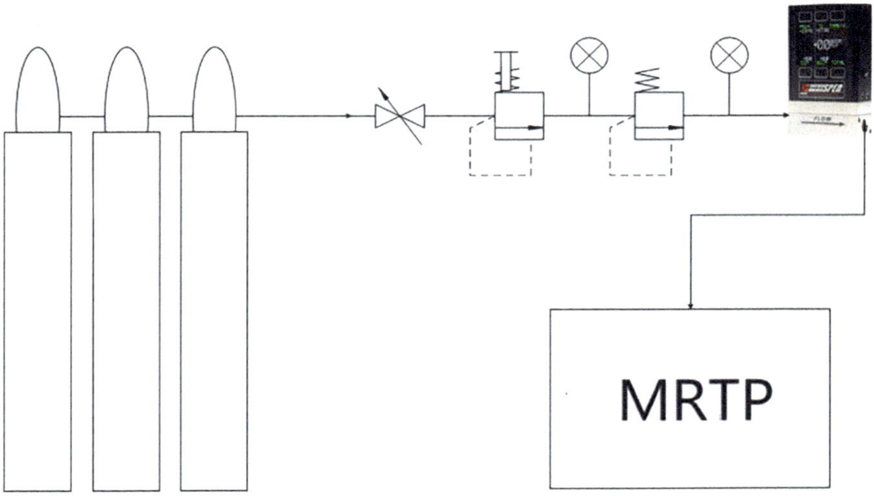

Fig. 20.10 Schematic diagram of gas supply system

Fig. 20.11 Chiller

20.3.5 Infrared Temperature Measurement

The schematic diagram of vacuum chamber and device temperature measurement is shown in the following figure. Real time temperature measurement of the target is achieved by placing optical, infrared, and other comprehensive measurement systems outside the chamber, in order to calculate the temperature of the graphitized sample inside the crucible. The optical, infrared, and other comprehensive measurement systems are composed of a dedicated dual channel infrared thermometer, an optical CCD camera equipped with specific spectral filters, high transmittance glass, dedicated processing software, and fiber optic signal connection lines (Fig. 20.12).

20.4 Experimental Study on MCMB Plasma Treatment

The experimental system for treating MCMB using MRTP plasma is shown in the following figure. Install an MRTP plasma device and a multi-layer high-temperature crucible carrying MCMB in the vacuum chamber, and measure the temperature at the crucible in real-time through an infrared camera to obtain the peripheral temperature of the sample (Fig. 20.13).

The average temperature of the crucible is over 3000 °C when the MRTP operates at a power of 50 kW. Due to testing conditions, only the average temperature outside the MCMB can be detected. Through the characteristics of the MCMB, it can be understood that the internal temperature of the MCMB is at least 400–500 °C higher than the average temperature outside, which is 3400–3500 °C or above (Fig. 20.14).

Conduct conductivity tests on the samples before and after the test. Before graphitization, they are all insulation materials. After high-temperature graphitization, the conductivity of the materials gradually improves, indicating that the degree of graphitization gradually increases (Fig. 20.15).

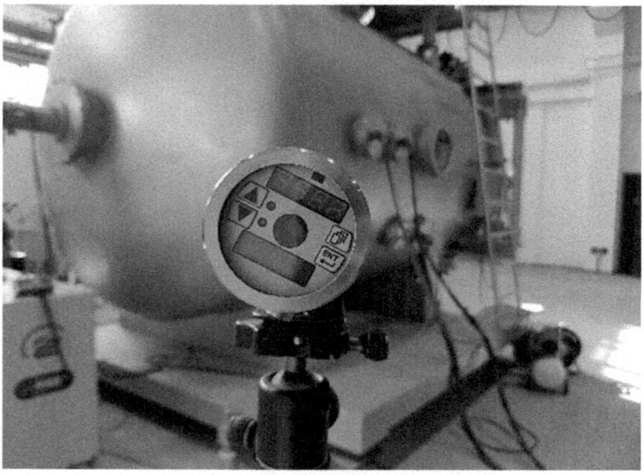

Fig. 20.12 Infrared thermometer

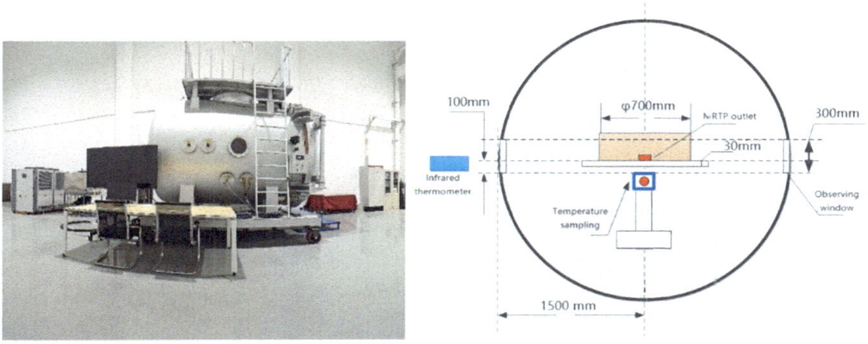

Fig. 20.13 Schematic diagram of vacuum chamber and device temperature measurement

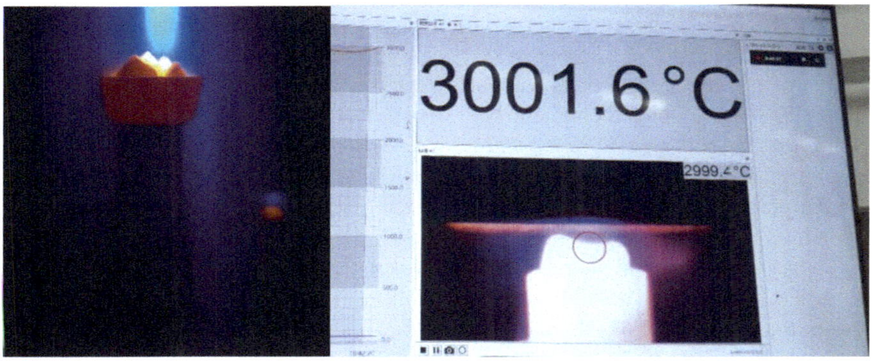

Fig. 20.14 MCMB high temperature plasma treatment process and outer infrared temperature measurement

Fig. 20.15 Comparison of MCMB before and after MRTP plasma treatment

20.5 Comparative Analysis of MCMB Treated Experiments

20.5.1 Study on the Effect of Plasma on the Microstructure of MCMB

The MCMB samples tested in this study were provided by Hebei Qinhuangdao Huarui Carbon Co., Ltd. During the experiment, a graphite crucible (uncovered) was used to carry MCMB raw material samples. The plasma generated by MRTP directly bombarded the surface of the material, reaching a temperature of over 2800 °C and undergoing a 40-min treatment. SEM testing and analysis were conducted on the materials before and after treatment. The comparison between the original and processed samples tested by SEM is shown in the following figure. From the figure, it can be seen that the original sample is mainly spherical particles, with small flake like particles attached to the sphere, which can be regarded as impurities and affect the material properties; After treatment, the sample has a clear layered structure, with thin layers in some areas. The layered structure is coated with spherical particles and exhibits obvious graphene characteristic structures (Fig. 20.16).

In addition, bulk density tests were conducted on the samples before and after plasma treatment. The density of the original MCMB sample was 1.425 g/cm^3, and after temperature treatment above 2800°, the sample volume density reached 1.99 g/cm^3, which is higher than the 1.91 g/cm^3 of the "Toyo ISO80" product produced by the Japanese Toyo carbon plant. Further microscopic imaging and ICP purity testing (as shown in the table below) showed that ultra-high temperature can effectively remove impurities in MCMB, especially for elements such as Na, Mg, Zn, Fe in carbon materials, with a purity increase from 70% to over 94% (Table 20.1).

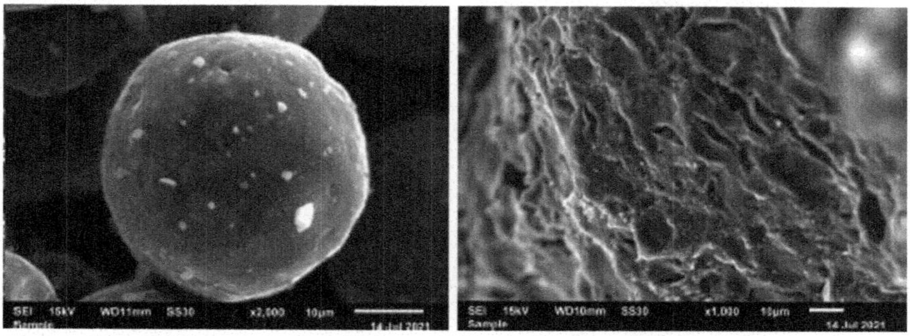

Fig. 20.16 MCMB original sample (left) and processed sample (right)

Table 20.1 Comparison of impurity element content before and after MCMB treatment

Sample	Elements in ppm						
	Na	Mg	Mn	Ni	Zn	Li	Fe
MCMB	51.5	30.3	3.1	2.2	409	0.16	271
MCMB (MRTP processing)	< 0.5	2.6	0.3	0.1	0.6	0.02	30

20.5.2 Study on the Effect of Plasma on the Graphitization Degree of MCMB

XRD analysis was conducted on MCMB treated at different temperatures and times to compare and analyze the effects of graphitization degree and plasma treatment. XRD analysis was performed on MCMB before and after treatment using an X-ray diffractometer. Coherent diffraction peaks with crystal characteristics will appear in the XRD image, and no other miscellaneous peaks will appear. The main diffraction peak is located in (002). The cell constants such as a and c can be calculated from the data in the XRD diffraction pattern, and the spacing d between each crystal plane can be calculated using the spacing formula (where h, k, l are the crystal plane indices):

$$\frac{1}{d^2} = \frac{4\left(h^2 + k^2 + hk\right)}{3a^2} + \frac{l^2}{c^2}$$

The formula for graphitization degree g is:

$$g = \frac{0.3440 - d_{002}}{0.3440 - 0.3354}$$

Samples 4, 5, 6, and 8 were subjected to plasma treatment at 1500, 2000, 2500, and 3000 °C for 45 min, respectively. The resulting Raman curves and SEM patterns are shown in the following figure. In sample 4, the irregular spherical surface is coated with sheet-like substances. Under this treatment temperature and time, the particles deform and the size tends to decrease; The sphere scanned in sample 5 is coated with a sheet-like structure, with a large volume of about 15–20 μm. After ultrasound, the transmission can reveal the peeling of the sheet-like substance. The sphere is small, 50–200 nm, and sheet-like objects can also be seen on the surface. After high-temperature heat treatment, the microspheres form a sheet-like structure from the outside to the inside. After ultrasound, the sheet-like structure is peeled off, leaving behind the core of the microspheres that form the sheet-like structure; In sample 6, the microstructure begins to exhibit irregular shapes, and at this treatment temperature and time, the sample may be dispersed by high-energy plasma; In sample 8, there are microspheres with irregular shapes and sheet-like structures, and the microspheres have deformation. Through testing, Raman shows that the degree of graphitization increases with the increase of temperature. When treated under the same conditions, the degree of graphitization of the powder is greater than that of the bulk. SEM shows that the surface of the high-temperature treated microspheres is coated with sheet-like substances, which roughens the surface. The higher the temperature, the rougher the surface, and the higher the degree of graphitization (Fig. 20.17).

4. The analysis of the graphite degree characteristic parameters of samples 5, 6, and 8 is shown in the Table 20.2. From the table, it can be seen that as the heat treatment temperature gradually increases, the ID/IG gradually decreases and the degree of graphitization also increases. Graphene features I2D/IG = 0.7, reduced interlayer spacing, reduced defects, and significantly increased graphitization degree.

20.5.3 Plasma Treatment MCMB Specific Capacity Test

Make MCMB material samples treated with plasma into batteries and conduct charge discharge tests. After 100 cycles of preliminary testing, the capacity of the graphitized MCMB material battery can still reach 360 mAh/g. By mixing the plasma treated sample with nanosilicon, the graphitized MCMB/Si composite material battery capacity can reach 1200 mAh/g. The specific capacity of the cyclic test is shown in the following Fig. 20.18.

20.6 Subsequent Research and Planning

The further study mainly focuses on the relationship between carbon-based materials and plasma high-temperature treatment, conduct treatment and analysis on carbon-based materials such as graphite. Continue charging and discharging cycle tests on the batteries made of processed MCMB samples, and test the composite of MCMB and silicon treated at different temperatures to explore materials and processes with high specific capacity and stability. On this basis, we will carry out research on the manufacturing process of vacuum ultra-high temperature and high efficiency mesophase carbon microsphere graphitization, as well as the standardized intelligent control technology for efficient, stable and continuous operation, to explore the application prospects of plasma in the field of carbon-based material modification and treatment.

20.7 Conclusion

This article introduces the principle and development status of magnetic rotary thermal plasma (MRTP) treatment technology, MCMB plasma treatment system, experimental research, and comparative analysis of material treatment. Research has shown that MCMB can be quickly heated to over 3000 °C within a few minutes. The vacuum environment is conducive to the precipitation of volatile substances in MCMB. After heat treatment, its internal structure is reordered and the degree of

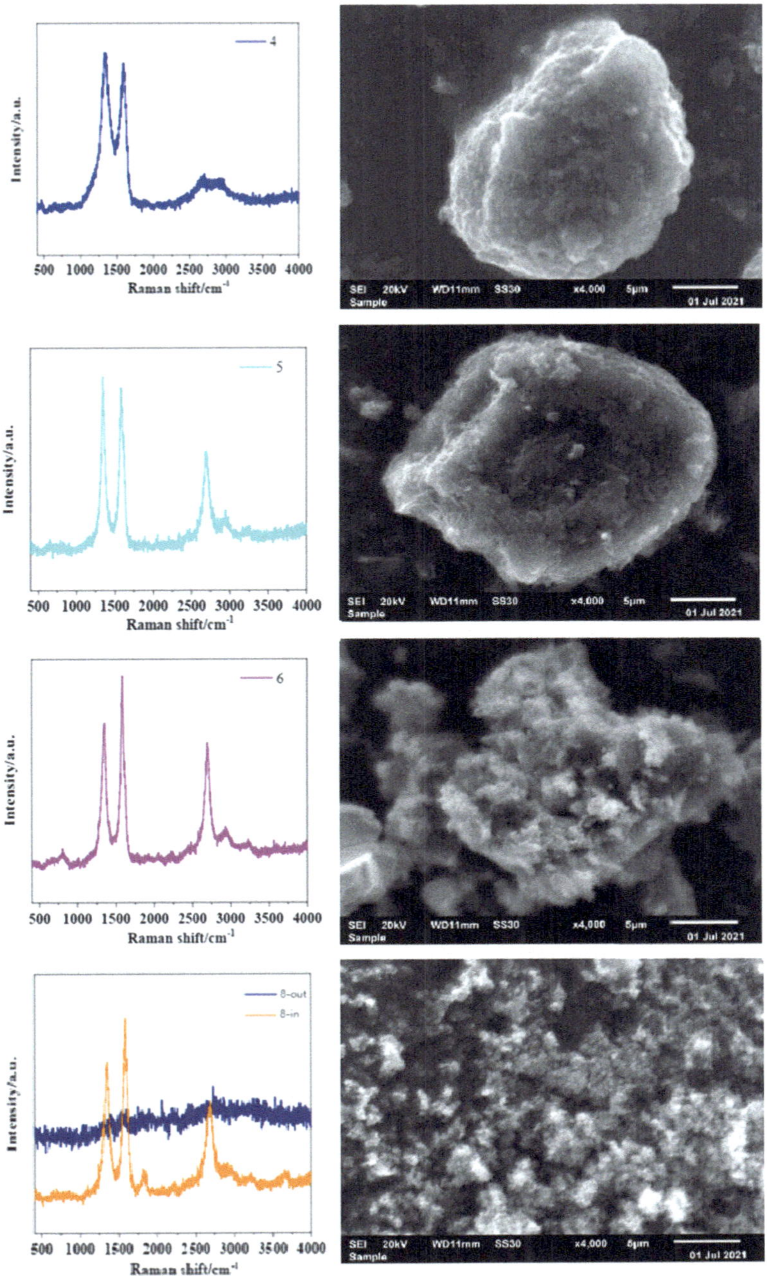

Fig. 20.17 Raman curves and SEM images of MCMB treated samples 4, 5, 6, and 8

Table 20.2 Graphite parameters after MCMB treatment

Samples	D position/ cm^{-1}	D strength	G position/ cm^{-1}	G strength	2D position/ cm^{-1}	D strength	D + G position/ cm^{-1}	D + G strength	ID/IG
4	1334.44	570.95	1587.22	530.15	2650.89	75.1	2906.51	59	1.0769594
5	1348.64	382.17	1585.8	359.11	2694.91	184.04	2947.69	52.09	1.0642143
6	1354.32	660.72	1585.8	909.62	2697.75	501.48	2932.07	87.55	0.7263693
8	1344.38	342.08	1584.38	442.92	2694.91	201.6	2915.03	53.47	0.7723291

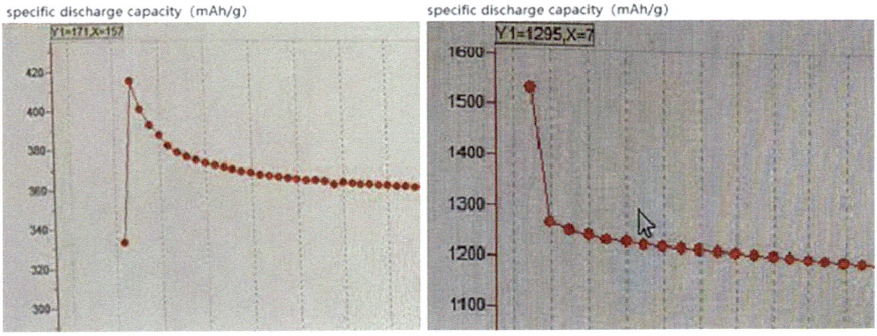

Fig. 20.18 Specific capacity test of MCMB processed samples

graphitization is significantly improved. The volume of MCMB samples treated with MRTP decreased, and the density and purity increased, especially the purity increased from 70% to over 94%. The processed MCMB sample was made into a battery and subjected to a charge discharge cycle test, with a capacitance of up to 360 mAh/g. By further mixing with silicon powder, the capacitance can be increased to 1200 mAh/g. MRTP treatment has changed the microstructure of MCMB and improved its thermal stability and electrochemical properties, giving it potential advantages for application in energy storage materials.

References

1. Z. Zhang, Study on the process and properties of self sintering mesophase carbon microspheres to prepare carbon graphite materials. National University of Defense Science and Technology (2023)
2. D. Le, Preparation of high-density and high-strength carbon/graphite materials using mesophase carbon microspheres (MCMB). Hunan University (2023)
3. T. Li, C. Wang, C. Guo, et al, Structural evolution characteristics of mesophase carbon microspheres during medium to low temperature carbonization. National Symposium on New Carbon Materials. Shanxi Institute of Coal Chemistry, Chinese Academy of Sciences (2005)
4. S. Dongliang, Plasma preparation and modified carbon materials. Suzhou University (2016). 10.7666/d. D0104065
5. S. Samal, Thermal plasma technology: the prospective future in material processing. J. Clean. Prod. **142**, 3131–3150 (2016). https://doi.org/10.1016/j.jclepro.2016.10.154
6. W. Ebeling, Thermal plasma fundamentals and applications. Zeitschrift f ü r Physikalische Chemie **190**, 152–153 (1995). https://doi.org/10.1524/zpch.1995.190.Part_1.152a

Chapter 21
Synthesis of Alloyed Nanoparticles from Perovskite Oxide by Plasma

Qinghao Li, Zihe Zhu, Jiaming Yang, Lei Fu, Jun Zhou, and Kai Wu

Abstract Bimetallic alloy nanoparticles (NPs) exsolved on the surface of perovskite oxides (ABO_3) alter the structural and electronic properties of the primarily monometallic counterparts. These synergistic effects make the whole material present superior electrolytic and catalytic performances. The strong bonding interaction between alloy NPs and perovskite substrate enables the stability and durability of the catalysis. However, traditional thermal exsolution requires a long-time treatment at a high temperature. Researchers have been pursuing more time-saving and energy-efficient alternative routes. Herein, a low-temperature and fast pathway based on the dielectric barrier discharge (DBD) reactor is provided to *in-situ* exsolve alloy NPs. A favorable activity of the catalyst for photothermal catalysis is demonstrated. Additionally, we conclude that the factor that restricts metallic agglomeration is the morphology of the substrate. Even after an overtime plasma treatment, the exsolved alloy NPs are still embedded in the porous structure of perovskite oxides.

21.1 Introduction

Over the past few decades, there has been a significant increase in energy consumption attributed to global population growth and the notable improvement in people's quality of life [1]. Unfortunately, the excessive use of fossil fuels has resulted in the release of greenhouse gases (primarily CO_2), leading to global warming, energy depletion, and environmental pollution. To address these challenges and promote sustainability, converting CO_2 into industrial feedstock offers a promising solution. Among the various methods available, the reverse water gas conversion reaction (RWGS) involves the conversion of CO_2 and H_2 into syngas (CO and H_2O), serving as a fundamental and crucial intermediate step in numerous CO_2 hydrogenation

Q. Li · Z. Zhu · J. Yang · L. Fu · J. Zhou (✉) · K. Wu
State Key Laboratory of Electrical Insulation and Power Equipment, Xi'an Jiaotong University, Xi'an 710049, Shaanxi, China
e-mail: zhoujun@mail.xjtu.edu.cn

Z. Fang et al. (eds.), *Proceedings of the 5th International Symposium on Plasma and Energy Conversion*, Springer Proceedings in Physics 398,
https://doi.org/10.1007/978-981-97-2245-7_21

reactions. The RWGS reaction facilitates the production of raw materials for Fischer–Tropsch synthesis, enables the creation of high-value-added fuels, and plays a pivotal role in achieving effective CO_2 recycling [2].

To enhance the catalytic performance of catalysts, researchers have developed heterojunctions by depositing metal nanoparticles onto oxide surfaces, thereby generating more active sites on the catalyst [3]. This strategy significantly enhances both the activity and selectivity of the reaction. Traditional methods to fabricate metal–semiconductor heterojunctions such as impregnation, physical mixing, and chemical vapor deposition suffer from challenges including metal particle agglomeration, uneven distribution, and limited durability. Recently, the *in-situ* exsolution of oxides has aroused great interest. A-site deficient perovskite oxides ($A_{1-\alpha}BO_{3-\delta}$) are reduced at a high temperature, and the B-site cations are exsolved towards the surface. Unlike conventional deposition methods, the exsolved metal nanoparticles possess the unique ability to integrate with the oxide surface, resulting in excellent dispersion, stability, and catalytic properties [4]. This technique has been extensively applied as catalysts in various energy-related fields. Nevertheless, the current standard exsolution procedure refers to high energy consumption, a long treatment period (>8 h), and safety concerns associated with the use of hydrogen gas as the reducing agent.

In this work, the novel plasma exsolution at room temperature was achieved. The A-site deficient and B-site doped perovskite $La_{0.43}Ca_{0.37}Ti_{0.86}Ni_{0.08}Cu_{0.06}O_{3-\delta}$ was treated by reduced atmosphere under DBD plasma. The obtained hybrid metal–semiconductor catalyst exhibited favorable performance for RWGS reaction in photothermal conditions. In topography, the exsolved alloys were embedded in the pores of the semiconductors and the porous structure of the substrate constrains the size of the nanoparticles, preventing uncontrolled growth. Totally, the plasma-driven exsolution provides a time and energy-efficient approach for synthesizing high-performance catalysts and introduces an innovative route for the exsolution of metal nanoparticles.

21.2 Experimental

$La_{0.43}Ca_{0.37}Ti_{0.86}Ni_{0.08}Cu_{0.06}O_{3-\delta}$ was prepared via the sol–gel method. Stoichiometric amounts of lanthanum nitrate hexahydrate [$La(NO_3)_3 \cdot 6H_2O$, 99.9%], nickel nitrate [$Ni(NO_3)_2 \cdot 6H_2O$, 98%], copper nitrate hydrate [$Cu(NO_3)_2 \cdot xH_2O$], calcium acetate [$2(C_2H_3O_2)Ca$, 99.0%], tetrabutyl titanate ($C_{16}H_{36}O_4Ti$), and citric acid ($C_6H_8O_7$, 99.8%) were dissolved in deionized water. The molar ratio of citric acid to total metal ions was 3:1. Subsequently, the solution was heated to 200 °C by the magnetic stirrer and stirred for 5 h to dry the water and obtain the sol. Then we moved the LCuR sol into the oven for overnight heating at 150 °C to get the dry gel. After the ball milling, the precursor underwent the subsequent calcination process. The sample was heated to 400 °C from room temperature by 1 °C/min and held at 400 °C for 1 h. Then the sample was heated from 400 to 850 °C in 100 min and

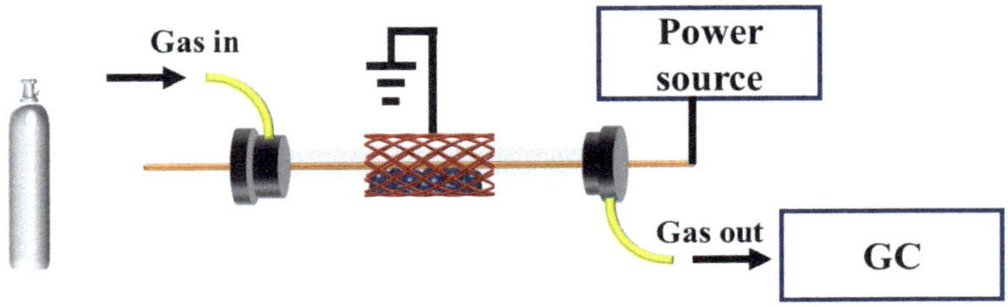

Fig. 21.1 Schematic diagram of the experimental setup

calcined at 850 °C for 6 h. The same experimental method was finished to make a non-doped sample $La_{0.43}Ca_{0.37}TiO_3$ (denoted as LCT).

As depicted in Fig. 21.1, the entire DBD device was assembled by two sealed valves within a 15 cm long glass tube. A 1 mm diameter copper rod served as the high-voltage electrode of the DBD, while a 5 cm wide copper mesh was wrapped around the exterior of the glass tube, serving as the grounding electrode. The length of the plasma discharge gap was 4 mm with a 5 mL discharge volume. We put 0.05 g sample at the bottom of the discharge area and inserted quartz cotton to prevent the catalyst from dispersing outside the discharge zone. To exsolve Ni–Cu alloy nanoparticles, a 5% H_2 gas with a flow rate of 25 mL/min was introduced. The DBD reactor was powered by a Suman CTP-2000 K, and the sample was treated at 30 W discharge power for 30 min.

XRD patterns of all the samples were obtained by X-ray diffraction (XRD, Bruker D8 Advance X-ray) and the step (2θ) was set from 10 to 85° with a step of 0.2°. The nanostructure and morphology of the LCuRpl samples were monitored by a scanning electron microscope (SEM, Zeiss Gemini SEM 500) and transmission electron microscope (TEM, Thermo Fisher Talos F200X). The RWGS reaction was conducted on Aulight OPTH platform. We placed 0.15 g catalyst in the center of a quartz tube and fully inlet 7.5 mL/min CO_2 and 30 mL/min H_2 into the system. Then after the platform was heated to 300 °C, the full-spectrum Xe light source was activated. The output gas of the photothermal reaction was directed into GC (Agilent) for analysis, and the concentration of the output gas was measured at a 30-min interval.

21.3 Results and Discussion

Figure 21.2 illustrates the XRD patterns of LCuR and LCuRpl. The peak intensities of these samples correspond well with the $CaTiO_3$ (JCPDS 22-0153). The highly crystalline nature of orthorhombic perovskites is evident, indicating that the 30-min plasma treatment has a minimal impact on the crystal structure. The peak positions of LCuR and LCuRpl exhibit significantly larger angle shifts compared to $CaTiO_3$. This can be attributed to oxygen vacancies introduced by A-site deficiencies, resulting in reduced cell volume [5]. Due to the small volume and wide distribution of exsolved alloy nanoparticles, the peaks of these nanoparticles cannot be recognized. After the introduction of Cu^{2+} (0.73A) and Ni^{2+} (0.69A) to replace Ti^{4+} (0.60A), the cell volume will increase but in the plasma environment cations are converted to the nanoparticles on the surface, weakening the effect of lattice expansion. Consequently, the significant peak shift to a larger angle in LCuRpl reflects the material's structural changes.

Figure 21.3 depicts the porous structure of the LCuRpl. The substrate bulks with pores of 50 ~ 200 nm are fabricated by the sol–gel method, which has large surface areas, providing more active sites on the surface of the catalyst [6]. We can observe that the alloy nanoparticles with an approximate diameter of 100 nm are embedded in the pores. This distinctive metal–semiconductor mosaic morphology is rare when compared to other samples produced via the sol–gel method and *in-situ* exsolution. In the patterns obtained from backscattered electron imaging (BSD mode), the contrast difference between the pores and the particles is readily discernible, signifying their distinct elemental composition.

Fig. 21.2 XRD patterns of LCuRpl and LCT

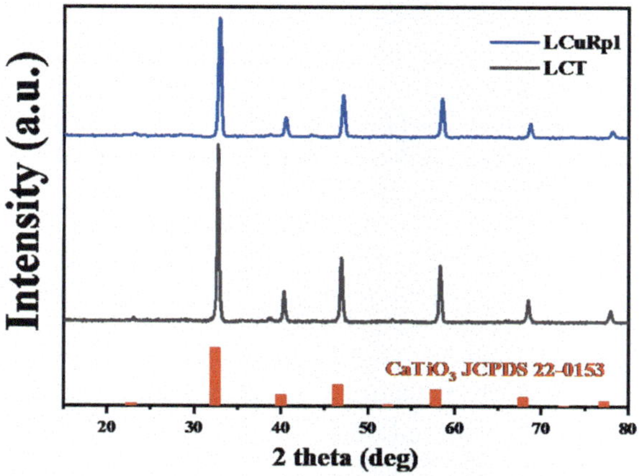

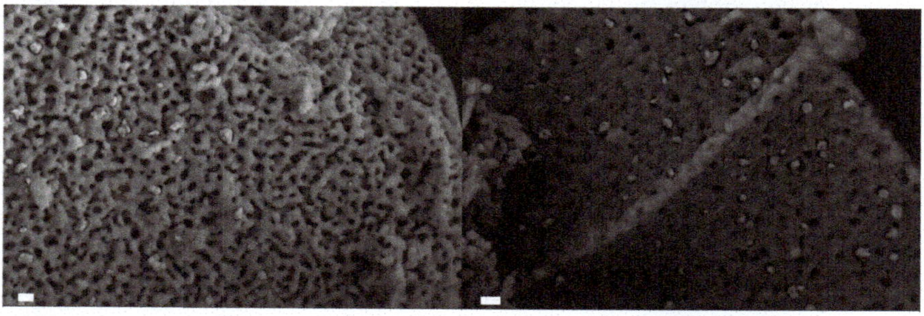

Fig. 21.3 SEM images of LCuRpl (left) and the SEM image in BSD mode (right)

The high-resolution TEM images of LCuRpl provide clear evidence that the exsolved CuNi alloys are embedded in the pores of $CaTiO_3$ (Fig. 21.4). The spacing of lattice fringes of 0.179 nm is in accordance with the (2 0 0) crystal plane of CuNi and the diameters of these NPs are ~ 110 nm. The characteristic spacing of 0.274 nm corresponds to the (1 2 1) lattice planes of $CaTiO_3$. Furthermore, HAADF and EDS mappings were carried out to detect the distribution of elements (Fig. 21.5). The continuous and uniform distributions of Ca, La, O, and Ti elements are ascribed to the substrate. The presence of some aggregations of Ni and Cu elements serves as further evidence of the *in-situ* growth of CuNi alloys, embedded within the pores. The exsolved nanoparticles are dipersely distributed, explaining the reason why CuNi peaks are not found in the XRD patterns of LCuRpl. In conclusion, LCuRpl synthesized via the sol–gel method exhibits favorable crystalline characteristics and a porous structure and the exsolved alloys are readily observable.

The fine-scanned XPS spectra of La 3d, Ti 2p, O 1s, and Cu 2p for LCT, LCuR, and LCuRpl are presented in Fig. 21.6. As shown in Fig. 21.6a, the La 3d spectra illustrate that B-site doping leads to a shift in the La 3d peaks towards higher binding energy. This shift is attributed to alterations in the stoichiometric ratio [7]. Additionally, the doping of Ni and Cu also affects the valence state of La. As depicted in Fig. 21.6b, a reduction in the B-site element Ti is evident. The sharp peaks of Ti^{4+} at 458.2 eV in LCuR and LCuRpl can be observed, and the peaks of Ti^{3+} at 462.6 eV are apparent in LCuRpl. The proportions of Ti^{4+} and Ti^{3+} are 95.75 and 4.25%, indicating a conversion of a certain amount of Ti^{4+} to Ti^{3+}, resulting in an increase in oxygen vacancies.

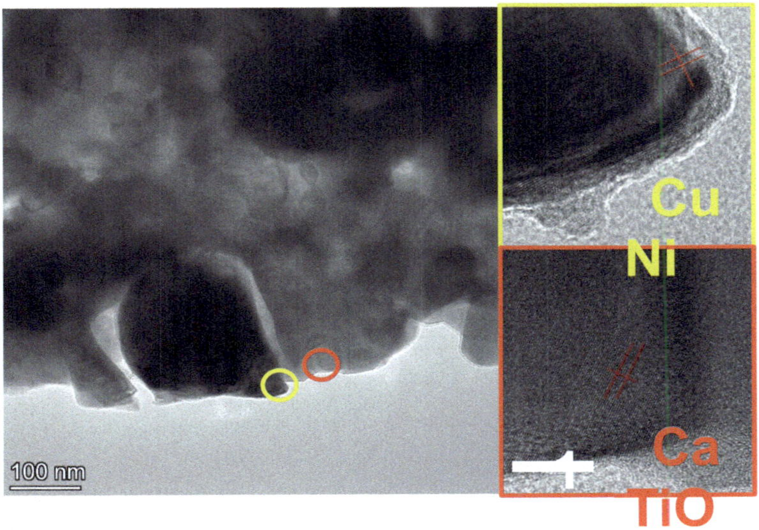

Fig. 21.4 TEM patterns of LCuRpl

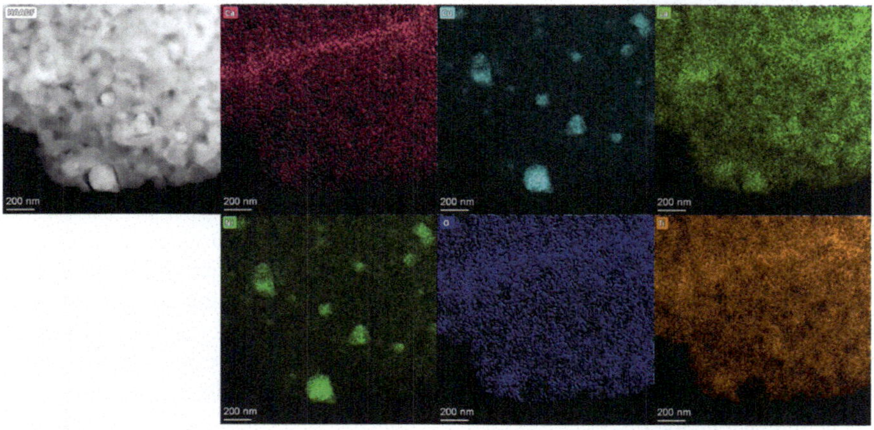

Fig. 21.5 HAADF and EDS mappings of LCuRpl

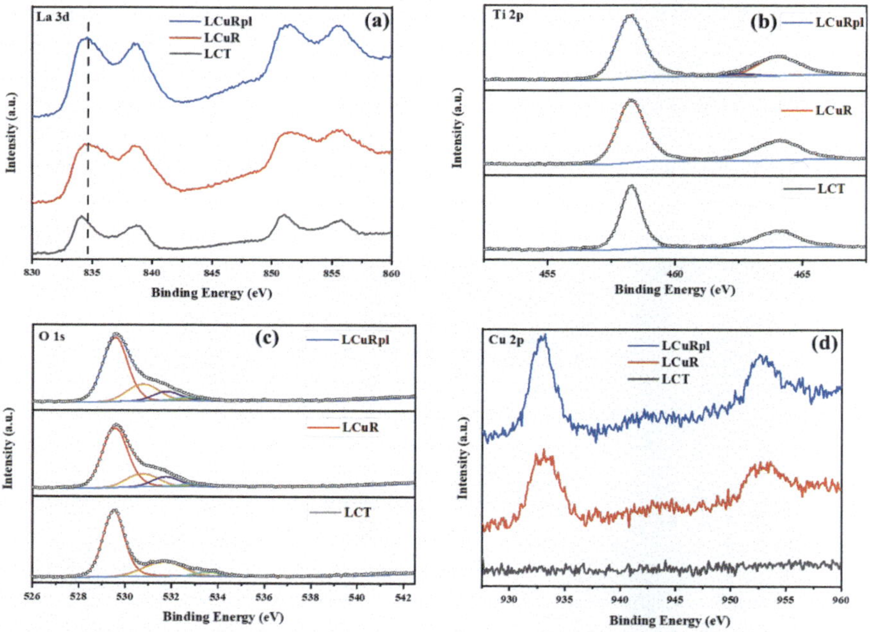

Fig. 21.6 XPS spectra of **a** La 3d, **b** Ti 2p, **c** O 1 s, and **d** Cu 2p in LCuRpl

In the high-resolution O 1s patterns (Fig. 21.6c), peaks at 529.5, 530.7, 531.7, and 532.8 eV correspond to four oxygen species: lattice oxygen, oxygen vacancies, adsorbed surface molecular oxygen, and absorbed H_2O, respectively [8]. The content of lattice oxygen presents a decrease from 62.53 (LCuR) to 65.18% (LCuRpl), due to the absence of perovskite structure after reduction. The content of oxygen vacancies increases from 17.16 (LCuR) to 21.18% (LCuRpl). Additionally, as demonstrated in Fig. 21.6d, the Cu 2p spectrum confirms the successful incorporation of Cu elements into the LCuR sample.

The photothermal and thermal RWGS reactions of LCuRpl were conducted at various temperatures. The experimental results are presented in Fig. 21.7. In terms of CO production, the RWGS reaction can be sustained at a very low gas production rate until 200 °C in the absence of light. Under dark conditions, the gas production rates at 250 and 300 °C can reach half of those under photothermal conditions, 0.112 and 0.281 mmol/g/h, respectively. In comparison to the thermal reaction, the photothermal reaction exhibits a significant increase in yield, which further intensifies with rising temperatures, reaching its maximum at 300 °C (0.530 mmol/g/h). The CH_4 production results are shown in Fig. 21.7. Below 250 °C, the peak of CH_4 is not observable in the chromatography. However, at 300 °C, in both thermal and photothermal reactions, there are CH_4 with values of 2.623 and 3.272 μmol/g/h, respectively. Throughout the RWGS process, the selectivity of CO remains consistently above 99%.

The enhanced performance of the RWGS reaction under light can be attributed to the origin of LCuRpl. The doping introduces an intermediate energy level within the sample, facilitating the migration of valence band electrons. This is evident in the broadening of the visible light absorption band [9]. Additionally, the interaction between CuNi alloy particles and the semiconductor substrate allows electrons to traverse the Schottky barrier between the alloy and semiconductor, leading to their reaction with CO_2 on the alloy surface [10]. This process hinders the recombination of photogenerated charges, consequently increasing the reaction rate.

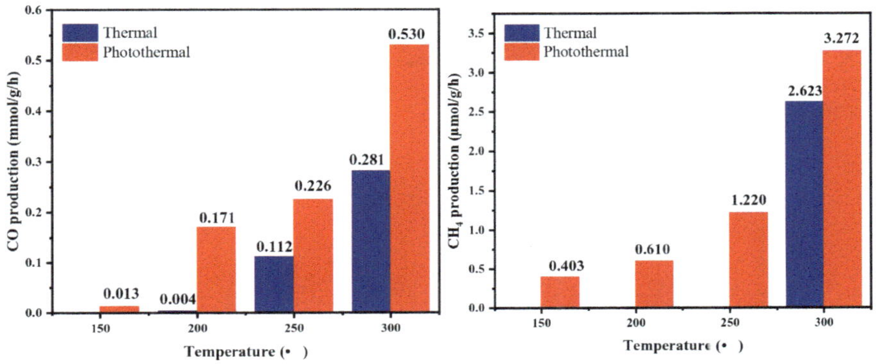

Fig. 21.7 CO (left) and CH_4 (right) production rates of LCuRpl in thermal and photothermal conditions

21.4 Conclusion

In this study, we achieved the *in-situ* exsolution of CuNi alloy on a semiconductor substrate through plasma, completing the entire process rapidly at room temperature. This approach significantly reduces preparation time and energy consumption in comparison to traditional high-temperature exsolution processes. With the *in-situ* exsolved alloy embedded in semiconductor substrates, the pore diameter restricts the unrestricted growth of alloy particles. The resulting alloy-perovskite catalyst exhibits favorable performance in the RWGS reaction, yielding a CO output of 0.503 mmol/g/h.

References

1. Y. Wang, Q. Zhou, Y. Zhu, D. Xu, High efficiency reduction of CO_2 to CO and CH_4 via photothermal synergistic catalysis of lead-free perovskite $Cs_3Sb_2I_9$. Appl. Catal. B Environ. **294**, 120236 (2021)
2. A. Kamkeng, M. Wang, J. Hu, W. Du, F. Qian, Transformation technologies for CO_2 utilisation: current status, challenges and future prospects. Chem. Eng. J. **409**, 128138 (2021)
3. N. Xiao, S. Li, X. Li, L. Ge, Y. Gao, N. Li, The roles and mechanism of cocatalysts in photocatalytic water splitting to produce hydrogen. Chin. J. Catal. **41**, 642–671 (2020)
4. J. Zhang, M. Gao, J. Luo, In situ exsolved metal nanoparticles: a smart approach for optimization of catalysts. Chem. Mater. **32**, 5424–5441 (2020)
5. H. Miao, X. Wu, B. Chen, Q. Wang, F. Wang, J. Wang, C. Zhang, H. Zhang, J. Yuan, Q. Zhang, A-site deficient/excessive effects of $LaMnO_3$ perovskite as bifunctional oxygen catalyst for zinc-air batteries. Electrochim. Acta **333**, 135566 (2020)
6. K. Zhuang, P. Jin, L. Yang, J. Yao, L. Yu, Z. Sheng, X. Chu, Z. Zhuang, X. Chen, Different morphologies on Cu–Ce/TiO_2 catalysts for the selective catalytic reduction of NO_x with NH_3 and DRIFTS study on sol-gel nanoparticles. RSC Adv. **13**, 25989–26000 (2023)
7. J. Faye, A. Baylet, M. Trentesaux, S. Royer, F. Dumeignil, D. Duprez, S. Valange, J. Tatibouët, Influence of lanthanum stoichiometry in $La_{1-x}FeO_{3-\delta}$ perovskites on their structure and catalytic performance in CH_4 total oxidation. Appl. Catal. B Environ. **126**, 134–143 (2012)
8. Q. Ji, L. Bi, J. Zhang, H. Cao, X. Zhao, The role of oxygen vacancies of ABO_3 perovskite oxides in the oxygen reduction reaction. Energy Environ. Sci. **13**, 1408–1428 (2020)
9. N. Guo, Y. Zeng, H. Li, X. Xu, H. Yu, Crumpled and flexible cotton-fiber-like TiO_2 with Pt anchored and its notable photocatalytic activity facilitate by Schottky junction interface. Mater. Lett. **221**, 183–186 (2018)

Chapter 22
Effect of PDMS Concentration on the Flashover Voltage Improvement of Epoxy Resin by Plasma Modification

Jincheng Chu, Zhengbing Tian, Jiatao Ding, Zhaorui Luo, Pengfei Wei, Zijian Zhou, Xi Zhu, and Zhi Fang

Abstract With the increase of voltage level in power systems, epoxy resin, as a commonly used insulating material, is prone to surface flashover under strong electric field. Research has shown that the use of low-temperature plasma to deposit silicon-containing films on the material surface could effectively improve the flashover voltage. In this paper, through treating the epoxy resin (EP) surface with Ar/Polydimethylsiloxane (PDMS)/Surface Dielectric Barrier Discharge (SDBD), the influence of PDMS addition (5, 15, 25 mL/min) on the surface properties of EP was studied. The results showed that with the increase of PDMS concentration, the morphology of the deposited silicon-containing film on EP surface changed from discrete particles to agglomerated coral-like structures, and then to stacked particles. Surface water contact angle and electrical insulating properties both showed a trend of increasing first and then decreasing, and the best surface performance appeared at PDMS addition of 15 mL/min, where the flashover voltage and water contact angle were increased by up to 75.2% and 61.5%, respectively. The improvement of surface properties is associated with the deposited silicon-containing film, with the changing of physical morphology and chemical compositions.

22.1 Introduction

In recent years, with the continuous operation and upgrading of power systems and electrical equipment, key insulating components have experienced a gradual decline in their electrical resistance and weather fastness due to aging. To improve

---

J. Chu · Z. Luo · P. Wei · Z. Zhou · X. Zhu (✉) · Z. Fang
College of Electrical Engineering and Control Science, Nanjing Tech University, Nanjing 211816, China
e-mail: zhuxi210825@163.com

Z. Tian · J. Ding
ZTT Cable Accessories Co., Ltd, Nantong 226010, Jiangsu, China

Z. Fang et al. (eds.), *Proceedings of the 5th International Symposium on Plasma and Energy Conversion*, Springer Proceedings in Physics 398,
https://doi.org/10.1007/978-981-97-2245-7_22

the surface characteristics of insulations and enhance their service life, insulation modification techniques have been employed. Currently, plasma, as an emerging technology, is widely utilized in the field of insulating material modification due to its advantages such as excellent modification effects, environmental friendliness, diverse structures and high controllability. Plasma modification of insulating materials often involves methods such as etching and deposition. Etching by plasma is to tailor the morphology of the material surface, effectively roughening the material surface; by contrast, deposition method is to pass into a medium containing specific elements during the discharge process, destroy internal molecular structure through electrical energy, recombine molecules, produce sedimentary substances and eventually form a film with specific elements on the material surface. Both etching and deposition methods can effectively enhance electrical insulation and weathering performance, herein reducing insulation safety hazards.

Currently, plasma modification offers a variety of discharge structures, different precursor substances and the ability to treat various types of insulating materials. Chen et al. utilized CF_4/He atmospheric pressure plasma jet to treat epoxy resin, and results showed that the flashover voltage on the surface of the treated epoxy resin is increased by 25–30% [1]. Cui et al. used octamethylcyclotetrasiloxane (OMCTS) as a reactant medium and employed plasma jet to generate silicon-containing films on the surface of epoxy insulation layers, adjusting their surface energy, flashover voltage and surface conductivity [2]. Shao et al. performed etching treatment on the surface of epoxy resin using Dielectric Barrier Discharge (DBD) in a pure oxygen environment, and it demonstrated that the plasma treatment increased the surface roughness of the epoxy resin, enhancing its surface electrical characteristics [3]. Furthermore, researchers conducted DBD plasma treatment on PMMA in the Ar/CF_4 environment, and the results showed that the surface hydrophobicity of PMMA and the flashover voltage both increased [4].

It has been discovered that Surface Dielectric Barrier Discharge (SDBD), as a form of DBD, possesses advantages such as fast response speed and the ability to achieve high efficient plasma discharge under normal experiment conditions, making it more advantageous for material treatment. However, there is currently insufficient exploration of the modification mechanisms for SDBD-treated materials. Therefore, this research focuses on SDBD as the chosen discharge form for film deposition on the surface of epoxy resin, and Ar/PDMS is selected as the precursor substance to explore the changes in various surface properties of epoxy resin (EP) at different PDMS concentrations.

22.2 Experiment Set Up

22.2.1 Sample Preparation

The pre-treated material chosen in this research is a solid epoxy resin (EP with dimensions of 100 × 100 × 1 mm, obtained from Beijing Xin Furunda Insulation Material Co., Ltd.) After the treatment, the samples' surfaces were subjected to cutting and processing to investigate their surface properties.

22.2.2 SDBD Discharge Experimental Platform

The atmospheric pressure Ar/SDBD system with PDMS addition was used to treat the surface of EP, as shown in Fig. 22.1. The Ar/PDMS gas mixture first enters through a 5-hole inlet and then compressed in the reactor before being blown out from the front end of the high-voltage electrode. The high-voltage electrode (100 mm long, 10 mm wide) is placed near the material, while the ground electrode (100 × 100 mm) is placed below a glass plate. The EP material and the glass plate are applied as barrier media. The experimental discharge is powered by a homemade nanosecond pulse power supply, with adjustable frequency (0–10 kHz) and voltage amplitude (0–20 kV). Based on previous experimental data, the optimal parameters for stable and highly active plasma discharge were determined as follows: 9 kV voltage, 1 kHz frequency, 100 ns rise/fall time, and 800 ns pulse width. Controlled by an ADW-20k10a flow meter, discharges is performed for 5 min using Ar (1.75 L/min) and PDMS as the precursor material at flow rates of 5, 15, and 25 mL/min.

22.2.3 Characterization Methods

To ensure the accuracy of the testing, all measurement points are set at a position 50 mm away from the high voltage electrode. The following characteristics will be tested.

(1) Material hydrophobicity

With the aim of obtaining the hydrophobicity of materials before and after treatment, the water contact angle (WCA) test is employed. A droplet of 3μL deionized water is carefully placed on the material surface, and the angle formed between the water droplet and the material is automatically generated by a computer.

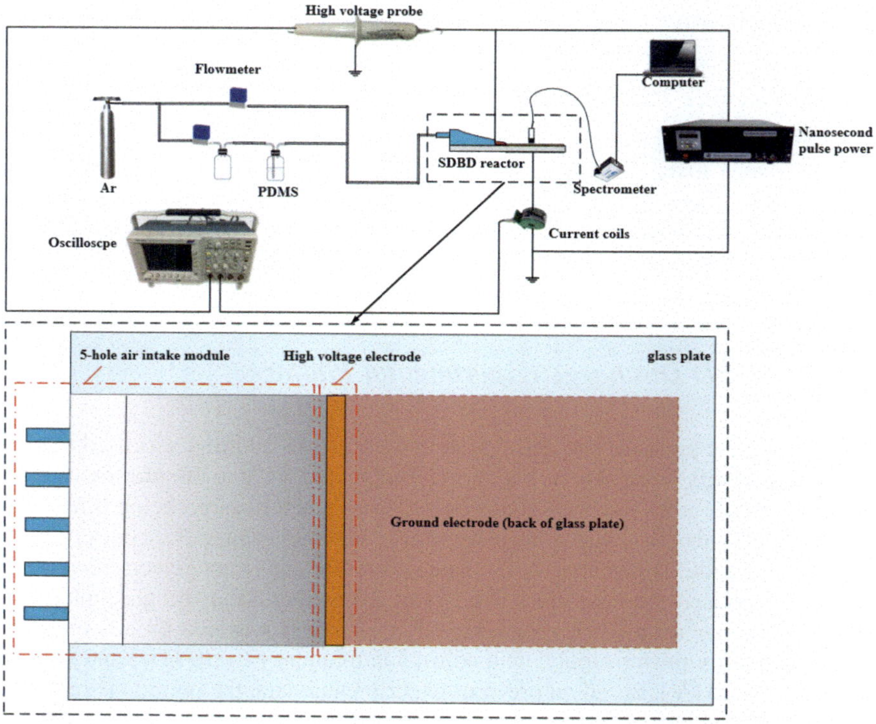

Fig. 22.1 Ar/PDMS discharge experimental system based on SDBD

(2) Electrical insulating properties

Testing the electrical insulation performance of EP requires the essential measurement of flashover voltage. The flashover voltage test primarily consists of a high-voltage power supply, current coil, high-voltage probe and oscilloscope. The platform is composed of polytetrafluoroethylene (PTFE), wherein a finger-shaped electrode with a radius of 10 mm and a spacing of 5 mm is placed.

The voltage is gradually increased until flashover occurs, and the voltage value at that moment is recorded. Finally, the Weibull distribution is used to analyze the flashover data (U) using Eq. (22.1) [5].

$$P(U) = 1 - \exp\left(-\left(\frac{U}{U_f}\right)^{\beta}\right) \quad (22.1)$$

where $P(U)$ is the probability of flashover, β is the shape parameter and U_f is the characteristic flashover voltage.

(3) Physical and chemical properties

For the purpose of investigating the surface morphology of EP before and after treatment, the scanning electron microscope (SEM) method is utilized. Magnifications of 0.5 kx and 30 kx are chosen to observe the deposition effect at low magnification and to investigate the morphology of the substances at high magnification.

To characterize the chemical functional groups on the surface, fourier transform infrared spectroscopy (FTIR) is employed. The measurements are made with a resolution of 0.1 cm^{-1} and a testing range of 500–4000 cm^{-1}.

The electron spectrometer is utilized to measure the energy distribution of photoelectrons and Auger electrons emitted from the surface of the sample under X-ray photon irradiation. This analysis is conducted to qualitatively analyze the elemental composition and molecular structure of the EP surface.

22.3 Experiment Results

22.3.1 Spectral Characteristics of SDBD

The emission spectrum can effectively reflect the concentration of active particles involved in the discharge process. As shown in Fig. 22.2, the emission spectrum of SDBD discharge under different PDMS concentrations primarily exhibits variations in the peaks of Ar (696.16–851.32 nm) and Si (514.79–524.23 nm). It can be observed that with an increase in PDMS concentration, the spectral intensity of Ar particles gradually decreases, while the concentration of Si particles initially increases and then decreases. Among them, the emission intensity of Si is highest at 15 mL/min. It can be seen that an increase in PDMS concentration inhibits the ionization intensity of Ar particles, but within an appropriate concentration range, it effectively enhances the concentration of Si active particles.

22.3.2 Water Contact Angle

After treatment with different PDMS concentrations, a hydrophobic silicon film is deposited on the surface of EP. The results of the water contact angle (WCA) test are shown in Fig. 22.3. It can be seen that the untreated EP surface has a WCA of 89.2°. After treatment with different PDMS concentrations, the surface WCA shows a significant improvement compared to the untreated surface, achieving hydrophobic to superhydrophobic properties. With an increase in PDMS concentration, the WCA initially increases and then decreases. It can be observed that at a PDMS concentration of 15 mL/min, the WCA on the EP surface reaches a superhydrophobic level of 156.3°, which represented a 75.2% increase compared to the untreated EP. The enhancement of surface hydrophobicity is beneficial for improving the weather resistance of insulating materials and extending the lifespan of insulators.

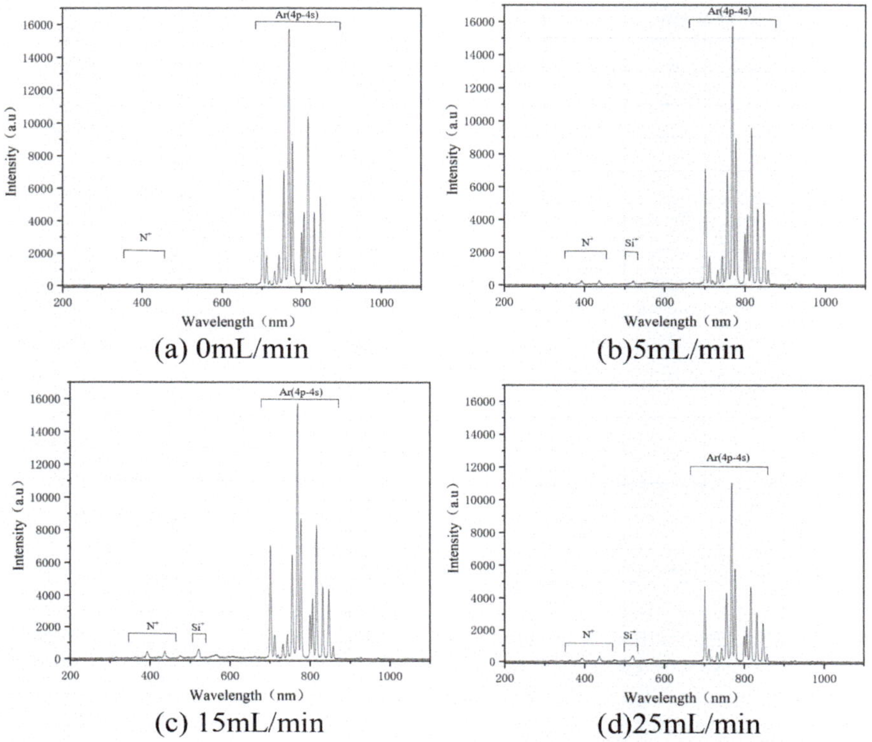

Fig. 22.2 Emission spectra at different concentrations

Fig. 22.3 WCA of EP surface under different PDMS additions

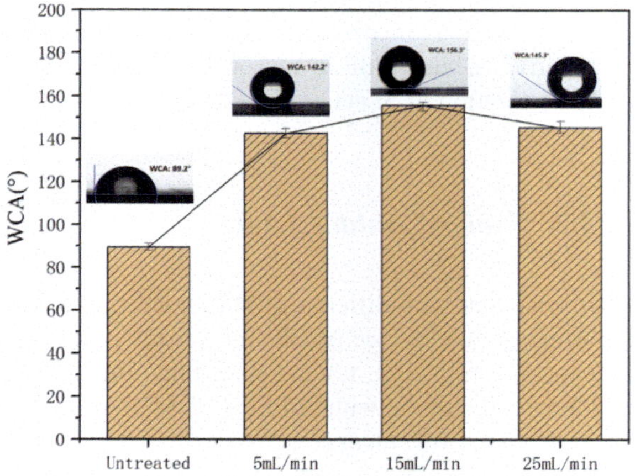

22.3.3 Flashover Voltage

The Weibull distribution is used to analyze the flashover voltage on the surface of EP with different PDMS concentrations, as shown in Fig. 22.4 and Table 22.1. First of all, it can be seen that the flashover voltage of the untreated EP sample is only 7.8 kV, which corresponds to the distribution line on the far left in the figure. After SDBD treatment, the surface flashover voltage distribution line moves to the right, and the flashover voltage changes in the range of 10.8–12.6 kV. As the PDMS concentration increases, the surface flashover voltage increases first and then decreases. At a concentration of 15 mL/min, the maximum value of 12.6 kV is reached, which corresponds to the distribution line on the far right in the figure, which is 61.5% higher than the untreated sample. The flashover voltage is one of the parameters that reflects the insulation properties of the sample. Therefore, in order to obtain a larger flashover voltage, the adjustment of PDMS concentration is essential.

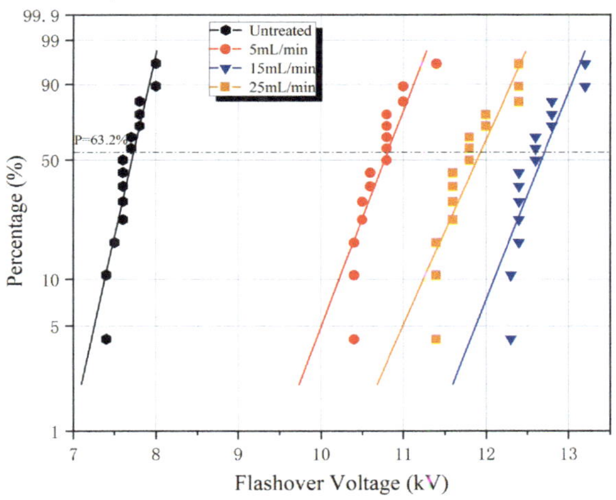

Fig. 22.4 Weibull distribution of flashover voltage

Table 22.1 Flashover voltage value table

Sample	Untreated	5 mL/min	15 mL/min	25 mL/min
U_f (kV)	7.8	10.8	12.6	11.9

22.4 Discussion

22.4.1 Effect of Plasma Conditions on Surface Morphology

The surface morphology of the EP, both untreated and treated with different PDMS concentrations, is observed using SEM at magnifications of 0.5 and 30 kx, as shown in Fig. 22.5. The untreated sample exhibited a smooth and flat surface. After being treated with different PDMS concentrations, the surface morphology changes significantly. At a PDMS concentration of 5 mL/min, the surface material appears as irregular particles and is relatively sparse. At a PDMS concentration of 15 mL/min, the surface material accumulated in a coral-like structure with a high density of deposition. At a PDMS concentration of 25 mL/min, the surface material appeared as particles with a higher deposition density. With an increase in the PDMS concentration, the degree of surface deposition initially increases and then decreases. From the emission spectrum (Fig. 22.2), it is evident that at lower PDMS concentrations, there is limited participation of Si elements in the discharge reaction, resulting in an inability to form dense film deposition on the material surface. Additionally, excessive addition of the medium not only affects the ionization intensity of Ar gas but also reduces the ionization intensity of Si elements, thereby reducing the surface deposition density.

22.4.2 Effect of Plasma Conditions on Chemical Components

From Fig. 22.6, it can be observed that the untreated EP surface primarily contains organic peaks such as C–O–C (1000–1050 cm^{-1}), O–CH_3 (1100–1260 cm^{-1}), and C–H (780–975 cm^{-1}). After SDBD treatment, the surface features of EP exhibit silicon-containing groups such as Si–CH_3 (1200–1260 cm^{-1}), Si–O–Si (1025–1220 cm^{-1}), and Si–H (866 cm^{-1}). It can be seen that under SDBD treatment, regardless of the variation in the medium concentration, it is not possible to quantitatively determine the content of these functional groups. For comparison, the two groups of functional groups with the highest absorption peaks, Si–O–Si and Si–CH_3, are selected. At a PDMS concentration of 5 mL/min, the Si–O–Si/Si–CH_3 ratio is 125.6%, while at a concentration of 15 mL/min, the ratio is 107.9%. At a concentration of 25 mL/min, the ratio becomes 97.5%. This is due to the molecular formula of PDMS fragments, $CH_3[Si(CH_3)_2O]_nSi(CH_3)_3$, of which Si–$CH_3$ occupies the largest proportion. As the concentration of PDMS increases, the degree of dissociation of Si–CH_3 gradually decreases, and the PDMS molecules remain on the surface of the material in their most primitive state, resulting in a gradual decrease in the ratio of Si–O–Si/Si–CH_3.

Use XPS to observe the relative element composition of the EP surface as shown in Table 22.2. The Si2p sub-peak fitting diagram is shown in Fig. 22.7. First of all, it can be seen from the elemental composition that on the surface of untreated EP, element C accounts for 73.2%, element O accounts for 21%, and there is a trace

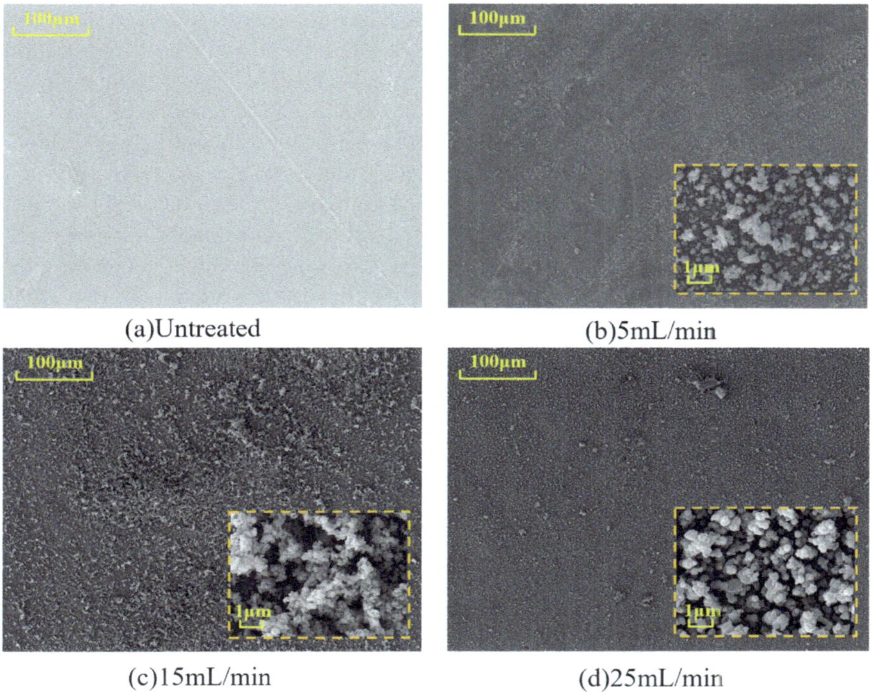

Fig. 22.5 Surface morphology of EP before and after SDBD treatment

Fig. 22.6 FTIR results of EP before and after SDBD treatment

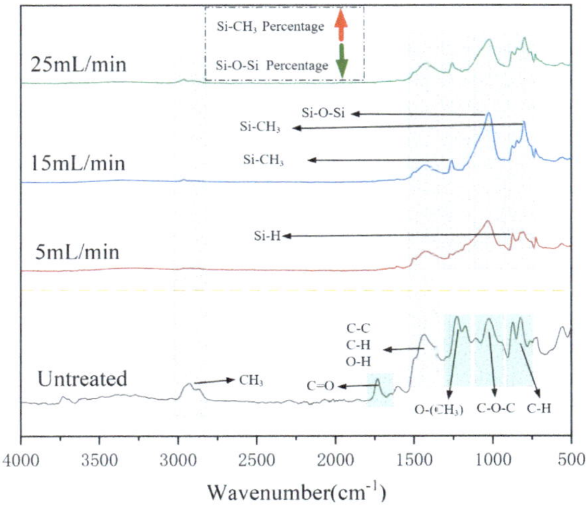

amount of Si element (5.8%). After SDBD discharge treatment with different PDMS concentrations, the EP surface element changes significantly. As the concentration increases, the C element gradually increases from 28.1 to 53.7%, and element O and Si gradually decreases from 45.0 and 26.9% to 25.5 and 20.8%, respectively. It can be seen from the peak fitting of Si2p that after treatment with different PDMS concentrations, the number of different chemical bonds on its surface changes significantly. At first, when the PDMS concentration is 5 mL/min, the surface chemical bond has Si–O_4 as the dominant, accounting for 59.6%; when the concentration is increased to 15 mL/min, the surface chemical bond is Si–O_3 as the dominant, accounting for 57.5%; when the concentration is 25 mL/min, the highest proportion of its surface is the Si–O_2 bond, accounting for 59.2%. It can be seen from the changes in the content of surface chemical bonds that the increase in the concentration of PDMS will reduce the fracture of the Si–CH_3 bond in its molecule, and the insufficient binding of the O atom is deposited on the EP surface in different fragmented forms.

It is noted that in the SDBD, the working gas Ar and reactive medium PDMS are excited by the pulsed electric field, leading to the generation of high-energy active particles (Si^+) and excited state particles Ar*. PDMS is ionized in the discharge space and undergoes chemical reactions with energy transfer, resulting in the formation of new molecular structures. The chemical bond of Si–CH_3 is broken, and O atoms replace the positions of $-CH_3$. Different chemical molecular structures are formed based on the degree of ionization. Subsequently, the recombined new components will deposit on the surface of EP, introducing a large number of hydrophobic groups such as SiO_x and $-CH_3$. This is the main reason for the increase in water contact angle on the EP surface. From the surface morphology, it can be observed that the surface roughness of EP significantly increases after treatment. The change in surface roughness effectively inhibits and hinders electron movement, increasing the creepage distance of electrons. Therefore, the flashover voltage on the EP surface also increases, and its trend is consistent with the changes in physical morphology.

Table 22.2 Surface chemical composition of EP under plasma treatment

Sample	Chemical composition (%)			Atomic ratio (%)	
	C	O	Si	Si/O	Si/C
Untreated	73.2	21.0	5.8	27.6	7.9
5 mL/min	28.1	45.0	26.9	59.8	95.7
15 mL/min	39.9	36.0	24.1	66.9	60.4
25 mL/min	53.7	25.5	20.8	81.6	38.7

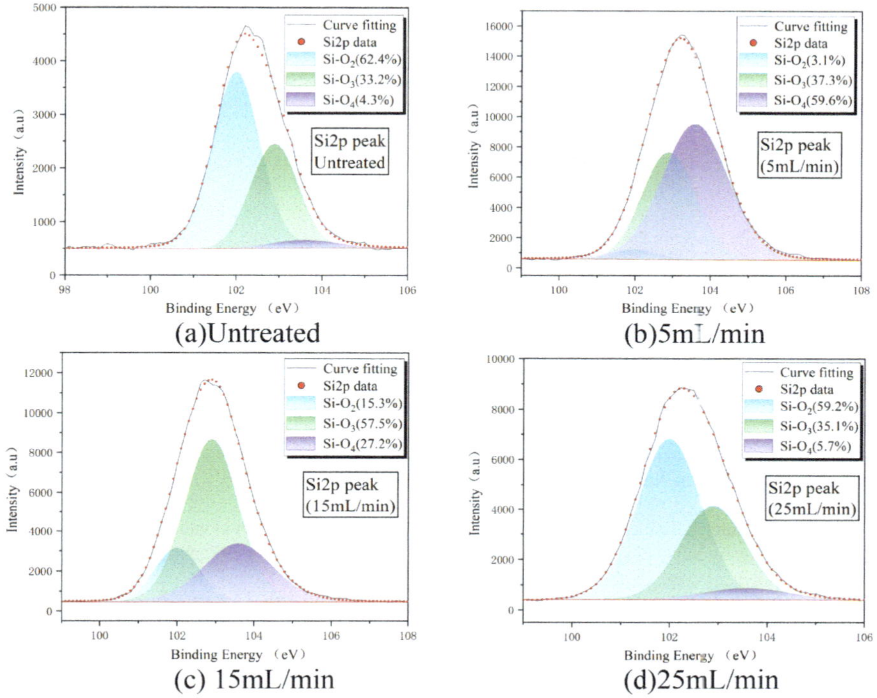

Fig. 22.7 Si2p results of EP before and after SDBD treatment

22.5 Conclusion

In this research, a Ar/PDMS/SDBD was used to modify EP surfaces for better surface performance. Through various surface property tests, the effect of different PDMS addition on the deposition properties of EP was analyzed. The main conclusions are as follows:

(1) After Ar/PDMS treatment, the hydrophobicity of the EP surface increases first and then decreases with the increase of PDMS concentration. The optimal hydrophobic surface is fabricated at PDMS of 15 mL/min, where WCA increases from 89.2° to 156.3°.
(2) After Ar/PDMS treatment, the flashover voltage of the EP surface increases first and then decreases with the increase of PDMS concentration. Similar to the WCA results, the best insulating surface appears at PDMS of 15 mL/min, where flashover voltage increases from 7.8 to 12.6 kV.
(3) After Ar/PDMS treatment, silicon-containing groups are obviously introduced on the surface of the EP. When the PDMS addition is small, most of the surface Si and O molecular structures are present in SiO_4; after more PDMS addition, the surface structure becomes SiO_2-dominated.

Acknowledgements This research was funded by Nature Science Foundation of China (Grant Nos. 52207160, 52037004), Natural Science Foundation of Jiangsu Province (BK20220341), Natural Science Foundation for Colleges in Jiangsu Province (22KJB470003), China Postdoctoral Science Foundation (2023M740642) and Postdoctoral Research Project in Zhejiang Province (ZJ2023119).

References

1. S. Chen, S. Wang, Y. Wang et al., Surface modification of epoxy resin using He/CF4 atmospheric pressure plasma jet for flashover withstanding characteristics improvement in vacuum. Appl. Surf. Sci. **414**, 107–113 (2017)
2. X. Cui, J. Shen, Y. Zhou et al., Nanosecond pulse-driven atmospheric-pressure plasmas for polymer surface modifications: wettability performance, insulation evaluation and mechanisms. Appl. Surf. Sci. **597**, 153640 (2022)
3. T. Shao, F. Liu, B. Hai et al., Surface modification of epoxy using an atmospheric pressure dielectric barrier discharge to accelerate surface charge dissipation. IEEE Trans. Dielectr. Electr. Insul. **24**(3), 1557–1565 (2017)
4. T. Shao, W. Yang, C. Zhang et al., Enhanced surface flashover strength in vacuum of polymethylmethacrylate by surface modification using atmospheric-pressure dielectric barrier discharge. Appl. Phys. Lett. **105**(7), 071607 (2014)
5. W. Li, X. Shi, P. Su et al., Direct-current breakdown test method based on electrode-embedded fully extruded molding crosslinked polyethylene sample. IEEE Trans. Dielectr. Electr. Insul. **29**(1), 247–254 (2022)

Chapter 23
Application of Magnetic Rotating Thermal Plasma Method in Solid Waste Treatment of Fly Ash

Ge Wang, Cheng Zhou, Yong Li, Yuntian Cong, Chuncai Kong, Zhimao Yang, Baojun Wang, Peng Wu, and Fenglin Ding

Abstract Fly ash is a hazardous solid waste. After plasma treatment, heavy metals are transferred to glassy slag, greatly reducing the leaching rate of heavy metals and achieving the goal of eliminating the toxicity of fly ash. This article conducted a 100 kW plasma incineration fly ash melting experiment in a low vacuum system. The experimental material is the fly ash and the mixture of fly ash and sludge, and the plasma source is a magnetic rotating thermal plasma (MRTP) device., the medium gas is argon, which can make the test object reach a melting temperature exceeding 1500°C within 2 min. The experimental results indicate that the solid waste treated with MRTP presents an opaque vitreous state. Randomly selected experimental samples were tested for element content in the mixture of vitreous water/acid leaching. The results showed that the leaching toxicity after treatment was much lower than the standard limit specified in GB/T41015-2021 "Technical Requirements for Solid Waste Vitrification Treatment Product".

23.1 Introduction

In recent years, plasma melting technology has been applied in solid waste treatment [1]. Compared with traditional heat treatment technologies, plasma technology has the advantages of excellent heating performance, high treatment temperature, effective promotion of chemical reactions, and high efficiency of destruction and removal. It integrates pyrolysis, gasification, and molten glass, and has been widely applied.

G. Wang (✉) · C. Zhou · Y. Li · Y. Cong · B. Wang · P. Wu · F. Ding
Beijing Institute of Control Engineering, Beijing 100190, China
e-mail: wangge522@sina.com

C. Kong · Z. Yang
Ministry of Education Key Laboratory for Non-Equilibrium Synthesis and Modulation of Condensed Matter, Shaanxi Province Key Laboratory of Advanced Functional Materials and Mesoscopic Physics, School of Physics, Xi'an Jiaotong University, Xi'an 710049, China

Z. Fang et al. (eds.), *Proceedings of the 5th International Symposium on Plasma and Energy Conversion*, Springer Proceedings in Physics 398,
https://doi.org/10.1007/978-981-97-2245-7_23

Beijing Institute of Control Engineering is committed to the research of high-power electric propulsion technology for spacecraft, and has successfully developed the ground engineering prototype of China's first 100 kW level high-power magnetic plasma electric thruster and completed relevant experimental testing work [2, 3]. On the basis of this research, a 100 kW level strong magnetic rotating thermal plasma (MRTP) device was developed, which has the ability to quickly heat up the target to 3000 K. This article uses this device to melt fly ash and analyzes the treated glass body.

23.2 Composition of Assessment System

23.2.1 MRTP Device

The basic working principle of the high-power MRTP device is shown in Figs. 23.1 and 23.2. The high-temperature arc generated by a large current ionizes the propellant and injects energy, using the Lorentz force of the magnetic field and current to accelerate the plasma, producing a tail flame containing high-density high-energy particles. The coupling of multiple acceleration mechanisms has achieved plasma particle velocities of up to tens of thousands of meters per second, resulting in extremely high temperatures for electrons and ions themselves, and continuously moving and colliding to achieve an overall gas temperature of several thousand degrees. Under high intensity heat sources, the activity energy of basic particles is much greater than that of intermolecular chemical bonding. The microscopic motion of matter is mainly atomic thermal motion, and the original material is shattered by high-energy particles, thereby transforming complex substances into simple harmless substances.

23.2.2 Infrared Temperature Measurement System

The temperature measurement system consists of temperature sensors, infrared thermometers, PLCs, industrial control computers, and industrial configuration software (see in Fig. 23.3). The temperature sensor is used to detect the ambient temperature in the vacuum chamber, away from heat sources. The infrared thermometer can measure temperatures ranging from 1000 to 3800 °C.

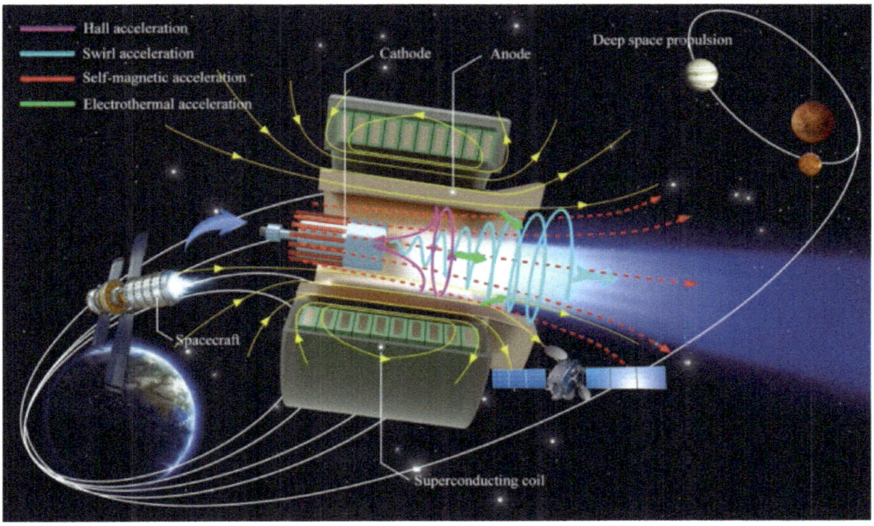

Fig. 23.1 The diagram of magnetic rotating thermal plasma (MRTP) device

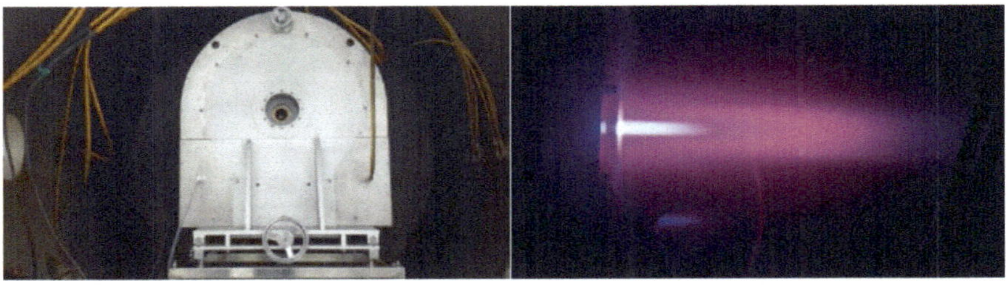

Fig. 23.2 The picture of magnetic rotating thermal plasma (MRTP) device

23.2.3 Vacuum and Power Supply System

The vacuum system is composed of various systems such as vacuum chamber, exhaust system, filtration system, cooling system, etc. The vacuum chamber can reach a vacuum degree of 0.1 Pa. The size of the vacuum chamber is 2.5 m in diameter and 3 m in length, with one on each side φ300 quartz glass observation window (see in Fig. 23.4). The power supply consists of an arc starting power supply, anode power supply, excitation power supply, etc., with an adjustable output power of 10–150 kW.

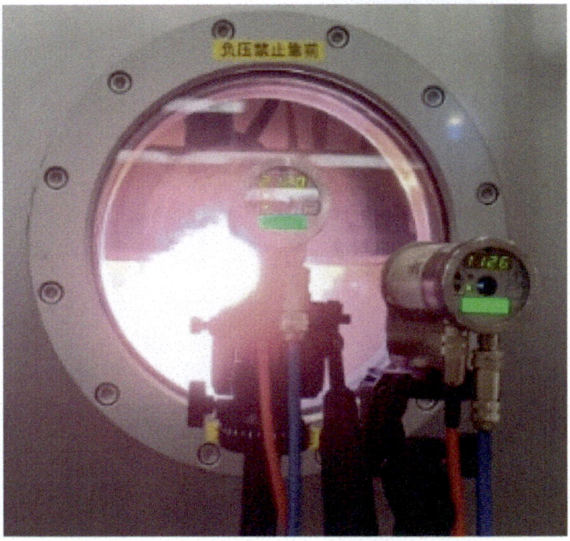

Fig. 23.3 The picture of infrared temperature measurement system

Fig. 23.4 The picture of vacuum and power supply system

23.3 Fly Ash Treatment Experiment

23.3.1 Fly Ash Samples

Considering the requirements for handling items, a mixture of 15% sludge and 85% fly ash was selected as the experimental treatment sample for this experiment. The morphology and composition of the sample are shown in Fig. 23.5. The main components of the sample measured by SEM are Si, Ca, Al, O, C, Mg, Cl, etc. The appearance of fly ash is a matte block structure, 2 μ The M-scale SEM image shows that the interior is composed of loose particle clusters, forming larger particles together μ The M-scale SEM image shows that its internal structure is very loose.

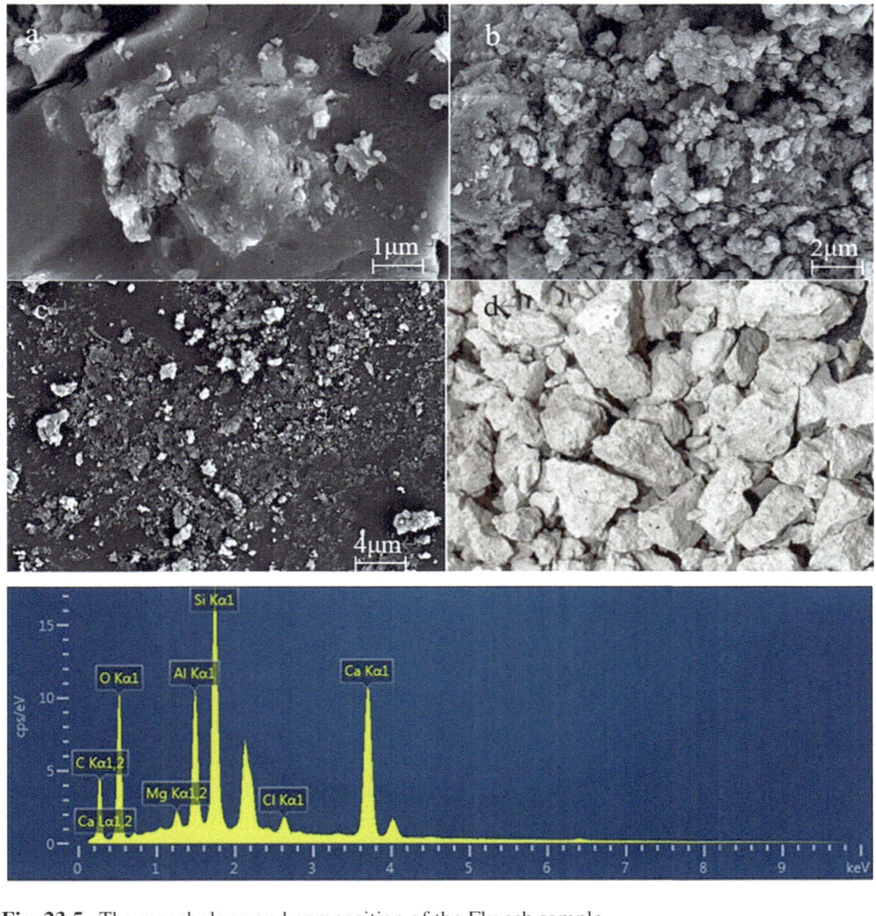

Fig. 23.5 The morphology and composition of the Fly ash sample

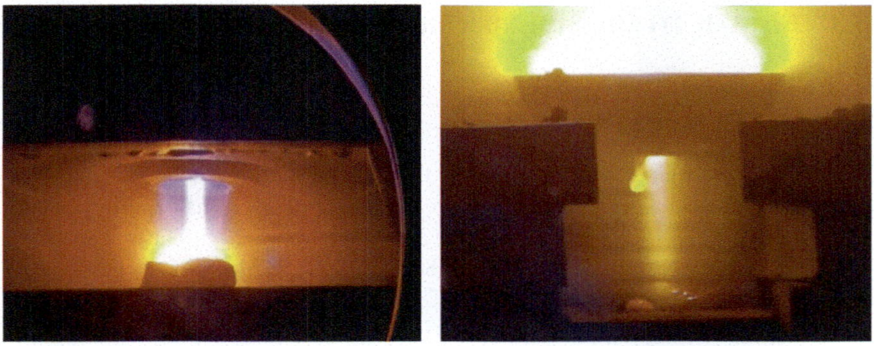

Fig. 23.6 The photos of MRTP device handling fly ash process

23.3.2 Experiment Process

The MRTP power of this test is not greater than 15 kW. The test sample is placed at a distance of 100 mm from the anode outlet. The test sample is placed in a crucible with a bottom hole, with a thickness of no more than 30 mm and a weight of 150–300 g.

Firstly, before the experiment, place the sample in the designated position of the vacuum system and close the vacuum chamber to vacuum to 10 Pa;
Secondly, after the vacuum level stabilizes, start the MRTP device and adjust the output power to 15 kW, with a process of about 1 min;
Thirdly, maintain a power output of 15 kW from the MRTP device until all fly ash melts and drips, with a process of about 2 min and a vacuum of 100 Pa;
Finally, reduce the power and shut down the machine, maintain the vacuum state until all the slag has cooled, and open the cabin to take out the slag sample.

The process of fly ash test and droplet drop of molten liquid is shown in Fig. 23.6.

23.3.3 Plasma Action Temperature

During the experiment, the temperature rise of plasma treated samples was measured using an infrared thermometer. After 2 min of processing at a power of 15 kW, the sample can reach 1500 °C and the temperature can be stably maintained (see in Fig. 23.7). Compared with traditional plasma processing devices, under low vacuum conditions, the heating efficiency of the plasma beam on the processed samples is significantly improved, the temperature rise rate is far higher than that of the traditional plasma torch, and the high-temperature area can be concentrated in the limited processing area.

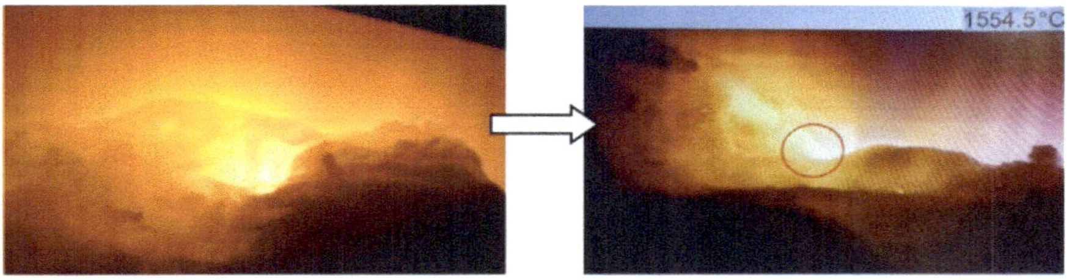

Fig. 23.7 The photos of plasma action temperature measurement

23.4 Results and Analysis

23.4.1 Appearance and Crystal Phase Analysis

After cooling the molten body, slag is obtained (see in Fig. 23.8). From the outside, the cooled sample is an opaque glass body with a smooth surface, a flat cross-section, and no pores with the naked eye. 2 μ The M-scale SEM image shows that the interior of the slag is a large and flaky structure, starting from 1 μ The M-scale SEM image shows that the large block structure is formed by the stacking of small sheet structures. This is because during the experiment, viscous molten material drips in the form of droplets through the small hole in the center of the crucible, and the dripping molten material is covered layer by layer. After cooling, it forms a vitreous slag.

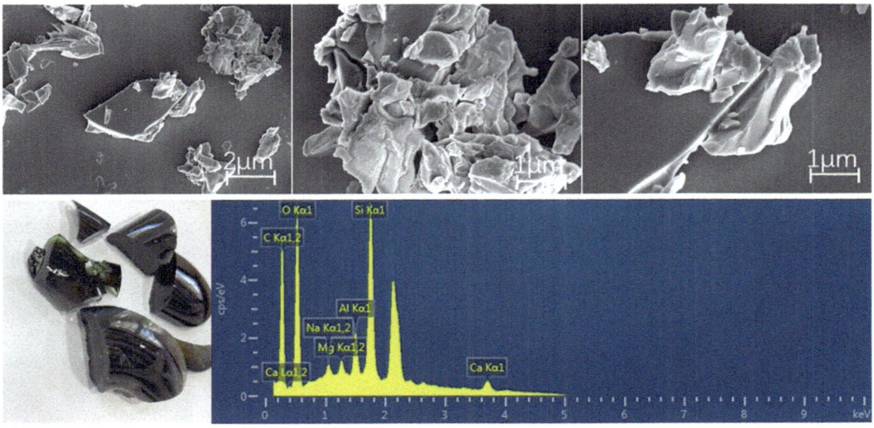

Fig. 23.8 The photos of slag and SEM images

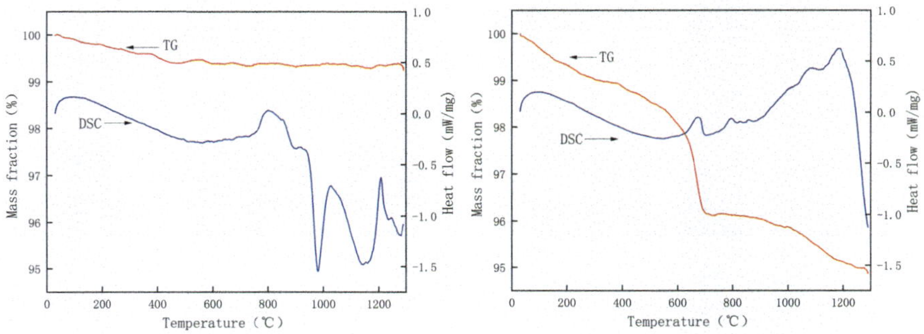

Fig. 23.9 The results of nitrogen domain TG test

23.4.2 Thermogravimetric Analysis

The relationship between the mass fraction and heat flux of ash and temperature can be understood through thermogravimetric analysis. Complex physical and chemical changes occurred in fly ash and slag within the range of 0–1300 °C (see Fig. 23.9). According to the DSC curve, both fly ash and slag have a less obvious absorption peak in the range of 0–200 °C, which may be caused by the evaporation of water adsorbed by the ash. After 600 °C, the TG curve of fly ash shows a significant decrease, resulting in a mass loss of 94.9% at 1300 °C. This is mainly due to the pyrolysis and combustion of residual organic matter in the fly ash. The TG curve of the slag has always been relatively flat, with 99.3% remaining at 1300 °C, indicating a good degree of vitrification of the slag.

23.4.3 Vitreous Content Analysis

The fly ash is treated by the MRTP device to form vitreous slag, which is tested by XRD (Fig. 23.10). The results show that the proportion of vitreous body reaches 96%, far exceeding the national standard of 85% (GB/T 41015-2021 Technical Requirements for Solid Waste Vitrification Treatment Products).

Glass content: The fly ash sample is about 79%, and the glass sample is about 96%.

23.4.4 Environmental Characteristics of Slag

The heavy metal leaching characteristics of slag directly determine its behavior in the environment. According to GB 5086 2-2010 "Horizontal Oscillation Method for Toxicity Leaching of Solid Waste" measured the leaching concentrations of several

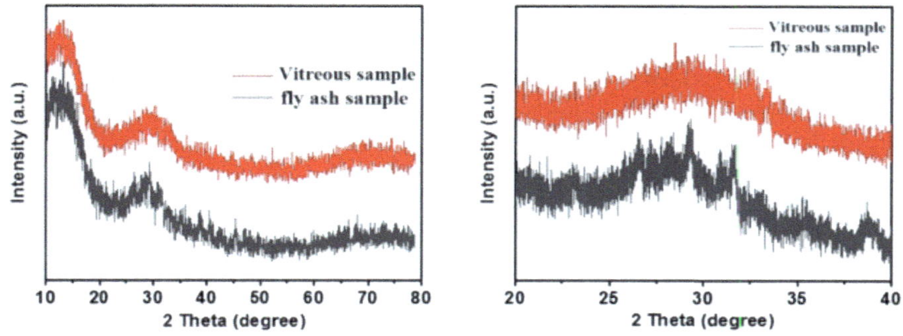

Fig. 23.10 Comparison of vitreous content before and after treatment with MRTP device

commonly concerned heavy metals in the environment, and compared them with the standard limits specified in GB/T41015-2021 "Technical Requirements for Solid Waste Vitrification Treatment Products" (Table 23.1). Randomly selected experimental samples were tested for element content in the mixture of vitreous water/acid leaching. The results showed that the leaching toxicity after treatment was much lower than the standard limit specified in GB/T41015-2021 "Technical Requirements for Solid Waste Vitrification Treatment Products".

Table 23.1 Testing and analysis of element content in vitreous water/acid leaching

Toxic substances project	Water leaching		Acid leaching	
	GB/T 41015-2021 standard limits (mg/L)	Test values (mg/L)	GB/T 41015-2021 standard limits (mg/L)	Test values (mg/L)
Cu	≤ 1.0	1.77×10^{-2}	≤ 1.0	0.108
Zn	≤ 1.0	$< 4.42 \times 10^{-5}$	≤ 1.0	1.15×10^{-2}
Cd	≤ 0.005	$< 4.02 \times 10^{-6}$	≤ 0.03	2.24×10^{-5}
Pb	≤ 0.01	$< 2.01 \times 10^{-6}$	≤ 0.3	5.05×10^{-4}
Cr	–	–	≤ 0.2	0.0620
Cr6+	≤ 0.05	4.51×10^{-3}	–	–
Hg	≤ 0.001	7.55×10^{-5}	–	–
Be	≤ 0.002	8.14×10^{-5}	–	–
Ba	≤ 0.7	0.134	–	–
Ni	≤ 0.02	1.66×10^{-3}	≤ 0.2	2.56×10^{-3}
As	≤ 0.01	4.01×10^{-4}	≤ 0.1	1.55×10^{-4}
Se	≤ 0.01	2.18×10^{-4}	–	–
Mn	≤ 0.1	1.38×10^{-3}	≤ 1.0	0.0494

Note Indicates that the project do not require testing

The heavy metal leaching concentration of slag is far below the corresponding national standard limits (see Table 23.1). Due to the small particle size and porous nature of fly ash, the leaching solution is easily immersed into the interior of the particles, promoting the dissolution and leaching of heavy metals; and the slag is more dense, with a glassy network structure that can solidify heavy metals and prevent their leaching.

23.5 Conclusion

Firstly, under a vacuum degree of 100 Pa, the MRTP device outputs a power of 15 kW (1500 °C) to process fly ash samples for 2 min, which shows continuous melting and dripping of the samples.

Secondly, the fly ash slag treated by the MRTP device exhibits an opaque vitreous state. Randomly selected experimental samples were tested for element content in the mixture of vitreous water/acid leaching. The results showed that the leaching toxicity after treatment was much lower than the standard limit specified in GB/T41015-2021 "Technical Requirements for Solid Waste Vitrification Treatment Products".

References

1. D. Changming, ed., *Plasma Treatment Technology for Solid Waste* (Chemical Industry Press, 2017)
2. T. Haibin, W. Yibai, W. Yanming, Review and understanding of magnetic plasma power thruster. J. Propul. Technol. **39**(11), 2401–2414 (2018)
3. C. Yuntian, L. Yong, T. Haibin, Z. Cheng, W. Baojun, Z. Boqiang, W. Ge, Design of 100 kW high-power magnetic plasma power propulsion system. Solid Rocket Technol. **44**(3), 304–310 (2021)

Chapter 24
Study on SF_6 Degradation Scheme and Products Based on Thermal Plasma

Yunshun Guo, Yutong Ding, Hao Sun, and Yi Wu

Abstract SF_6 gas is widely used in the power industry, especially in switchgear, due to its unique insulation performance and arc extinguishing ability. However, due to the extremely strong greenhouse effect of SF_6, its exhaust gas can cause significant harm to the environment. At present, SF_6 gas emissions are mainly reduced through recycling and degradation. In the above methods, thermal plasma is suitable for large-scale centralized processing of SF_6 due to its high power and easy control characteristics. In this work, a SF_6 degradation platform was established using the method of thermal plasma degradation. Explored the effect of different carrier gas flow rates on the degradation rate and energy efficiency of SF_6 in the $SF_6/H_2/Ar$ reaction system. At the same time, changing the flow rate of H_2 in the experiment and analyzing the product components under different conditions have certain guiding significance for the design of harmless treatment schemes for degradation products.

24.1 Introduction

SF_6 gas molecules themselves are colorless, odorless, non-toxic, non corrosive, and basically insoluble in water, with very stable chemical properties. SF_6 gas was first artificially synthesized by Moissan and Lebeau in the laboratory in 1900 [1]. Due to its excellent physical and chemical properties, it began industrial application in 1937 and is widely used in the power industry, magnesium metal smelting, semiconductor processing, and other fields. 90% of SF_6 gas is used in the power industry [2]. In the power industry, SF_6 gas is usually used to replace air as insulation and arc extinguishing gas, injected into circuit breakers, gas insulated metal enclosed transmission lines, and gas insulated metal enclosed switches.

Although SF_6 has been widely used in industrial applications, it is a strong greenhouse gas and its enormous harm to the environment is gradually being exposed. The

Y. Guo · Y. Ding · H. Sun (✉) · Y. Wu
School of Electrical Engineering, Xi'an Jiaotong University, Xi'an 710049, China
e-mail: sunhao1031@xjtu.edu.cn

Z. Fang et al. (eds.), *Proceedings of the 5th International Symposium on Plasma and Energy Conversion*, Springer Proceedings in Physics 398,
https://doi.org/10.1007/978-981-97-2245-7_24

global warming potential (GWP) value of SF_6 gas is 23500 times that of CO_2, and its atmospheric lifespan also exceeds 3200 years.

In order to limit the use and emissions of SF_6, countries signed the Kyoto Protocol in Japan in December 1997, which for the first time restricted greenhouse gas emissions in the form of regulations and included sulfur hexafluoride gas as one of the greenhouse gases that needed strict control. Therefore, it is necessary to study an efficient and harmless technical method for treating SF_6 gas that will be decommissioned in large quantities.

In terms of experiments, scholars have also attempted various methods to treat SF_6 gas to reduce emissions, commonly used methods include recycling, purification, reuse, thermal catalytic degradation, and plasma degradation.

Hao et al. [3] used zirconium phosphate, aluminum phosphate, and calcium phosphate as catalysts to reduce the reaction temperature to 800 °C, and the degradation rate of SF_6 gas was close to 100% within 8 h; Shih et al. [4] demonstrated the decomposition of SF_6 in radio frequency (RF) plasma. In a SiO_2 reactor, the degradation rate of SF_6 and the distribution of product mole fractions were studied as a function of input power and feed O_2/SF_6 ratio; Shen Yan et al. [5] used a dielectric barrier discharge cold plasma method to treat SF_6 waste gas, and quantitatively analyzed the degradation effect of SF_6 gas using gas chromatography. At the same time, infrared spectroscopy was used to analyze the gas degradation products. The results showed that under the conditions of auxiliary active gases with different contents such as N_2, Ar, H_2O, O_2, and air, the degradation rate of SF_6 gradually increased with the increase of air humidity. After mixing SF_6 gas at a pressure of 2.0 kPa with air at a relative humidity of 51% at a pressure of 28.2 kPa, the degradation rate of SF_6 gas is 92%, and the degradation products contain SO_2F_2, SOF_2, SOF_4, SF_4, SiF_4, etc.

Although there are many methods for treating SF_6 waste gas at present, most of them have their own insurmountable problems and cannot achieve efficient and harmless treatment of SF_6 gas. In contrast, due to the extremely high reaction temperature of thermal plasma, the average temperature can reach over 5000 K. Under high-energy particle bombardment, SF_6 molecules can be effectively decomposed into atomic products, which is conducive to improving degradation efficiency. At the same time, this method has the advantages of simple principle, easy control of reaction conditions, and large processing capacity. In addition, due to the fact that the hot plasma method generates extremely high reaction temperatures in a small area and can quickly degrade high concentration and large flow of SF_6 waste gas, the significant improvement in its processing capacity actually leads to higher energy efficiency than existing cold plasma methods and thermal catalytic decomposition methods.

Therefore, our team proposed the use of thermal plasma for the degradation of SF_6. This article mainly introduces the latest progress of our team in the degradation of SF_6 by thermal plasma, explores the impact of different carrier gas flow rates on the degradation effect and the product composition at different active gas concentrations, which has certain guiding significance for the design of schemes for the degradation of SF_6 using thermal plasmons.

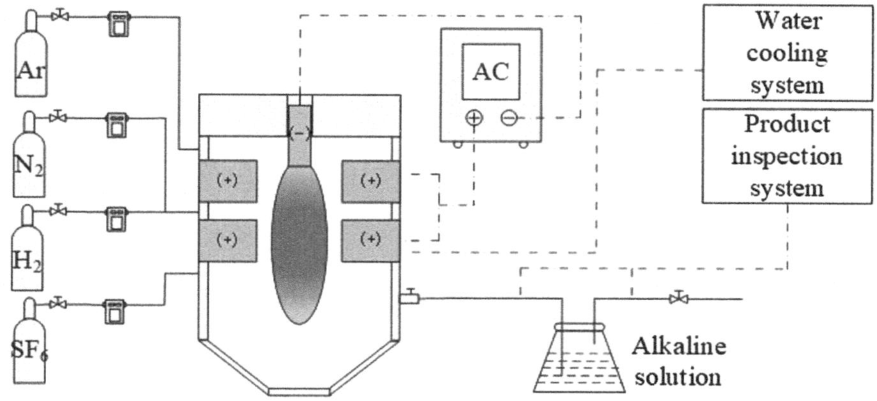

Fig. 24.1 Degradation platform diagram

24.2 Experimental Platform and Methods

24.2.1 Experimental Platform

Based on the principle of thermal plasma torch reaction, an experimental platform for plasma torch discharge degradation of SF_6 gas was established. As shown in the system diagram and structural diagram in Fig. 24.1, the experimental platform used in this article is mainly divided into three subsystems: intake control system, plasma torch reaction chamber, and harmless treatment system.

The intake control system mainly achieves quantitative control of the input gas, the plasma torch is used for discharge degradation of SF_6 gas, and the harmless treatment system is used to treat the solid and gaseous products obtained after SF_6 gas degradation, and to detect the exhaust gas composition.

24.2.2 Product Evaluation Standards

In order to evaluate the degradation effect of plasma torch discharge on SF_6 gas, this paper takes the destruction removal efficiency (DRE) and energy yield (EY) as the main evaluation indicators.

The concentration of SF_6 gas before and after the reaction was measured using an Agilent 8860 gas chromatograph. Considering the significant change in the molar fraction of the input gas and the output gas, in order to ensure consistency in the calculation of SF_6 gas concentration before and after the reaction, the inert gas Ar with the same total number of moles before and after the reaction was selected as the reference gas, and the output SF_6 gas concentration was converted; Meanwhile, by selecting 22.4 L/mol as the molar volume of the gas, the mass flow rate of the degraded SF_6 gas can be obtained. Based on the integration of discharge power and time, the energy yield EY of SF_6 gas degradation can be further calculated.

24.3 Results and Discussion

In the latest experimental exploration, we investigated the impact of different carrier gas flow rates on degradation efficiency and the product composition under different active gas concentrations by controlling variables.

24.3.1 The Effect of Different Carrier Gas Flowrates

In order to further investigate the impact of carrier gas flow rate on degradation efficiency, this article maintains a discharge current of 100 A and a SF_6: H_2 gas mixture ratio of 8:30, increasing the Ar gas flow rate from 30 to 40 and 50 L/min. At this point, the total flow rate of the mixed gas increases, while the gas path size remains unchanged, resulting in an proportional increase in gas flow rate and a corresponding reduction in reaction time through the hot plasma region. It can be approximated that the increase in Ar gas flow rate leads to a corresponding increase in the rate of temperature rise and fall during the particle reaction process. The experimental results are shown in Fig. 24.2.

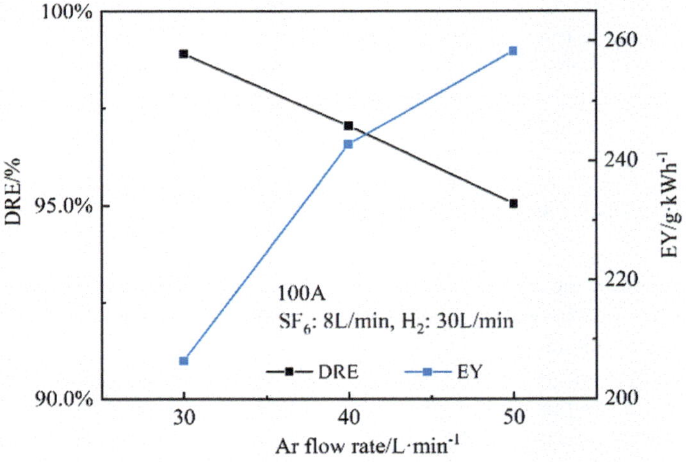

Fig. 24.2 Experimental results under different carrier gas flow rates

It can be seen that with the increase of Ar gas flow rate, the destruction removal efficiency (DRE) of SF_6 gas decreases from 99 to 95%, with a difference of less than 4%, indicating that the effect of temperature rise and fall rate on the degradation effect of SF_6 gas is not significant. The main reason for the slight decrease in the destruction removal efficiency (DRE) of SF_6 gas, is that as the total gas flow rate increases, the input SF_6 molecules and H_2 molecules are gradually diluted, reducing the probability of the F and H radicals they release colliding with each other and reacting in the hot plasma region, resulting in some F radicals recombining into SF_6 molecules.

On the contrary, with the increase of Ar gas flow rate, the energy yield (EY) of SF_6 gas degradation significantly increases, from 206 to 260 g/kWh, an increase of about a quarter. This is because the increase in Ar gas flow rate increases the proportion of Ar in the mixed gas, while the proportion of SF_6 molecules relatively decreases. Among them, SF_6 molecules are very difficult to ionize and easily capture free electrons. They are often used as arc extinguishing gases in circuit breakers, and their presence significantly increases the equivalent impedance of the arc. In order to maintain the average discharge current basically unchanged, the average arc voltage significantly increases. On the contrary, Ar is more easily excited and ionized to maintain arc discharge, so an increase in Ar ratio leads to a decrease in the equivalent impedance of the arc, a decrease in the average arc voltage, and a corresponding decrease in the average discharge power of the hot plasma torch from 15 kW to around 12 kW. However, if the DRE of SF_6 gas does not significantly decrease, the EY of SF_6 gas degradation correspondingly increases.

24.3.2 Exploration of Degradation Products When H_2 is Present

In order to further analyze the final degradation product components of SF_6 gas in excess of H_2, gas degradation products were sampled at a mixture ratio of 7.5:30 and 3:30 for SF_6: H_2 gas, and analyzed using infrared spectroscopy (FTIR). The detection results are shown in Figs. 24.3 and 24.4.

It can be seen that when H_2 gas is sufficient, the final gas degradation products are mainly a small amount of remaining SF_6 gas and HF gas. HF itself has strong corrosiveness and can react with the gas sample pool in the spectrometer to generate SiF_4 gas. Due to Ar and H_2 being non-polar gases, there are no absorption peaks in the infrared spectrum. When H_2 gas is further increased to a far excess, the remaining SF_6 gas in the final gas degradation product significantly decreases. However, it is expected that no significant characteristic peaks were detected for the H_2S gas generated under this condition, as the concentration of the generated H_2S gas is low and can corrode the metal components of the sampling probe and the plastic film of the sampling bag, resulting in most of the H_2S gas being consumed and unable to be detected.

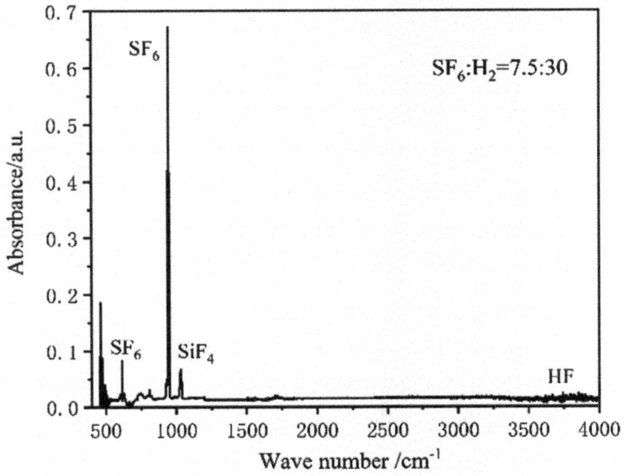

Fig. 24.3 Infrared spectroscopy analysis at SF_6: $H_2 = 7.5:30$

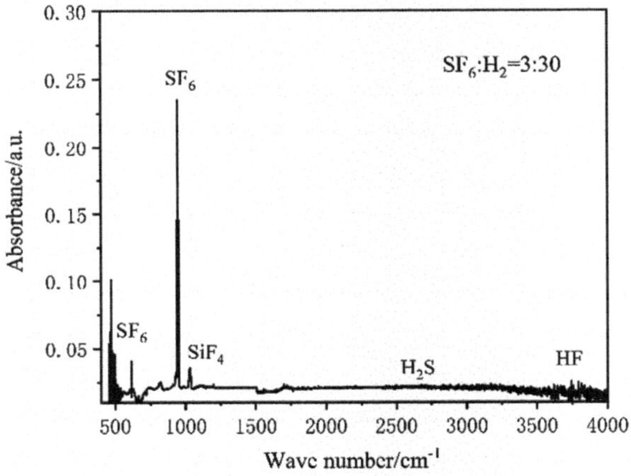

Fig. 24.4 Infrared spectroscopy analysis at SF_6: $H_2 = 3:30$

24.3.3 Exploration of Degradation Products When H_2 is Insufficient

In order to verify the degradation product components under this condition, the precipitates in the alkaline solution pool were first collected for drying and grinding treatment, and then analyzed using X-ray diffraction (XRD) method. The detection results are shown in Fig. 24.5.

According to the principle of three strong peaks in X-ray diffraction (XRD) analysis, the precipitation in the alkaline solution pool mainly consists of three components: S elemental, CaF_2, and $CaSO_4$. Among them, elemental S refers to the part of S powder generated in the reaction chamber after long-term operation of the device, which is carried into the alkaline solution pool by the airflow. CaF_2 is the precipitate

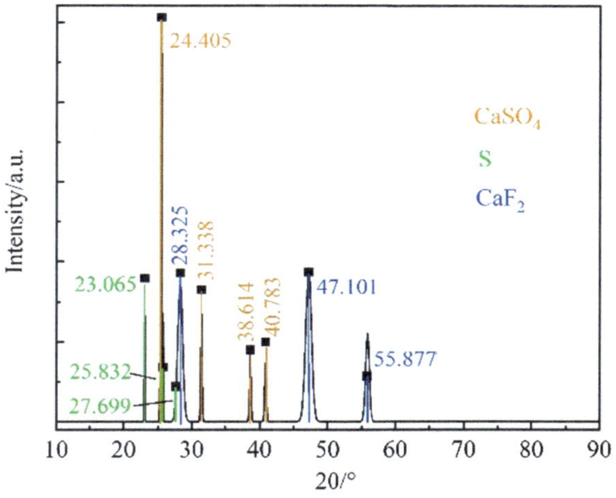

Fig. 24.5 XRD test results of solid products

generated by Ca (OH) 2 alkaline solution absorbing HF gas, and $CaSO_4$ is the final product obtained by the incomplete decomposition of SF_6 and the reaction of SFx with alkaline solution to generate sulfite ions, which are further oxidized to sulfate ions.

In order to further investigate the components of low fluorine sulfide SFx generated when there is a small amount of H_2, gas degradation products were sampled at a SF_6: H_2 gas mixture ratio of 15:30 and analyzed using infrared spectroscopy (FTIR). The detection results are shown in Fig. 24.6.

It can be seen that the remaining SF_6 gas content is relatively high at this time, and S_2F_{10} and SF_4 gases appear in the final degradation products, both of which are acidic and toxic gases. Although they can be absorbed by alkaline solution and

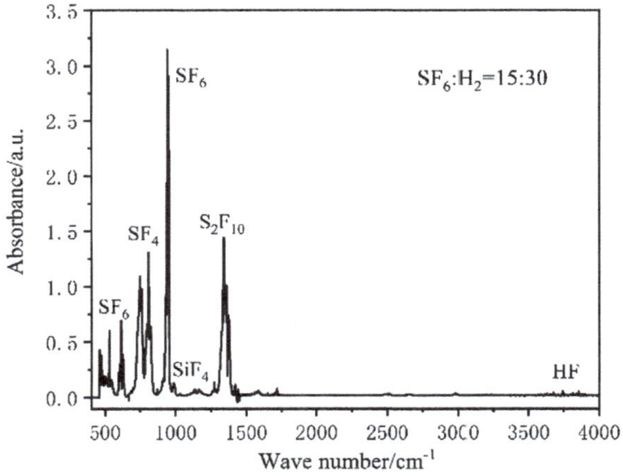

Fig. 24.6 Infrared spectroscopy analysis at SF_6: H_2 = 15:30

generally do not enter the environment to cause harm to the human body, it is still necessary to avoid their generation as much as possible.

24.4 Conclusion

As the carrier gas flow rate increases, the energy efficiency will gradually increase, but the degradation rate of SF_6 will gradually decrease. To achieve a sufficiently high degradation rate, the carrier gas flow rate should not be too high;

By comparing the effect of H_2 flow rate on the product, when the concentration of H_2 is too high, S element will react with H_2 to generate H_2S, and when the H_2 flow rate is low, SF_6 will decompose incompletely to form low fluorine sulfides;

In order to avoid the incomplete decomposition of SF_6 gas and the generation of low fluorine sulfide SFx, it is necessary to maintain an excess of H_2. At this time, the final degradation product almost only contains S elemental and HF.

Acknowledgement This work was supported by the National Key Research and development Program (2021YFB2400604), the National Natural Science Foundation of China under Grant No. 52077170.

References

1. L.G. Christophorou, J.K. Olthoff, R.J. Van Brunt, Sulfur hexafluoride and the electric power industry. IEEE Electr. Insul. Mag. **13**(5), 20–24 (1997)
2. X.X. Zhang, H.Y. Xiao, Y.J. Huang, Review of low-temperature plasma treatment of SF_6 waste gas. Trans. China Electrotech. Soc. **31**(24), 16–24 (2016)
3. X. Hao, C. Tie, L. Hang et al., Study on thermal catalytic degradation of SF_6 waste gas based on phosphate. High Volt. Apparat. **57**(03), 180–185 (2021)
4. M. Shih, W.J. Lee, C.H. Tsai et al., Decomposition of SF_6 in an RF plasma environment. J. Air Waste Manag. Assoc. **52**(11), 1274–1280 (2002)
5. S. Yan, H. Li, Z. Renxi et al., Study on the degradation of SF_6 by dielectric barrier discharge. Environ. Chem. **03**, 275–279 (2007)

Chapter 25
Experimental Study on the Effect of Glass Beads on DBD Degradation of Sulfuryl Fluoride

Jianben Liu, Yalong Li, Zhaodi Yang, Yan Liu, Ni Li, Xiaoxing Zhang, Kun Wan, and Peng Gao

Abstract SO_2F_2 is an inorganic compound that is colorless and odorless gas at room temperature and gas pressure. Because of its strong diffusion permeability and fast insecticidal speed, it is widely used in termite and pest control in containers, buildings, gardens and other places. However, SO_2F_2 is a toxic gas with strong greenhouse effect, and its emission after degradation can reduce the harm to the atmosphere and organisms. In this paper, a dielectric barrier discharge (DBD) plasma was used to degrade SO_2F_2, and the influence of filled glass beads on its degradation was explored. The results show that with the increase of input power, the degradation rate increases, and the energy efficiency decreases gradually. In the empty tube system without filling dielectric, the DRE of SO_2F_2 with a concentration of 2% reached 61.5% and the EY was 7.2 g/kWh under the condition of flow rate of 150 ml/min and power of 90 W. After filling the glass beads, the amount of charge transfer increased significantly, and the degradation rate and energy efficiency were also significantly improved. Under the condition that SO_2F_2 concentration was 2%, flow rate was 150 ml/min and power was 70 W, the DRE and EY of SO_2F_2 were 91.48% and 10.71 g/kWh, respectively. The main decomposition products of SO_2F_2 are SOF_2, SO_2, SOF_4, OF_2, SiF_4, etc. The addition of glass beads promotes the decomposition of SO_2F_2 to produce more SO_2 that is easy to handle, which provides a research direction for the harmless degradation of SO_2F_2.

J. Liu · Y. Li · Y. Liu · N. Li
State Key Laboratory of Power Grid Environmental Protection, China Electric Power Research Institute, Wuhan 430074, P. R. China

Y. Li · Z. Yang · X. Zhang (✉) · K. Wan · P. Gao
Hubei Engineering Research Center for Safety Monitoring of New Energy and Power Grid Equipment, Hubei University of Technology, Wuhan 430068, P. R. China
e-mail: xiaoxing.zhang@outlook.com

Z. Fang et al. (eds.), *Proceedings of the 5th International Symposium on Plasma and Energy Conversion*, Springer Proceedings in Physics 398,
https://doi.org/10.1007/978-981-97-2245-7_25

25.1 Introduction

Sulfuryl fluoride (SO_2F_2) is an inorganic compound that is a colorless and odorless gas at room temperature and pressure [1]. It is widely used in termite and pest control in containers, buildings, gardens and other places due to its advantages of strong diffusion and penetration, fast insecticidal speed, and ease of use at low temperatures [2, 3]. However, SO_2F_2 is a strong greenhouse effect gas, its global warming potential (GWP) is 4780 times that of CO_2, and its lifetime in the atmosphere is 36 years [4, 5]. Since 1978, SO_2F_2 has been accumulating in the global atmosphere at a rate of 5% per year [6]. With the proposal of double carbon goal, the efficient and harmless degradation of strong greenhouse gases has become an urgent problem to be solved. In addition, SO_2F_2 has a certain degree of toxicity, inhalation of human body will inhibit the human nervous system and respiratory system, higher concentrations of SO_2F_2 may lead to obvious lung and kidney damage, and may even lead to death [7], which is not negligible to the human body.

At present, domestic and foreign treatment methods for SO_2F_2 exhaust gas are lye treatment method [8], physical adsorption method [9], bio-based solvent absorption method [10], non-thermal plasma method (NTP) [11] and so on. Among them, the lye treatment method in lye is a highly corrosive chemical, improper handling will cause harm to human beings and the environment, and the treatment efficiency of lye absorbing SO_2F_2 is low. Physical adsorption method is more efficient for SO_2F_2 treatment, but the disposal of adsorbent after treatment has become a new secondary problem. Bio-based solvent absorption method has the advantages of low cost and low toxicity, but the treatment efficiency is low. NTP with the advantages of high treatment efficiency and rate, easy to operate and so on has received wide attention.

NTP technology utilizes a large number of active particles and high-energy electrons in the plasma and gas pollutant molecules undergo frequent collisions, so that they are decomposed and transformed into harmless products or easy-to-treat products. According to the generation method of NTP, it can be divided into corona discharge, sliding arc discharge, dielectirc barrier discharge (DBD), etc. [12–14]. DBD technology has the advantages of high energy efficiency, simple device, stable and reliable discharge, etc., and has been applied to the conversion of CO_2 [15], degradation of SF_6 [16], and volatile organic compounds (VOCs) treatment, etc. [17]. Our group used a DBD reactor to investigate the effect of additional H_2 and H_2O concentration on the degradation of SO_2F_2, and it was found that when the concentration of SO_2F_2 is 2% (Volume fraction), the destruction and removal efficiencies (DRE) of SO_2F_2 could reach a maximum of 86.26 and 80.29% with the addition of 1% H_2O or H_2, while excessive H_2O and H_2 would inhibit the degradation of SO_2F_2 instead [11].

In addition, the addition of catalysts can also significantly improve the degradation of SO_2F_2.In 2012, Nie Yong et al. explored the influence of quartz glass spheres and alumina on the efficiency of SO_2F_2 removal. At a discharge voltage of 11.4 kV, a discharge frequency of 8.0 kHz, and a concentration of 0.7% of SO_2F_2 treated by a plasma reactor filled with quartz glass spheres, the DRE of SO_2F_2 reached 100%, and the degradation products were SO_2, SiF_4, and S. This study mainly examined the DRE of SO_2F_2, and there was no detailed analysis of the decomposition products and decomposition mechanism of SO_2F_2 [17].

In this paper, a filled-bed DBD experimental platform was constructed to carry out the research on the influence law of glass bead-filled media on the DRE and energy yield (EY) of SO_2F_2, and to analyze the distribution law of decomposition products and the possible decomposition path of SO_2F_2. The research results in this paper provide experimental and theoretical basis for the efficient and harmless degradation of SO_2F_2.

25.2 Experimental Platform and Calculation Method

25.2.1 Experimental Platform

The filled-bed DBD experimental platform built in this paper is shown in Fig. 25.1, which mainly includes the gas distribution system, DBD power supply and reaction system, and tail gas detection and treatment system. In this paper, GC500 four-channel intelligent dynamic gas distribution instrument is used to control the input SO_2F_2 gas concentration and flow rate. The concentration of SO_2F_2 gas used in the experiments of this paper is 2%, the background gas is Ar, and the total gas flow rate is 150 ml/min. A fast Fourier Transform Infrared Spec-trophotometer (FTIR) was used to qualitatively analyze the decomposition products of SO_2F_2 gas to determine the types and trends of the decomposition products. The quantitative detection of SO_2F_2 and its main decomposition products SO_2, SOF_2 and SiF_4 using a gas chromatography-mass spectrometer (GC–MS) was used to analyze the main decomposition pathways of SO_2F_2.

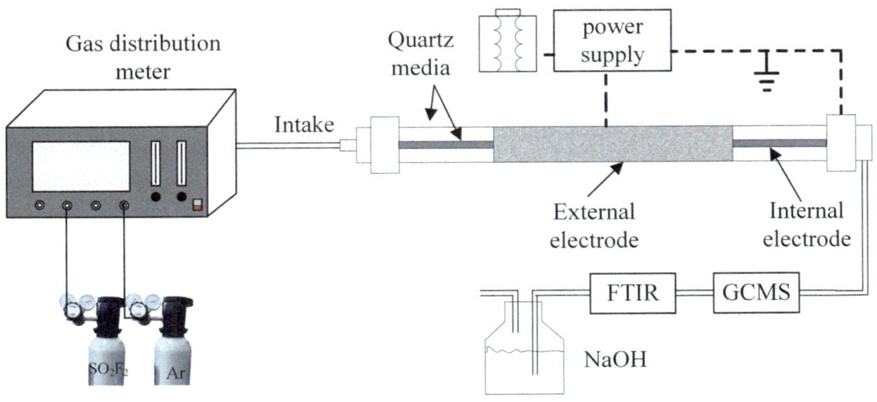

Fig. 25.1 Experimental platform

25.2.2 Calculation Method

The DRE of SO_2F_2 is calculated as follows:

$$\mathrm{DRE}(\%) = \frac{C_{\mathrm{in}} - C_{\mathrm{out}}}{C_{\mathrm{in}}} \times 100\% \tag{25.1}$$

where C_{in} is the concentration of SO_2F_2 before degradation and C_{out} is the concentration of SO_2F_2 remaining after the degradation reaction.

The EY of SO_2F_2 is defined as the mass of SO_2F_2 solved per unit input energy drop and is calculated as shown in Eq. (25.2):

$$\mathrm{EY}(\mathrm{g/kWh}) = \frac{\mathrm{M}_{SO_2F_2}}{P \times \mathrm{t} \times 3.6} \times 1000 \tag{25.2}$$

where, is the mass of SO_2F_2 gas degraded at time t, in g. P is the input power, in W. t is the time required to degrade SO_2F_2, in h, and EY is the energy yield, in g/kWh.

25.3 Experimental Results

25.3.1 Electrical Parameters

In order to investigate the variation of electrical parameters under the glass bead filled system, the discharge voltage and current were examined under this system and the Lissajous graph of the discharge were analyzed as shown in Figs. 25.2 and 25.3. From Fig. 25.2, it can be seen that the discharge voltage and current waveforms are sinusoidal and the discharge system is capacitive. The empty tube without filling medium has more filament discharge generated by uneven discharge. After filling glass beads due to the change of reactor capacitance leads to a significant increase in discharge voltage and current, which reaches 14.4 kV, 217 mA, respectively. After filling glass beads the discharge is more uniform, and the filament discharge is significantly reduced. Figure 25.3 shows the Lissajous graph of reactor discharge before and after glass bead filling, the area of the Lissajous graph increased significantly after filling glass beads, and the power of the power supply discharge also increased, and more energy was used to stimulate the decomposition reaction of SO_2F_2 molecules. In addition, the amplitude of the charge transfer also increased, indicating that more energy is involved in the reaction in each discharge cycle.

Fig. 25.2 Voltage and current waveform diagram

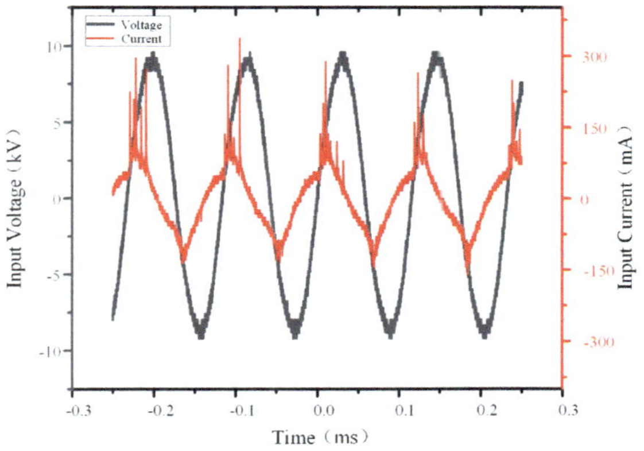

(a) Glass bead filling

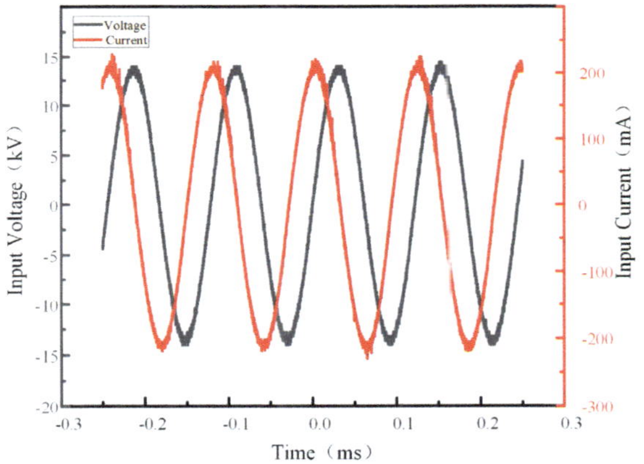

(b) Unpopulated

25.3.2 DRE and EY

Figures 25.4 and 25.5 show the DRE and EY of SO_2F_2 before and after glass bead filling at different input powers, respectively. The DRE and EY were significantly increased after filling glass beads. Under the empty tube system, the degradation rate showed a linear increase with the increase of input power, and the DRE of SO_2F_2 was 73.25% when the power supply was 100 W, while the energy efficiency showed a gradual decrease. Under the glass bead filling system, the DRE of SO_2F_2 showed a trend of increasing and then decreasing with the increase of input power, and the highest DRE of SO_2F_2 was observed when the power input power was 70 W. At this time, the DRE and EY of SO_2F_2 were 91.48% and 10.71 g/kWh, respectively. After filling the glass beads, the amount of charge transfer was increased, and more energy was used for excitation of electrons, and a large number of High-energy electrons and

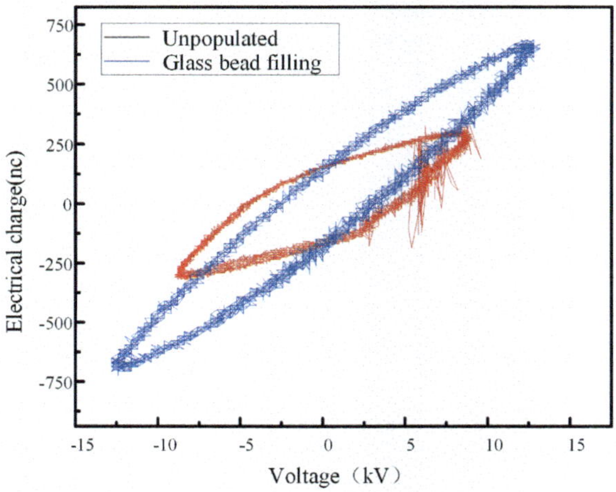

Fig. 25.3 Lissajous figure under different systems

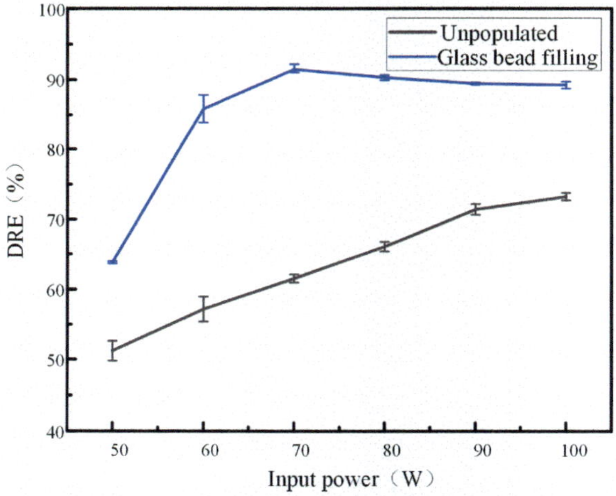

Fig. 25.4 DRE of different systems

active particles are excited by collision with SO_2F_2, which gradually breaks the bond and decomposes it, and then improves the DRE of SO_2F_2; and with the increase of the input power of the power supply, the DRE of SO_2F_2 shows a trend of increasing and then decreasing.

25.3.3 Product Analysis

In order to further clarify the decomposition process of SO_2F_2, FTIR was used to qualitatively detect and analyze the SO_2F_2 decomposition products. Figure 25.6

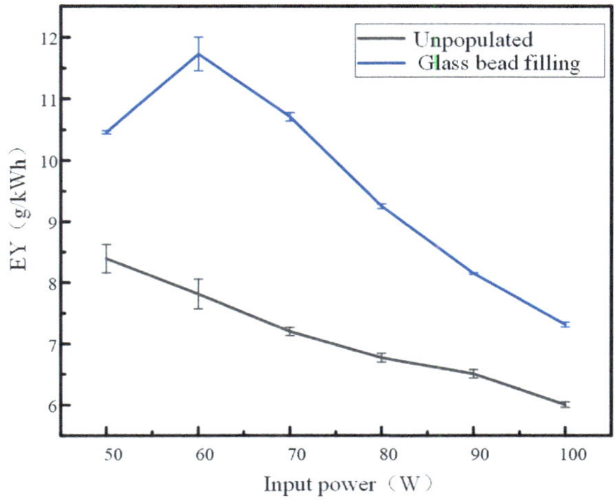

Fig. 25.5 EY of different system

shows the infrared spectra of SO_2F_2 decomposition products under different systems. It can be seen that the main decomposition products of SO_2F_2 are SOF_2, SO_2, SiF_4, OF_2 and so on. Comparison of Fig. 25.6a and b shows that the absorption peaks of SO_2F_2 under the glass bead filling system are relatively weak, while the absorption peaks of SiF_4 and SO_2 are stronger, which indicates that the degradation effect of SO_2F_2 under the glass bead filling is more excellent and the content of SiF_4 and SO_2 in the products is higher. Under the empty tube system, the SOF_2 absorption peak was stronger, indicating that the SOF_2 content was higher under the empty tube. The FTIR spectrum of the experimental group without glass bead filling showed the absorption peak of SF_6 at a wave number of 945.91 cm^{-1}, which was attributed to the combination of S and F ions generated during the bond-breaking decomposition of SO_2F_2 to generate SF_6 with a more stable structure, while no SF_6 absorption peak was found when glass beads were filled, which was attributed to the combination of S and F ions. This is S and F ions combined to form SF_6 reaction is a reversible reaction, glass bead filling system in the glass bead material is SiO_2, in the plasma discharge process will be involved in the decomposition of SO_2F_2 reaction to block the S and F ions combined to form SF_6 reaction. The presence of SiF_4 indicates that the glass surface etching reaction, and the glass bead filled with a moment of etching reaction is more intense, reaction process as Eq. (25.3), this may be due to the reaction of F ions with SiO_2 to generate SiF_4.

$$4F^- + SiO_2 \rightarrow SiF_4 + 2O^{2-} \tag{25.3}$$

In order to detect the content of the main gaseous products after the decomposition of SO_2F_2, the three main decomposition products of SO_2, SOF_2 and SiF_4 were quantitatively detected and analyzed by GCMS in this paper. Figure 25.7a and b show the contents of the main decomposition products of SO_2F_2 under different systems, respectively. In the empty tube system, SOF_2 is the main decomposition product,

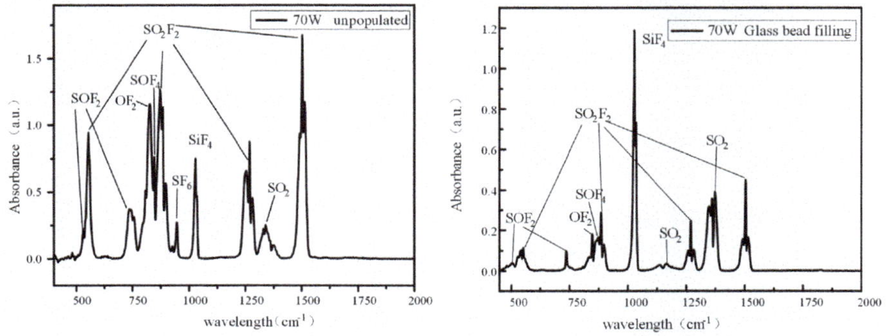

(a) Unpopulated (b) Glass bead filling

Fig. 25.6 Infrared spectroscopy

with the increase of input power, its content gradually increased, SO_2 content in the power greater than 70 W before there is a significant increase in the concentration of SiF_4 does not change with the input power, glass bead filling system, the degradation reaction is more complete, the main product is SO_2, the decomposition products have been detected in SOF_2 gas. The highest SO_2 content was 9421.71 ppm when the power supply was 100 W. The SiF_4 content under the empty tube system was stable at about 2000 ppm, while under the glass bead filling system, the SiF_4 gradually increased with the increase of power, and reached 8500 ppm when the power supply was 100 W, which indicated that the etching reaction of the glass reacted more strongly with the increase of power supply.

According to Fig. 25.7b, it can be seen that the concentration of SO_2 and SiF_4 gradually increased with the increase of power input from the power supply, and the SOF_2 content was almost zero. Comparing with Fig. 25.7a, it is possible that part of

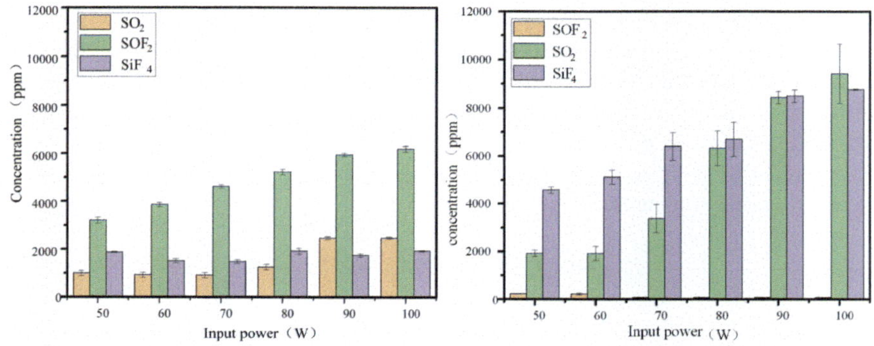

(a) Unpopulated (b) Glass bead filling

Fig. 25.7 Product content

the energy under the glass bead-filled system was used to stimulate the decomposition of SOF_2, which led to the decomposition of SO_2 into SO_2. And the energy used for degrading SO_2F_2 was reduced, and this may be the reason for the degradation rate of SO_2F_2 under the power of 80 W. This may also be the reason for the decrease of DRE of SO_2F_2 under 80 W power. It can be seen that the DRE and EY are not a single criterion for judging the degradation effect, but also consider the products after degradation.

25.4 Conclusion

(1) The DRE and EY of SO_2F_2 increased significantly after glass bead filling, and there was the highest DRE at 70 W. At this time, the DRE and EY of SO_2F_2 were 91.48% and 10.71 g/kWh. Respectively, which were significantly higher than those of the empty tube system (61.5% and 7.2 g/kWh).
(2) Under the glass bead filling system, with the increase of input power, the DRE showed a trend of increasing and then decreasing, probably because when the input power reaches a certain value, part of the energy of the system is used to stimulate the decomposition of SOF_2, so that the energy used for degrading SO_2F_2 is reduced, and the DRE of SO_2F_2 decreases gradually.
(3) Under the glass bead filling system, the main decomposition products of SO_2F_2 are SOF_2, SO_2, SOF_4 and OF_2. Accompanied by a large number of SiF_4, indicating that the etching reaction has occurred.

Acknowledgements The work was supported by Open Fund of State Key Laboratory of Power Grid Environmental Protection (No. GYW51202301432).

References

1. X.F. Jiang, Preparation of sulfuryl fluoride and its application in organic synthesis. Chem. Agent **43**(04), 454–457 (2021)
2. Y.A. Xu, Overview of application and development of sulfuryl fluoride fumigation technology. Grain, Oil Food Technol. **29**(04), 50–56 (2021)
3. X.P. Yan, Z.Y. Mu, D.D. Li et al., Research and application progress of sulfuryl fluoride on the control of stored grain pests. Grain Storage **47**(4), 15–19 (2018)
4. C. Zhang, X.J. Xu, Synthesis and application of sulfamoyl fluoride. Organo Fluorine Ind. **02**, 39–41 (2006)
5. B. Zhang, R. Zhang, L. Zhu et al., Effect of sulfuryl fluoride on ozone attenuation and research on tail gas treatment technology. Cereals, Oils Food Sci. Technol. **29**(04), 79–84 (2021)
6. J. Muhle, J. Huang, R.F. Weiss et al., Sulfuryl fluoride in the global atmosphere. Geophy. Res.: Atmos. **114**(D5), 1–13 (2009)
7. X.P. Shen, First aid and nursing of sulfuryl fluoride poisoning patients. General Nurs. **10**(22), 2070–2071 (2012)

8. Y. Nie, X.J. Liang, J.B. Ji et al., Harmless treatment of sulfuryl fluoride by chemical absorption. Ecol. Environ. Conser. **32**(9), 789–795 (2015)
9. G.Z. Cao, Z.J. Hua, X.Y. Zhu et al., A method for eliminating sulfuryl fluoride: CN104324704A.P. 2015.02.04
10. Z.J. Zhang, Study on the dissolution characteristics of sulfruyl fluoride in bio-based solvents. D. Zhejiang University of Technology (2020)
11. X.X. Zhang, C. Zhou, Y.J. Wu et al., Effect of H_2O and H_2 on degradation of sulfuryl fluoride by Argon dielectric barrier discharge. High Volt. Eng. **49**(01), 382–390 (2023)
12. S.Z. Li, S.H. Xie, Y. WU et al., Decomposition of greenhouse gas of SF6 and CF_4 by means of atmospheric pressure microwave plasma torch. High Volt. Eng. **47**(8), 3012–3019 (2021)
13. M.S. Li, S.L. Li, D. Li et al., Experimental study of the synergistic effect of catalyst and pulsed discharge plasma on removal of dimethyl sulfide. High Volt. Eng. **42**(2), 462–467 (2016)
14. N. Lu, M.J. Liu, Y.D. Liu et al., Research progress of hydrogen production by rotating gliding arc plasma reforming. High Volt. Eng. **47**(8), 3001–3011 (2021)
15. Y. Gao, L.G. Dou, J.W. Li et al., Progress of low-temperature plasma-catalyst synergistic catalytic CO_2 conversion. High Volt. Eng. **48**(04), 1607–1619 (2022)
16. Z.L. Chui, Y.P. Hao, L. Yang et al. Review on harmless degradation of SF_6 exhaust gas, in *Proceedings of the CSEE* (2023), pp. 1–16
17. Q.F. Zheng, Y. Nie, D.Y. Gu et al., Study on removal of sulfuryl fluoride from filled dielectric barrier discharge plasma. High Volt. Eng. **38**(05), 1157–1162 (2012)

Chapter 26
Pulsed Gas–liquid Discharge Synergistic $CuFe_2O_4$ Stimulated Fenton Reaction for Tetracycline Degradation

Pengju Chen, Yuankun Ye, Sen Wang, and Zhi Fang

Abstract Excessive used antibiotics in various industries has caused significant impacts on both the human health and ecological environment. Low-temperature plasma, has gradually gained attentions in wastewater treatment. To improve the degradation performance of antibiotics in water, plasma coupled with catalysts are usually conducted. In this study, pulsed gas–liquid discharge coupled with $CuFe_2O_4$ was utilized for the degradation of tetracycline in water. The results showed that $CuFe_2O_4$ dosage, pulse voltage, initial concentration of tetracycline, and working gas had important effects on the degradation of tetracycline. Compared to plasma treatment alone, the plasma-catalytic system had a significant enhancement in the degradation rate. The catalytic mechanism of gas–liquid discharge synergistic $CuFe_2O_4$ is proposed. Radical quenching experiments showed that H_2O_2 is one of the most key active species in the plasma/$CuFe_2O_4$ degradation. The results of the study provide an effective way to treat antibiotics in water.

26.1 Introduction

Antibiotics have been extensively employed in the medical and agricultural fields in recent years due to their antibacterial and antimicrobial properties, which have contaminated the aqueous environment [1]. The stable and complex molecular structure of antibiotics makes them difficult to be degraded by microorganisms in the nature environment, leading to their long-term presence in water bodies. Bacterial resistance thus develops and "superbugs" inevitably arise, endangering ecosystems and human health. To date, a variety of advanced oxidation techniques (AOPs) have been attempted in the antibiotic degradation, including photocatalysis, persulfate reaction, and Fenton reaction [2]. However, most AOPs suffer from long treatment

P. Chen · Y. Ye · S. Wang (✉) · Z. Fang
Nanjing Tech University, Nanjing 211816, China
e-mail: Wang_sen@njtech.edu.cn

Z. Fang et al. (eds.), *Proceedings of the 5th International Symposium on Plasma and Energy Conversion*, Springer Proceedings in Physics 398,
https://doi.org/10.1007/978-981-97-2245-7_26

times, low energy efficiency and secondary pollution. An environmentally friendly and energy-efficient technology is urgently needed to remove antibiotics from water.

Recently, non-thermal plasma (NTP) has gradually attracted attentions in the water treatment, especially for organic pollutants that are difficult to biodegrade [3, 4]. Pulsed gas–liquid discharge plasma (PGLDP) as a NTP can produce reactive species ·OH, O_3, and H_2O_2, as well as ultraviolet irradiation, shock waves, and electric fields [5]. The antibiotic molecules can be degraded in a short time due to the combined effects of the above-mentioned physicochemical processes. Moreover, PGLDP has a broad development prospect in the field of antibiotic degradation for the advantages of high efficiency, no secondary pollution, and easy operation. However, the physical effects in PGLDP are usually wasted, which may be the reason for the unsatisfactory energy efficiency [6]. Many studies have pointed out that physical effects like ultrasound and light radiation can enhance the Fenton-like catalysts to convert aqueous H_2O_2 produced in PGLDP to ·OH, which has a stronger oxidizing property, and thus optimizing the degradation [7–9]. Therefore, combining PGLDP with Fenton-like catalysts can enhance the energy utilization of the system.

$CuFe_2O_4$ (CFO), as a magnetic nanospinel ferrite, exhibits easy separation, excellent chemical stability and high electrocatalytic activity [10]. Its activation of various oxide species (H_2O_2, persulfate, etc.) also makes it promising for non-homogeneous Fenton reactions. For example, Gao et al. [11] used the $CuFe_2O_4$ electro-Fenton process for the degradation of semi-coking wastewater. Thayane et al. [12] researched the removal of dyes by $CuFe_2O_4$ with the radiation of visible light. However, CFO has never been combined with PDP to degrade antibiotics in water.

Herein, the photo-Fenton catalyst CFO was prepared by sol–gel method, and pulsed gas–liquid discharge plasma synergistic with CFO for tetracycline degradation was proposed. The performance of PGLDP combined with CFO on tetracycline degradation was investigated. Working conditions, including catalyst amount, pulse voltage, and working gas were also optimized. The mechanism of tetracycline degradation by PGLDP/CFO was analyzed by radical trapping experiment and ROS concentration measurement.

26.2 Experimental Setup

26.2.1 Plasma Generation System for Degradation

The experimental system of PDLDP/CFO for tetracycline(TC) degradation is illustrated in Fig. 26.1, which consists of a power supply (Xi'an HV-2015), a plasma reactor, air supply system and diagnosis devices. A needle-to-water electrode is adopted, with a 2 mm-diameter stainless steel needle fixed 1 mm above the liquid surface and connected to the power supply. 50 mL TC solution is contained in a quartz vessel as the treated solution and a stainless-steel plate is mounted at the bottom of the vessel as the ground electrode. A nanosecond pulsed power source is used to

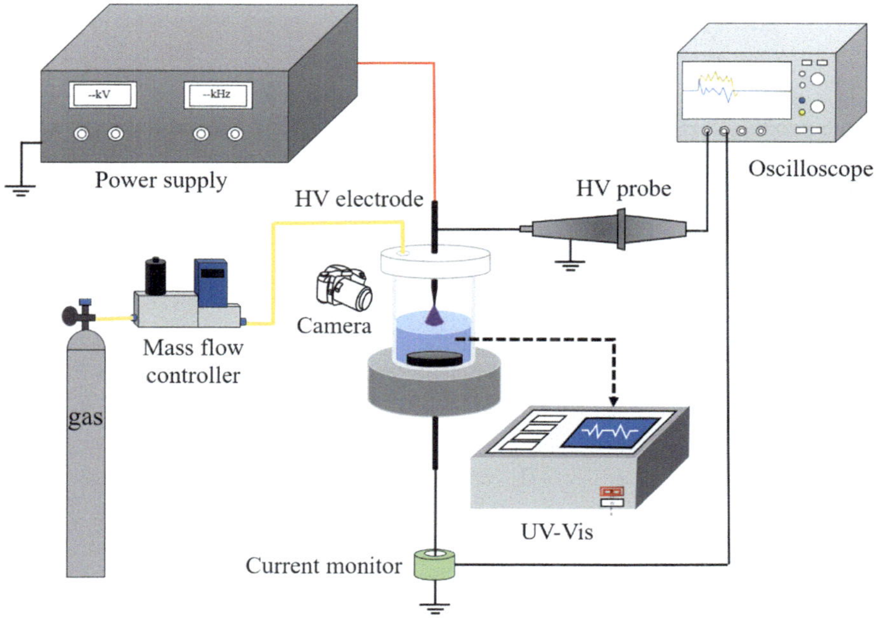

Fig. 26.1 Schematic diagram of the experimental system

excite the discharge, with an adjustable voltage amplitude and frequency range of 0–20 kV and 0–10 kHz, respectively. Voltages-currents waveforms were measured by a high voltage probe (Tektronix, P6139A) and a current coil (Pearson, 4100), and the data is recorded by an oscilloscope (Tektronix, TDS-3054c).

26.2.2 $CuFe_2O_4$ Preparation

The sol–gel method was employed to synthesized the non-homogeneous catalyst $CuFe_2O_4$. $Cu(NO_3)_2{\cdot}3H_2O$, $Fe(NO_3)_3{\cdot}9H_2O$ and urea were first mixed into deionized water. Next, the mixture was placed in a water-bath thermostatic magnetic stirrer with the stirrer parameters set at 70 °C and 200 r/min. A green suspension was obtained after 0.5 h of treatment, which can convert to gel with further heating. The gel was treated in a muffle furnace at the constant temperature of 600 °C for 20 h. The product obtained was finally milled to form the non-homogeneous catalyst $CuFe_2O_4$.

26.2.3 *Analytic Methods*

H_2O_2 concentration was measured by detecting the absorbance at wavelength 407 nm of the product formed by the reaction between H_2O_2 and titanium sulfate under acidic conditions by a UV–Visible Spectrophotometer (Jinghua UV 1800). The O_3 concentration was then determined by the indigo method, in which O_3 was used to discolor sodium indigo disulfate in a phosphate buffer environment and absorbance was obtained at 610 nm [13]. Isopropanol, uric acid, potassium dihydrogen phosphate, and pyruvate were used as scavengers for ·OH, O_3, e^*, and H_2O_2, respectively to analyze the role of reactive species in TC degradation [12].

TC concentration was calculated from the absorbance at 354 nm measured by UV 1800 [14]. The TC degradation performance was evaluated by degradation rate and reaction rate. The degradation rate and reaction rate were calculated by the Eq. (26.1) and (26.2):

$$\eta = \frac{c_0 - c_t}{c_0} \tag{26.1}$$

$$\ln(c_0/c_t) = kt \tag{26.2}$$

where C_0 represents the initial concentration, C_t means the concentration after treatment time t; k is the reaction rate.

26.3 Results and Discussions

26.3.1 *Performance of PGLDP with $CuFe_2O_4$ Under Different Conditions*

26.3.1.1 Effect of CFO Dosage

Figure 26.2 presents the TC degradation rate in PGLDP+CFO system with different CFO dosages. As illustrated in Fig. 26.2a, the degradation rate was significantly improved by CFO addition. With the increase of CFO dosage, the degradation rate also increased gradually. At a CFO dosage of 1.0 g/L, the highest degradation rate of 86.62% was achieved at the treatment of 10 min, while the degradation rate of plasma alone was only 66.75%. When the CFO dosage is further elevated, the degradation performance decreases. In Fig. 26.2b, the variation of reaction rate constants shows that the best degradation performance was obtained at 1.0 g/L CFO with a reaction rate of 0.170 min^{-1}. The higher degradation rate of PGLDP+CFO system is due to the photo-Fenton reaction after CFO is introduced into the solution. Moreover, the reactive sites at the catalysts surface gradually increase with the CFO dosage, which also provides a larger reactive specific surface area. Excess CFO hinders light

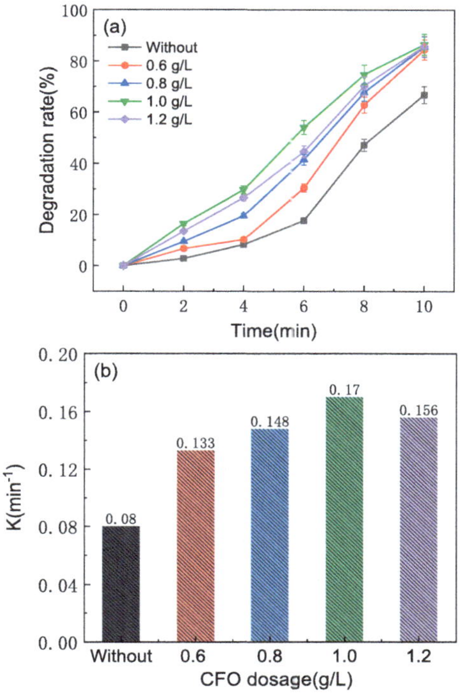

Fig. 26.2 Effect of CFO dosage on TC degradation: **a** degradation rate; **b** reaction rate

scattering, leading to fewer photogenerated electrons, and the active sites overlap, resulting in weakened degradation performance [15].

26.3.1.2 Effect of Voltage

Figure 26.3 demonstrates the effect of pulse voltage on TC degradation rate in the PGLDP+CFO system. As shown in Fig. 26.3a, the increasing of pulse voltage promoted the degradation rate of TC. The degradation rate increases from 34.50 to 86.62% when the voltage increases from 6.5 to 8.0 kV. In particular, when the voltage exceeds 7 kV, the degradation rate increases rapidly in 8 min. The reaction rates at different voltage amplitudes are shown in Fig. 26.3b. The reaction rate increases with pulse voltage, which is observed to have the highest value of 0.170 min^{-1} with the pulse voltage of 8.0 kV. The increase in pulse voltage boosts the electric field at the discharge gap, allowing the electrons to gain more energy and producing more active species. The reactions at the gas–liquid interface are thus more intense and the production of aqueous ROS is elevated. Additionally, the stronger physicochemical effect (high-energy electrons, UV radiation, shock waves, etc.) from the more intense discharge further enhances the degradation [16, 17].

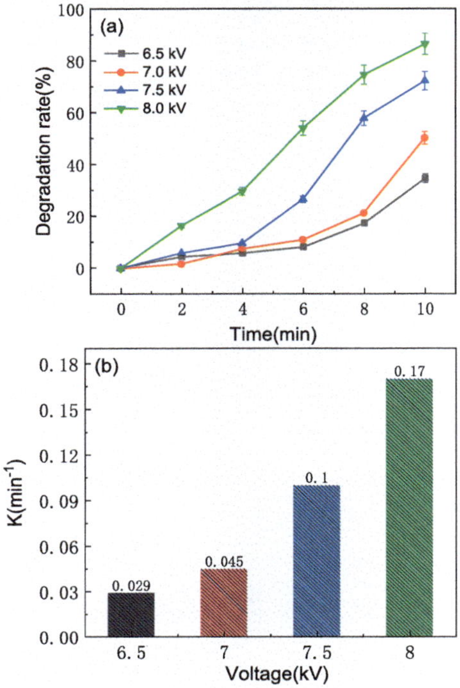

Fig. 26.3 Optimization of pulse voltage for TC degradation: **a** degradation rate; **b** reaction rate

26.3.1.3 Effect of Initial TC Concentration

Degradation rate and reaction rate under different initial TC concentrations are shown in Fig. 26.4. It is suggested from Fig. 26.4a that either too low or too high initial TC concentration is not favorable for the degradation. While in a initial concentration of 80 mg/L, the highest degradation rate of 84.63% can be achieved in 10 min. It can also be found from Fig. 26.4b that the reaction rate reachs 0.133 min^{-1} at the initial concentration of 80 mg/L, which is much higher than 0.036 and 0.069 min^{-1} for 10 or 120 mg/L. When the TC concentration is 40–80 mg/L, it is relatively dispersed in the solution and has a low probability of collision or reaction with the reactive species, resulting in a low degradation rate. As the initial concentration exceeds 80 mg/L, there is a decrease in the percentage of active species in each tetracycline molecule. Furthermore, degradation intermediates at high TC concentrations may also react with reactive species and outcompete TC molecules. Although the degradation rate is unsatisfactory at high TC concentrations, the absolute removal of TC is still substantial at an initial concentration of 120 mg/L, reaching 75.36 mg. This phenomenon can be attributed to a large amount of charge in the pollutant solution at high concentrations, which increases the probability of collision with free radicals [18].

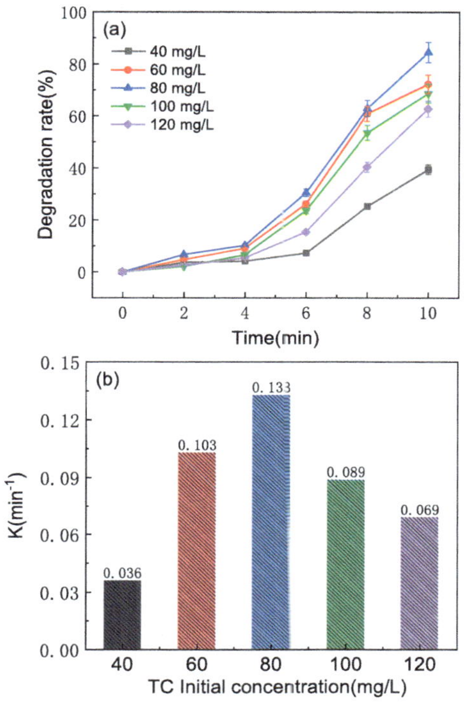

Fig. 26.4 Effect of initial concentrations on TC degradation: **a** degradation rate; **b** reaction rate

26.3.1.4 Effect of Working Gas

Figure 26.5 illustrates the impact of different working gases on degradation performance of PGLDP+CFO system. Under Ar conditions, the degradation effect is optimal, and it is the worst under O_2 conditions. The degradation of TC is more powerful in Air than in N_2. The degradation rate under Air, N_2, and O_2 discharges first increases slowly in ~~at~~ the first 4 min, and then rapidly increases. After 10 min of treatment, the degradation rate achieves 30.75%(O_2), 64.00%(N_2), 86.13%(Ar), and 84.63%(Air), respectively. Figure 26.5b shows the reaction rate in different working gases. The reaction rate under Ar is nearly six and two times higher than those of O_2 and N_2, while being 20% higher than those under Air. Differences in degradation performance under different operating gases are associated with different discharge characteristics. The lowest degradation rate in O_2 discharge is due to the weak discharge intensity [19]. Compared to other gases, Ar discharge has a higher electron density and energy, which promotes the production of active species in the liquid phase [20]. The degradation performance in N_2 discharge is weaker than that in Air, mainly because ROS can only be produced by the reactions between high-energy electron and H_2O, resulting in a lower concentration of ROS.

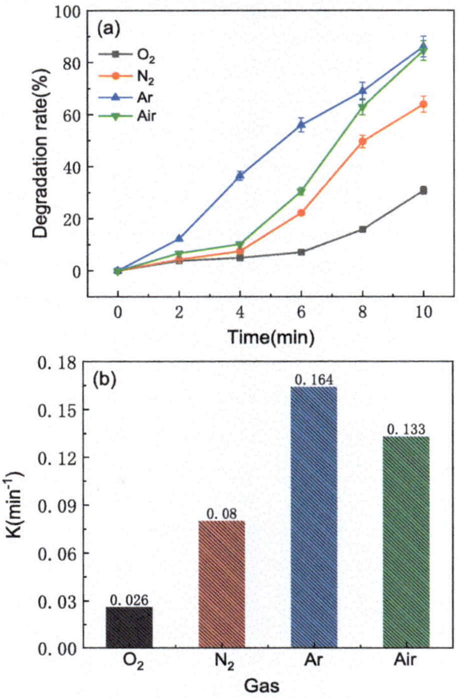

Fig. 26.5 Effect of working gas on TC degradation: **a** degradation rate; **b** reaction rate

26.3.2 Synergetic Mechanism of PGLDP/CFO

26.3.2.1 Concentration of Long-Lived Species H_2O_2 and O_3

Figure 26.6 illustrates the distribution of aqueous H_2O_2 and O_3 concentrations over time in the PGLDP and PGLDP+CFO systems. It is clearly shown that the concentration of H_2O_2 in both systems shows an increasing trend over discharge time. However, the H_2O_2 concentration in the PGLDP+CFO system is lower compared to the PGLDP system, due to the Fenton reaction between a portion of the H_2O_2 and the CFO of the PGLDP+CFO system [21]. O_3 concentration in the PGLDP+CFO system is higher than in the PGLDP system, which suggests that the PGLDP+CFO system facilitates the generation of O_3 (R(3)- R(5)) [22].

The complex reaction of ·OH and the bombardment of H_2O by high-energy electrons are the main pathways to generate H_2O_2 (R(6)-R(7)) [23]. The reaction between ·O and O_2 in the gas phase is the main pathway for O_3 generation. H_2O_2 and O_3 can also be indirectly involved in the degradation of tetracycline [24], converted to more oxidizing reactive species by further reaction and then reacted with TC (R(8)-R(10)) [25].

$$\equiv \mathrm{Fe(II)} + H_2O_2 \rightarrow \mathrm{Fe(III)} + \cdot OH + OH^- \quad (26.3)$$

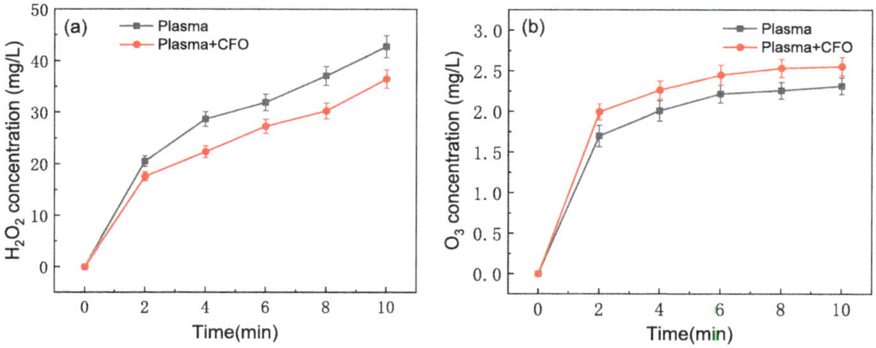

Fig. 26.6 Variation of H_2O_2 (**a**), and O_3 (**b**) in the PGLDP and PGLDP+CFO systems

$$\equiv Fe(III) + H_2O_2 \rightarrow Fe(II) + \cdot HO_2 + H^+ \tag{26.4}$$

$$\cdot OH + O_2 \rightleftarrows O_3 + e^- + H^+ \tag{26.5}$$

$$\cdot OH + \cdot OH \rightarrow H_2O_2 \tag{26.6}$$

$$2H_2O \xrightarrow{e} H_2O_2 + H_2 \tag{26.7}$$

$$H_2O_2 + 2O_3 \rightarrow 2 \cdot OH + 3O_2 \tag{26.8}$$

$$H_2O_2 \xrightarrow{hv} 2 \cdot OH \tag{26.9}$$

$$O_3 \xrightarrow{hv} O_2 + \cdot O \tag{26.10}$$

26.3.2.2 Role of Active Species

Radical scavengers (isopropanol (IPA), uric acid (UA), monopotassium phosphate (MP) and pyruvate (SP)) were added to remove ·OH, O_3, e^-, and H_2O_2 to investigate the role of active species in the degradation. The degradation rates with different radical scavengers are plotted in Fig. 26.7. The degradation rate decreases from 84.63 to 43.25%, 56.62, 65.62 and 50.25% with the addition of IPA, SP, UA, and MP, with the corresponding reaction rates decreasing from 0.133 min^{-1} to 0.043 min^{-1}, 0.059 min^{-1}, 0.070 min^{-1} and 0.089 min^{-1}, respectively. It is indicated that ·OH contributes the most to the degradation of tetracycline, which is due to its high oxidation potential ($E_0 = 2.8$ V) and reaction rate (1×10^9 L/(mol·s)) [13], which can

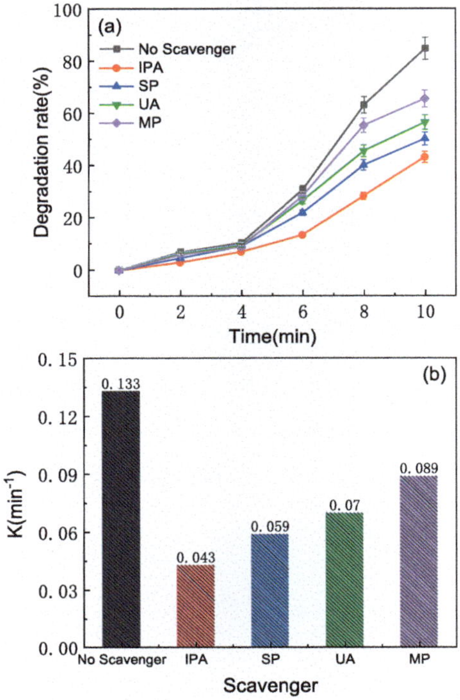

Fig. 26.7 Degradation rate (**a**) and reaction rate (**b**) in the PGLDP+CFO system with radical scavengers added in the solution

rapidly participate in the degradation of tetracycline. Moreover, H_2O_2 also plays an important role in generating stronger oxidizing active substances through the Fenton reaction between H_2O_2 and CFO, in addition to its direct or indirect participation in the degradation [26]. O_3 can interact with tetracycline directly and also by conversion to the more oxidizing $\cdot O_2^-$ and $\cdot OH$ radicals(R(11)-R(13)) [27, 28].

$$3O_3 + H_2O \rightarrow 2 \cdot OH + 4O_2 \tag{26.11}$$

$$O_3 + \cdot HO_2 \rightarrow \cdot OH + O_2 + \cdot O_2^- \tag{26.12}$$

$$O_3 + \cdot OH \rightarrow \cdot HO_2 + \cdot O_2^- \tag{26.13}$$

26.3.2.3 Proposed Synergetic Mechanism of PGLDP and CFO

Figure 26.8 shows the synergetic mechanism of PGLDP and CFO. UV radiation generated by PGLDP activates CFO, which forms photogenerated electrons and electron holes (R(14)-R(15)) [10]. High-energy electrons produced by PGLDP bombard

H_2O molecules, producing ·OH radicals. The reaction of ·OH generates H_2O_2, which then reacts with the active sites Fe(III) and Cu(II) on CFO to generate $\cdot HO_2$ and $\cdot O_2^-$ radicals. Fe(III) and Cu(II) can be reduced to Fe(II) and Cu(I) during the reaction of H2O2 with the active site(R(16)-R(18)) [11]. Metal-to-metal reactions of varying valence levels produce oxidizing radicals (·OH) In addition, reactions between metals of different valence levels promote the generation of oxidizing radicals(·OH) [29]. Since there is some leaching of Fe and Cu in the liquid phase, some homogeneous Fenton reactions also occur in the liquid phase. Moreover, the presence of Lewis acidic copper in CFO promotes the interaction between Fe and Cu iron and copper, indirectly accelerating the production of strong oxidizing radicals [30]. Finally, the physical and chemical effects of PGLDP can also degrade TC directly or indirectly.

$$\mathrm{CuFe_2O_4} + hv \rightarrow e^- + h^+ \tag{26.14}$$

$$h^+ + \mathrm{H_2O} \rightarrow \cdot\mathrm{OH} + \mathrm{H^+} \tag{26.15}$$

$$\equiv \mathrm{Cu(II)} + \mathrm{H_2O_2} \rightarrow \equiv \mathrm{Cu(I)} + \cdot\mathrm{O_2^-} + 2\mathrm{H^+} \tag{26.16}$$

$$\equiv \mathrm{Cu(I)} + \mathrm{H_2O_2} \rightarrow \equiv \mathrm{Cu(II)} + \cdot\mathrm{OH} + \mathrm{OH^-} \tag{26.17}$$

$$\equiv \mathrm{Fe(III)} + \mathrm{Cu(I)} \rightarrow \equiv \mathrm{Fe(II)} + \mathrm{Cu(II)} \tag{26.18}$$

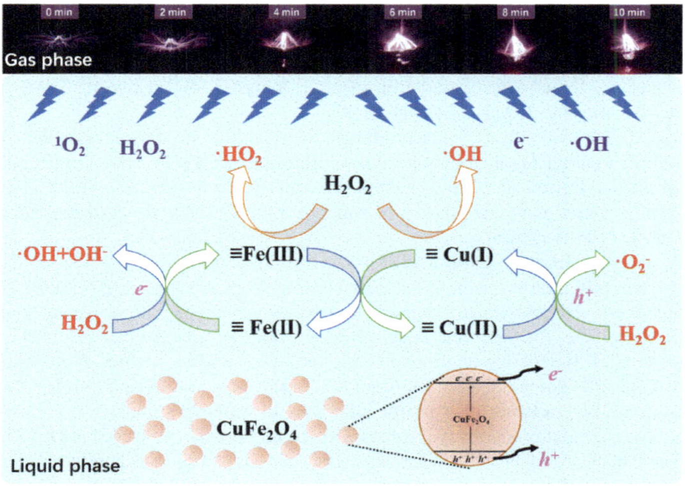

Fig. 26.8 Schematic diagram of the mechanism of TC degradation by the PGLDP+CFO system

26.4 Conclusions

In this research, PGLDP coupled with CFO was employed for the TC degradation. The performance of PGLDP/CFO system under different working conditions was analyzed. The study revealed that pulse voltage, initial TC concentration, CFO amount, and working gas are important factors affecting TC degradation. The degradation rate in the PGLDP+CFO system reached 86.62% after 10 min treatment, which is only 66.75% was obtained by plasma alone. It is also founded that H_2O_2 concentration in the PGLDP+CFO system is lower than in the PGLDP system, suggesting that H_2O_2 was consumed by CFO induced Fenton reaction. It is also proved that ·OH radical acts the most contribution to the TC degradation in PGLDP+CFO system, followed by H_2O_2. This research provides an effective way for plasma synergistic Fenton reaction to deal with antibiotic wastewater.

References

1. S. Obregón et al., Electrophoretic deposition of $PbMoO_4$ nanoparticles for photocatalytic degradation of tetracycline. Appl. Surf. Sci. **457**, 501–507 (2018)
2. M. Magureanu et al., A review on non-thermal plasma treatment of water contaminated with antibiotics. J. Hazard Mater. **417**, 125481 (2021)
3. M.A. Clement et al., Degradation of glucocorticoids in water by a synergistic system of peroxymonosulfate, microbubble and dielectric barrier discharges. J. Water Process Eng. **42**, 102175 (2021)
4. M. Hatzisymeon et al., Highly energy-efficient degradation of antibiotics in soil: extensive cold plasma discharges generation in soil pores driven by high voltage nanopulses. Sci. Total Environ. **786**, 147420 (2021)
5. G. He et al., Degradation of flumequine in water by pulsed discharge plasma coupled with reduced graphene oxide/TiO_2 nanocomposites. Sep. Purif. Technol. **218**, 206–216 (2019)
6. S. Kefeng et al., Synergetic degradation of acid orange 7 (AO7) dye by DBD plasma and persulfate. Chem. Eng. J. **311**, 378–384 (2017)
7. C. Yong et al., An evaluation of metronidazole degradation in a plasma-assisted rotating disk reactor coupled with TiO_2 in aqueous solution. Engineering **11**(7), 1603–1610 (2021)
8. H. Weijie et al., Pulsed discharge plasma on water surface coupled with $CaFe_2O_4/Bi_2O_3$ composites for synergistic degradation of aqueous tetracycline hydrochloride. Separat. Purifi. Technol. **279**, 119691 (2021)
9. G. He et al., Accelerated fenton reaction for antibiotic ofloxacin degradation in discharge plasma system based on graphene-Fe_3O_4 nanocomposites. Vacuum **185**, 110022 (2021)
10. W. Zhang et al., Degradation of tetracycline by activating persulfate using biochar-based $CuFe_2O_4$ composite. Environ. Sci. Pollut. Res. Int. **44**(29), 67003–67013 (2022)
11. W.W. Gao et al., Efficient degradation of semi-coking wastewater in three-dimensional electro-fenton by $CuFe_2O_4$ heterocatalyst. Environ. Sci. Pollut. Res. Int. **49**(29), 74163–74172 (2022)
12. Z. Tianqi et al., Degradation of cefixime antibiotic in water by atmospheric plasma bubbles: performance, degradation pathways and toxicity evaluation. Chem. Eng. J. **421**, 127730 (2021)
13. A.R. Ribeiro et al., An overview on the advanced oxidation processes applied for the treatment of water pollutants defined in the recently launched directive 2013/39/EU. Environ. Int. **75**, 33–51 (2015)
14. C. Fang et al., Degradation of tetracycline by atmospheric-pressure non-thermal plasma: enhanced performance, degradation mechanism, and toxicity evaluation. Sci Total Environ **812**, 152455 (2022)

15. Z. Xin et al., A facile heterogeneous system for persulfate activation by $CuFe_2O_4$ under LED light irradiation. RSC Adv. **55**(9), 32328–32337 (2019)
16. W. Chao et al., Removal of tetracycline antibiotics from wastewater by pulsed corona discharge plasma coupled with natural soil particles. Chem. Eng. J. **346**, 159–170 (2018)
17. M.M.E. Salomon et al., Degradation of sulfamethoxazole by double cylindrical dielectric barrier discharge system combined with Ti /C-N-TiO_2 supported nanocatalyst. J. Hazard. Mater. Adv. **5**, 100051 (2022)
18. L. Zhen et al., Insights into water film DBD plasma driven by pulse power for ibuprofen elimination in water: performance, mechanism and degradation route. Separat. Purif. Technol. **277**, 119415 (2021)
19. P. Marija et al., Monte Carlo simulation of RF breakdown in oxygen—the role of attachment. Euro. Phys. J. D **4**(74), 72–80 (2020)
20. L. Chandana et al., Atmospheric pressure non-thermal plasma jet for the degradation of methylene blue in aqueous medium. Chem. Eng. J. **282**, 116–122 (2015)
21. H. Guo et al., Multi-catalysis induced by pulsed discharge plasma coupled with graphene-Fe_3O_4 nanocomposites for efficient removal of ofloxacin in water: mechanism, degradation pathway and potential toxicity. Chemosphere **129089**(265), 129089 (2021)
22. D. Ruoyu et al., Enhanced synergistic performance of nano-Fe^0-CeO_2 composites for the degradation of diclofenac in DBD plasma. Chem. Eng. J. **406**, 126884 (2021)
23. Z. Ke et al., Degradation of sulfamethoxazole in aqueous solution by dielectric barrier discharge plasma combined with Bi_2WO_6-rMoS_2 nanocomposite: mechanism and degradation pathway. Chemosphere **222**, 872–883 (2019)
24. G. Shi et al., Degradation of levofloxacin in aqueous solution by non-thermal plasma combined with Ag_3PO_4/activated carbon fibers: mechanism and degradation pathways. Separat. Purif. Technol. **250**, 117264 (2020)
25. W. Sen et al., Highly synergistic effect for tetracycline degradation by coupling a transient spark gas–liquid discharge with TiO_2 photocatalysis. Chem. Eng. J. **450**, 138409 (2022)
26. O.T. Portela et al., Synthesis, characterization, and photocatalytic investigation of CuFe2O4 for the degradation of dyes under visible light. Catalysts **6**(12), 623 (2022)
27. H. Chunjing et al., Degradation of pharmaceutical contaminant tetracycline in aqueous solution by coaxial-type DBD plasma reactor. IEEE Trans. Plasma Sci. **2**(48), 471–481 (2020)
28. G. He et al., A comprehensive insight into plasma-catalytic removal of antibiotic oxytetracycline based on graphene-TiO_2-Fe_3O_4 nanocomposites. Chem. Eng. J. **425**, 130614 (2021)
29. D. Rong-Rong et al., Oxygen vacancy on hollow sphere $CuFe_2O_4$ as an efficient Fenton-like catalysis for organic pollutant degradation over a wide pH range. Appl. Catal. B: Environ. **291**, 120069 (2021)
30. W. Yanhong et al., Photo-catalytic oxidation for pyridine in circumneutral aqueous solution by magnetic Fe-Cu materials activated H_2O_2. Chem. Eng. Res. Des. **163**, 1–11 (2020)

Chapter 27
Characteristics of Degradation and Product Precipitation of Phenolic Wastewater Enhanced by Ultrasonic Photocatalysis Combined with Discharge

Jiayu Wang, Libo Zhu, Boyang Lin, Qianwen Li, Danyue Jiang, Xiang He, and Bingyan Chen

Abstract The total amount of phenolic wastewater is large, difficult to degrade, and traditional treatment methods can't recover and utilize phenolic substances. In this study, air plasma jet and photocatalyst were used to treat phenolic wastewater and produce reactive species in the liquid, and then ultrasonic combined with composite photocatalyst CQDs/TiO_2 was further used to enhance the subsequent reaction and produce solid recyclable precipitates. The experimental results show that when the flow speed of gas–liquid two-phase flow is 4.20 m/s, the plasma jet has the best effect of generating reactive species and degrading phenol in water. The continuous flow of liquid phase after discharge treatment is conducive to the degradation of phenolic wastewater, and the best effect is when the liquid phase flow rate is 0.60 m/s. The degradation efficiency of p-nitrophenol solution was increased 18.23% by using ultrasonic coordinated CQDs/TiO_2 supported aluminum oxide pellets. This study has certain reference value for the application of plasma combined with photocatalysis in the treatment of phenolic wastewater with high toxicity and difficult degradation.

27.1 Introduction

Phenolic wastewater is a kind of highly toxic and refractory wastewater, which exists in the wastewater of all walks of life. Such as oil refineries (6–500 mg/L), coking operations (28–3900 mg/L), coal processing (9–6800 mg/L) and petrochemical plants (2.8–1220 mg/L). Other sources containing 0.1–1600 mg/L of phenol

J. Wang · L. Zhu · B. Lin · Q. Li · D. Jiang · B. Chen (✉)
School of Mechanical and Electrical Engineering, Hohai University, Changzhou 213200, Jiangsu, China
e-mail: cbyan@foxmail.com

X. He · B. Chen
School of Science, Hohai University, Changzhou 213200, Jiangsu, China

Z. Fang et al. (eds.), *Proceedings of the 5th International Symposium on Plasma and Energy Conversion*, Springer Proceedings in Physics 398,
https://doi.org/10.1007/978-981-97-2245-7_27

include the pharmaceutical, plastics, wood products, paint, pulp and paper industries [1]. Domestic and foreign experts and scholars have conducted numerous studies on phenolic wastewater treatment methods, which can be divided into physical methods [2–4], biological methods [5–7] and chemical methods [8–10] in principle. The traditional chemical method makes it difficult to meet the discharge standard when treating high-concentration wastewater, and it is easy to produce secondary pollution. The physical method only changes the existence state of phenolic organic compounds, the consumption of macroporous resin is large, and phenolic substances cannot be recovered. The biological method has high environmental requirements and poor adaptability and colonization in open environment. With the development of science and technology and the improvement of environmental requirements, these traditional methods are gradually unable to meet all the requirements for the treatment of phenolic wastewater, so it is urgent to develop some treatment processes to degrade phenolic wastewater.

The principle of atmospheric pressure plasma to achieve water treatment is that the reactive species such as electron collision, ultraviolet ray (UV), free radicals (·OH, ·O, ·HO_2), ozone (O_3), hydrogen peroxide (H_2O_2) generated in the discharge environment interact with water or toxic substances in water, resulting in decomposition, oxidation and other synergistic reactions. At present, it mainly focuses on: corona discharge, dielectric barrier discharge (DBD), sliding arc discharge, plasma jet and so on. Compared with other methods, atmospheric pressure plasma jet (APPJ) water treatment has the characteristics of fast, efficient and no secondary pollution [11].

Focus on the field of phenolic wastewater treatment, many plasma reactors have been developed for the treatment of wastewater according to the plasma jet process at home and abroad. However, the treatment efficiency of these reactors is generally not high, so it is difficult to popularize and use in a large area. Many researchers have made many improvements in improving the treatment efficiency. One is to use other advanced oxidation technologies to jointly enhance the treatment efficiency [12]; the other is to increase the contact area between plasma and wastewater by atomizing wastewater to improve the treatment efficiency [13]. However, these studies do not make full use of the energy in the reaction process, such as ultraviolet ray, and the harmful substances in phenolic wastewater are eliminated by degradation, and there is no way to recycle.

Therefore, this study is based on a set of ultrasonic photocatalytic combined discharge system for the treatment of phenolic wastewater. In an atmospheric pressure plasma discharge environment, the reaction with ultrasonic and photocatalyst is coordinated. By adjusting the power supply of the reactor, the gas flow rate, the liquid flow rate after discharge, and the increase of the degradation amount of the overall photocatalyst and ultrasonic coordination, etc. The law of degradation and product precipitation of phenolic wastewater was studied.

27.2 Experimental Device and Methods

27.2.1 Experimental Device

27.2.1.1 Mechanism Analysis of Reaction Rate of Discharge System

As shown in the experimental system established in this paper, it is mainly divided into two units, a discharge processing unit and a circulation flow unit. In this study, 500 ml of p-nitrophenol solution with a concentration of 20.0 mg/L was used as a sample solution to investigate the degradation of phenolic wastewater and the law of product precipitation. The p-nitrophenol solution enters the discharge treatment unit through the water pump. The CQDs/TiO_2 composite catalytic material synthesized by carbon quantum dots (CQDs) [14] and titanium dioxide (TiO_2) is added to the gas–liquid–solid three-phase discharge device and combined with ultrasound to improve the degradation efficiency of wastewater. At the same time, an air source is connected to provide a gas–liquid–solid three-phase continuous environment for the discharge reaction. The reactive species generated by the discharge is used to oxidize p-nitrophenol in water. After the discharge, the nitrophenol solution entered the circulation flow unit for cyclic treatment, and CQDs/TiO_2 was added to the reactor, and combined with ultrasound, the combined enhancement effect of secondary ultrasonic photocatalysis was performed.

A stable discharge APPJ was used to treat a flowing solution of p-nitrophenol (PNP). The APPJ reactor is discharged in the flowing liquid phase to produce hydroxyl free radical (·OH), oxygen free radical (·O), hydrogen superoxide (HO_2), nitrogen oxide (NO_x), hydrogen peroxide (H_2O_2), ozone (O_3) and other active substances to react with phenolic organic compounds. The reaction formation mechanism is shown in the formula [15–17]:

$$e + H_2O \rightarrow e + H + OH \tag{27.1}$$

$$e + O_2 \rightarrow O + O + e \tag{27.2}$$

$$O + H_2O \rightarrow OH + OH \tag{27.3}$$

$$OH + OH \rightarrow H_2O_2 \tag{27.4}$$

$$O + O_2 \rightarrow O_3 \tag{27.5}$$

At the same time, the added high-efficiency catalyst (CQDs/TiO_2) is in full contact with the circulating liquid, and continues to produce a synergistic effect, showing more active catalytic groups and increasing catalytic efficiency. Due to the interaction between the fast-flowing solution and the gas and solid, the gas, liquid and solid three-phase reaction system is formed, which increases the average concentration of the active reactants (such as ·OH, O_3 and NO_x) in the solution.

As shown in Fig. 27.1, the gas–liquid–solid three-phase discharge device is generally a cuboid, whose length, width and height are 195.0 mm, 55.0 mm and 40.0 mm, respectively. The bottom plate consists of a 2.0 mm thick nickel plate, while the sides and top plate are made of 10.0 mm thick quartz glass. Above the reaction device is an APPJ generator for generating APPJ, which enters the flowing liquid to simulate PAW generation. A piezoelectric transducer is bonded to the nickel plate at the bottom of the reaction device, and CQDs/TiO_2 pellets are added to the lower part of the reactor.

As shown in Fig. 27.2, the circulation flow unit device adopts quartz glass with length, width and height of 215 mm, 75.0 mm and 45.0 mm respectively, on which a special light source that can simulate natural light is placed in a close distance as the excitation light source of the photocatalyst. The catalyst in the reaction pool works cooperatively with ultrasonic waves to continuously contact with the flowing liquid and accelerate the catalytic rate. A vessel containing a highly efficient catalyst (CQDs/TiO_2) is ultrasonic-treated to achieve a micro-perturbation effect, thereby

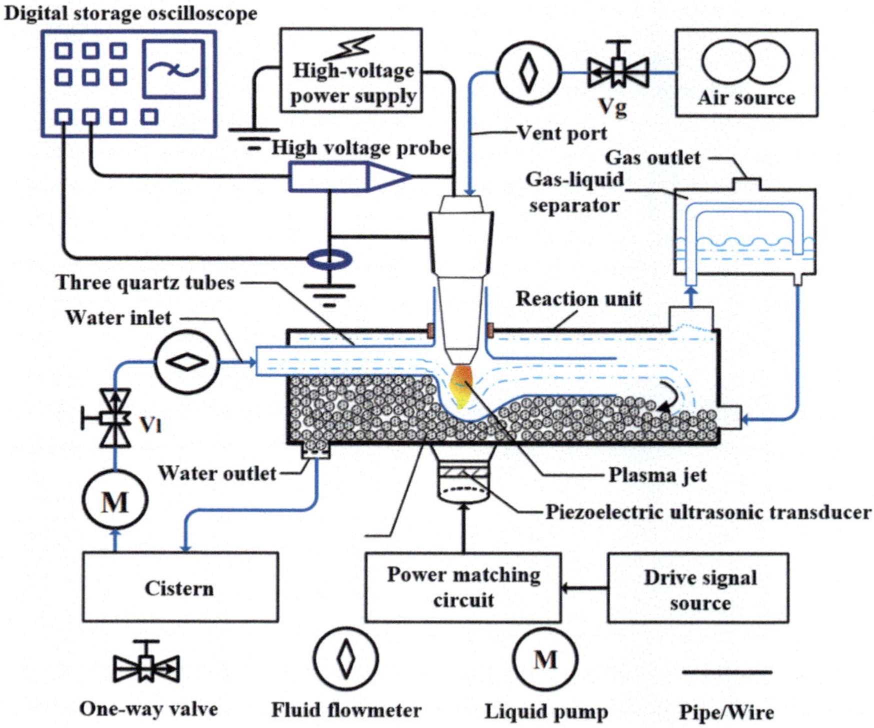

Fig. 27.1 Gas–liquid–solid three-phase discharge device

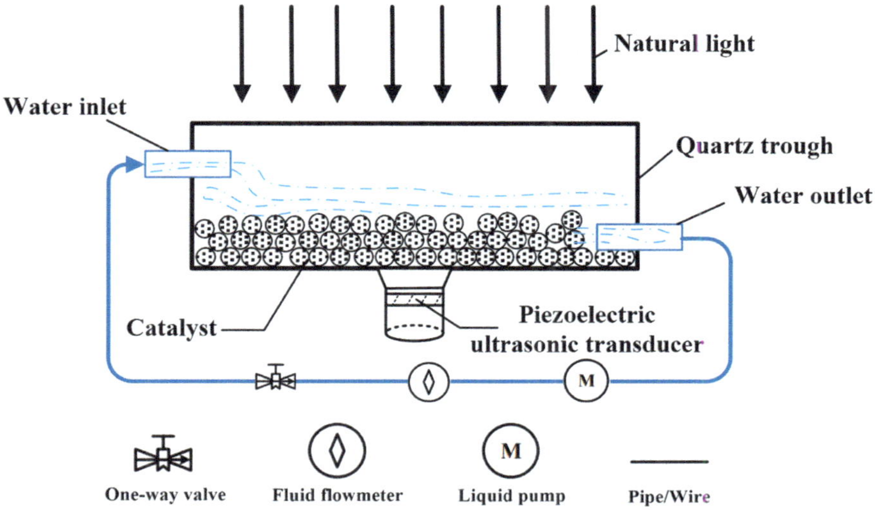

Fig. 27.2 Circulation flow device

utilizing its mechanical effects to enhance the mass transfer and diffusion of the reactive species in water. Secondly, during the experiment, the peristaltic pump was used for circulating treatment, and the degradation rate of the p-nitrophenol solution was investigated under the optimal liquid flow rate after discharge treatment by adjusting different liquid flow rate parameters.

27.2.2 Analytical Methods

The optimal PNP degradation efficiency and maximum precipitation recovery can be obtained by adjusting electrical parameters, gas flow rate parameters, liquid flow rate parameters, ultrasonic parameters catalysts and so on. At the microscopic level of the underwater APPJ discharge reaction system, there are complex physical and chemical effects such as microflow, microturbulence, acoustic (shock) wave and catalysis. These complex effects will change the diffusion rate between the gas and liquid phase and the activation energy of the reaction, thus changing the reaction rate of the system. In the gas–liquid–solid three-phase reaction system, fluid diffusion characteristics are regulated by physical parameters such as ultrasonic factor, fluid gas content and flow rate, and the reaction expression is as follows [18]:

$$k_{d+} = \frac{2RT}{3\mathrm{N_A}\left[\beta^2\eta_g + (1-\beta)^2\eta_l + \beta(1-\beta)\left(\eta_g\frac{\rho_l}{\rho_g} + \eta_l\frac{\rho_g}{\rho_l}\right)\right]} \cdot \frac{(r_\mathrm{A} + r_\mathrm{B})^2}{r_\mathrm{A} r_\mathrm{B}} \cdot f_u f_v \tag{27.6}$$

where, k_{d+} is the forward diffusion rate, which has a decisive influence on the gas–liquid reaction rate. $\boldsymbol{f}_v$ is the fluid flow factor, $\boldsymbol{\beta}$ is the aeration rate and $\boldsymbol{f}_u$ is the ultrasonic factor, $\boldsymbol{\eta}_g$ is the viscosity coefficient of air and $\boldsymbol{\eta}_l$ is the viscosity coefficient of water, $\boldsymbol{rA}$ and $\boldsymbol{rB}$ are the reactant molecular radius, $\boldsymbol{\rho}_g$ is the density of air and $\boldsymbol{\rho}_l$ is the density of water.

The catalyst can effectively reduce the reaction threshold energy, thereby increasing the reaction rate. The reaction rate ratio m varies exponentially with the activation energy difference $E_a - E_{ac}$:

$$m = \exp\left(\frac{E_a - E_{ac}}{RT}\right) \quad (27.7)$$

Therefore, the addition of the catalyst in full contact with the circulating liquid can continue to produce a synergistic effect, show more active catalytic groups, and increase the catalytic efficiency. Due to the interaction between the fast-flowing solution and gas and solid, the gas, liquid and solid three-phase reaction system is formed, which increases the diffusion rate of the active reactants in the solution (such as ·OH, O_3 and NO_x) and effectively improves the reaction rate.

27.2.2.1 Circulation System Mechanism Analysis and Catalyst Selection

According to formula (27.7), the catalyst can effectively reduce the reaction threshold energy and increase the reaction rate. The nano-catalyst has a valence band (VB) and conduction band (CB) structure, and the region between CB and VB is called a band gap (E_g). When light with energy equal to or higher than the semiconductor bandgap energy is injected into the semiconductor surface, electrons and holes (e^- and h^+) with reducing capacity and oxidation capacity are generated, and then the resulting e^-/h^+ will migrate to the catalyst surface and participate in the chemical reaction.

Formula (27.6) shows that the forward diffusion rate is also affected by ultrasonic factors. Under the mechanical disturbance of ultrasound, the collision frequency between reactant molecules and reactive species can also be increased, and the reaction rate can be increased. Based on this theory, in order to maximize the degradation efficiency of organic matter, ultrasonic combined with photocatalysis was used to improve the degradation efficiency.

In this study, titanium dioxide (TiO_2) and carbon quantum dots (CQDs) were used to synthesize composite catalytic materials as photocatalysts, and the catalytic mechanism was shown in Fig. 27.3. TiO_2 has the advantages of low cost, non-toxicity and high chemical stability. After absorbing visible or near-infrared light, carbon quantum dots can stimulate the excellent characteristics of visible or ultraviolet light with shorter wavelength through the anti-Stokes effect, making up for the defects of TiO_2's large band-gap width (3.2 eV) resulting in small wavelength range of absorption spectrum (less than 387.5 nm) and low utilization rate of natural light. At

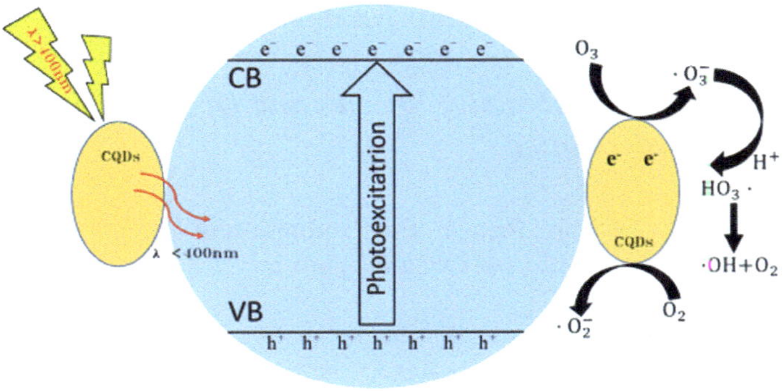

Fig. 27.3 Catalytic mechanism

the same time, the electron transient effect of carbon quantum dots can effectively prevent the recombination of electrons and holes and improve the catalytic activity.

27.2.2.2 Inspection Methods

The data detection part of this study includes the detection of reactive species O_3, H_2O_2, nitrogen-containing active substances (NO_X) and the detection of degradable substances in the liquid to be treated. For the detection of reactive species O_3, H_2O_2 and nitrogen-containing active matter (NO_X), O_3 detector (Jida Cygnet, GDYS-101SC2, measuring range 0–2.50 mg/L, resolution:0.01 mg /L), nitrite nitrogen detector (Jida Cygnet, GDYS-101SX, measuring range: 0–0.45 mg/L, resolution:0.01 mg/L), H_2O_2 detector (Kyoritsu chemical-check Lab. Corp., DPM-H_2O_2, measuring range: 0–2.00 mg/L, resolution: 0.01 mg/L) to detect O_3, NO_2^- and H_2O_2 in solution. The concentration of degradable substances in phenolic wastewater was detected by ultraviolet–visible spectrophotometer (Persee, T6U, wavelength 190–1100 nm, wavelength accuracy 1 nm, absorbance −0.3–3.0 Abs, transmittance 0–200%).

By controlling the reactor power supply, gas flow rate, circulating liquid flow rate and other parameters, the contents of reactive species O_3, H_2O_2, nitrogen-containing active matter (NO_X) and phenol in phenolic wastewater under different treatment times were detected, and the images with time were drawn, so as to analyze the degradation and product precipitation laws of phenolic wastewater.

27.3 Experimental Results and Analysis

27.3.1 *Generation of Reactive Species and Degradation of Organic Matter*

27.3.1.1 Effect of Gas Flow Rate on the Production of Reactive Species and the Degradation of Organic Matter

According to formula (27.6), fluid flow factor and gas content have an effect on the forward diffusion rate. On the one hand, changing the gas flow rate can change the gas content, so as to change the number of gaseous particles passing through the high-pressure electric field in a unit time, which will have a certain effect on the yield of high-energy particles. On the other hand, the flow factor of the fluid can be changed, and then the diffusion rate of high-energy particles can be changed. In view of this theoretical significance, the gas flow rate is used as a regulatory parameter to study the influence of gas flow rate on the production of reactive species and the degradation of organic matter.

PNP solution with a volume of 400 ml and a concentration of 100 mg/L was selected for this experiment. As shown in Fig. 27.4, with the increase of gas flow rate, the energy efficiency ratio of O_3 and H_2O_2 yield in the liquid phase increases logarithmically, increasing rapidly at the low flow rate stage (gas flow rate < 4.20 m/s) and slowly at the high flow rate stage (gas flow rate > 4.20 m/s). Finally, the maximum values are 7.57 and 30.59 μg/kJ at 6.30 m/s, respectively. The energy efficiency ratio of NO_2^- yield increases first and then decreases with the gas flow rate increasing, and reaches the maximum value of 284.13 μg/kJ at the gas flow rate of 4.20 m/s. As can be seen from Fig. 27.5, with the increase in gas flow rate, the residual rate of PNP presents a linear decline trend, which is because a higher gas flow rate brings higher reactive species yield, which promotes the degradation of PNP. The energy efficiency ratio of PNP degradation rate also increased logarithmically, reaching 51.80 μg/kJ at the gas flow rate of 6.30 m/s. The overall trend of PNP degradation rate energy efficiency ratio is similar to the energy efficiency ratio of O_3 and H_2O_2 yield in the liquid phase, because O_3 and H_2O_2 have higher oxidation potential and stronger oxidation ability, and can play a greater degradation role compared with NO_2^-.

Since the energy efficiency ratio of reactive species production and PNP degradation was not much improved by the subsequent increase of gas, the most suitable gas flow rate was 6.30 m/s.

27.3.1.2 Effect of Power Supply on Production of Reactive Species and Degradation of Organic Matter

The discharge energy is the main supplier of the reaction threshold energy, which has an impact on the yield of high-energy particles. Therefore, based on the gas flow

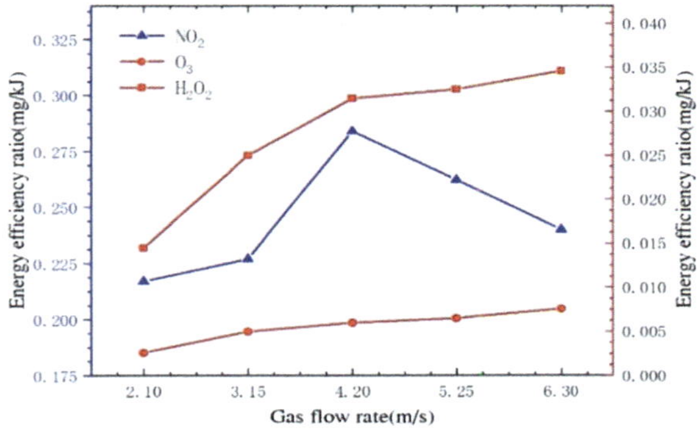

Fig. 27.4 Energy efficiency ratio of active ingredients produced at different gas flow rates

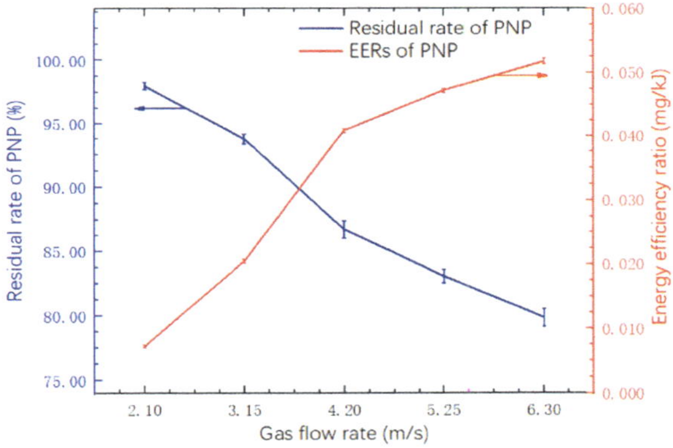

Fig. 27.5 Energy efficiency ratio of organic matter degradation at different gas flow rates

rate, the influence of the power supply on the production of reactive species and the degradation of organic matter will be further investigated in this section.

Firstly, whether the power supply affects the types of reactive species produced was investigated, and emission spectra of APPJ discharge in air under different power supplies were collected. As can be seen from Fig. 27.6, the change of power supply does not affect the position of major spectral peaks, that is, does not affect the type of reactive species produced, but the relative intensity of each spectral peak increases with the increase of power supply. This is because the relative intensity of the emission spectrum is proportional to the electron density and electron energy in the discharge environment. When the power supply on the reactor is increased, the electron density in the discharge region will also increase, thus increasing the probability of particle collision in the reaction region, which will increase the discharge reaction rate, bring higher concentrations of reactive species, and eventually lead to the relative spectral intensity of each spectral peak.

Then, the influence of power supply on the reaction rate was further explored. The control experimental conditions were as follows: gas flow rate of 6.30 m/s, power

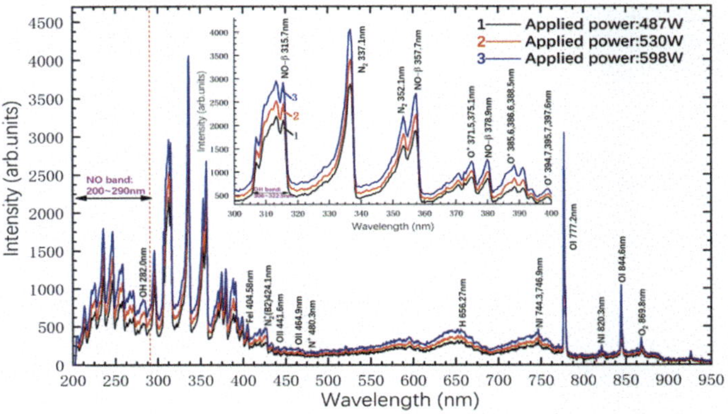

Fig. 27.6 Emission spectra of APPJ discharging in the air under different power supply

supply frequency of 23.50 kHz, gas solution flow rate of 1.575 m/s, treated solution volume of 400 mL, and PNP solution to be treated of 100 mg/L.

Figure 27.7 shows the energy efficiency ratio of reactive species under different power supplies. It can be seen that as the power supply increases from 487 to 598 W, the energy efficiency ratio of O_3 yield in the liquid phase shows a trend of first increasing and then unchanged, and an inflection point of 9.99 μg/kJ is reached when the power supply is 530 W. The energy efficiency ratio of H_2O_2 yield reaches a maximum of 35.61 μg/kJ when the power supply is 530 W, and then decreases with the increase of power. This is because the increase in power supply enhances the ultraviolet ray at the APPJ nozzle, which will cause the photolysis of H_2O_2 and consume it. The energy efficiency ratio of NO_2^- yield decreased gradually with the increase in power supply.

As can be seen from Fig. 27.8, with the increase of power supply, the residual rate of PNP presents a linear decline trend, which is because in the process of discharge,

Fig. 27.7 Energy efficiency ratio of reactive species under different power supply

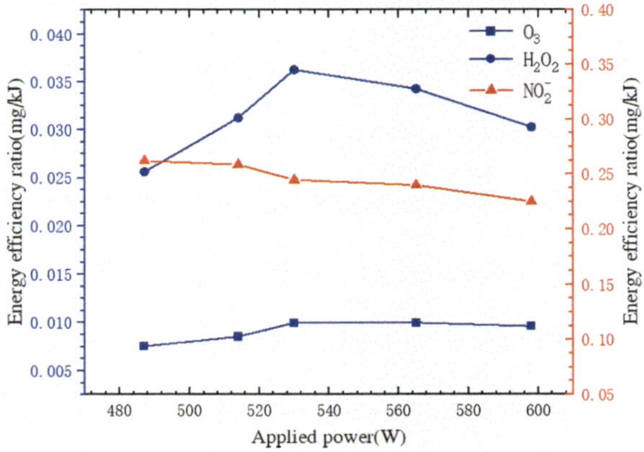

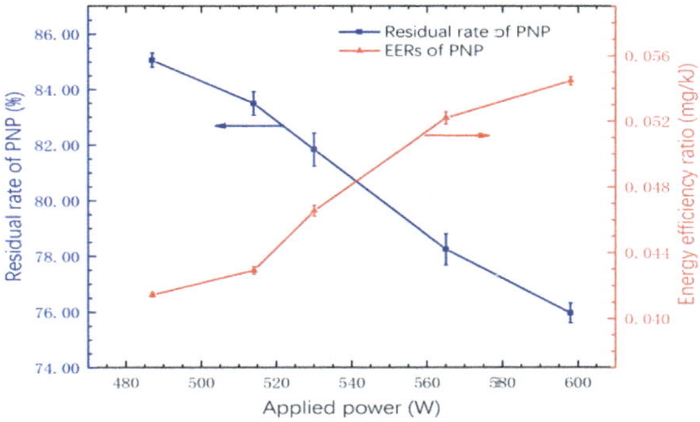

Fig. 27.8 PNP degradation rate and energy efficiency ratio under different discharge power

not only O_3, NO_2^- and H_2O_2 and other reactive species participate in the reaction, but the increase of power supply also enhances the degradation of PNP by free radicals and ultraviolet ray. The energy efficiency ratio of PNP degradation rate also increased linearly and was 54.47 μg/kJ at a 598 W power supply.

It can be concluded that although the increase in power supply will reduce the energy efficiency ratio of reactive species, it has a positive correlation with the increase of PNP degradation rate and energy efficiency ratio of degradation.

27.3.2 Death of Reactive Species and Degradation of Organic Matter After Discharge

27.3.2.1 Effect of Liquid Velocity on Organic Matter Degradation

In the pre-experiment, it was found that whether the treated solution flowed or not had a significant difference to the degradation result of PNP. This phenomenon can be explained by formula (27.6). After treatment, there are still unreacted reactive species in the solution, and flowing the solution can change the fluid flow factor, increase the collision frequency between reactant molecules and reactive species, and increase the reaction rate.

As shown in Fig. 27.9, the residual rate of PNP solution after discharge treatment decreased to 78.4% after standing for 3.0 h, and the concentration remained basically unchanged in the following time. The discharge-treated PNP solution decreased by 6.3% after 12.0 h, and the residual rate decreased by 19.4% after 72.0 h. Therefore, the experiment chose to let the liquid continue to flow after discharge to obtain the best treatment effect. Fig. 27.10 shows that the concentration of PNP is the lowest when the solution flow rate is 0.60 m/s at the same time, and the ranking is 0.60 > 0.90 > 0.30 m/s according to the degradation effect. With the progress of flow, the concentration of PNP has been low, and the degradation effect is not obvious, so the

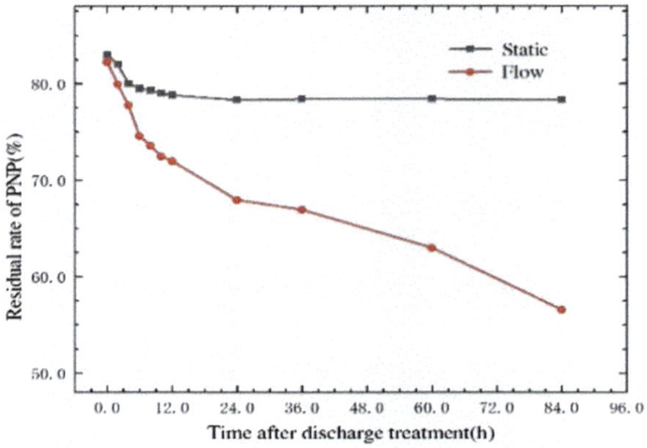

Fig. 27.9 Change of PNP residual rate with time during static and flow

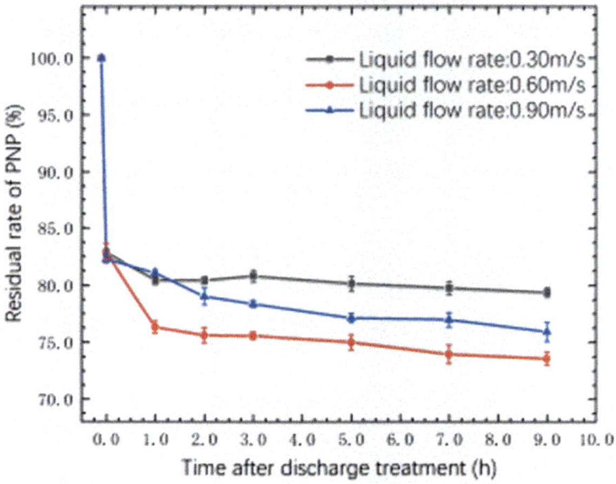

Fig. 27.10 Changes of PNP residual rate with time at different solution flow rates

subsequent curve tends to be smooth. Based on the above conclusions, the optimal liquid flow rate is 0.60 m/s.

27.3.2.2 Effect of Photocatalysis on the Extinction Rule of Reactive Species and the Degradation Rate of PNP

In order to further explore the influence of ultrasound combined with photocatalysis on PNP degradation, the changes of reactive species concentration and PNP participation rate over time in the presence of matt catalyst CQDs/TiO_2 were studied, and the results were shown in Fig. 27.11.

Figure 27.11a shows that without catalyst addition, ozone concentration decreases from 6.20 mg/L to about 1.80 mg/L within 0.0–4.0 h, and the overall extinction rate is

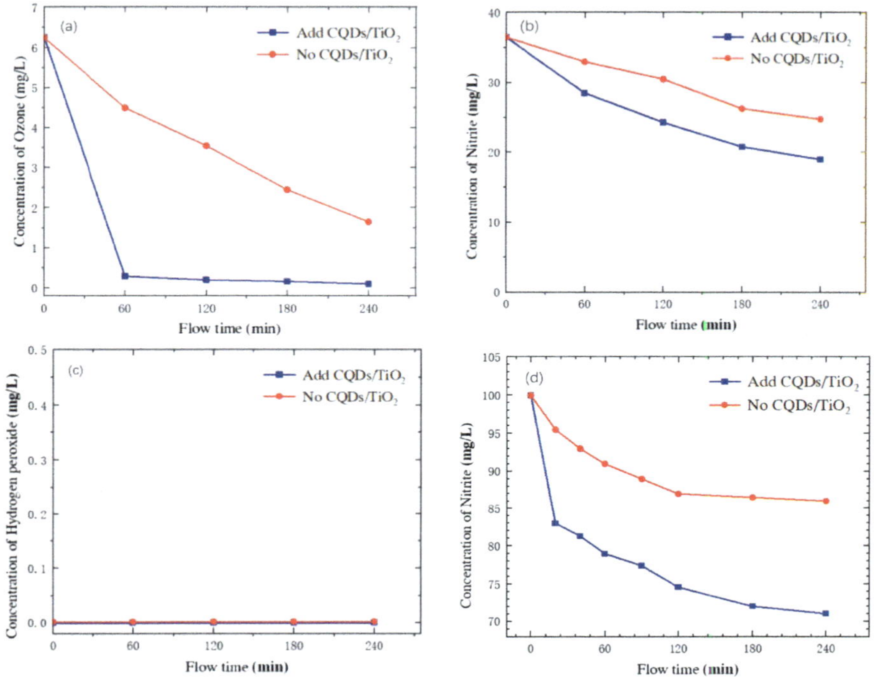

Fig. 27.11 Changes of reactive species concentration and PNP participation rate over time with or without catalyst **a** O_3, **b** NO_2^-, **c** H_2O_2, and **d** PNP

relatively small. When the catalyst is added in the flow process, the ozone extinction rate is greatly increased. When the solution flows for 1.0 h, the ozone concentration has been reduced to a very low value of about 0.24 mg/L, which saves more than 4.0 h compared with the non-catalyst group. Therefore, the collaboration between photocatalysis and fluid flow will significantly accelerate the extinction of ozone, which is conducive to reducing the flow time of treatment liquid, saving the time of overall sewage treatment, and speeding up the treatment rate.

Figure 27.11b shows that the concentration of NO_2^- decreases from the initial 36.00–18.00 mg/L when the liquid flows for 4.0 h without the addition of a catalyst. In the presence of the catalyst, when the liquid flow is 4.0 h, the NO_2^- extinction rate is relatively small, decreasing from the initial 36.00–24.00 mg/L, and the extinction trend of NO_2^- with or without a catalyst is relatively flat.

Figure 27.11c shows that the concentration of H_2O_2 detected in the flow process is always 0.0 mg/L, so when the discharge time is greater than or equal to 3.0 min, that is, when H_2O_2 completely disappears, no new H_2O_2 will be generated in the subsequent flow process.

Figure 27.11d shows that without the addition of a catalyst, the concentration of PNP gradually decreases to about 90.00% in the flow time of 0.00–120.0 min, and

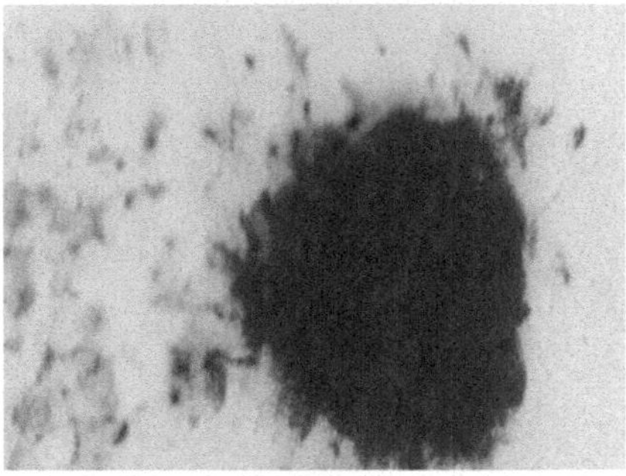

Fig. 27.12 The yellow–brown precipitation from PNP

the concentration of PNP hardly decreases during the flow time of 120.0–240.0 min. The results showed that the reactive species in the solution basically disappeared and could no longer degrade PNP. After the addition of the catalyst, the total degradation amount of PNP in the photocatalytic collaborative fluid flow within 0.0–240.0 min was 28.23%, and the degradation amount of PNP increased by 18.23% compared with that without the addition of the catalyst. This can be attributed to the high reduction potential of photogenerated electrons on $CQDs/TiO_2$ can promote the reaction with O_3, thus improving the utilization efficiency of O_3. The production of ·OH was increased, which eventually led to the increase of PNP degradation.

27.3.2.3 Precipitation Generation Rule

After being treated in the flow system for 36.0 h, the yellow–brown precipitation as shown in Fig. 27.12 began to appear in the PNP solution after discharge. Combined with the above experimental principles, the further degradation of PNP during the flow process after discharge is influenced by other factors besides the effect of reactive species. The cause of the phenomenon may be that more oxygen dissolves in water during the flow of the solution, resulting in greater contact space between PNP and oxygen for oxidation, and the reaction forms quinones. The type of this substance is not yet clear, and the specific formation mechanism needs to be further studied. In future studies, mass spectrometry and nuclear magnetic resonance can be used to detect the specific composition and type of brown precipitation generated, so as to further the law of ultrasonic photocatalytic degradation of organic matter.

27.4 Conclusion

(1) In the process of APPJ discharge with photocatalytic materials, the formation rate of active substances changes with the change of gas velocity.

(2) When the gas flow rate at the plasma jet nozzle is 4.20 m/s, the energy efficiency ratio of the reactive species is the highest.
(3) In the process of circulation flow after discharge, the degradation effect is best when the liquid flow rate is 0.60 m/s.
(4) Through the combination of photocatalyst and ultrasound to enhance the effect, the degradation of p-nitrophenol solution increased by 18.23%.
(5) After being treated for 36.0 h in the flow cycle, a substance suspected to be quinone will be oxidized to form. The type of this substance is not yet clear, and the specific formation mechanism needs to be further studied. In future studies, mass spectrometry and nuclear magnetic resonance can be used to detect the specific composition and type of brown precipitates generated. Thus, the law of organic matter degradation by ultrasonic photocatalysis can be further studied.
(6) This study has a certain reference value for the application of plasma combined with photocatalysis in the treatment of phenolic wastewater with high toxicity and difficult degradation. In the process of the flow cycle, the ultrasonic cooperative liquid flow parameter regulation model and photoelectric precipitation separation technology are used to transform phenolic substances into easily precipitated substances with low solubility at a high rate and recover them, so as to realize resource utilization.

Acknowledgements Natural Science Foundation of Jiangsu Province (BK20191162), Changzhou Key Research and Development Program (CJ20190046), Central University Basic Research Service Project (B210203006), National Innovation and Entrepreneurship Training Program for College Students (202310294047Z).

References

1. S. Mohammadi, A. Kargari, H. Sanaeepur et al., Phenol removal from industrial wastewaters: a short review. ChemInform **46**, 2215–2234 (2015)
2. P. Ariana, F. Catarina, P. Joana et al., Use of cork powder and granules for the adsorption of pollutants: a review. Water Res. **46**(10), 3152–3166 (2012)
3. M. Brígida, E. Graziele, C. Deivson et al., Mono and binary component adsorption of phenol and cadmium using adsorbent derived from peanut shells. J. Cleaner Prod. **201**, 219–228 (2018)
4. L. Ren, Y. Eman, X. Fan et al., Novel electrospun TPU/PDMS/PMMA membrane for phenol separation from saline wastewater via membrane aromatic recovery system. Sep. Purif. Technol. **212**, 21–29 (2019)
5. S. Muñoz, R. García, G. Cerqueda et al., Anaerobic conversion of saline phenol-containing wastewater under thermophilic conditions in a membrane bioreactor. Front. Bioeng. Biotech. **8**, 565311 (2020)
6. S. Deng, L. Jothinathan, Q. Cai et al., FeOx@GAC catalyzed microbubble ozonation coupled with biological process for industrial phenolic wastewater treatment: catalytic performance, biological process screening and microbial characteristics. Water Res. **190**, 116687 (2021)
7. Y. Rouphael, G. Colla, Editorial: biostimulants in agriculture. Front. Plant Sci. **11**, 40 (2020)
8. X. Lin, W. Kong, X. Lin, Degradation of high-concentration p-nitrophenol by Fenton oxidation. Water Sci. Technol. **81**(10), 2260–2269 (2020)

9. A. Verma, S. Pal, J. Kuntail et al., Visible light enhanced p-nitrophenol reduction by glycerol over Ag/Cu core-shell bimetallic nanocatalysts. J. Environ. Chem. Eng. **9**, 105655 (2021)
10. K. Huang, J. Wang, D. Wu et al., Copper hydroxyl sulfate as a heterogeneous catalyst for the catalytic wet peroxide oxidation of phenol. RSC Adv. **5**(11), 8455–8462 (2015)
11. X.L. Hao, Z.Y. Shi et al., Progress in the application of atmospherie pressure plasmaiets in environmental field[J]. J. Environ. Eng. Technol. **13**(4), 1415–1424 (2023)
12. Institute of Electrical Engineering, Chinese Academy of Sciences.Plasma collaboration with ultrasonic cavitation effect wastewater treatment reactor: CN201410337343.7[P] (2014)
13. Dalian University of Technology, Jinzhou Sihai Biochemistry Co., LTD. Device and method for atomizing plasma treatment of wastewater: CN202210951134.6[P] (2022)
14. F.X. Zhao, X.Y. Li et al., Preparation of photocatalysts decorated by carbon quantum dots (CQDs) and their applications: a review. J. Environ. Chem. Eng. **2**, 109487 (2023)
15. K.F. Liu, H.Y. Zhao et al., Degradation of Organic Wastewater by Pulsed Discharge Plasma. High Voltage Eng. **34**, 12–16 (2009)
16. B.Y. Chen, C.P. Zhu et al., Reaction kinetics of phenols and p-nitrophenols in flowing aerated aqueous solutions generated by a discharge plasma jet. J. Hazard. Mater. **363**, 55–63 (2019)
17. J.H. Yan, S.L. Dai et al., Emission Spectroscopy Diagnosis of the Radicals Generated in Gas-Liquid Phases Gliding arc Discharge. Spectrosc. Spectral Anal. **8**, 1851–1855 (2008)
18. W.C. Wang, F. Liu et al., Study of Radicals in Pulsed Corona Discharge by Emission Spectroscopy[J]. Spectrosc. Spectral Anal. **11**, 1288–1292 (2004)

Chapter 28
Study on Plasma-Activated Medium Adjuvant Cisplatin Chemotherapy in Human Endometrial Cancer Cells

GuoWen Xiong, Haoyu Zhan, Yanze Zhang, Xiaoyue Chen, Lei Lan, and Xishan Wen

Abstract Cancer cells with lower differentiation levels can exhibit higher levels of aldehyde dehydrogenase (ALDH) activity, which will inhibit the efficacy of chemotherapeutic drugs. Cisplatin, a commonly used chemotherapeutic drug that can be used to treat human endometrial cancer cells, but the treatment effect of it was not obvious for HEC cells. Recent research results show that plasma-activated medium (PAM) is widely used in the field of plasma medicine research, and has great potential in cancer treatment. In this paper, under the high frequency and high, pressure dielectric barrier discharge plasma generator is used to prepare PAM, and relevant cell experiments are conducted on human endometrial cancer cells. The effect of PAM adjuvant chemotherapy was explored experimentally, and the results verified the effectiveness of PAM in inducing apoptosis in human endometrial cancer cells. Compared with traditional cisplatin alone, PAM could significantly inhibit ALDH activity, thus contributing to adjuvant cisplatin to improve the efficacy of cisplatin chemotherapy. The research results of this work can provide new methods and ideas for the anticancer therapy.

28.1 Introduction

Low temperature plasma can be obtained by discharging on the electrode surface through high frequency excitation of external power supply at atmospheric pressure or low pressure (vacuum). Its important feature is the ability to produce a large number

G. Xiong · H. Zhan (✉) · Y. Zhang · X. Chen (✉) · L. Lan · X. Wen
School of Electrical Engineering and Automation, Wuhan University, Wuhan 430072, Hubei, China
e-mail: haoyuzhan@whu.edu.cn

X. Chen
e-mail: chenxiaoyue@whu.edu.cn

State Key Laboratory of Power Grid Environment Protection, School of Electrical Engineering and Automation, Wuhan University, Wuhan 430072, Hubei, China

Z. Fang et al. (eds.), *Proceedings of the 5th International Symposium on Plasma and Energy Conversion*, Springer Proceedings in Physics 398,
https://doi.org/10.1007/978-981-97-2245-7_28

of chemically active substances, such as reactive oxygen species (ROS) and reactive nitrogen species (RNS), against a background close to the ambient temperature (30–60°C). In recent years, low temperature plasma has been intensively used in material processing, energy conversion, biomedicine, and agricultural food field [1, 2]. Among them, plasma biomedicine has been achieving fruitful results in the fields of pathogen inactivation, hemostasis and coagulation, cancer and tumor treatment [3–5], which has a relatively broad application prospect.

Plasma applications in biomedicine are mainly divided into two categories: the first type is used in site treatment, that is, biological target directly contacts with low temperature plasma jet [6], which can be used for wound infection treatment and other biomedical scenarios [7–10]; the second type is used for off-site treatment, using plasma sources to treat biological solutions, so that the latter produces various long life (from several hours to several days) of active substances (RONS) [11, 12]. This solution is called plasma-activated medium (PAM), which can further act on biological organs and chambers where plasma sources are difficult to act directly [13].

The PAM has the same fluidity and homogeneity as the original biological solution medium. In addition to its rich RONS, it also has many advantages, such as being stable at room temperature and harmless to healthy tissues at effective doses [14]. The results of recent years show that the use of PAM is becoming one of the most promising means in the field of plasma medicine research [15]. It has been studied and verified in many medical application scenarios, such as living tissue sterilization and blood coagulation [16]. In studies aiming at the inhibition of cancer cells, PAM can selectively induce apoptosis in cancer cells [17]. In recent years, studies have shown that for some cancer cells with no obvious chemotherapy effect (such as human endometrial cancer cells), PAM can achieve the purpose of adjuvant chemotherapy drug therapy by inhibiting the activity of aldehyde dehydrogenase.

Clinically used cisplatin can exert anticancer effects through various mechanisms, including DNA cross-linking, induction of apoptosis, cell cycle arrest, and anti-angiogenesis. Under the synergistic action of the above mechanisms, the survival and proliferative capacity of cancer cells will be destroyed, which plays an important role in inhibiting the tumor growth and dissemination. Aldehyde dehydrogenase (ALDH) is an important catalytic substance in the human body, which can be converted from the latter into the corresponding acid through catalyzing the oxidation reaction of aldehyde, so ALDH plays an important role in cell metabolism. It has recently been shown that some members of the ALDH family of enzymes can act against protection during apoptosis. Specifically, cells with high expression of ALDH may inhibit the occurrence of apoptosis through mechanisms such as activating cell survival signaling pathway, up-regulating the expression of apoptosis inhibitory proteins, and enhancing DNA repair capacity [18]. In cancer treatment, these mechanisms precisely attenuate the apoptotic effect of cancer cells induced by chemotherapeutic drugs, leading to the reduced sensitivity of cancer cells to chemotherapeutic drugs. Because it is rich in active substances, PAM may affect the ALDH expression level by changing the intracellular oxidative environment or directly modifying the ALDH enzyme protein to change the conformation and function of the enzyme protein.

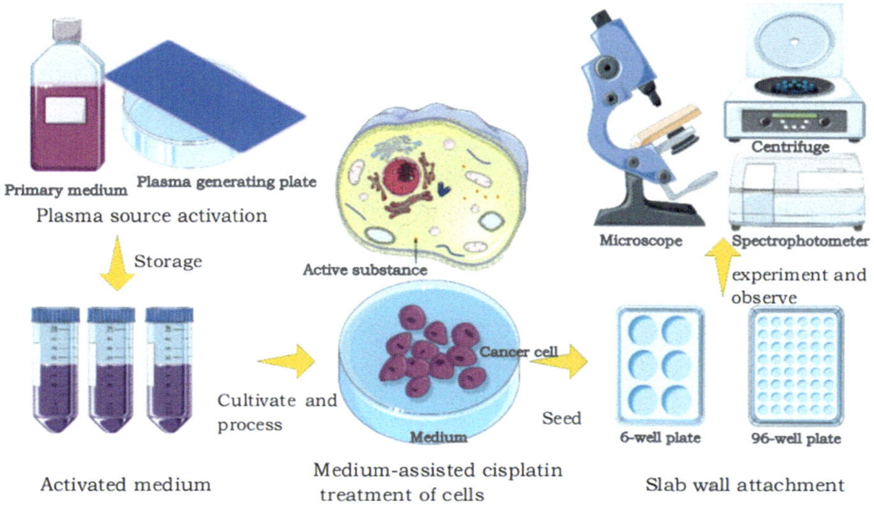

Fig. 28.1 Concept diagram of the research route of the work

In this paper, using high frequency and high pressure plasma generator for activating cancer cell medium to get PAM. After acting PAM on human endometrial cancer cells, designing multiple cell experiments to observe the effect of PAM on cancer cells. Then, studying the effect of PAM on adjuvant chemotherapy drugs cancer treatment by promoting cancer cell apoptosis and inhibiting ALDH expression in chemotherapy-insensitive cancer cells (Fig. 28.1).

28.2 Experiment Setting

28.2.1 Surface Dielectric Barrier Discharge Plasma Generator

Plasma is generated by dielectric barrier discharge. The dielectric barrier discharge plate is composed of metal grid, metal plate, ceramic panel and metal reinforcing plate. In order to protect the generating plate from mechanical impact damage, a metal reinforcing plate is added at the bottom of the generating plate. When the high frequency and high pressure excitation is applied between the metal net and the metal plate, the ceramic plate acts as the blocking medium, and a large area of microdischarge of the metal grid occurs, as shown in Fig. 28.2. The air around the metal grid produces a large amount of active substance, which can be used to activate the culture medium.

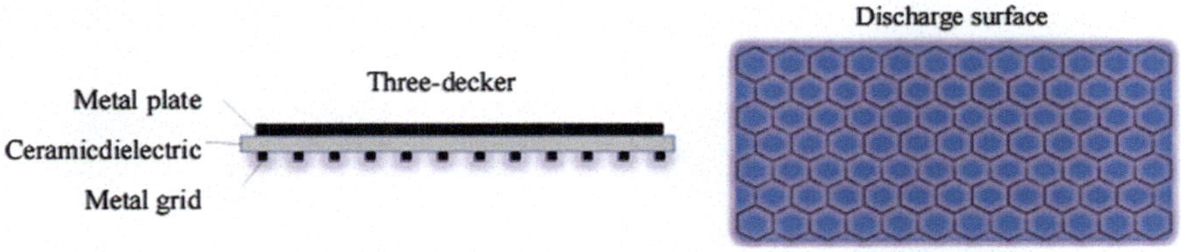

Fig. 28.2 Schematic diagram of dielectric barrier discharge plasma generating plate metal grid discharge

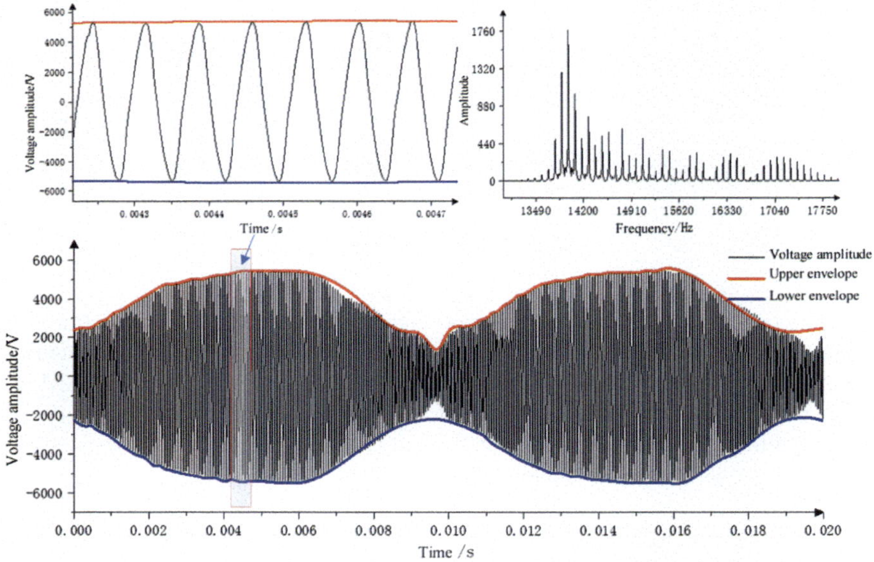

Fig. 28.3 The voltage amplitude waveform of the discharge plate during power frequency AC input

Using the power frequency 220 V power supply as the input, the high-frequency sinusoidal output wave with the peak value between 1440–5520 V and the main frequency at 14 kHz can be obtained. As measured by the oscilloscope, the voltage waveform of the two ends of the discharge plate is shown in Fig. 28.3.

28.2.2 Activation Treatment

The activation treatment was conducted by making an indirect contact between the low-temperature plasma and the culture medium to be treated [19]. As shown in Fig. 28.4, the medium-blocking discharge plate was placed above the culture medium, and the distance between the cell culture medium and the discharge surface

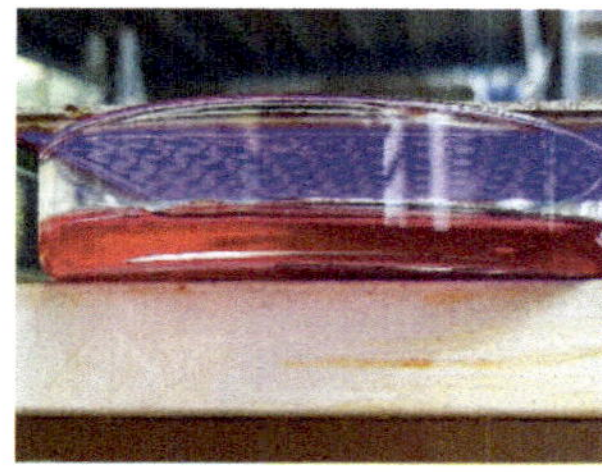

Fig. 28.4 Activation of the medium

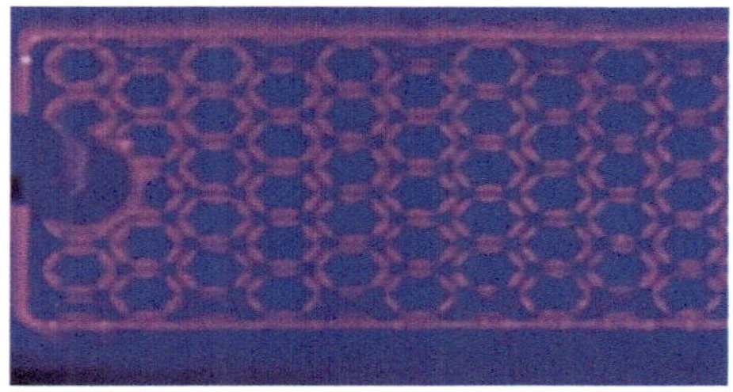

Fig. 28.5 A plasma generating plate being energized

was approximately 1 cm. Figure 28.5 shows the plasma generator plate undergoing discharge. At this time, a large area of discharge on the metal network, so that the surrounding air is rapidly ionized, producing a large number of active substances. When energized, the active substances produced by ionizing air will be partially dissolved in the culture medium to activate the treatment effect. The volume of medium for one treatment was about 20 ml and the discharge treatment time was about five minutes.

28.2.3 Cell Experimental Design

Cancer-Initiating Cells(CICs) are a small population of cells with tumor initiation and regeneration capacity, which can initiate tumor formation and maintain tumor growth and progression. CICs have some specific features and functions, and the high level of ALDH expression is one of them [20]. Studies have shown that less differentiated cancer cells prefer to have a higher content of CICs because they retain certain stem cell characteristics and capabilities, such as self-renewal and multi-directional differentiation. In this paper, human endometrial cancer HEC cells (intermediate differentiation) and ISK cells (high differentiation) were selected for experimental study. Due to the low differentiation, ALDH in HEC cells was expressed at higher levels than that in ISK cells. HEC cells were cultured using DMEM medium, and ISK cells were cultured using RPMI 1640 medium. In cell experiments, HEC cells were mainly studied for the effect of PAM treatment, and ISK cells were used to verify the effect of ALDH expression level on the experimental results.

In this paper, the differences in the efficacy of chemotherapeutic drugs on HEC and ISK cells were first investigated: **Firstly**, HEC and ISK cells were treated with cisplatin drugs alone, and the cell activity was detected to study the difference in cell sensitivity to chemotherapy, and the ALDH expression level was verified to detect the difference of intracellular ALDH enzyme activity. **Secondly**, the effect of PAM treatment on chemotherapy-insensitive HEC cells was studied: HEC cells were treated with PAM alone, and the cell activity was tested to study the effect of PAM directly inducing apoptosis to inhibit cell activity, and cell ALDH levels were measured to study the effect of PAM inhibiting cell ALDH expression level. **Finally**, the effect of PAM and chemotherapy drugs in treating HEC cells was studied: combining different concentrations of PAM and cisplatin to detect the cell activity to study the effect of PAM adjuvant chemotherapy, and the detection of apoptosis confirmed that a large number of cells underwent apoptotic process after PAM adjuvant chemotherapy.

28.2.4 Evaluation of Biochemical Indicators

(1) **Determination of Active Substances in the Activated Culture Medium**

The active substance dissolved in the culture medium will change the oxidation environment of the culture medium. The reactive oxygen species exist mainly as H_2O_2. Applying H_2O_2 Oxidation of divalent iron ions to produce ferric iron ions, and then using a microplate reader to detect purple products (240 nm wavelength), which were produced in a specific solution by ferric iron ions and xylenol orange, can achieve the determination of hydrogen peroxide concentration.

(2) **Measurement of Cell Viability and Counting**

The effect of each group on apoptosis was assessed by measuring the percentage of apoptotic cells, using the CCK8 test agent. WST-8 in the reagent can be reduced by live cells to generate orange-yellow products. When light pass the detected object, the difference in the energy is the energy absorbed by the detected object. At a specific wavelength, the concentration of the same detected object is quantitative to the absorbed energy. Measurement the corresponding wavelength (absorbance) of the product can reflect the number of living cells. The absorbance values of the experimental group were corresponding to those of the untreated cells, and the results were expressed as the survival index (the ratio of the activity of the experimental group to the control cells).

(3) **Detection of Cell Apoptosis**

Apoptotic status was assessed by measuring the percentage of apoptotic cells, using annexin V-APC/PI, an apoptotic detection reagent. Apoptotic cells were determined by using a Beckman Coulter flow cytometer, and the results were analyzed.

(4) **ALDH Enzyme Activity Detection**

Using the ALDH enzyme activity detection reagent, the band absorbance value of the ALDH enzyme catalytic product was measured to assess ALDH expression levels, and the amount of 1 nmol product by catalytic reduction per minute per 10,000 cells was defined as 1 enzyme viability unit.

28.2.5 Data Processing

The experimental data were analyzed for the average value using the independent sample t-test method. Differences were considered statistically significant at $P < 0.05$.

28.3 Experimental Results

28.3.1 Study of the Effects of Cisplatin Treatment of HEC and ISK Cells

Measuring the expression level of ALDH in HEC cells and ISK cells. It showed that the **ALDH enzyme activity in ISK cells was very low**. Under the same number, ALDH enzyme activity detection value in HEC cells was about 42 times that in ISK cells (0.0337/0.008).

To analyze the efficacy of chemotherapeutic agents on human endometrial cancer cells at medium and high differentiation levels, the viability of cancer cells treated with cisplatin was measured using CCK8 reagent, and the survival indices of the two cells were compared. The cisplatin concentration values ranged from 0 to 250 μM.

From Fig. 28.6, under cisplatin, ISK cells showed a higher chemosensitivity compared to HEC cells. The survival index of ISK cells was significantly different when the cisplatin concentration is higher than 25 μM, while HEC cells were significantly different when the cisplatin concentration is higher than 75 μM.Besides, at the maximum cisplatin concentration 250 μM, cisplatin still did not achieve the half inhibition of HEC cells. Therefore**, cisplatin alone has achieved better survival inhibition for ISK cells.** However, **for HEC cells, study using PAM for adjuvant treatment is needed.**

28.3.2 Study of the Effect of PAM Treatment of HEC Cells

(1) **Inhibition of PAM Activity in HEC Cells**

DMEM medium after activation treatment, Reactive oxygen species concentration represented by H_2O_2 is increased by 122.78 μM, which is rich in active substances

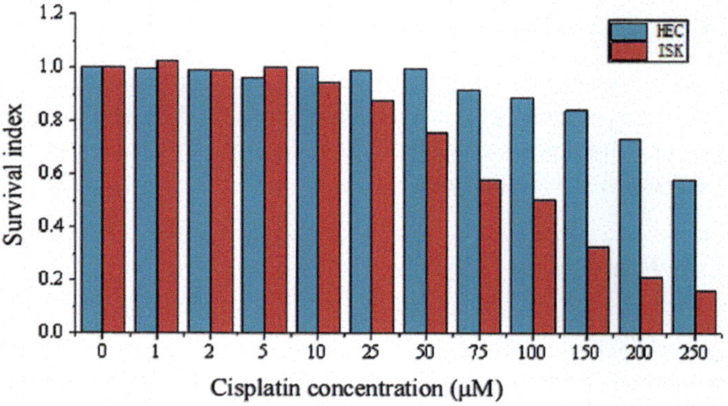

Fig. 28.6 Survival index of HEC and ISK treated with different concentrations of cisplatin

and can directly cause damage to cancer cells. Active species can trigger oxidative stress and cause DNA damage within cancer cells. These chemicals react with biomolecules such as DNA, proteins and lipids of cancer cells to disrupt their structure and function, thereby inhibiting their growth and proliferation. Oxidative stress and DNA damage may trigger the apoptotic mechanisms in cancer cells. Apoptosis is a form of programmed cell death that is capable of effectively eliminating abnormal cells. Active substances may induce apoptosis in cancer cells by interfering with cell signaling pathways and promoting apoptosis-related protein expression. PAM reduced the survival index of HEC cells to about 58% compared to untreated cells.

(2) **Inhibition of ALDH Expression in HEC Cells by PAM**

As shown in Fig. 28.7, without activation treatment, PAM after two times dilution and undiluted PAM, **the mean absorbance of the enzyme catalytic product** was 0.01467, 0.00233 and 0.00133 respectively. Taking the enzyme catalytic products in cells without activation as the benchmark value, the corresponding enzyme activity index of the latter two were 0.0159 and 0.0909, respectively, indicating that **PAM played a significant inhibitory effect on the ALDH expression level in HEC cells.**

28.3.3 PAM Combined with Cisplatin Treating HEC Cells

(1) **Cell Survival Index Was Measured Under the Combination of Different Concentrations of PAM and Cisplatin**

To study the effect of chemotherapy after PAM adjuvant cisplatin, in this paper, **the cell survival index assay experiment** after the combination of PAM and different concentrations of cisplatin was performed, and the result is shown in Fig. 28.8. Where the vertical axis represents cisplatin with different concentrations ranging from 0 to 300 μM, increasing from top to bottom, and the horizontal axis represents

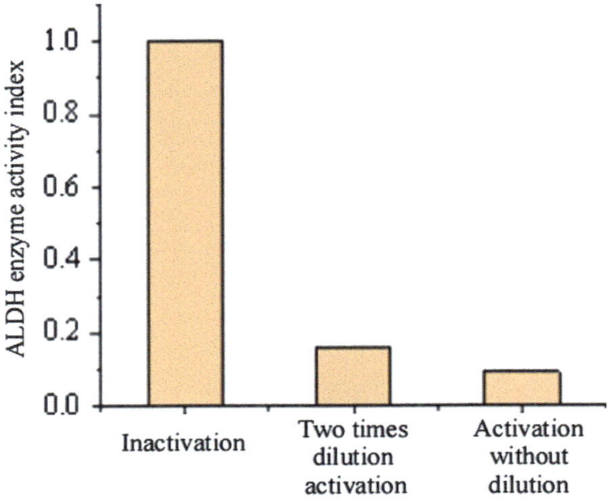

Fig. 28.7 Enzyme activity index of HEC cells treated with PAM

PAM with different dilutions, increasing from left to right. **Cisplatin concentration was 0 and the medium was not activated** as the base value, and the cell survival index characterized by the data of each table cell is the ratio of the cell activity to the benchmark cell activity under the combination of different concentrations. The lower the HEC cell survival index is, the lighter the color of the corresponding table cell is. Figure 28.8 clearly shows that **with the increase of cisplatin concentration and the decreasing dilution times of PAM, the inhibition of HEC cells with cisplatin or PAM alone was gradually increased, and the combination of the two will further enhance the inhibition.**

(2) **Apoptosis Detection Under the Combined Treatment**

Apoptosis was measured using FACS on HEC cells in the presence of cisplatin with PAM and without PAM, and the comparative result is shown in Fig. 28.9. In particular, LL regions characterize normal cells, LR region early apoptotic cells, UR region late apoptotic cells, and UL region necrotic cells. According to the result, **in contrast to cisplatin alone, a large number of HEC cells underwent apoptosis after treatment with PAM-assisted cisplatin.**

28.4 Discussion and Analysis

As an emerging research field in recent years, atmospheric pressure low temperature plasma has attracted wide attention in many fields, especially in biomedical applications, because to its characteristics of atmospheric pressure, low gas temperature and high particle activity. At present, the main discharge mode of low-temperature plasma is DBD, which is mainly used in plasma jet and surface air plasma. However, various research teams have differences in the structural design, gas introduction, and

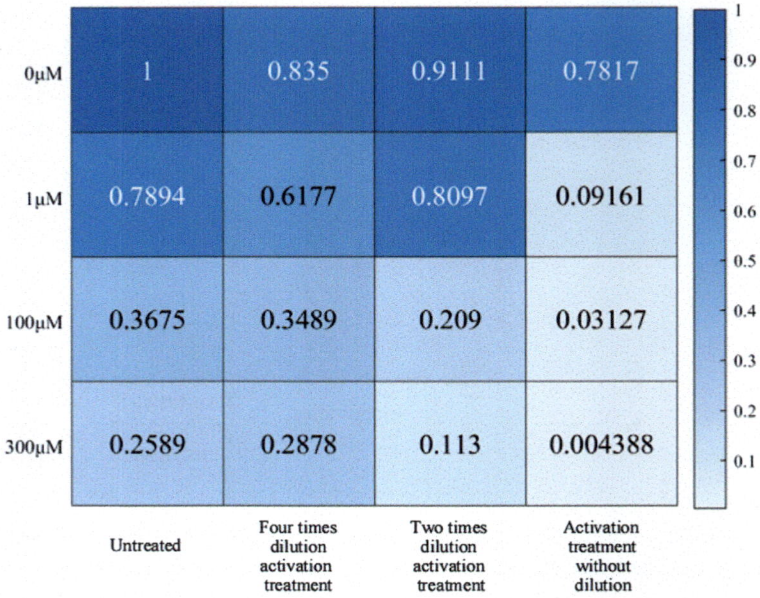

Fig. 28.8 Cell survival index of HEC cells after treatment

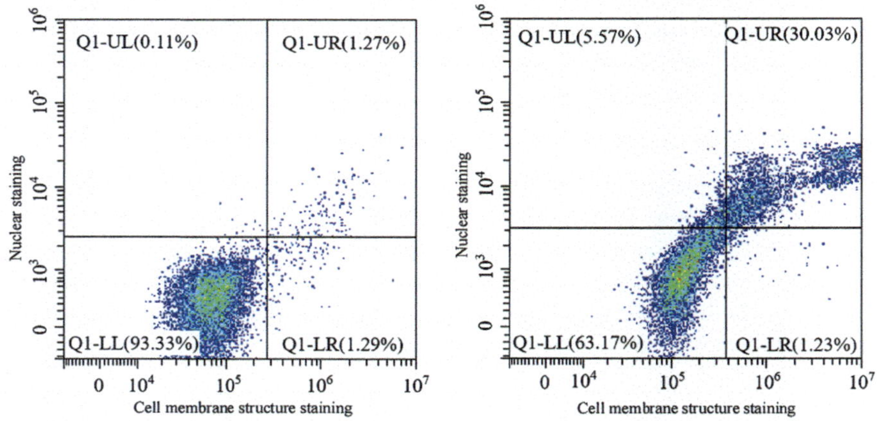

Fig. 28.9 Results of apoptosis detection after cisplatin without/with PAM

parameter adjustment of the CAP power generation device. In this study, we used the surface air plasma treatment device with simple and efficient characteristics such as high frequency, high pressure and low power. Plasma generation is characterized by surface microdischarge, a high frequency and high voltage AC power supply with a main frequency of 14 kHz. This device was used to generate discharge at 1 cm from the level of the medium, the treatment time was 5 min and 20 ml of medium at one time. Represented by H_2O_2, the active material in the treated medium was increased by 122.78 μM compared with the original medium.

The results of cell experiments showed that highly differentiated ISK cells showed high drug sensitivity to the action of cisplatin, and cisplatin alone has achieved better survival inhibition for ISK cells. However, due to the resistance effect of ALDH on cisplatin, moderately differentiated HEC cells with higher levels of ALDH expression showed lower chemotherapy sensitivity when treated with cisplatin alone. After plasma source activation treatment, abundant active substances are produced in PAM, which can not only induce the apoptosis of cancer cells, but also affect the expression of ALDH in cancer cells, thus playing a role in improving the efficacy of chemotherapy. In cell experiments, the results showed that PAM alone had a significant inhibitory effect on HEC cells, and a significant inhibitory effect on ALDH expression in HEC cells.

After treating HEC with PAM and cisplatin, survival index measurement and apoptosis detection of HEC cells were further conducted. The results of cell survival index assay experiment showed that the activity of HEC cells after PAM treatment decreased. With the increase of cisplatin concentration and the dilution of PAM, the inhibitory effect of cisplatin or PAM alone is gradually enhanced, and the combination of the two will further enhance the inhibitory effect, and the activity of HEC cells decreased more significantly; the results of apoptosis test confirmed that after the combination of PAM and cisplatin, a large number of HEC cells became apoptotic, and the efficacy of chemotherapy was significantly improved.

The results of the above cell experiments showed that **PAM can not only directly inhibit HEC cell activity, but also assist cisplatin for chemotherapy by inhibiting the expression level of ALDH.**

28.5 Conclusion

In conclusion, we first designed a plasma-generating device that can be used for PAM preparation, then we proposed a chemotherapy approach using PAM as adjuvant cisplatin for human endometrial cancer. Human endometrial cancer HEC cells have some resistance to chemotherapy drugs and poor chemotherapy effect due to their high ALDH level. After a series of cellular experiments, we reach a conclusion that PAM treatment could effectively reduce the ALDH expression in HEC cells, in the same time had a direct inhibitory effect on its cellular activity. The results illustrate the possible mechanism of PAM adjuvant chemotherapy on cancer cells, which has a certain development prospect. After combining PAM with cisplatin, the effect of chemotherapy was significantly improved, and a large number of cells developed the process of apoptosis. The results confirm the effect of PAM as adjuvant cisplatin in chemotherapy, and provide a new idea to solve the problem of chemotherapy-insensitive cancer treatment.

References

1. M.G. Kong, G. Kroesen, G. Morfill et al., Plasma medicine: an introductory review. New J. Phys. **11**, 115012 (2009)
2. G. Fridman, G. Friedman, A. Gutsol, A.B. Shekhter, V.N. Vasilets, A. Fridman, Applied plasma medicine. Plasma Process. Polym. **5**, 503–533 (2008)
3. D. Liu, Y. Zhang, M. Xu et al., Cold atmospheric pressure plasmas in dermatology: sources, reactive agents, and therapeutic effects. Plasma Process. Polym. **17**(4), 1900218 (2020)
4. K. Miyamoto, S. Ikehara, H. Sakakita et al., Low temperature plasma equipment applied on surgical hemostasis and wound healings. J. Clin. Biochem. Nutr. **60**(1), 25–28 (2017)
5. M. Keidar, R. Walk, A. Shashurin et al., Cold plasma selectivity and the possibility of a paradigm shift in cancer therapy. Br. J. Cancer **105**(9), 1295–1301 (2011)
6. S. Bekeschus, A. Schmidt, K.-D. Weltmann, T. von Woedtke, The plasma jet kINPen—a powerful tool for wound healing. Clin. Plasma Med. **4**, 19–28 (2016)
7. J.-W. Lackmann, J.E. Bandow, Inactivation of microbes and macromolecules by atmospheric-pressure plasma jets. Appl. Microbiol. Biot. **98**, 6205 (2014)
8. K. Wende et al., Atmospheric pressure plasma jet treatment evokes transient oxidative stress in HaCaT keratinocytes and influences cell physiology. Cell Biol. Int. **38**, 412 (2014)
9. H.J. Ahn et al., Atmospheric-pressure plasma jet induces apoptosis involving mitochondria via generation of free radicals. PLoS ONE **6**, e28154 (2011)
10. J.F. Kolb, A.A.H. Mohamed, R.O. Price, R.J. Swanson, A. Bowman, R.L. Chiavarini, K.H. Schoenbach, Cold atmospheric pressure air plasma jet for medical applications. Appl. Phys. Lett. **92**, 241501 (2008)
11. T. Io, G. Uchida, A. Nakajima, K. Takenaka, Y. Setsuhara, Control of reactive oxygen and nitrogen species production in liquid by nonthermal plasma jet with controlled surrounding gas. Jpn. J. Appl. Phys. **56**, 01AC06 (2016)
12. N.J. Jha, J.J. Ryu, E.H. Choi, N.K. Kaushik, Generation and role of reactive oxygen and nitrogen species induced by plasma, lasers, chemical agents, and other systems in dentistry. Oxid. Med. Cell Longev. 7542540 (2017)
13. D.B. Graves, The emerging role of reactive oxygen and nitrogen species in redox biology and some implications for plasma applications to medicine and biology. J. Phys. D Appl. Phys. **45**, 263001 (2012)
14. M. Horiba, T. Kamiya, H. Hara, T. Adachi, Cytoprotective effects of mild plasma activated medium against oxidative stress in human skin fibroblasts. Sci. Rep. **7**, e42208 (2017)
15. J. Heinlin, G. Isbary, W. Stolz, G. Morfill, M. Landthaler, T. Shimizu, B. Steffes, T. Nosenko, J. Zimmermann, S. Karrer, Plasma applications in medicine with a special focus on dermatology. J. Eur. Acad. Dermatol. Venereol. **25**, 1–11 (2011)
16. K.N. Kumar, G. Bhagirath, L. Ying et al., Biological and medical applications of plasma-activated media, water and solutions. Biol. Chem. **400** (2018)
17. N. Kumar, J.H. Park, S.N. Jeon, B.S. Park, E.H. Choi, P. Attri, The action of microsecond-pulsed plasma-activated media on the inactivation of human lung cancer cells. J. Phys. D. **49**, 115401 (2016)
18. I. Ma, A.L. Allan, The role of human aldehyde dehydrogenase in normal and cancer stem cells. Stem Cell Rev. **7**, 292–306 (2011)
19. D. Xu, Q. Cui, Y. Xu et al., Systemic study on the safety of immuno-deficient nude mice treated by atmospheric plasma-activated water. Plasma Sci. Technol. (2018)
20. N. Rahadiani, J. Ikeda, S. Mamat et al., Expression of aldehyde dehydrogenase 1 (ALDH1) in endometrioid adenocarcinoma and its clinical implications. Cancer Sci. **102**, 903–908 (2011)

Chapter 29
Cold Atmospheric Plasma Activated Hydrogel and Its Biotoxicity

Yinglong Li, Xiaojiao Zhang, Ruonan Ma, and Ruixue Wang

Abstract Objective: Cold atmospheric plasma activated hydrogel (CAP-Gel) is an effective controlled-release method of the reactive species. The present study evaluated its physicochemical properties, acute oral toxicity and genetic toxicity. Methods: The plasma hydrogel was prepared by dielectric barrier discharge (DBD) plasma device. The reactive species in the hydrogel were evaluated by optical emission spectroscopy (OES), and the pH and oxidation reduction potential (ORP) were measured. A total of 70 mice of SPF grade were selected, and 20 mice were used for acute oral toxicity test by one-time maximum test method. Oral gavage method was applied and the toxicity symptoms of animals were observed and recorded for 14 days consecutively. To evaluate the genetic toxicity, 50 mice were used for micronucleus test of bone marrow polychromatic erythrocytes. Results: The emission spectrum results showed that the main components of the plasma were reactive oxygen species and reactive nitrogen species. After 10 min of treatment, the pH decreased from 6.86 $\pm$ 0.26 to 1.57 $\pm$ 0.34, and the ORP increased from 161 $\pm$ 0.12 to 570 $\pm$ 10.54. Acute oral toxicity test indicated that there was toxicity symptom was observed. The micronucleus test of bone marrow polychromatic erythrocytes showed that no significant difference between CAP-Gel and negative control group ($p > 0.05$). Conclusion: Cold atmospheric plasma activated hydrogel is a safe, no obvious acute oral toxicity and genetic toxicity.

Y. Li · X. Zhang
Department of Stomatology, Beijing Chaoyang Hospital, Capital Medical University, Beijing 100020, China

R. Ma (✉)
Henan Key Laboratory of Ion-Beam Bioengineering, School of Agricultural Sciences, Zhengzhou University, Zhengzhou 450052, China
e-mail: maruonan@zzu.edu.cn

R. Wang (✉)
College of Mechanical and Electrical Engineering, Beijing University of Chemical Technology, Beijing 100029, China
e-mail: wrx@mail.buct.edu.cn

Z. Fang et al. (eds.), *Proceedings of the 5th International Symposium on Plasma and Energy Conversion*, Springer Proceedings in Physics 398,
https://doi.org/10.1007/978-981-97-2245-7_29

29.1 Introduction

Different with ceramic veneers and porcelain crowns, tooth bleaching with chemical agents is increasingly the chosen means of modifying the appearance of teeth for cosmetic reasons. Besides, veneers, crowns and other methods have the disadvantages of destroying normal tooth tissue, high cost, and complex operation. More important, tooth bleaching agents do not lead to macroscopic structural changes of the tooth surface [1]. It is worth noting that high concentrations of H_2O_2 (30–35%) contact with enamel surface over 20 min are commonly used in in-office tooth bleaching, which will lead to enamel etching and cause tooth sensitivity and potential soft tissue damage [2, 3]. At the same time, the duration of whitening is limited, and it is necessary to work with home whitening measures to achieve lasting whitening effect [3]. Thus, more secure, comfortable and high efficiency tooth bleaching methods is strongly needed in clinic.

Plasma is regarded as the fourth fundamental state of matter after solid, liquid and gas. It is an electrically conducting medium, rich in unbound positive and negative ions, numerous reactive radicals, quanta of electromagnetic radiation and strong electric fields [4, 5]. Different with thermal plasma, cold plasmas generated under atmospheric conditions (CAP), which can be produced under mild conditions, such as corona discharges, glow discharges, dielectric barrier discharges and plasma jets [6]. Nowadays, CAP technologies had been applied in biomedical, material processing and food industry and dentistry.

The water or other solutions activated by cold plasma was named plasma activated water (PAW), the water molecules interact with plasma-generated particles and generate many reactive species, which exhibit remarkable biological activities in agricultural and biomedical applications [7, 8]. The biochemical activity of PAW is derived from a series of active species, which results in various primary and secondary reactive species being dissolved in water [9, 10]. These reactive species, including reactive nitrogen species (RNS) and reactive oxygen species (ROS), are responsible for the biochemical effects of PAW [11].

Besides, PAW and plasma-activated hydrogels were regarded as a green disinfectant because of the low drug resistance than traditional chemical agents [12]. Thus, PAW exhibited great potential in dentistry, previous studies indicated that periodontal diseases related bacteria (*Porphyromonas gingivalis*) and caries related biofilm (*Streptococcus mutans* and *Actinomyces viscosus*) can be killed by PAW, demonstrating the antimicrobial effects of PAW [13]. Besides, the hydrogel prepared using plasma-activated water (CAP-Gel) was regarded as a better form in some occasions, such as medical dressing and sustained-release drug carrier. The H_2O_2 gel treated with cold plasma or plasma activated water gel was endowed with antibacterial and better tooth bleaching [14]. However, the biotoxicity of the hydrogel prepared using plasma-activated water (CAP-Gel) is also important before use.

In this study, the CAP-Gel was synthesized and the active species of cold plasma was evaluated as well as the physicochemical properties were determined. The biotoxicity of the CAP-Gel was tested with acute oral toxicity and micronucleus test of bone marrow polychromatic erythrocytes.

29.2 Materials and Methods

29.2.1 Cold Plasma Device and Fabrication of CAP-Gel

Figure 29.1a shows the schematic diagram of an air plasma generator, which was designed based on a dielectric barrier structure with hollow electrodes (HEDBS). A more detailed description of the device is provided in our previous studies [15, 16]. The plasma jet was mainly composed of copper electrodes and a quartz dielectric tube with an inlet diameter of 1.5 mm. Compressed air with a flow rate of 260 L/h was pumped into the quartz tube as the working gas. The copper foil, operating as a high-voltage electrode, was connected to an excitation voltage with a peak voltage of 25 kV and a power source of 20 kHz. The input power was approximately 25 W. Twenty milliliters of H_2O were treated with 20 min of cold plasma treatment. As shown in Fig. 29.1b, continuous plasma was generated at the end of the plasma jet. As shown in Fig. 29.1c, The CAP-Gel was generated and kept at 4 °C [3].

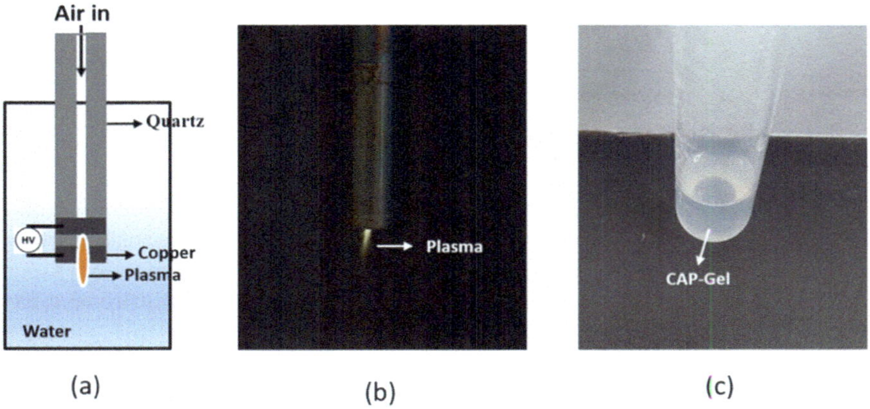

Fig. 29.1 **a** Schematic diagram of the plasma microjet and **b** photograph of the plasma plume, **c** CAP-Gel

29.2.2 Main Reactive Species Measurement

To identify the main reactive species generated by the cold plasma, optical emission spectrometry (OES) was applied using a fiber optic spectrometer (AvaSpec-2048-8-USB2, Avantes, Netherlands), the resolution: 0.04–20 nm, sensitivity up to 5000 counts/(μW ms). Measurements were performed in a dark room in order to reduce the interference of the external impurity light source. The dispersed emission spectra was recorded range from 250 to 1000 nm by a 2048-pixel charge-coupled device (CCD) detector array. The fiber probe was perpendicular to the plasma plume. Details of the OES measurement could be found in our earlier study [15].

29.2.3 Temperature Measurement

The plasma treatment process is accompanied by a rise in temperature. In order to measure the liquid temperature change, a thermal imager FLIR E50 (Danderyd, Sweden) was used to record the temperature change. This instrument is easy to use and has a determination sensitivity of 0.07 °C. The plasma device was placed below the liquid level, and the FLIR E50 was installed at a distance of 30 cm away from the cold plasma device. The measurements were performed three times at the unexcited state, as well as at the excited plasma generation state for 5 min, 10 min, and 20 min, respectively.

29.2.4 Measurement of pH Value and ORP

ORP reflects the active part of the oxidation property and reflects the potential value of the active part. The higher the ORP value represents the stronger oxidation of the water, which is related to the bactericidal ability. The pH/ORP instrument (Xima, AR8401, China) was used for the measurement in this study. The instrument can measure the PH range from 0.0 to 14.00 with an accuracy of 0.01pH, the ORP measurement range from − 1999 to + 1999 mV. Therefore, measurements were performed at 0, 5, 10, 15, and 20 min, and each experiment was repeated three times.

29.2.5 Concentration of H_2O_2

H_2O_2 is an important chemical signal. According to the Chinese disinfection technical specification (2002), the content of H_2O_2 in CAP-Gel was determined. Solutions such as 2 mol/L sulfuric acid and 100 g/L manganese sulfate were prepared. In addition, 0.02 mol/L potassium permanganate titration solution was prepared and calibrated. An appropriate amount of the sample was precisely sucked, which was equivalent to about 0.3 g of H_2O_2 and diluted to 100 mL with distilled water. Taking 10.0 mL of H_2O_2 diluent, put it into 100 mL iodine volumetric flask, add 20 mL of 2 mol/L sulfuric acid and 3 drops of 100 g/L manganese sulfate, and shake well. The solution was titrated with 0.02 mol/L potassium permanganate titrant (contained in a 25 mL titrant tube) until pink, and the amount of potassium permanganate titrant was recorded. The measurements were repeated two times, and the average of the two times was used for the following calculation. Since 1 mL of 1 mol/L potassium permanganate titration solution is equivalent to 0.08505 g of H_2O_2, the content (ρ) can be calculated according to the following formula:

$$\rho = \frac{(V_1 - V_0) \times c \times 0.08505 \times 1000}{V} \quad (29.1)$$

(ρ represents the H_2O_2 content, g/L; c is the concentration of standard solution, mol/L; V blank titration consumption, V0 = 0.02 mL; V1 sample titration consumption of standard solution, mL; V Volume of H_2O_2-like liquid contained in iodine volumetric flask, mL).

29.2.6 Acute Oral Toxicity Test

Acute oral toxicity test can be used to determine whether plasma-activated gel can be absorbed through skin, and the toxic reaction caused by short-term action is one of the important bases for determining sub-chronic toxicity. The experimental protocols were developed on the basis of above regulations and approved by Ethics Committee of Beijing Chaoyang Hospital of Capital Medical University (Registration number: 2020-ke-28). In this study, specific pathogen free (SPF) Kunming (KM) mice were used, which has the characteristics of strong reproductive ability, disease resistance and environmental adaptability, is suitable for acute oral toxicity experiments. Twenty KM mice (10 male and 10 female) were selected, and the dose was 5000 mg/kg body weight. 5.0001 g of CAP-Gel was weighed, and purified water was added to 20 mL.

The specific experimental procedures were as follows: (1) mice were fasted overnight without restriction of water before drug administration; (2) The volume of gavage was 0.2 mL/10 g body weight, and the frequency was 24 h/ time; (3) The poisoning symptoms and death of mice were observed and recorded for 14 consecutive days, and the body weight was measured once a week. If death occurred, the dead animals were dissected, and if abnormal organs were found, pathological examination was performed.

29.2.7 In Vivo Bone Marrow Cell Chromosome Aberration Test

Micronucleus test is mainly used to evaluate whether CAP-Gel can break chromosomes or damage the association between chromosomes and spindles. In this study, 50 km mice (25 male and 25 female) of SPF grade, weighing 25.2–28.0 g, were selected and randomly divided into 5 groups (n = 10/group): The negative control group was treated with pure water, and the positive control was treated with 0.0403 g cyclophosphamide (CTX) and sterile normal saline to 20 mL. The experimental group weighed 25.0002 g of CAP-Gel and add sterile water to 100 mL, according to the weight of 5000, 2500, and 1000 mg/kg.

All tested materials were exposed to 0.2 mL of body weight by gavage twice within 30 h. At 6 h after the second gavage, all rats were sacrificed by cervical dislocation method, and the femoral bone marrow was mixed with calf serum, fixed by routine smear, and stained with Giemsa. 1000 complete bone marrow polychromatic erythrocytes (PCE) of each animal were observed, the number of cells with micronucleus was recorded, and the micronucleus frequency of cells was calculated. The Poinsson distribution u test was used for statistical analysis. At the same time, the number of mature cells (NCE) seen in 200 PCE was counted, and the PCE/NCE ratio was calculated.

29.2.8 Statistical Analysis

All data were presented as mean standard deviation (M ± Std) using a completely randomized design. The graphs in this study were plotted by Origin (OriginLab Inc., version 2022, USA). SPSS Statistics 22.0 (SPSS Inc., USA) was used to process the data, and One-Way ANOVA was used to analyze the data. The $p < 0.05$ was considered as the significant difference level.

29.3 Results

29.3.1 Characterization of Free Radicals

As shown in Fig. 29.2, the emission spectra are dominated by N_2, as N_2 is the major component of air (78%). The figure displays the N_2^+ first negative system at 400–520 nm, the N_2 first positive system at 550–730 nm, and the second positive system at 300–420 nm. Additionally, N and O emission lines can be detected at 730–900 nm. Moreover, two oxygen atom transition lines are detected at 777 nm and 845 nm, respectively. The dissociation of O_2 and N_2 generates O and N, respectively, which are excited from the ground state by electron impact:

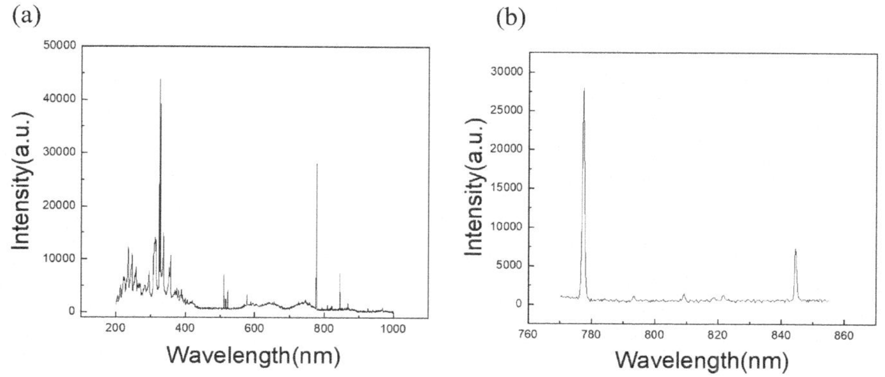

Fig. 29.2 Optical emission spectroscopy results in the range 250–1000 nm. **a** The emission spectra are dominated by N2. **b** Oxygen atom transition lines are detected at 777 nm and 845 nm

$$e + O_2 \rightarrow O(^3P) + O(^1D) + e \tag{29.2}$$

$$e + N_2 \rightarrow N(^4S) + N(^2D) + e \tag{29.3}$$

29.3.2 Temperature Changes

Figure 29.3a plots typical IR imaging photographs for four processing times. Figure 29.3b shows the photos after three repeated experiments. It can be observed that at 0 min, the water temperature was 15.07 ± 0.40 °C, at 5 min, at 10 min, the water temperature was 23.53 ± 0.57 °C, at 10 min, and at 20 min, the temperature reached 30.60 ± 1.05 °C.

29.3.3 pH/ORP Value

pH is a common index to evaluate the plasma activation solution. The pH results of different experimental groups are shown in Fig. 29.4. The results show that after plasma treatment, the pH value gradually decreased, and the decrease was most obvious within the first 1 min, from 6.86 ± 0.26 to 4.34 ± 0.20 at 0 min. At 10 min, it was basically stable at 1.59 ± 0.06, and the pH did not change significantly after 20 min of treatment ($p > 0.05$).

As shown in Fig. 29.4b, ORP also increased most rapidly in the first 3 min, from 161.00 ± 12.12 (0 min) to 382.33 ± 2.52 (1 min) and 528.33 ± 4.04 (3 min), and the

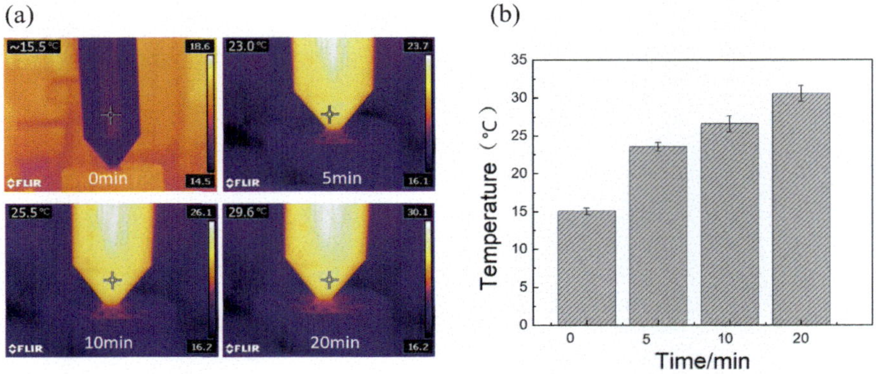

Fig. 29.3 Plasma temperature variation. **a** Infrared imaging photographs; **b** temperature change

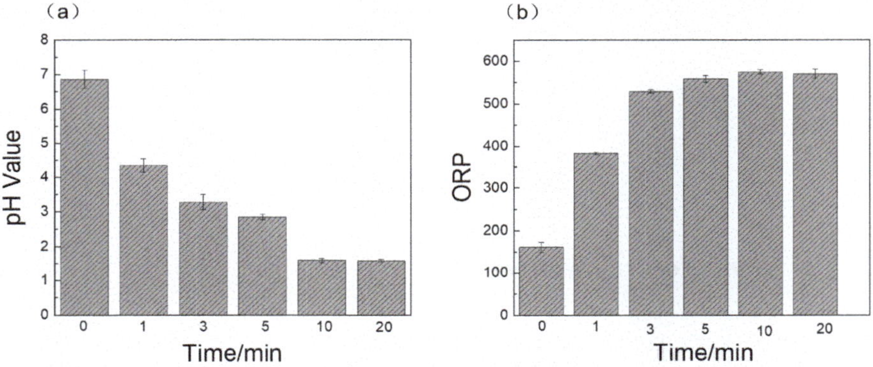

Fig. 29.4 The results of pH and ORP. **a** pH value, **b** ORP value

increase of ORP was not obvious after 3 min ($p > 0.05$). The ORP value increased to 574.67 ± 5.13 at 10 min and 570 ± 10.54 at 20 min.

29.3.4 H_2O_2 Concentration in CAP-Gel

As shown in Table 29.1, the concentration of H_2O_2 detected in the CAP-Gel prepared in this study can be observed that the concentration of H_2O_2 in the four samples is 16.5, 16.6, 16.1, and 16.3 g/L, and the average is 16.375 g/L.

Table 29.1 H_2O_2 concentration in CAP-Gel

Sample	Sample volume V (mL)	Standard solution dosage V1 (mL)	Concentration (g/L)
1	0.5	4.62	16.5
2	0.5	4.64	16.5
3	0.5	4.52	16.1
4	0.5	4.58	16.3

Table 29.2 Results of the acute oral toxicity test for CAP-Gel

Dosage (mg/kg)	Gender	Number	Weight (g, M ± SD)			Death
			0d	7d	14d	
5000	Male	10	18.80 ± 0.60	26.40 ± 0.65	35.90 ± 0.84	0
5000	Female	10	19.30 ± 0.65	26.90 ± 1.06	36.40 ± 1.02	0

29.3.5 Results of Acute Oral Toxicity Test

As shown in Table 29.2, during the observation period, all mice had normal activity, good hair gloss, and no toxic symptoms were observed. The body weight of mice in each dose group changed normally, no abnormal body weight occurred, and no mice died (0).

29.3.6 Results of Bone Marrow Polychromatic Erythrocyte Micronucleus Test

As shown in Table 29.3, the results of the CAP-Gel bone marrow polychromatic erythrocyte micronucleus test showed that compared with the negative control group (H_2O), the higher the dose (5000, 2500, and 1000 mg/kg) of the experimental group as early as possible, the higher the dose, the higher the PCE containing micronucleus and the PCE/NCE ratio also increased. The number of PCE with micronucleus in female and male were 7 and 6, respectively, which were slightly higher than those in negative control group (4 and 5), but there was no significant difference between them and negative control group ($p > 0.05$). Compared with positive control group, both of them were significantly different ($p < 0.05$).

Table 29.3 CAP-Gel bone marrow polychromatic erythrocyte micronucleus test

Group	Dosage (mg/kg)	Gender	No	Detected PCE	Micronucleus (PCE)	PCE/NCE	Ratio of micronucleus
EG	5000	Female	5	5000	7	0.97 ± 0.05	1.40 ± 0.89
		Male	5	5000	6	1.05 ± 0.06	1.20 ± 0.45
	2500	Female	5	5000	6	1.03 ± 0.08	1.20 ± 0.45
		Male	5	5000	5	1.01 ± 0.09	1.00 ± 0.71
	1000	Female	5	5000	4	1.02 ± 0.08	0.80 ± 0.45
		Male	5	5000	5	0.97 ± 0.07	1.00 ± 0.71
NC	H_2O	Female	5	5000	4	1.04 ± 0.08	0.80 ± 0.45
		Male	5	5000	5	1.05 ± 0.07	1.00 ± 0.71
PC	CTX	Female	5	5000	105	1.03 ± 0.13	21.00 ± 2.74
		Male	5	5000	103	0.99 ± 0.06	20.60 ± 3.36

EG—experimental group, NC—negative control, PC—positive control, CTX—cyclophosphamide

29.4 Discussion

Microbial infection, such as chronic periodontitis, periodontitis, and tooth discoloration are common diseases in dentistry. Plasma activated hydrogel has been reported to act as reactive species carrier that allow good storage and controlled slow-release of reactive species to achieve long-term antibacterial effects and tooth bleaching [7, 8]. The reliance on the use of high concentrations of H_2O_2 has potential biosafety problems in tooth bleaching [9]. The present study indicated that the CAP-Gel is relatively non-toxic and does not cause significant chromosome aberration of bone marrow cells.

The OES results of HEDBS cold plasma device indicated that the main reactive species are reactive nitrogen species (RNS) and reactive oxygen species (ROS). In general, N_2 (78.084%), O_2(20.9476%), Ar (0.934%), and CO_2 (0.04%) account for about 99% of the composition of air. The plasma jet was immersed in water to generate PAW, this could help generating more reactive species in water. The production of these plasma-induced RNS/ROS depends on different typical parameters such as types of plasma power supply, type of working gas, electrode configuration, applied voltage, voltage polarity, treatment time, gas flow, the volume of solution, the distance between electrode and liquid surface, the composition of the solution, pulse duration and so on [16].

When plasma contacts with water, the plasma interacts with itself and the water molecules to generate a series of primary and secondary species [17]. The species formed in PAW include short-lived reactive oxygen and nitrogen species such as singlet oxygen ($^1O_2$), superoxide (O_2^-), hydroxyl radical (·OH), peroxynitrite ($ONOO^-$), nitric oxide (NO·), and peroxynitric ($OONO_2^-$) as well as long-lived species such as nitrite (NO_2^-), nitrate (NO_3^-), hydrogen peroxide (H_2O_2), and ozone (O_3). These species had many biomedical functions, such as antibacterial, tooth bleaching and food sterilization. It is reported that superoxide anion radical ($\bullet O_2^-$) played a critical role in the inactivation of bacterial biofilms [18]. Meanwhile, the hydrogen peroxide (H_2O_2) is the key component of tooth bleaching.

As Fig. 29.4 shows, pH value was decreased with processing time, and kept a relatively stable state was reached after 10 min. The pH value decreased with treatment time. Because of the production of nitric acid and an acid that consists of a hydrogen cation (H+) and a superoxide anion ($\bullet O_2^-$) [19]:

$$H_2O^+ + H_2O \rightarrow H^+(H_2O) + OH\cdot \quad (29.4)$$

Meanwhile, ORP increased significantly within 1 min and basically stabilized at 3 min, with no significant difference between 5 and 20 min. At the same time, many studies have shown that only when ORP value is greater than 700 can stable sterilization effect be guaranteed [20, 21]. The high ORP value.

Compared with PAW, CAP-Gel have stronger sterilization activity because of their unique chemical composition and gel structure, which can be better activated by the reactive species [22]. The revealed mechanisms of the antibacterial ability of the PAH identifies the unique combination of short-lived species ($^1O_2$, $\bullet OH$, $ONOO^-$ and O_2^-) stored in hydrogels [23, 24]. Meanwhile, previous study has confirmed that CAP-GEL has a significant tooth bleaching effect [14]. However, biotoxicity should also be considered in practical applications, as it is difficult to avoid non-contact with human soft tissues.

The temperature change of cold plasma was measured, the temperature increased from 15.07 ± 0.40 °C to 30.60 ± 1.5 °C. In general, increasing the reaction temperature will accelerate the percentage of activated molecules, the number of activated molecules increases, the effective collision between molecules increases, and the chemical reaction rate accelerates. Thus, this is helpful to generate more reactive species in PAW. Main reasons lead to temperature increase is that air is harder to excite than noble gases such as He and Ar [25]. Nevertheless, the final temperature (20 min) basically stabilizes around 30 °C, which is a comfortable temperature for human tissues.

It is suggested that the concentration of 3% containing H_2O_2 is relatively safe in theory. Besides, cold plasma treatment could increase the activity of related free radicals in H_2O_2. As to CAP-Gel, the average concentration of H_2O_2 is around 16.375 g/L (1.634%), which is far lower than usual bleaching products (35% H_2O_2). Thus, CAP-Gel is safe for humans.

Among all these common long-lived species, H_2O_2 is regarded as a fundamental molecule in anti-microbial, tooth bleaching. In present study, the concentration of H_2O_2 in the four samples is 16.5 g/L (1.634%), which is much lower than 30% H_2O_2 (1122 g/L). The H_2O_2 could be generated with the treatment of cold plasma:

$$OH_g \rightarrow OH_{aq} + OH_{aq} \rightarrow H_2O_{2aq}$$

$$H_2O \rightarrow H_{aq} + OH_{aq}$$

Meanwhile, the small amount of H_2O_2 could provide multiple reaction paths for the generation of·OH, which is helpful to improve the bleaching effect and antimicrobial effects.

$$H_2O_2 + O_3 \rightarrow \cdot OH + \cdot OOH + O_2$$

$$H_2O_2 + O \rightarrow \cdot OH + \cdot OOH$$

$$H_2O_2 \rightarrow 2 \cdot OH$$

Acute toxicity tests provide preliminary information on the toxic nature of a material for which no other toxicology information is available. This study was the first to investigate the acute toxicity of CAP-Gel. In this study, the acute toxicity showed no obvious effects on body weight. Male and female mice were not found to be more sensitive to CAP-Gel. All mice are live healthy after 14d. *Dehui Xu* reported that the maximum dose of plasma-activated water (PAW) for 15 min had no lethal effect and other acute toxicity [26]. *Dingxin Liu* reported that there were no related side effects in rabbits receiving the maximum dose of 700 μL plasma-activated liquid [27]. The above results indicate that CAP-Gel treatment had no acute toxicity and basically consistent with previous studies.

29.5 Conclusion

This work provides a foundation for safety and toxicity studies of CAP-Gel. There was no toxicity symptom.

Acknowledgements This work was supported by National Natural Science Foundation of China (No. 81901064 to Li) and Beijing Hospitals Authority Youth Programme (QML20210304 to Li).

References

1. Y. Al-Tarakemah, B.W. Darvell, On the permanence of tooth bleaching. Dent. Mater. **32**(10), 1281–1288 (2016)
2. A. Reis, S. Kossatz, G.C. Martins, A.D. Loguercio, Efficacy of and effect on tooth sensitivity of in-office bleaching gel concentrations: a randomized clinical trial. Operative Dent. **38**(4), 386–393 (2013)
3. C. Peng, S. Park, F.B. de Sousa, H. Gan, S.J. Lee, W. Wang, S. Lavender, S. Pilch, J. Han, Enhanced teeth whitening by nanofluidic transport of hydrogen peroxide into enamel with electrokinetic flows. Dent. Mater. **35**(11), 1637–1643 (2019)
4. X. Lu, G. Naidis, M. Laroussi, S. Reuter, D. Graves, K. Ostrikov, Guided ionization waves: theory and experiments. Phys. Rep. **540**(3), 123–166 (2014)

5. X. Lu, G. Naidis, M. Laroussi, K. Ostrikov, Reactive species in non-equilibrium atmospheric-pressure plasmas: generation, transport, and biological effects. Phys. Rep. **630**, 1–84 (2016)
6. K. Ashok, H.K. Dwivedi, N. Vijay, Atmospheric non-thermal plasma sources. Int. J. Eng. **2**(1), 53–68 (2008)
7. Q.-Y. Han, X. Wen, J.-Y. Gao, C.-S. Zhong, Y.-Y. Ni, Application of plasma-activated water in the food industry: a review of recent research developments. Food Chem. **405**, 134797 (2023)
8. Z. Qian, R. Ma, Y. Tian, B. Su, K. Wang, S. Yu et al., Sterilization efficiency of a novel electrochemical disinfectant against staphylococcus aureus. Environ. Sci. Technol. **50**(6), 3184 (2016)
9. R. Zhou, R. Zhou, P. Wang, Y. Xian, A. Mai-Prochnow, X. Lu, P. Cullen, K. Ostrikov, K. Bazaka, Plasma-activated water: generation, origin of reactive species and biological applications. J. Phys. D **53**, 303001 (2020)
10. V.A. Kolikov, V.E. Kurochkin, L.K. Panina, A.F. Rutberg, F.G. Rutberg, V.N. Snetov, A.Y. Stogov, Prolonged microbial resistance of water treated by a pulsed electrical discharge. Tech. Phys. **52**, 263–270 (2007)
11. D. Qiao, Y. Li, J. Pan, J. Zhang, Y. Tian, K. Wang, Effect of plasma activated water in caries prevention: the caries related biofilm inhibition effects and mechanisms. Plasma Chem. Plasma Process. **42**, 801–814 (2022)
12. J. Chen, Z. Wang, J. Sun, R. Zhou,* L. Guo, H. Zhang, D. Liu, M. Rong, K. (Ken) Ostrikov, Plasma-activated hydrogels for microbial disinfection. Adv. Sci. 2022074007 (2023)
13. Y. Li, J. Pan et al., In vitro studies of the antimicrobial effect of non-thermal plasma activated water as a novel mouthwash. Eur. J. Oral Sci. **125**(6), 463–470 (2017)
14. X. Yang, K. Sun, W. Zhu, Y. Li, J. Pan, Time-dependent efficacy and safety of tooth bleaching with cold plasma and H_2O_2 gel. BMC Oral Health **22**(1), 1–9 (2022)
15. R.N. Ma, H.Q. Feng, Y.D. Liang, Q. Zhang, Y. Tian, B. Su, J. Zhang, J. Fang, J. Phys. D Appl. Phys. **46**, 285401 (2013)
16. H. Xu, C. Chen, D. Liu et al., Contrasting characteristics of aqueous reactive species induced by cross-field and linear-field plasma jets. J. Phys. D Appl. Phys. **50**(24), 245201 (2017)
17. N. Bolouki, W.-H. Kuan, Y.-Y. Huang, J.-H. Hsieh, Characterizations of a plasmawater system generated by repetitive microsecond pulsed discharge with air, nitrogen, oxygen, argon gases species. Appl. Sci. **11**(13), 6158 (2021)
18. B. Xia, H.K.N. Vyas, R. Zhou, T. Zhang, J. Hong, J.G. Rothwell, S.A. Rice, D. Carter, K. (Ken) Ostrikov, P. Cullen, A. Mai-Prochnow, The importance of superoxide anion for Escherichia coli biofilm removal using plasma-activated water. J. Environ. Chem. Eng. **11**, 109977 (2023)
19. G. Park, Y.H. Ryu, Y.J. Hong et al., Cellular and molecular responses of Neurospora crassa to non-thermal plasma at atmospheric pressure. Appl. Phys. Lett. **100**, 063703 (2012)
20. S.-R. Yoon, J.Y. Lee, J.-S. Yang, J.-H. Ha, Bactericidal effects of diluted slightly acidic electrolyzed water in quantitative suspension and cabbage tests. LWT **152**, 112291 (2021)
21. T. Chen, H. Yao, D.-Y. Liu, Study on the bactericidal effect of electrolytic ozone water and its influencing factors of ORP value. Chin. J. Disinfection **10**, 721–723 (2019)
22. J. Chen, Z. Wang, J. Sun, R. Zhou, L. Guo, H. Zhang, D. Liu, M. Rong, K. (Ken) Ostrikov, Plasma-activated hydrogels for microbial disinfection. Adv. Sci. 2207407 (2023)
23. H. Zhang, S. Xu, J. Zhang, Z. Wang, D. Liu, L. Guo, C. Cheng, Y. Cheng, D. Xu, M.G. Kong, M. Rong, P.K. Chu, Biomaterials **276**, 121057 (2021)
24. C. Sun, X. Zeng, S. Zheng, Y. Wang, Z. Li, H. Zhang, L. Nie, Y. Zhang, Y. Zhao, X. Yang, Chem. Eng. J. **427**, 130843 (2022)
25. L. Zhang, D. Zhuang, M. Zhao, Q. Gong, L. Guo, L. Ouyang, R. Sun, Y. Wei, S. Zhan, The effects of selenium content on u(InGa)Se2 thin film solar cells by sputtering from quaternary target with Se-free post annealing. Vacuum **137**, 205–208 (2017)
26. Xu D, Cui Q, Xu Y et al., Systemic study on the safety of immuno-deficient nude mice treated by atmospheric plasma-activated water. Plasma Sci. Technol. **20**(4), 044003 (2018)
27. Xu Y, Peng S, Li B et al., Systematic safety evaluation of cold plasma-activated liquid in rabbits **9**, 659227 (2021)

Chapter 30
Grape Preservation by Plasma Synthesized NO Fumigation

Chunyang Zhang, Xu Li, Rongchen Shang, Qi Liu, Yongfeng Jiang, Xiang He, and Bingyan Chen

Abstract In this paper, the high concentrations of NO are produced under different gas flow speeds and enhanced by coconut shell charcoal. Further, thus were used to fumigation fresh grapes (shine muscat). With the applied voltage and frequency loaded on the discharge reactor are controlled at 17.0 kV and 8.0 kHz. When the gas flow speeds increased from 0.4 to 2.0 m/s, the concentration ratios of NO and total nitrogen oxides (NO/NO_x) have presented firstly increased and then decreased, reaching a maximum value of 61% at 1.2 m/s. In order to obtain higher concentrations of NO, fully contacting the discharged gas with 200 °C coconut shell charcoal, which can reduce O_3 and NO_2 by 85%. Stored at 23 °C, grapes were fumigated with different concentrations of NO. The experimental results indicate that as the concentrations of NO increased, the respiration rates of grapes are inhibited. Further, the concentrations of CO_2 and the weight loss rate decrease. Simultaneously, the decline in the soluble solids of grapes was inhibited. Compared to unprocessed group, the treated grapes with the concentration of NO for 600 ppm showed an overall decrease of about 20% in respiratory rate and a decrease in weight loss rate from 2.8 to 1.9% on the 15th day.

30.1 Introduction

After fruits and vegetables are picked, lack of oxygen supply from the roots and branches of plants, they will accelerate aging and decay, in order to ensure the quality of them, which will be preserved before being shipped to the market [1]. For preservation, the exogenous gas fumigation technologies on the market currently include SO_2 fumigation technology [2] and O_3 preservation technology [3], but both

C. Zhang · X. Li · R. Shang · Q. Liu · Y. Jiang · B. Chen (✉)
School of Mechanical and Electrical Engineering, Hohai University, Changzhou 213200, Jiangsu, China
e-mail: cbyan@foxmail.com

X. He · B. Chen
School of Science, Hohai University, Changzhou 213200, Jiangsu, China

Z. Fang et al. (eds.), *Proceedings of the 5th International Symposium on Plasma and Energy Conversion*, Springer Proceedings in Physics 398,
https://doi.org/10.1007/978-981-97-2245-7_30

of them are harmful to human health and will also cause pollution, and there will be SO_2 residues after using SO_2 treatment to cause certain effects on consumers; If the concentration of O_3 is too high, it will cause damage to the surface plasma membrane of fruits and vegetables, and even accelerate the aging and corruption of fruits and vegetables.

Air discharge plasma is widely used in the synthesis of NO, and the reaction process includes the collision dissociation, displacement and association of particles, which are in a state of thermal equilibrium, so that nitrogen and oxygen in the air are thermally ionized to generate N and O free radicals, which are then combined to form NO [4]. The sliding arc plasma has a large processing capacity and high energy density, which is widely used in the activation process of inert small molecules [5]. Therefore, we adopt the sliding arc discharge reactor discharge with ball-helical electrode structure to realize the excitation and activation of nitrogen to produce high concentration NO gas.

As a small molecule signaling substance, NO participates in many physiological and biochemical processes of plants, regulates the growth and development of plants [6], can also reduce the respiration of fruits and vegetables, inhibit ethylene biosynthesis, effectively delay the color change and ripening of fruit peels, and is a convenient treatment, time-saving and labor-saving, green and safe preservation method [7]. However, most of the NO gas comes from industrial production and storage of high-purity NO through cylinders. As a result, the market application of NO fumigation process has been greatly reduced. At this stage, it has been found that NO fumigation technology can be combined with cold chain logistics and pre-cooling treatment, which can reduce the postharvest respiratory intensity of fruits and vegetables to a greater extent, inhibit the production of enzymes and the release of ethylene, reduce the physiological metabolic rate of fruits and vegetables, and reduce physiological diseases [8].

In this project, the high concentrations of NO are produced under different gas flow speeds and enhanced by coconut shell charcoal. Further, thus were used to fumigation fresh grapes (shine muscat). The effect of NO fumigation on inhibiting the decay of grapes and deterioration was studied by plate counting method and detection of CO_2 content, weight loss rate, soluble solids content and grain drop rate. According to the above, this project can be used to reduce the loss of fruit and vegetable products, improve the national cold chain system, and ensure national food safety [9], which has high economic and social value.

30.2 Methods and Equipment

30.2.1 Experimental Materials and Equipment

In this project, fresh grapes (shine muscat) for fumigation. In the experiment, coconut shell charcoal, columnar carbon, spherical carbon, and ALK-Z203 NaOH modified columnar carbon (Carbon ernuo New Materials Co, Ltd.) were used to carry out the high-temperature activated carbon adsorption experiment, and the types of activated carbon that could be used for the experiment were further determined according to the comparison of adsorption effects.

The main equipment and instruments used in the experiment include programmable AC/DC power supply (GW Instek, GPK-2302); Imakoshi high-precision electronic scale (Wuyi Takuetsu Industry and Trade Co., Ltd., A-05); Brix meter (Zhejiang Lianxiang Hardware Technology Co, Ltd. TD10); Hardness tester (ARRIP, GY-3 type); Portable composite gas analyzer (Shenzhen Jishunan Technology Co., Ltd.; JK90-M4-GD); CO_2 sensor (Shenzhen Kenuo Electronic Technology Co, Ltd. GT2000-LZ2); Nutrient agar medium (Bichman Biotech Ltd.).

30.2.2 Research Protocol

In this experiment (see Fig. 30.1), Discharge plasma was used to synthesize mixed gas containing NO. After the high-temperature activated carbon adsorption device is introduced, NO_2 and O_3 in the mixed gas are removed to obtain purer NO gas, and when the gas is cooled and mixed with nitrogen to a suitable concentration, it is introduced into the fumigation bag, and the concentration of NO gas is monitored in real time by using a portable composite gas analyzer to achieve the effect of controlling the concentration of NO fumigation; Stop aeration after fumigation for 5 min; After fumigation, each group of grapes is sealed and packaged in food-grade ziplock bags.

30.2.3 Principle of Plasma Synthesis of NO

In order to obtain a larger plasma action area, an discharge reactor with spherical-helical electrode structure is designed. The high-voltage electrode adopts a ball electrode to maintain the uniformity of the spatial distance between the high-voltage electrode and the grounded electrode; The grounding electrode is a spiral electrode, on the one hand, the arc generated between the two electrodes can slide along the spiral electrode under the action of air flow, and on the other hand, it is convenient for the experimenter to observe the morphological characteristics of the discharge plasma plume and the measurement of the discharge spectrum.

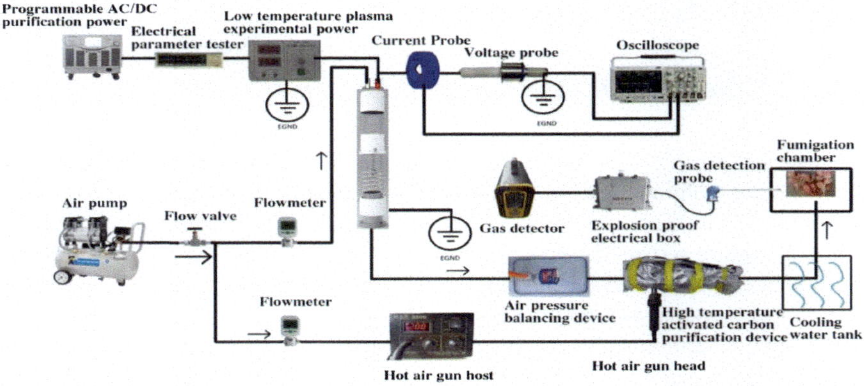

Fig. 30.1 Process flow diagram

When a high voltage is connected to the ball electrode, a bright arc is formed by breakdown between the shortest distance between the ball electrode and the spiral electrode, and under the action of the incoming airflow, the end of the arc located on the spiral electrode slides along the spiral electrode, thus forming a sliding arc discharge.

Plasma is a material aggregate state composed of many charged particles and neutral particles, and the whole is electrically neutral. When energy is introduced into matter, the substance undergoes the transformation of solid, liquid, and gaseous states, and then continues to introduce energy to gaseous matter, and the atoms that make up gas molecules obtain enough energy to be excited or dissociated.

In an air discharge[10–12], the gas between the electrodes is broken down to form a plasma plume, and a large number of free electrons, positive and negative ions, excited atoms or gas molecules exist in the discharge region. Interactions such as collisions between active ingredients cause the kinetic energy of particles to be converted into the internal energy of the collided particles, causing more particles to be excited, dissociated or ionized. Chemical reactions or mutual conversion of active ingredients to form gas-phase products such as O_3 and NO_x [13–15]. The formation of O_3 comes from O_2 molecules in the air, and 0–5 eV electrons can decompose O_2 to obtain O radicals, and when the heat carrier molecule M (N_2 or H_2O) is involved, O radicals and O_2 molecules form O_3 [16, 17].

$$e + O_2 \rightarrow O(^1D) + O(^1D) + eT \approx 0 - 5eV \tag{30.1}$$

$$O + O_2 + M \rightarrow O_3 + M \tag{30.2}$$

N_2 molecules in the air are easily excited by free electrons to form N_2^*, and further react with O_2 [18, 19]:

$$e + N_2 \rightarrow N_2^* + e \tag{30.3}$$

$$N_2^* + O_2 \rightarrow N_2 + 2O(^1D) \tag{30.4}$$

The excited oxygen atom $O(^1D)$ can react with N_2 to form NO and N radicals as follows [18, 20, 21]:

$$O(^1D) + N_2 \rightarrow NO + N \tag{30.5}$$

$$N + O_2 \rightarrow NO + O \tag{30.6}$$

OH can react with N atoms to form NO:

$$N + OH \rightarrow NO + H \tag{30.7}$$

The gas phase product in air discharge also includes NO_2, which is derived from the oxidation process of NO, and the O_3 and O produced during the discharge oxidize NO to NO_2 [21–23].

$$NO + O_3 \rightarrow NO_2 + O_2 \tag{30.8}$$

$$NO + O + M \rightarrow NO_2 + M \tag{30.9}$$

There is a reciprocal conversion reaction between NO and NO_2, and NO_2 is reduced by H to produce NO and OH, which are reversible. The specific reaction process is as follows (Fig. 30.2):

$$NO_2 + H \rightarrow NO + OH \tag{30.10}$$

30.2.4 High-Temperature Activated Carbon Enhances NO Conversion

Activated carbon is a commonly used adsorbent, catalyst or catalyst carrier with selective adsorption properties [24]. Using the selective adsorption characteristics of activated carbon, a NO purification device can be made to avoid nitrite from NO_2, O_3 and other gases. After the mixed gas is introduced into the copper box containing activated carbon, the activated carbon is first adsorbed, and then desorbed as the temperature rises. The equation is:

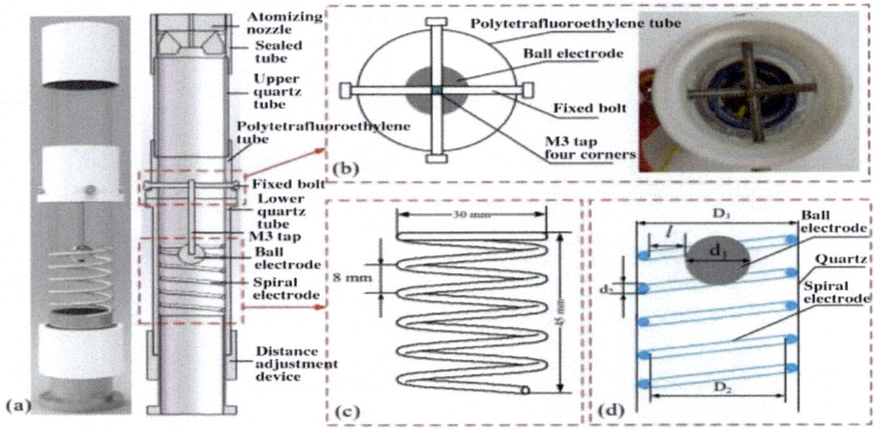

Fig. 30.2 Discharge reactor. **a** Structural diagram; **b** ball electrode and its fixing; **c** spiral electrode; **d** relative position of ball-spiral electrode

$$C^* + NO_2 \rightarrow C(NO_2) \quad (30.11)$$

$$C(NO_2) \rightarrow C(O) + NO \quad (30.12)$$

$$C(O) + NO_2 \rightarrow C(ONO_2) \quad (30.13)$$

The active center of activated carbon reacts with NO_2, and $C(ONO_2)$ is finally generated with $C(NO_2)$ and $C(O)$ as intermediate products, and NO_2 is released. At 250 °C, the pyrolysis of $C(ONO_2)$ is the mainstay, and the desorption equation is:

$$C(ONO_2) \rightarrow C(O) + NO_2 \quad (30.14)$$

$$C(ONO_2) \rightarrow CO_2 + NO \quad (30.15)$$

$C(ONO_2)$ may appear as an intermediate $C(O_2)$ before decomposition, while surface compounds containing two oxygen atoms easily decompose into CO_2:

$$C(ONO_2) \rightarrow CO + NO_2 \quad (30.16)$$

When the temperature is higher than 300 °C, the following reactions may occur:

$$2C(O) \rightarrow CO_2 + C^* \quad (30.17)$$

$$C(O) \rightarrow CO \quad (30.18)$$

Among them, C^* represents the active center sites on the surface of activated carbon, C(O), $C(O_2)$, $C(NO_2)$, $C(ONO_2)$ are the surface complexes of activated carbon, where $C(ONO_2)$ is the chemically adsorbed product of NO_2 after adsorption. During the desorption process, the escape of CO_2 and CO causes the loss of activated carbon. When the temperature is higher than 300°C, the desorbed intermediate products generate active center sites while releasing CO_2.

The highly acidic functional groups in activated carbon can increase the adsorption capacity of NO_2. Many literature shows [25], coconut shell charcoal has more acidic functional groups in activated carbons such as columnar carbon, spherical carbon, modified columnar carbon, coconut shell charcoal, etc., so coconut shell charcoal was used to selectively adsorb NO_2 (see Fig. 30.3).

After heating coconut shell charcoal, the NO/NO_x content is high, and compared with columnar activated carbon, spherical activated carbon and modified activated carbon, coconut shell carbon can achieve the best adsorption effect of NO_2 at 23 °C. Under the condition of the same quality of activated carbon, the heating temperature of coconut shell charcoal was adjusted, and under the different temperatures (23, 100 and 200 °C) when the best adsorption effect can be achieved when the coconut shell charcoal is heated to 200–220 °C (see Table 30.1), so that the NO_2 content is reduced to a concentration that cannot produce toxic substances such as nitrite during the fumigation process.

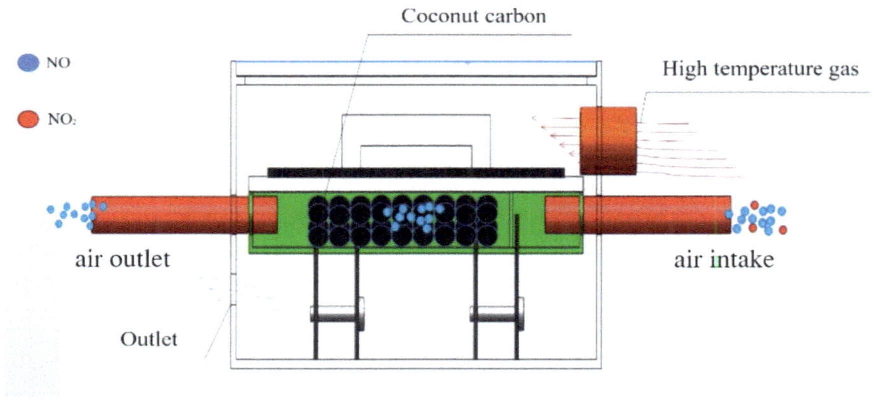

Fig. 30.3 Schematic diagram of NO_2 removal experiment

Table 30.1 Heating effect of activated carbon at different temperatures

Temperature (°C)	NO (ppm)	NO_2 (ppm)	NO_x (ppm)
23	1755	377	3066
100	1500	210	2508
200	1579	184	2604

30.2.5 Measurement Items and Methods

In this experiment, the soluble solids content of grapes was observed by Brix meter. To verify whether NO fumigation inhibits the growth of epidermal microorganisms, nutrient agar medium was used to cultivate grape epidermal colonies after 15 days of fumigation. The grapes used a CO_2 sensor to detect the CO_2 content every three days and used it to reflect the respiration rate of the grapes to verify NO whether inhibited the respiration of grapes; The weight loss rate(a) of grapes before fumigation and after three days of storage (19) was used to calculate the change of weight loss rate, quality before storage(m_0) minus quality after storage(m_t) then divided by quality before storage. Change law of grape weight in the experimental process was explored. The particle drop rate of different concentration fumigation groups was used to verify the freshness preservation effect.

$$a(\%) = \frac{(m_0 - m_t)}{m_0} \times 100\% \tag{30.19}$$

30.2.6 Power Supply Energy Detection and Calculation

In the application of gas discharge plasma, in order to obtain the operating parameters of efficient discharge, it is often necessary to evaluate the effect of the discharge treatment. For the calculation of power supply energy during high-voltage discharge, it is necessary to obtain the waveform data of the power supply voltage and current on the reactor during the discharge process, and in the actual experimental research process, high-voltage voltage probe, high-frequency current probe and digital oscilloscope and other equipment are used to collect it. The calculation of the power supply energy E during high voltage discharge can be obtained by integrating the voltage and current product over time(t) [26–28]:

$$E = \int_0^T u(t)i(t)\mathrm{d}t \tag{30.20}$$

where $u(t)$ is the function of the power supply voltage on the discharge reactor, $i(t)$ is the function of the power supply current in the discharge circuit, and T is the discharge time of the reactor. Typically, $u(t)$ and $i(t)$ are complex functions that change over time, and the power supply energy is difficult to obtain directly by integral. In arc discharge and spark discharge, the power supply voltage and current waveform is more regular and the noise interference is small, and the power supply energy of sliding arc discharge is calculated by instantaneous power method.

Figure 30.4a shows the supply voltage current waveform in sliding arc discharge at a power supply frequency of 5.0 kHz and a gas flow speed of 1.6 m/s, when the supply voltage waveform is close to the sine wave and the supply current waveform

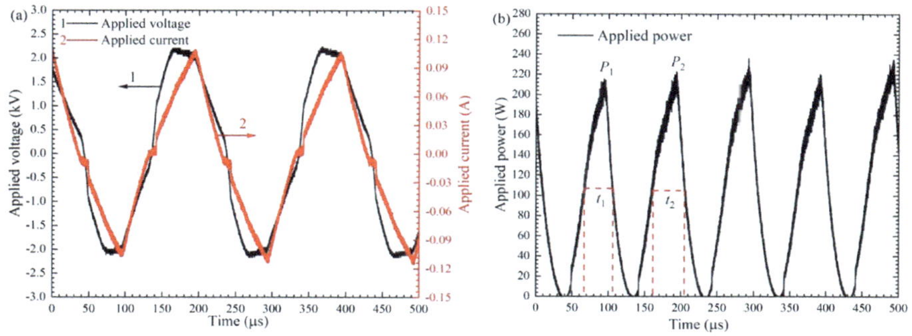

Fig. 30.4 Applied waveform on reactor with 5.0 kHz. **a** Applied voltage and current waveform; **b** applied power waveform

is close to the sawtooth wave; Multiply the supply voltage and supply current data to obtain an instantaneous power spectrum, see Fig. 30.4b. The full-width at half-maximum: FWHM method was used to calculate the power supply curve and the area around the abscissa to obtain the power supply energy [26–28]. The calculation expression is as follows:

$$E = P_1 t_1 + P_2 t_2 \tag{30.21}$$

Among them, P_1 is the power peak of the first half of the power supply cycle, P_2 is the peak of the second half of the power supply cycle, and t_1 and t_2 are the length of time occupied by the power half-peak. In sliding arc discharge, due to changes in experimental parameters and parasitic parameters on the reactor, a secondary voltage and current peak appears on the supply voltage and current waveform, which leads to a secondary peak in the power supply waveform obtained by the product of supply voltage and current (see Fig. 30.5), the supply voltage and current waveform is obtained under experimental conditions with a supply frequency of 7.0 kHz and a gas flow speed of 4.8 m/s. When calculating the power supply energy when there is a secondary peak, take the trough of the two power peaks in half a power supply cycle as the boundary, divide it into two parts, and then calculate its area by the half-peak full width method.

30.2.7 *Effect of Gas Flow State on NO Content*

In order to obtain the parameter regulation characteristics of discharge mode conversion and active ingredient generation in discharge, the gas flow regulation experiment was carried out under the experimental conditions of fixed voltage of 17.0 kV and frequency of 8.0 kHz. The gas flow range of the reactor is 0–180.0 L/min, and the

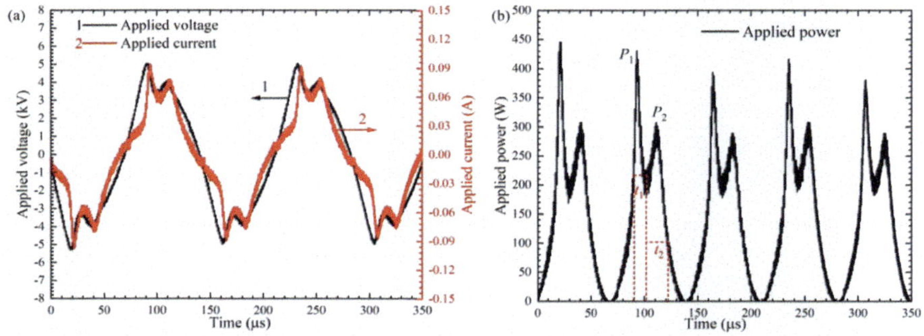

Fig. 30.5 Applied waveform on reactor with 7.0 kHz. **a** Applied voltage and current waveform; **b** applied power waveform

plasma plumes generated by the discharge have different morphological characteristics under different gas flow speeds. According to the experimental results, when the control gas flow speed is 1.6 m/s into the ball-spiral discharge reactor for sliding arc discharge, the mixed gas containing NO, NO_2, O_3 and other components can pass through the pipeline at a small flow speed, and the process can reach full contact with material. The principle is as follows:

When the gas flow speed is 1.6 m/s, the plasma plume formed between the ball electrode and the spiral electrode is a narrow and bright arc, and the discharge mode is arc discharge. When the gas flow speed increases to 4.0 m/s, the discharge pattern changes from an elongated arc to a dispersed plasma plume, the area of the discharge area increases, the color changes to orange-red and there are a small number of discharge filament, and the discharge mode is arc-spark transition mode. When the gas flow speed continues to increase to 7.2 m/s, the plasma plume evolves into a large number of bright white filament, the orange-red diffuse state disappears with a popping sound, and the discharge mode is spark discharge.

Table 30.2, the three flow states corresponding to flow velocities of 1.6, 4.0 and 7.2 m/s. In the flow regulation experiment, in order to study the relationship between the flow field and the change of discharge mode, the fluid velocity at the shortest horizontal distance is used to characterize the flow field change, and the fluid velocity calculation formula here is as follows:

$$u = \frac{4Q_t}{\pi({D_2}^2 - {d_1}^2)} \tag{30.22}$$

Table 30.2 Corresponding table of air discharge flow rate and velocity

Gas flow rate (L/min)	10.0	20.0	30.0	40.0	50.0	60.0	70.0	80.0	90.0	100.0
Gas velocity (m/s)	0.4	0.8	1.2	1.6	2.0	2.4	2.8	3.2	3.6	4.0

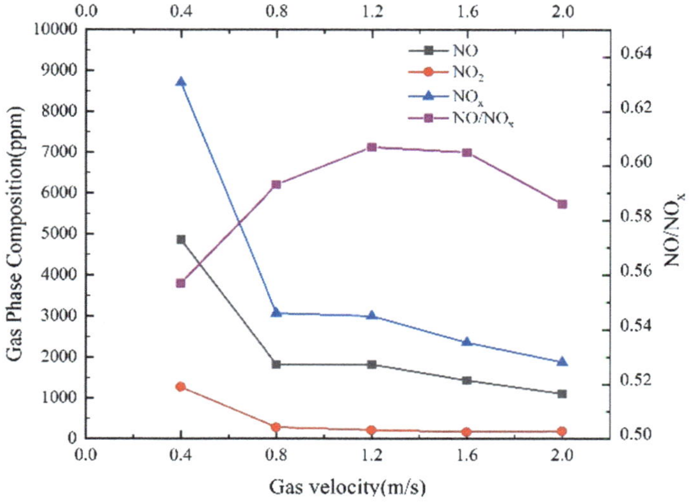

Fig. 30.6 Product concentrations at different gas flows

where Q_t is the total volume flow through the discharge area, d_1 is the diameter of the ball electrode, and D_2 is the diameter of the fluid channel (spiral electrode inner diameter) in the plasma action area, which can be calculated by the following formula:

$$D_2 = D_1 - 2d_2 \tag{30.23}$$

Among them, d_2 is used to make the diameter of the titanium wire of the spiral electrode (Figs. 30.6 and 30.7).

30.3 Results and Discussion

30.3.1 Fumigation Preservation Effect

Six sets of experiments were set up according to NO concentration (see Fig. 30.8), which were 0, 300, 600, 900, 1200, and 1500 ppm, respectively. Fumigation at different concentrations has different fresh-keeping effects. Experimental data showed that the best fumigation effect was the concentration of NO was 600 ppm, and the growth of epidermal mold had been significantly different 15 days after fumigation. The control group had a complete decay cycle of 23 days, after fumigation extended the complete decay cycle of grapes to 32 days. One of the main reasons for the decay of grapes is the rapid growth and abundant number of bacteria in the epidermis. After 15 days of fumigation, the skin of each group of grapes was

Fig. 30.7 Phenomenon of sliding arc discharge

Fig. 30.8 Effect of grape spoilage

cultured with bacteria (see Fig. 30.9). It was verified that the NO fumigation could greatly inhibit the growth of grape epidermal microorganisms and delay its decay rate, especially the 600 ppm NO concentration may be the best.

30.3.2 Respiration Rate Analysis

The respiration of grapes is closely related to storage, the photosynthesis of grapes stops after harvest, respiration becomes the dominant process of metabolism, and as a result, the organic matter of the fruit is consumed, the water is reduced, resulting

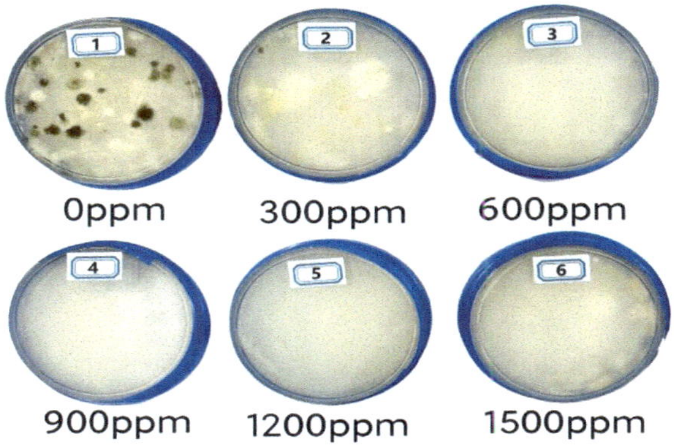

Fig. 30.9 Culture of common bacteria

in a gradual decline in quality, and the change of grape respiration rate is one of the main physiological indicators during grape storage [29]. After the grape fumigation is sealed and protected from light for a period of time, the CO_2 concentration (volume ratio) in the fumigation chamber was determined by carbon oxide detector. After the measurement, the naturally fallen fruit is removed, the air in the bag is drained and sealed for storage.

In this experiment, 6 groups of experiments were set up according to the concentration of NO, each group of 5 bags of grapes, each bag of grapes mass of 130 $\pm$ 5 g:0 ppm control group, 300, 600, 900, 1200 and 1500 ppm under the same conditions of grape species and storage environment. The experiment showed that the NO gas fumigation inhibited the respiration of grapes, so that the exhaled CO_2 content was significantly lower than that of the control group, and the CO_2 content of the experimental group with NO concentrations of 600 and 900 ppm was significantly lower than that of other groups at last. (see Fig. 30.10).

30.3.3 Weight Loss Rate Analysis

The weight loss rate was detected, and the weighing method [30] was used to detect the quality of the grapes in the bag before fumigation and then fumigated, sealed and protected from light for a period of time. Before the CO_2 concentration was determined in the next measuring cycle, the quality of the grapes in the fumigation bag was measured again and the weight loss rate was calculated by formula (30.19) above.

The weight loss rate of grapes gradually increased with the increase of storage time during the spoilage process (see Fig. 30.11).The NO fumigation delayed the reduction in the weight loss rate, and the loss rate of the experimental group with NO concentrations of 600 and 900 ppm was significantly lower than that of other groups at last.

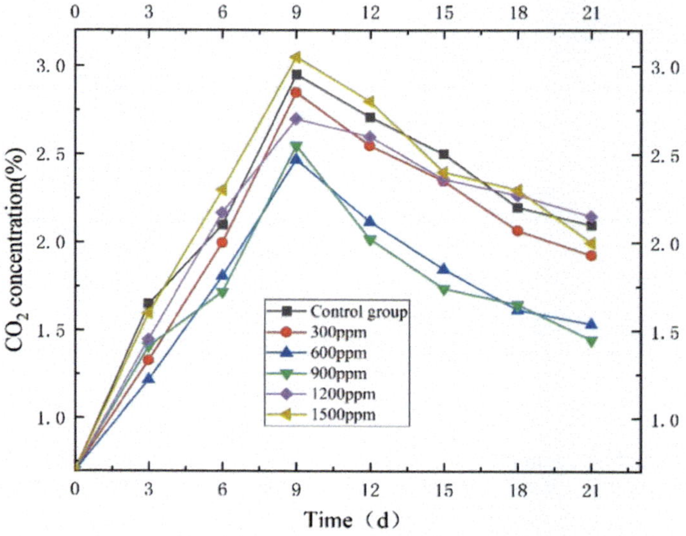

Fig. 30.10 CO_2 volume ratio after fumigation of NO gas

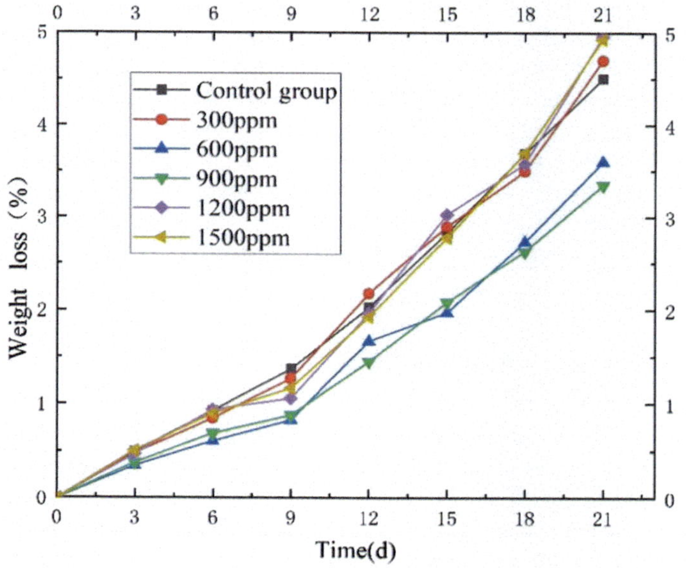

Fig. 30.11 Change in weight loss rate

30.3.4 Soluble Solids Content Analysis

In the early stage of grape preservation, with the continuous transformation of organic matter inside, the soluble solids content will gradually increase, and in the later stage, due to the continuous respiration, the grapes have no external nutrient supply, the internal organic matter will continue to decrease, and the soluble solids content will gradually decrease. Use a handheld refractometer when detecting, draw *10uL* of distilled water drops with a dropper to the sampling area of the refractometer, observe the reading of the internal ruler through the eyepiece in a bright place and zero it, and wipe the sampling area clean after completion; The must is poured into a small beaker and the *10uL* of the must is drawn into a clean dropper to observe and read the soluble solids content of the grapes.

The experiment showed that NO as a signaling molecule inhibits respiration (see Fig. 30.12), resulting in a delay in the occurrence of the peak of soluble solids content and a decrease in the rate of later decline, which confirmed that the use of NO-containing gas fumigation effectively slowed down the rate of grape decay, effectively delayed the ripening process of grape fruit, and prolonged its freshness time.

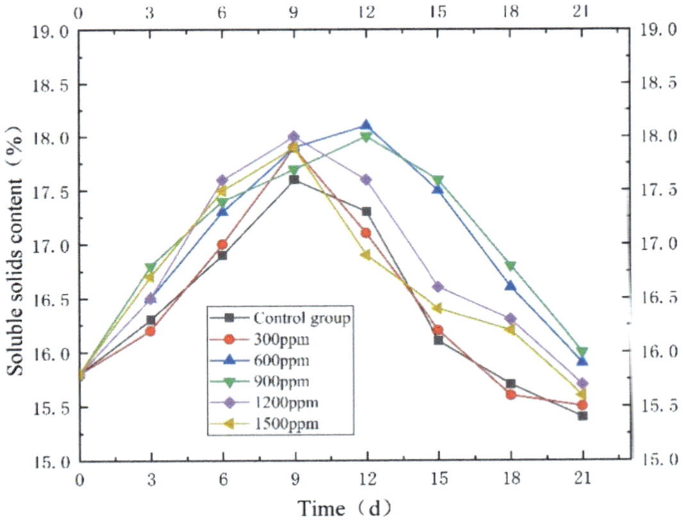

Fig. 30.12 Change in soluble solids content after NO fumigation

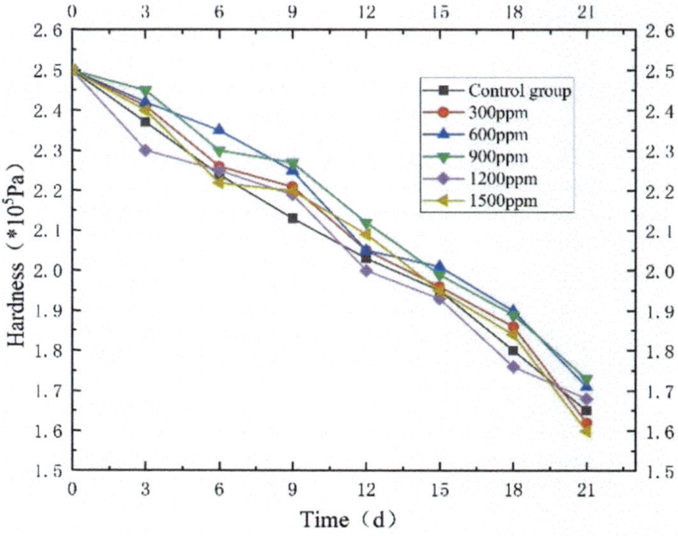

Fig. 30.13 Change in hardness

30.3.5 Hardness Analysis

Experiments showed (see Fig. 30.13) that the hardness of the grapes decreases significantly with accelerated spoilage over a longer period of storage. Under different NO concentrations, the experimental group with a concentration of 600 ppm had a relatively gentle decrease in hardness due to the physiological effect of low concentrations of NO inhibiting grapes, that is, the rate of decay was slower. Conversely, the experimental group with a high concentration of NO fumigation at a concentration of 1500 and the experimental group with a low concentration of 300 ppm had a greater decrease in hardness.

30.4 Conclusion

Experimental data and observation phenomena show that discharge plasma combined with high-temperature activated carbon adsorption to generate NO in the fumigation and preservation process, low concentration of NO can reduce the ethylene produced by plant tissues during growth and water stress. Conversely, when the fumigation concentration of NO is lower or higher, the fresh-keeping effect may be worse and even accelerate decay. NO can effectively reduce the respiration of grapes, delay the reduction of soluble solids content and inhibit the growth of epidermal bacteria. This method needs further research in terms of food safety and public acceptance. Compared with sulfur dioxide, O_3 and other preservation processes, this experiment is greener and safer, and has certain reference value for the research on fruit and vegetable preservation.

Acknowledgements Natural Science Foundation of Jiangsu Province (BK20191162).
Changzhou Key Research and Development Program (CJ20190046).
Central University Basic Research Service Project (B210203006).
National Innovation and Entrepreneurship Training Program for College Students (202310294047Z).

References

1. X. Wu, S. Ren, J. He et al., Current situation and prospect of fruit and vegetable preservation technology. Shandong Chem. Ind. **51**(12), 111–114 (2022)
2. Z. Zhang, *The Mechanism of Sulfur Dioxide (SO_2) Fumigation to Maintain the Postharvest Quality of Munag Grapes* (Xinjiang Agricultural University, 2022)
3. N. Bambalele, A. Mditshwa, N. Mbili et al., The antifungal effect of gaseous O_3 on Lasiodiplodia theobromae causing stem-end rot in 'Keitt' mangoes. Foods **12**, 195 (2023)
4. Q. Li, W. Zhang, H. Wang et al., Research progress in preparation of NO for medical purposes. Int. J. Biomed. Eng. **42**(5), 430–435 (2019)
5. W. Wang, B. Patil, Heijkers et al., Nitrogen fixation by gliding arc plasma: better insight by chemical kinetics modelling. Chem. Sus. Chem. **10**(10), 2145–2157 (2017)
6. G. Dong, Y. Qiao, C. Liu et al., Effects of calcium chloride combined with different packaging materials on postharvest aging and reactive oxygen species metabolism of Cocoa alba. Food Ferment. Ind. **48**(18) (2022)
7. M.S. Aghdam, Z. Luo, L. Li et al., Melatonin treatment maintains nutraceutical properties of pomegranate fruits during cold storage. Food Chem. **303** (2020)
8. C. Li, M. Feng, L. Wei et al., Research progress on the application of 1-methylcyclopropene in broccoli storage and preservation. Food Ferment. Ind. **47**(04) (2021)
9. K. Azizm, Natural antimicrobial/anti oxidantagents in meatand poultry products as well as fruits and vegetables: a review. Crit. Rev. Food Sci. Nutr.Nutr. **12**(3), 486–511 (2016)
10. K. Chen, L. Wan, B. Chen et al., Characteristics of water volatilization and oxides generation by using positive and negative corona. Plasma Sci. Technol **24**, 044007 (10 pp) (2022)
11. C. Bingyan, L. Changyu, W. Lianghao et al., Surface morphology and wetting properties of white pine modified by positive and negative corona discharge. High Voltage Technol. **47**(3), 786–795 (2019)
12. Y. Li, J. Jiang, M. Hinshelwood et al., Characterization of plasma catalytic decomposition of methane: role of atomic O and reaction mechanism. J. Phys. D Appl. Phys. **55**(15), 155204 (2022)
13. H. Li, D. Yu, W. Sun et al., Review on research progress of atmospheric pressure discharge plasma. High Voltage Technol. **42**(12), 3697–3727 (2016)
14. Z. Qing, L. Shuzhang, T. Honghui et al., *Plasma Technology and Application* (National Defense Industry Press, Beijing, 2009)
15. X. Xu, D. Zhu, *Gas Discharge Physics* (Fudan University Press, Shanghai, 1996)
16. Y.D. Li, J.K. Jiang, M. Hinshelwood et al., Characterization of plasma catalytic decomposition of methane: role of atomic O and reaction mechanism. J. Phys. D: Appl. Phys. **55**(15), 155204 (16pp) (2022)
17. A.T. Marnix, R.V. Hardeveld, E.M. Veldhuizen, Reactions of NO in a positive streamer corona plasma. Plasma Chem. Plasma **17**, 371–391 (1997)
18. P.J. Bruggeman, R.R. Frontiera, U.R. Kortshagen et al., Plasma-driven solution electrolysis. J. Phys. D Appl. Phys. **129**(20), 200902 (2021)

19. V. Gamaleev, N. Iwata, M. Hiramatsu et al., Tuning of operational parameters for effective production of NO using ambient air rotating glow discharge jet. Jpn. J. Appl. Phys.. J. Appl. Phys. **59**(4), 1–8 (2020)
20. G. Nayak, G. Oinuma, Y. Yue et al., Plasma-droplet interaction study to assess transport limitations and the role of OH, O, H, $O_2(a1\Delta g)$, O_3, He(23S) and Ar(1s5) in formate decomposition. Plasma Sources Sci. Technol., **30**, 115003(17pp) (2021)
21. J. Pawłat, P. Terebun, M. Kwiatkowski et al., Evaluation of oxidative species in gaseous and liquid phase generated by mini-gliding arc discharge. Plasma Chem. Plasma **39**, 627–642 (2019)
22. R.J. Wandell, H.H. Wang, R.K. Bulusu et al., Formation of nitrogen oxides by nanosecond pulsed plasma discharges in gas-liquid reactors. Plasma Chem. Plasma **39**(3), 643–666 (2019)
23. A. Moldgy, G. Nayak, H.A. Aboubakr et al., Inactivation of virus and bacteria using cold atmospheric pressure air plasmas and the role of reactive nitrogen species. J. Phys. D Appl. Phys. **53**, 434004 (2020)
24. X. Lu, *Study on the Characteristics of Highly Selective Purification of* NO_2 *in Medical NO Using Activated Carbon* (Huazhong University of Science and Technology, 2011)
25. G. Shoushan, D. Jingqi, L. Wenchuan, Chemical modification of activated carbon fiber. J. Anshan Iron Steel Inst. **23**(6), 406–409 (2000)
26. M.A. Lieberman, A.J. Ridenberg, *Discharge Plasma Principle and Material Treatment* (Science Press, Beijing, 2007)
27. H. Bluhm, *Principle and Application of Pulse Power System* (Tsinghua University Press, Beijing, 2008)
28. B.Y. Chen, Y.L. Gan, C.P. Zhu et al., Oxides yield comparison between DBD and APPJ in water-gas mixture. IEEE T. Plasma Sci. **44**(12), 3369–3378 (2016)
29. Y. Zhu, *Study on Improvement of Measurement Method of Respiration Rate of Fruits and Vegetables and Design Method of Air Conditioning Package* (Jiangnan University, 2012)
30. W. Chuanzeng, D. Fei, Z. Xuedan et al., Research progress of fruit and vegetable preservation under pressure. Acta Agriculturae Sinica **6**(3), 68–71 (2016)

Part III
High Voltage Insulation

Chapter 31
Effects of Cross-Linking Density on the Breakdown Characteristics of Silicone Gels for Power Device Encapsulation

Kexin Zhu, Fuping Zeng, Rirong Chen, Langlang Lv, Hua Jiang, and Guangzhi Chen

Abstract Silicone gels are widely used as a high-performance polymer insulating material in the packaging of high-voltage, high-power devices. However, under long-term thermal stress conditions, silicone gels undergo cross-linking reactions, and the degree of cross-linking increases with the aging degree. In order to study the potential relationship between the aging degree and the crosslink density, this paper builds the AC and DC withstand voltage test platform, by changing the ratio of the specimen preparation, to obtain the silicone gel specimens with different crosslink density, and carries out the negative polarity under the role of the rod-plate electrode DC breakdown test, and draws the Weibull distribution diagram. The study shows that the increase of cross-linking density leads to the decrease of the breakdown voltage of the organosilicon gel, and the change of the cross-linking density of the organosilicon gel due to thermal aging is an important means by which it affects the breakdown voltage of the organosilicon gel.

31.1 Introduction

Power devices are the core components for realizing the conversion of electric energy and its process control, which makes the use of electric energy more efficient, energy-saving, and environmentally friendly. In the twenty-first century, many semiconductor materials with superior performance have brought new impetus to the development of power devices, and their material properties make them more

K. Zhu · F. Zeng (✉) · R. Chen · L. Lv · H. Jiang · G. Chen
Hubei State Key Laboratory of Power Grid Environmental Protection, School of Electrical Engineering and Automation, Wuhan University, Wuhan 430072, Hubei, China
e-mail: Fuping.Zeng@whu.edu.cn

Hubei Key Laboratory of Power Equipment & System Security for Integrated Energy, School of Electrical Engineering and Automation, Wuhan University, Wuhan 430072, Hubei, China

Z. Fang et al. (eds.), *Proceedings of the 5th International Symposium on Plasma and Energy Conversion*, Springer Proceedings in Physics 398,
https://doi.org/10.1007/978-981-97-2245-7_31

suitable for high-temperature and high-voltage operation. A new generation of wide bandgap (WBG) semiconductor materials, such as silicon carbide (SiC) and gallium nitride (GaN), have gradually become a research hotspot due to their better electrical, mechanical, and thermal properties than Si [1–4]. However, at the same time, it also puts forward higher requirements for guaranteeing the reliable operation of the power system and puts forward new challenges for the insulation performance of power devices. At present, the encapsulation technology at the end of the power device production technology is still not perfect enough to keep pace with the development of the chip, so it is necessary to carry out in-depth and detailed research on the encapsulation technology of power devices.

Power device encapsulation is a highly comprehensive discipline, including material research and form process and so on many fields [5], in which the performance of the insulating material for encapsulation will directly affect the insulation performance of the encapsulation module. The main insulating materials for encapsulation include potting insulating materials and substrate insulating materials [6]. Among them, the basic purpose of potting materials is to protect chips and metal components from harsh environments such as moisture, chemicals, gases, etc., and at the same time provide additional insulation protection and mechanical protection between conductors and components [7]. Silicone gel, as a polymer insulating material with excellent performance, is widely used in the encapsulation of high-voltage and high-power devices, such as the typical power module Insulated Gate Bipolar Transistor (IGBT) encapsulation, the soldered silicone-based IGBT module is often used in silicone gel as a potting material.

In actual operating conditions, the chip in the high-voltage high-power devices through the continuous switching state transition in the thermal cycle often due to conduction losses, switching losses, etc., a large amount of heat, and through heat conduction will be transferred to the surrounding packaging materials. In the limited heat dissipation conditions, this will directly lead to an increase in the internal temperature of the module. As the number of power devices used and the number of years of operation increases, the encapsulation material used for module packaging will be in a long-term high-temperature environment. Under the conditions of long-term thermal stress, the accelerated aging of the material will lead to changes in its chemical structure, which in turn will lead to changes in the relevant characteristics. Therefore, based on the accelerated thermal aging experiments on the breakdown characteristics of silicone gel under different aging cycles to analyze the mechanism of the relevant thermal and electrical properties of silicone gel under the influence of thermal aging, not only to improve the theoretical study of power module packaging in polymer materials but also to enrich the relevant content of the application of power device technology for the power device packaging of the current insulation design and reliability checking. Provide reference basis.

At present, domestic and foreign research on silicone gel is mainly focused on mechanical properties, dielectric properties, partial discharge, space charge characteristics, etc., mostly considering the changes and mechanisms of these physical and chemical properties under the action of temperature and frequency. However, the research on the effect of thermal aging on the insulating properties of silicone gel is still not systematic enough, lacking the breakdown test data of silicone gel under different types of voltages, and the mechanism analysis of the effect of thermal aging on the insulating properties of silicone gel is not perfect. For this reason, this paper builds the AC and DC voltage test platform, studies the negative polarity DC breakdown test of silicone gel specimens with different cross-linking densities under the action of rod-plate electrodes, obtains its breakdown voltage change rule under different aging cycles and different voltage types, and thus analyzes the mechanism of thermal aging on its breakdown characteristics.

31.2 Preparation of Specimens with Different Crosslinking Densities

The silicone gel used in this paper is BeGel 8606 A/B dielectric insulating silicone gel produced by Shanghai Bayco Polymers Co. A component is a polysiloxane polymer, and B component is a hydrogen-containing polysiloxane polymer, which is colorless and odorless oily liquids. A and B components are cured to produce silicone gel through the addition reaction that occurs in the process of mixing, and the molecular formulas of the A and B components are shown in Fig. 31.1. The molecular formula of components A and B is shown in Fig. 31.1. By controlling the mass ratio of components, A and B when mixing to control the crosslinking density of the finished silicone gel obtained, that is, by changing the ratio to artificially change the crosslinking density. First use ethanol to clean the utensils in addition to impurities, and then calculate the A, and B components of the mass ratio of 0.6:1 when the components of the required mass, weighing the two components of the colloid poured into a beaker, stirring at a uniform speed for 30 min until the mixing of the colloid in the fewer bubbles, it can be poured into a mold for curing, curing 24 h, demolding can be obtained from the organosilicon gel specimen. After that, the distribution of 1.0:1, 0.9:1, 0.8:1, and 0.7:1 for the A, and B components of the mass ratio for the repetition of the preparation, and finally can obtain five different crosslinking densities of the silicone gel sample.

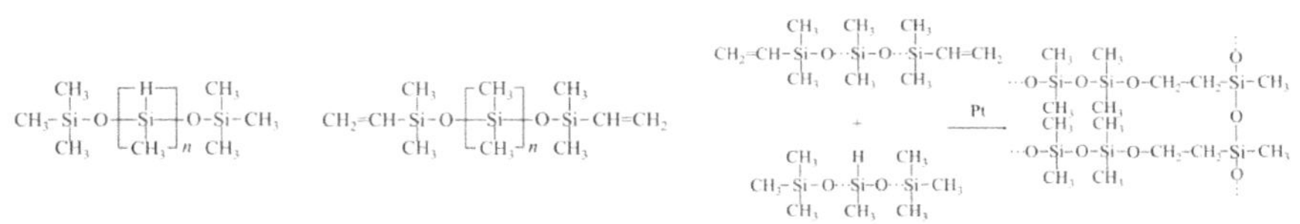

(a) Molecular formula of the main components of component A (b) Molecular formula of the main components of component B (c) Molecular formulae for the addition reactions of components A and B

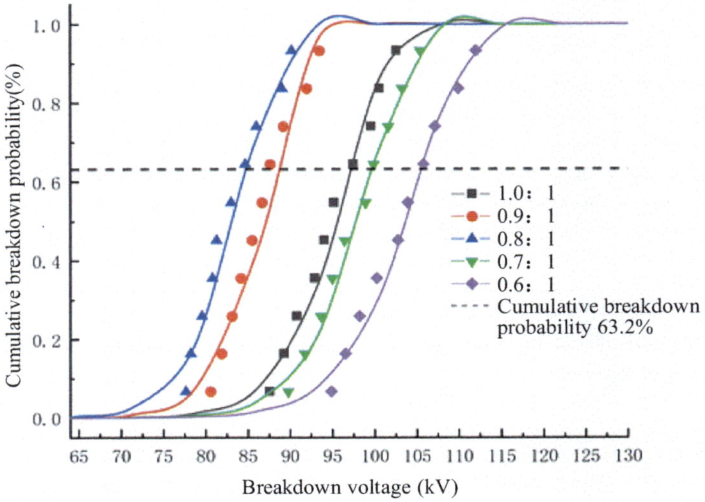

Fig. 31.1 Weibull distribution of negative DC breakdown voltage of rod-plate with different ratios of specimens

31.3 Test Method for Crosslink Density

Rubber-based polymers are characterized by both flexibility and large reversible deformability. Flexibility is a common property of all amorphous polymers, but to have both flexibility and reversible deformability requires the introduction of cross-linking points using chemical reactions or physical interactions. The biggest role of cross-linking points is to inhibit the sliding between polymer chains, and the general three-dimensional network structure in polymers is generally formed through the addition of binder through the cross-linking reaction. This crosslinked three-dimensional network structure is usually characterized by two parameters: crosslink bond type and crosslink density. The crosslink density can be described by physical quantities such as the density of the web chains (υ_e), the density of the crosslinking points (μ_e), and the average molecular mass of the chain segments between the crosslinking points (M_c). Among them, the web density refers to the number of web chains contained per unit volume of the elastomer, the cross-link density refers to the number of cross-linking points per unit volume of the elastomer, and the average molecular mass of the chain segments between cross-linking points refers to the length of the molecular chains between cross-linking points.

The general methods of determining crosslink density include chemical determination based on the mechanism of thermosetting polymers and crosslinking agents, physical determination based on the statistical theory of rubber elasticity such as the equilibrium swelling method, the stress–strain method, the Dynamic Mechanical and Thermal Analysis (DMA) method, and the rheological method, and so on. With the advancement of theory and technology, new methods such as Nuclear Magnetic Resonance (NMR), Small Angle Neutron Scattering (SANS), Atomic Force Microscopy (AFM), and Gas Chromatography (GC) have emerged.

While comprehensively analyzing the above methods, the equilibrium dissolution method is the most suitable test method for this test. Cross-linked or reticulated polymers cannot be dissolved in good solvents until the cross-linking bonds are broken, but the cross-linked polymers can still be dissolved by absorbing a large amount of solvent. During the swelling process, the forces of the solvent trying to penetrate the polymer to expand its volume compete with the forces of elastic retraction of the crosslinked network in the polymer resulting from the expansion and stress. When the two-reach equilibrium, i.e., the swelling forces of the solvent and the contraction forces of the polymer cancel each other out, swelling reaches equilibrium, at which point the volume of swelling no longer changes. The degree of swelling Q can be defined in terms of the mass or volume of solvent absorbed by the crosslinked polymer:

$$Q = \frac{m - m_0}{m_0} = \frac{V - V_0}{V_0} \tag{31.1}$$

where m_0 and V_0 denote the mass and volume of the cross-linked polymer before swelling, and m and V denote the mass and volume of the cross-linked polymer after swelling.

The change of Gibbs free energy of the system during swelling consists of the free energy of mixing of the polymer with the solvent and the elastic free energy of the crosslinked network deformation, which adds up to zero when swelling equilibrium is reached. Then, according to the statistical theory of high elasticity of rubber and the lattice model theory of Flory–Huggins, the formula for the cross-linking density can be obtained:

$$v_e = \frac{\rho_b}{M_c} = \frac{\left[\ln(1 - v_2) + v_2 + \chi v_2^2\right]}{v\left(v_2^{1/3} - 2v_2/f\right)} \tag{31.2}$$

$$\rho_b = \frac{m_b}{V_b} \tag{31.3}$$

where v_e is the crosslinking density of the elastomer, mol/cm^3; ρ_b is the density of the elastomer; M_c is the relative average molecular mass between crosslinking points; v_2 is the volume fraction of the polymer in the solubilized sample; χ is a parameter of the interaction between the polymer and the solvent; v is the molar volume of the solvent; and f is the functionality of the polymer network. Therefore, the crosslink density v_e can be derived from Eq. (31.2) by simply measuring the v_2 of the specimen.

In this experiment, the cross-linking density of silicone gel specimens prepared in different ratios was estimated based on the equilibrium swelling method. In the previous measurement, a piece of silicone gel of a certain proportion was removed after demolding, two small pieces of silicone gel samples were cut off completely using a cutting tool, and the weights of the two samples were measured and recorded

Table 31.1 The crosslink density of silicone gels with different ratios

Ratio	1.0:1	0.9:1	0.8:1	0.7:1	0.6:1
Crosslink density (10^{-5} mol/cm^3)	1.21	1.66	1.71	2.21	2.36

using a precision mass measuring instrument, one piece was put into a measuring cylinder containing water to measure its volume, and the other was put into a beaker containing a toluene solvent for dissolution. Using the mass and volume measured by the former can be calculated according to Eq. (31.3) the density of the proportion of the organosilicon gel sample, the latter dissolved 7d after the removal of the toluene solution left on the surface of the absorbent after the measurement of the mass of the dissolved, according to Eq. (31.1) to calculate the degree of dissolution of the proportion of the organosilicon gel. The silicone gel specimens with ratios of 1.0:1, 0.9:1, 0.8:1, 0.7:1, and 0.6:1 were sequentially subjected to the above measurements, and the cross-linking densities of different ratios of the organosilicon gel specimens were obtained by calculation as shown in Table 31.1. It can be seen that it is feasible to change the crosslink density of the specimen by changing the ratio.

31.4 Effect of Crosslink Density on Breakdown Voltage of Silicone Gels

By applying negative DC voltage to the specimens with different ratios under the action of rod-plate electrodes for the DC breakdown test, the negative DC breakdown voltage data of the specimens with different crosslinking densities were obtained, and the corresponding Weibull distribution graphs were plotted as shown in Fig. 31.1, and the values of the two characteristic parameters of the Weibull distributions corresponding to the specimens with different crosslinking densities were listed in Table 31.2.

As can be seen from Fig. 31.1, the breakdown voltages of the specimens with different ratios show a decreasing and then increasing trend with the increase of crosslink density due to the decrease of the ratio, and a sudden increase occurs before the beginning of the increase. When the ratio decreased in the interval of 1.0 to 0.8, the crosslink density of the specimen increased from 1.21×10^{-5} mol/cm^3 to 1.71×10^{-5} mol/cm^3, and its breakdown voltage decreased from 97.3 kV in the 1.0 ratio to 88 kV in the 0.9 ratio, and then to 85.10 kV in the 0.8 ratio; while when

Table 31.2 Weibull distribution parameters of negative polarity DC breakdown voltage with different proportions

Ratio	1.0:1	0.9:1	0.8:1	0.7:1	0.6:1
η/kV	97.32	88.38	85.10	99.90	105.75
β	20.75	22.04	20.35	20.95	19.74

the ratio decreased in the interval of 0.7 to 0.6, the crosslink density of the specimen showed a decreasing and increasing trend. When the ratio decreases from 0.7 to 0.6, the crosslink density increases from 2.21×10^{-5} mol/cm^3 to 2.36×10^{-5} mol/cm^3, and the breakdown voltage rises from 99.0 kV in the ratio of 0.7 to 105.75 kV in the ratio of 0.6, in which the breakdown voltage of the specimen with a ratio of 0.7 exceeds that of the specimen with a ratio of 1.0. It can be seen, that when the ratio drops to less than 0.8–0.7 this interval in a range, the breakdown characteristics of the specimen will change the law of change.

Before this range, the breakdown characteristics of organosilicon gel specimens decrease with the increase of crosslink density, which is caused by the increase of free volume in the specimen. Free volume in polymers can be understood to some extent as caused by the loose arrangement of molecular chains and irregular thermal movement of molecules, which exists between molecular chains in the amorphous region of the polymer, and usually refers to voids that are not occupied by the chain segments (intrinsic static voids between the molecular chains) or a small amount of unoccupied volume at the end of the chain segments (dynamic voids). According to the free volume breakdown theory, as the cross-linking density increases, the number of cross-linking points in the organosilicon gel specimen increases, and the structure of the cross-linking system thus becomes neater, with the degree of aggregation of the molecular chains increasing, while the small molecular side-chain structure is consumed, which, in the overall scheme of things, results in an increase in the free volume of the specimen. When the free volume increases, the average free travel of the charged protons in the specimen is increased, and the energy obtained under the same electric field is also increased. Therefore, when the cross-linking density of the organosilicon gel specimen increases, the energy obtained by the charged protons therein during the migration process is more likely to exceed the threshold barrier corresponding to the breakdown, thus increasing the breakdown voltage of the specimen to a certain extent.

When the ratio exceeds the above interval range, the breakdown characteristics of the specimen rise with the increase of the crosslink density, which is caused by the change of the crosslinking network of the specimen due to the decrease of the ratio. When the ratio is reduced to a certain extent, the preparation of the beaker of mixed colloids within the B component is much more than the A component, this time the reaction generated by the cross-linking network has been with the ratio of 1.0 (actual encapsulation of silicone gel prepared with the ratio used in the preparation of the specimen has a great difference, its performance is similar to the liquid silicone rubber, the study shows that the breakdown voltage of liquid silicone rubber does increase with the increase in cross-linking density and rise. That is to say that the crosslinking grid of the specimen under this ratio has been fundamentally changed.

Therefore, when considering the effect of crosslinking density on the breakdown characteristics of silicone gel, the main reference is the breakdown characteristics of silicone gel specimens with a ratio of 1.0–0.8. According to the above analysis can be known, by changing the ratio obtained by different crosslinking density silicone gel specimens of negative DC breakdown voltage change rule and different aging cycle specimens of negative DC breakdown voltage change rule is similar. Thus, this group of specimens verifies the thermal aging by affecting the crosslinking density of silicone gel and thus has an impact on the breakdown characteristics of silicone gel.

31.5 Conclusion

(1) By changing the ratio of A and B components during preparation, silicone gel samples with different crosslinking densities were obtained. A total of 5 sets of samples were prepared with 5 different ratios of 1.0, 0.9, 0.8, 0.7, and 0.6, and the crosslinking density of 5 groups of samples was estimated by equilibrium expansion method, which proved that with the decrease of the proportion, the crosslinking density of the specimens was indeed increasing.
(2) The negative polarity DC breakdown test under the action of the rod-plate electrode was carried out on the prepared silicone gel samples with different crosslinking densities, and the Weibull distribution map was drawn by processing the data by the same statistical method, and it was observed that the breakdown voltage decreased with the increase of crosslinking density without changing the fundamental crosslinking network structure of the silicone gel sample.
(3) The main mechanism of thermal aging affecting the breakdown characteristics of silicone gels is: increasing the crosslinking density of silicone gels to increase their free volume, further increasing the general insulation defects inside the material, so that the breakdown characteristics of cross-linked polymer silicone gels decrease.

References

1. L. Li et al., Analysis of thermal aging characteristics of silicone gel for high-voltage IGBT packaging, in *2023 IEEE 4th International Conference on Electrical Materials and Power Equipment (ICEMPE)*, Shanghai, China (2023), pp. 1–4
2. L. Li et al., Infrared spectrum analysis and aging mechanism of high temperature vulcanized silicone rubber under ultraviolet irradiation. Spectrosc. Spectral Anal. **40**(04), 1063–1070 (2020)
3. X. Wang et al., Study on thermal decomposition behavior of phenyl silicone rubber based on thermogravimetric analysis. Rubber Ind. **68**(09), 699–704 (2021)
4. S. Seal, M. Glover, 3D wire bondless switching cell using flip-chip bonded silicon carbide power devices. IEEE Trans. Power Electron. **33**(10), 8553–8564 (2017)
5. N. Iwamuro, T. Laska, IGBT history, state-of-the-art, and future prospects. IEEE Trans. Electron Devices **64**(03), 741–752 (2017)
6. P. Fu et al., Partial discharge measurement and analysis in high voltage IGBT modules under DC voltage. CSEE J. Power Energy Syst. **4**(04), 513–523 (2018)
7. L. Coppola, D. Huff, F. Wang et al., Survey on high-temperature packaging materials for SiC-based power electronics modules, in *2007 IEEE Power Electronics Specialists Conference*. Orlando, FL, USA (2007), pp. 2234–2240

Chapter 32
AC Surface Flashover Voltage Prediction of Epoxy Resin by Tree-Based Model

Chuanjie Lin, Longyang Zhu, Xiaosi Lin, Yuwei Chen, Hao Chen, Hui Chen, and Yuesheng Zheng

Abstract AC surface flashover is a prevalent insulation fault in gas-insulated equipment. Given the complex dependency of this fault on various factors and the limitations of destructive experimental methods, this paper proposes the use of tree-based machine learning models as an economical and efficient alternative to obtain the flashover voltage. The analysis concentrates on salient features impacting AC surface flashover, such as electroluminescence (EL), water contact angle (WCA), and percentage of silicon atoms (PSA). The extra trees algorithm was employed and optimized by feature selection. The model, utilizing six features—EL, WCA, PSA, Polished status, Ra, and surface resistivity—yielded excellent performance with MAE, MSE, MAPE, and R^2 scores of 0.50, 0.37, 1.47%, and 0.93, respectively. The study highlights the potential of tree-based models in providing insights and prediction of surface flashover voltage for the insulation in gas-insulated equipment.

32.1 Introduction

32.1.1 A Subsection Sample

Surface flashover, occurring at the interface between the epoxy resin (EP) insulator and SF6 gas, represents a prevalent insulation fault in AC gas-insulated equipment [1–5]. Analyzing the surface flashover process is challenging due to its dependency

C. Lin · L. Zhu · Y. Chen · H. Chen · H. Chen
Fuzhou Power Supply Company, State Grid Fujian Electric Power Co., Ltd., Fuzhou 350009, China

X. Lin
State Grid Fujian Technical Training Centre, State Grid Fujian Electric Power Co., Ltd., Quanzhou 362000, China

Y. Zheng (✉)
College of Electrical Engineering and Automation, Fuzhou University, Fuzhou 350108, China
e-mail: yueshengzheng@fzu.edu.cn

Z. Fang et al. (eds.), *Proceedings of the 5th International Symposium on Plasma and Energy Conversion*, Springer Proceedings in Physics 398,
https://doi.org/10.1007/978-981-97-2245-7_32

on numerous factors at the gas–solid interface. Consequently, AC surface flashover experiments serve as the primary method for studying this phenomenon. However, these destructive experiments demand high-quality samples and rigorous control over experimental conditions, resulting in substantial consumption of manpower, materials, and financial resources. Hence, employing machine learning algorithms as an alternative to actual experiments to predict the AC surface flashover voltage holds significant value for EP insulator development and gas-insulated equipment design [6, 7].

Previous studies on AC flashover prediction have predominantly focused on experiments and analysis conducted under conditions of insulation material surface pollution or icing. Mahdjoubi et al. [8] proposes an efficient and rapid method that utilizes artificial intelligence techniques, such as support vector machines and genetic algorithms, to optimize the maintenance of outdoor insulators. Arshard et al. [9] emphasizes the effective application of machine learning algorithms in predicting flashover parameters of polymeric insulators under contaminated conditions, achieving a prediction error reduction of approximately 5% or less. Cui et al. [10] introduces a novel electric field measurement technology for predicting flashover voltage on various types of insulators, exploring the relationship between the conductivity of the contamination layer and the environmental relative humidity. Gencoglu et al. [11] develops a dynamic model using the least square support vector machine regression method to estimate AC flashover voltages of polluted insulators. Despite these advancements, there is a notable scarcity of research on the use of different interface characteristics on the gas–solid interface to predict the AC surface flashover voltage, particularly for EP insulators.

In this study, we conduct experiments to obtain the water contact angle, electroluminescence, surface conductivity, and other interface characteristics of EP samples. By performing correlation coefficient analysis, we identify key characteristics influencing AC surface flashover. Further, we employ machine learning algorithms based on tree-based models to predict flashover voltage, and we validate the effectiveness of these algorithms through testing.

32.2 Experimental Setup

32.2.1 Experimental Data Acquisition

The experimental samples in this study were prepared by blending EP with micro-sized alumina particles, as previously detailed in our earlier works [12, 13]. To achieve diverse surface micro morphologies, the alumina particles underwent sand-blasting and polishing treatments. The gas–solid interface characteristics and corresponding test methods employed in this research are summarized in Table 32.1. Except for the surface electroluminescence test and AC flashover test, the rest are common tests, which will not be described in detail [14]. The reported values for water contact angle, surface chemical composition, surface chemical bonding, surface conductivity, dielectric constant, dielectric loss, and partial discharge represent average measurements obtained from multiple samples. In the case of surface morphology assessment, the most representative morphology was selected after observing multiple samples. The surface trap density and flashover voltage were determined by directly analyzing the distribution of results obtained from testing multiple samples.

The luminescence phenomenon comes from the process that excited electrons emit photons to realize the deexcitation. Electroluminescence is initializing the excitation of electrons by electric field. EL can be used to study the characteristics of traps on the surface or inside of materials and the characteristics of charge storage and transport [15–17]. The surface electroluminescence test platform based on the above principle is shown in Fig. 32.1. A dark chamber, containing a pair of finger electrodes, was utilized in the luminescence test. The distance between the electrodes was 20 mm. The voltage − 8 kV in steps of 300 V per second was employed, and the earth strap passed through a high frequency current transformer, which was connected to an oscilloscope, to ensure that luminescence was not from measurable PD and surface flashover. The luminescence was measured by a counting unit (Hamamatsu C8855-01) combined with a photon counting head (Hamamatsu H8259-01). The spectral response of the photon-counting unit was in range of 185–850 nm with the dark count of less than 20 counts per second after 30 min stored in darkness.

The finger electrode structure used is shown in Fig. 32.2. The distance between the two electrodes was 20 mm. In the platform, a pressing knob was set to apply pressure in the vertical direction to keep the surface of the sample having a close contact with the metal electrode. The process of AC flashover test was as follows: (1) Wiped the surface of EP samples and electrodes with alcohol. (2) Installed a sample on the plat-form, used the knob to make a close contact between the sample and electrodes. (3) Applied gradually increased AC voltage on the electrodes until flashover occurs. (4) Recorded the flashover voltage. The whole process of pressurization was carried out in the atmosphere of SF6 at 20 °C and 1 atmospheric pressure. Each experiment group included 20 samples, and each sample carried out 3 times of AC flashover.

Table 32.1 Test methods for gas–solid interface characteristics

No.	Interface characteristics	Test methods
1	Surface micro morphology	Scanning electron microscope (SEM)
2	Water contact angle	Contact angle meter
3	Interface chemical composition	X-ray photoelectron spectroscopy (XPS)
4	Interface chemical bond	Infrared spectrum (IR)
5	Surface trap density	Surface electroluminescence
6	Surface conductivity	High electric field conductivity test
7	Dielectric constant and dielectric loss	Dielectric spectrometer
8	Partial discharge	Pulse current method
9	AC flashover voltage	AC surface flashover test

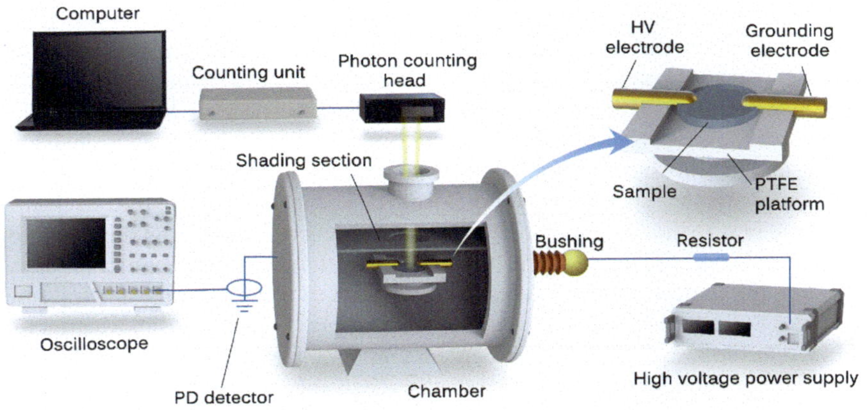

Fig. 32.1 Luminescence test platform

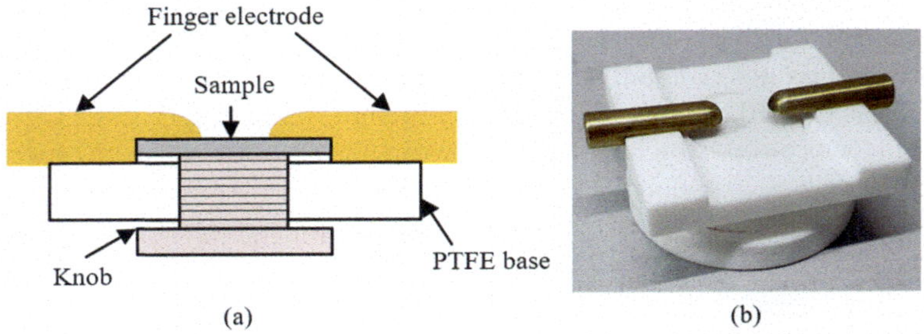

Fig. 32.2 Finger electrodes. **a** Cross section diagram; **b** practical diagram

32.2.2 Machine Learning Algorithms

In this study, multiple tree-based models, including Decision Tree (DT), Adaptive Boost (AB), Extra Trees (ET), and Random Forest (RF), are used to predict flashover voltage. These tree-based models are compared with Linear Regression (LR) and Support Vector Machine (SVM) models to identify the optimal algorithm [18, 19].

Decision Tree. Decision Tree (DT) is a supervised learning algorithm which is characterized by its use of a tree structure to make predictions. In supervised learning, we have a sample set, each with a distinct set of attributes and a classification result [20]. DT builds a tree, where each internal node signifies an attribute judgement. Each branch indicates a judgement result, and each leaf node symbolizes a classification result. The process of building a decision tree involves two main steps: node splitting and threshold determination. Node splitting occurs when an attribute represented by a node cannot deliver a final judgment, requiring the node to split into children. Determining the threshold involves selecting an appropriate value to minimize the classification error rate. Commonly used decision trees include ID3, C4.5, and CART. Owing to CART's superior classification effect, it is the method employed in this study. CART utilizes the Gini coefficient to select division attributes and measures the purity of dataset D with the Gini value:

$$Gini(D) = 1 - \sum_{k=1}^{N} p^2(k) \tag{32.1}$$

where $p(k)$ denotes the probability of drawing the kth category of samples. The Gini coefficient responds to the probability that two samples are randomly drawn from dataset D with inconsistent category markers, so a smaller Gini (D) indicates a higher purity of dataset D.

Adaptive Boost. Adaptive Boost (AB) is a prominent integrated learning technique that transforms a weak learner with prediction accuracy slightly above random guesses into a strong learner with significantly improved prediction accuracy. It increases the weights of the misclassified samples of the previous basic classifier, and it decreases the weights of the correctly classified samples, using this information to train the next basic classifier. A new weak classifier is added in each iteration until a predetermined error rate or a maximum number of iterations is reached [21]. In this study, DT is utilized as the weak learner for Adaboost.

Random Forest. Random Forest (RF) is an algorithm that combines multiple trees using the bagging concept of integrated learning, where decision trees are the basic unit. To classify an input sample, it needs to be input into each tree for classification, and the results are aggregated by voting to form a strong classifier. Random forest introduces two types of randomness to improve performance: randomness in sample selection and randomness in feature selection. This introduction of randomness aids in mitigating overfitting and enhances noise immunity [22].

Extra Trees. Extra Trees (ET) extends the concept of random forest by eliminating the randomness of sample selection and introducing the randomness of tree branching. Instead of using evaluation indicators like the Gini coefficient in dividing samples, ET randomly selects a characteristic, thus augmenting the randomness. ET reduces the variance relative to RF, but it increases the bias. However, under certain conditions, the generalization capability of ET outperforms that of RF [23].

Metrics. In this study, an objective of comparing various algorithms necessitates the utilization of four distinct evaluation metrics, namely mean absolute error (MAE), mean square error (MSE), mean absolute percentage error (MAPE), and coefficient of determination (R^2). These metrics have been selected to quantify and analyze the disparities among the algorithms under investigation. To facilitate a comprehensive evaluation, the formulas corresponding to each of these evaluation metrics are presented in the Table 32.2.

Table 32.2 Formulas and explanation of the evaluation metrics

Metrics	Formula	Explanation
MAE	$MAE = \frac{1}{n}\sum_{i=1}^{n}\lvert y_{true} - y_{pred}\rvert$	The range is [0, + ∞). The predicted value is equal to 0 when it exactly matches the true value, i.e., the perfect model. The larger the error, the larger the value
MSE	$MSE = \frac{1}{n}\sum_{i=1}^{n}(y_{true} - y_{pred})^2$	The range is [0, + ∞). The predicted value is equal to 0 when it exactly matches the true value, i.e., the perfect model. The larger the error, the larger the value
MAPE	$MAPE = \frac{100\%}{n}\sum_{i=1}^{n}\left\lvert\frac{y_{true}-y_{pred}}{y_{true}}\right\rvert$	The range is [0, + ∞). A MAPE of 0% indicates a perfect model, and a MAPE greater than 100% indicates an inferior model
R^2	$R^2 = 1 - \frac{\sum_{i=1}^{n}(y_{true}-y_{pred})^2}{\sum_{i=1}^{n}(y_{true}-\overline{y_{true}})^2}$	The range is (− ∞, 1]. The closer the value of R^2 is to 1, the better the regression curve fits the observed values. Conversely, the smaller the value of R^2 is, the worse the regression curve fits the observed values

32.3 Results and Discussion

32.3.1 Data Visualization and Analysis

As shown in Fig. 32.3, the correlations between each feature and the AC surface flashover voltage are ranked according to their Pearson coefficient.

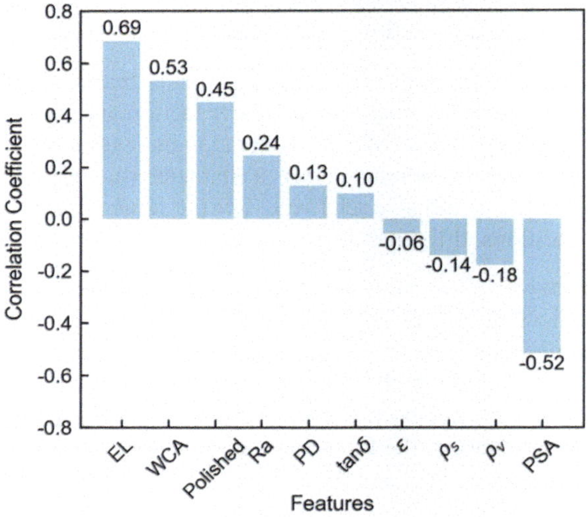

Fig. 32.3 Correlation between the features and the flashover voltage

The absolute values of the correlation coefficients indicate that four features possess the strongest correlation with the AC surface flashover voltage, which are electroluminescence (EL), water contact angle (WCA), percentage of silicon atoms (PSA), and polish status, respectively.

Figure 32.4a, b reveal a positive correlation between both EL and WCA and the AC surface flashover voltage. For EL, most data points lie below the fitted curve before 200 ms^{-1} and after 300 ms^{-1}, while between 200 ms^{-1} and 300 ms^{-1}, most data points are located above the curve. For WCA, a high concentration of points is observed around 100°. Conversely, Fig. 32.4c illustrates that PSA negatively correlates with the AC surface flashover voltage, with a uniform scattering of data around the fitted curve.

To address the binary variable of polish status, the distributions of EL, WCA, and PSA for polished and unpolished samples are plotted separately, as presented in Fig. 32.4d–f. Compared with unpolished samples, the distribution densities of EL and WCA in polished samples shift significantly to the right, while PSA distribution density moves to the left.

EL indicates the local micro-discharge phenomenon on the sample surface [14], and the photon count in the measurements corresponds to the number of shallow traps on the surface. An increased steady-state photon count suggests a higher number of surface shallow traps. It is hypothesized that the process of carrier capture by surface shallow traps reduces the average free range of carriers, thereby obstructing collisional ionization. Consequently, a higher steady-state photon count makes it more challenging for AC surface flashover to occur.

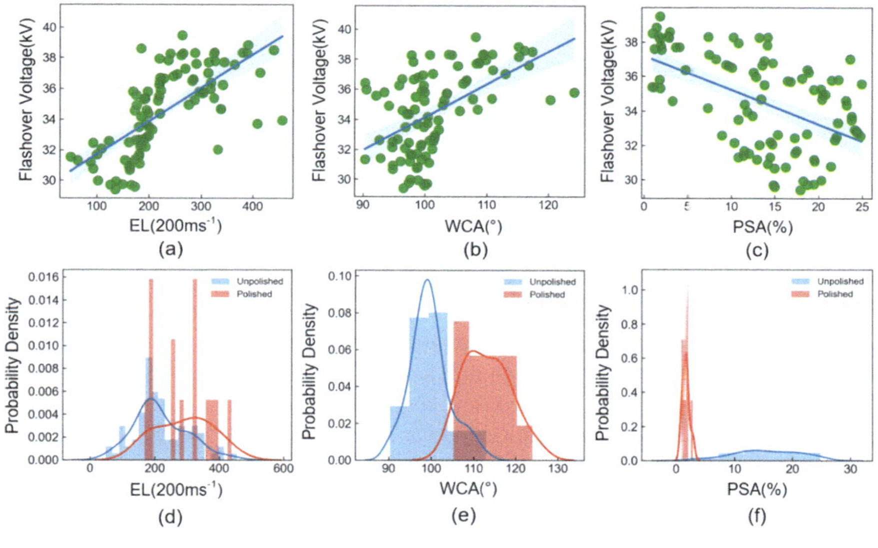

Fig. 32.4 Distribution of the top four features. **a** Flashover voltage of EL; **b** flashover voltage of WCA; **c** flashover voltage of PSA; **d** distribution of EL; **e** distribution of WCA; **f** distribution of PSA

WCA is associated with the hydrophobic of the material. A larger contact angle signifies increased hydrophobicity on the material surface, reducing the likelihood of water vapor forming a path. Hence, fewer water carriers are present on the material surface, making it less prone to violent carrier collisions that lead to AC surface flashover.

PSA reflects the thickness of the silicon release agent layer on the material surface. A higher percentage denotes a thicker release agent layer, which generally has lower electrical strength compared to the EP. As a result, a thicker release agent layer could potentially be a vulnerability for AC surface flashover. This explains the negative correlation of PSA with AC surface flashover voltage.

Changes in distribution presented in Fig. 32.4d–f show that EL and WCA slightly increase after polishing, while PSA increases after the release agent layer is removed by polishing. Consequently, the overall impact is an increase in the AC surface flashover voltage.

32.3.2 Prediction Results

The performance metrics of various machine learning algorithms are depicted in Fig. 32.5. In terms of MAE, MSE, MAPE, and R^2, the data reveals that the overall tree model outperforms both Linear Regression (LR) and Support Vector Machine (SVM). Within the test set, the lowest MAE, MSE, MAPE, and the highest R^2 values are achieved by ET and RF, with the highest R^2 value reaching 0.91. This high score suggests the feasibility of surface flashover voltage prediction based on the tree model.

A further comparison between ET and RF indicates that ET significantly surpasses RF on the training set, suggesting a severe overfitting issue. In contrast, RF's performance on the training and test sets is more balanced, indicating that RF possesses stronger generalization capabilities and has greater practical utility. Principally, while ET employs all samples, RF employs bagging to extract samples, resulting in ET's inferior generalization capabilities compared to RF during the integration process.

The actual and predictive values of the surface flashover voltage for the samples obtained by ET and RF are shown in Figs. 32.6. The maximum single error for ET is found for sample NO. 3, and the error between the predictive and actual values is 0.9 kV, with a relative error of 2.7%. The minimum error is found in the sample NO. 7, with an error value of 0. The maximum single error of RF is found in the sample NO. 3, with an error value of 1.2 kV between the predictive and actual values, with a relative error of 3.6%. The minimum error was found for sample NO. 8, with an error value of 0.1 kV and a relative error of 0.3%. It can be seen that the actual prediction results of both ET and RT reach a high level, and it is feasible to use the calculation of predictive values to replace the actual values for the test acquisition.

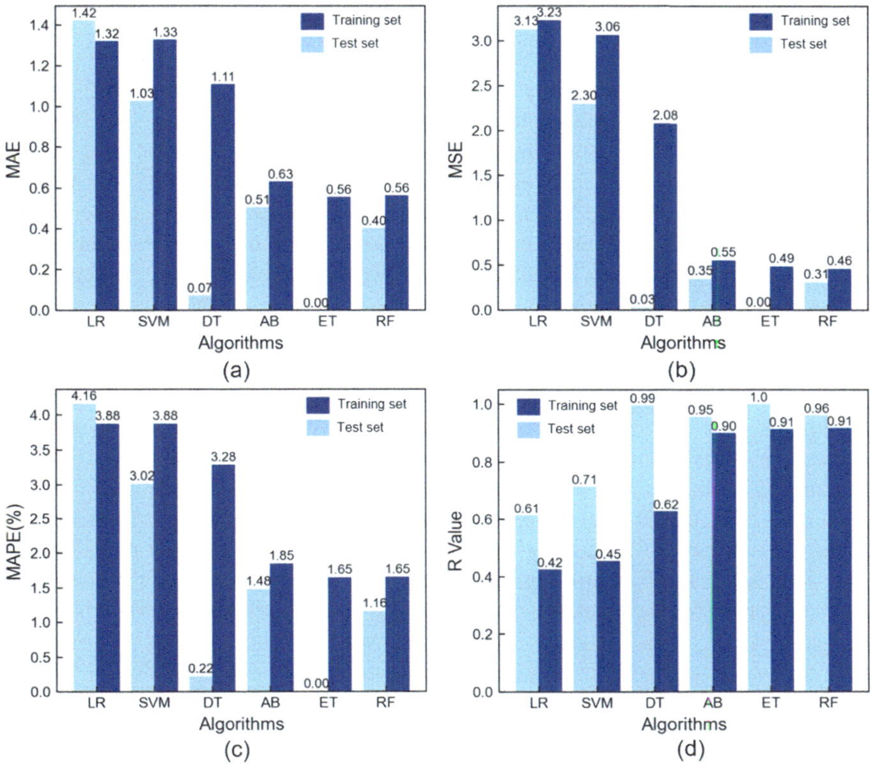

Fig. 32.5 Comparison of different machine learning algorithms for flashover voltage. **a** MAE; **b** MSE; **c** MAPE; **d** R value

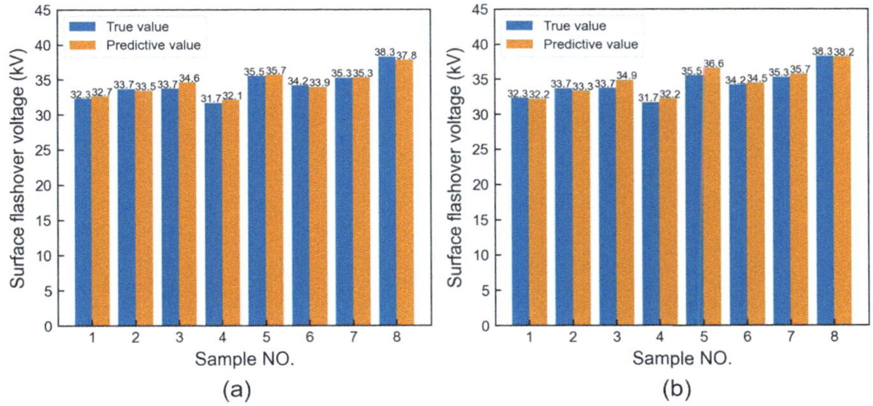

Fig. 32.6 Comparison of true and predicted values using ensemble techniques. **a** ET; **b** RF

32.3.3 Prediction Results with Feature Selection

In practical applications, the ability to use minimum features to predict the AC surface flashover voltage is desirable, as fewer features imply lower testing costs. For the tree-based model, feature importance can be represented by information gain. Lower information gain indicates less critical features. Based on this, a flowchart for feature selection can be constructed, as shown in Fig. 32.7. The crux of this process is to determine whether the feature count is less than or equal to one after completing the test set prediction. If not, the feature with the lowest feature importance is removed, the model is trained and predicted again, and the evaluation score is obtained.

The relationship curve between the score and the number of removed features from the above process is presented in Fig. 32.8. It shows that the MAE, MSE, and MAPE score curves for all algorithms, excluding SVM, display an increasing trend, while the R^2 score curve exhibits a decreasing trend. When the number of removed features is less than four, the evaluation index fluctuates only within a small range, indicating the model's prediction accuracy does not decrease significantly. When the number of removed features exceeds four, the MAE, MSE, and MAPE scores begin to escalate rapidly with the increase of removed features, and the R^2 score curve decreases sharply, implying a significant decrease in the model's prediction accuracy. For SVM, the reduction in the number of features does not significantly

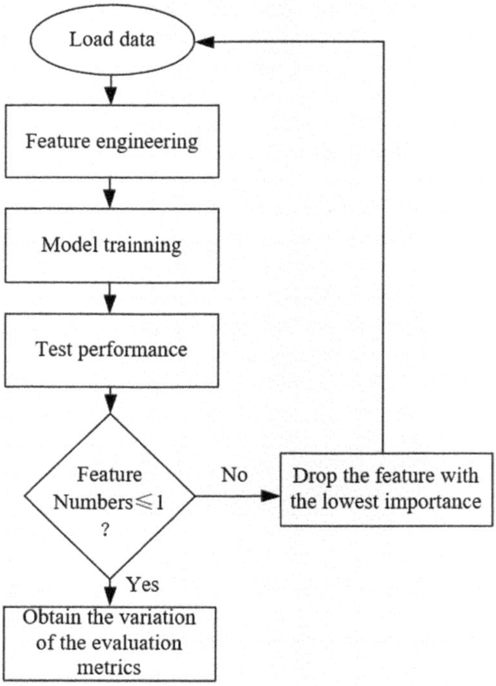

Fig. 32.7 Feature selection process

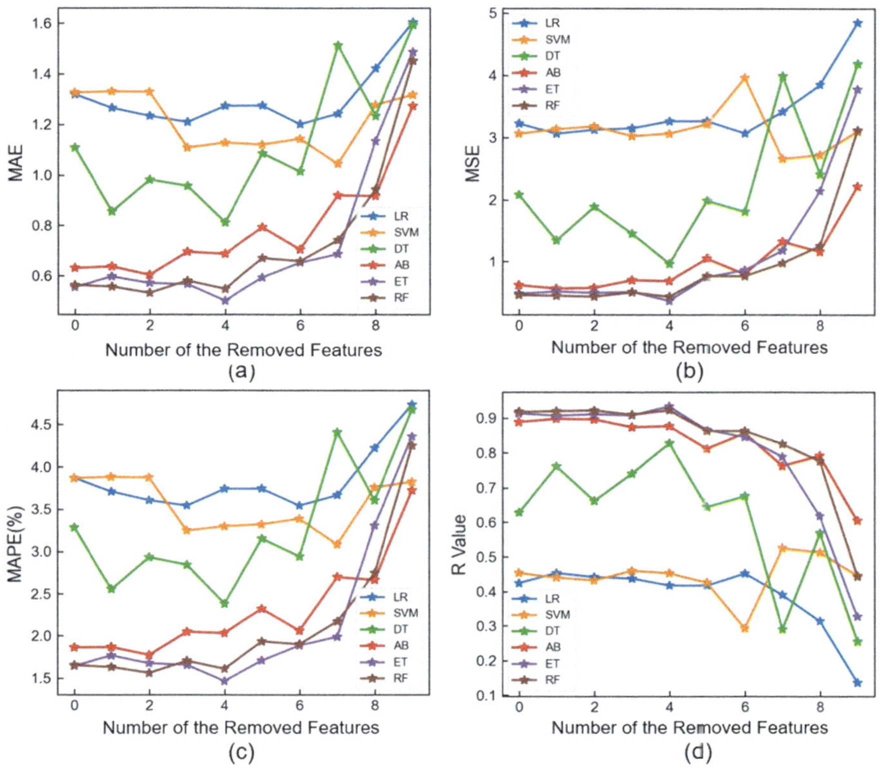

Fig. 32.8 Variation of the evaluation metrics with the change of the number of the removed features. **a** MAE; **b** MSE; **c** MAPE; **d** R value

impact its score, suggesting that SVM maintains certain model accuracy with fewer features and is appropriate for modeling environments with limited data.

Among the various algorithms, LR displays the lowest prediction accuracy due to the absence of features with strong linear correlation with the AC surface flashover voltage, resulting in poor fitting using the linear formula. The three DT-based ensemble methods AB, ET, and RF significantly outperform DT, with ET and RF demonstrating exceptional performance when the number of removed features is less than four. The best prediction performance is exhibited by the ET algorithm with four features removed. At this point, the algorithm's MAE, MSE, and MAPE scores are 0.50, 0.37, and 1.47% respectively, with a decrease of 10.7%, 24.5%, and 10.9% compared to the ET algorithm without feature removal (0.56, 0.49, and 1.65%). The R^2 score stands at 0.93, marking a 2.2% increase from 0.91, indicating an overall improvement in performance.

In light of these findings, it can be inferred that the most optimal predictive solution is to utilize the ET algorithm with a reduced feature set, specifically retaining EL,

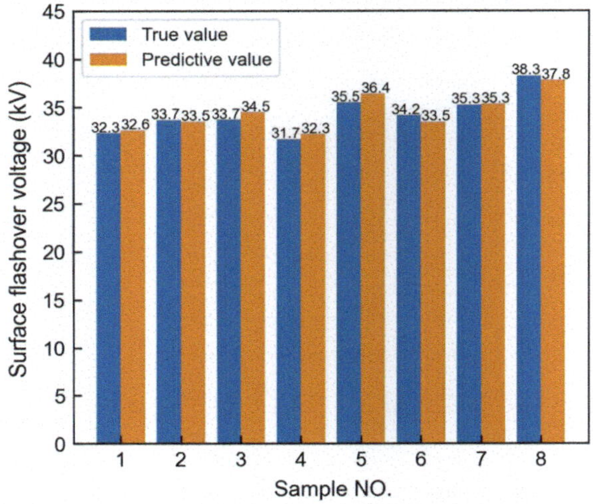

Fig. 32.9 Comparison of true value and predictive value obtained by ET with feature selection

WCA, PSA, Polished status, Ra, and surface resistivity. This combination will most effectively predict the AC surface flashover voltage.

The actual and predictive values of surface flashover voltages for the samples obtained by ET are shown in Fig. 32.9 after removing 4 features and calculating based on 6 features, including EL, WCA, PSA, Polished status, Ra, and surface resistivity. The maximum single error appears in the sample NO. 5, and the error value between the predictive and actual values is 0.9 kV, with a relative error of 2.5%. The minimum error is found in sample NO. 7, with an error value of 0. It is clear that the actual ET-based prediction after feature selection is at a high level, and it is feasible to use the calculation of predicted values instead of the experimental acquisition of actual values. In practical applications, only 6 features, including EL, WCA, PSA, Polished status, Ra, and surface resistivity, need to be obtained for the prediction of surface flashover voltage.

32.4 Conclusions

This study has conducted an analysis and prediction of the AC surface flashover voltage of epoxy resin materials using tree-based models. The following conclusions can be drawn from the results:

1. The prediction model utilizing the ET algorithm showed commendable performance with MAE, MSE, MAPE, and R^2 scores of 0.50, 0.37, 1.47%, and 0.93, respectively. This effectively demonstrates the viability of machine learning in predicting out-comes of destructive tests such as surface flashover voltage based

on non-destructive tests. These findings offer novel insights and potentially cost-effective alternatives for future research pertaining to the destructive testing of insulating materials.

2. Based on both the absolute values of correlation coefficients and the feature importance within the tree-based model, the six predominant features influencing the surface flashover voltage were identified as EL, WCA, PSA, Polished status, Ra, and surface resistivity of the material.
3. The results suggest that enhancements to the AC surface flashover voltage of epoxy resin can be achieved by increasing the EL photon number, increasing the WCA, reducing the PSA on the surface, and employing sanding techniques to remove the surface silicon release agent.

It is important to note that the dataset employed in this study is relatively small. As such, the division of training and test sets, as well as the training of the machine learning model, may have introduced some error. Therefore, the conclusions drawn here necessitate validation with a larger experimental sample size in future studies.

References

1. H.C. Miller, Surface flashover of insulators. IEEE Trans. Electr. Insul. **24**(5), 765–786 (1989)
2. E. Volpov, Dielectric strength coordination and generalized spacer design rules for HVAC/DC SF/sub 6/gas insulated systems. IEEE Trans. Deilectrics Electr. Insul. **11**(6), 949–963 (2004)
3. Y. Yamano, S. Kobayashi, Y. Takahashi, Reduction of surface charge-induced electric field enhancement and increase in ac surface flashover voltage. IEEE Trans. Electr. Insul. **EI-20**(3), 529–536 (1985)
4. T.S. Sudarshan, R.A. Dougal, Mechanisms of surface flashover along solid dielectrics in compressed gases: a review. IEEE Trans. Electr. Insul. **E1–21**(5), 727–746 (1986)
5. J.R. Laghari, A.H. Qureshi, Surface flashover of spacers in compressed gas insulated systems. IEEE Trans. Electr. Insul. **E1–16**(5), 373–387 (1981)
6. M.T. Gencoglu, M. Uyar, Prediction of flashover voltage of insulators using least squares support vector machines. Exp. Sys. Appl. **36**(7), 10789–10798 (2009)
7. S.A. Bessedik, H. Hadi, Prediction of flashover voltage of insulators using least squares support vector machine with particle swarm optimization. Electr. Power Sys. Res. **104**, 87–92 (2013)
8. A. Mahdjoubi, B. Zegnini, M. Belkheiri, T. Seghier, Fixed least squares support vector machines for flashover modelling of outdoor insulators. Electr. Power Sys. Res. **173**, 29–37 (2019)
9. Arshad et al., Forecasting flashover parameters of polymeric insulators under contaminated conditions using the machine learning technique. Energies **13**(15), 3889 (2020)
10. L. Cui, M. Ramesh, Prediction of flashover voltage using electric field measurement on clean and polluted insulators. Int. J. Electr. Power Ene. Sys. **116**, 105574 (2020)
11. M.T. Gençoğlu, M. Uyar, Prediction of flashover voltage of insulators using least squares support vector machines. Exp. Sys. Appl. **36**(7), 10789–10798 (2009)
12. C. Li et al., Novel HVDC spacers by adaptively controlling surface charges–part II: experiment. IEEE Trans. Deilectrics Electr. Insul. **25**(4), 1248–1258 (2018)
13. C. Li et al., Surface charge migration and dc surface flashover of surface-modified epoxy-based insulators. J. Phys. D Appl. Phys. **50**(6), 065301 (2017)
14. C. Lin et al., Luminescence reveals micro discharge as a potential triggering factor for surface flashover. J. Phys. D Appl. Phys. **53**(44), 445103 (2020)

15. V. Griseri, L.A. Dissadol, J.C. Fothergill, Photoluminescence, recombination induced luminescence and electroluminescence in epoxy resin. J. Phys. D: Appl. Phys. **34**(16), 2534–2540 (2001)
16. V. Griseri et al., Electroluminescence excitation mechanisms in an epoxy resin under divergent and uniform field. Proc. IEEE Int. Conf. Solid Dielectrics **9**(1), 150–160 (2002)
17. D. Mary, D. Malec, Electroluminescence measurements to detect accumulated charge at the electrode-insulator interface. IEEE Trans. Deilectrics Electr. Insul. **8**(5), 771–775 (2001)
18. R. Jiang, Y. Zheng, Series arc fault detection using regular signals and time-series reconstruction. IEEE Trans. Ind. Electron. **70**(2), 2026–2036 (2023)
19. W.N. Venables, B.D. Ripley, Tree-Based Methods. Modern Applied Statistics with S-Plus (1999), pp. 303–327
20. J.R. Quinlan, Induction of decision trees. Mach. Learn., pp. 81–106 (1986)
21. C. Ying, M. Qi-Guang, L. Jia-Chen, G. Lin, Advance and prospects of AdaBoost algorithm. Acta Automatica Sinica **39**(6), 745–758 (2013)
22. L. Breiman, Random forests. Mach. Learn. **45**, 5–32 (2001)
23. O. Maier et al., Extra tree forests for sub-acute ischemic stroke lesion segmentation in MR sequences. J. Neurosci. Meth. **240**, 89–100 (2015)

Chapter 33
Active Noise Control of Power Transformer Based on Time Series Forecasting

Chuanjie Lin, Longyang Zhu, Xiaosi Lin, Yuwei Chen, Hao Chen, Hui Chen, and Yuesheng Zheng

Abstract As urbanization accelerates, substations have become a part of residential and downtown areas. The low-frequency noise generated by transformers has increasingly become a major source of complaints from neighboring residents. Due to the long wavelength of low-frequency noise, traditional passive noise reduction technology is bulky, expensive and limited in effectiveness. Active noise control (ANC) has gained attention due to its compact size, light weight, low cost, and excellent ability to reduce low-frequency noise. However, the precision of ANC relies on the control algorithm, and conventional least mean square-based algorithms still face challenges such as sluggish convergence, susceptibility to dispersion, and poor stability. In light of the time-varying characteristics of transformer noise, this paper proposes a machine learning-based algorithm for generating inverse noise waveform through time-series forecasting, and this approach aims to achieve rapid and stable noise reduction. The experimental result shows that the algorithm achieves noise reduction effects of 49.67 dB, and the time used for training and prediction is 1.78 s and 1.64 s, which shows a promising solution for transformer noise reduction.

C. Lin · L. Zhu · Y. Chen · H. Chen · H. Chen
Fuzhou Power Supply Company, State Grid Fujian Electric Power Co., Ltd., Fuzhou 350009, China

X. Lin
State Grid Fujian Technical Training Centre, State Grid Fujian Electric Power Co., Ltd., Quanzhou 362000, China

Y. Zheng (✉)
College of Electrical Engineering and Automation, Fuzhou University, Fuzhou 350108, China
e-mail: yueshengzheng@fzu.edu.cn

Z. Fang et al. (eds.), *Proceedings of the 5th International Symposium on Plasma and Energy Conversion*, Springer Proceedings in Physics 398,
https://doi.org/10.1007/978-981-97-2245-7_33

33.1 Introduction

The power industry serves as a cornerstone of national economy. However, the issues of noise pollution have persisted alongside the nation's ongoing economic development. In recent years, urban electricity demand has surged, leading to a gradual rise in the number of large transformers within cities. This increase is exacerbated by the expanding urban landscape and the necessity for urban grid transformation. Consequently, some substations are constructed near residential areas, significantly disrupting the daily lives and work routines of residents. Therefore, urgent actions are required to address transformer noise, and active noise control (ANC) stands out as one of the promising solutions [1, 2].

Compared with traditional passive noise reduction technology, the main advantages of ANC technology are as follows. Firstly, it is suitable for controlling low-frequency and medium-frequency noise. Additionally, the control system is characterized by its real-time responsiveness, lightweight construction, and compact size. Furthermore, it can adapt its strategies based on the specific characteristics of the noise it aims to control, resulting in a clearer noise control objective. This adaptability enhances its value for engineering applications. With the rapid development of digital signal processor technology, digital filtering technology is widely used in the field of noise reduction. Currently, the prevailing trend in noise reduction technology is the development of digital noise control systems, with the core focus on digital signal processors and their associated adaptive control algorithms [3, 4].

ANC falls under the category of noise reduction methods applied within the noise propagation path. In the field of ANC, many countries have conducted research and obtained certain results. However, certain limitations are still present, including challenges related to core winding and system stability [5–7].

The control algorithm stands at the core of any control system, drawing significant attention from researchers. There are many kinds of adaptive algorithms, and the appropriate selection of these adaptive algorithms directly impacts the feasibility and performance of the adaptive control system. Currently, the Least Mean Square (LMS) algorithm and its improved algorithms are commonly used as adaptive control algorithms. The LMS algorithm, proposed by Hoff and Widrow in 1960, is one of the most widely used adaptive control algorithms and belongs to linear adaptive algorithms. There is no integral, average and square operations in the algorithm, rendering it both simple and efficient. However, in time-varying systems, the algorithm will be unstable, slow to converge, and even divergent [8].

To address the limitations of the LMS algorithm, researchers have proposed several improved algorithms based on the LMS algorithm, such as Normalized Least Mean Square (NLMS), Block Least Mean Square (BLMS) and Filter-x Least Mean Square (FxLMS). Compared to the LMS, the NLMS exhibits a moderate increase in computational complexity, but it offers significant improvements in algorithm performance, notably enhancing convergence speed. However, the algorithm is likely to exhibit divergence when the estimated gradient is disturbed by a large amount of noise. The BLMS algorithm possesses similar performance to the LMS algorithm.

The gradient vector estimation of the BLMS algorithm is more accurate. However, this algorithm suffers from the conflicting problems of low steady state error and high convergence speed. The FxLMS algorithm is derived from the LMS algorithm, but takes into account the influence of the secondary channels on the control system in real systems. In order to compensate for the influence of the secondary channel, the reference signal must be filtered by the secondary channel identification, and the secondary channel identification has a direct influence on the anechoic effect and the stability of the system. If the phase error between the reference signal and the secondary channel identification is greater than 90°, the algorithm may diverge. In summary, the existing active noise control algorithms are mainly based on mathematical iterative calculations, and it is difficult to solve the problems of algorithm convergence speed and dispersion. Therefore, a new control algorithm is urgently required to achieve the tracking of noise waveforms [9].

This paper introduces an active noise feed-forward control algorithm based on waveform prediction, derived from the mechanism and characteristics of noise generation in power transformers. The basic principles of time series prediction algorithms and their application in active noise reduction systems are introduced. The impact of the presented algorithm on noise reduction performance in transformers is studied through simulations and theoretical analysis. Furthermore, a software system is developed using LabVIEW virtual instrument technology. When integrated with hardware equipment, an active noise control system based on adaptive theory is designed and implemented. The experimental result shows that the algorithm achieves noise reduction effects of 49.67 dB, and the time used for training and prediction is 1.78 and 1.64 s.

33.2 Experimental Setup

33.2.1 Transformer Noise Analysis

The noise is generated by a 220 kV main transformer SFSZ11-240,000/220 with an operating life of 11 years in the downtown area of Fuzhou City, China, as shown in Fig. 33.1. The time-domain image of the noise amplitude is shown in Fig. 33.2, and the noise amplitude is up to 104.42 dB. The frequency-domain amplitude of the noise is shown in Fig. 33.3, and it mainly distributes in the low-frequency band and concentrates on the integer multiple of the low frequency of 50 Hz.

33.2.2 Structure of Active Noise Control System

In this paper, a feed-forward active noise reduction system is used, and the structure is shown in Fig. 33.4. This type of system takes the ANC algorithm as the core,

Fig. 33.1 Site photograph of the transformer that is the object of noise collection

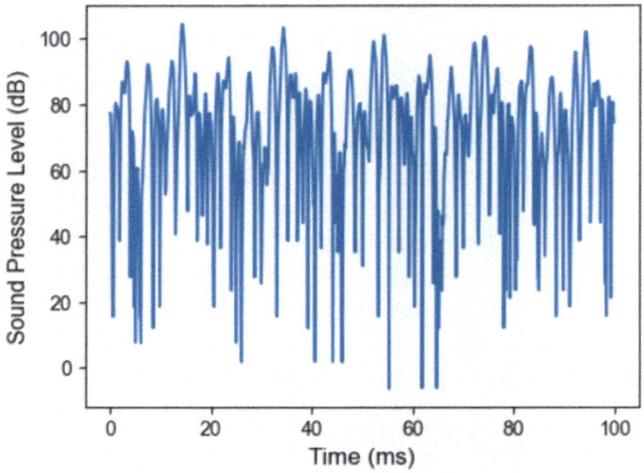

Fig. 33.2 Time domain image of the noise

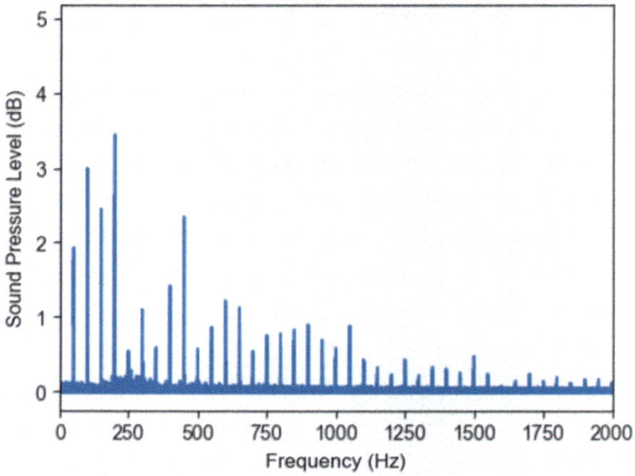

Fig. 33.3 Frequency domain image of the noise

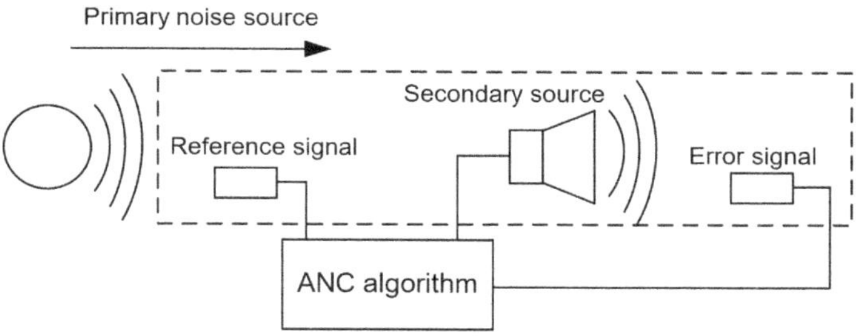

Fig. 33.4 Structure of feed-forward active noise reduction system

through the acquisition of the reference signal, to generate the reversed-phase sound wave to offset the primary noise in the secondary source. The acquisition of the error signal is used as further feedback to adjust the output sound wave of the secondary source for the effect of noise reduction.

33.3 Time Series Forecasting Algorithms

A time series is a series of data generated sequentially at fixed intervals, where data at later moments usually correlate with data at earlier moments. The time series is created after correcting transformer noise through the reference microphone. Essentially, the time series reflects the trend of transformer noise over time. The crux of time series prediction lies in extracting patterns from this data, enabling the use of collected transformer noise data to forecast future noise levels [10].

33.3.1 Time Series Decomposition and Noise Reduction Filtering

This paper combines time series decomposition and filtering processing to process the noise signal, and the logic flow is shown in Fig. 33.5. The transformer noise signal is simplified by the noise reduction processing of the trend component and the direct removal of the random component, which can achieve a reduction in the complexity of the algorithm and improve the signal quality while increasing the computing speed.

For the time series $\{x_t\}$, the sequence decomposition method is to decompose the transformer noise S_0 into a trend component S_1, a periodic component S_2, and a random component S_3 as follows:

$$S_0 = S_1 + S_2 + S_3. \tag{33.1}$$

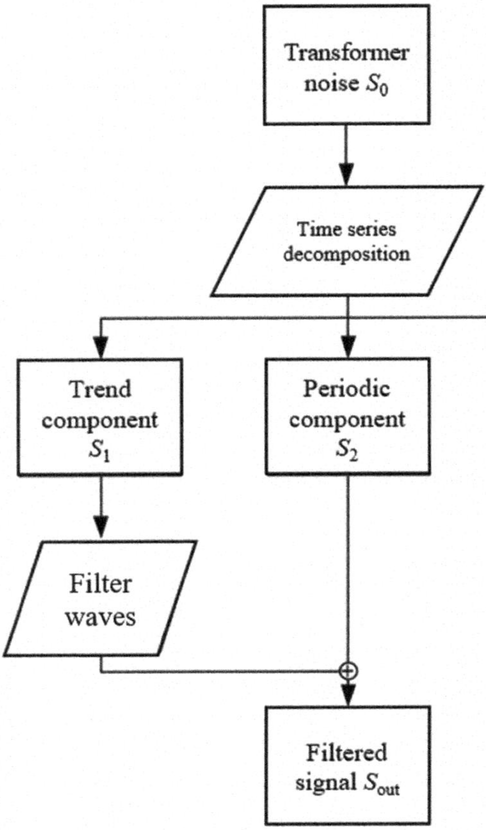

Fig. 33.5 Time series noise reduction and filtering process

The decomposition of S_1 is extracted using the centered moving average method, which is calculated as follows:

$$S_1 = \frac{0.5x_{t-\frac{f}{2}} + x_{t-\frac{f}{2}+1} + x_{t-\frac{f}{2}+2} + \cdots + x_{t+\frac{f}{2}-1} + 0.5x_{t+\frac{f}{2}}}{f} \tag{33.2}$$

where f is the periodic frequency of the time series, which is 50 Hz for power-transformer noise. The decomposition of S_2 is calculated using the average of the values at the same frequency as follows:

$$x'_t = x_t - S_1 \tag{33.3}$$

$$S_2 = \sum_{i=0}^{n} x'_i / f \tag{33.4}$$

where n is obtained by dividing the actual sequence length N by the cycle frequency f and then rounding:

$$n = [N/f] \tag{33.5}$$

The decomposition of S_3 comes directly from S_0 minus S_1 and S_2:

$$S_3 = S_0 - S_1 - S_2. \tag{33.6}$$

For the filtering of the trend component S_1, the common methods, median filtering, mean filtering, weighted filtering, Kalman filtering, etc., can be used to eliminate the spark part of the trend signal, making the trend signal smooth.

The transformer noise signal S_0 collected in the field is decomposed into the trend component S_1, the periodic component S_2 and the random component S_3 as shown in Fig. 33.6. The median filter is used to reduce the noise of S_1, and the random component S_3 is removed to obtain the input signal required for machine learning. A comparison of the signal before and after processing is shown in Fig. 33.7. It can be seen that the re-synthesized signal after noise reduction and filtering is significantly more stable than the original signal, which is more suitable for the model construction of machine learning.

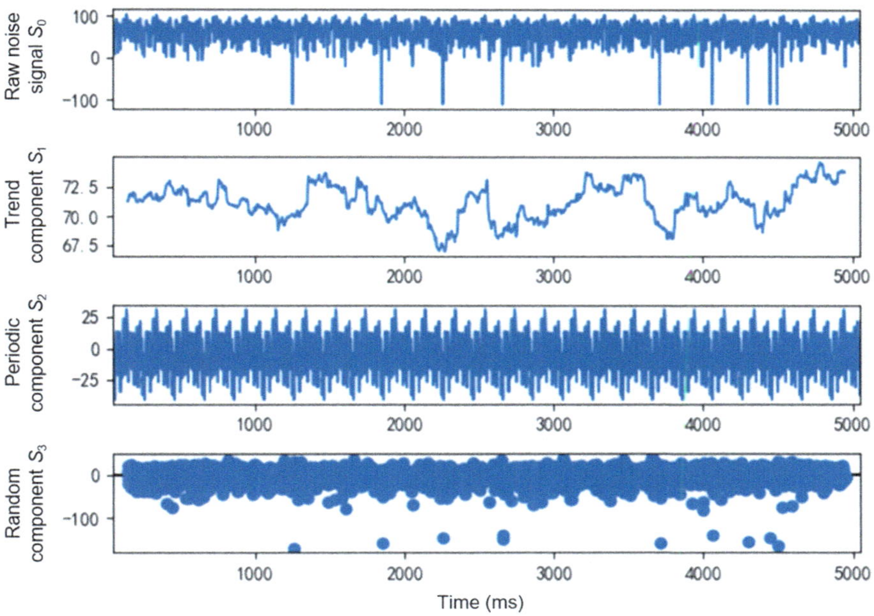

Fig. 33.6 Transformer noise decomposition

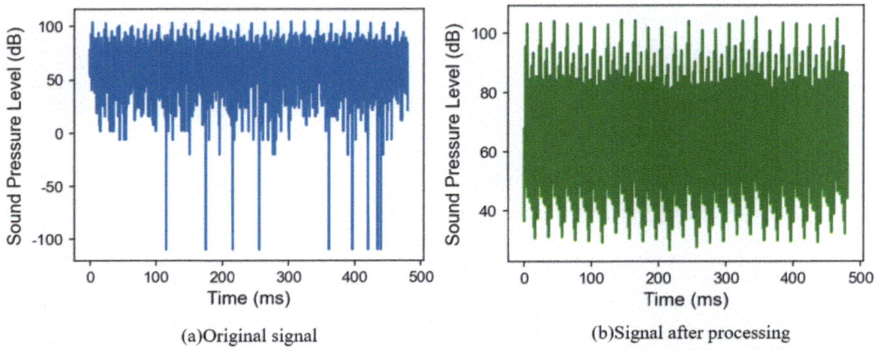

Fig. 33.7 Comparison of transformer noise before and after noise reduction and filtering

33.3.2 Time Series Prediction Model Construction

The prediction process using a time series prediction model for transformer noise is as follows (Fig. 33.8).

The time series prediction model requires four parameters: hyperparameter, training sample, test sample, and sample to be predicted. In this paper, two machine learning methods, a tree model and a LSTM model, are used for the prediction of noise waveforms. The hyperparameters of the tree model include the number of trees, the depth of the tree, the minimum number of samples of the division node, the minimum number of samples of the result node, etc. The hyperparameters of the LSTM model include the number of neurons, the depth of the network, and the learning rate, etc.

The flowchart is divided into four main parts: the grid search part, the training part, the testing part and the prediction part. These four parts work in the following way: firstly, the grid search and cross validation are performed to obtain the optimal combination of hyperparameters by giving smaller values of the initial hyperparameters and the training samples. Secondly, the optimal hyperparameter combinations are used as inputs and combined with the extreme random tree algorithm to train the training samples into a time series prediction model. Then the accuracy of the model is verified by inputting test samples. If the model does not reach the required accuracy, the process of the first two parts is repeated, and the operation is repeated until the model reaches the required accuracy. Finally, we only need to input the samples to be predicted, and then we can achieve the prediction of the predicted samples and get the prediction results through the trained time series prediction model. In this paper, only the first four parts of the flowchart are carried out to complete the training of the time series prediction model.

The model training samples are derived from actual noise acquisition with a sampling frequency of 10 kHz and a sampling time of 2 h. For the tree model, its prediction process is to map from input X to output y, and there is no long and short-term memory mechanism similar to LSTM, so it is necessary to shift and splice the

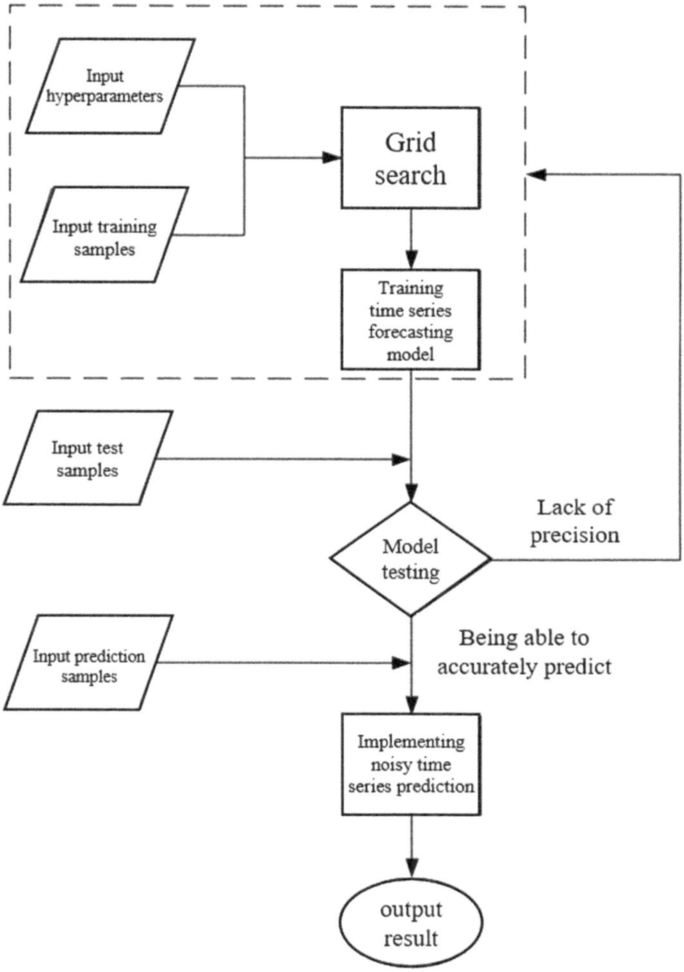

Fig. 33.8 Flowchart of timing prediction model construction

collected noise sequences. As shown in Fig. 33.9, the original sequence is shifted by 1, 2, …, n steps to form a matrix, and each row of the matrix forms the input value an of the model, which is calculated by the model algorithm to obtain the predicted value y_n at the next moment. The specific number of shifting steps n needs to be selected by cross validation.

The selection of the shift amount is based on the basic decision tree (DT). The training set constructed with different shift amounts is scored, and its R2 score results are shown in Fig. 33.10. The higher the value is, the more accurate the prediction will be. From the results, it can be seen that when the shift amount is 60, the prediction

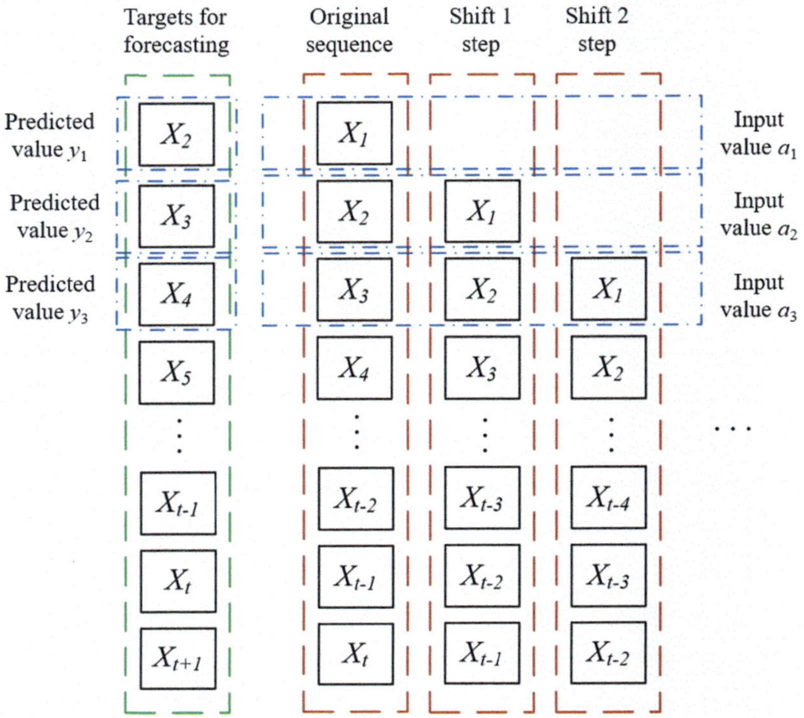

Fig. 33.9 Schematic diagram of sequence shift splicing

effect is optimal, so this shift amount value is chosen as the feature generation parameter.

Fig. 33.10 DT prediction scores for different shift amounts

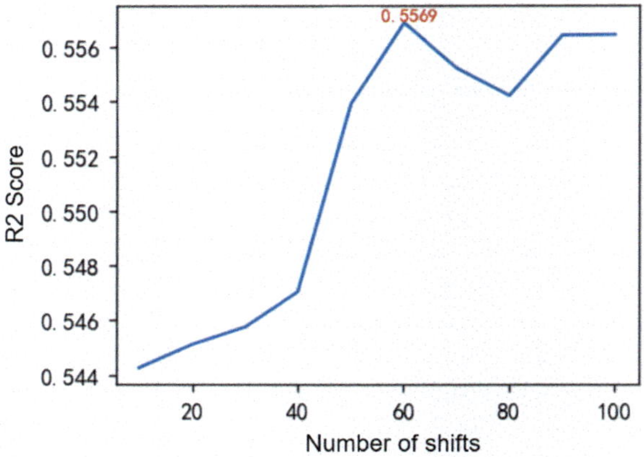

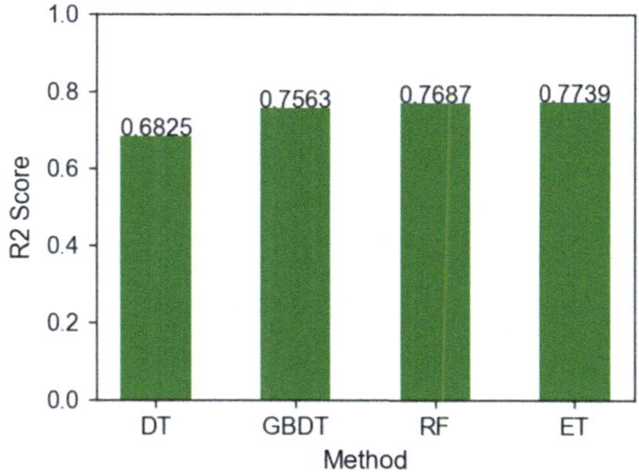

Fig. 33.11 Predicted scores of different tree model methods

In terms of dividing the training test set, the sampling frequency of the training set is 10 kHz, and the sampling time is 1 h and 50 min, totaling 66,000 k data points. The sampling frequency of the test set is 10 kHz, and the related time length is 10 min, totaling 6000 k data points.

Time series forecasting based on tree model. The original decision tree (DT), gradient boosting decision tree (GBDT), random forest (RF), and extra tree (ET) are the four main tree-based models used for time series forecasting. In the model establishing process, the optimal hyperparameters are firstly obtained by grid search and cross validation. The optimal hyperparameters are used to train the model, and the R2 scores are shown in Fig. 33.11. It can be seen that the ET has the highest prediction accuracy, and the R2 score can reach 0.7739.

The comparison between the predicted value and the real value is shown in Fig. 33.12. It can be seen that the yellow prediction curves of all four algorithms basically cover the blue curve, with better prediction results. This is because the accuracy of the R2 score of the algorithms exceeds 0.68 (i.e., the prediction accuracy exceeds 68%). It can be seen that the noise waveform prediction through the tree model is feasible and effective.

LSTM-based time series forecasting. Tree model-based time series prediction requires human adjustment and screening of parameters due to the large range of hyperparameters. LSTM is a special type of neural network, which tends to have only one layer, so the most critical parameter is the number of neurons. The number of neurons is 128, 64, 32 and 16, and the results are shown in Fig. 33.13. It can be seen that as the number of neurons decreases, the prediction accuracy first increases and then decreases, reaching a maximum value of 0.7978 at 64 neurons.

The comparison between the predicted value and the real value is shown in Fig. 33.14. It can be seen that the yellow prediction curves of all four algorithms overlap with the blue curves to such an extent that it is difficult for the naked eye to distinguish the differences. Since LSTM has a capability of long and short-term memory, it may show better results for longer periods of time other than 2 h.

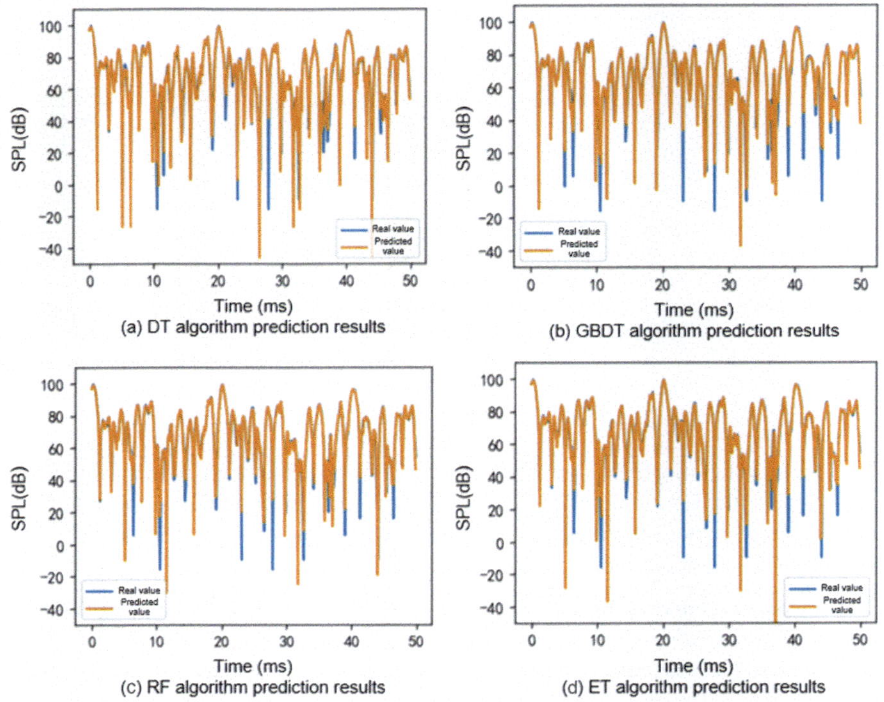

Fig. 33.12 Predictions from different tree model methods

Fig. 33.13 Scores predicted by LSTM with different number of neurons

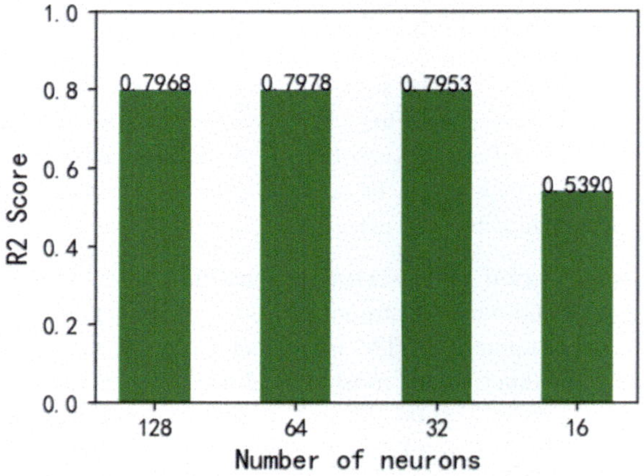

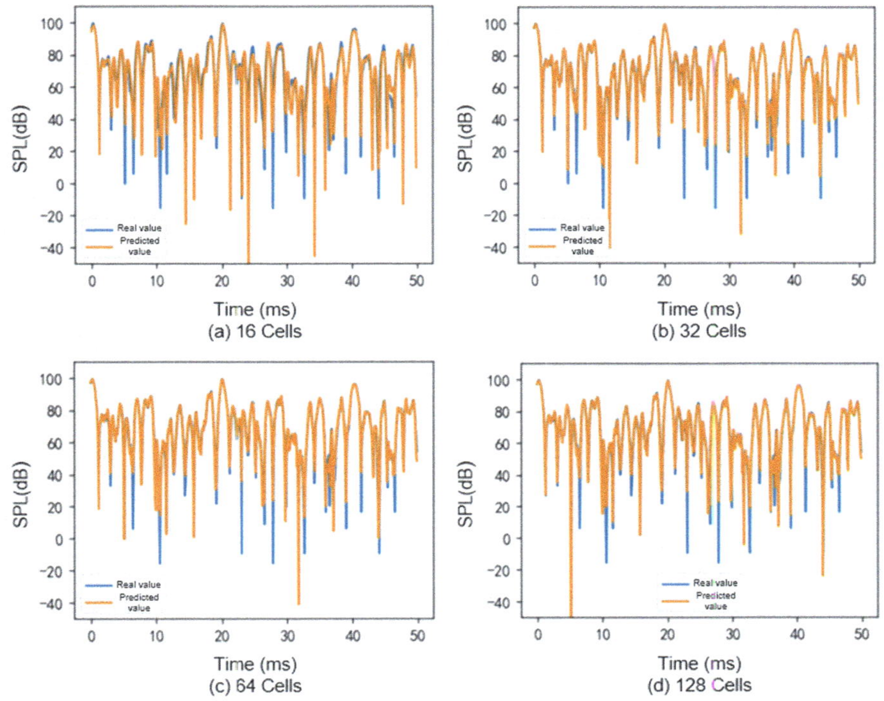

Fig. 33.14 Prediction of LSTM with different number of neurons

The prediction effect, time consumed and theoretical optimal noise reduction capacity of all the above methods are compared and shown in Table 33.1.

Considering the time consumption and noise reduction effect, the GBDT, which has better effect and fastest computation speed among the tree models, is chosen as the core algorithm for time series prediction.

Table 33.1 Performance comparison of different time series prediction methods

Method	R2 score	Training time	Predicting time	Suppression ability (dB)
DT	0.6799	10.2 s	1.03 s	44.78
GBDT	0.7563	1.78 s	1.64 s	49.67
RF	0.7689	10 min 41 s	1 min 2 s	51.96
ET	0.7747	4 min 19 s	1 min 22 s	52.59
LSTM-128 Cells	0.7930	40 min 9 s	2 min 23 s	52.67
LSTM-64 Cells	0.7954	46 min 50 s	1 min 52 s	52.33
LSTM-32 Cells	0.7904	16 min 46 s	3 min 3 s	52.81
LSTM-16 Cells	0.4718	6 min 54 s	3 min 13 s	31.71

33.4 Active Noise Control System Based on Time Series Forecasting

In this paper, on the basis of the mechanism of power transformer noise generation and its characteristics, an active noise feed-forward control algorithm based on time series prediction waveform is proposed, and a set of active noise control system based on adaptive theory is designed and developed for reducing low-frequency noise of power transformers.

33.4.1 Software System Design

The active noise control software system is mainly composed of three modules: a data acquisition module, an active noise control module and a data output module. The software system design diagram is shown in Fig. 33.15.

Data Acquisition Module. This module can realize real-time acquisition and storage of transformer noise signals.

LabVIEW can drive the NI data acquisition card according to the set sampling frequency. The microphone is connected through the analogue input channel to collect the noise signals and store the noise data.

Active Noise Control Module. The main body of the module is to program the time series prediction algorithm described in the previous section, which calculates the secondary sound source signal based on the acquired reference and error signals.

Data Output Module. This module can generate the inverse acoustic wave to eliminate the original noise.

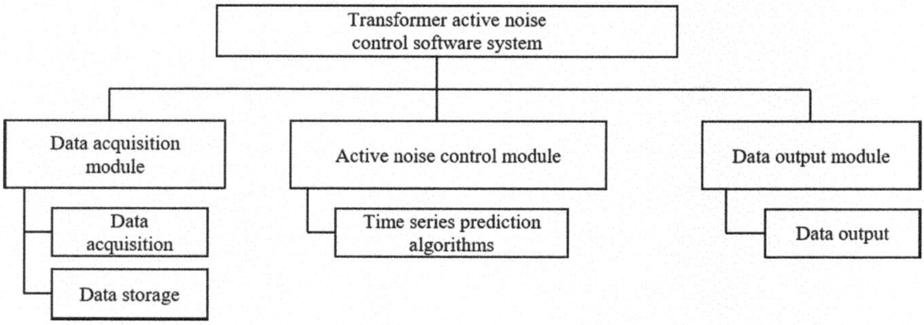

Fig. 33.15 Software system design diagram

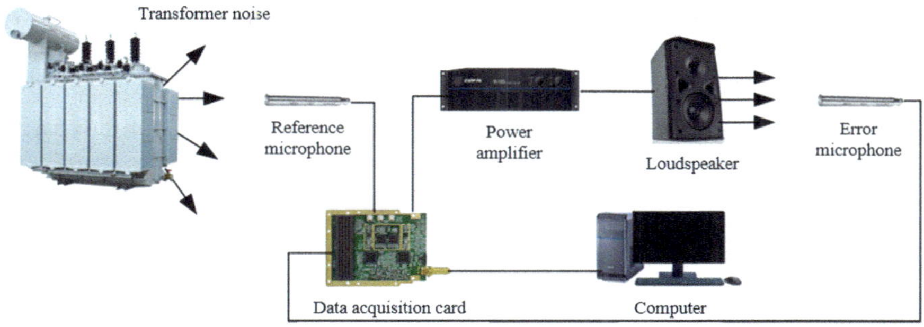

Fig. 33.16 Hardware system design diagram

33.4.2 Hardware System Design

In Fig. 33.16, the active noise control hardware includes a reference microphone, an error microphone, an NI data acquisition card, a computer, a power amplifier, and a loudspeaker. Among them, the reference microphone collects the reference signal, and the error microphone collects the error signal. The NI acquisition card inputs the collected signals into the computer, and then the software system processes the signals and calculates the secondary source signals by the time series prediction algorithm. Finally, the NI acquisition card drives the power amplifier and the loudspeaker to emit the secondary source signal for achieving noise reduction.

Microphone and Preamplifier. The AWA14423 microphone is manufactured by Hangzhou Aihua Instrument Company and has a nominal sensitivity of 50 mV/Pa, an upper measurement limit of 140 dB, and a frequency range from 10 to 20,000 Hz. The AWA14604 preamplifier is manufactured by the same company. Its frequency range is 10 Hz–50 kHz, and the gain is − 0.3 dB.

NI Data Acquisition Card. This experiment uses the NI USB-6211 data acquisition card, which has 16 analogue input channels and 2 analogue output channels. The highest single-channel sampling rate is 250 kS/s, and the voltage resolution is 16 bits.

Power Amplifier and Loudspeaker. The employed HC-720 power amplifier has an output power of 35 W and 50 W, respectively, and a frequency response range from 40 Hz to 20 kHz. The VX5-C loudspeaker has an output frequency range from 40 Hz to 20 kHz.

33.5 Conclusion

According to the mechanism and characteristics of noise generation in power transformers, this study proposes an active noise feed-forward control algorithm based on waveform prediction. The GBDT algorithm achieves superior noise reduction performance and faster computation speed, with a theoretical optimal noise reduction capability of 49.67 dB. Furthermore, an active noise control system based on adaptive theory is designed and developed using this algorithm. This system enables real-time acquisition, storage, and output of noise signals. Hardware devices, whose sound pressure level is larger than 120 dB, were selected to accommodate the specific characteristics of transformer noise.

References

1. S.C. Zhong, L.H. Zhu, Q.C. Wang et al., Analysis of power transformer vibration noise and its active noise control. J. Electrotechnol. **37**(S1), 11–21 (2022)
2. D.S. Liu, Z.S. Li, P. Li et al., A review on the research of transformer vibration characteristic analysis and its noise reduction technology. Transformer **60**(7), 18–27 (2023)
3. S.C. Ji, Y.H. Shi, F. Zhang et al., Research status and prospect of power transformer vibration and noise and its control measures. High Voltage Electr. Apparatus **55**(11), 1–17 (2019)
4. W.D. Xu, L.F. Wang, J. Wang et al., A review of research on power transformer noise analysis and control technology. Guangdong Elect. Power **33**(8), 86–94 (2020)
5. C.L. Li, G.Q. Liu, H.Y. Chen et al., Research on active noise control method of extra-high voltage transformer based on dipole equivalent sound source. J. Electrotechnol. **33**(1), 221–226 (2018)
6. X.H. Yan, P.C. Cha, G.Q. Liu et al., Study on the optimal placement of secondary sound sources for dual-channel active noise control. High Voltage Electr. Apparatus **56**(3), 150–155 (2020)
7. J.Y. Wang, B.D. Bai, H.L. Liu et al., Effect of DC bias on transformer vibration noise. J. Electrotechnol. **30**(8), 56–61 (2015)
8. T. Zhao, J.B. Liang, X. Xia et al., Active noise control system for power transformer based on LMS adaptive filtering algorithm. High Voltage Technol. **42**(7), 2299–2307 (2016)
9. T. Zhao, J.B. Liang, H. Liu, et al., Variable structure FXLMS adaptive transformer active noise control algorithm based on multi-frequency trap filtering and signal power variable parameter. Chin. J. Electr. Eng. **37**(22), 6719–6729+6785 (2017)
10. R. Jiang, Y. Zheng, Series arc fault detection using regular signals and time-series reconstruction. IEEE Trans. Industr. Electron. **70**(2), 2026–2036 (2023)

Chapter 34
High-Temperature Energy Storage Dielectric with Double-Layer Deposition Structure

Yu Feng, Weiye Cheng, Hao Yang, and Dong Yue

Abstract The lower energy density and decreasing insulation performance at high temperatures of energy storage polymer dielectric limit their application in military and civilian fields such as electromagnetic weapons and new energy vehicles. In order to further improve the breakdown strength and energy density at high temperatures, silica (SiO_2) and boron nitride (BN) with wide band gaps were selected as the inorganic layers to in-situ grow and construct a double-layer barrier layer on the surface of polycarbonate (PC) films. The total deposition thickness with optimal performance was determined by adjusting the single-layer SiO_2 thickness. Subsequently, BN was introduced to construct a heterojunction structure on this basis to further improve breakdown and energy storage performance. Achieving a higher energy density of 8.57 J/cm^3 at 780 kV/mm and room temperature in PC composite dielectric, and an excellent energy density of 4.94 J/cm^3 at 600 kV/mm and 150 °C. The PC composite dielectric with heterojunction structures can effectively improve breakdown and energy storage performance by constructing an internal reverse electric field. This work provides more optimization pathways for research and development on high-temperature energy storage dielectric.

Y. Feng · W. Cheng · H. Yang · D. Yue (✉)
The Key Laboratory of Engineering Dielectrics and Its Application, Ministry of Education, Harbin University of Science and Technology, Harbin 150080, China
e-mail: yuedong@hrbust.edu.cn

The School of Electrical and Electronic Engineering, Harbin University of Science and Technology, Harbin 150080, China

Y. Feng
e-mail: fengyu@hrbust.edu.cn

Z. Fang et al. (eds.), *Proceedings of the 5th International Symposium on Plasma and Energy Conversion*, Springer Proceedings in Physics 398,
https://doi.org/10.1007/978-981-97-2245-7_34

34.1 Introduction

Electricity, as the key to a low-carbon economy, is assuming the role of energy source for more and more devices, and the large-scale application of new energy is the foreseeable future [1–4]. Capacitors as electromagnetic equipment, new energy generation and other areas of the core devices, generally divided into ceramic capacitors and polymer capacitors, although ceramic capacitors are used in large quantities and many areas because of high energy storage density (U_e), [5] but the polymer film capacitors have unique high flexibility and high operating voltage, that ceramic capacitors do not have, and the future miniaturization of electrical equipment and high efficiency have an important impact, so in recent years it got more and more attention from researchers [6–8]. However, polymer film capacitors still have many difficulties that need to be solved, the most important of which is that most film capacitors on the market have a low temperature tolerance [9, 10]. For example, the most widely used biaxially oriented polypropylene film (BOPP) has an operating temperature of up to 105 °C [11, 12]. Applications in high ambient temperatures, such as inverters for new energy vehicles and a variety of pulsed-power devices, require additional cooling devices. In addition, the U_e of BOPP is only 1–2 J/cm^3, the low energy density leads to a large volume and weight of the device, which is a greater limitation on the use in limited space [10].

Existing high temperature resistant polymers such as polycarbonate (PC, $T_g \approx 150$ °C), polyeterimide (PEI, $T_g \approx 217$ °C), fluorinepolyester (FPE, $T_g \approx 330$ °C) etc. [13, 14] have high temperature resistance because of aromatic functional groups or ether-bonded structures in main chain contains, but there are still some drawbacks that need to be overcome. On the one hand the conductivity of the material cannot be neglected in relation to the concentration of carriers and their transport, and high temperatures lead to an increase in leakage currents, resulting in a serious degradation of the energy storage efficiency (η) and breakdown strength (E_b). According to Schottky emission theory (Schottky emission) and Thermionic-field emission theory (Thermionic-field emission), in a high temperature environment, electrons in the vicinity of the metal electrodes tend to gain enough energy to become excited, thus crossing the potential barrier to become free electrons to increase the leakage current density, resulting in an increase in the loss of the film and leading to thermal runaway [15–17]. On the other hand, due to the presence of benzene rings in the main chain of the high temperature resistant polymer molecules, their not easily rotatable characteristics can hinder the generation of steering polarization, resulting in low dielectric constants (ε_r) of energy storage dielectric applied in high temperature fields, which limits the enhancement of U_e as well as its application in practice [18, 19]. Considering that conventional polymers usually cannot satisfy the simultaneous enhancement of dielectric constant ε_r and E_b, in addition to searching for materials with better temperature tolerance, certain measures must be taken to suppress the synergistic enhancement of carrier injection and augment ε_r together at high temperatures, and to optimize energy storage performance of the polymer at high temperatures [7, 20, 21].

In order to suppress the increase of leakage current of polymers at high temperatures, wide-band inorganics are widely used to prepare various fillers and coatings, and their higher band gap width and stable chemical properties have a significant effect on enhancing the overall barrier of the composite dielectric [22–24]. In recent years, in order to overcome the inherently low U_e of high-temperature-resistant polymers and the severe degradation of E_b at high temperatures and high field strengths, the most popular means adopted by researchers is to add coatings containing broad-banded inorganic nanoparticles or nanofibers on the surface of polymer substrates or to construct inorganic barrier layers on the surface of the substrates in a variety of ways to mitigate the degradation of the E_b at high temperatures. For example, Wang et al. [25] constructed montmorillonite (MMT) coatings by making polyamideimide films (PAI) impregnated with MMT nanosheet solution to improve the high-temperature electrical insulation and energy storage properties of the composite films, and achieved a E_b of 551 kV/mm, a high U_e of 2.9 J/cm^3, and a charging/discharging efficiency > 80% at 150 °C. Using simulations, Thomas et al. [26] confirmed the significant effect of 2D hexagonal boron nitride (*h*-BN) nanocoatings on the blocking effect of carriers at high temperatures and on the improvement of the E_b of thin films by surface potential attenuation and dielectric breakdown measurements, and elucidated the related breakdown mechanisms. In addition, Zhang et al. [27] also grew SiO_2 barrier layers on both sides of PEI by magnetron sputtering, which effectively reduced the conductivity loss and achieved 2.96 J/cm^3 and more than 90% charging/discharging efficiency at 150 °C. The above studies show that by introducing BN, SiO_2, and other wide bandgap inorganic materials with similar ε_r to the polymer matrix on the surface of the polymer matrix in recent years has been one of the mainstay ways to inhibit carrier injection and internal development at high temperatures and to optimize the overall dielectric properties of the dielectrics by using the substrate-barrier interfacial polarization at the same time [28–30]. In addition, multilayer structures have been shown to be effective in enhancing the energy storage performance of thin films [31]. Not only can more interfaces be created between multilayers to increase the extra polarization strength, but heterojunctions are formed when two inorganic materials with different bandgap structures are combined together. Through the presence of different energy band structures, a built-in electric field is formed during electron transport, which has a significant hindering effect on the development of breakdown paths [32, 33].

Plasma-enhanced chemical vapor deposition (PECVD) is an emerging method applied to the preparation of composite films in recent years. Compared with traditional multilayer structure preparation processes such as electrostatic spinning and hot pressing, PECVD can be used to directly prepare the target deposition layer on the polymer surface by in-situ growth in a room temperature environment. Since the deposition occurs directly on the surface of the polymer substrate, it solves the problem of poor densification and the existence of pores between layers in the traditional hot-pressing transfer preparation method. In this work, PC was selected as the matrix, and BN and SiO_2 barrier layers were sequentially grown on both sides of the matrix by PECVD to construct a heterojunction structure, and the inhibitory effects of different thickness ratios of the barrier layers on the carrier injection and transport processes at high temperatures were investigated. The results show that under the condition of depositing total thickness of 150 nm, the two inorganic layers (BN/SiO_2) deposited at 50 nm and 100 nm respectively have the best enhancement on the performance of the composite dielectric, the highest U_e and E_b reached 8.57 J/cm^3 and 780 kV/mm at room temperature, which is 35% higher than pure PC. Even at 150°C, the highest U_e and E_b reached 4.94 J/cm^3 and 600 kV/mm, which is 26.6% higher than that of pure PC, effectively improving the E_b of composite film at high temperature. The composite films prepared in this work exhibit excellent high-temperature tolerance along with good energy storage properties, which provides more references for the subsequent structural design of dielectric films at high temperatures.

34.2 Experimental Materials and Test Methods

34.2.1 Materials and Instruments

PC, PolyK Technology Co., Ltd., $T_g = 180$ °C. Dimethylaminomethylborane (DMAB), Shanghai Aladdin Reagent Co., Ltd. Ethyl orthosilicate (TEOS), Shanghai Aladdin Reagent Co., Ltd. Tetrahydrofuran (THF), Shanghai Aladdin Reagent Co., Ltd.

34.2.2 Preparation of Composite Films

The composite films were prepared as shown in Fig. 34.8. An appropriate amount of PC particles was taken in a beaker, and THF solvent was added in the ratio of 1.8 g:10 ml, sealed and stirred at room temperature for 10 h until the PC was well dispersed in THF solvent and continued to stand for 2 h to remove air bubbles from the mixed solution. The mixed solution was uniformly coated on the glass plate using a coating spatula and then the wet film was transferred to a coating oven and kept at 60 °C for 8 h to fully evaporate the solvent, followed by a vacuum drying oven at 80 °C for 10 h to obtain the desired PC film. Finally, different thicknesses of BN and SiO_2 inorganic layers were deposited in the PECVD tube furnace according to different deposition time requirements to obtain the corresponding composite films. The schematic diagram of the PECVD equipment is shown in Fig. 34.8.

34.2.3 Testing Instruments

X-ray physical characterization was carried out using an EMPYREAN Sharp Shadow X-ray Diffractometer (XRD) manufactured by Panacor, Netherlands. The surface characterization of the composite films was carried out using a SU8020 scanning electron microscope (SEM) manufactured by Hitachi, Japan. Elemental characterization was performed using X-ray photoelectron spectroscopy (XPS) with Helium light source irradiation and simultaneous application of a − 5V bias test. The test was carried out by Axis Ultra^{DLD} X-ray photoelectron spectrometer, KRATOS, UK. Alpha-A broadband dielectric spectrum analyzer is used, the electrode is 9 mm diameter aluminum electrode. E_b and U_e measuring instrument for the United States Radiant Premier II company, measuring the unilateral hysteresis return line and unilateral current density curve, electrode for the 3 mm diameter aluminum electrode, breakdown test boost rate of 300 V/s, energy storage test each round of boosting 20 V/μm. The Ultraviolet ray (UV) diffuse reflectance pattern was obtained by UV-3600 PLUS UV–Vis spectrophotometer. Ultravioletphotoelectron spectroscopy (UPS) test was carried out by UV photoelectron spectrometer, ThermoFisher ESCALAB 250Xi model.

34.3 Results and Discussion

In order to study the effect of different heterojunction thickness ratios on the performance of composite films, the optimal total thickness of the PC matrix film for plasma deposition must be investigated [34], so the first step of the experiment is to use only SiO_2 as the deposition material to investigate the effect of different deposition thicknesses of plasma on the surface of composite films during the deposition process. As shown in Fig. 34.9, in order to characterize the presence of SiO_2 in the composite film, the composite film was first tested by XPS. The characteristic peaks of the three elements (Si, C, O) can be seen in the XPS spectrum, which proves the existence of SiO_2 barrier layer in the composite film. In order to investigate the effect on the composite films at different deposition thicknesses, the changes in the surface morphology and thickness of the deposited layers were observed using SEM, as shown in Fig. 34.1a–d. By observing the electron microscope cross-section of the composite film, it can be found that with the increase of the deposition time, thickness of the inorganic layer on the layer surface is gradually increasing, which indicates that the thickness of the deposited layer is directly proportional to the deposition time. In addition, by observing the electron microscope surface map of the composite film, it is clearly observed that the number of pores on the surface of the composite film is increasing. In Fig. 34.1a_1, a_2, it can be seen that the deposited layer on the surface of the composite film is relatively uniform and does not show defective areas that can be clearly observed. As the deposition thickness continues to increase reaching about 75 nm, more pronounced pores appear in Fig. 34.1b_1, b_2. When the deposition thickness reaches 220 nm or more, as shown in Fig. 34.1c_1, c_2, obvious defects appear in the deposited layer on the surface of the composite film. This indicates that as the deposition thickness increases, the deposition layer will inevitably lead to gradually larger defects, even the stripping of part of the deposition layer as shown in Fig. 34.1d_1, d_2. This is also confirmed in the AFM image (Fig. 34.1a_3–d_3), the

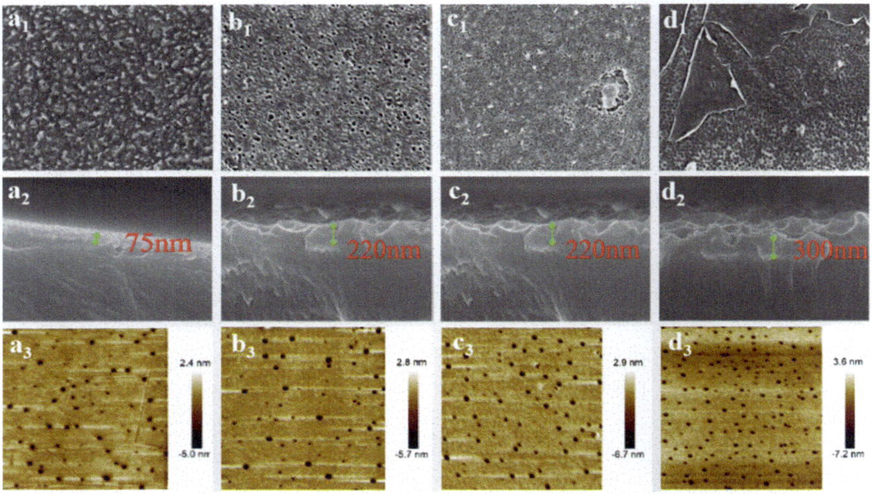

Fig. 34.1 SEM and AFM images of composite films with different deposition times, $\mathbf{a}_1$–$\mathbf{a}_3$ 75 nm; $\mathbf{b}_1$–$\mathbf{b}_3$ 150 nm; $\mathbf{c}_1$–$\mathbf{c}_3$ 220 nm; $\mathbf{d}_1$–$\mathbf{d}_3$ 300 nm

depth of the pores on the surface of the composite film gradually increases from the deepest 5 nm when the thickness of the deposited layer is about 75 nm to the deepest 7.2 nm when the thickness of the deposited layer is about 300 nm, which indicates that the densification of the composite film is deteriorating although the thickness of the deposited layer becomes thicker with the longer deposition time, and therefore, it is not possible to purely pursue the thickness of the deposited layer and also consider the performance degradation caused by the deterioration of densification and find the optimal point.

In order to more accurately investigate the effects of different deposition thicknesses on the composite films, the E_b and U_e were tested at room temperature, 100 °C and 150 °C, respectively, and the results were obtained as shown in Fig. 34.2. In Fig. 34.2a, it can be found that with the increase of the deposition thickness, the E_b at different temperatures shows a trend of increasing and then decreasing, in which the E_b of 710 kV/mm at room temperature and 564 kV/mm at 150 °C are obtained when the deposition thickness reaches 150 nm, which show the optimal insulating properties. However, as the deposition thickness continues to increase, the E_b deteriorates and even eventually falls below that of pure PC film (575 kV/mm). In addition, as shown in Fig. 34.2b, U_e and η also gained enhancement along with the increase of the deposition thickness, and the composite film deposited with 150 nm SiO_2 reached 7.17 J/cm^3 and 95% at room temperature, and it also maintained a higher performance of 4.32 J/cm^3 and 89.2% at 150 °C, which was the same as the breakdown trend. When the thickness of deposited SiO_2 exceeds 150 nm, the energy storage performance of the composite film starts to decrease. This trend may be partly due to the fact that the increase in the thickness of the surface SiO_2 deposition layer with the increase in deposition time leads to an increase in

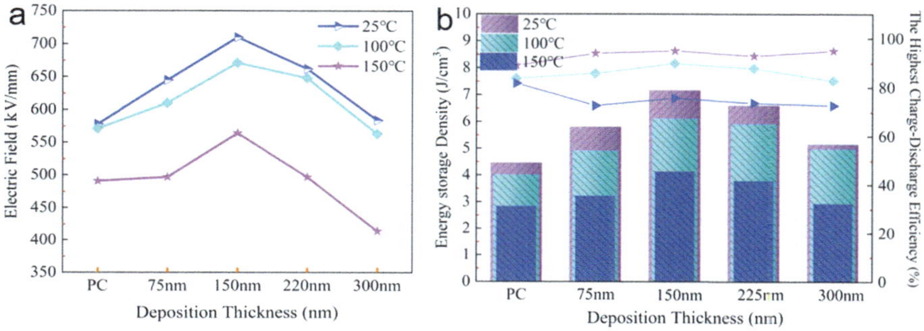

Fig. 34.2 **a** Breakdown strength at different sedimentation thicknesses; **b** energy storage density and efficiency at different sedimentation thicknesses

the overall potential barrier of the composite film, which enhances the inhibition effect on carrier injection. However, at the same time, with the prolongation of the deposition time precursor in the high-frequency electric field generated ions are also constantly impacting the surface of the composite film, resulting in an increase in the number of surface pores and defect depth expansion, so that the degradation of composite film energy storage properties of the rate is increasing [35]. The deposition of 150 nm can be considered to be the deposition layer for the composite film to bring the carrier suppression effect and the defects caused by plasma to reach a balance.

At room temperature when the deposition thickness reaches 300 nm, the U_e and η of the composite film is still better than that of the pure PC film, although the degradation degree is great compared with the deposition of 150 nm, whereas the energy storage performance of the composite film deposited for 60 min at 150 °C is seriously lower than that of the pure PC film, in particular, the charging and discharging efficiency is only about 70%. This suggests that at high temperatures, surface defects can exacerbate the effect on the overall deterioration of the composite film, and that defects caused by excessively long deposition times result in the composite film being more easily injected into the interior of the composite film at high temperatures, making the composite film more susceptible to breakdown [36]. Due to the fact that the conductive loss dominates the dielectric loss at high temperatures, the leakage current density due to the electric field inhomogeneity increases the conductive loss dramatically, which affects the charging and discharging efficiency [37]. Tests have demonstrated that the deposition of 150 nm is the optimal deposition duration for PC films, and that other structural adjustments are needed to further improve the breakdown strength and energy storage properties of the composite films under this total deposition duration. The bilayer heterostructure, due to the different bandgaps of different inorganic materials, will show energy band bending at the interface, forming an electric field pointing from the wide bandgap inorganic layer to the narrower bandgap inorganic layer, which has an inhibitory effect on the internal transport of carriers in the composite film due to the direction opposite to the applied electric field [38, 39]. In this work, BN and SiO_2 are used as different inorganic layers to construct heterojunctions, and the effects of different deposition thicknesses on the composite films are investigated under the total thickness of 150 nm deposition.

The XPS images of PC films after deposition of BN and SiO_2, respectively, are shown in Fig. 34.10, and the elemental characteristic peaks of Si, B, N, and O can be seen in the test maps, indicating the presence of BN and SiO_2 deposited layers. In order to more visually indicate the presence of a double deposition layer, the changes in the surface and cross-section of the composite film for different deposition thicknesses of the deposition layer were observed using SEM, as shown in Fig. 34.3. In the surface diagram, it can be seen that the composite film deposition surface is more uniform, and there is no obvious defects and inorganic layer peeling phenomenon. The cross-section diagram shows that the growth rate of BN and SiO_2 inorganic layer is close to the growth rate per unit time, and the interlayer fits neatly and densely, and the thickness is more uniform.

The dielectric properties of composite films with different thicknesses of BN and SiO_2 deposited sequentially at different temperatures are shown in Fig. 34.4. On the whole, the dielectric constant increases with the increase of SiO_2 deposition thickness, which can be attributed to the fact that SiO_2 dielectric constant is slightly higher than that of PC [40]. In the case of the same total thickness of the two substances deposited by the composite film, the larger the relative thickness of SiO_2 in the overall deposition layer of the composite film, the higher the overall dielectric constant of the composite film rises. And there is a small increase in the dielectric constant at high temperature of 150 °C compared to that at room temperature, which is due to the fact that the high temperature environment causes an intensification of the thermal motion of the carriers, which makes the steering polarization easier to establish [41].

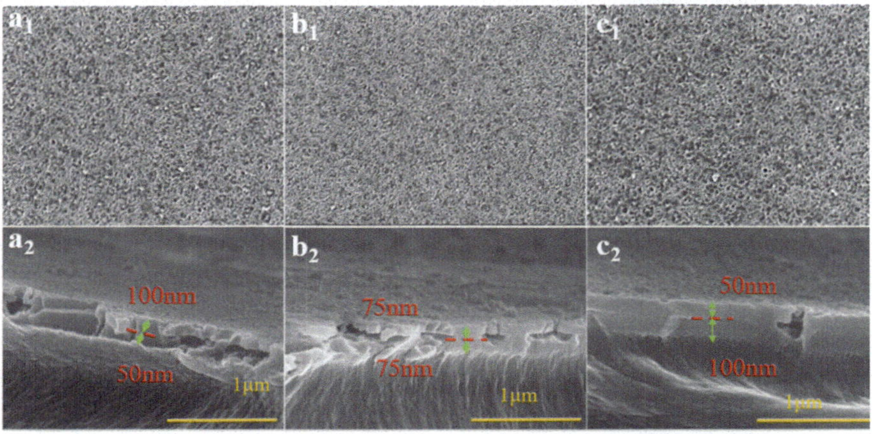

Fig. 34.3 Surface and cross-sectional SEM maps of composite films with different deposition times of BN-SiO_2; $\mathbf{a}_1$–$\mathbf{a}_2$ BN50 nm-$SiO_2$100 nm; $\mathbf{b}_1$–$\mathbf{b}_2$ BN75 nm-$SiO_2$75 nm; $\mathbf{c}_1$–$\mathbf{c}_2$ BN100 nm-$SiO_2$50 nm

It is worth noting that at the low frequency band of 150 °C, some samples showed an abnormal increase in dielectric constant, which may be caused by the space charge not being released in time, but it does not affect the film properties at high frequencies. In addition, it can be found that the dielectric loss is lower when depositing a single layer of SiO_2 or BN compared to depositing a double layer of inorganic material. This aspect may be due to the fact that more interfaces are formed during bi-layer deposition, resulting in more interfacial losses [42]. For the same deposition time, there is also a slight increase in dielectric loss with increasing temperature, which is again an increase in the value of loss due to increased thermal movement of molecules as a result of increasing temperature.

The E_b intuitively represents insulating ability of the dielectric in the electric field, and its performance directly affects the scope of application of the dielectric [43]. Increasing the E_b is an important part of the insulation dielectric development process. The Weibull plots of composite films deposited with different thicknesses of BN and SiO_2 at different temperatures are shown in Fig. 34.5a–c. The breakdown properties of different deposition layer thicknesses at different temperatures are summarized in Table 34.1.

The E_b of the bilayer structure is significantly better than that of the monolayer structure within a certain thickness ratio, which is attributed to the heterojunction structure formed at the interface of the two inorganics due to the difference in the height of the potential barriers between BN and SiO_2, constituting a built-in electric field. Where the SiO_2 bandgap is more ambassadorial the direction of the built-in electric field is the opposite direction of the applied electric field, and the presence of the built-in electric field under the applied electric field plays a certain degree of inhibition of the carrier injection, so that the breakdown field strength increases [43, 44]. It is worth noting that, with the gradual decrease in the proportion of SiO_2 in the deposition layer, the E_b is also decreasing, even lower than the single-layer SiO_2 deposition layer, which indicates that the heterojunction structure for the inhibition of carrier injection has a certain limit, the total thickness of the deposition is unchanged, when the increase in the thickness of the BN deposition, which inevitably results in the reduction of the thickness of the SiO_2 deposition to make the composite film of the overall bandgap width is reduced. This can be assumed that the increase in the thickness of the BN deposition layer does not allow the more effective built-in electric field established by the heterojunction to provide more suppression effects for the

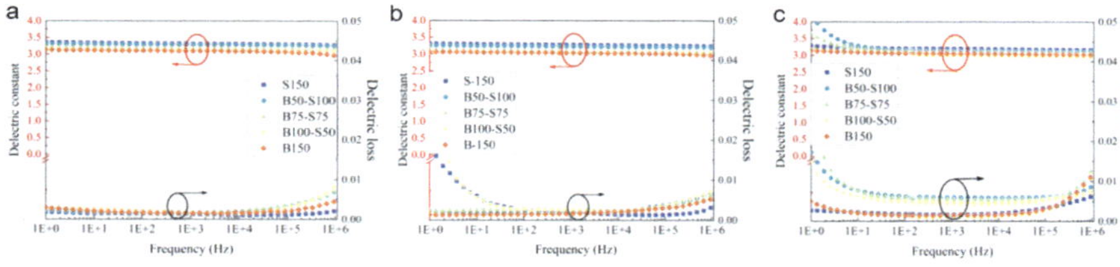

Fig. 34.4 The dielectric properties of composite films with different thicknesses of BN and SiO_2 **a** Room temperature; **b** 100 °C; **c** 150 °C

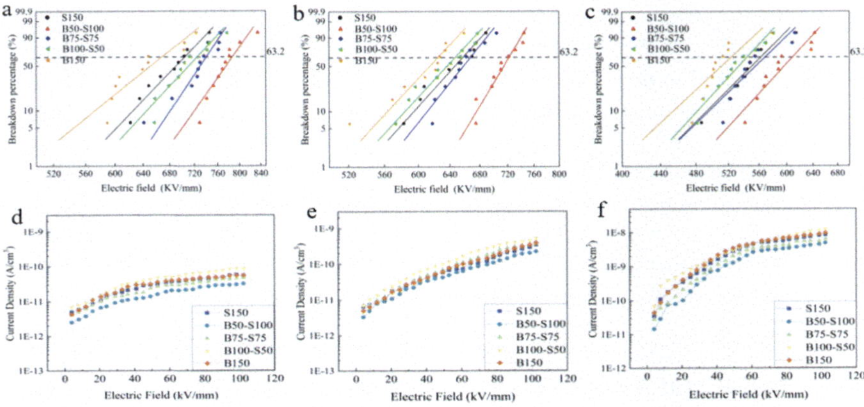

Fig. 34.5 **a–c** The Weibull plots of composite films deposited with different thicknesses of BN and SiO_2 at different temperatures; **a** Room temperature; **b** 100 °C; **c** 150 °C; **d–f** The leakage current performance curves of composite films deposited with different thicknesses of BN and SiO_2 at different temperatures; **d** room temperature; **e** 100 °C; **f** 150 °C

Table 34.1 Comparison of breakdown strength of different deposition layer thicknesses at different temperatures

Thicknesse (nm)	Temperature (°C)		
	25 (kV/mm)	100 (kV/mm)	150 (kV/mm)
S150	700	660	560
B50-S100	775	720	600
B75-S75	750	665	566
B100-S50	725	650	550
B150	675	625	525

carriers, while the corresponding decrease in the thickness of the SiO_2 deposition leads to weakening the overall barrier height of the composite film. The built-in electric field from the heterojunction structure between the thicker BN deposition layer and the thinner SiO_2 deposition layer cannot compensate for the decrease in the electron injection barrier due to the decrease in the thickness of the SiO_2 deposition layer, allowing for an increase in the number of carriers injected. The E_b reached the highest at the deposition of 50 nm BN and 100 nm SiO_2. In addition, with the increase of temperature, the E_b of the composite film is also gradually reduced, which is related to the thermal excitation of carriers, the high temperature makes the molecular thermal movement intensified to obtain more energy, which is more likely to cross the electrode-interface barriers into free electrons to increase the leakage current density [43].

The magnitude of the leakage current density of the dielectric film has a critical impact on the performance of energy storage films operating at high temperatures and high electric field environments. Larger leakage current densities can lead to localized temperature increases in the film, which can lead to breakdown of the composite film at lower electric fields. By performing leakage current test on the composite films, the leakage current performance curves were obtained as shown in Fig. 34.5d–f. In Fig. 34.5d, it can be seen that under the total deposition time of 150 nm, the leakage current density shows a trend of decreasing and then increasing with the increase of the thickness of the deposited BN, and reaches the lowest at the deposition of 50 nm BN and 100 nm SiO_2, which is the same trend as that of the Weibull plot. In addition, it can be seen in Fig. 34.5f that the leakage current densities of the composite films all increase as the temperature increases up to 150 °C, which is again related to the thermal excitation of the carriers.

In order to represent the energy storage performance of the composite film more intuitively, the storage property tests at different temperatures were carried out, as shown in Fig. 34.6. Overall, the changes in the energy storage properties of the composite films are consistent with the trends of E_b and leakage current density, both of which increase and then decrease with the increase in BN deposition time. The best storage performance was obtained when 50 nm BN and 100 nm SiO_2 were deposited, U_e of 8.57 J/cm^3 and η of 95.8% were achieved at 780 kV/mm and room temperature. Even in a 600 kV/mm electric field at 150 °C, the U_e of 4.94 J/cm^3 and η of 89.5% were obtained. It is worth noting that the bi-layer deposition structure has a significant enhancement effect on the energy storage performance of the composite film, compared with the single-layer deposition of 150 nm SiO_2, and the enhancement of the U_e by deposition of 50 nm BN and 100 nm SiO_2 reaches 19.5% and 14.3% at ambient temperature and 150 °C, respectively. This is mainly due to the carrier suppression effect of the heterojunction structure formed by the bilayer deposition that enhances the breakdown field strength, and the additional increase in the interface also optimizes the ε_r to a certain extent, which synergistically enhances the U_e.

The enhancement of η by the double-layer deposition structure is also very obvious. Since the heterojunction structure effectively hinders the carrier injection of the composite film under high electric field, it reduces the conductivity loss, so that the composite films containing the double-layer deposition structure maintain more than 93% η at ambient temperatures, and most of the components are better than those of the pure PC films. And as the temperature rises to 150 °C, the η decreases to about 90%, which is mainly due to the intensification of molecular thermal motion at high temperatures, which makes it easier for carriers to jump to higher energy levels, and the increase in leakage current density causes the conductivity loss to rise [45].

The bilayer deposited structured film in this experiment was able to withstand the highest E_b at high temperatures and obtained the maximum U_e, in order to illustrate the excellent energy storage properties obtained in this work, the maximum electric field as well as the energy storage density were compared with the current research work as shown in Table 34.2.

The above experiments show that by preparing wide bandgap inorganic layers and heterojunction structures on polymer films can effectively enhance the U_e by improving the E_b of the composite films. Especially in high-temperature environments, the synergistic action can effectively inhibit carrier injection from the metal electrode in the face of the enhanced Schottky effect. In Fig. 34.7, a mechanistic

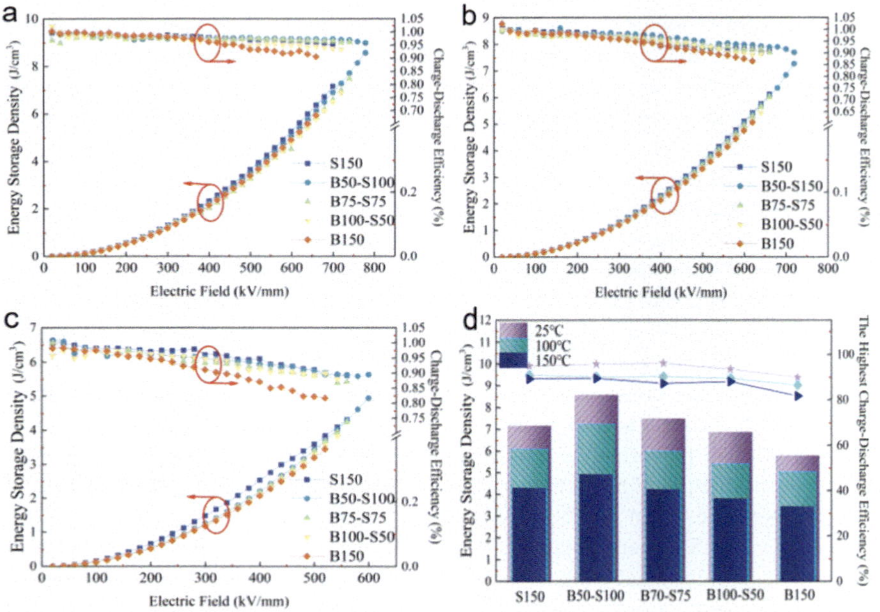

Fig. 34.6 The energy storage performance of composite films deposited with different thicknesses of BN and SiO_2 **a** Room temperature; **b** 100 °C; **c** 150 °C; **d** energy storage performance statistics bar chart

Table 34.2 Comparison of breakdown strength and discharge energy density for dieletric composites with different fillers at 150 °C

Materials	E_b (kV mm^{-1})	U_e (J cm^{-3})	References
PEI/SiO_2	450	2.96	Li et al. [25]
PI/BN	200	0.49	Chen et al. [46]
PI/BZT	350	1.83	Cheng et al. [47]
PI/$BaTiO_3$	550	2.10	Hu et al. [48]
PEI/Al_2O_3	450	4.31	Li et al. [49]
PC/BN	500	5.52	Liu et al. [50]
PEI/MgO	500	4.13	Zhang et al. [37]
PEI/BN	500	4.37	Liu et al. [51]
PAI/MMT	551	2.90	Zhang et al. [27]
PI/BN/TiO_2	500	5.00	Wang et al. [52]
PC/BN/SiO_2	600	4.94	This work

diagram of the composite film suppressing the carrier injection process is shown. Fig. 34.7a, b show the energy bands of BN and SiO_2 schematically, and the band gap width of SiO_2 is higher than that of BN as calculated by the UPS and UV, what is shown in Figs. 34.11, 34.12, 34.13 and 34.14. As shown in Fig. 34.7c, there will be a

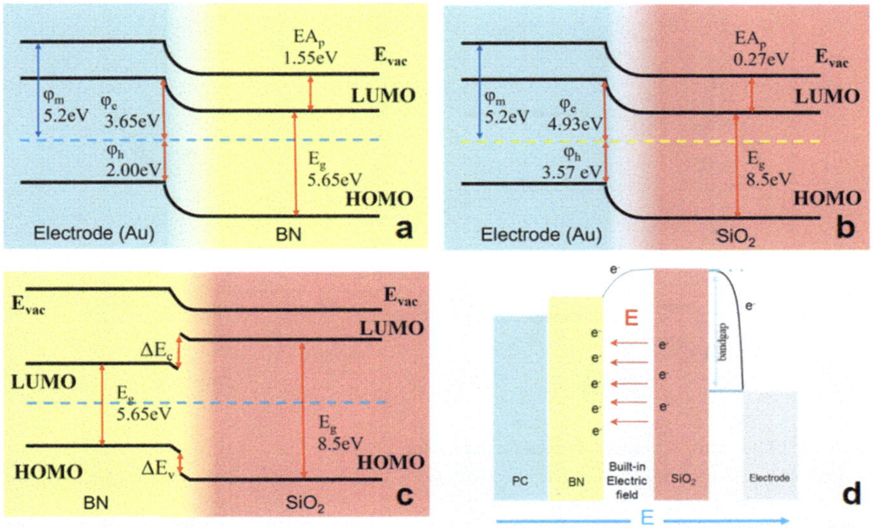

Fig. 34.7 **a** Band diagram of BN. **b** Band diagram of SiO_2. **c** Band diagram at the BN/SiO_2 interface. **d** Schematic of the intrinsic electric field

difference in the barrier height at the interface of the two materials, so in the process of electron injection will be due to the difference in the energy band structure so that the electrons will be gathered in the side close to the BN, the formation of the built-in electric field from the SiO_2 pointing to the BN, so as to achieve the role of inhibiting the development of carriers in the internal development of the composite film. Figure 34.7d shows the mechanism of the composite film to maintain a high E_b under the action of electric field. Firstly, the electrons at the electrode interface need to absorb energy in order to be able to overcome the high potential barrier of SiO_2 to enter into the interior of the composite film, which realizes a certain obstruction to carrier injection. Subsequently, as the free electrons progress from the first SiO_2 barrier layer to the second BN barrier layer, the reverse built-in electric field between the two will weaken the energy of the free electrons and inhibit their transport in the composite film.

34.4 Conclusion

In this work, BN and SiO_2 with different bandgap widths are deposited by PECVD to construct inorganic barrier layers on both sides of the PC film, and the effective enhancement of the energy storage performance of the composite film is realized by searching for the optimal total thickness of the deposition and adjusting the relative thicknesses of the double inorganic layer deposition. Experiments show that the

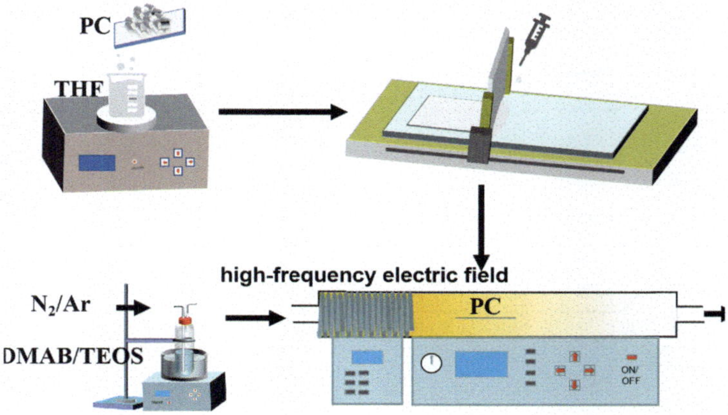

Fig. 34.8 Flow chart of composite film preparation

construction of the heterojunction structure is carried out on the basis of the optimal deposition thickness, and the synergistic effect of the high electronic barrier of the deposited inorganic material and the internally formed built-in electric field achieves an effective suppression of carrier injection and development, optimizes the problem of large loss value of the pure polymer due to the increase of the electrical conductivity at high temperatures, and confirms the validity of the structural tuning by the performance tests. The method of preparing composite films in this experiment is simple and convenient, and the inorganic layer structure constructed by the PECVD method can achieve further performance enhancement of the existing energy storage dielectric with versatility, which is important for the continued optimization of the formed materials in the future.

Appendix

See Figs. 34.8, 34.9, 34.10, 34.11, 34.12, 34.13 and 34.14

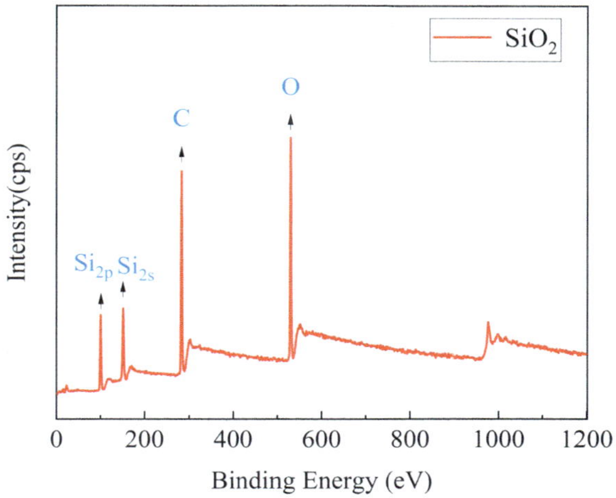

Fig. 34.9 XPS pattern of SiO_2

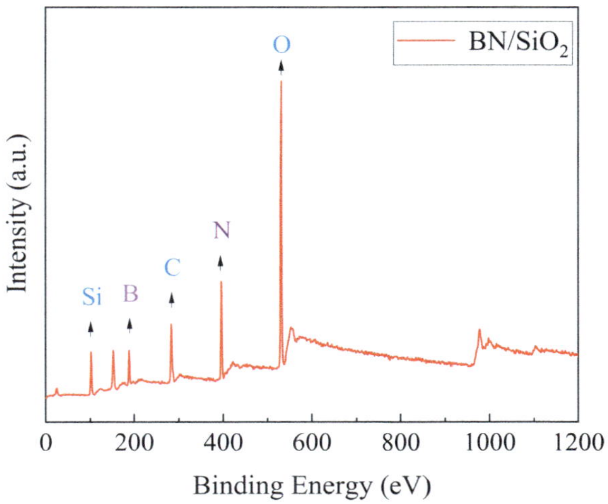

Fig. 34.10 XPS pattern of BN/SiO_2

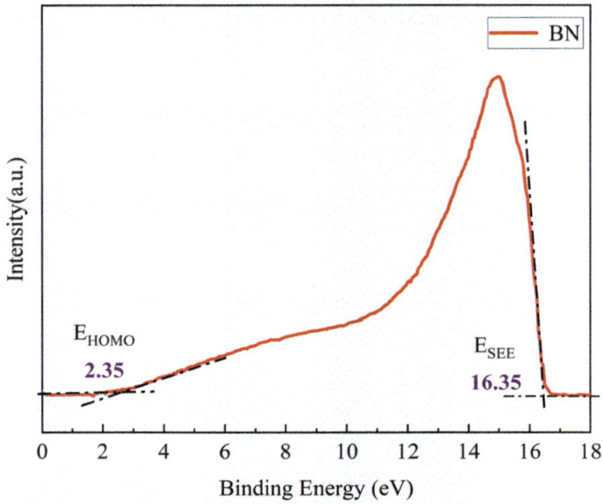

Fig. 34.11 UPS spectra of BN

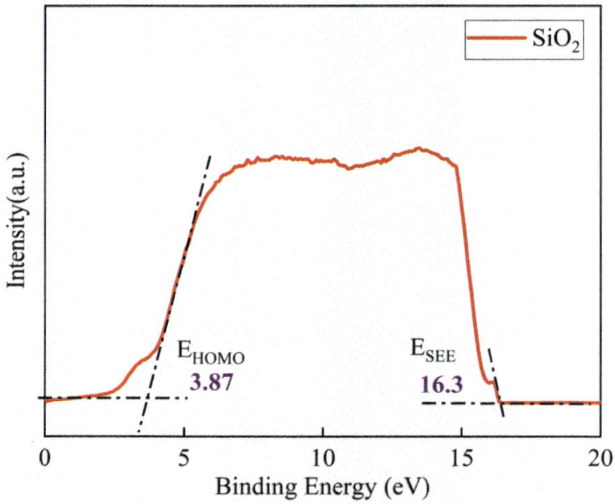

Fig. 34.12 UPS spectra of SiO_2

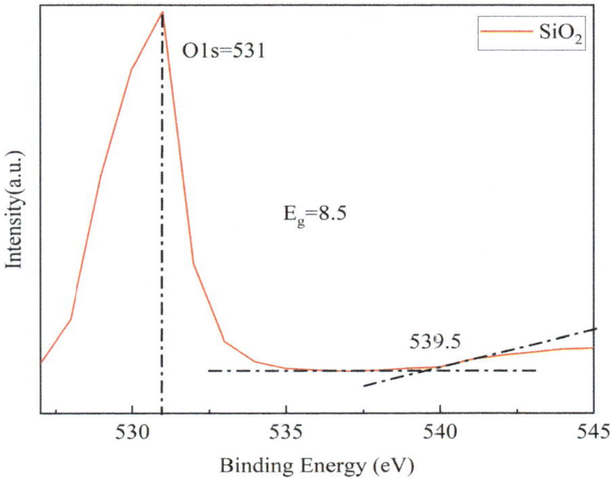

Fig. 34.13 XPS spectrum of O1S electrons in SiO_2

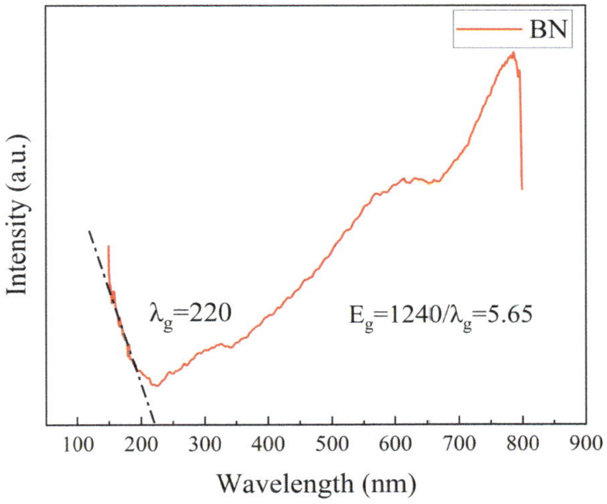

Fig. 34.14 UV–Vis spectra of BN

References

1. B. Fjijoe, E. Stip, Special section on harmonics stability and mitigation in power electronics. IEEE J. Emerg. Sel. Topics Power Electron. **4**(1), 1–2 (2016)
2. S. Mjajos, Mapping dynamics of low-carbon energy innovation for small hydropower technology in Nigeria. Afr. J. Sci. Technol. Innov. Dev.Innov. Dev. **10**(5), 569–578 (2018)
3. P. Zhao, L. Xiong, X. Li et al., Research and application progress of conductive films in energy storage devices. Adv. Mater. Technol. (2023). https://doi.org/10.7536/PC141201
4. J. Zhou, X. Meng, H. Wang et al., A novel DC transmission and distribution system with 100% new energy consumption. Proc. J. Phys.: Conf. Ser., F **2465**, 012026 (2023)
5. Y. Zhao, X. Meng, X. Hao, Synergistically achieving ultrahigh energy-storage density and efficiency in linear-like lead-based multilayer ceramic capacitor. Scripta Mater. **195** (2021)

6. C. Bin, X. Hou, Y. Xie et al., Ultrahigh energy storage performance of flexible BMT-based thin film capacitors. Small **18**(4), 2106209 (2022)
7. X. Li, Y. Yang, Y. Wang et al., Enhanced energy storage density of all-organic fluoropolymer composite dielectric via introducing crosslinked structure. RSC Adv. **11**(25), 15177–15183 (2021)
8. Y. Zhang, X. He, S. Li et al., Innovative all-organic dielectric composite for dielectric capacitor with great energy storage performance based on thermodynamic compatibility. J. Chem. Phys. **158**(21), 214902 (2023)
9. J. Mao, S. Feng, S. Wang et al., Improving the high-temperature energy storage performance of epoxy films: moderately reducing unsaturation for extremely high efficiency. Langmuir (2023). https://doi.org/10.1021/acs.langmuir.3c01211
10. J. Mao, S. Wang, Y. Cheng et al., High energy storage density and efficiency achieved in dielectric films functionalized with strong electronegative asymmetric halogen-phenyl groups. Chem. Eng. J. **444**, 136331 (2022)
11. Z. Bao, X. Du, S. Ding et al., Improved working temperature and capacitive energy density of biaxially oriented polypropylene films with alumina coating layers. ACS Appl. Energy Mater. **5**(3), 3119–3128 (2022)
12. T. Zhang, H. Yu, Y.H. Jung et al., Significantly improved high-temperature energy storage performance of BOPP films by coating nanoscale inorganic layer. Energy Environ. Mater. (2022). https://doi.org/10.1002/eem2.12549
13. Y. Liu, H. Luo, F. Wang et al., Enhanced energy density and efficiency of all-organic composites by designing a multilayer gradient structure. J. Mater. Chem. C **11**(32), 10985–10992 (2023)
14. N. Pfeiffenberger, F. Milandou, M. Niemeyer et al., High temperature dielectric polyetherimide film development. IEEE Trans. Dielectr. Electr. Insul.Dielectr. Electr. Insul. **25**(1), 120–126 (2018)
15. H.-M. Chang, K.-L. Fan, A. Charnas et al., Experimental analysis of the Schottky barrier height of metal contacts in black phosphorus field-effect transistors. J. Phys. D-Appl. Phys. **51**(13), 135306 (2018)
16. M. Grajower, U. Levy, J.B. Khurgin, The role of surface roughness in plasmonic-assisted internal photoemission Schottky photodetectors. ACS Photonics **5**(10), 4030–4036 (2018)
17. X. Tan, P. Rumbach, N. Griggs et al., Theoretical analysis of 1D resonant tunneling behavior in ion-enhanced cold field and thermo-field emission. J. Appl. Phys. **120**(21), 213301 (2016)
18. S. Chen, G. Meng, B. Kong et al., Asymmetric alicyclic amine-polyether amine molecular chain structure for improved energy storage density of high-temperature crosslinked polymer capacitor. Chem. Eng. J. **387**, 123662 (2020)
19. Y. Zhang, X. He, S. Li et al., Design of heterogeneous sandwich-structured dielectric composites with excellent charge-discharge efficiency and energy storage density. J. Mater. Chem. C (2023). https://doi.org/10.1039/d3tc01817b
20. S. Liu, W. Zhang, F. Guan et al., The simple synthesis of all-organic polymer bilayer films: achieving simultaneous enhancements in energy storage density and efficiency. J. Appl. Polym. Sci.Polym. Sci. **140**(26), e53991 (2023)
21. X.-J. Liu, M.-S. Zheng, Q. Wang et al., Enhancement of high temperature capacitive performance of sulfone-containing polyimide by suppressing carrier transport. Appl. Phys. Lett. **121**(24), 243902 (2022)
22. L. Gao, J. Zhang, L. Song et al., Low-content core-shell-structured TiO_2 nanobelts@SiO_2 doped with poly(vinylidene fluoride) composites to achieve high-energy storage density. J. Mater. Sci.-Mater. Electron. **33**(23), 18345–18355 (2022)
23. X. Wang, X. Zhang, B. Zhu, Influence of surface fluorination treatment on properties of polypropylene film. IEEE Trans. Dielectr. Electr. Insul.Dielectr. Electr. Insul. **28**(6), 2058–2064 (2021)
24. C. Zhang, H. Wang, Z. Liu et al., Enhanced breakdown strength and energy storage density of PMMA/PVDF blends by coating superficial layers and doping organic fillers. J. Mater. Sci.-Mater. Electron. **34**(23), 1657 (2023)

25. Y.F. Wang, Z.Z. Li, C. Wu et al., Polyamideimide dielectric with montmorillonite nanosheets coating for high-temperature energy storage. Chem. Eng. J. **437**, 135430 (2022)
26. T. Linker, Y. Wang, A. Mishra et al., Deep well trapping of hot carriers in a hexagonal boron nitride coating of polymer dielectrics. ACS Appl. Mater. Interfaces **13**(50), 60393–60400 (2021)
27. T. Zhang, L. Yang, J. Ruan et al., Improved high-temperature energy storage performance of PEI dielectric films by introducing an SiO_2 insulating layer. Macromol. Mater. Eng. **2021**, 2100514
28. A. Albo, Y.V. Flores, Q. Hu et al., Two-well terahertz quantum cascade lasers with suppressed carrier leakage. Appl. Phys. Lett. **111**(11), 111107 (2017)
29. M. Feng, Y. Feng, C. Zhang et al., Ultrahigh energy storage performance of all-organic dielectrics at high-temperature by tuning the density and location of traps. Mater. Horiz.Horiz. **9**(12), 3002–3012 (2022)
30. H. Zhao, C. Xi, X.-D. Zhao et al., Elevated-temperature space charge characteristics and trapping mechanism of cross-linked polyethylene modified by UV-initiated grafting MAH. Molecules **25**(17), 3973 (2020)
31. E. Wruss, L. Hoermann, O.T. Hofmann, Impact of surface defects on the charge transfer at inorganic/organic interfaces. J. Phys. Chem. C **123**(12), 7118–7124 (2019)
32. R. Balabai, A. Solomenko, Flexible 2D layered material junctions. Appl. Nanosci.Nanosci. **9**(5), 1011–1016 (2019)
33. X. Li, H. Luo, C. Yang et al., Enhancing high-temperature energy storage performance of PEI-based dielectrics by incorporating ZIF-67 with a narrow bandgap. ACS Appl. Mater. Interfaces **15**(35), 41828–41838 (2023)
34. J.-H. Park, S.H. Choi, J. Zhao et al., Thickness-controlled multilayer hexagonal boron nitride film prepared by plasma-enhanced chemical vapor deposition. Curr. Appl. Phys. **16**(9), 1229–1235 (2016)
35. Y. Gong, D. Chen, J. Duan et al., Largely enhanced energy density of BOPP-OBT@CPP-BOPP sandwich-structured dielectric composites. J. Mater. Chem. C **10**(36), 13074–13083 (2022)
36. D. Li, W. Li, R. Wang et al., High-temperature strength of zirconia and its sensitivity to surface defects. Mater. Res. Express **6**(7), 075511 (2019)
37. T. Zhang, L. Yang, C. Zhang et al., Polymer dielectric films exhibiting superior high-temperature capacitive performance by utilizing an inorganic insulation interlayer. Mater. Horiz.Horiz. **9**(4), 1273–1282 (2022)
38. T. Zhang, W. Li, Y. Zhao et al., High energy storage performance of opposite double-heterojunction ferroelectricity-insulators. Adv. Func. Mater. **28**(10), 1706211 (2018)
39. S. Zuo, X. Zhu, J. Feng et al., High-performance inverted Perovskite solar cells by reducing electron capture region for electron transport layers. Solar RRL **3**(9), 1900207 (2019)
40. J. Low, J. Yu, M. Jaroniec et al., Heterojunction photocatalysts. Adv. Mater. **29**(20), 1601694 (2017)
41. Q.-K. Feng, Y.-X. Zhang, D.F. Liu et al., Dielectric and energy storage properties of all-organic sandwich-structured films used for high-temperature film capacitors. Mater. Today Energy **29**, 101132 (2022)
42. Y. Zhang, Y. Ma, Y. Lin et al., Ultrahigh breakdown strength and discharge energy density of newly designed asymmetric sandwich-structured PVDF-based nanocomposite film. J. Mater. Chem. C (2023). https://doi.org/10.1039/d3tc02825
43. X. Ji, X. Liu, Y. Guo et al., Developing visible light responsive Z-scheme BN-PDI photocatalysts with good degradation performance for antibiotics. Chem. Eng. J. **425**, 131260 (2021)
44. S. Cheng, Y. Zhou, Y. Li et al., Polymer dielectrics sandwiched by medium-dielectric-constant nanoscale deposition layers for high-temperature capacitive energy storage. Energy Storage Mater. **42**, 445–453 (2021)
45. D. Yue, Y. Feng, X.-H. Liu, Prediction of energy storage performance in polymer composites using high-throughput stochastic breakdown simulation and machine learning. Adv. Sci. **9**(17), 2105773 (2023)

46. W.Z. Chen, M.H. Cao, H.Y. Wang et al., Energy storage performance of silica-coated $K_{0.5}Na_{0.5}NbO_3$-based lead-free ceramics. J. Mater. Sci.-Mater. Electron. **33**(13), 10121–10130 (2022)
47. S. Cheng, Y. Zhou, J. Hu et al., Polyimide films coated by magnetron sputtered boron nitride for high-temperature capacitor dielectrics. IEEE Trans. Dielectr. Electr. Insul.Dielectr. Electr. Insul. **27**(2), 498–503 (2020)
48. P.H. Hu, W.D. Sun, M.Z. Fan et al., Large energy density at high-temperature and excellent thermal stability in polyimide nanocomposite contained with small loading of $BaTiO_3$ nanofibers. Appl. Surf. Sci. **458**, 743–750 (2018)
49. H. Li, D. Ai, L.L. Ren et al., Scalable polymer nanocomposites with record high-temperature capacitive performance enabled by rationally designed nanostructured inorganic fillers. Adv. Mater. **31**(23), 1900875 (2019)
50. G. Liu, T.D. Zhang, Y. Feng et al., Sandwich-structured polymers with electrospun boron nitrides layers as high-temperature energy storage dielectrics. Chem. Eng. J. **389**, 124443 (2020)
51. X.J. Liu, S.L. Zhong, M.S. Zheng et al., Enhanced high-temperature capacitive performance of a bilayer- structured composite film employing a charge blocking layer. ACS Appl. Mater. Interfaces **15**(1), 1105–1114 (2023)
52. P. Wang, L.M. Yao, Z.B. Pan et al., Ultrahigh energy storage performance of layered polymer nanocomposites over a broad temperature range. Adv. Mater. **33**(42), 2103338 (2021)

Chapter 35
Electric Field Simulation Analysis of Typical Defects of Metal Particles in Main Insulation of Converter Transformer Based on Multi-physics Coupling

Yongpeng Liu, Kai Nie, Chengzhen Li, Ze Li, Wenhao Gao, Fanbo Meng, and Qingquan Li

Abstract With the continuous expansion of China's power grid and the ongoing improvement of transformer voltage levels, electrical insulation in converter transformers has become increasingly prominent as a critical component. The presence of typical external defects such as metal particles can easily lead to electric field distortion and partial discharge in micro-areas within converter transformers, resulting in significant insulation deterioration. Utilizing COMSOL finite element simulation software, we have developed a comprehensive multi-physics coupling simulation model to analyze the impact of metal particle defects on internal electric field distortion. Additionally, we investigate how different types, sizes, and positions of metal particles affect the distribution of main insulation electric fields and maximum

Y. Liu (✉) · K. Nie · C. Li · Z. Li · W. Gao · F. Meng · Q. Li
Shandong Provincial Key Laboratory of UHV Transmission Technology and Equipment, Shandong University, Jinan 250000, China
e-mail: grlyp4545@163.com

K. Nie
e-mail: 2072625131@qq.com

C. Li
e-mail: chengzhen_li@outlook.com

Z. Li
e-mail: lize20001229@163.com

W. Gao
e-mail: 2695401277@qq.com

F. Meng
e-mail: mengfanbo@sdu.edu.cn

Q. Li
e-mail: lqq@sdu.edu.cn

Z. Fang et al. (eds.), *Proceedings of the 5th International Symposium on Plasma and Energy Conversion*, Springer Proceedings in Physics 398,
https://doi.org/10.1007/978-981-97-2245-7_35

field strength in converter transformers. This research aims to provide guidance for relevant factory tests.

35.1 Introduction

Since the reform and opening up, the power system of our country is booming, the scale is expanding, however, with climate change, regional instability, the global energy industry chain has been seriously affected, therefore, to build a green low-carbon sustainable new power system has become the development direction of the new era of power industry. The completed UHV transmission projects have provided strong support for the optimization of national resource allocation and the construction of a new power system. Power transmission equipment represented by oil-immersed converter transformers is the hub of energy transfer and voltage conversion in new power systems, and it is very important to ensure their safe and stable operation [1, 2].

According to statistics, the partial discharge fault caused by typical transformer defects accounts for 30% of the transformer fault types. Converter transformers are prone to typical insulation defects such as metal particles, which induce partial discharge and electric field distortion during their delivery and operation. In complex operating conditions, the insulation system of converter transformers may undergo irreversible deterioration. The oil-immersed converter transformer has complex structure and harsh working environment. The material properties of the insulation system are greatly affected by temperature, involving multiple physical fields such as electric field, flow field and temperature field, and the coupling influence is significant and the relationship is complex.

Literature [3] built a two-dimensional model of the winding end of a power transformer, obtained the electric field distribution of the main insulation end, explored the position of the maximum field strength, and carried out numerical calculation and analysis. Literature [4] explored the influence of different temperatures on the properties of insulating materials, and obtained the influence of temperature on the electric field distribution at the end of the transformer. In Ref. [5], a simplified two-dimensional pumping symmetry model of oil-immersed transformer was established, and the fluid-temperature field coupling was carried out to explore the influence of oil flow velocity and oil flow channel width on its temperature rise. Literature [6] proposed a multi-physics field coupled finite element method for calculating winding temperature rise.

However, in the above studies, the mapping relationship between electricity and heat and the nonlinearity between the insulation material parameters and the electric field were not considered in the fluid-temperature field simulation. When studying the electric field of transformer winding end, the simulation only depends on the variation law of insulation material parameters with temperature, and the temperature field and electric field are not really coupled. In order to facilitate the calculation, the models built are simplified models, which are greatly different from the actual situation of transformer. In the current research, the typical defects of metal particles that may occur in the actual factory and operating conditions are not considered, and their size, shape and position have a huge impact on the field strength distribution at the end of the winding.

In order to truly and accurately reflect the influence of typical metal particle defects of converter transformers on the electric field at the winding end, this paper considers the coupling relationship between the involved multi-physical fields according to the actual operation situation of the transformer, and builds a multi-physical field coupling simulation model through the finite element simulation software. The effects of different metal particle sizes, positions and types on the electric field distribution and maximum field strength at the winding end are investigated.

35.2 Multi-Physics Field Coupling Model

35.2.1 Simulation Model of Converter Transformer

Taking a certain type of converter transformer as an example, this paper constructs an actual two-dimensional geometric model, and simplifies the model according to the principle of thin paper small oil gap between the high and low voltage windings of the transformer. In the simulation, a quarter of the model is used for calculation. Because of its symmetric structure, only an eighth model is shown, as shown in Fig. 35.1. The winding end of converter transformer has complex structure, which assumes the main role of transformer insulation. The temperature field and flow field are complex, and the electric field distortion is serious, so it needs to be analyzed emphases.

Among them, 1 is the voltage regulating winding, 2 is the net side winding, 3 is the insulating cardboard, 4 is the valve side winding, 5 is the electrostatic ring, 6 is the iron core, and 7 is the insulating oil.

35.2.2 Electromagnetic-Fluid-Temperature Field Coupling

During the operation of the transformer, the windings and the core produce the main heat source of the transformer, and the electromagnetic loss of the core is closely related to the distribution of the magnetic flux density [7, 8]. According to Maxwell's equations, the magnetic field governing equation of the transformer based on the magnetic vector potential is shown in Eq. (35.1).

$$\nabla \times \frac{1}{\mu}(\nabla \times A) = J_s - \sigma \frac{\partial A}{\partial t} \tag{35.1}$$

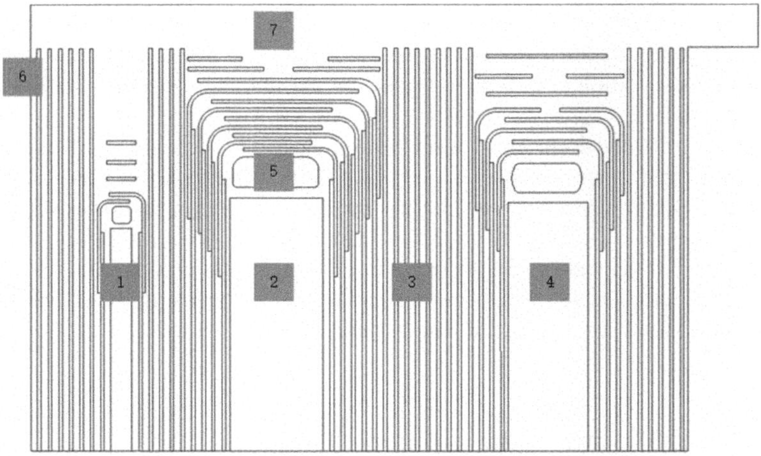

Fig. 35.1 Simulation model of converter transformer

where, A is the vector magnetic potential, J_s is the winding current density, μ is the magnetic permeability, σ is the electrical conductivity, and t is time.

The heat generated by the transformer core and winding is mainly dissipated by the flow and circulation of the transformer oil, and the heat transfer between the tank wall and the outside world. Transformer oil is an incompressible ideal fluid, and its flow inside the tank satisfies the laws of conservation of momentum, conservation of mass and conservation of energy [9, 10]. The governing equations are shown in Eq. (35.2).

$$\begin{cases} \dfrac{\partial \rho_1}{\partial t} + \nabla \cdot (\rho_1 U_1) = 0 \\ \dfrac{\partial (\rho_1 U_1)}{\partial t} + \nabla \cdot (\rho_1 U_1 \times U_1) = -\nabla p + \mu_1 (\nabla^2 U_1) + g(\rho_1 - \rho_{ref}) \end{cases} \tag{35.2}$$

where, p and U_1 are the unknown quantities for solving the governing equation, respectively representing the pressure and velocity of the oil flow; μ is the dynamic viscosity of transformer oil; ρ_1 and ρ_{ref} respectively represents the density and density reference value of transformer oil; g is the acceleration of gravity.

Therefore, the finite element method is used to analyze the electromagnetic field of the transformer, and the electromagnetic heat generated by the transformer core and winding is used as the heat source of the temperature and flow field coupling, and the internal temperature field and flow field distribution of the transformer are obtained. By consulting a large number of historical materials and data, it is found that the relative permittivity of insulating cardboard and transformer oil is not affected by temperature, but the conductivity shows obvious nonlinear characteristics by temperature, and then affects the distribution of electric field at the end of the winding.

Based on the temperature field distribution results obtained by the above simulation, the mapping method is used to transfer the data, and the interpolation method is used to transfer the data obtained by the temperature field simulation to the relevant grid of the electric field calculation, so as to realize the deep coupling calculation of electromagnetium-fluid-temperature field, which ensures the accuracy and authenticity of the electric field simulation of the winding end of the converter transformer.

35.3 Simulation Analysis

Based on the above multi-physical coupling model, in the comsol simulation software, material properties and boundary conditions are set, the model is divided into grids, and the converter transformer is simulated and analyzed to explore the influence of different metal particle sizes, locations and types on the electric field distribution and maximum field strength at the winding end.

35.3.1 Effect of Metal Particle Defects on Electric Field of Main Insulation of Converter Transformer

The ambient temperature was set as room temperature, and the simulation analysis was carried out under DC voltage. In order to better reflect the influence of typical defects of metal particles on the electric field at the winding end, the research was carried out along the path AB in Fig. 35.2a, and the change of field strength was analyzed, and the change law shown in Fig. 35.2b was obtained. Under the action of DC voltage, the internal electric field of the metal particle is zero, forming an electrostatic shield. Compared with no metal particle defects, the appearance of metal particle defects induces the electric field distortion of the oil gap and the paperboard between the electrostatic ring at the end of the winding and the paperboard, and the electric field intensity of the oil gap and the paperboard near the metal particle increases significantly. The maximum field strength of the transformer winding end appears at the oil gap of the metal particle near the electrostatic ring, and the maximum field strength is 79.34 kV/mm.

The simulation is performed at AC voltage and the electric field distribution shown in Fig. 35.3 is obtained. The presence of metal particles severely changes the electric field distribution of the oil gap between the electrostatic ring and the metal particles at the end of the winding, and has little effect on the insulating paperboard. The maximum electric field intensity appears in the oil gap where the metal particles are near the electrostatic ring, and the maximum electric field intensity is 78.82 kV/mm.

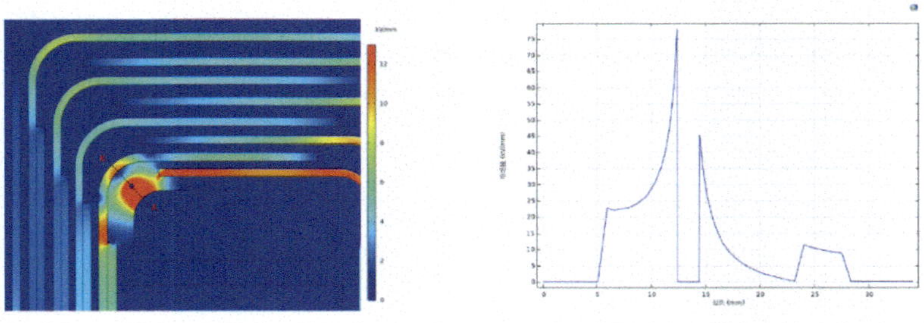

(a) The electric field distribution at the winding end (b) the electric field distribution along the path AB

Fig. 35.2 Electric field distribution of defects containing metal particles at the winding end under DC voltage

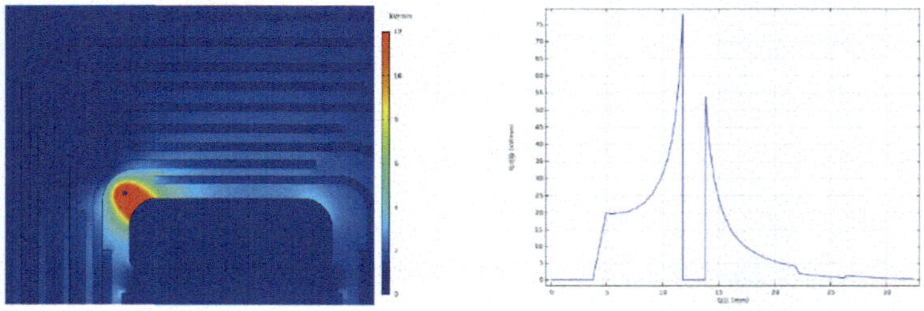

(a) The electric field distribution at the winding end (b) the electric field distribution along the path AB

Fig. 35.3 Electric field distribution of defects containing metal particles at the winding end under AC voltage

35.3.2 Effect of Different Metal Particle Radius on Electric Field of Winding End

By changing the size of the metal particle radius, the influence of different metal particle radius on the field strength of the transformer winding end under AC/DC voltage is explored. Figure 35.4 shows that the electric field distribution of the main insulation of the transformer has the same variation trend under DC and AC voltages. When the radius is less than 0.1 mm, the distortion effect of metal particles on the electric field is very significant, and the maximum field strength under DC voltage and AC voltage is 397.912 kV/mm and 402.875 kV/mm, respectively. When the radius is between 0.1 and 3.5 mm, the electric field intensity decreases with the increase of the radius, and the change is not large. When the radius is larger than 3.5 mm, the maximum field strength shows an increasing trend with the increase of the radius. When small metal particles appear in the transformer, that is, the radius is less than 0.5 mm, the effect of AC voltage is more obvious. When large metal

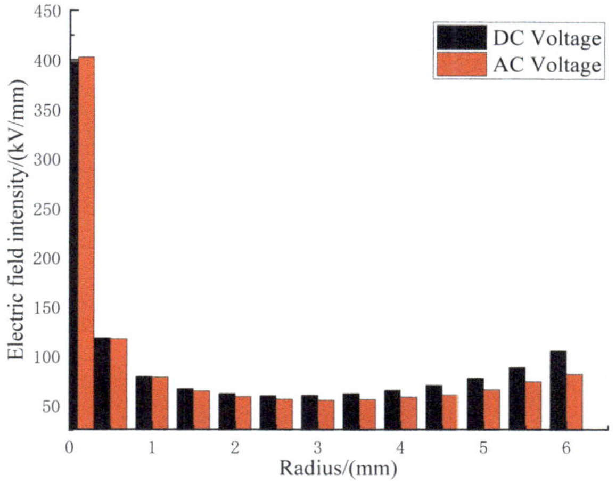

Fig. 35.4 Effect of different metal particle radii on the electric field

particles appear in the transformer, that is, the radius is greater than 4 mm, the effect of DC voltage is more significant.

35.3.3 Influence of Different Positions of Metal Particles on the Electric Field at the Winding End

As shown in Fig. 35.5, when the metal particles are located in the oil gap between the transformer winding and the electrostatic ring, the maximum value of the electric field appears in the oil gap of the metal particles near the winding side, and the electric field in the oil gap is seriously distorted, which is easy to induce partial discharge. For the part of the transformer end away from the metal particles, the electric field distribution changes little and is less affected by the metal particles.

As shown in Fig. 35.6, when the metal particles appear between the Angle ring at the end of the transformer and the Angle ring, there is a significant difference between the DC voltage and the AC voltage. Under the action of DC voltage, the metal particles have little effect on the electric field distribution in the nearby oil gap, and the electric field intensity on the insulating cardboard between the metal particles and the electrostatic ring increases significantly, and the insulating cardboard will bear greater insulation pressure. The maximum field intensity appears on the side of the insulating cardboard near the metal particles near the electrostatic ring. Under the action of AC voltage, the metal particles have a great influence on the electric field distribution of the nearby oil gap, and the maximum intensity appears in the oil gap where the metal particles are close to the electrostatic ring, and the influence on other parts is small.

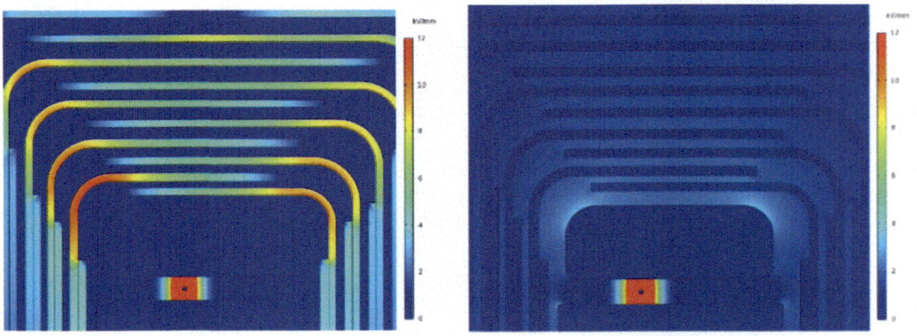

(a) DC voltage action (b) AC voltage action

Fig. 35.5 The metal particle is located between the winding and the electrostatic ring

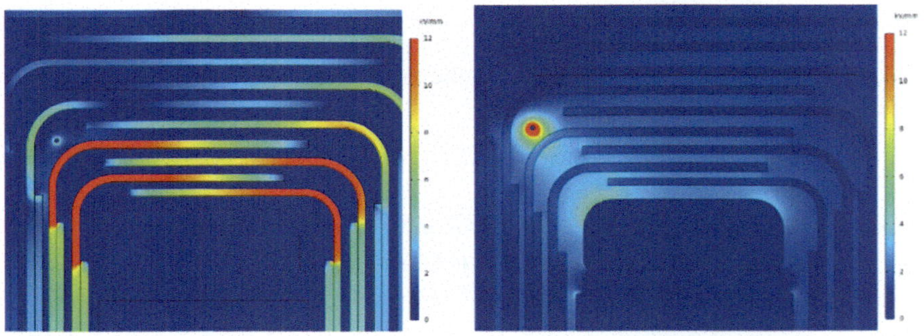

(a) DC voltage action (b) AC voltage action

Fig. 35.6 The metal particle is located between the Angle ring and the Angle ring

35.3.4 *Effect of Different Metal Particle Morphology on Electric Field of Winding End*

In the process of converter transformer leaving the factory and running, not only spherical metal particles may appear, but also metal particles in the form of strip wire defects. As shown in Fig. 35.7, compared with spherical metal particles, the presence of typical defects of strip metal wires has a more prominent effect on the distortion of electric field, and its influence range is wider, so it will seriously harm the performance of converter transformer insulation system in practical operation.

No matter what kind of metal particles exist, they will profoundly change the electric field distribution of the main insulation at the end of the winding, and produce a great electric field intensity distortion, as shown in Fig. 35.8. Under DC voltage, the maximum field strength caused by spherical metal defects and strip metal defects is 79.344 kV/mm and 83.979 kV/mm, respectively, which is 6.28 and 6.65 times

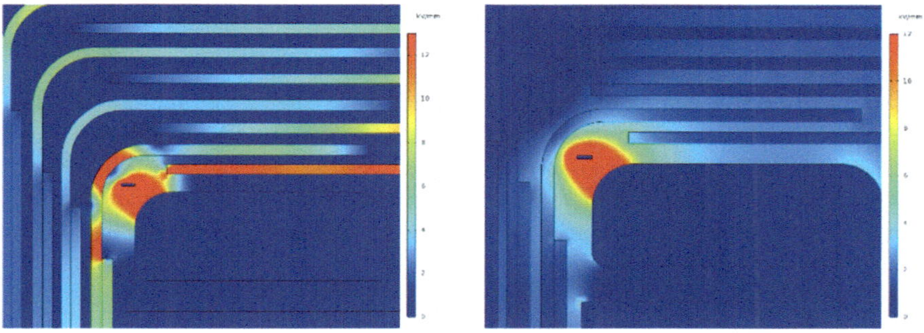

(a) DC voltage action (b) AC voltage action

Fig. 35.7 Effect of the wire on the electric field at the end of the winding

higher than that without defects. Under the action of AC voltage, the maximum field strength values caused by spherical metal defects and strip metal defects are 78.816 kV/mm and 84.557 kV/mm, respectively, which are 21.35 and 22.63 times higher than those without defects, respectively. Therefore, the strip wire defects have a more obvious distortion effect on the electric field, and should be protected to avoid partial discharge and insulation deterioration.

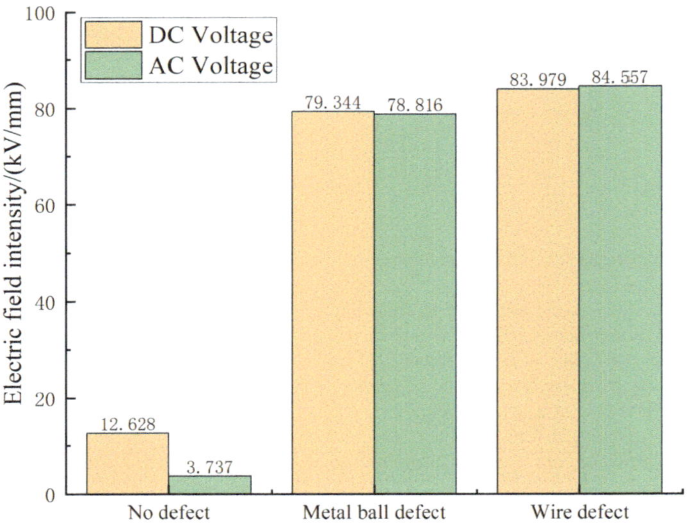

Fig. 35.8 Effect of different types of metal particles on the maximum field strength at the winding end

35.4 Conclusion

Based on COMSOL simulation platform, this paper builds a simulation model of converter transformer, summarizes the influence of temperature on the relevant material parameters of converter transformer, and transfers the simulation results of heat-flow coupling to the electric field through the mapping method, and analyzes the electric field distribution of the winding end with metal particle defects. The conclusions are as follows:

1. The existence of metal particle defects induces serious electric field distortion in the oil gap between the electrostatic ring at the end of the winding and the paperboard, and the transformer oil and the insulating paperboard will also bear greater insulation pressure. When there are small particles, the maximum field strength is inversely proportional to the radius. When there are extremely small metal particles, the maximum field strength will increase sharply, which seriously endangers the safety of the insulation system.
2. When the metal particles appear between the Angle rings, they have little effect on the nearby oil gap under the action of DC voltage, but cause serious electric field distortion in the adjacent insulating paperboard; At AC voltage, the electric field distribution of the oil gap near the metal particles will be seriously affected.
3. Strip metal wire is also a common type of metal particles, which will also change the electric field distribution at the end of the winding and induce serious electric field distortion. Compared with spherical metal particles, it has a stronger harmful effect, and should be protected to avoid inducing partial discharge and ensure the safety and stability of the insulation system.

References

1. Y.C. Wu, Optimized design of transformer main insulation structure. Integr. Circ. Appl. **39**(10), 158–159 (2022)
2. L. Li et al., Study on the influence of valve side winding defects on electric field distribution of converter transformer, in *2023 9th International Conference on Electrical Engineering, Control and Robotics (EECR)*, Wuhan, China (2023), pp. 1–6
3. H. Lei, *Numerical Analysis of Electric Field of Power Transformer Winding End Insulation Structure* (Hebei University of Technology, 2014)
4. F. Zhou, D. Zhang, H. Liu, K. Hu, L. Li, W. Miao, L. Wu, Effect of temperature on electric field distribution of converter transformer winding end. High Voltage Apparatus **56**(01), 80–86 (2020)
5. X. Zhu, Z. Liu, S. Li, Simulation analysis of temperature rise characteristics of oil-immersed transformer based on fluid-temperature field coupling. Electrotech. Tech. **15**, 46–50 (2022)
6. Y. Xie, L. Li, Y. Song, et al., Calculation method of winding temperature rise of oil-immersed power transformer based on multi-physical field coupling. Proc. CSEE **36**(21), 5957–5965+6040 (2016)
7. H. Liu, J. Du, Z. Zhang et al., Numerical modeling and analysis of electromagnetic, temperature and fluid field bidirectional coupling in pie windings of oil-immersed transformer. J. Beijing Jiaotong Univ. **44**(04), 141–149 (2020)

8. C. Liao, J. Ruan, C. Liu et al., Oil-immersed transformer 3d electromagnetic fluid-temperature field coupling analysis method. Electric Power Autom. Equipment **35**(9), 150–155 (2015)
9. L. Wang, K. Hou, X. Lu et al., Research on electric field distribution at winding end of converter transformer considering temperature gradient, in *2021 International Conference on Electrical Materials and Power Equipment (ICEMPE)*, Chongqing, China(2021), pp.1–4
10. R. Xiao, *Temperature Field Simulation Modeling and Real-Time Temperature Fast Prediction Method for Converter Transformer* (Chongqing University, 2022)

Chapter 36
Research on the Improved Installation of Pressure Relief Valves in Response to Internal Tank Pressure Increase

Qiang He, Yongxin Liu, Tianheng Chen, Long Xu, Tianjian Zhao, Qianle Cong, and Qianhang Ni

Abstract Oil-immersed transformers are crucial equipment in power transmission. The use of pressure relief valves in transformers is widespread as a response to the risk of tank rupture caused by a sharp rise in pressure. However, even with pressure relief valves equipped, transformer explosions occur from time to time, raising concerns about how to enhance the effectiveness of such valves in preventing explosions. The purpose of this research is to analyses pressure fluctuations inside a transformer tank by examining the installation method of pressure relief valves, using the finite element analysis and unidirectional fluid–solid interaction method. The installation method of pressure relief valves is crucial for ensuring the operational safety of transformers, and this study aims to provide relevant guidance in this regard.

Q. He · Y. Liu · T. Chen · L. Xu · T. Zhao · Q. Cong · Q. Ni (✉)
State Grid Zhejiang Electric Power Company, Ltd. Shaoxing Power Supply Company, Shaoxing 312000, Zhejiang, China
e-mail: ni_qianhang@zj.sgcc.com.cn

Q. He
e-mail: oline9@sina.com

Y. Liu
e-mail: liu_yomghxin@zj.sgcc.com.cn

T. Chen
e-mail: chen_tianheng@zj.sgcc.com.cn

L. Xu
e-mail: xu_long@zj.sgcc.com.cn

T. Zhao
e-mail: zhao_tianjian@zj.sgcc.com.cn

Q. Cong
e-mail: cong_qianle@zj.sgcc.com.cn

Z. Fang et al. (eds.), *Proceedings of the 5th International Symposium on Plasma and Energy Conversion*, Springer Proceedings in Physics 398,
https://doi.org/10.1007/978-981-97-2245-7_36

36.1 Introduction

Oil-immersed transformers, a crucial component of power transmission equipment, have experienced explosions and fires in recent years, despite being equipped with pressure relief valves. This has severely impacted the reliability of power supply and the operational stability of the power system [1].

The SERGI team based in France has developed a system that allows for the discharge of oil and the injection of nitrogen [2]. This system releases transformer oil by rupturing the rupture disc, triggered by excessive pressure within the transformer. Simultaneously, nitrogen is injected into the oil tank to cool, maintain constant pressure, and isolate the oxygen. According to research [3], different calibers of pressure relief valves were assessed for their venting effects under different energy faults using numerical methods. Calculations indicate that a pressure reduction of significance can be achieved by increasing the caliber size of the pressure relief valve to at least 100 mm.

To investigate the correlation between the pressure relief valve's pressure relief capacity and the installation method, this study utilizes finite element analysis and fluid–solid interaction to create simulations that closely replicate real-world conditions. Various installation scenarios for the pressure relief valve are examined in different cases.

36.2 Modeling Short Circuits in Transformer Tanks

36.2.1 Pressure Source Modelling

When a short-circuit fault occurs in a power transformer, the generation of large quantities of gas through evaporation and cracking reactions becomes the source of pressure inside the transformer. It should be noted that pressure waves propagate in liquids [4]. The equation for the pressure wave in transient form can be expressed as:

$$\rho \frac{\partial^2 P}{\partial t^2} = \nabla(\mathrm{K}\nabla P) - \frac{4}{3}\rho \frac{\partial \mathbf{u}}{\partial t} - \eta \nabla^2 P \tag{36.1}$$

where P is the pressure of the pressure wave, ρ is the density of the insulating oil, t is the time, K is the bulk modulus of elasticity of the insulating oil, ∇ is the gradient operator to describe the spatial gradient, $\mathbf{u}$ is the velocity field of the insulating oil, η is the viscosity of the insulating oil. and ∇^2 is the Laplace operator to describe the second order spatial derivative.

Fig. 36.1 Three-dimensional fuel tank model

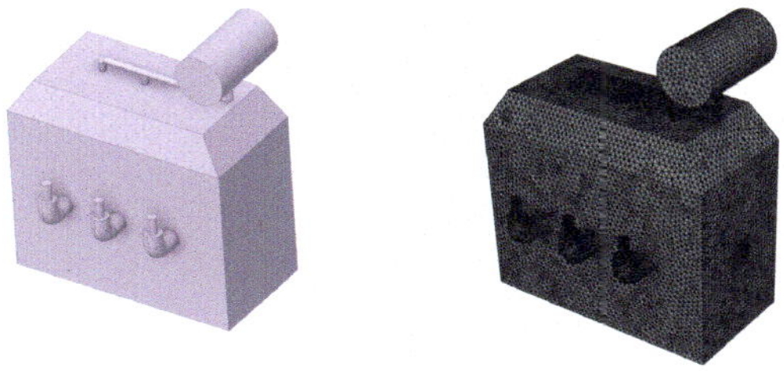

(a) Geometric model (b) Finite element model

36.3 Fuel Tank Modeling and Pressure Relief Settings

36.3.1 Fuel Tank Geometry Model

Based on the dimensions and material parameters of the actual transformer tank, a three-dimensional geometric model of the transformer was established. The structure of the transformer tank was simplified while disregarding parts that have minimal impact on the internal pressure. The resulting transformer tank geometry model is depicted in Fig. 36.1.

36.3.2 Relief Valve Operation Condition Setting

The paper's example employs a User-Defined Function (UDF) to demonstrate that the pressure relief valve opens at a designated pressure and streamlines the procedure for reaching the release pressure. The flow rate of the pressure relief valve is determined by the equation [5].

$$Q = uA\sqrt{2\frac{P}{\rho}} \tag{36.2}$$

where Q is the flow rate, μ is the flow coefficient and the shape of the valve, this example takes 0.9. A is the cross-sectional area of the flow diameter, P is through the valve before and after the pressure difference. ρ is the density of the fluid medium.

36.4 Simulation Results and Analysis

To investigate the pressure relief valve's installation conditions for its pressure relief capacity and the overall pressure inside the transformer tank, this paper utilizes five examples. The simulation time is set to 10 ms, simulation step size is set to 10 μs, and

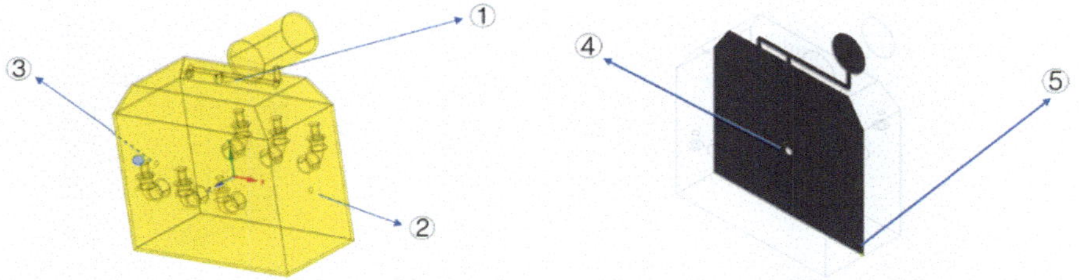

(a) Pressure relief valve installation location (b) Pressure measurement position

Fig. 36.2 Pressure relief valve installation location and pressure measurement location

the calculation is carried out for 1000 steps. Figure 36.2 depicts the mounting position of the pressure relief valve and the measuring position for pressure. Additionally, Table 36.1 displays the example set-up conditions.

The figures above illustrate how variations in the installation of the pressure relief valve impact the internal pressure of the transformer. Based on pressure distribution analyses from simulation calculations, the following conclusions can be drawn:

1. Based on Fig. 36.3, the gravitational force causes a slight difference in pressure between the center and bottom of the transformer tank. Additionally, internal discharge results in pressure fluctuations at the bottom of the tank.
2. According to Fig. 36.4, the pressure on the side of the transformer tank is higher than the pressure on the top due to gravity. As a result, the pressure on the side reaches the action conditions of the pressure relief valve first, and the moment of pressure relief on the top will occur later than on the side.
3. As depicted in Fig. 36.5, increasing the aperture of the pressure relief valve enhances the rate of pressure release from the transformer tank.
4. As shown in Fig. 36.6, the installation of pressure relief valves in multiple locations enables quicker pressure relief inside the transformer tank.

Table 36.1 Example setup conditions

Name	Pressure relief valve installation number	Pressure relief valve installation location	Pressure relief valve orifice	Pressure measurement point
Example 1	0	/	/	④, ⑤
Example 2	1	①	130 mm	④
Example 3	1	②	130mm	④
Example 4	1	②	80mm	④
Example 5	2	②, ③	80mm	④

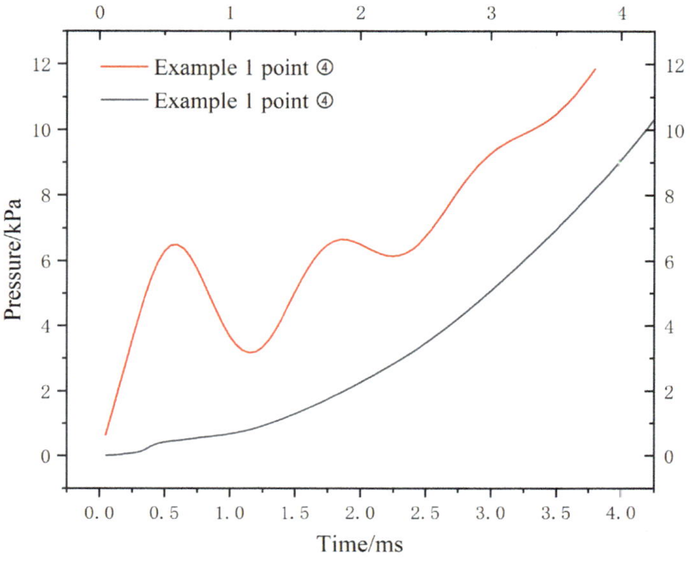

Fig. 36.3 Pressure at different locations inside the transformer

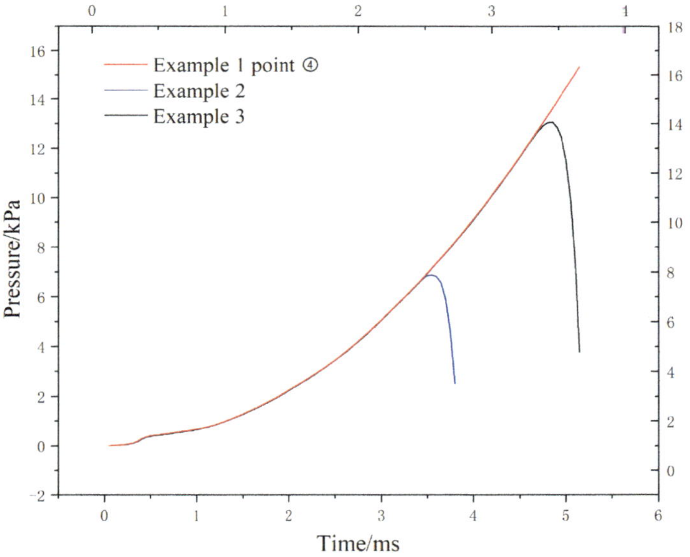

Fig. 36.4 Pressure relief valve internal pressure diagrams for different positions

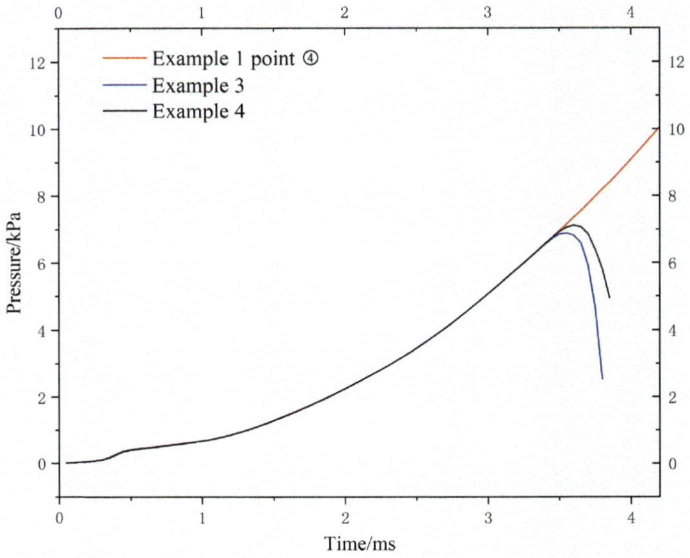

Fig. 36.5 Pressure relief valve internal pressure diagrams for different orifices

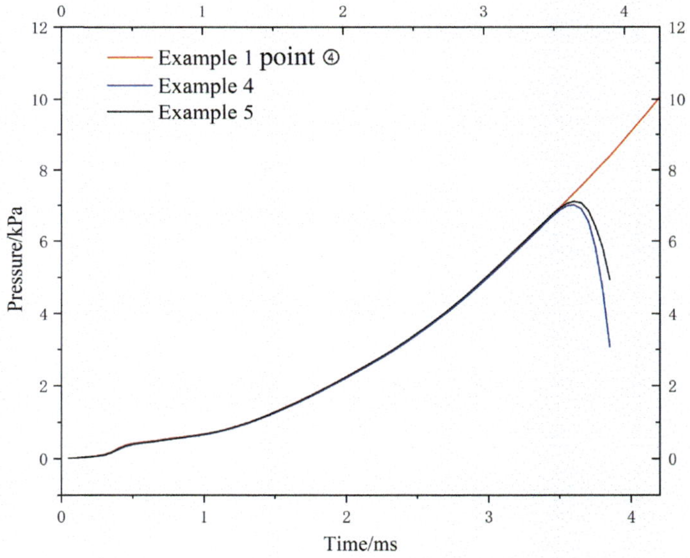

Fig. 36.6 Pressure relief valve internal pressure diagrams for different quantities

36.5 Conclude

In this study, a three-dimensional model of the tank was created, and the pressure relief valve was achieved by using User-Defined Function (UDF). The finite element simulation was employed to determine the pressure inside the transformer for various installation scenarios of the pressure relief valve. The findings of this analysis are presented below.

1. The pressure at the bottom of the transformer was slightly higher than the pressure in the center, caused by gravity. Pressure fluctuations occurred at the bottom of the tank due to internal transformer discharges.
2. Since the pressure on the side relief valve is slightly higher than that on the top, the pressure on the lateral side of the tank reaches the condition for relief valve action first, and the moment of pressure relief on the top will occur later than on the side.
3. Increasing the size of the pressure relief valve aperture can significantly enhance its pressure relief capacity, subsequently providing fast and efficient alleviation of pressure to avert overpressure in the transformer.
4. Installing pressure relief valves in multiple locations allows for quick pressure relief in emergencies and reduces the burden on individual valves.

The simulation results and conclusions can serve as a guide for installing pressure relief valves for transformers [6].

Acknowledgements This paper is solely funded by the Technology Project of State Grid Zhejiang Electric Power Company of China (B311SX230002).

References

1. M. El-Harbawi, Fire and explosion risks and consequences in electrical substations—A transformer case study. ASME Open J. Eng. **1** (2022)
2. S.B. Muller, Prevention of transformer tank explosion: Part 1—experimental tests on large transformers, in *ASME Pressure Vessels and Piping Conference* (Pressure Vessels and Piping Chicago, Illinois, USA, 2008), pp. 357–365
3. T. Roseberg, Transformer tank rupture—A protection engineer's challenge, in *2017 70th Annual Conference for Protective Relay Engineers (CPRE)* (IEEE, College Station, TX, USA, 2017), pp. 1–29
4. C. Yan, A novel digital non-electrical protection for power transformer based on internal pressure characteristics, in *2016 IEEE Power and Energy Society General Meeting (PESGM)*. (IEEE, Boston, MA, USA, 2016), pp. 1–5
5. P.K. Kundu, I.M. Cohen, D.R. Dowling, *Fluid Mechanics* (Academic Press, Cambridge, Massachusetts, USA, 2015)
6. J. Kolb, Streamers in water and other dielectric liquids. J. Phys. D: Appl. Phys. **41**(23), (2008)

Chapter 37
Modeling and Analysis of Deformation of Transformer Oil Tank Caused by Internal Increased Pressure

Qianhang Ni, Jia Ren, Zhiliang Wang, Qiang He, Feng Li, Dongdong Zhou, and Daiming Zhou

Abstract The internal failure of the transformer tank may lead to increased pressure and further deformation and rupture of the tank. It is of great significance to study the stress characteristics of the mechanical structure of transformer tank. In this paper, the deformation of the transformer tank after the internal pressure rises is simulated and analyzed by using the fluid–structure coupling calculation method and the modified evaporation–condensation model, The findings indicate that gravity causes a larger shape variation at the bottom of the tank compared to the top. The stress level at the oil pillow remains consistently low. Stress concentrations occur at the connection points between the connecting pipe and the tank's top, and between the turret and the tank's side. Enhancing the mechanical strength of these areas can enhance the tank's overall explosion-proof capability. The simulation outcomes in this study serve as a reference for designing transformer tanks.

Q. Ni · J. Ren · Z. Wang · Q. He (✉) · F. Li · D. Zhou · D. Zhou
State Grid Zhejiang Electric Power Company, Ltd. Shaoxing Power Supply Company, Shaoxing 312000, Zhejiang, China
e-mail: oline9@sina.com

Q. Ni
e-mail: ni_qianhang@zj.sgcc.com.cn

J. Ren
e-mail: ren_jia@zj.sgcc.com.cn

Z. Wang
e-mail: wang_zhiliang@zj.sgcc.com.cn

F. Li
e-mail: li_feng@zj.sgcc.com.cn

D. Zhou
e-mail: zhou_dongdong@zj.sgcc.com.cn

D. Zhou
e-mail: zhou_daiming@zj.sgcc.com.cn

Z. Fang et al. (eds.), *Proceedings of the 5th International Symposium on Plasma and Energy Conversion*, Springer Proceedings in Physics 398,
https://doi.org/10.1007/978-981-97-2245-7_37

37.1 Introduction

Transformer is a crucial component of the power system, and its safe and dependable operation is crucial for the reliable transportation and secure utilization of electrical energy [1]. Instances of transformer oil tank deformation, rupture, and explosion resulting from sudden pressure increases are frequent, and the ensuing fire poses a significant threat to the safety of operators and surrounding equipment.

Considering the serious harm of transformer tank rupture and explosion accident. Since the 1970s, scholars at home and abroad have studied the mechanism of transformer internal fault pressure and tank bearing capacity. Yan Chenguang, from Xi'an Jiaotong University, developed a transient mechanical model of the oil tank. A finite element simulation was conducted to observe the stress distribution changes of the oil tank wall under varying fault energies and positions. The results indicated a direct relationship between fault energy, oil pressure amplitude, and the stress magnitude of the oil tank [2]. Li Boyu, from Xi'an Jiaotong University, constructed a three-dimensional converter transformer model and observed a faster pressure increase in the narrow sections [3].

In this paper, we establish a three-dimensional model of a transformer tank and simulate the effect of changes in pressure fields on external tank deformation caused by gasification decomposition in the internal fluid domain using a fluid–structure coupling simulation method. We calculate the deformation characteristics of the transformer tank after the center of the tank fails. The findings indicate that gravity causes a larger shape variation at the bottom of the tank compared to the top. The stress level at the oil pillow remains consistently low. Stress concentrations occur at the connection points between the connecting pipe and the tank's top, and between the turret and the tank's side. Enhancing the mechanical strength of these areas can enhance the tank's overall explosion-proof capability. The simulation outcomes in this study serve as a reference for designing transformer tanks.

37.2 Mathematical Model

37.2.1 Pressure Source Model

The modified evaporative condensation model is used instead of arc decomposition transformer oil to produce gas During the cavitation process, liquid–gas mass transfer is controlled by the vapor transport equations:

$$\frac{\partial}{\partial t}(\alpha_v \rho_v) + \nabla . \left(\alpha \rho_v \vec{V_v} \right) = \dot{m}_{lv} - \dot{m}_{vl} \tag{37.1}$$

where v represents gas phase, α_v is the steam accumulation score, ρ_v is the density of steam, V_v is the velocity of gas phase, m_{lv}, m_{vl} which is caused by evaporation and condensation, respectively. The unit of these rates is kg/(s m^3).

The condensing mass transfer rate $\dot{m}_{vl}$ is changed to zero to simulate the process of arc decomposition transformer oil. A heating wall is arranged inside the oil tank, and arc power is used as heat source input instead of arc energy transfer.

37.2.2 Mechanical Model of Oil Tank

The mechanical model of oil tank is composed of transient equilibrium equation, constitutive equation and compatibility equation [4].

The transient equilibrium equation is used to describe the load-stress relationship,

$$\rho_\iota \frac{\partial^2 u}{\partial t^2} - \nabla \cdot \sigma_\iota = F_p \tag{37.2}$$

Where, ρ_ι is the wall density of the tank, and u is the position shift of the element; $\boldsymbol{\sigma}_\iota$ is its stress tensor; F_p is the density of the fault pressure acting on the element and its magnitude is equal to the ratio between the pressure p_ι on the box wall and the thickness d of the box wall.

According to Hooke-Duhamel's law, the constitutive equation used to describe the stress–strain relationship of the oil tank is

$$\sigma - \sigma_0 = C_{:}\left(\varepsilon - \varepsilon_0 - \varepsilon_{\mathrm{p}} - \varepsilon_{\mathrm{lh}}\right) \tag{37.3}$$

where, $\boldsymbol{\sigma}_0$ and $\boldsymbol{\varepsilon}_0$ are the initial stress and initial strain, respectively; C is the elasticity tensor of order 4; $\boldsymbol{\varepsilon}_p$ Is the deformation tensor of the material.

In the constitutive Eq. (37.3), the relationship between the total strain tensor and the position shift of the element is expressed by the compatibility equation

$$\varepsilon = \frac{1}{2}[(\nabla u) + \left(\nabla u\right)^T] \tag{37.4}$$

37.2.3 Mechanical Properties of the Oil Tank

The transformer oil tank material used in the simulation is carbon structural steel, which has multilinear isotropic properties. The elastic modulus of the iron shell is 200 GPa, Poisson's ratio is 0.3, the yield stress is 250 MPa, the corresponding elastic strain limit is 0.1775%, and the hard tensile strength is 460 MPa.

(a) geometric model (b) finite element model

Fig. 37.1 Three-dimensional model of transformer

37.3 Transformer Simulation Model

Based on the actual size of the transformer, a three-dimensional geometrical model is established. Figure 37.1 displays the geometrical model and the finite element mesh model of the transformer. A heating wall is implemented within the transformer to substitute arc heat transfer. Using the finite element method, 1,457,411 tetrahedral elements with 2,197,949 vertices are divided in this paper. The time step in this example is 10 μs.

37.4 Simulation Analysis

The modelled point of failure is placed at the tank's center, with experimental arc power used as the heat flux input rather than bubble-generating arc decomposition of the surrounding liquid. As presented in Fig. 37.2, the deformation diagram of the oil tank displays a conspicuous bulging effect throughout the whole tank, accompanied by outward expansion of the surrounding walls, bending of the turret, and the oil pipeline section. It is evident that, owing to gravity, the extent of shape variance at the lower region of the tank is larger than that at the upper end, whereas the variable's magnitude at the tank's side is markedly greater than that at the front side.

As shown in Fig. 37.3, the stress distribution in the oil tank at the end of the simulation, the maximum stress reaches 8.4 MPa. Due to the obstructing and attenuating effect on the propagation of the pressure wave by the connecting pipeline between the oil pillow and the oil tank, the value of the stress at the oil pillow is at a lower level during the whole process. The stress in some characteristic positions of the tank is significantly greater than others, that is, the stress concentration point of the tank. The stress at the connection between the turret and the tank wall and the connection between the pipe and the top of the tank is significantly greater than other places. Because the stress concentration will more easily lead to mechanical fatigue of the tank, so that the area is more prone to fracture, so strengthening the mechanical

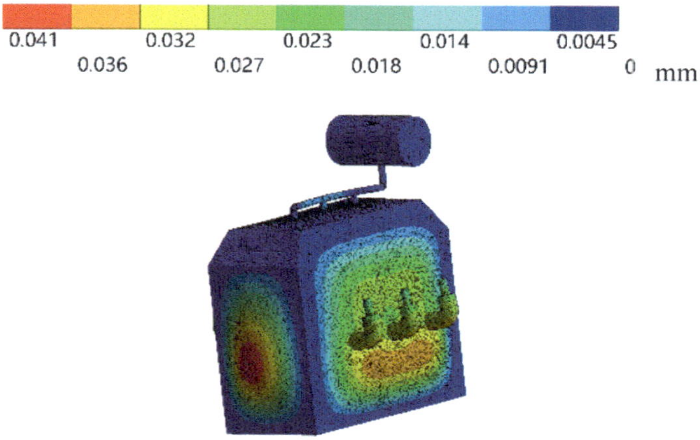

Fig. 37.2 Total deformation of transformer oil tank

strength of the area can effectively improve the overall explosion-proof ability of the transformer tank.

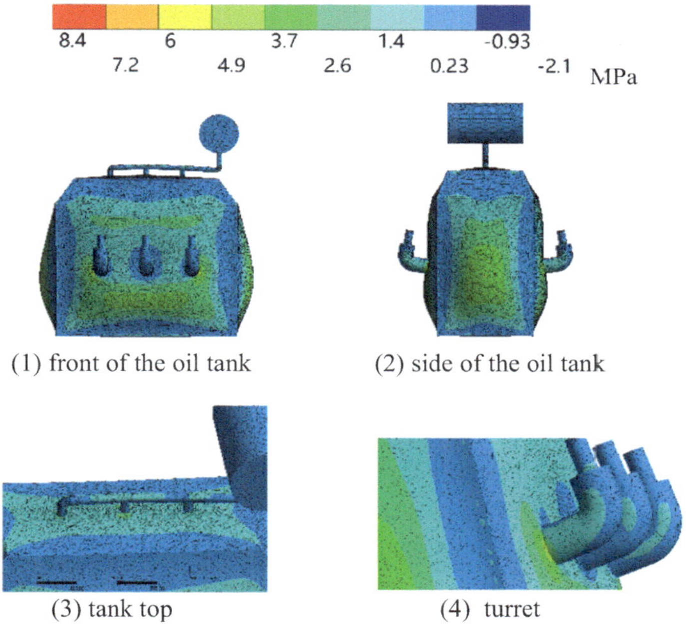

Fig. 37.3 Stress distribution of oil tank

37.5 Conclusion

The simulation results show that.

1. Under the influence of gravity, the shape variable at the bottom of the tank is larger than that at the top of the tank. The shape variable and stress value of the side of the tank are obviously greater than that of the front of the tank.
2. Due to the connection pipe between the oil pillow and the oil tank, the propagation of the pressure wave is hindered and attenuated. In the whole process, the stress value of the oil pillow is at a low level;
3. The connection between the turret and the wall is a stress concentration point, strengthening the mechanical strength of this area can effectively improve the overall explosion-proof capability of the tank.

Acknowledgements This paper is solely funded by the Technology Project of State Grid Zhejiang Electric Power Company of China (B311SX230002).

References

1. L. Zehong, Findings in development of ± 1100 kV UHVDC transmission. Proc. CSEE **40**(23) (2020)
2. C.G. Yan, Modeling and simulation of deformation rupture of power transformer tank. Electric Eng. **31**(3), 180–187 (2016)
3. B. Li et al., Modeling and analysis of tank rupture due to internal electrical fault in oil-filled power transformer, in *2021 International Conference on Advanced Electrical Equipment and Reliable Operation (AEERO)* (IEEE, Beijing, China, 2021), pp. 1–5
4. R. Courant, On the partial difference equations of mathematical physics. IBM J. Res. Dev. **11**(2), 215–234 (1967)

Chapter 38
A Reduced Order Model for Arc Heat Transfer in Oil Based on POD-LSTM

Shuwen Zhang, He Ren, and Linlin Zhong

Abstract In real-time simulation applications, the full-order model often necessitates lengthy computation times, rendering it inadequate to meet the demands of real-time performance. In pursuit of establishing a digital twin for predicting the arc temperature field within transformer oil, this paper introduces a reduced-order model capable of rapidly solving the temperature distribution. Initially, a full-order model of arc heat transfer within transformer oil is developed employing the finite volume method. Subsequently, Proper Orthogonal Decomposition (POD) is employed for feature analysis, followed by dimensionality reduction based on an energy ratio criterion to capture the predominant modes. Finally, a prediction model for each primary modal coefficient is constructed using a Long Short-Term Memory (LSTM) network. The effectiveness of this approach is demonstrated through numerical simulations of a two-dimensional arc heat transfer multiphase flow model. The results indicate that the calculation outputs from the reduced-order model meet the specified error criteria. Concurrently, the computational time is dramatically reduced from hours to seconds. This research underscores the accuracy and timeliness of the reduced-order model, enhancing computational efficiency while maintaining the precision of the digital twin model.

38.1 Introduction

The digital twin can map the spatial scale and life cycle of physical entities to the digital realm [1]. In power system, the digital twin emerges as a key enabler for comprehensive life cycle management of power equipment [2, 3]. Multiphysics modeling is one of the essential technologies in constructing digital twin system. While traditional numerical methods (such as finite element and finite volume

S. Zhang (✉)
School of Software, Southeast University (Suzhou), Suzhou 215127, China
e-mail: zhangshuwen@seu.edu.cn

H. Ren · L. Zhong
School of Electric Engineering, Southeast University, Nanjing 210096, China

Z. Fang et al. (eds.), *Proceedings of the 5th International Symposium on Plasma and Energy Conversion*, Springer Proceedings in Physics 398,
https://doi.org/10.1007/978-981-97-2245-7_38

methods) have been widely used in multi-physics simulation, they fall short when applied to digital twin modeling due to their real-time performance and high calculation speed requirements [4]. For instance, the modeling of arc heat transfer in transformer oil, with its involvement in multiphase flow, poses a complex challenge of multi-physics simulation. Hence, a reduced-order model with reduced computational requirements becomes essential for real-time simulation operation. This is particularly significant for constructing a digital twin model describing the arc heat transfer in transformer oil.

Consequently, the research landscape has evolved towards developing corresponding reduced-order models (ROM). ROMs utilize simplified mathematical models to express the input/output relationship of computational fluid dynamics (CFD) systems, capturing essential physical characteristics. Proper Orthogonal Decomposition (POD), derived from statistical analysis, is widely used in ROM for various dynamic systems [5]. The combination of POD and the finite element method allows for the rapid calculation of multi-physics models [6]. In the field of heat transfer, scholars have extensively studied the reduced-order model of transient heat conduction. For instance, Fic et al. effectively improved computational efficiency by applying the classical Fourier law to model the thermal diffusion and using the POD method to extract classical orthogonal functions [7]. In the realm of power equipment, Xie et al. studied the transient temperature rise in oil-immersed power transformer windings, and used the POD method to reduce the order of the model and achieve efficiency improvements [8].

Despite these advancements, there remains a gap in research on model reduction in the field of arc heat transfer in transformer oil. This paper introduces the Long Short-Term Memory (LSTM) surrogate model based on the basic idea of POD. The model focuses on establishing a reduced-order model for arc heat transfer in oil. Results from a two-dimensional multiphase flow model demonstrate the effectiveness of the ROM. While ensuring calculation accuracy, the proposed method significantly reduces calculation time from hours to seconds. The POD-LSTM method and ROM hold promise for practical engineering applications for digital twins.

38.2 Methodology

38.2.1 Proper Orthogonal Decomposition (POD)

Proper Orthogonal Decomposition (POD) serves as an analysis method for high-dimensional data. Leveraging projection technology, data in a high-dimensional space is condensed into a low-dimensional subspace, facilitating order reduction [9]. The proper method processes the data to extract a set of optimal basis functions ϕ from a vast array of known system solutions X. This extraction ensures that elements in the data set X project most significantly onto this basis set.

$$\max_{\phi} \sum_{k=1}^{m} \frac{(X_k, \phi)^2}{\|\phi\|^2}, \ \Phi^T \Phi = I \tag{38.1}$$

where $(\cdot,\cdot)$ denotes the inner product operation, and vector ϕ represents the orthogonal basis of POD. The matrix Φ comprises the set of POD orthogonal bases. In practice, the physical attribute values (e.g., pressure, speed, temperature) at each node over m discrete time steps, obtained through simulation, form a snapshot matrix—a column vector in n dimensions, where n is the number of grid nodes in the simulation. Singular value decomposition of this matrix allows for model order reduction by truncating the number of feature vectors. The contribution of each POD basis to the matrix is assessed through its eigenvalues.

$$\sum_{i=1}^{r} \left\| \overline{\Phi}^{\mathrm{T}} x^i \right\|_2^2 = \sum_{j=1}^{r} \lambda_j \tag{38.2}$$

where $\overline{\Phi}$ denotes the truncated POD vector, λ_j is the corresponding eigenvalue, and the subscript 2 indicates the distance norm.

38.2.2 Long Short-Term Memory (LSTM)

Long Short-Term Memory (LSTM) [10], a specialized type of recurrent neural network (RNN), addresses the challenge of gradient disappearance and explosion encountered by traditional RNN models when dealing with lengthy sequences. Its design incorporates a complex internal structure comprising three pivotal gating units: the forget gate, input gate, and output gate. These units empower LSTM to selectively forget, store, and read information, effectively managing the state of cells. Each LSTM module takes the input of the current time step and the hidden state from the previous time step. The output is computed through the fully connected layer with the Sigmoid activation function.

38.2.3 POD-LSTM

The fundamental concept behind the POD method is to utilize the results of a full-order model's calculations to establish an optimal set of orthogonal bases that comprehensively capture the dynamic characteristics of the system. The model is then reduced by selecting an appropriate number of these POD bases. Once the suitable POD basis modes are obtained, snapshots of the full-order model at different time instances are projected into the reduced subspace. The overall process architecture of POD-LSTM is illustrated in Fig. 38.1.

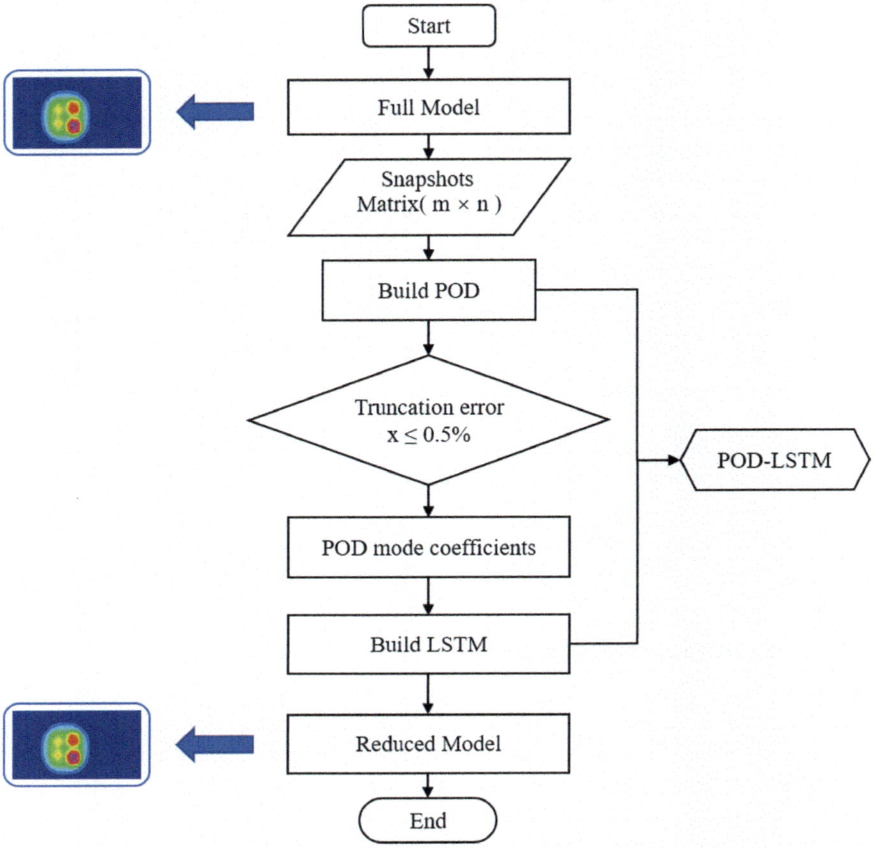

Fig. 38.1 Construction process of reduced-order model

For the snapshot matrix X, composed of the full-order model, the initial step involves taking the average value of snapshots at different times, resulting in the residual snapshot.

$$\overline{x} = \sum_{k=1}^{m} X_k / m \tag{38.3}$$

$$x_k = X_k - \overline{x} \tag{38.4}$$

The feature vector ϕ_0 is derived through singular value decomposition of the snapshot matrix. As the initial portion of the vector basis encapsulates the primary features of all samples, the features of the entire sample space can be approximated by truncating the appropriate number of POD bases ϕ. The truncation follows a principle calculated based on the energy capture ratio. The total energy $E = \sum_{i=1}^{m} \lambda_i$ is defined, allowing the energy contribution rate of a single ϕ_i calculated as $k_i = \lambda_i/E$. After intercepting the first r POD bases according to the desired energy ratio, the snapshot matrix is projected into the subspace to obtain.

$$X \approx [\phi_1, \phi_2, \ldots, \phi_r][\alpha_i^1, \alpha_i^2, \ldots, \alpha_i^m]^{\mathrm{T}} + \overline{x} \tag{38.5}$$

Next, LSTM is employed to model the time series of modal coefficients obtained through POD decomposition. The constructed model satisfies

$$\breve{\alpha}^{m+k} = [\breve{\alpha}_1^{m+k}, \breve{\alpha}_2^{m+k}, \ldots, \breve{\alpha}_r^{m+k}] = N_{lstm}(k) \tag{38.6}$$

where $N_{lstm}(k)$ denotes the kth snapshot prediction result obtained by N_{lstm}. Therefore, the predicted results can be calculated as follows.

$$\breve{X}^{r+k} = [\phi_1, \phi_2, \ldots, \phi_r]\breve{\alpha}^{m+k} + \overline{x} \tag{38.7}$$

38.3 Case Study of 2D Arc Heat Transfer Model

38.3.1 Full Model of 2D Arc Heat Transfer

In this paper, a mathematical model for the two-dimensional compressible gas–liquid two-phase flow process is established based on the Navier–Stokes equations. The finite volume method is employed for solving the model, with a volume function model selected for gas–liquid two-phase flow calculations. The computational domain is rectangular, measuring 1.5 m in length and 0.5 m in width. A uniform grid division approach is adopted, resulting in 7500 grids. The fault discharge position is located at 0.26 m from the x-axis and 0.03 m from the y-axis, with a discharge energy of 6 MJ and a discharge time of 60 ms. The governing equations are as follows.

Mass conservation:

$$\frac{\partial \rho}{\partial t} + \frac{\partial}{\partial x}(\rho V) = 0 \tag{38.8}$$

Momentum conservation:

$$\frac{\partial}{\partial t}(\rho \vec{v}) + \nabla \cdot (\rho v v) = -\nabla p + \nabla \cdot \left[\mu\left(\nabla v + \nabla v^T\right)\right] + \rho g + F \tag{38.9}$$

Energy conservation:

Table 38.1 Medium properties

Properties	Oil	Gas
Density (kg/m^3)	895	0.08189
Thermal conductivity (W/(m K))	0.128	0.1672
Viscosity (kg/(m s))	0.97	8.411×10^{-6}
Specific heat (kJ/(kg K))	1.8	$6.88 + 0.000066\text{T} + 0.279 \times 10^{-6}\text{T}^2$

$$\frac{\partial}{\partial t}(\rho E) + \nabla \cdot [v(\rho E + p)] = \nabla \cdot \left(k_{eff} \nabla T\right) + S_h \tag{38.10}$$

where ρ is the fluid density; t is time; V is the volume; x is displacement; ν is speed; p is pressure; μ is fluid dynamic viscosity; g is the acceleration of gravity; F is the additional volume force (buoyancy, electromagnetic force, etc.); E is the total energy; k_{eff} is the effective thermal conductivity; T is temperature; S_h is the volume heat source. In addition, the solution model includes the gas state equation and the liquid state equation, as shown in Eqs. (38.11) and (38.12), respectively.

$$P = \rho RT \tag{38.11}$$

$$\left(\frac{\rho}{\rho_0}\right)^n = \frac{\kappa}{\kappa_0} \tag{38.12}$$

where P is pressure; ρ is fluid density; R is the molar gas constant; T is temperature; ρ_0 is the liquid density under the reference pressure; n is the density index; κ is the bulk modulus under pressure; κ_0 is the bulk modulus under the reference pressure. The physical parameters of gas–liquid two-phase flow are listed in Table 38.1.

38.3.2 Comparison Between Full- and Reduced-Model

The total simulation duration is 500 ms with a time step of 1 ms. Results from 0 to 450 ms constitute the training set, while the remaining 50 ms forms the test set. Figure 38.2 displays the continuous arc discharge calculation results for the initial 60 ms from the full-order model, capturing a multiphase flow process. Within this timeframe, the oil temperature in proximity to the arc steadily increases, followed by the onset of insulating oil vaporization.

The POD-LSTM method is employed to construct the reduced-order model. Figure 38.3 illustrates the energy ratios of POD bases at various orders. The abscissa represents the POD base, and the ordinate denotes the energy ratio. Notably, the energy proportion for the first three modes surpasses 0.8, followed by a sharp decline in subsequent modal energies. This energy decline aligns with prior POD research.

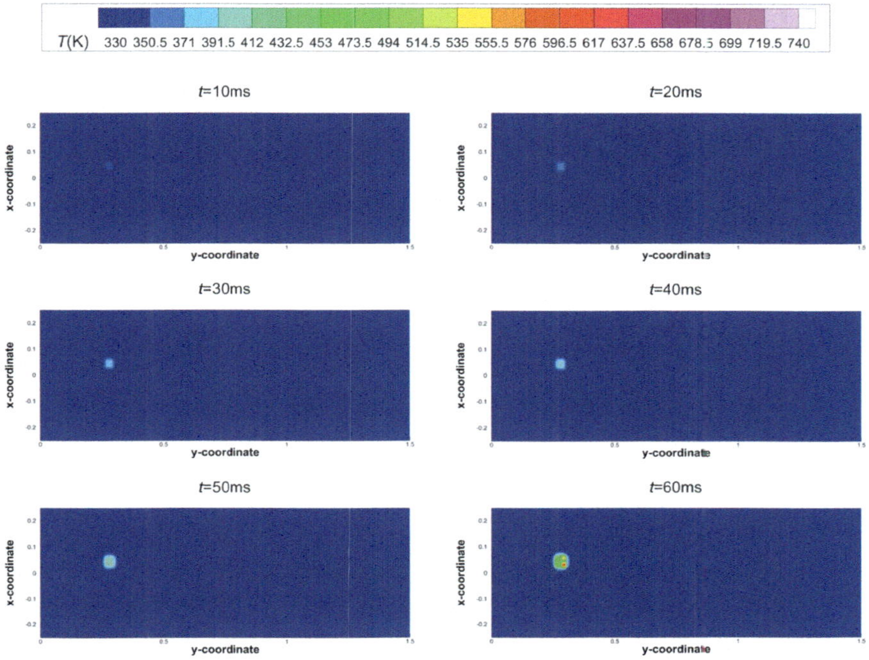

Fig. 38.2 Calculated temperature field of full-model

For this study, given that the first 15 basis modes encapsulate the majority of temperature field characteristics, 15 is chosen as the number of POD bases for model reduction.

To quantitatively assess the disparity between the full-order model and the reduced-order model's calculation results, this article presents the reduced-order model's outcomes within 60 ms (Fig. 38.4). Figure 38.5 illustrates the absolute errors between the full-order and reduced-order models across different time periods. Notably, the reduced-order model exhibits a calculation error within $\pm$ 2 K, aligning with POD's specified error criteria.

Table 38.2 provides insights into the calculation time for reduced and full models of varying orders. Notably, the full-order model necessitates approximately 9746 s, while the reduced-order model achieves computations at a second-level pace. In conclusion, leveraging the reduced-order model for transient temperature field problem-solving significantly reduces calculation time while maintaining accuracy requirements.

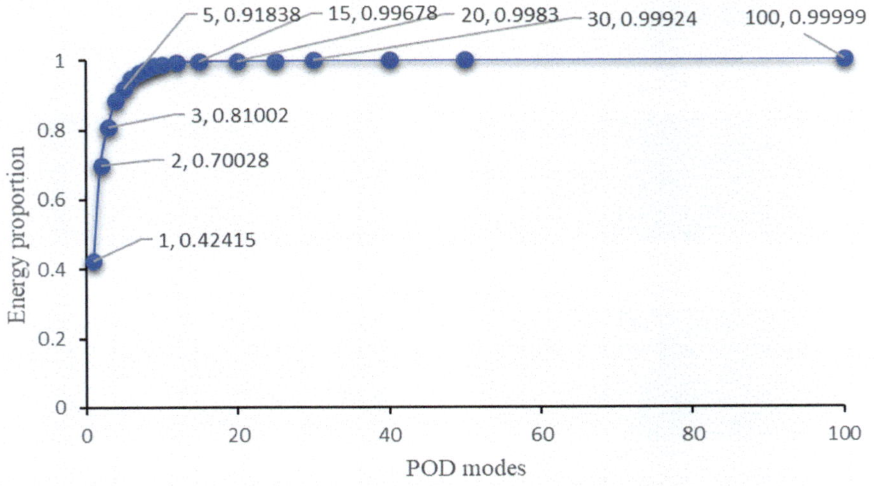

Fig. 38.3 Energy proportion of POD modes

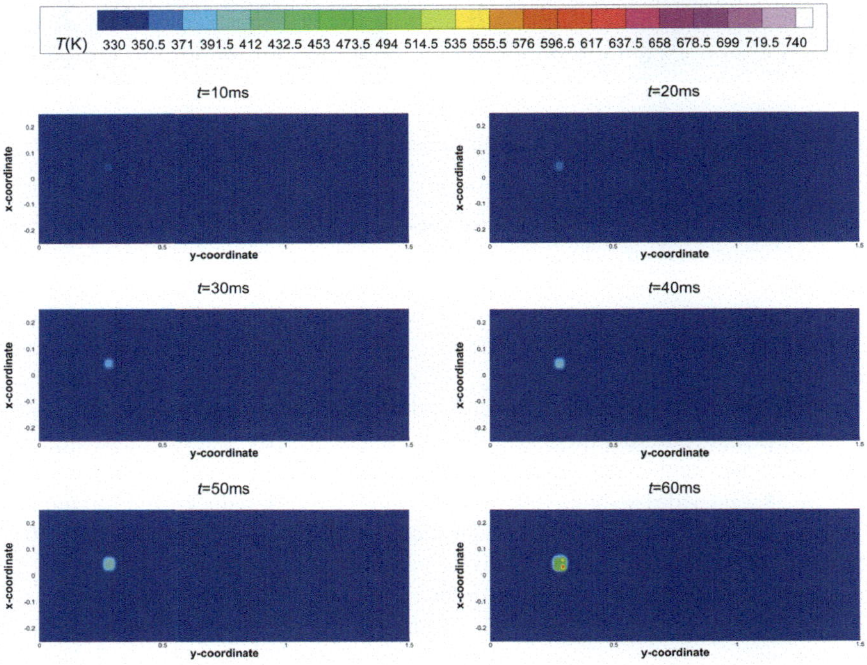

Fig. 38.4 Calculated temperature field of reduced-model

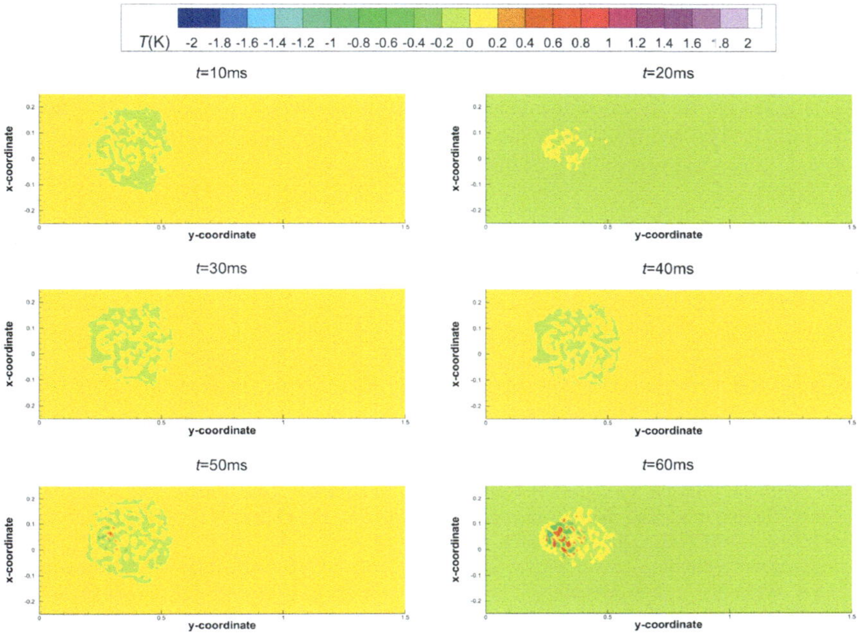

Fig. 38.5 Calculation error between full and reduced model

Table 38.2 Computation time of different models

Method	Number of selected POD modes	E/pa	Training time/s	Calculating time
Full model	/	6740	0	9746.1
POD-LSTM-1	5	6190	496	3.2
POD-LSTM-2	10	6660	987	5.8
POD-LSTM-3	15	6719	1435	8.9
POD-LSTM-4	20	6729	1973	12.4

38.4 Conclusions

This paper establishes a 2D arc heat transfer multiphase flow reduction model using the POD-LSTM method. Initially, a finite volume full-order model for the transient temperature field is developed, followed by the proposal of a POD-LSTM digital twin model reduction calculation method. The energy distribution of POD bases with different orders is analyzed, and an optimal order is selected to construct the reduced-order model. The conclusions are as follows:

1. A minimal number of POD bases capture the majority of energy in the flow field, meeting accuracy requirements. Increasing the number of orthogonal bases correlates with a higher energy ratio, approaching that of the full-order model.
2. Comparative analysis of the calculation time for models of different orders is provided. The reduced-order model significantly reduces calculation time from hours to seconds compared to the full-order model.

References

1. F. Tao, H. Zhang, A. Liu, A.Y.C. Nee, Digital twin in industry: state-of-the-art. IEEE Trans. Ind. Inf. **15**(4), 2405–2415 (2019). https://doi.org/10.1109/TII.2018.2873186
2. P. Moutis, O. Alizadeh-Mousavi, Digital twin of distribution power transformer for real-time monitoring of medium voltage from low voltage measurements. IEEE Trans. Power Delivery **36**(4), 1952–1963 (2021). https://doi.org/10.1109/TPWRD.2020.3017355
3. M. Fahim, V. Sharma, T.-V. Cao, B. Canberk, T.Q. Duong, Machine learning-based digital twin for predictive modeling in wind turbines. IEEE Access **10**, 14184–14194 (2022). https://doi.org/10.1109/ACCESS.2022.3147602
4. Science and technology development service center of China Electric Power Enterprises Federation. White paper on Application of digital twin technology in electric power industry [R/OL] [2022-07-02]. https://www.thepaper.cn/newsDetail_forward_19289460
5. P. Frame, A. Towne, Space-time POD and the Hankel matrix. PLoS ONE **18**(8), e0289637 (2023)
6. G. Jiang et al., Online reconstruction of 3D temperature field fused with POD-based reduced order approach and sparse sensor data. Int. J. Thermal Sci. **175**, 107489 (2022)
7. A. Fic, R.A. Bialecki, A.J. Kassab, Solving transient nonlinear heat conduction problems by proper orthogonal decomposition and the finite-element method, in *CHT-04 - Advances in Computational Heat Transfer III. Proceedings of the Third International Symposium 2004.*
8. Xie et al., Multi-physics coupling calculation method for winding temperature rise of oil-immersed power transformer. Chin. J. Electr. Eng. **36**(21), 5957–5965+6040 (2016). https://doi.org/10.13334/j.0258-8013.pcsee.160806
9. K.Z. Wu, Congchang, Snapshot POD analysis of transient flow in the pilot stage of a jet pipe servo valve. J. Turbulence **19**(10) (2018)
10. S. Hochreiter, J. Schmidhuber, Long short-term memory. Neural Comput. **9**, 1735–1780 (1997)

Chapter 39
Research on Surface Charge Dissipation Characteristics of Epoxy Resin Insulation Materials Under Different Aging Conditions

Zhaohui Song, Xin Li, Min Liu, Zehong Chen, Feng Wang, and She Chen

Abstract Epoxy resin is a commonly used insulating material in gas-insulated metal-enclosed switchgear (GIS), recognized for its commendable insulation, mechanical, thermal, and adhesive characteristics. The insulators of GIS circuit breakers experience varying degrees of deterioration due to different load, overvoltages, and environmental factors, which significantly affect the equipment replacement cycles. Currently, it is challenging to achieve non-destructive determination of the aging degree of epoxy resin insulators. This paper proposes a method for evaluating the insulation degradation by measuring the surface charge dissipation rate. Experiments were conducted using a four-axis motorized rotation platform for epoxy resin specimens under different thermal aging conditions. Surface charge dissipation tests were carried out after pre-charging, investigating the surface charge dissipation before and after aging. Combining SEM and FTIR tests on specimens, the results indicate that thermally aged epoxy resin specimens experience more severe degradation, with a significant increase in surface charge dissipation rate.

39.1 Introduction

Gas Insulated Switchgear (GIS) offers significant advantages, including a high transmission capacity, low transmission losses, and enhanced safety performance [1, 2]. Within GIS, the circuit breaker plays a pivotal role in control and protection. Its operational status directly impacts the stability and reliability of the entire power system. However, GIS features a complex structure, which translates into substantial routine maintenance work and high maintenance costs [3]. In the event of major repairs,

Z. Song · X. Li · M. Liu
Northwest Branch of State Grid Corporation of China, Xi'an 710001, China

Z. Chen · F. Wang · S. Chen (✉)
Hunan University, Changsha 410082, China
e-mail: 280777544@qq.com

Z. Fang et al. (eds.), *Proceedings of the 5th International Symposium on Plasma and Energy Conversion*, Springer Proceedings in Physics 398,
https://doi.org/10.1007/978-981-97-2245-7_39

complete disassembly is required, incurring both time and substantial expenses, and often leading to various new challenges.

Epoxy resin material plays a critical role in the insulation of GIS circuit breakers [4]. The degree of insulation degradation is intimately tied to the circuit breaker's lifespan. Over years of operation, during which it experiences various short-term voltage surges and prolonged periods of operating voltage, the surface resistivity of epoxy resin insulation components diminishes, and molecular chains break [5]. Severe degradation can lead to a significant reduction in flashover voltage and breakdown voltage, thereby substantially increasing the operational risks within the power system.

In recent years, the research focus on GIS flashovers has largely centered on the accumulation of surface charges on bushing insulators [6–9]. It is believed that surface charges are a significant factor contributing to the decline in flashover voltage [10–12]. Some researchers have manufactured epoxy resin samples from the same material as bushing insulators and subjected them to aging processes, including thermal aging, electrical-thermal combined aging, and humidity-thermal aging [13, 14]. Subsequent studies covered electrical, physical, and chemical properties, including activation energy, dielectric constant, breakdown voltage, and others, allowing for the analysis of the lifespan of the aged samples [15–17].

This article presents the establishment of an experimental platform using a four-axis motorized slide module for measuring surface charge on insulating components. The study focuses on the surface charge dissipation rate of insulating support components and insulating pull rods made from epoxy resin within circuit breakers. It encompasses the production of specimens using the same epoxy resin material, subjecting these specimens to thermal aging, and conducting studies on the surface charge dissipation rate, as well as SEM and FTIR analysis before and after aging.

39.2 Test Platform

39.2.1 Test Platform

The four-axis stepper motor rotation measurement platform, as illustrated in Fig. 39.1a, comprises four distinct components. The first component includes a rotating table controlled by a motor, on which the test subject is positioned, allowing continuous rotation in either a clockwise or counterclockwise direction within the XY plane. The second component consists of a passive capacitive probe and its fixture, which can be controlled by the motor to rotate in the XZ plane, either clockwise or counterclockwise, facilitating the vertical measurement of the curved surface of the test subject. The third component involves a control platform capable of moving along the Z-axis, enabling measurements at various heights of the supporting insulators. Lastly, the fourth component consists of a control platform that moves along

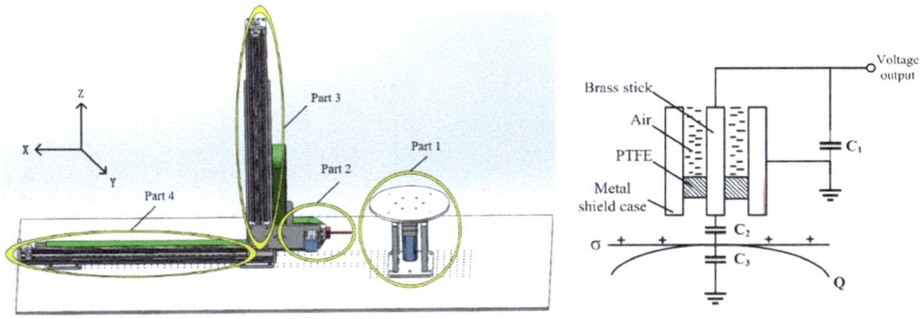

(a) Stepper motor slide module measurement platform

(b) Principle diagram of the linear scaling method for the capacitive probe

Fig. 39.1 **a** Stepper motor slide module measurement platform. **b** Principle diagram of the linear scaling method for the capacitive probe

the X-axis, regulating the distance between the passive capacitive probe and the insulator's surface.

In this study, a custom-made passive capacitive probe is utilized to measure the potential on the insulator's surface. By applying linear scaling to the probe, the measured potential can be converted into the magnitude of surface charge [18]. The results presented in this article are depicted as a curve illustrating the decrease in potential, which fundamentally reflects the dissipation of surface charge. The equivalent circuit diagram of the probe is provided in Fig. 39.1b.

39.2.2 Object and Method

Support Insulator The insulators chosen for this study are the supporting insulators of a 750 kV GIS, as depicted in Fig. 39.2. In this image, the upper position is located near the side close to the central high-voltage conductor, where it operates under consistently high electric field intensity. The lower end is connected to the housing, experiencing lower electric field intensity during operation. The measurement point is indicated by the yellow circle.

To prepare the insulator for measurement, non-woven fabric soaked in anhydrous ethanol is used to thoroughly clean the surface of the test insulator. It is left to sit for 12 h to ensure complete evaporation of the anhydrous ethanol. Then, the custom-made passive capacitive probe is used to measure the surface, ensuring that the initial potential is set to zero.

There are two methods to measure the surface charge dissipation of this supporting insulator:

The first method involves the direct application of voltage. Place the end of the supporting insulator with lower electric field intensity on the rotating table, with

Fig. 39.2 Schematic diagram of the measurement point location

the table grounded. Connect the metal fixture at the end with higher electric field intensity to a negative high-voltage DC power source. Apply voltage continuously for a specified period, then turn off the negative high-voltage DC power source. Initiate the slide stage module to measure the surface potential of the supporting insulator, thus obtaining the potential decline curve at that point.

The second method employs the preset charge approach. Apply voltage to a brass rod with a curved tip of 0.5 mm radius at one end, causing discharge at the tip. This action precharges a certain amount of charge on the surface of the supporting insulator at the measurement point. Measure the surface potential in that area to obtain the potential decline curve at that point.

Epoxy Resin Specimen In this study, epoxy resin specimens, composed of the same material as the supporting insulator, are utilized. Each specimen takes the form of a circular, thin sheet with a thickness of 1 mm and a diameter of 100 mm. There are three unaged specimens designated as ①, ②, and ③, alongside three specimens that have been subjected to aging for 25 days at 115 °C, also marked as ①, ②, and ③.

To prepare the specimens, they are placed on a 10 mm-thick, 150 mm-diameter aluminum metal disc, and the aluminum metal disc is grounded. In the center of each epoxy resin specimen, a certain amount of charge is precharged using the brass rod by applying voltage. The subsequent step involves measuring the curve depicting the potential decline on the surface of the specimens.

39.3 Result

39.3.1 Surface Charge Dissipation Rate

Support Insulator Align the tip of the brass rod precisely with the measurement point at a distance of 10 mm, and apply a voltage of − 10 kV to the brass rod for 5 min. Following the power shutdown, initiate the slide stage module to measure the measurement point. The results of this procedure are depicted in Fig. 39.3.

To precharge a specific charge quantity at the measurement point on the supporting insulator, voltage applications of − 30 and − 60 kV to the brass rod for 30 min are carried out. The outcomes of measuring the potential at this point are presented in Fig. 39.4a.

It is noteworthy that, owing to the application of a negative high-voltage DC power supply, the precharged charge on the surface in the test area is negative, leading to a negative surface potential. As the applied voltage increases, the quantity of charge at this point also increases. However, with higher charge amounts, the dissipation rate also accelerates.

At the 1000-s mark, the residual charge amounts for both applied voltage levels are nearly identical. This similarity persists until the 2000-s mark. Subsequently, the charge at − 30 kV applied voltage dissipates slowly, while the charge at the − 60 kV voltage level essentially ceases to dissipate. This phenomenon is likely due to the − 60 kV applied voltage, during the application process, providing the charges with more energy, trapping them in deeper traps, which makes it more challenging for them to escape from these traps.

Applying voltages of − 30 and − 60 kV to the metal fixture of the supporting insulator for a duration of 3 h yields the potential decline at the measurement point, as

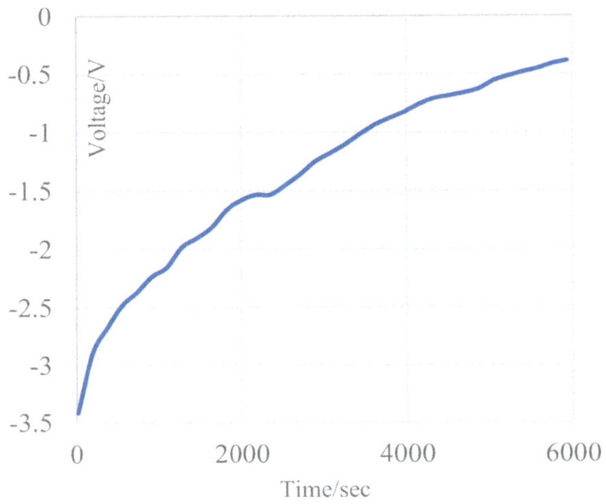

Fig. 39.3 Variation of surface potential at the measuring point of supporting insulator in the preset charge mode

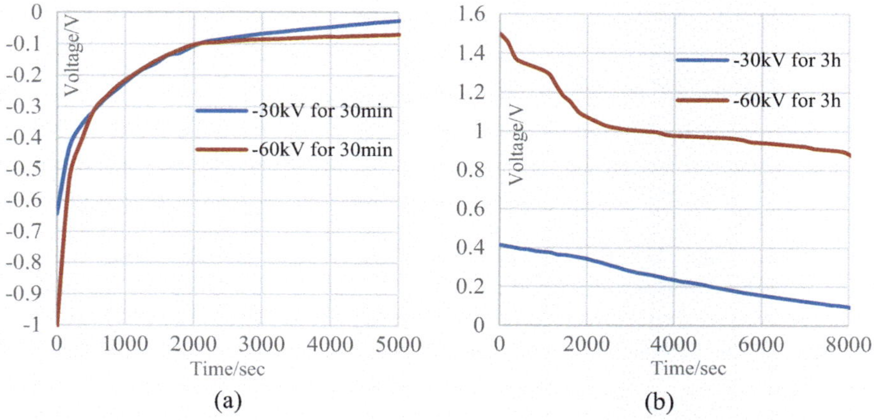

Fig. 39.4 Change curve of amplitude of applied voltage on surface potential

depicted in Fig. 39.4b. It's important to note that, despite the application of a negative voltage, the measured potential eventually becomes positive [8]. This phenomenon is attributed to the prolonged duration of voltage application, which results in a charge reversal effect.

Similarly, higher applied voltages lead to the accumulation of more charge. Under the − 30 kV applied voltage level, the charge dissipation exhibits a nearly linear declining trend. However, when subjected to the − 60 kV applied voltage level, the charge dissipation initially occurs rapidly, followed by a slowing trend.

Epoxy Resin Specimen Position the tip of the brass rod precisely at the center of the epoxy resin specimen, with the distance set at 10 mm. Apply pressure to the brass rod at 7.5 kV for 5 min. It can be observed that, whether the specimens are unaged or aged for 25 days, the static electric potential at the center of the specimens is approximately − 3 kV. After normalizing the data, the potential decline curve on the surface is presented in Fig. 39.5.

From the data, it can be concluded that unaged specimens exhibit a slower rate of surface potential decline, reducing by approximately ten percent over the course of three hours. In contrast, aged specimens experience a surface potential decrease of approximately fifty percent after four hours.

39.3.2 SEM

As shown in Fig. 39.6, the circled areas correspond to SEM results magnified 2000 times, while the areas without circles are results magnified 20,000 times. It is evident that unaged specimens exhibit the fewest surface defects, the supporting insulator

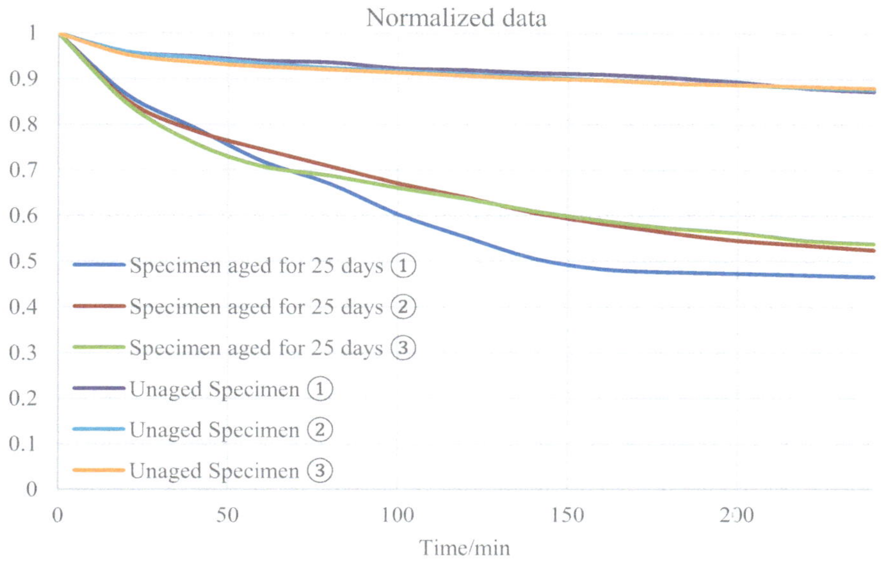

Fig. 39.5 Normalized curves of voltage drop changes for unaged specimen and specimen aged for 25 days

for 750 kV shows a relatively greater number of surface defects, and the specimens aged for 25 days have the highest number of surface defects.

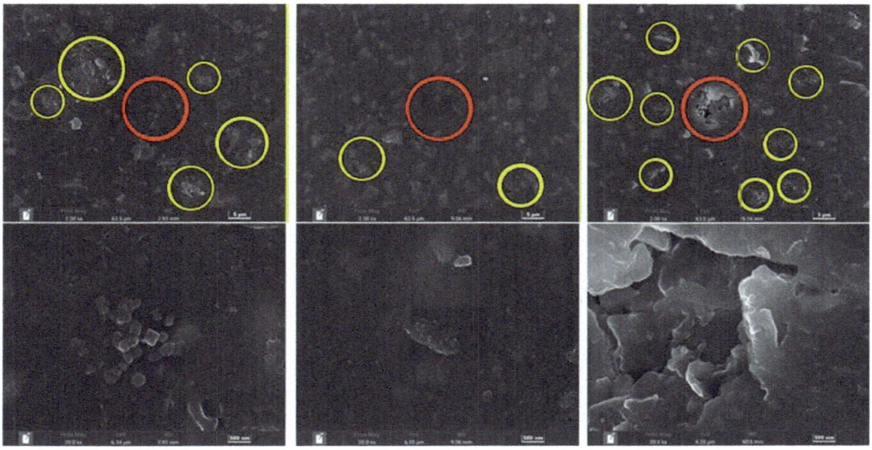

Fig. 39.6 SEM photo

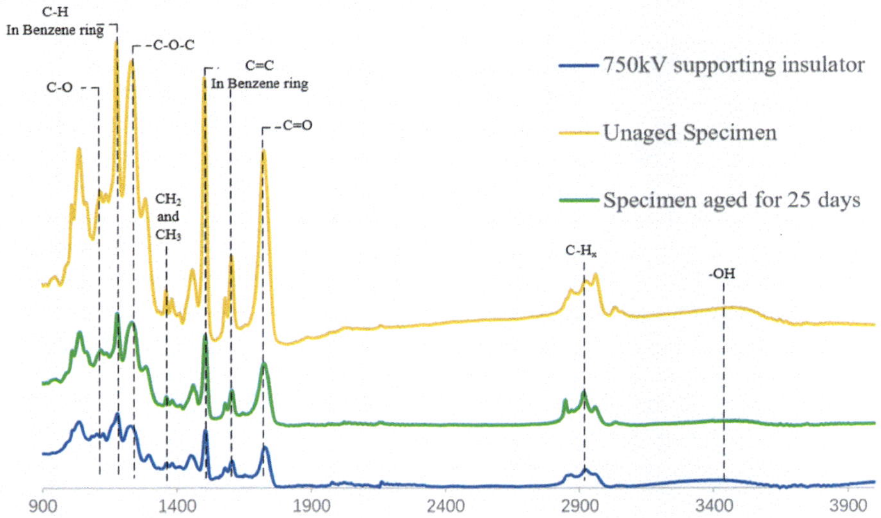

Fig. 39.7 FTIR line spectrum

39.3.3 FTIR

FTIR analysis was conducted on the 750 kV supporting insulator, unaged specimens, and specimens aged for 25 days, and the results are presented in Fig. 39.7.

Epoxy resin being an organic compound, all characteristic peaks contain C groups, particularly the C–H bond characteristic peak, which is most prominent. For instance, the strong characteristic absorption peak at 1180 cm^{-1} corresponds to the C–H bonds in the phenyl group. Multiple C-H characteristic peaks appear in the range of 2850–3037 cm^{-1}. Additionally, a characteristic peak for bending vibrations of CH_2 and CH_3 is observed at 1386 cm^{-1}. A C–O–C bending vibration characteristic peak is detected at 1237 cm^{-1}, and a characteristic peak for the C=O bond is observed at 1732 cm^{-1}, indicating the presence of ester groups in the specimens. Peaks corresponding to C=C bonds appear at 1506 and 1605 cm^{-1} due to para-substituted benzene rings. In summary, the primary chemical groups in the epoxy resin specimens consist of C, H, and O.

It can be observed that the unaged specimens exhibit the highest peak values for various chemical bonds. In contrast, the specimens aged for 25 days display a significant decrease in the peak values for all characteristic peaks. The results for the supporting insulator closely resemble those of the specimens aged for 25 days.

39.4 Conclusions

This study investigates the surface charge dissipation of supporting insulator and specimens made of epoxy resin before and after aging. The aged specimens exhibit molecular chain breakage and an increased number of surface defects. Both the 750 kV supporting insulator and the insulating pull rod, after being in operation, display a relatively faster rate of surface charge dissipation. Similarly, the aged epoxy resin specimens also demonstrate a significantly higher surface charge dissipation rate compared to their unaged counterparts. This perspective provides a novel approach to the assessment of insulator insulation degradation.

References

1. K. Hermann, *Gas Insulated Transmission Lines (GIL)* (Wiley, Chichester, UK, 2011)
2. J. Fang, Gas insulated metal enclosed transmission line (GIL) and its application. Electric Power **41**(8), 38–43 (2008)
3. K. Huang, Research on the evaluation of remaining life of high-voltage circuit breakers. MS dissertation, North China Electric Power University, China (2013)
4. Y. Zhang, Exploration on the characteristics of new ceramic materials for insulators used in SF6 gas insulated HVDC equipment. Proc. CSEE **41**(1), 174–182 (2021)
5. W. Zhang, An improved dielectric spectroscopy method to obtain activation energy of epoxy resin composite insulation. Insul. Mater. **53**(7), 98–105 (2020)
6. X. Qing, Distribution of polymer surface charge under DC voltage and its influence on surface flashover characteristics. IEEE Trans. Dielectr. Electr. Insul. **25**(6), 2157–2168 (2018)
7. M. Jingtan, Quantitative analysis on the influence of surface charges on flashover of insulators in SF6. IEEE Trans. Dielectr. Electr. Insul. **28**(1), 274–281 (2021)
8. D. Qiandong, Surface charge distribution on DC basin-type insulator. IEEE Trans. Dielectr. Electr. Insul. **26**(1), 17–25 (2019)
9. G. Zhang, Experiment study of surface charge accumulation and decay on a cone-type insulator in HVDC GIL High Voltage Eng. **41**(5), 1430–1436 (2015)
10. B. Lutz, Surface charge accumulation on cylindrical polymeric model insulators in air: simulation and measurement. IEEE Trans. Dielectr. Electr. Insul. **18**(6), 2040–2048 (2011)
11. A. Winter, Transient field distribution in gas-solid insulation systems under DC voltages. IEEE Trans. Dielectr. Electr. Insul. **21**(1), 116–128 (2014)
12. M. Schueller, Role of ion sources for spacer charging in SF6 Gas insulated HVDC system. IEEE Trans. Dielectr. Electr. Insul. **21**(1), 352–359 (2014)
13. Y. Liu, Aging characteristics of epoxy resin in hygrothermal environment. Conf. Proc. ISEIM **1**(24), 283–286 (2017)
14. Y. Wang, The effect of electrothermal aging on the properties of epoxy resin in dry-type transformer. Trans. China Electrotech. Soc. **33**(16), 3906–3916 (2018)
15. J. Huang, Experiment and analysis of permittivity curve transition of epoxy resin used in power equipment. Southern Power Syst. Technol. **17**, 1–6 (2023)
16. P. Ren, Residual lifetime expectancy prediction of the gis basin-type insulators based on thermal dynamics analysis. Proc. CSEE **39**(22), 6774–6782 (2019)
17. A. Winter, Stationary resistive field distribution along epoxy resin insulators in air under DC voltage. IEEE Trans. Dielectr. Electr. Insul. **19**(5), 1732–1739 (2012)
18. W. Feng, Insulator surface charge accumulation under impulse voltage. IEEE Trans. Dielectr. Electr. Insul. **11**(5), 847–854 (2004)

Chapter 40
Insulation Prediction and Descriptor Selection Based on Random Forest Algorithm

Nian Tang, Yanze Pang, Dongwei Sun, and Boya Zhang

Abstract To find more efficient and environmentally friendly substitutes for sulfur hexafluoride (SF_6), it is still necessary to screen more potential insulating gases. However, the commonly used experimental methods are inefficient. In this article, random forest (RF) algorithm are chosen to predict gas insulation strength with gas molecular descriptors. The origin of descriptors includes those used in existing journals (such as electric dipole moment μ (IEEE Trans Dielectr Electr Insul 20(3):Jun [1]), positive surface area, and surface area (Neurocomputing 216 [2])A_s) (J Comput Chem 38(10):721–729 [3]), and the highly weighted ones in our pre-experiment with RF (such as Hall-Kier alpha connected with charge distribution and Kappa to describe the degree of molecular curvature). The descriptors sorted by RF reveals more correlations between the molecular properties and insulation, and the trained model provides important insights to design and to screen other SF_6 substitutes.

40.1 Introduction

Sulfur hexafluoride (SF_6), a gas with good insulating properties, has been wildly used in the power system, which has brought great revolution in high voltage power equipment since 1967 [4]. However, SF_6 can terribly absorb the infrared radiation, and its global warming potential (GWP) is evaluated up to 23,500 times of CO_2. Worse still, the lifetime of SF_6 reaches 3200 years. In 1995, high GWP value of SF_6 had been noticed [5], and it was listed in Kyoto Protocol [6] to strictly limit. The Paris

N. Tang · D. Sun
Electric Power Research Institute of Guangdong Power Grid Co., Ltd, Guangzhou 510080, China

Sulfur Hexafluoride Key Lab of China Southern Power Grid, Guangzhou 510080, China

Y. Pang · B. Zhang (✉)
State Key Laboratory of Electrical Insulation and Power Equipment, Xi'an Jiaotong University, Xi'an 710049, China
e-mail: zhangby@xjtu.edu.cn

Z. Fang et al. (eds.), *Proceedings of the 5th International Symposium on Plasma and Energy Conversion*, Springer Proceedings in Physics 398,
https://doi.org/10.1007/978-981-97-2245-7_40

Agreement of 2016 [7] also explicitly called for zero net greenhouse gas emissions in the second half of this century.

Lately, many fluorinated gases were introduced as the insulation medium with low GWP. These gases have similar or even superior electrical strength to SF_6. Since 2015, the synthetic gases announced by 3M company, known as C_4F_7N and $C_5F_{10}O$, have drawn close attention [1, 8]. However, at present, no gas can perfectly replace SF_6. It indicates that we still should invent potent gases to replace SF_6.

As mentioned by Rabie [1], the breakdown experiment could be quite expensive, due to the expensive gas manufacturing. Meanwhile, directly theoretical analysis is unfeasible because the breakdown process includes the electron diffusion, drifting and reactions, each of which also needs the experimental data of electron impact cross sections [12]. After all, it is necessary to find a feasible scheme to predict the properties of the gases before manufacture efficiently.

Previous work indicates that some of the molecular properties are closely related to the electrical strength, such as electric dipole moment μ [1], positive surface area [2], and surface area A_s [3]. The potential properties could be the integrated optical absorption polarizability, ionization energy [1], spectrum [9], chemical hardness [10]. Rabie's work [1] is quite promised by testing 67 molecules and the best correlation coefficients R^2 were 0.71 for polar molecules and 0.92 for nonpolar molecules respectively. In Yu's work [10], 43 molecules were tested and the correlation coefficient R^2 could reach 0.985. In Wu's work [11, 12], positive surface area is also taken into consideration.

Considered factors are often limited, and the form of the expression can be subjectivity if only one empirical formula is selected to describe the relationship between the electrical strength and the descriptor of a gas. By contrast, by using machine learning algorithms, it could provide better efficiency and more diverse connections [12].

In the past few years, the nonlinear statistical methods based on machine learning algorithms have already been founded in the field of electrical engineering such as artificial neural network (ANN), support vector machine (SVM) and random forest (RF). The RF algorithm is finally selected in this paper thanks to such advantages [13]: reduced overfitting due to the law of large number of the decision trees, accuracy, robustness, flexibility, and variable importance. Especially, random forest algorithm provides variable importance, which makes the RF explicable and helps to identify the influential features to find out the underlying relationships in the data [14].

Overall, random forest is a good regression prediction algorithm, with the ability to handle complex data sets and provide accurate and robust predictions.

40.2 Method

40.2.1 Why Need to Filter Descriptors

To predict electrical strength by descriptors of gases, experimental data are collected in Table 40.1. After that, the molecular properties are calculated. In this project, the descriptors are got with rdkit and tkinter, two open source python libraries. Twenty percent were randomly selected as the test set and the rest as the training set.

Table 40.1 Table captions should be placed above the tables

Molecule_name	Er (rel SF_6)	Molecule_name	Er (rel SF_6)	Molecule_name	Er (rel SF_6)
c-C_4F_6	1.6	$CF_3CH = CF_2$	0.94	CHF_3	0.27
c-C_4F_8	1.25	$CF_3CH = CH2$	0.8	$CHF{=}CHCF_3$	0.4
c-C_4F_8O	1.38	CF_3Cl	2.41	CHFC12	0.92
c-C_5F_8	2.1	CF_3CN	1.34	CO	0.4
c-C_6F_{10}	1.9	$CF_3N{=}SF_2$	1.94	CO_2	0.35
c-C_6F_{12}	2.35	CF_3NO_2	1.5	$F_2C{=}CFCl$	0.72
c-C_7F_{14}	2.24	$CF_3OCF(CF_3)CN$	1.43	H_2	0.22
c-$CF_3(C_2F_2O)$ CF_3	1.6	CF_3OCF_3	1	HCCH	0.6
c-$CF_3(C_4F_6)$ CF_3	2.3	CF_3SCF_3	1.5	i-C_3F_7CN	2.2
$C_2F_{50}C_2F_5$	1.51	CF_3SO_2F	1.43	i-C_3F7COC_2F_5	2.8
C_2F_5CN	2	CF_4	0.43	i-$C_3F_7COCF_3$	2.1
C_2F_6	0.8	$CFCl_2CFCl_2$	2.37	N_2	0.38
C_3F_7CN	2.4	$CFCl_3$	1.84	N_2O	0.46
C_3F_8	0.98	$CH_2{=}CHCF_3$	0.8	NSF_3	1.4
C_4F_{10}	1.32	CH_2C_{12}	0.68	O_2	0.33
C_5F_{12}	1.75	CH_2F_2	0.27	OCS	0.9
C_6F_{14}	2.26	CH_2FCl	0.39	SeF_6	1.21
C_6F_6	1.4	CH_3Br	0.45	SF_3N	1.37
CCl_4	2.36	CH_3CF_2Cl	1.04	SF_5CF_3	1.53
CF_2ClBr	1.32	CH_3CF_3	0.41	SF_6	1
CF_2ClCF_2Cl	1.71	CH_3CH_2Cl	0.85	SO_2	0.52
$CF_3OC_2F_5$	1.24	CH_3CHCl_2	1.01	SO_2F_2	0.73
CF_3Br	0.76	CH_3Cl	0.32	SOF_2	1.42
CF_3CCCF_3	2.19	CH_3CN	0.8	CH_3I	1.12
CF_3CCl_3	2.47	CH_4	0.43	CF_3I	1.21
$CF_3CF{=}CF_2$	1.01	$CHCl_3$	1.77		

RandomForestRegressor in sklearn is called for training a RF model. All of the descriptors are sent into RF, so that potentially useful descriptors can be filtered out.

We tried to use this model directly as the final result, but when it is evaluated by Pearson correlation coefficient (PCC), we gave up.

PCC is a way to measure a linear relationship between two variables x and y. It is formally defined as in (40.1):

$$r = \frac{\sum (x_i - \overline{x}) \sum (y_i - \overline{y})}{\sqrt{\sum (x_i - \overline{x})^2} \sqrt{\sum (y_i - \overline{y})^2}} \tag{40.1}$$

In this project, PCC between electrical strength obtained by experiment and by this RF model is calculated by functions in sklearn. In fact, even though there is a high PCC in the training set (higher than 0.9), PCC of test set is only 0.5–0.7. It means that the model is overfitting, which may be caused by too few electrical data and too many descriptors. To be specific, there are only 77 kinds of gases, while the number of descriptors reach 250.

When carefully analyzing these descriptors, it will be found that many descriptors are not useful. For example, many descriptors are used to describe whether the organic matter contains certain groups (such as carboxylic acids, benzene ring, etc.). However, due to the boiling point limitations of environmentally friendly gases, most gases do not have these groups. Therefore, it is necessary to screening gas descriptors.

40.2.2 Determine Which Descriptors

Two approaches are taken to filter descriptors in this project. On the one hand, refer to the empirical formula constructed earlier, and select ones that can be easily calculated. On the other hand, select descriptors that occur more frequently with high weights according to the RF algorithm in Sect. 40.2.1. As in Fig. 40.1, compare these descriptors with those found in previous empirical formulas, and then screen out descriptors that may potentially affect the electrical strength of the gas.

In [1], the predictors were more straightforward and included those can be directly derived based on the molecular structure, which were: polarizability α, dipole moment μ, molecular mass m, highest occupied molecular orbital energy HOMO and lowest unoccupied molecular orbital energy LUMO. In [2], the surface area A_s has been shown to correlate with insulation strength. In [3], positive surface area seems more important than negative surface area. In [10], the electrostatic potential surface V_s, together with the molecular properties (including polarizability α and electronegativity χ), is used to drive further parameters, such as s_{tot}^2 (the statistical variance of the surface potential) and ν (degree of balance between the positive and negative surface potentials), which are defined in formula (40.2).

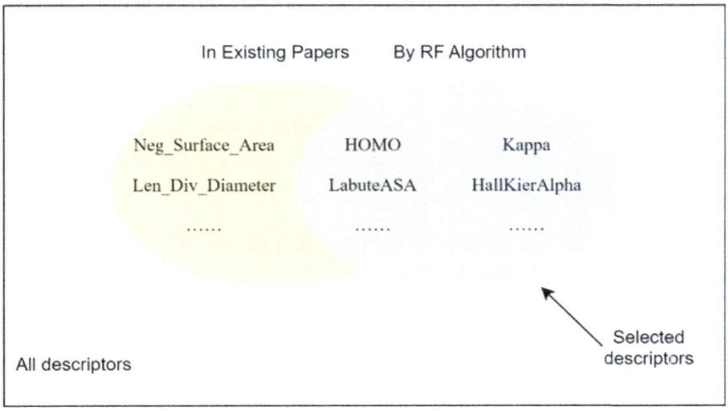

Fig. 40.1 Selected descriptors are from both existing papers and RF algorithm

$$\begin{cases} s_{tot}^2 = s_+^2 + s_-^2 \\ \quad = \frac{1}{n}\sum_{i=1}^{n}\left[V^+(r_i) - \overline{V_s^+}\right]^2 + \frac{1}{m}\sum_{i=1}^{m}\left[V^-(r_i) - \overline{V_s^-}\right]^2 \\ \Pi = \frac{1}{n+m}\sum_{i=1}^{m+n}\left|V_s(r_i) - \overline{V}_s\right| \\ \nu = \frac{s_+^2 + s_-^2}{\left(s_{tot}^2\right)} \end{cases} \tag{40.2}$$

where the indexes m and n are numbers of positive (V_s^+) and negative potentials (V_s^-) on the surface, respectively. $\overline{V_s^+}$ and $\overline{V_s^-}$ are the positive and negative average potentials, respectively. Parameter $\prod$ is the average deviation of V_s(r), corresponding to the local polarity of the molecule.

Based on the above literature, the following descriptors in Table 40.2 are selected. Due to differences in the names used for the same descriptors in different literature. Considering that our calculation is based on two open source python libraries, we choose the names from rdkit and tkinter to present here.

Table 40.2 Descriptors from earlier empirical formula

Selected descriptors	Selected descriptors	Selected descriptors
Weight	Len_Div_Diameter	Overall_Variance
HOMO	Dipole_Moment	Nu
LUMO	Volume	Pi
HOMO_LUMO_Gap	Density	MPI
Mol_Radius	Pos_Surface_Area	
LabuteASA	Neg_Surface_Area	

Table 40.3 Descriptors from RF algorithm

Selected descriptors	Selected descriptors
Kappa1	HallKierAlpha
Kappa2	MolLogP
Kappa3	MolMR

Meanwhile, excluding the descriptors appearing in the above table, here are ones from RF algorithm in Table 40.3.

Kappa1-3 describe the curvature of the space around atoms or groups in a molecule, which means that the shape of molecules is also a factor affecting electrical strength. HallKierAlpha is used to represent the polarity and charge distribution of molecules. MolLogP, the ratio of lipid solubility to water solubility of molecules. MolMR, the ratio of relative molecular weight to polarization of molecules.

40.2.3 *Establishing the RF Model*

With these descriptors, the RF for electrical strength can be established. Considering the data size, 20% of the gas is randomly selected as the test set, and the rest is the training set. The number of estimators (n_estimators) is 70, and with this number increasing, the result tends to converge; max_features is not limited, because the descriptors have been heavily screened in 2.2 and by comparison, not limiting the maximum number of features per tree performs better; max_depth is 17.

Due to the randomness of the RF algorithm and the fact that there are still few gases with known electrical strength, the output results are not stable. Especially the small number of test set samples further aggravates the randomness. While the test set usually has a correlation coefficient above 0.9, the training set has an unstable correlation coefficient. Even so, compared with no screening descriptors in 2.1, the scheme in this section still performs much better, both in terms of average performance and maximum correlation coefficient. Still, There is a solution to this instability. As long as you save the model that performs well on the test set, the good model can be used later.

40.3 Result

Figure 40.2 shows the result of one of the random forest models.

In Fig. 40.2, "exp" on the X-axis is the experimental value (true value) of the electrical strength of the gas; The Y-axis "pre" is the predicted value. The blue dots are the training set and the orange dots are the test set. The closer the data point is to the "y = x" line, the better the fit.

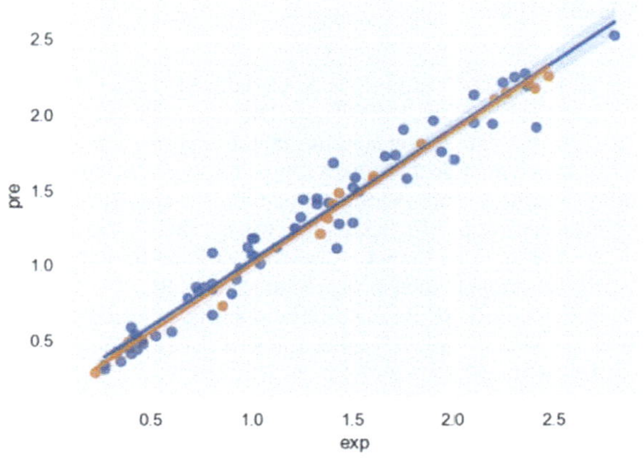

Fig. 40.2 The relationship between experimental values (true values) and predicted values in the test and training sets

In this model, the correlation coefficient for the training set is 0.929, while the correlation coefficient for the test set is 0.818. Although the correlation coefficient on the test set was slightly lower than on the training set (as expected), the data on the training set did not significantly deviate from the true value.

40.4 Conclusion

The relationship between gas electrical strength and gas descriptor is established in this paper. Firstly, all the calculated descriptors are put into the random forest algorithm, and some descriptors with high correlation are obtained. Subsequently, to avoid terribly overfitting, this paper searched the previous fitting formula on the electrical strength of gases, searched the descriptors used therein, merged them with the descriptors obtained by the method in this paper, and screened the final descriptors used. The random forest model is trained with the selected descriptor and the final model is obtained.

In this paper, by filtering descriptors and limiting the number of descriptors, overfitting caused by too little known data is avoided. Furthermore, Kappa1~3, HallKier-Alpha, MolLogP, and MolMR are proposed as related descriptors that affect the insulation of gases. In addition, the algorithm in this paper is cross-verified by test sets to ensure reliability.

References

1. M. Rabie, D.A. Dahl, S.M.A. Donald, M. Reiher, C.M. Franck, Predictors for gases of high electrical strength. IEEE Trans. Dielectrics Electr. Insul. **20**(3), Jun
2. H. Zhou, et al., A new sampling method in particle filter based on Pearson correlation coefficient. Neurocomputing **216** (2016). https://doi.org/10.1016/j.neucom.2016.07.036
3. Y. Xiaojuan, et al. Prediction on dielectric strength and boiling point of gaseous molecules for replacement of SF_6. J. Comput. Chem. **38**(10), 721–729 (2017)
4. M. Rabie, C.M. Franck, Assessment of eco-friendly gases for electrical insulation to replace the most potent industrial greenhouse gas SF_6
5. P. O'Connell, SF_6 in the electric industry, status 2000. Electra **200**, 16–25 (2002)
6. K. Protocol, 'United Nations framework convention on climate change. Kyoto Protocol Kyoto **19**, 497 (1997)
7. J. Rogelj, M. Den Elzen, N. Höhne, T. Fransen, T.H. Fekete, H. Winkler, R. Schaeffer, F. Sha, K. Riahi, M. pt Meinshausen, Nature **534**, 631 (2016)
8. J.D. Mantilla, N. Gariboldi, S. Grob, M. Claessens, 'Investigation of the insulation performance of a new gas mixture with extremely low GWP, in *Proceedings of IEEE Electrical Insulation Conference (EIC)*, Jun. 2014, pp. 469–473
9. N. Meurice, E. Sandre, A. Aslanides, D.P. Vercauteren, Simple theoretical estimation of the dielectric strength of gases. IEEE Trans. Dielectrics Electr. Insul. **11**(6), 946–948 (2004)
10. X. Yu, H. Hou, B. Wang, Prediction on dielectric strength and boiling point of gaseous molecules for replacement of SF_6. J. Comput. Chem. **38**(10), 721–729 (2017)
11. Y. Wu, C. Wang, H. Sun, M. Rong, A.B. Murphy, T. Li, J. Zhong, Z. Chen, F. Yang, C. Niu, Evaluation of SF_6-alternative gas C5-PFK based on arc extinguishing performance and electric strength. J. Phys. D, Appl. Phys. **50**(38) (2017), Art. no. 385202
12. H. Sun et al., Prediction of the electrical strength and boiling temperature of the substitutes for greenhouse gas SF_6 using neural network and random forest. IEEE Access **99**, 1–1 (2020)
13. L. Breiman, Random forests. Mach. Learn. **45**(1), 5–32 (2001)
14. Y. Amit, G. Blanchard, Multiple randomized classifiers: MRCL (2000)

Chapter 41
Molecular Electrical Strength Prediction Method Based on Machine Learning

Nian Tang, Jiaxin Tan, Dongwei Sun, and Boya Zhang

Abstract Sulfur hexafluoride (SF_6) has excellent insulation properties and is widely used in the field of electrical equipment. However, SF_6 exhibits an extremely strong greenhouse effect, with a global warming potential of 25,200 times that of carbon dioxide. In response to environmental concerns, finding alternative gases to SF_6 is a crucial issue that needs to be addressed in the field of electrical power equipment. The use of machine learning algorithms in this context is instrumental in accelerating the development of eco-friendly alternatives to SF_6. In this paper, we propose a method using machine learning and neural network techniques to predict the electrical strength of alternative gases. By establishing a neural network model, which learns the complex relationship between the known molecular descriptors and the electrical insulation strength, the electrical properties of various gases can be accurately predicted. The flexibility and versatility of this method can be used to guide the design of novel, environmentally friendly gases, and can quickly and efficiently screen a large number of molecular candidate materials with unknown insulation electrical strength.

41.1 Induction

Sulfur hexafluoride (SF_6) is widely used in the field of electrical equipment because of its good insulating properties [1]. However, SF_6 has a strong greenhouse effect, and its global warming potential is 25,200 times that of carbon dioxide (CO_2), so reducing the use of SF_6 is an important issue that needs to be solved in the field of power equipment to achieve green development, and the research on finding alternative gases of sulfur hexafluoride has attracted much attention [2].

N. Tang · D. Sun
Electric Power Research Institute of Guangdong Power Grid Co. Ltd, Guangzhou 510080, China

J. Tan · B. Zhang (✉)
Key Laboratory of Electrical Insulation and Power Equipment, Xi'an Jiaotong University, Xi'an, China
e-mail: zhangby@xjtu.edu.cn

Z. Fang et al. (eds.), *Proceedings of the 5th International Symposium on Plasma and Energy Conversion*, Springer Proceedings in Physics 398,
https://doi.org/10.1007/978-981-97-2245-7_41

Electrical strength is one of the key properties in screening alternative gases. Electrical strength is a measure of the properties of gas insulation. It can represent the maximum uniform electric field strength that the gas can withstand, and the gas will lose its insulation when the electric strength is exceeded. In power equipment, insulating gas or gas mixture with high electrical strength is required, and there have been many studies on this [3]. There are many precise ways to study electrical breakdown in gases: namely, breakdown experiments for the study of the breakdown phenomenon [4], or dynamic modeling of the breakdown process based on microscopic electron-molecular cross sections [5]. However, breakdown experiments are time-consuming, and dynamic modeling of breakdown processes requires the measurement of cross sections of all relevant processes. These methods consume time and resources, and cannot be quickly screened and evaluated in the face of large quantities of gases with unknown electrical strength. Therefore, a fast and accurate prediction method is needed to replace the traditional experimental method, and it is of great practical significance to develop a method based on computational model to predict the molecular insulation strength.

Machine learning as a powerful tool can accelerate the development of environmentally friendly gas alternatives. This selection can be based on transparent and cost-effective methods that can quickly identify candidate gases. In the past few decades, nonlinear statistical methods based on machine learning algorithms, such as artificial neural networks, support vector machines and random forests, have been developed and applied in the field of electrical engineering [6–8]. Because of its multi-layer structure and nonlinear activation function, neural network has strong flexibility and expression ability, and can learn complex nonlinear relationships. In contrast, the QSPR (Quantitative Structure–property Relation) equation regression prediction method [9] usually models the linear relationship between the feature and the target, and its expression ability is relatively weak. The method described in this paper promises to provide an innovative solution to this challenge by using machine learning and neural network techniques to predict the electrical insulation strength of gases.

In this paper, we propose a method using machine learning and neural network techniques to predict the electrical insulation strength of alternative gases. Molecular descriptors with high correlation with insulation strength are selected and a convolutional neural network model is established. By learning the complex relationship between known molecular descriptors and electrical insulation strength, the model can accurately predict the electrical properties of various gases.

41.2 Methods

41.2.1 Machine Learning

Machine learning is a subfield of the field of artificial intelligence that involves developing algorithms and models that can learn patterns and make predictions based on data. In the context of the search for alternatives to SF_6, machine learning algorithms can be a powerful tool for predicting the electrical insulation properties of gases, thus guiding the selection of gases in power equipment.

Machine learning algorithms can be used to learn the relationship between electrical strength and molecular properties, which can achieve better efficiency and accuracy, and have good stability. The neural networks used in this paper are a key technology in machine learning, which can handle various complex tasks by simulating the information transmission mode between neurons in the human brain. Neural networks are made up of artificial neurons that mimic biological neurons, the neurons receive the input, apply the weights, and generate the output through the activation function (Fig. 41.1). The input layer receives raw data, the hidden layer performs feature extraction and representation learning, and the output layer generates the final output of the model. The backpropagation algorithm is used to reduce errors by comparing the output of the network with the target value, calculating the gradient, and updating the weights in the network according to the gradient.

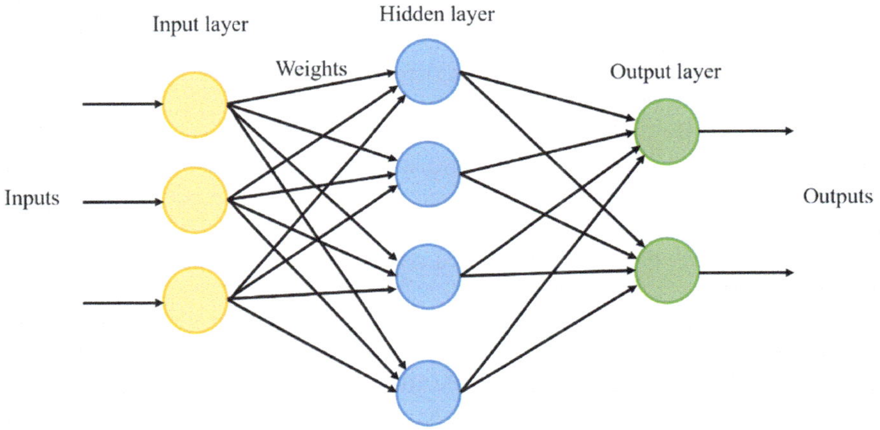

Fig. 41.1 Neural network structure diagram

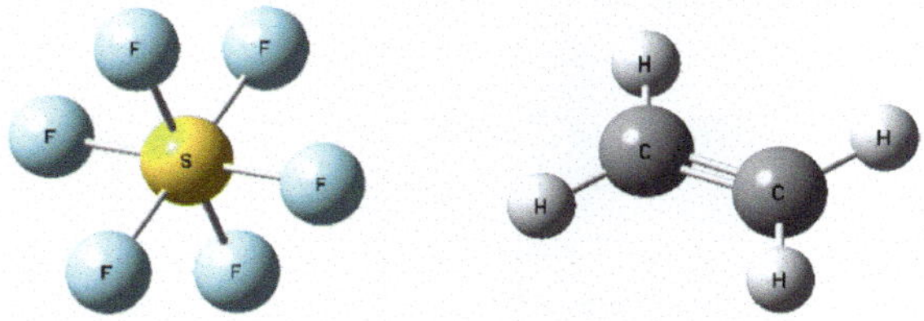

Fig. 41.2 Diagram of molecular models, with SF_6 and C_2H_4 as examples

41.2.2 Quantitative Calculation

The calculation is based on 80 SF_6 substitute gases. The three-dimensional structure of the molecule was constructed by GaussView, a molecular modeling software. The types, connections and spatial arrangements of atoms and other relevant molecular parameters are preliminarily determined. The diagram of molecular models as shown in Fig. 41.2. Subsequently, the molecular structure usually needs to be optimized to ensure that the molecular geometry is the lowest energy state. For this purpose, the quantum chemistry calculation software Gaussian16 [10, 11] was used to apply the density functional theory (DFT) method, and the molecular structure optimization was performed using the M06-2X method with the 6-311G++ (d,p) base set. Through a numerical optimization algorithm, the system adjusts the coordinates of the atoms to reduce the total energy of the molecules until the most stable geometry is reached.

Molecular descriptors are obtained by calculating an optimized molecular model. Use RDKit, a powerful open source chemical information toolkit widely used in computational chemistry and pharmaceutical research, to compute structural class molecular descriptors. It can calculate a variety of descriptors such as structural class molecules, physicochemical properties, molecular conformation, etc., such as molecular mass, molecular atomic composition, polar surface area. Molecular descriptors such as surface electrostatic potential correlation and conceptual density functional theory correlation can be obtained through molecular wave function calculation, which can help us understand the electronic structure and properties of molecules more deeply. In this paper, Multiwfn [12] is used to calculate the molecular wave function results.

Molecular insulation strength can describe the response ability of molecular electrons in external electric field, and is an important parameter to understand the electronic properties of molecules. The molecular insulation strength is obtained by consulting existing databases and recognized literature.

41.2.3 Molecular Descriptor Filtering

Molecular descriptors are numerical representations of molecular structure and chemical properties. In 2.1, a large number of molecular descriptors were obtained through quantitative calculation, but not all of them are related to insulation properties, so it is necessary to filter the descriptors.

The Lasso regression model was used to screen important molecular descriptors. Lasso regression is a commonly used feature selection technique that can be used to screen for molecular descriptors associated with target variables. Lasso regression is very useful for fitting sparse models, and it tends to compress some coefficients to zero, enabling feature selection. Cross-validation and other techniques are used to select the appropriate regularization parameter alpha value to ensure that the appropriate feature subset is found. Descriptors with non-zero coefficients are considered important, so descriptors with zero coefficients are eliminated. By observing the coefficients of the Lasso regression model, it is preliminarily determined that a group of molecular descriptors are important for the prediction of target variables.

In the context of studying the insulation strength of gas molecules, specific molecular descriptors are closely related to the electrical insulation strength of gas. Common descriptors related to insulation include ionization potential, electronegativity, etc. Some descriptors related to insulation properties are added when constructing neural network model. Ionization potential is the energy required to remove an electron from a molecule in a gas. The ionization potential of a gas is usually related to the strength of its electrical insulation, because a high ionization potential means that more energy is required to remove the electron, indicating that the molecule has a strong binding force on the electron. Electronegativity describes how strongly an atom or molecule attracts electrons. Molecules with high electronegativity tend to attract electrons, which may help maintain the electrical insulation of the molecule since electrons are not easily removed from the molecule. For electrical insulation materials, they usually have low polarizability, because they are not easy to deform the electron cloud under the external electric field. The electron density distribution obtained by wave function analysis describes the distribution of electrons in a molecule. For electrical insulation materials, usually the electron density is evenly distributed within the molecule and does not easily accumulate at the interface or surface, which helps to maintain electrical insulation.

The machine learning model presented in this paper analyzes these molecular descriptors and establishes an association between these factors and electrical insulation strength. Through a process of training and validation, the model can predict the electrical insulation properties of gases not included in the initial data set. This predictive power is a key resource when looking for alternatives to SF_6.

41.3 Result

When screening the importance of molecular descriptors in the Lasso regression model, as alpha increases, the curves in the figure continue to decrease from left to right to 0, and the descriptors are correspondingly compressed smaller. When alpha reaches a certain value, some unimportant variables will be compressed to 0,

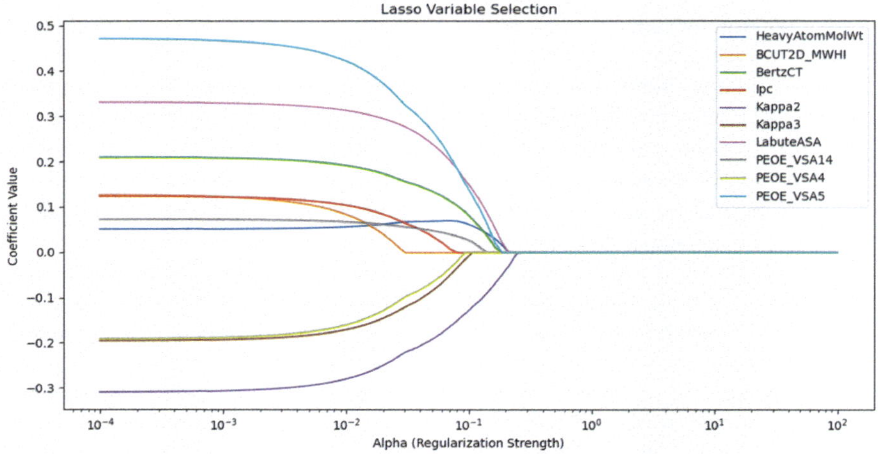

Fig. 41.3 The Lasso regression model filters the dynamic process diagram of molecular descriptors

indicating that the variable has been removed from the model. Figure 41.3 shows how some of the descriptors change with alpha in the lasso regression.

In this paper, 80 molecular models of SF_6 substituted gases are constructed. Based on the results of 80 molecular descriptors and the electrical strengths, a neural network model was established, and a model that can predict the insulation strength of new molecules was obtained. The 80 molecules were randomly divided into the training set and the test set according to the ratio of 4:1. In each machine learning, 64 samples are input as training, and the network is trained 1000 times.

In this paper, mean absolute error (MAE) is used as the evaluation coefficient. MAE is an intuitive metric that measures the average absolute error between a model's predictions and actual observations. It is less sensitive to outliers and outliers, and can be used to describe the performance of the model intuitively. In the initial prediction of insulation strength, the smaller error is more critical, and MAE can reflect the absolute error more accurately. At the same time, correlation coefficients R^2 is also given as a reference.

After training, a neural network model with good prediction effect is obtained. The training set R^2-value is 0.96 and MAE-value is 0.1. The test set R^2-value is 0.85, and the MAE-value is 0.23. The model training results are satisfactory, as shown in Fig. 41.4. Since the objective of this paper is to design SF_6 replacement molecules, the relative dielectric strength (E_r) of SF_6 is used.

A key advantage of the proposed method is its flexibility and versatility. It is not limited to a specific gas or molecular structure, instead, it can be applied to a variety of different gases, making it possible to quickly and accurately predict the electrical insulation strength of a large number of molecules during initial mass screening.

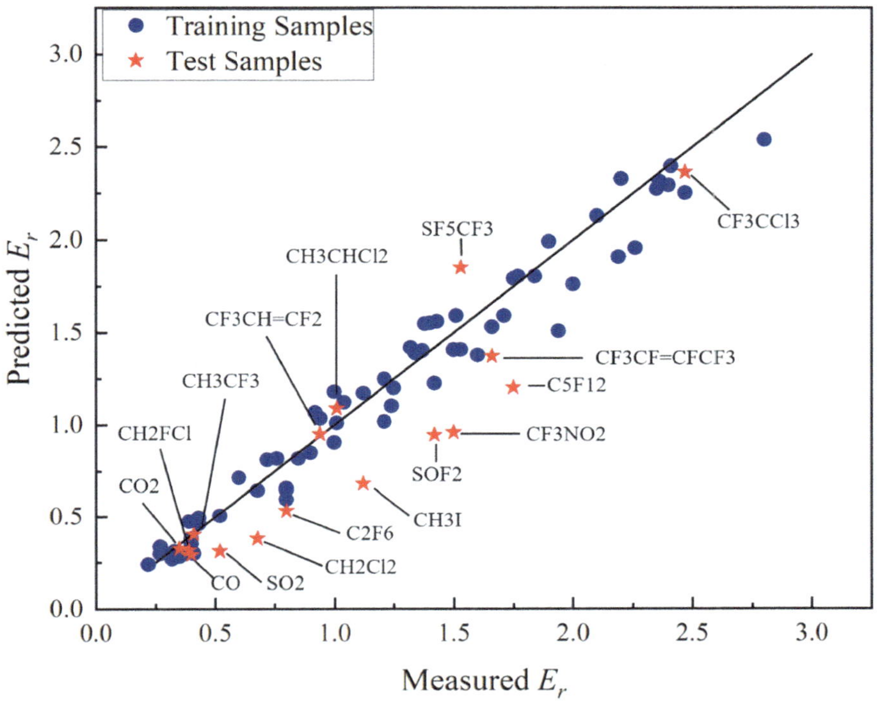

Fig. 41.4 The comparison between predicted and measured electrical strength E_r by the training samples and the test samples

41.4 Conclusion

In this paper, we propose a method using machine learning and neural network techniques to predict the electrical insulation strength of alternative gases. Molecular descriptors with high correlation with insulation strength are selected and a convolutional neural network model is established. By learning the complex relationship between known molecular descriptors and electrical insulation strength, the model can accurately predict the electrical properties of various gases. The method proposed in this paper is flexible and universal, and can be used to guide the design of new and environmentally friendly gases. A large number of molecular candidate materials with unknown insulation strength can be quickly and efficiently screened.

References

1. M. Rabie, C.M. Franck, Assessment of eco-friendly gases for electrical insulation to replace the most potent industrial greenhouse gas SF6. Environ. Sci. Technol. **52**(2), 369–380 (2018)
2. L. Xingwen, Z. Hu, A.B. Murphy, SF6-alternative gases for application in gas-insulated switchgear. J. Phys. D Appl. Phys. **51**(15), 153001 (2018)
3. J.C. Devins, Replacement gases for SF6. IEEE Trans. Electr. Insul. **2**, 81–86 (1980)
4. G. Biasiutti, Neue isoliergase für gekapselte hochspannungsanlagen. Diss. ETH Zurich (1985)
5. R.E. Robson et al., Kinetic theoretical and fluid modelling of plasmas and swarms: the big picture. Plasma Sources Sci. Technol. **17**(2), 024020 (2008)
6. D.J. Sobajic, Y.H. Pao, Artificial neural-net based dynamic security assessment for electric power systems. IEEE Trans. Power Syst. **4**(1), 220–228 (1989)
7. Z. Linlin, Fast prediction of electron-impact ionization cross sections of large molecules via machine learning. J. Appl. Phys. **125**(18) (2019)
8. M. Rabie, D.A. Dahl, S.M.A. Donald, Predictors for gases of high electrical strength. IEEE Trans. Dielectr. Electr. Insul. **20**(3), 856–863 (2013)
9. X. Yu, H. Hou, B. Wang, Prediction on dielectric strength and boiling point of gaseous molecules for replacement of SF6. J. Comput. Chem. **38**(10), 721–729 (2017)
10. F. Ogliaro, M. Bearpark, J.J. Heyd, E. Brothers, K.N. Kudin, *Gaussian09, Revision a. 02. Gaussian* (Wallingford, U.K., 2009), p. 28
11. J.C.W. Bauschlicher, H. Partridge, 'A modification of the Gaussian-2 approach using density functional theory. J. Chem. Phys. **103**(5), 1788–1791 (1995)
12. L. Tian, C. Feiwu, Multiwfn: a multifunctional wavefunction analyzer. J. Comput. Chem. **33**(5), 580–592 (2012)

Chapter 42
Study on the Requirements for the 10 kV Voltage Transformers Considering Live Working

Qi Han, Yongqiang Gao, and Yongjian Sun

Abstract Voltage transformers can explode during live disconnection, jeopardizing the safety of operating personnel. To address this issue, this paper constructs an ATP-EMTP simulation model for the connection of 10 kV distribution network voltage transformers to the primary system. It analyzes the impact of live disconnection transient processes and evaluates the safety of transformer operation under various conditions. Additionally, safety measures for live disconnection of voltage transformers are proposed. Research findings reveal that electronic voltage transformers exhibit no hazardous overvoltages or overcurrents during live disconnection. On the other hand, electromagnetic voltage transformers experience transient overvoltages and overcurrents upon live connection. The method of connecting transformers to the primary system and the grounding approach on the high-voltage side significantly affect the safety of transformer operation. During live operations, mitigating measures can involve the insertion of damping resistors between the transformer and the primary system or grounding the high-voltage side of the transformer through linear resistors to mitigate these effects.

42.1 Introduction

In recent years, distribution network automation and non-stop operation technology have developed rapidly and are widely used in various power companies at home and abroad [1]. As an important part of the distribution network automation device, the voltage transformer has a very small load current during normal operation, and the secondary side is equivalent to no-load. However, in the actual disconnection process, due to the operation mode of the distribution network, the quality of the transformer itself, and Various factors such as the construction level of the operators may lead to abnormal operation of the transformer, which may cause the transformer

Q. Han · Y. Gao (✉) · Y. Sun
State Grid Shandong Electric Power Company Dongying Power Supply Company, Dongyin 257000, China
e-mail: sdddgyq@163.com

Z. Fang et al. (eds.), *Proceedings of the 5th International Symposium on Plasma and Energy Conversion*, Springer Proceedings in Physics 398,
https://doi.org/10.1007/978-981-97-2245-7_42

to overheat or even explode, including Shandong and Jiangsu, have experienced such problems, posing a significant threat to the safety of personnel and equipment [2].

Qi Zheng from North China Electric Power University analyzed the generation mechanism of multi-PT ferromagnetic resonance, and proposed the suppression technology of ferromagnetic resonance at the same time [3]. Cui Juyong of State Grid Liaoning Electric Power Company proposed a new test standard for the excitation characteristics of electromagnetic voltage transformers through actual cases and combined with simulation calculations [4]. State Grid Xinjiang Electric Power Company He Long et al. conducted a quantitative comparison of the conditions for the generation of ferromagnetic resonance and the effects of its influencing factors to determine the key influencing factors that cause ferromagnetic resonance [5]. State Grid Qinghai Electric Power Company Liu Shengnan et al. analyzed the mechanism and hazards of various faults of electromagnetic voltage transformers, and introduced harmonic elimination measures and their advantages and disadvantages [6].

To sum up, the existing research focuses on the ferromagnetic resonance fault of the transformer, and seldom studies the influence of the transient process and the corresponding safety measures when the voltage transformer is disconnected in live working. In this paper, aiming at the most commonly used electromagnetic and electronic voltage transformers in 10 kV distribution network, a simulation model of voltage transformer ATP-EMTP is established to analyze its electromagnetic transient process in the process of live disconnection and connection, and to study the possible Factors that cause harm to workers, and finally put forward safety measures in live work, and illustrate the feasibility of measures through simulation results.

42.2 Transient Process Analysis of Disconnected and Connected Voltage Transformer

With the comprehensive advancement of the automation transformation of urban power grid distribution lines, energy-harvesting Voltage Transformer that provide power for load switch controllers and operating devices have been widely used [7], among which electromagnetic Voltage Transformer and electronic Voltage Transformer are the most widely used. The analysis of the electromagnetic transient characteristics of the live disconnection and connection of the voltage transformer can deeply understand the difference in the transient characteristics of the two types of voltage transformers in the disconnection lead, and provide a basis for the safety research of the live disconnection and connection of the transformer.

42.2.1 Analysis of Transient Characteristics During Live-Line Disconnection and Connection of Electronic Voltage Transformers

The electronic voltage transformer, typically integrated and sealed within the column-mounted intelligent switch, mainly consists of a capacitive voltage divider and a power supply sensor [8]. The ATP-EMTP simulation model for live-line connection of a 10 kV distribution network electronic voltage transformer is illustrated in Fig. 42.1.

By controlling the time interval between the closing and opening of the time-controlled switch to simulate the connection and disconnection of the three-phase voltage transformer of the system, the three phases are respectively connected to the system at 0.1, 0.2 and 0.3s, and disconnected at 0.7s. Taking phase A as an example, the voltage and current changes during the connection and disconnection of the transformer are shown in Figs. 42.2 and 42.3

From the simulation waveforms, it can be observed that there is no significant change in the voltage waveform of the voltage transformer at the moment of live-line connection and disconnection. The peak current in the lead of Phase A voltage transformer does not exceed 3 mA during the connection and disconnection process. This indicates that the process of live-line connection of electronic voltage transformers with lead wires carries no risk.

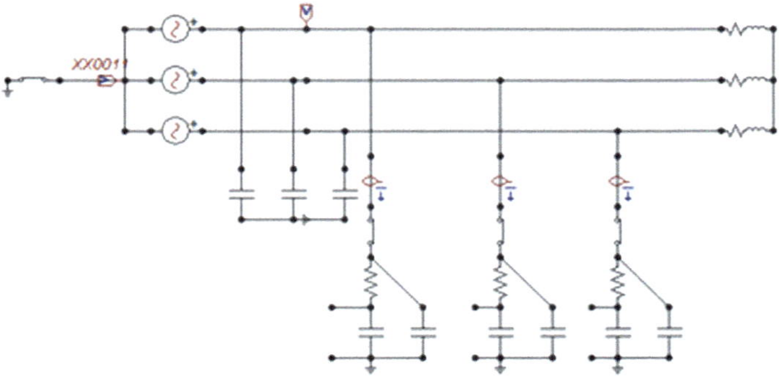

Fig. 42.1 Model of live-line connection and disconnection of 10 kV distribution network electronic voltage transformers

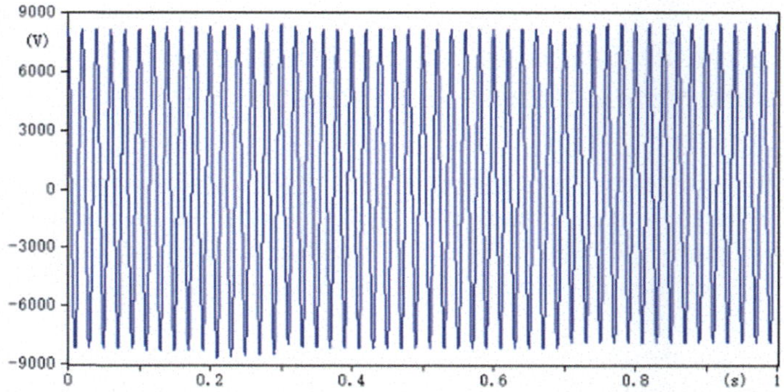

Fig. 42.2 Phase A bus voltage

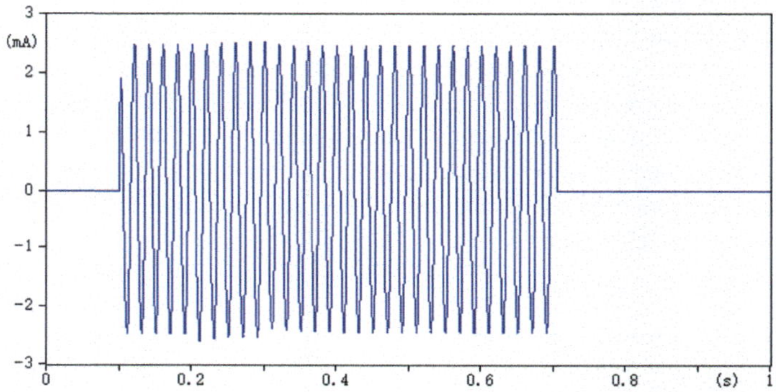

Fig. 42.3 Current waveform of phase A voltage transformer lead

42.2.2 Analysis of Transient Characteristics During Live-Line Disconnection and Connection of Electromagnetic Voltage Transformers

The electromagnetic voltage transformer is used in conjunction with the switch on the pole on the distribution line. When the live wire is lapped, the transformer will explode and burn due to overvoltage, overcurrent or secondary side wiring errors [9]. In the electromagnetic transient calculation, the JMarti model is usually used for the line, and the load of the distribution network is replaced by an R-L-C element whose neutral point is not grounded. Build the ATP-EMTP simulation model of the live connection of the automatic terminal of the 10 kV ungrounded system electromagnetic voltage transformer (Fig. 42.4).

During the simulation, it is considered that the three-phase parameters are symmetrical, and the connection and disconnection of the three-phase voltage transformer of the system are simulated by controlling the time interval between the closing and

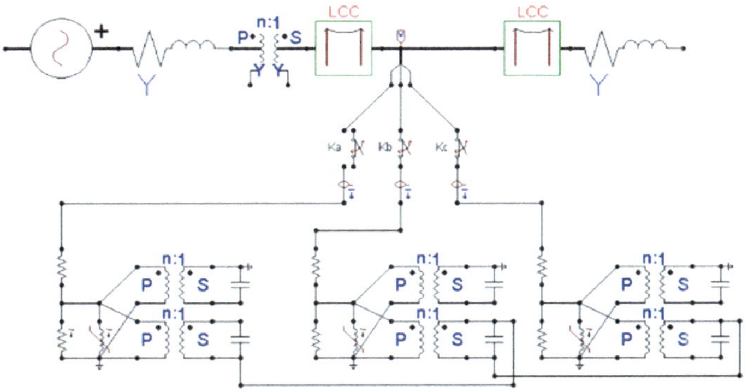

Fig. 42.4 ATP simulation model

opening of the time-controlled switch. The three phases (A, B, and C) are successively connected to the system at 0.1, 0.2, and 0.3s, and then disconnected at 0.7s. Taking phase A as an example, after one phase is connected, the voltage and current waveforms of the voltage transformer are as follows (Figs. 42.5 and 42.6).

From the above simulation results, it can be known that the electromagnetic voltage transformer resonates when it is charged into the primary system, accompanied by transient overvoltage and overcurrent, and there is no obvious resonance, overvoltage and overcurrent when it is disconnected. On the one hand, the overvoltage will make the excitation process of the transformer violent and saturate the iron core; on the other hand, it will cause the insulation breakdown of the transformer, resulting in a short circuit between internal components. Overcurrent may blow the fuse on the primary side of the system, and more serious transformers may explode due to overvoltage and overcurrent, which poses a huge hidden danger to the safety of workers. The research shows that the connection method of the transformer, the grounding method of the high-voltage side of the transformer, the system overvoltage, and the

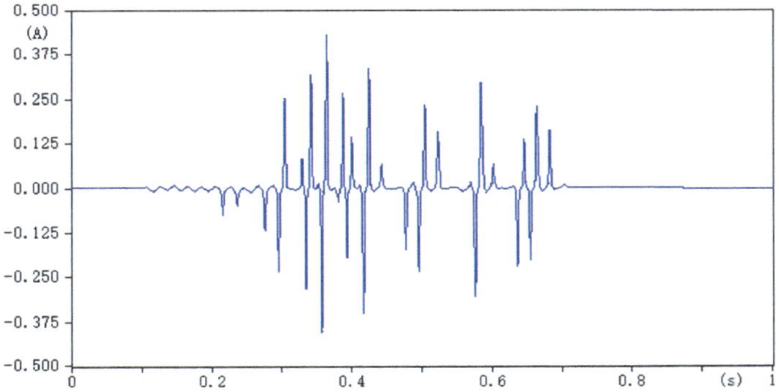

Fig. 42.5 Excitation current of phase A on high voltage side of voltage transformer

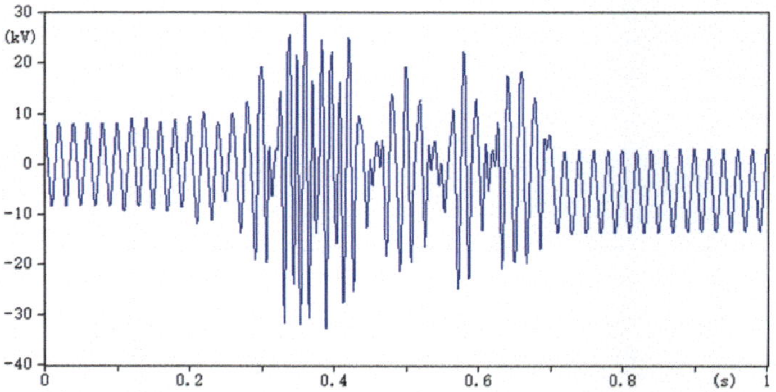

Fig. 42.6 Voltage waveform of phase A on the high voltage side of the voltage transformer

grounding method of the distribution network may affect the transient process when the transformer is connected. It cannot be changed during live work. Therefore, here we focus on the transient impact of the connection method of the transformer and the grounding method of the high-voltage side of the transformer on the connection of the transformer and the corresponding safety measures.

42.2.3 Ways of Connecting Voltage Transformers to the Primary System

In order to reduce the risk of electric shock, reduce the scope of faults and improve the continuous power supply capacity of the system, the 10 kV distribution network usually chooses the operation mode of the neutral point of the system not being grounded. When the transformer is directly connected to the primary system, if the neutral point of the distribution network is not grounded, the exciting inrush current generated when the transformer is connected cannot be released and dissipated in time, which prolongs the duration of the exciting inrush current and increases the impact on the transformer and the system. At the same time, the lack of a low impedance path limits the flow of the inrush current, resulting in a larger amplitude of the inrush current, which increases the temperature rise of the transformer and affects the stability and service life of the transformer [10].

42.2.4 Grounding Method of High Voltage Side of Voltage Transformer

The neutral point of the high-voltage side of the voltage transformer is generally directly grounded. When a short-circuit current is generated on the high-voltage side due to a fault, it can be quickly eliminated through the grounding system. At the same time, it can improve the fault detection capability and simplify the system design. However, when the electromagnetic voltage transformer is charged into the primary system, the direct grounding of the neutral point on the high voltage side is likely to cause resonance, because the direct grounding of the neutral point will form a resonance composed of transformer capacitance and inductance, and other capacitance and inductance in the system The loop produces a coupling effect. At the same time, due to the lack of damping in the system, it will also cause the resonance oscillation to be difficult to attenuate, which will continue to affect the system.

42.3 Optimization Scheme Design of Transformer Live Disconnection and Connection

To reduce the impact of transient overvoltage and overcurrent on equipment and operators when the electromagnetic voltage transformer is disconnected, there are two main approaches. Firstly, considering the transformer and the operators themselves, it involves using transformers with better excitation characteristics or manufacturing transformers using materials with improved insulation and heat resistance properties. Operators should also enhance their personal protection and adhere to standardized operating procedures. Secondly, from the perspective of live working, efforts can be made to shorten the duration of the transient process or reduce the amplitude of transient overvoltage and overcurrent to mitigate this impact.

42.3.1 Inserting a Damping Resistor Between the Voltage Transformer and the Primary System

Due to the ungrounded neutral point operation mode in the 10 kV distribution network, the excitation inrush current generated when the electromagnetic voltage transformer is connected to the primary system cannot dissipate in a timely manner, causing continuous damage to the equipment. Impedance elements can accelerate the release of excitation inrush current to reduce the duration of the transient process and the amplitude of overvoltage and overcurrent. By inserting damping resistors with different resistance values in series between the transformer and the primary system and then performing transient process simulations, the influence of damping resistors on the transient process is analyzed [11] (Fig. 42.7).

When there is no damping resistor connected between the transformer and the primary system, the excitation current waveform at the high voltage side of the transformer is shown in Fig. 42.5. When the damping resistor is connected between the transformer and the primary system, the excitation current decays rapidly and the transient process gradually disappears At the same time, with the increase of

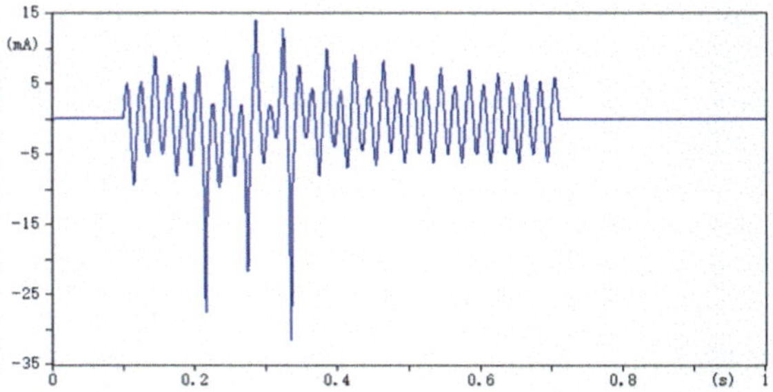

(a) Incorporate a 2.5kΩ damping resistor in series between the voltage transformer and the primary system

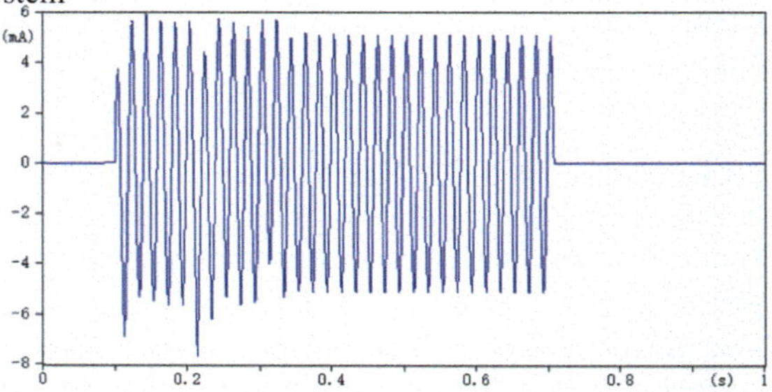

(b) Incorporate a 25kΩ damping resistor in series between the voltage transformer and the primary system

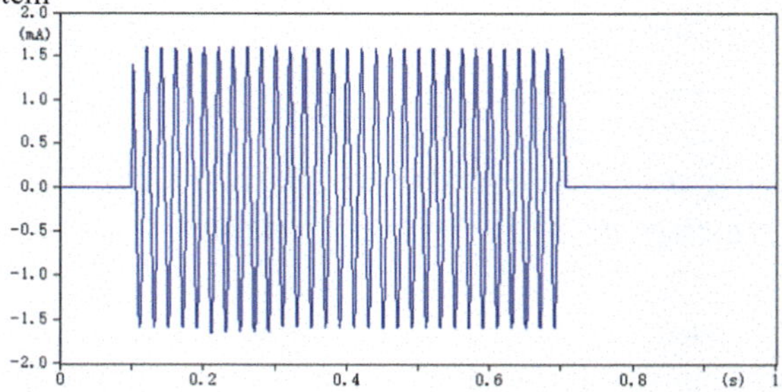

(c) Incorporate a 250kΩ damping resistor in series between the voltage transformer and the primary system

Fig. 42.7 Excitation current of phase A on primary side of voltage transformer under different damping resistance

the damping resistance, the overcurrent generated by the transient process is also gradually decreasing until R = 250 kΩ, the voltage transformer is connected to the primary system and the transient process will not occur. From this, it can be concluded that the damping resistor can effectively slow down the impact of the transient process when the electromagnetic voltage transformer is connected to the primary system.

42.3.2 The High Voltage Side of the Transformer is Grounded Through a Resistor

The neutral point of the voltage transformer's high-voltage side is directly grounded, which can lead to the formation of a resonance loop and result in coupling effects. In order to shorten the electromagnetic transient process and reduce the time for resonance damping, it is possible to increase the system's damping. Therefore, considering a linear resistor grounding connection at the neutral point of the voltage transformer's high-voltage side can enhance system damping. This approach not only restricts the current through the transformer windings and limits the saturation of the transformer core but also accelerates energy dissipation and suppresses resonance. Grounding the high-voltage side of the voltage transformer through a resistor is equivalent to introducing the resistor into the zero-sequence loop of the system. Under normal operating conditions, it does not affect the operation of the system. However, when connecting live terminals with voltage transformer -type automation terminals, it increases the zero-sequence impedance of the voltage transformer's primary side, thereby limiting the current on the voltage transformer's primary side. The schematic diagram of grounding the high-voltage side of the voltage transformer through a resistor is shown in Fig. 42.8.

The neutral point of the high-voltage side of the transformer is grounded through linear resistors with different resistance values, followed by transient process simulations to analyze the impact of grounding resistance on the transient process during the energization and disconnection of the transformer. In a 10 kV distribution network, the typical grounding resistance ranges from 20 to 30 kΩ. For this simulation experiment, linear resistors with grounding resistances of 10, 25, and 40 kΩ were selected to simulate the transient process. The simulation results obtained are as follows (Fig. 42.9).

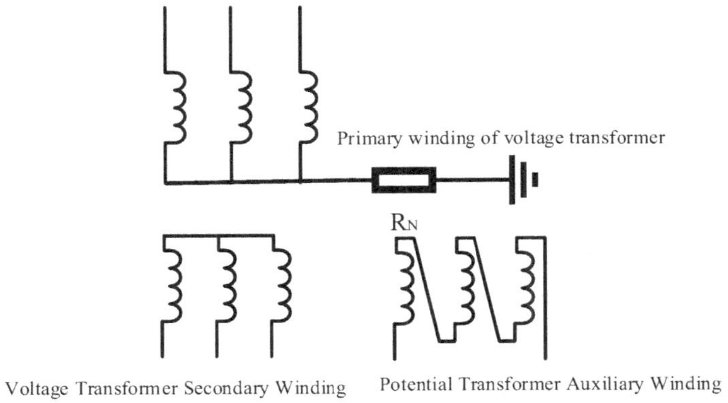

Fig. 42.8 Schematic diagram of the voltage transformer high-voltage side grounded through a resistor

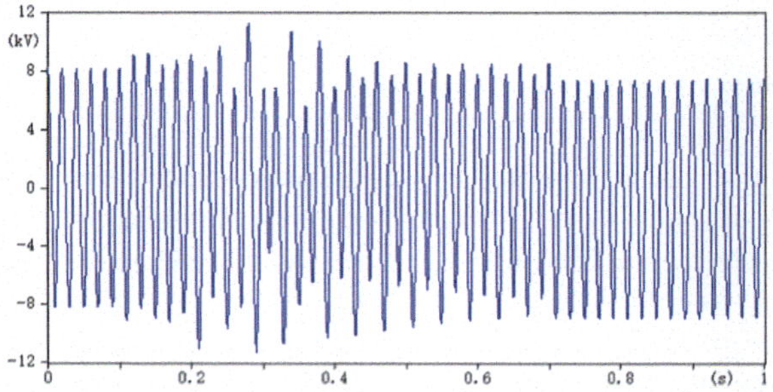

(a) The neutral point on the high voltage side is grounded through a 10kΩ linear resistor

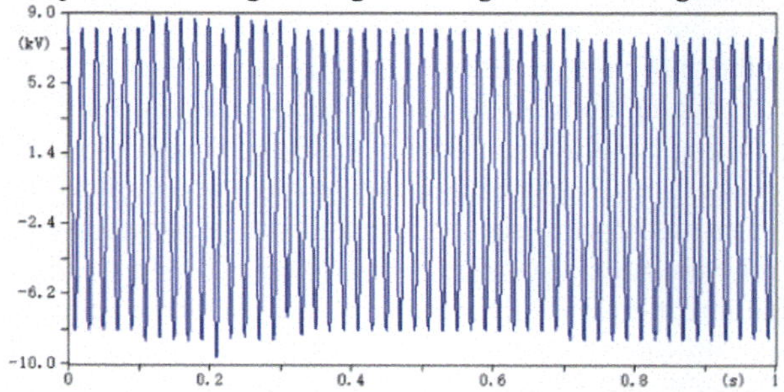

(b) The neutral point on the high voltage side is grounded through a 25kΩ linear resistor

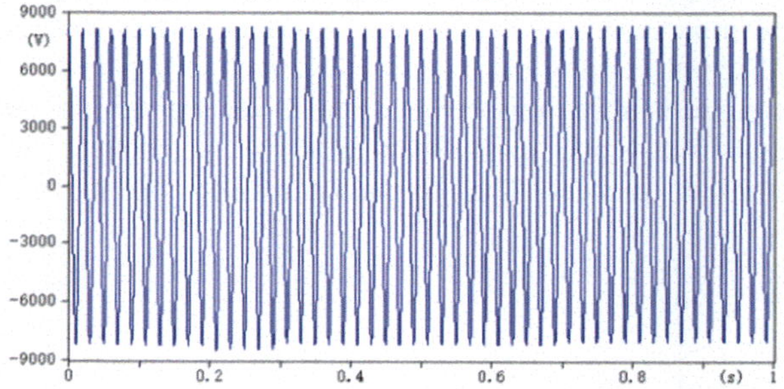

(c) The neutral point on the high voltage side is grounded through a 40kΩ linear resistor

Fig. 42.9 Voltage waveform of phase A on the high voltage side of the voltage transformer

The simulation waveform of the high-voltage side neutral point of the voltage transformer directly grounded is shown in Fig. 42.6. The A-phase voltage waveform on the high-voltage side exhibits sustained resonance, resulting in overvoltage. When the high-voltage side of the voltage transformer is grounded through a linear resistor, the simulation results show that as the resistance value of the linear resistor increases, the amplitude of the transient overvoltage gradually decreases, and the duration of the transient process becomes shorter. When the resistance value of the linear resistor is 40 kΩ, the suppression effect on the transient process is similar to that of 20 kΩ. From this, it can be concluded that grounding the high-voltage side of the voltage transformer through a linear resistor can mitigate the impact of transient processes when an electromagnetic voltage transformer is connected to the primary system.

It can be seen from Figs. 42.7 and 42.9 that the transient process during the disconnection process of the transformer can be suppressed by connecting the damping resistor in series between the transformer and the primary system or by grounding the high voltage side of the transformer through a linear resistor. And the larger the resistance value of the damping resistance, the better the suppression effect on the amplitude and attenuation speed of the voltage and current in the transient process; but there is not a simple linear relationship between the resistance of the high-voltage side of the transformer and the suppression effect. Consider both economy and efficiency.

42.4 Conclusion

In order to study the transient process of the live disconnection and connection voltage transformer and its influence on the voltage transformer and the operators, and improve the safety of the disconnection and connection transformer in the live work, an ATP-EMTP simulation model of the live connection primary system of the transformer is built in this paper. The following conclusions were obtained:

1. When the electromagnetic voltage transformer is connected to the primary system with electricity, there will be a transient process, resulting in overvoltage and overcurrent; before and after the electronic voltage transformer is disconnected and connected from the primary system with electricity, the voltage waveform does not change significantly, and the peak current does not exceed 3mA, indicating that there is no risk in the process of electronic voltage transformer live lap lead.
2. The transient process when the electromagnetic voltage transformer is charged into the primary system is mainly related to the way the transformer is connected to the primary system and the grounding method of the high voltage side of the transformer.
3. By connecting the damping resistor in series between the transformer and the primary system or grounding the high-voltage side of the transformer through a linear resistor, the impact of the transient process on the transformer and the operator can be effectively reduced, and the damping resistor and linear resistor are 250 kΩ and When 25 kΩ, the suppression effect is more obvious.

References

1. Y. Hu, K. Liu, Y. Peng et al., Research status and development trend of live working key technology. High Voltage Eng. **40**(07), 1921–1931 (2014)
2. S.H. Wang, X.J. Wang, Y. Xia et al., Ferromagnetic resonance characteristics and identification of PT for distribution network switchgear under uninterruption working condition. Power Syst. Technol. **46**(11), 4534–4540 (2022)
3. Z. Qi, H.Y. Li, Research on multi-PT ferroresonance in island microgrid. Power Syst. Protect. Control **47**(22), 52–58 (2019)
4. J.Y. Cui, S. Wang, Z.J. Zhao et al., Analysis of electromagnetic voltage transformer ferroresonance characteristics. Electr. Energy Manage. Technol. **01**, 62–65+77 (2023)
5. L. He, J.C. Ma, L.J. Du et al., Research on key influencing factors of ferroresonance fault in ineffective grounding system. J. Electr. Eng. **15**(02), 61–69 (2020)
6. S.G. Liu, Y.M. Zhang, Y. Yang et al., Review on common faults of electromagnetic potential transformers and the countermeasures. J. Xihua Univ. (Nat. Sci. Edn.) **42**(03), 47–58 (2023)
7. X.Q. Chen, Planning and construction of 10 kV distribution network automation. Heilongjiang Sci. **14**(04), 151–153 (2023)
8. Z.G. Tian, M.D. Zhu, L.J. Chi et al., Research on transient characteristics of resistor-capacitor voltage dividing electronic transformers. Process Autom. Instrum. **43**(04), 27–32 (2022)
9. Y.W. Tu, T. Deng, H. Shen et al., Fault analysis and treatment of electromagnetic potential transformer. Electr. Eng. **09**, 92–95 (2022)
10. M. Zhou, Y.Z. Sun, Eliminating ferro-resonance for distribution network with different neutral grounding modes. High Voltage Apparatus. **51**(01), 80–85 (2015)
11. J. Liu, H. Liu, C.Y. Jiang et al., Live replacement technology of voltage transformer for distribution network. Sci. Technol. Eng. **19**(26), 244–248 (2019)

Chapter 43
Comparison of Insulation Performance Between PP and XLPE at 110 kV High Voltage Cable Joints Based on Electro-Thermal-Stress Coupling Simulation

Xinyu Chen, Xinkui Wang, Baolei Zhang, Xiaoye Bai, Kai Cao, Runhua Li, Xinglei Cui, Zhi Fang, Pengfei Xu, Yu Ding, and Deyan Li

Abstract Currently, high-voltage cables with cross-linked polyethylene (XLPE) as the main insulation are widely used in the power industry. Nevertheless, polypropylene-based (PP) materials have a broad application prospect due to their advantages such as higher temperature stability and good recyclability, offering options as an alternative to XLPE cables for global electrical industry. In this paper, key material parameters such as relative dielectric constant, Young's modulus, Poisson's ratio and coefficient of thermal expansion of XLPE and PP materials at different temperatures were measured. The coupling simulation model of 110 kV high voltage cable joint was established to obtain the distributions of electric, thermal and stress fields when XLPE and PP were used as the main insulation of the joints, respectively. The results show that the intensity of electirc field at composite interface is lower for XLPE/SR (silicone rubber) due to its larger relative dielectric constant, leading to better insulation performance. Moreover, the thermal deformation of XLPE cable is smaller than that of PP, as its smaller heat capacity of constant pressure. As a result, the degree of electric field distortion is lower, bringing about better working stability at high temperature.

X. Chen · R. Li · X. Cui (✉) · Z. Fang
College of Electrical Engineering and Control Science, Nanjing Tech University, Nanjing 211816, Jiangsu, China
e-mail: cxl00@njtech.edu.cn

X. Wang · B. Zhang · X. Bai · K. Cao · P. Xu · Y. Ding · D. Li
ZTT Cable Accessories Co., Ltd., Nantong, China

Z. Fang et al. (eds.), *Proceedings of the 5th International Symposium on Plasma and Energy Conversion*, Springer Proceedings in Physics 398,
https://doi.org/10.1007/978-981-97-2245-7_43

43.1 Introduction

Cross-linked polyethylene (XLPE) cable and its accessories are widely used in power systems due to their convenient installation and high reliability. At present, many XLPE cables have already exhausted the lifespan, thus facing the difficulty of recycling and renewing. Given the fact that the environmental protection concept has become a global trend in today's society, developing brand new insulation materials has become a main topic of global power industry.

In recent years, polypropylene (PP)-based thermoplastic materials have raised enormous concern due to their high operating temperature and good recyclability [1, 2]. As cable materials work in electro-thermal-stress coupling conditions, their performance is determined by the electrical, thermodynamical and mechanical properties of the materials. Therefore, a study for comparing the performance of XLPE cables and PP cables is required. In order to examine the feasibility of replacing XLPE with PP, a comprehensive comparison of interfacial properties between the main insulation and stress cone is indispensable when the material of the main insulation changes from XLPE to PP.

At present, due to the difficulty of measuring the distribution of the composite interfacial stress in the operating high-voltage cable joint, multi-field coupling simulation has become an effective mean to solve the problem. Song etc. calculated the electric field distribution of cable joints with different defects [3]. Zhao etc. established a coupled electric-temperature-fluid model to analyze the effect of different laying environments on the thermal field of the cable joint [4]. Although the simulation around the cable composite insulation interface has been already carried out, the study related to the comparison between XLPE and PP considering the coupling process has not been implemented.

Focusing on the above problems, this paper analyzes the electro-thermal-stress coupling effect under the operating condition of high-voltage cable joints on the basis of multi-field simulation. A three-dimensional simulation model is established and the distribution of electric field is simulated at the composite interface when XLPE and PP are selected as main insulation material, as well as thermal and thermal stress field. The reasons for the difference of electric field distortion between the two materials are further analyzed.

43.2 Simulation Model

In this paper, the 110 kV high-voltage cable joint is taken as the target, and a three-dimensional simulation model is established in the multi-physics field simulation software, as shown in Fig. 43.1a. For concisely showing the model structure, the 1/2 axial section of the three-dimensional simulation model is given, as shown in Fig. 43.1b.

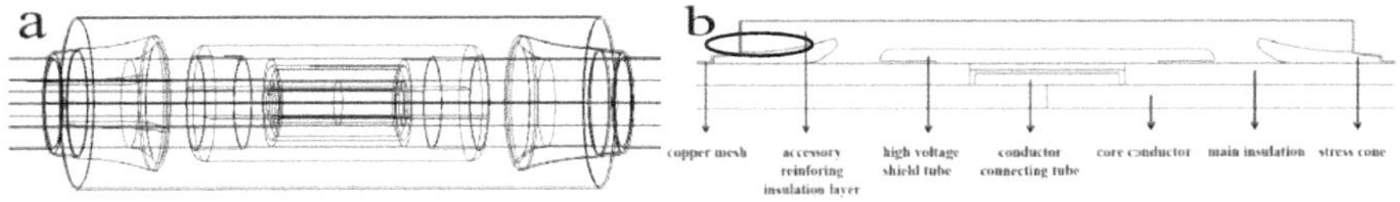

Fig. 43.1 Simulation model of cable joint. **a** Simulation model in three dimensions, **b** axial section of the simulation model

In Fig. 43.1b, the path in the dashed box is the computational path for the coupled electro-thermal-stress simulation performed in this paper. The governing equations of the electric field are shown as follows:

$$\nabla \cdot D = \rho_v \tag{43.1}$$

$$E = -\nabla V \tag{43.2}$$

where ∇ is vector differential operator; D is electric flux; ρ_v is charge volume density; E is electric field intensity; V is electric potential.

The governing equation of thermal field is shown in Eq. (43.3):

$$\rho C_P \frac{\partial T}{\partial t} + \rho C_P u \cdot \nabla T = \nabla \cdot (k \nabla T) + Q \tag{43.3}$$

where ρ is material density;k is thermal conductivity;C_p is constant pressure heat capacity; Q is heat source, and u is displacement。

The governing equation of stress field is shown in Eq. (43.4):

$$\rho \frac{\partial_U^2}{\partial t} = \nabla \cdot s + F_V \tag{43.4}$$

s is stress tensor; F_v is volumetric force.

According to the actual operating conditions, the convective heat transfer parameter h of the copper wire mesh shielding layer exposed to air directly is defined to 10 W/($m^2 \cdot$ C) [5]. Besides, the copper wire mesh ensures 0 potential. The rated current of core conductor is defined to 2 kA, and its potential is defined to 160 kV, considering the peak potential during the actual operation of the cable. The radial section away from the high-voltage cable joint is set to be electrically insulated. Moreover, the axial heat transfer is ignored. The axial displacement component is 0, and the constraint condition on the joint surface is no displacement.

43.3 Comparison of Physical Parameters Between PP and XLPE

In this paper, significant temperature rise of the cable joint is generated under rated current, which directly affects the physical parameters of the insulating material. Therefore, it is necessary to acquire the variations of these parameters with temperature. In the preliminary work, the values of each physical parameter of the two materials at 25, 75, 90 and 120 °C were measured.

Combined with the governing equation of thermal field, the value of thermal conductivity, constant pressure heat capacity and thermal expansion coefficient mainly determine the thermal field distribution. Figure 43.2a illustrates that the thermal expansion coefficient of XLPE is larger than that of PP in the range of 25–120 °C. As the temperature increases, the thermal expansion coefficients of both XLPE and PP show an evident augment trend. The thermal expansion coefficient of XLPE increases by 614% and peaked at 120 °C with a value of 0.002, while PP increases by 140% with a value of 0.0018.

The relative dielectric constant is the determining physical parameter to electric field. From Fig. 43.2b, relative dielectric constant of XLPE is obviously higher in the range of 25–120 °C. Moreover, it decreases evidently with temperature for XLPE, while that of PP remains relatively stable, and the gap between the two descends gradually with the increasing temperature. The lowest value of relative dielectric constant of PP is 2.64, which is 5.2% lower than that of 25 °C.

The physical parameters related to the stress field distribution in this study are Young's modulus and Poisson's ratio. Figure 43.2c clearly illustrates that the Poisson's ratio of PP hardly fluctuates by temperature and has a slight decreasing tendency in the range of 25–120 °C, while XLPE has a large fluctuation with the increase of temperature, reaching a minimum value of 0.29 at 47 °C and a maximum value of 0.45 at 105 °C, which is an increase of 30% compared with the room temperature. Figure 43.2d shows that the Young's modulus of PP is larger than that of XLPE in the range of 25–120 °C, and the difference between the two increases significantly with the growth of temperature. PP reaches the bottom with a value of 17.846 at 120 °C, which is a decrease of 83.2%, while the Young's modulus of XLPE decreases by 99.7%, with the lowest value of 0.31.

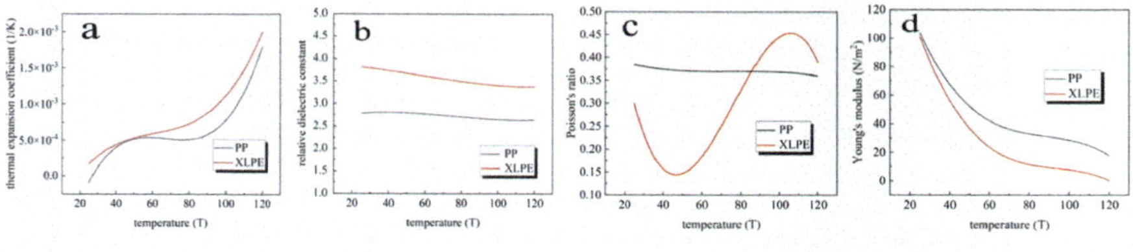

Fig. 43.2 Variations of four parameters with temperature. **a** Thermal expansion coefficient, **b** relative dielectric constant, **c** Poisson's ratio, **d** Young's modulus

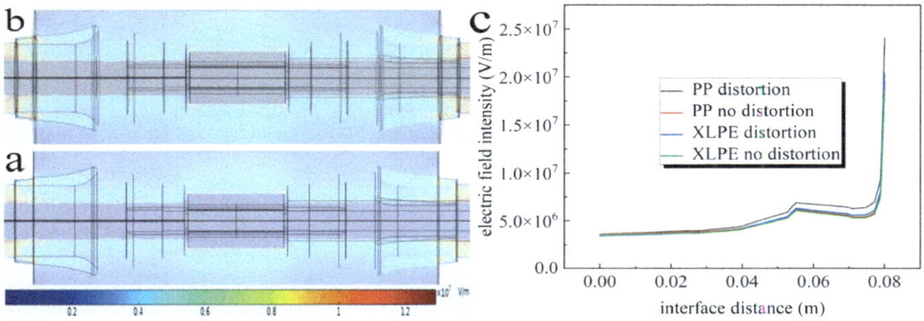

Fig. 43.3 Electric field distribution of the cable joint. **a** PP, **b** XLPE, **c** composite interface

43.4 Simulation Results

Based on the material parameters given above, the electro-thermal-stress coupling simulation is carried out. The distributions of the three physical fields are obtained when XLPE and PP are used as the main insulation material, respectively.

43.4.1 Electric Field Distribution

Figure 43.3a, b illustrate the electric field distributions. Since this study focuses on the electric field distortion of the cable at the composite interface, Fig. 43.3c shows the influence of thermal expansion to the electric field at composite interface. It shows that the electric field intensity at the corresponding point of PP cable is greater than that of the XLPE both before and after the thermal stress distortion on the composite interface. The maximum electric field intensity on the composite interface after distortion appears at the end of the path, known as the separation of the stress cone from main insulation. Compared with distortion-free circumstance, the electric field of PP increases by 23.4% at the separation point, while it of the XLPE cable increases by 9.7%.

43.4.2 Thermal Field Distribution

Figure 43.4a, b give the thermal field distribution. Figure 43.4c gives the variation of temperature at the composite interface. High voltage cable joints under normal operation show a significant temperature rise after the operation reaches stability due to the heating of the conductor cable core. Figure 43.4a, b show that the temperature rise of PP is evidently larger than that of XLPE and reaches 354 K at the cable core, equivalent to a temperature rise of 56 °C. This phenomenon results from the fact that

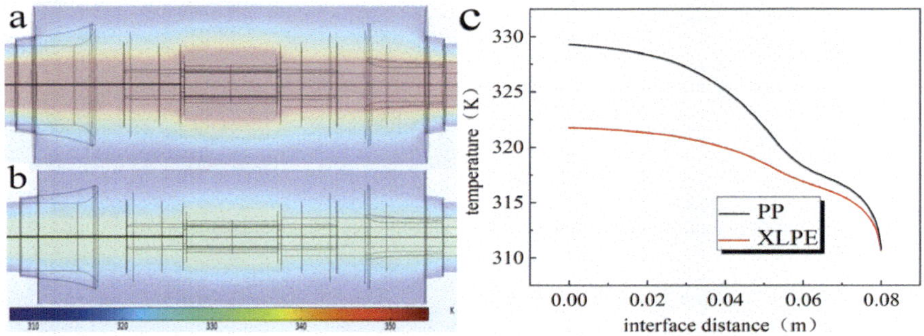

Fig. 43.4 Thermal field distribution of the cable joint. **a** PP, **b** XLPE, **c** composite interface

the constant pressure heat capacity of PP material is 28.7% lower than that of XLPE. Figure 43.4c concisely illustrates that the temperature of PP main insulation cable is higher than that of XLPE in all cases at composite material interface. Moreover, the difference between the two descends along the path with the lowest temperature rise at the separation of the stress cone from the main insulation of the cable at 311 K, equivalent to a temperature rise of 13 °C.

43.4.3 Thermal Stress Distribution

Figure 43.5a, b compare the overall thermal stress field of the cable joint with PP and XLPE as the main insulation materials, respectively. Figure 43.5c gives the thermal stress distribution at the interface of the composite material. It illustrates that due to the increasing temperature at composite material interface, different degrees of thermal expansion in each component of the joint is stimulated, resulting in radial and axial deformation and displacement. At the beginning and end of the path, intensive stress leads to deformation, which is severer in PP cable due to the higher temperature.

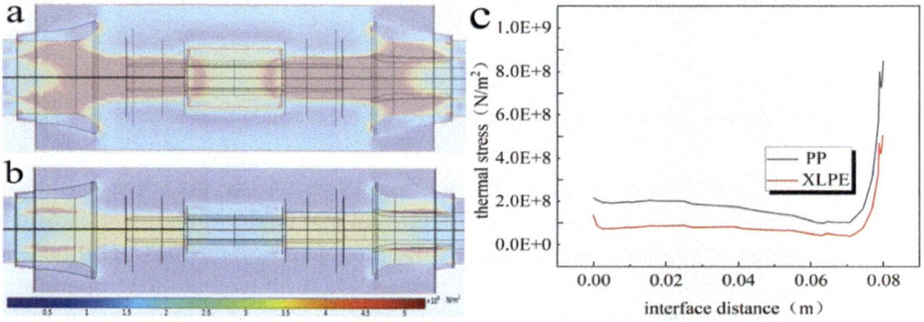

Fig. 43.5 Thermal stress distribution of the cable joint. **a** PP, **b** XLPE, **c** composite interface

43.5 Conclusion

This paper analyzes the electro-thermal-stress coupling process of the high voltage cable joint, establishes corresponding three-dimensional simulation model, and obtains the distributions of electric, thermal and thermal stress field of the cable joints when the main insulation materials are XLPE and PP respectively.This study obtains the following conclusions:

1. The overall electric field of the PP main insulated cable is stronger resulting from the obviously higher relative dielectric constant of XLPE than that of PP. At the composite material interface, electric field intensity increase by 23.4% at the separation point due to thermal stress distortion.
2. The temperature of the PP main insulation cable is significantly higher than that of the XLPE main insulation cable, which results from lower constant pressure heat capacity of the PP material, 28.7% lower compared with XLPE.
3. The thermal stress in the PP main insulation cable is overall larger than that in the XLPE main insulation cable, due to its the higher operating temperature.

References

1. X. Gu, S. He, Y. Xu et al., Partial discharge detection on 320 kV VSC-HVDC XLPE cable with artificial defects under DC voltage. IEEE Trans. Dielectr. Electr. Insul.Dielectr. Electr. Insul. **25**(3), 939–946 (2018)
2. Y. Chen, B. Hui, Y. Cheng et al., Failure investigation of buffer layers in high-voltage XLPE cables. Eng. Fail. Anal. **113**, 104546 (2020)
3. S. Xu, X. Huang, S. Yang et al., Three-dimensional simulation analysis of electric field distribution at the middle joint of 110 kV cable with typical defects. IOP Conf. Ser.: Earth Environ. Sci. **647**(1), 012034 (2021) (IOP Publishing)
4. X. Zhao, P. Meng, J. Hu et al., Simulation and design of 500 kV DC cable terminal accessory based on ZnO varistor microsphere composites. IEEE Trans. Dielectr. Electr. Insul.Dielectr. Electr. Insul. **27**(1), 10–16 (2020)
5. Y. Xin, L. Zhen, L. Zhen et al., Simulation calculation of thermal stress distribution of composite interface in operation of high voltage cable joint under electro-thermal coupling. Electric Mach. Control (Chinese) **24**(10) (2020)

Chapter 44
Lifetime Estimation of Metalized Film Capacitor Based on Self-healing Properties

Cheng Yao, Lei Jia, Zhiyuan Li, Lu Cheng, and Wenfeng Liu

Abstract Metalized film capacitors (MFC) are widely applied in power system, military weapons and railway traffics, etc. The lifetime of MFC is closely related to the self-healing (SH) process, which causes the loss of electrode area and thus leads to the capacitance reduction. As a result, a lifetime estimation method based on the SH properties and statistics was proposed in the present study. To establish the model, the artificial accelerate ageing experiments were firstly performed on the MFC elements under elevated DC voltages, and the capacitance of each element was monitored during the whole ageing periods. At the end of the life with 5% capacitance lost, the aged sample elements were disassembled to achieve the metalized films. Further, the SH parameters including the area of SH points and the points distribution were obtained with the help of image recognition technology. Based on the statistic model of SH parameters and the monitored capacitance reduction during ageing, the duration time of metalized films under certain DC voltage and the SH parameters could be correlated, and in turn the lifetime of metalized films could be deduced. Consequently, an efficient method for the lifetime evaluation of MFC based on the SH properties was proposed.

C. Yao · L. Jia
Electric Power Research Institute, United Laboratory of Advanced Electrical Materials and Equipment Support Technology, China Southern Power Grid Company Limited Guangzhou, Guangdong 510663, China

C. Yao · Z. Li · L. Cheng (✉) · W. Liu
State Key Laboratory of Power Equipment and Electrical Insulation, Xi'an Jiaotong University, Xi'an 710049, Shaanxi, China
e-mail: chenglu@xjtu.edu.cn

Z. Fang et al. (eds.), *Proceedings of the 5th International Symposium on Plasma and Energy Conversion*, Springer Proceedings in Physics 398,
https://doi.org/10.1007/978-981-97-2245-7_44

44.1 Introduction

44.1.1 A Subsection Sample

Capacitors, with the applications of energy storage, harmonics filtering, reactive power compensation and so on, are widely used in many areas. Especially with the development of green energies, such as solar energy, wind energy, tidal energy, et al., capacitors are facing a large challenge with much severer working condition including higher electric field, higher temperature and higher stability. For instance, the rated field of DC-link capacitors in VSC-HVDC can be more than 200 kV/mm, almost three times higher than shunt capacitors and twice higher than DC filter capacitors. To achieve such performances, metalized film capacitors (MFC) are adopted. Comparing with conventional liquid impregnant capacitors, the high energy density of MFCs is derived from the self-healing properties of metalized film. During the operation, breakdown continuously happens, which is followed by the SH process, resulting in the loss of electrode area. When the total lost area reaches 5%, i.e., the capacitance reduces to 95% of the original value, it is regarded as the end of the MFC's lifetime. As a result, the lifetime of MFC is largely dominated by the SH process.

Currently, investigations on the MFC lifetime are mainly categorized into the mathematic model and the ageing mechanism. Hereinto, the lifetime models were usually built by fitting the capacitance reduction curve with certain statistic models, such as Gamma distribution [1] and Birnbaum–Saunders distribution [2]. The obtained models should rely on large quantity of test results, and strongly associated with the ageing conditions, like voltage, temperature, waveforms and so on. The mechanism studies were usually derived from the artificial accelerated ageing tests [3–5]. It was found that the SH process was dominated by several parameters, such as the charged voltage, interlayer pressure, and temperature, etc. [6, 7], which in turn affected the lifetime of MFC.

The development of VSC-HVDC technology raised the high requirements of DC-link capacitors, while the working condition it, especially the operation field was far beyond the current studies. Consequently, in the present study a lifetime prediction model is proposed based on the operation condition of DC-link capacitors. Artificial ageing test on MFC elements was performed to achieve the evolutions of capacitance reduction under varied DC voltage. At the life ends of MFC elements, the characteristic parameters of more than 3000 SH points were obtained with the method of image recognition. Accordingly, the frequency of happened SH during certain period could be deduced, which finally led to the lifetime prediction method of MFC with higher efficiency.

44.2 Samples and Tests

44.2.1 Ageing Test

The artificial accelerated ageing test was performed on the MFC element with the capacitance ranging about 13 μF. The thickness of the base films was 6 μm, and the square resistance of the metalized films was 30Ω/□. The elements were aged under DC voltages of 1.4 U_N, 1.5 U_N, 1.6 U_N and 1.7 U_N, abbreviated as C1, C2, C3 and C4 respectively, where the U_N was the rated voltage of DC-link capacitors that was 2900 V in the present study. During the ageing, the capacitance of each sample element was measured at every 24 h. When the capacitance dropping to 95% of the original value, the test was ended.

44.2.2 SH Parameters

After the ageing, sample elements were teared down to obtain the metalized films. The length of film in sample element could be more than 60 m. To balance the accuracy and efficiency, films from three regions were selected as shown in Fig. 44.1.

From outer layer to the axis, the interface pressure decreased which caused the different SH behavior [6], so sample films were selected from outer layer, middle layer and inner layer respectively. At least films with the length of 30 cm of each region were tested. To obtain the density of SH spots (the amount of SH spots per unit area) and the area of each SH spot were obtained with the method of image recognition which has been calibrated.

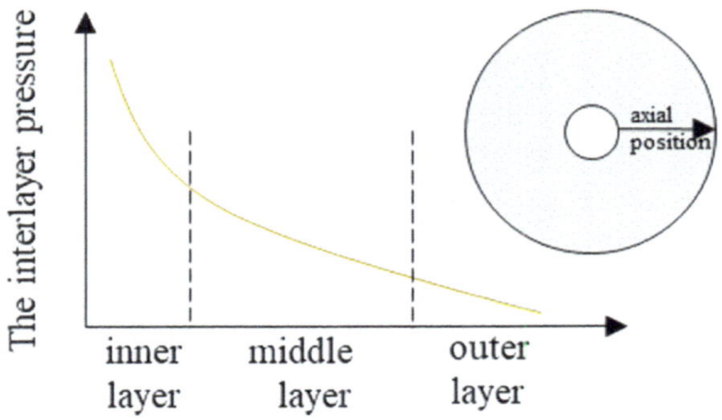

Fig. 44.1 Illustration of the sampling locations

44.3 Test Results

The capacitance of each sample element was measured offline every 24 h, and Fig. 44.2 shows the monitored capacitance evolution. It was noticed that with the ageing time going by, the capacitance of all the sample elements decreased, while the reduction rate went up both with the increase of ageing voltage and ageing duration time. When the ageing voltage was relatively low, the capacitance first maintained a stable level at the initial stage and then started the pronounced degradation from a critical ageing time. Just like C1 and C2, the capacitance exhibited few changes until about 200 h. Contrary to C1 and C2, the capacitance of C3 and C4 immediately dropped at the beginning of the ageing test and kept the fast reduction till the end of the life. Accordingly, at the end of life, i.e., the capacitance loss reaching 5% of original value, the ageing test duration of C1, C2, C3 and C4 was 955 h, 504 h, 264 h and 192 h respectively.

After the ageing test, all the sample elements were teared down for the further SH properties examination. Previous studies [8] indicated that the layer location of metalized film had obvious influences on the SH properties. Along the radical direction that from the axis to the outer layer, the interlayer pressure decreased which led to the increased probability of breakdown as well as SH. As a result, to make the results comparable, the metalized films at the fifth layer from the outside of the aged samples were selected for examination as shown in Fig. 44.3.

Basically, all the SH points showed the round shape with irregular edge, which was originated from the evaporation of the electrode. Once the weak point failed, the discharged energy evaporated the metalized layer, giving rise to the self-healing. At relatively low voltages of $1.4U_N$ and $1.5U_N$, the SH points were separated from each

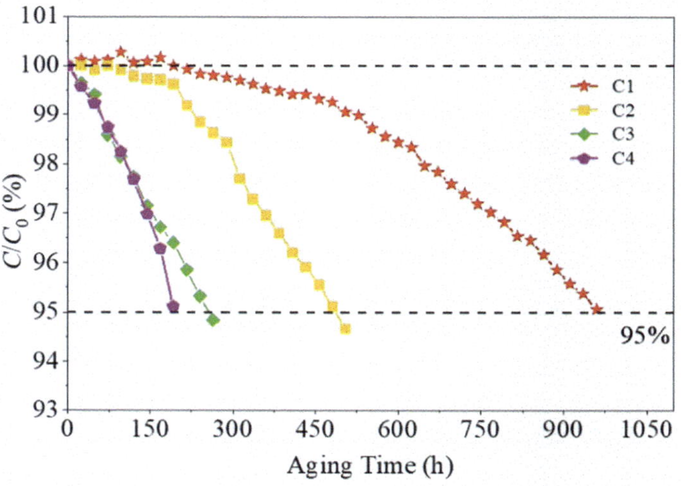

Fig. 44.2 Capacitance evolution aged under different voltage

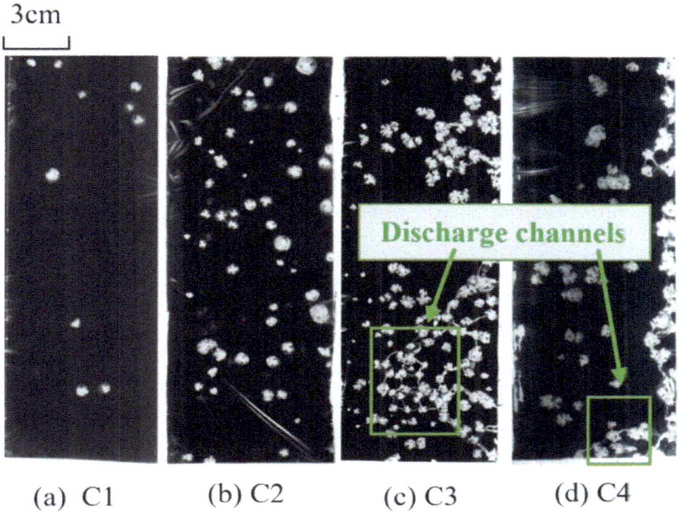

Fig. 44.3 Metalized films of capacitor element aged under different voltage

other, while with the voltage rising to $1.6U_{\mathrm{N}}$ and $1.7U_{\mathrm{N}}$, parts of the SH points overlapped and became connected by discharging channels due to the boost discharging energy.

To characterize the difference of SH points brought by the ageing voltage, the area of more than 3000 SH points was extracted by image recognition. The lognormal distribution was employed to fit the obtained area, which is appropriate for the statistical analysis of incomplete data caused by restricted measuring, testing and so on [9]. The function of lognormal distribution was shown as Eq. (44.1).

$$f(x,\mu,\sigma)=\begin{cases}\dfrac{1}{\sqrt{2\pi}\sigma x}\exp\left(-\dfrac{(\ln x-\mu)^2}{2\sigma^2}\right), & x>0\\ 0, & x\leq 0\end{cases} \tag{44.1}$$

The obtained SH area as well as the fitted distribution curves were depicted in Fig. 44.4. On the rising of the voltage, the area of SH point increased and the data distributed more scattered. For instance, about 80% data of C1 located at the range of 0–5.1 mm^2, while for C4, this interval increased to 0–49.5 mm^2.

According to the probabilistically distributed model, the expected SH area (S) of the sample elements aged under $1.4U_{\mathrm{N}}$, $1.5U_{\mathrm{N}}$, $1.6U_{\mathrm{N}}$ and $1.7U_{\mathrm{N}}$ was 3.08 mm^2, 8.33 mm^2, 22.03 mm^2 and 36.16 mm^2 respectively. The relation between S and the voltage could be expressed by the empirical equation of SH energy (W_{sh}) as below [8].

$$S\propto W_{sh}=\frac{k(U^xC_s)}{R^y}f(P) \tag{44.2}$$

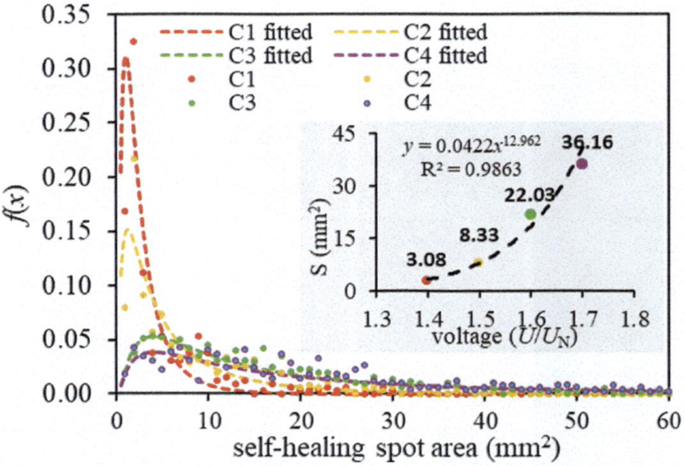

Fig. 44.4 Area distribution characteristics of self-healing under different voltages

where k, x and y are the coefficients, f(P) is the function of interlayer pressure, U is the charging voltage, and C_s is the capacitance.

The empirical equation indicated that the increase of voltage, capacitance and interlayer pressure could all lead to the rising of SH energy, which in turn caused the increase of SH area due to the energy was released by the evaporation of metalized layer. So, the SH area could also be expressed by Eq. (44.2). In the present study, the fitted x was 12.96 with the good fitting coefficient R^2 as high as 0.9863.

44.4 Lifetime Estimation Model

The test results indicated the dominant position of electric stress on the lifetime of MFC, since the rising of the voltage could lead to the decreased ageing test duration time and the increased SH area. As a result, it is of great necessity to propose the lifetime estimation model of MFC versus voltages. The accuracy of lifetime prediction model relies on the sample capacity, that the larger quantity of tested results achieved, the higher accuracy of lifetime model obtained. However, the large amount of test results was time- and cost-consuming, so how to balance the efficiency and the accuracy was a challenge for researchers.

The end of the lifetime of MFC was brought by a series of SH points, so it was possible to transform the test results of one capacitor into numerous SH points. With this hypothesis, the number of failed points (N) at every test period (24 h) could be calculated by the monitored results of capacitance reduction (ΔC) in Fig. 44.1 and the deduced expected SH area in Fig. 44.3 as shown in equation below.

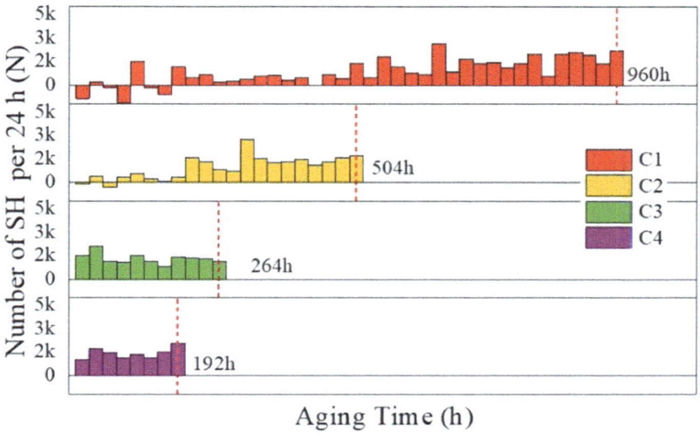

Fig. 44.5 Number of self-healing failures per specified time

$$N = \frac{d\Delta C}{\varepsilon_0 \varepsilon_r S} \tag{44.3}$$

where d is the thickness of BOPP film thickness, which in the present study was 6 μm. ε_0 and ε_r are the dielectric constant of vacuum and PP respectively. The calculated N of each sample were given in Fig. 44.5. The evolution of N in Fig. 44.5 was in accordance with variation trend in Fig. 44.2, that the capacitance of C1 and C2 kept stable at the initial stage of the ageing test, while C3 and C4 experienced a continuous capacitance reduction.

The SH points and their corresponding test duration time (t) should obey the Weibull distribution, which was firstly proposed to describe the failure rate of a series system caused by the weakest link [10, 11]. Nowadays, the Weibull distribution was widely used to estimate the lifetime of insulation materials [12]. Equation (44.4) showed the expression of Weibull distribution.

$$F(t) = 1 - \exp\left(-\left(\frac{t}{\beta}\right)^{\beta}\right) \tag{44.4}$$

where t is the lifetime and F(t) is the accumulated failure rate. α is the scale parameter and β is the shape parameter. The fitted results were shown in Fig. 44.6.

All the samples exhibited excellent fitting of Weibull distribution with the coefficient all above 0.9. The scaled parameters and shape parameters of C1, C2, C3 and C4 were shown in Table 44.1. Based on the two parameters, the duration time at any failure rate could be calculated with Eq. (44.4). Here we set the duration time with the failure rate of 95% as the lifetime, which were also given in Table 44.1.

The correlations between calculated lifetime (t) and the ageing voltage of four samples were plotted in Fig. 44.7, which could be fitted by the inverse power law function as shown below.

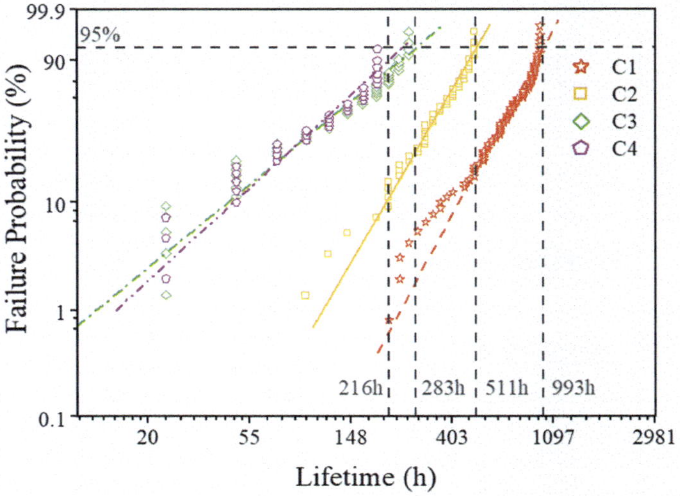

Fig. 44.6 Weibull probability distribution of failure times

Table 44.1 Weibull distribution parameters

Parameter	C1	C2	C3	C4
α	757	382	154	128
β	3.37	3.25	1.58	1.77
Calculated lifetime (h)	993	511	283	216

$$t = \frac{19781}{(U/U_N)^{-8.92}} \tag{44.5}$$

The inverse power law was widely used to describe the relations between the stability of insulation material determined by the electrical ageing [13]. Consequently, the excellent fitting indicated that the lifetime of MFC in the present study was dominated by the electrical strength. The fitted inverse power law of Eq. (44.5) provided the lifetime model of MFC under different voltages, by which the lifetime could be predicted. Under the rated voltage, the lifetime of the MFC was supposed to be about 20,000 h. To achieve the lifetime of 30 years that was the commonly required lifetime of power equipment, the rated voltage was suggested to be 75% of the current value.

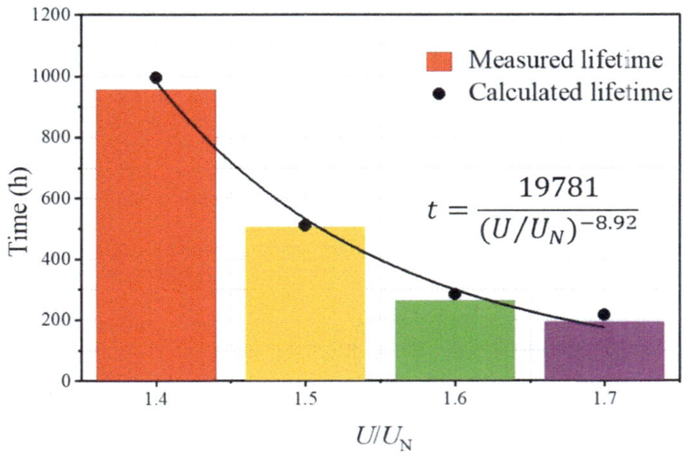

Fig. 44.7 Inverse power theorem life prediction

44.5 Conclusion

In the present study, artificial accelerated ageing tests of MFC elements were performed under elevated DC voltages with the range of 1.4UN to 1.7UN. With the monitoring of the capacitance decay and the statistics of SH points of metalized films, the correlations between capacitance loss and SH parameters were connected, which brought about the number SH points at every test cycle. The number of SH points was fitted by the Weibull distribution and thus the lifetime of MFC under certain voltages could be obtained with the failure rate reaching 95%. Consequently, a lifetime model with inverse power law of voltages was built, which could be applied for the lifetime prediction. Results showed that based on the ageing condition in the present study, the lifetime of MFC element running under rated voltage at the ambient temperature of 60 °C was deduced to be 2.35 year. The study provided an efficient and economical method for reliability evaluation of DC-link capacitors for practical application.

Acknowledgements The work was supported by United Laboratory of Advanced Electrical Materials and Equipment Support Technology, CSG (Program No. CSGULAEMEST-2021-KF-04).

References

1. X. Liu, H. Li, W. Zhang, Bootstrap method of metallized film capacitor gamma process degradation modeling. Hig. Vol. Eng. **37**, 198–202 (2011)
2. S. Xiong, J. Lu, Y. Zheng, Lifetime prediction of metallized film capacitor based on particle filter. Hig. Vol. Eng. **44**, 2378–2384 (2018)

3. V.O. Belko, O.A. Emelyanov, Self-healing in segmented metallized film capacitors: experimental and theoretical investigations for engineering design. J. Appl. Phys. **119** (2016)
4. J.H. Tortai, A. Denat, N. Bonifaci, Self-healing of capacitors with metallized film technology:: experimental observations and theoretical model. J. Electrostat. **53**, 159–169 (2000)
5. H. Li, M. Zhang, F. Lin, Study on theory and influence factors of self-healing in metallized film capacitors. Trans. China Electrotech. Soc. **27**, 218–223+230 (2012)
6. L. Jia, L. Cheng, W. Liu, in *Self-healing Properties of Metalized Polypropylene Film with Elevated Sheet Resistance*. 2020 IEEE International Conference on High Voltage Engineering and Application (ICHVE) (2020)
7. B. Peng, F. Lin, H. Li, Calculation and measurement of metalized film capacitor's inner pressure and its influence on self-healing characteristics. IEEE Trans. Dielectr. Electr. Insul. **17**, 1612–1618 (2010)
8. L. Cheng, Z. Li, J. Wang, Degradation behavior and mechanism of metalized film capacitor under ultrahigh field. IEEE Trans. Dielectr. Electr. Insul. **30**, 509–517 (2023)
9. Z. Zhao, M. Lai, L. Song, Estimation and testing of population parameters of two lognormal distributions with partially missing data. Statist. Decis. **20**, 7–9 (2009)
10. D.G. Shaw, S.W. Cichanowski, A. Yializis, A changing capacitor technology—failure mechanisms and design innovations. IEEE Trans. Electr. Insul. **EI-16**, 399–413 (1981)
11. X. Yang, D. Feng, C. Li, Life prediction model for oil-impregnated paper bushing based on Weibull distribution. Hig. Vol. App. **51**, 103–109 (2015)
12. Q. Sun, J.L. Zhou, Z. Zhong, Gauss-Poisson joint distribution model for degradation failure. IEEE Trans. Plasma Sci. **32**, 1864–1868 (2004)
13. IEC/TS 61251, *Electrical Insulating Materials-A.C. Voltage Endurance Evaluation-Introduction*. (B. S. Institution, 2008)

Chapter 45
Development and Evaluation of a Polarization/Depolarization Current Measurement Device for High-Voltage Cable Insulation Analysis

Haozhou Wang, Xiangnan Cao, Shuyu Chen, and Chengyan Ren

Abstract To measure the polarization/depolarization current (PDC) of cross-linked polyethylene (XLPE) power cables in service, a micro-current measurement device was developed, which can be placed on the high-voltage side of the cable and other electrical equipment. Traditional PDC testing methods are not suitable for field measurements and are significantly affected by field interference. By comparing several kinds of common PDC measurement methods, the measuring circuit in which a micro-current measurement unit was placed between the high-voltage power supply output and the high-voltage side of the test object was chosen. Firstly, the micro-current measurement device was developed according to the requirement of high-voltage side measurement. Then the steady-state current measuring result of self-designed device was contrasted and verified with that of Keithley 6517B. For a segment of 110 kV XLPE cable, PDC tests were also conducted using both the self-designed device and 6517B. The measuring results of steady-state current and cable by the self-designed device are close to the results by 6517B. Meanwhile the self-designed device can avoid low-frequency signal interference introduced by the ground wire. The PDC measurement and corresponding analysis can provide a more accurate assessment for the active cable insulation condition.

H. Wang · X. Cao · S. Chen · C. Ren (✉)
Beijing International S&T Cooperation Base for Plasma Science and Energy Conversion, Institute of Electrical Engineering, Chinese Academy of Sciences, Beijing 100190, China
e-mail: rcy@mail.iee.ac.cn

H. Wang · S. Chen · C. Ren
University of Chinese Academy of Sciences, Beijing 100049, China

Z. Fang et al. (eds.), *Proceedings of the 5th International Symposium on Plasma and Energy Conversion*, Springer Proceedings in Physics 398,
https://doi.org/10.1007/978-981-97-2245-7_45

45.1 Introduction

Cross-linked polyethylene (XLPE) cables are extensively used in power transmission and distribution lines due to their superior mechanical and electrical properties [1]. The insulation of power cables will age to varying degrees due to manufacturing processes, mechanical stress, high temperatures, and humidity during production, installation, and long-term operation [2]. This aging can severely affect the safe and stable operation of power systems. Therefore, it is necessary to diagnosis the condition of XLPE cable insulation, analyze the characteristics of insulation aging, and ensure the safety and reliability of the power system.

The polarization and depolarization current test, based on the principle of dielectric response [3], is a non-destructive method that offers simplicity, rich diagnostic information, and broad applicability. However, the methods for PDC measurement in laboratory are not suitable for field conditions where cable ground wires are fixed, and these methods can be subject to significant interference during testing, leading to inaccurate assessments of cable insulation conditions [4]. Devices capable of conducting PDC measurements in the field are rare at present [5], and there is an urgent need to develop devices suited for field measurement conditions.

This paper introduces a microcurrent measurement device designed for field environments that can be placed on the high-voltage side of cables. We compared this device with traditional low-voltage side measurements, verifying the testing method and the detection system.

45.2 Design of the PDC Measurement Device

Three types of circuits, as shown in Fig. 45.1, can be used to measure the cable's PDC. HV denotes the high-voltage DC power supply that provides the polarization voltage to the cable. A is the ammeter used to measure the PDC. Relay refers to the high-voltage relay that switches the cable between polarization and depolarization states.

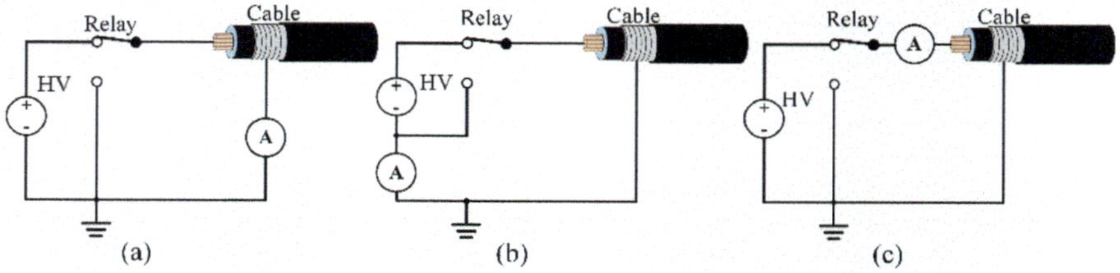

Fig. 45.1 Three PDC measurement circuits for cables: **a** ammeter connected between the cable armor and the ground wire, **b** ammeter connected between the low-voltage side of HV and the ground wire, **c** ammeter connected between the high-voltage side of HV and the cable conductor

In Fig. 45.1a, the PDC flows into the measurement device from the cable armor. However, since the cable armor is grounded in service, this method is not effective [6]. In Fig. 45.1b, the ammeter may inadvertently capture the current from HV output to the ground and the current from the high-voltage side to the low-voltage side of the relay in the results, leading to an overestimation of the polarization current [7]. This error increases with higher polarization voltages. The ammeter in Fig. 45.1c can directly measure the PDC of cables while avoiding interference from the currents of other circuit components. Additionally, stray currents on the ground wire are blocked by the cable's capacitive characteristics.

Using the circuit shown in Fig. 45.1c, the most important component is the microcurrent measuring device placed on the high-voltage side. This device requires a measurement range of ±10 mA to ±1 pA and is powered by a lithium battery to maintain complete isolation from the high-voltage and the ground wire. Measurement data is wirelessly transmitted using Zigbee technology and processed for statistical analysis and fitting on a host computer. The device uses a relay-controlled multi-stage I/V conversion circuit and data is collected via a 24-bit Analog to Digital Converter (ADC). The multi-stage I/V conversion amplification circuit is depicted in Fig. 45.2a, where I_{IN+} and I_{IN-} represent the positive and negative inputs of the microcurrent measuring device, and V_{OUT} is the output of the I/V conversion connected to the ADC. This circuit, controlled by a Microcontroller Unit (MCU), switches resistors of appropriate values to change the amplification factor of the amplifier. The relationship between input and output is expressed as:

$$V_{IN} = I_{IN} R_f \tag{45.1}$$

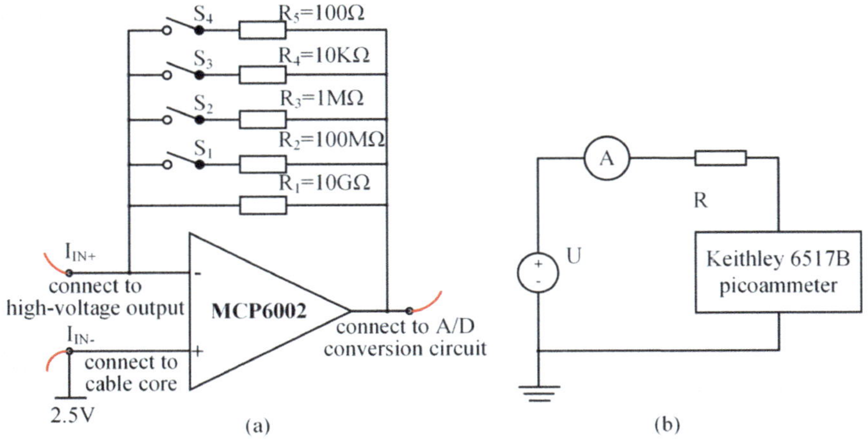

Fig. 45.2 **a** Multi-stage I-V amplifier circuit, **b** steady-state current test circuit

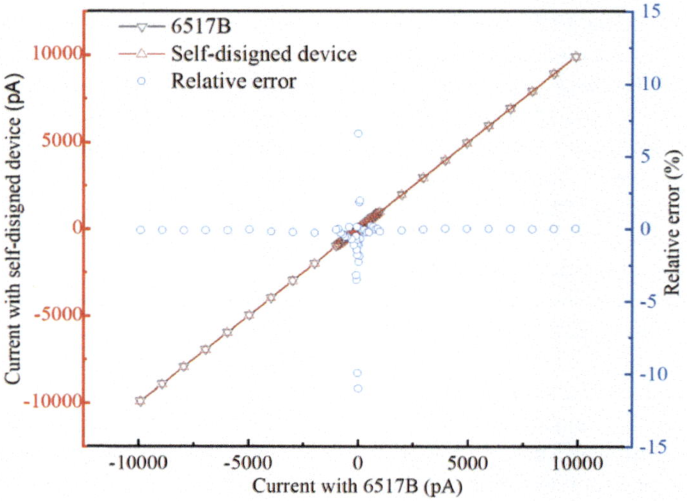

Fig. 45.3 Steady-state current comparison test

where V_{IN} is the converted voltage, I_{IN} is the current passing through I_{IN+} and I_{IN-}, and R_f is the feedback resistance between the amplifier's negative terminal and output.

To ensure the self-designed device can accurately measure the PDC of cables, we devised a steady-state current test for comparison as shown in Fig. 45.2b. In this setup, U represents a high precision DC voltage source (0–±1000 V) of 6517B, A is the self-designed device, and R is a high precision non-inductive resistor. The commonly used range for PDC measurement is from −10 to 10 nA, with a gain of 10^8 V/A for this range. By setting R at 10 GΩ and varying the voltage from −1000 to 1000 V, the test results are presented in Fig. 45.3.

Comparing the measurement results of the self-designed device with the 6517B, the relative error is less than 3% when the measurement range is above 10 pA or below −10 pA. When the current approaches 1 pA, the relative error increases. The absolute error remains below 700 fA, which is generally acceptable for the PDC measurement of XLPE cables.

45.3 PDC Measurement in XLPE Cable

Using the circuit shown in Fig. 45.4, PDC tests by the self-designed device and 6517B were simultaneously conducted on a new 110 kV XLPE cable, with the length of 0.5 m, the type of ZC-YJLW02-64/110 $kV^{-1} \times 800\ m^2$.

The polarization voltage was set at 6 kV, with a polarization time of 1500 s and a depolarization time of 1500 s. R_0 is a current-limiting resistor with a resistance of 2 MΩ, preventing high pulse currents from damaging the self-designed device and

Fig. 45.4 PDC test circuit for 110 kV cable

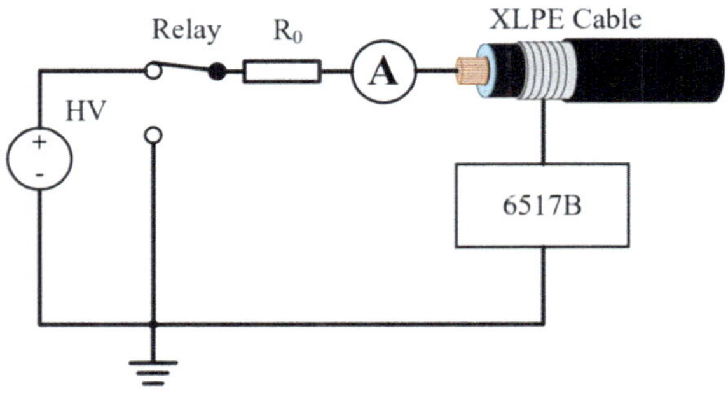

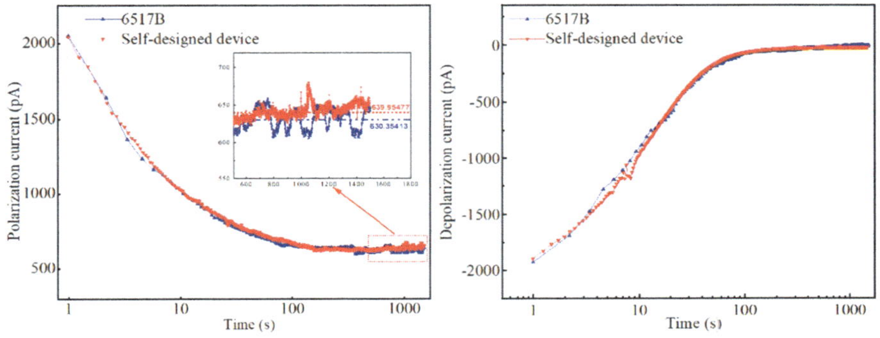

Fig. 45.5 PDC test results of 110 kV cable

6517B. The relay is used to switch the polarization and depolarization circuit modes. The measured PDC results are shown in Fig. 45.5.

The insulation resistance of the cable was calculated using the following formula:

$$R_0 = \frac{U_C}{I_{p0}} \tag{45.2}$$

where R_0 is the cable's insulation resistance, U_c is the polarization voltage, and I_{p0} is the steady-state polarization current.

The insulation resistances by the self-designed device and 6517B are respectively 9.38 TΩ and 9.52 TΩ, with a relative error of 1.4%, so the measurement accuracy of the self-designed device is good. Based on the isothermal relaxation current theory [8], the cable's depolarization current can be fitted with a third-order exponential function to calculate the cable insulation's aging factor.

$$I_d(t) = I_0 + \sum_{j=1}^{3} \alpha_j e^{-\frac{t}{\tau_j}} \tag{45.3}$$

Table 45.1 Depolarization current fitting results of different devices

Device	I_0	α_1	τ_1	α_2	τ_2	α_3	τ_3	A
6517B	0.88	909.19	4.44	1226.80	21.93	79.63	177.02	1.76
Self-designed	16.59	808.77	7.11	1089.23	22.19	69.37	136.67	1.60

where $I_d(t)$ is the depolarization current, t is the depolarization time, and I_0 is the steady-state value of the depolarization current. The parameters α_j and τ_j are related to the dielectric properties of the cable insulation, reflecting the trap density and trap depth of the cable insulation respectively [9].

Table 45.1 shows that the self-designed device and the 6517B provide similar results for the three-stage exponential fit of the depolarization current, with a slight difference in the calculated aging factors. This discrepancy is due to the introduced low-frequency interference from ground connection in 6517B measurement, and the results have a significant deviation in the τ_3 parameter. According to the German aging factor criterion, when the aging factor A is less than 1.75, the cable insulation is in good condition. This result corresponds to its unused status

45.4 Conclusions

A PDC measurement method for the high-voltage side test of cables was proposed, and it is more suitable for the field test. along with a microcurrent measurement device designed to measure cable PDC at high voltage. The comparison test results with the Keithley 6517B picoammeter indicate that the self-designed device has high accuracy for both steady-state current and PDC. The measurement range is from ±1 pA to ±10 mA. When the current is high to 10 pA, the relative error is below 3%, near 1pA, the absolute error is less than 700 fA. The device can avoid low-frequency interference from ground connections, and the analysis results can accurately assess the insulation condition of the cable.

References

1. T. Kubota, Y. Takahashi, S. Sakuma et al., Development of 500 kV XLPE cables and accessories for long distance underground transmission line-Part I: insulation design of cables. IEEE Trans. Power Delivery **9**(4), 1741–1749 (1994)
2. S. Boggs, J. Densley, J. Kuang, Mechanism for impulse conversion of water trees to electrical trees in XLPE. IEEE Trans. Power Delivery **13**(2), 310–315 (1998)
3. H. Wolfgang, L. Eberhard, *High-Voltage Test and Measuring Techniques* (Springer, Berlin, Heidelberg, 2014)
4. J. Wu, Y. Shihu, H. Jianping, et al., in *Study on the Time Characteristics of Polarization/Depolarization Currents in High Voltage Cables*. Proceedings of the 4th International Symposium on Plasma and Energy Conversion, pp. 87–91 (2023)

5. F. Yang, Z. Kai, A. Shakeel et al., Improving the polarization and depolarization current measuring method to avoid ground wire interference. High Vol. **7**(5), 847–855 (2022)
6. L. Tiandi, Z. Cheng, L. Hao et al., Insulation state evaluation of 110 kV cable based on isothermal relaxation current. Adv. Tech. Elect. Eng. Energy **39**(04), 43–52 (2020). ((in Chinese))
7. T. Jin, in *Study on the Insulation Diagnosis System of Cable Based on the Polarization/ Depolarization Current Method*. Huazhong University of Science and Technology (2018) (in Chinese)
8. J.G. Simmons, M.C. Tam, Theory of isothermal currents and the direct determination of trap parameters in semiconductors and insulators containing arbitrary trap distributions. Phys. Rev. B **7**(8), 3706–3713 (1973)
9. P. Birkner, Field experience with a condition-based maintenance program of 20 kV XLPE distribution system using IRC analysis. IEEE Trans. Power Delivery **19**(1), 3–8 (2004)

Chapter 46
Influence of Different Square Resistances of DC Support Capacitor Components on Their Self-healing and Withstand Voltage Performance

Shaowei Guo, Zhaoliang Xing, Guiyou Fan, Chong Zhang, Xin Chen, Yaotian Su, Juyang Liu, Wenjing Zhang, and Yankun Fan

Abstract As the key equipment of DC converter valve in flexible DC transmission, DC support capacitor mainly plays the role of voltage support and harmonic filtering, which requires it to be safe, reliable and resistant to high breakdown voltage. As an important parameter of capacitor, the square resistance of metallized film is related to the self-healing and withstand voltage performance of capacitor. In this paper, three kinds of square resistance metallized film capacitor components are selected to carry out ESR, self-healing test, limit voltage withstand test and other experimental studies. The experimental results show that the high square resistance capacitor has good self-healing and withstand voltage performance, but its ESR will also increase. It is necessary to comprehensively consider the selection of capacitor square resistance range.

46.1 Introduction

As an energy storage component, DC support capacitor is applied to the DC side rectifier arm module in flexible DC transmission, which mainly plays the role of voltage support, harmonic filtering and so on [1]. At present, metallized film capacitors are widely used in DC support capacitors [2]. Metallized film capacitors are mainly composed of three parts: metal electrode, dielectric film and gold spray. Compared with the traditional oil immersed full film capacitor, the advantages of metallized film capacitor are mainly reflected in the following points: (1) metallized film has

S. Guo · Z. Xing (✉) · C. Zhang · X. Chen · Y. Su · J. Liu
State Key Laboratory of Advanced Power Transmission Technology, State Grid Smart Grid Research Institute Co. Ltd, Beijing, China
e-mail: xingzhaoliang007@163.com

G. Fan · W. Zhang · Y. Fan
EHV Branch Company of State Grid Fujian Electric Power Co. Ltd, Fuzhou, China

Z. Fang et al. (eds.), *Proceedings of the 5th International Symposium on Plasma and Energy Conversion*, Springer Proceedings in Physics 398,
https://doi.org/10.1007/978-981-97-2245-7_46

self-healing performance and high reliability; (2) The capacitor unit has small energy volume, light weight, compact structure and simple process; (3) Dry oil-free, high fire rating, environmentally friendly; (4) Low noise [3].

The dielectric film of DC support capacitor is mostly polypropylene film, with a thickness of 2–10 μm. The typical thickness is 5.8 μm. The metallized polypropylene film was formed by evaporating Al or Zn-Al composite with a thickness of 10–100 nm on the polypropylene film [4, 5] the square resistance of the metallized film refers to the resistance of the planar conductor of a unit block, that is, the planar resistivity P of the planar conductor. Square resistance is an important parameter of metallized film. Theoretically, the thicker the coating on the active surface (which forms the electrode plate of the capacitor), the lower the loss tangent of the capacitor. However, the thicker the coating, the more thermal damage the base film will suffer during evaporation, and the self-healing point will increase during energization, which will reduce the pressure resistance of the metallized film. Therefore, it is of great significance for the development of metallized film capacitor manufacturing technology to study the self-healing and withstand voltage performance of DC support capacitor components under different square resistances.

In this paper, three kinds of typical square resistance metallized films for DC support capacitors are selected as the basis, and they are wound into components. Experimental studies such as equivalent series resistance test, self-healing test and ultimate withstand voltage test are carried out, and the influence of different square resistance on the self-healing and withstand voltage performance of DC support capacitor components is studied.

46.2 Samples and Tests

46.2.1 Sample Preparation

Biaxially oriented polypropylene film (BOPP) is made from extruded polypropylene melt into sheet, and then manufactured through a series of processes, such as biaxial stretching, heat treatment, cooling treatment, etc. The film material used in this test is BOPP film produced by Quanzhou Jiadeli Electronic Material Co., Ltd., and the polypropylene is Borealis HC300BF electrical grade high temperature resin. The performance test data of BOPP and polypropylene resin are shown in Tables 46.1 and 46.2.

Table 46.1 Polypropylene resin properties

Product	Oxidation induction time (200 °C)/min	Melt massflow rate /g/10 min	Yellowness index	Tacticity /%	Ash content / ppm
HC300BF	36.0	3.1	2.0	99.6	15

Table 46.2 BOPP performance

No.	Test item		Unit	Average value
1	Thickness		μm	5.81
2	Tensile strength	MD	Mpa	173
		TD		339
3	Surface roughness	Ra	μm	0.074
4	DC breakdown strength		V/μm	637
5	Electrical weak spots (450 V/μm)		count/m^2	< 0.1
6	Dissipation factor (23 °C /50 Hz)		/	2.3×10^{-4}
7	Permittivity		/	2.12

The BOPP film was metallized, which was biaxial stretched, was metallized by double edge coating process. Three groups of metallized film were designed, namely group **a** (3–8.5 Ω/□), group **b** (3–20 Ω/□) and group **c** (3–50 Ω/□). Then type DMR-1C device was used to test the square resistance of the metallized film. Each group of metallized film tested 1–11 square resistance points, as shown in Fig. 46.1a. Figure 46.1b shows the measured square resistance of each group of metallized films. The square resistance test results of the 3–8.5 Ω/□ group fluctuated between 2.2 Ω/□ and 8.2 Ω/□, while the square resistance test results of the 3–20 Ω/□ group fluctuated between 2.6 Ω/□ and 17.9 Ω/□. The square resistance test results of the 3–50 Ω/□ group fluctuated between 2.5 Ω/□ and 55.7 Ω/□ were also considered to be within the coating error range. At the same time, the breakdown strength of the metallized films in each group was measured. The breakdown field strength of the high square resistance metallized film in group **c** was the highest, As shown in Table 46.3.

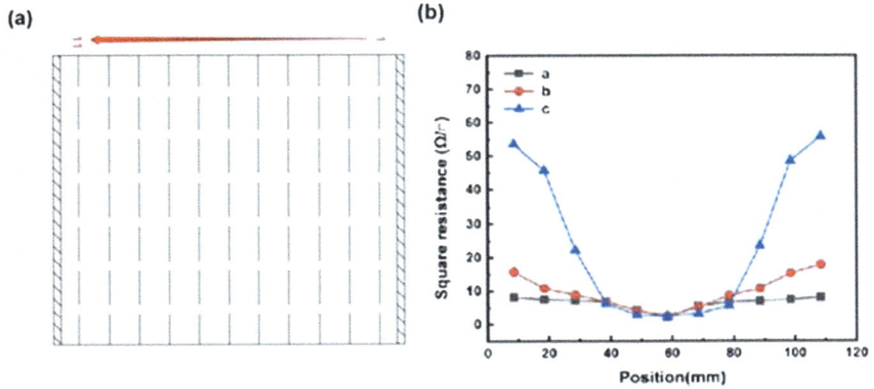

Fig. 46.1 **a** Schematic diagram for measuring the square resistance of metallized films; **b** measurement results of square resistance of metallized films

Table 46.3 Breakdown strength of metallized films

Product	Square resistance Ω/□	DC breakdown strength V/μm
a	3–8.5	612
b	3–20	621
c	3–50	630

Fig. 46.2 Capacitor components prepared with different square resistance metallized films

The evaporated metallized film is made into the test component by the same process. According to the metallized film evaporation resistance from low to high, it is divided into three groups: group A, group B and group C, with 10 components in each group. The element diameter d n is 86 mm, the height *h* is 161 mm, and the element mandrel diameter d_0 is 5 mm, as shown in the Fig. 46.2. Then, ESR, self-healing test, high temperature breakdown voltage, limit withstand voltage were carried out on the components.

46.2.2 *Test Method*

ESR test: at 20 °C, use LCR tester to measure the ESR of all components. The measuring voltage is 1 V and the measuring frequency is 50 Hz.

Self healing test: select 2 components from each group for test, as shown in Fig. 46.3a. Keep the whole body temperature of the test object at 70 °C. The capacitor element shall withstand a DC voltage of 1.5 U_{NDC} between terminals for 10 s. If there are less than 5 times of breakdown during this period (10 s), the voltage shall be slowly increased until 5 times of breakdown have occurred since the start of the test, or until

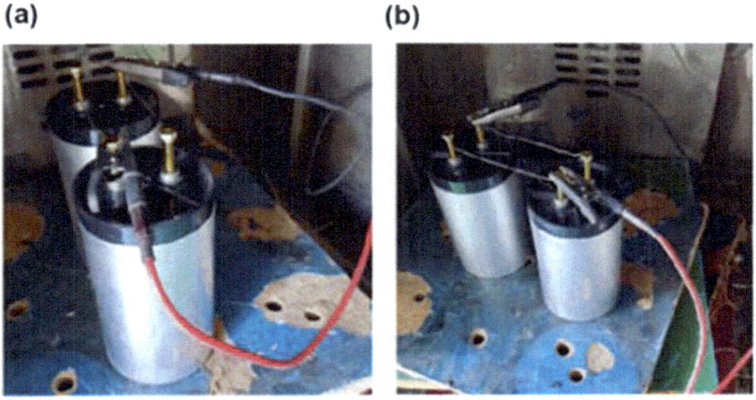

Fig. 46.3 **a** Self-healing test.; **b** ultimate withstand voltage test

the voltage reaches 2.5 U_{NDC}. If the voltage reaches 2.5 U_{NDC}, the breakdown is still less than 5 times, and the test time is more than 10 s, the test is over.

Limit withstand voltage test: select 2 components from each group for test. Each capacitor element shall be baked in the oven for more than 4 h to reach 80 °C of the whole body and applied with 1.6 U_{NDC} for 10 min, as shown in Fig. 46.3b. Breakdown and flashover shall not occur during the test. Self healing breakdown is allowed.

High temperature breakdown voltage test: select 3 components from each group of components for test. Place the element in an oven at 80 °C, set the voltage to climb at the speed of 100 V/s, and record the voltage value when the capacitor element is breakdown or its leakage current exceeds 50 mA.

46.3 Results and Discussion

In the actual production of the capacitor, because the electrode material used has resistance and the insulating medium has loss, the capacitor itself will have certain resistance and loss, which is called equivalent series resistance (ESR) [6]. Figure 46.4 shows the ESR measurements of different square resistance capacitor elements. The ESR of group a low square resistance elements is significantly lower than that of group B and C. This is because for metallized film materials, the equivalent resistance is directly proportional to the length of the material and inversely proportional to the cross-sectional area of the material, that is, when the material and length are constant, the larger the cross-sectional area, the smaller the resistance. Reducing the square resistance of metallized film, that is, increasing the thickness of metal coating, can effectively reduce the equivalent series resistance of capacitor.

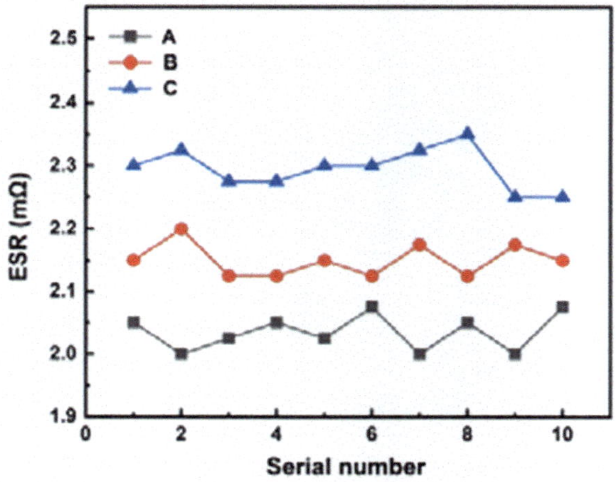

Fig. 46.4 ESR of different square resistance capacitor components

When the internal medium of metallized film capacitor breaks down, an arc will be generated to evaporate the metal layer at the breakdown point and "repair" the local part. This process is called "self-healing" [7]. Figure 46.5 shows the capacitance changes of different square resistance capacitor elements before and after self-healing test. With the increase of the square resistance, the capacity loss of the components before and after the experiment is gradually reduced, and the capacitance of group C components is even positively biased. This is because the self-healing energy of the metallized film is high under low square resistance, and the arc at the breakdown point cannot be extinguished, resulting in the breakdown and sintering of the capacitor dielectric film, which increases the self-healing area of the component and leads to capacity attenuation [8]. The use of high electrode square resistance can effectively reduce the self-healing energy and ensure the effective reduction of capacitance loss on the basis of successful self-healing.

The flexible DC support capacitor has high field strength under working conditions and can withstand DC superimposed harmonic voltage for a long time, so it has higher requirements for breakdown performance at high temperature. Figure 46.6a shows the change rate of capacitance of different square resistance capacitor elements before and after the limit withstand voltage test. It can be found that after the limit withstand voltage test, three groups of elements have positive capacitance bias, of which group C has the largest positive capacitance bias. It shows that the three groups of capacitor elements perform well in the limit voltage withstand test. Figure 46.6b is the limit value of DC withstand voltage of capacitor elements with different square resistances at 85 °C. With the increase of square resistances, the limit withstand voltage level of the element also increases. The average limit withstand voltage of group C element is 4.66% and 3.06% higher than that of group A and group B respectively. It shows that the withstand voltage level of capacitor element is high under high square resistance, which is consistent with the change of breakdown strength of metallized films with different square resistance, as shown in Table 46.3.

Fig. 46.5 Self healing test of different square resistance capacitor components

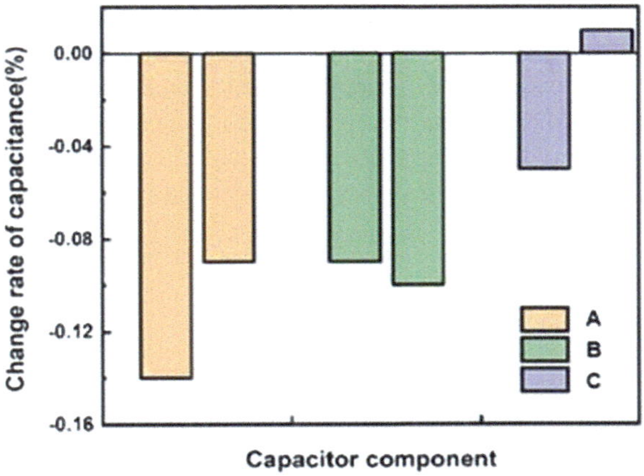

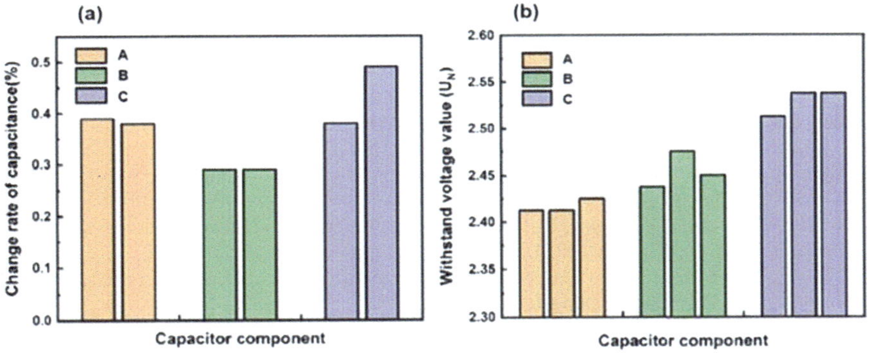

Fig. 46.6 **a** Ultimate withstand voltage experiment of different square resistance capacitor components; **b** high temperature ultimate breakdown voltage of different square resistance capacitor components

46.4 Conclusion

This study systematically tested and studied the influence of different square resistances on the self-healing and withstand voltage performance of DC support capacitor components. The results show that the ESR of capacitor element increases with the increase of the square resistance of metallized film, and the heating of high square resistance capacitor element will increase under working conditions; With the increase of the square resistance of the metallized film, the capacity loss of the capacitor element gradually decreases after the self-healing experiment, indicating that the high square resistance element has good self-healing performance and high safety; The final high temperature limit withstand voltage test results show that there is little difference in the high temperature withstand voltage test results of different square resistance capacitor components, but the limit withstand voltage level of high square resistance is higher than that of low square resistance components. Therefore, in the design of DC support capacitor, increasing the square resistance can improve the self-healing and withstand voltage performance of the capacitor, and it is also necessary to comprehensively consider the problem of electrode heating caused by the increase of ESR caused by high square resistance.

Acknowledgements This work was funded by the State Grid Corporation of China Headquarters Science and Technology Project (No. 5500-202340161A-1-1-ZN).

References

1. L. Wei, G. Qi, L. Feng, Discussion on the development of DC link capacitor for flexible HVDC transmission project. Power Capacit. Reactive Power Comp. **38**(3), 95–98 (2017)
2. B. Peng, F. Lin, H. Li, Y. Chen, M. Zhang, F. Lv, Calculation and measurement of metalized film capacitor's inner pressure and its influence on self-healing characteristics. IEEE Trans. Dielectr. Electr. Insul. **17**, 1612–1618 (2010)
3. O.G. Gnonhoue, A. Velazquez-Salazar, É. David, I. Preda, Review of technologies and materials used in high-voltage film capacitors. Polymers **13**, 766 (2021)
4. S. Qin, S. Ma, S.A. Boggs, in *The Mechanism of Clearing in Metallized Film Capacitors*. Proceedings of the IEEE International Symposium on Electrical Insulation, pp. 592–595 (2012)
5. A.R. Shete, A.R. Thorat, in *An Optimal Technique for Manufacturing of H.T. Capacitor by Using Metallized Polypropylene Film*. 2013 International Conference on Power, Energy and Control (ICPEC), India, pp. 177–180 (2013)
6. C. Song, L. Zhaolin, Square resistance of metallized film for capacitor and its test method. Shaanxi Elect. Pow. Qual. Saving Tech. Soc. **4** (2018)
7. Y. Xin, W.H. Ying, Metallized film capacitor self-healing effect factors and control measures. World Nonferr. Metals **23**, 95–96 (2016)
8. L. Hua, Y. Bosi, L. Li et al., Study on breakdown and self-healing properties of polypropylene film for DC-link capacitors. Power Cap. & React. Power Compens. **40**(3), 0079–0083 (2019)

Chapter 47
Influence of Phase of Multi-frequency Harmonics on Breakdown Characteristics of Polypropylene Film for High Voltage Shunt Capacitors

Qiming Sun, Zhaoliang Xing, Yingbin Li, Junjie Liu, Shaowei Guo, Yiming Zhang, Jingyu Yang, and Lingyu Zhu

Abstract Harmonics in the power grid distort the voltage at both ends of the high voltage shunt capacitor and affect the insulation of the polypropylene film inside the capacitor. In order to study the effect of the phase of harmonics on the insulation characteristics of polypropylene films, an experimental platform for breakdown of polypropylene films for high voltage shunt capacitors was established. The phase of the harmonics voltage was changed and record the breakdown field strength of the polypropylene film under different conditions. The effect of the phase of the harmonics voltage on the breakdown voltage of the polypropylene film was analyzed using the two-parameter Weibull distribution. The results show that the breakdown field strength of polypropylene film decreases with the increase of temperature, and the harmonics leads to the decrease of the breakdown field strength of polypropylene film. Moreover, the phase of harmonic voltages affect the breakdown characteristics of polypropylene films by changing the peak and rise rates of voltage.

Q. Sun (✉) · Y. Li · J. Liu · L. Zhu
State Key Laboratory of Electrical Insulation and Power Equipment, Xi'an Jiaotong University, Xi'an 710049, Shaanxi, China
e-mail: sunqiming@stu.xjtu.edu.cn

Z. Xing · S. Guo
State Key Laboratory of Advanced Power Transmission Technology, State Grid Smart Grid Research Institute Co., Ltd., Beijing 102209, China

Y. Zhang · J. Yang
State Grid Tangshan Power Supply Company, Tangshan 063000, China

Z. Fang et al. (eds.), *Proceedings of the 5th International Symposium on Plasma and Energy Conversion*, Springer Proceedings in Physics 398,
https://doi.org/10.1007/978-981-97-2245-7_47

47.1 Introduction

High voltage shunt capacitor is an important reactive power compensation equipment in power system. The gradual increase of harmonic content in the power grid leads to a gradual increase in the harmonic content of high-voltage shunt capacitors, which leads to the increase in the defect and failure rate of capacitors [1, 2]. The existence of harmonic increases the effective value and peak value of the voltage applied at both ends of the capacitor, so that the voltage waveform distortion is peaked, and affect the insulation characteristics of the dielectric materials. However, most of the research mainly focus on the production and manufacture of high temperature resistant polypropylene films and their modification [3, 4], and the breakdown test of polypropylene film only carries out AC and DC voltage, ignoring the influence of power frequency superimposed multi-frequency harmonic voltage conditions during operation on the film medium. Therefore, in order to improve the operational reliability of high voltage shunt capacitors, it is necessary to study the influence of harmonic phase on the breakdown characteristics of polypropylene films under the condition of multi-frequency harmonics.

At present, researches on the breakdown characteristics of polypropylene film used in high voltage shunt capacitors mainly focus on the change in breakdown strength caused by the change of the structural parameters of the film itself [5], or the change of the structural parameters of the film after electrical or thermal aging [6]. The applied breakdown voltage is also only DC and power frequency sine voltage. There are few researches on the effect of superimposed harmonics on the breakdown field strength of dielectric insulation in capacitor. Reference [7] measures the DC breakdown strength of a 4–15 μm thick polypropylene film at room temperature using five electrodes with an area of 0.045 to 9.3 cm^2, and the results show that the DC breakdown strength decreases with the decrease of electrode area. In order to test the properties of the films, reference [8, 9] carries out DC electrothermal aging and DC breakdown test. It is found that the breakdown field strength of 63.2% probability increases with the aging time, and the insulation properties of films with similar breakdown properties at room temperature are different after high temperature and thermal aging.

The phase of the harmonics affects the breakdown voltage of the polypropylene film medium by changing the voltage rise rate and the maximum voltage. The study of the effect of voltage waveform and temperature on the breakdown characteristics of composite dielectric insulation is more mature in the oil-paper insulation structure [10–12], and the study of polypropylene film for high voltage shunt capacitors is less. In this paper, the harmonic loading platform of polypropylene film for capacitors is set up. At 20 °C, 40 °C, 60 °C and 80 °C power frequency superimposed multi-frequency harmonic voltages, polypropylene film breakdown experiments are carried out to explore the relationship between the breakdown field strength and the harmonics by changing the phase of the harmonics. The method provides a reference for revealing the mechanism of the influence of the phase of harmonics on the breakdown field strength of polypropylene film for high voltage shunt capacitors.

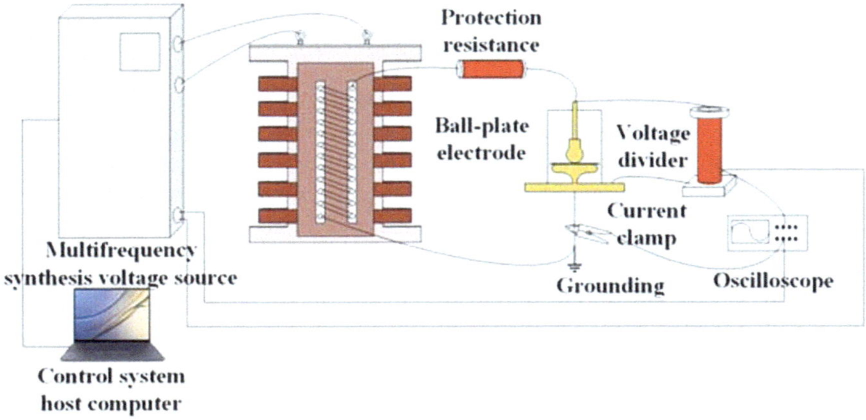

Fig. 47.1 Schematic diagram of polypropylene film breakdown test platform

47.2 Experimental Setup and Methods

47.2.1 Experimental Platform

Polypropylene film breakdown test platform mainly includes harmonic loading system, oil cup and measurement system. The platform diagram is shown in Fig. 47.1. In this test platform, the test film needs to be continuously loaded with a specific frequency combination voltage, and the power supply needs to be cut off quickly after the test product breakdown to avoid system overload. Therefore, in the process of operation, the control system needs to access two measurement signals, use the voltage divider to measure the voltage at both ends of the test for feedback adjustment, and use the current clamp to measure the current of the test as the basis for test interruption. At the same time, voltage divider and oscilloscope are used to read the voltage at the breakdown time of the film, and the rising edge of the branch current at the breakdown time is used as the trigger condition to record the breakdown time waveform of the test.

47.2.2 Experimental Platform

Before the test, the test product and the liquid insulating medium should be conditioned. The film stands at 23 °C and the relative humidity is 50% for more than 12 h, and the oil stands at 80 °C and the vacuum environment for more than 12 h, so that the water in the oil can be discharged, and then transferred to the oven and the film test product to stand together to achieve the test temperature conditions. In the experiment, the continuous voltage boost method was used to increase the voltage uniformly from zero until the breakdown of the test product at a voltage boost rate of 300 V/s. Each set of conditions contained ten samples, and each breakdown occurred between 10 and 20 s.

The voltage distortion can be calculated by using the maximum stable overcurrent that the capacitors can withstand during operation. According to the provisions of "DL/T 840–2016 Specification of high-voltage shunt capacitors for service", capacitors should be able to operate under a stable overcurrent with a square mean root value of $1.3I_n$. For capacitors with the largest positive capacitance deviation, this overcurrent is allowed to reach $1.37I_n$ [13]. The voltage distortion rate of each harmonic voltage can be expressed as:

$$\frac{0.83}{\sqrt{2}h_i} \leq U_{hi} \leq \frac{0.92}{\sqrt{2}h_i} \tag{47.1}$$

The distortion rates of each harmonic voltage in the test are determined as shown in Table 47.1. The combination of harmonics, initial phase of harmonics and temperature condition are set as shown in Table 47.2.

Under each group of experimental conditions, the film samples were broken down 10 times, and the peak voltage measured by the oscilloscope at the time of breakdown was recorded as the breakdown voltage. The breakdown field strength was calculated by dividing the peak value by the thickness of the film, and the data were analyzed and processed by using the two-parameter Weibull distribution function. The probability distribution function of the two-parameter Weibull distribution is:

Table 47.1 Distortion rate of each harmonic voltages

Order of harmonics	Distortion rate of harmonics (%)
3	20
5	12
7	9
9	7
11	6
13	5
15	4
17	3.5

Table 47.2 Experimental conditions of film breakdown

Order of combination harmonics	Initial phase of harmonics	Experimental temperature (°C)
$U_1 + U_3 + U_5$	0	20
$U_1 + U_7 + U_9$	π/6	40
$U_1 + U_{11} + U_{13}$	π/3	60
$U_1 + U_{15} + U_{17}$	π/2	80

$$F(E) = 1 - e^{-\left(\frac{E}{\alpha}\right)^{\beta}} \tag{47.2}$$

where E is the breakdown field strength of the film; $F(E)$ is the breakdown probability of the film at the field strength E; α is the scale parameter, representing the characteristic breakdown field strength of the film when the breakdown probability is 63.2%; β is the shape parameter, reflecting the dispersion of the breakdown field strength.

47.3 Results and Discussions

47.3.1 Waveform of Experimental Voltage

In order to study the influence of the initial phase of harmonics on the breakdown field strength of polypropylene films under the condition of power frequency superimposed harmonic voltage, the breakdown data of films under different voltage conditions in the film breakdown test are divided according to the initial phase of harmonics, and different voltage waveforms were shown in Fig. 47.2. The effective values of each harmonic are reduced according to Table 47.1, and the initial phase of the harmonics and temperature conditions are set as shown in Table 47.2. The purple line is 50 Hz power frequency sine waveform, and its effective value is equal to the effective value of the power frequency superposed harmonic voltage.

The 63.2% probability breakdown field strength of polypropylene film at different temperature, different initial phase of harmonics and different harmonic combinations conditions are shown in Fig. 47.3.

As shown in Fig. 47.3a and Table 47.2, when polypropylene film is applied with power frequency plus 3rd harmonic plus 5th harmonic combination voltages, at different temperature, the breakdown field strength is the largest when the harmonic initial phase is 30°. When the initial phase of harmonic is 0° or 90°, the breakdown field strength is the lowest. As shown in Fig. 47.2a, when the power frequency is superimposed with the 3rd and 5th harmonics, the voltage contains three peaks in the positive half cycle. When the initial phase of the harmonic is 0°, the rise rate of the waveform is the highest after the zero crossing, and the peak value of the first wave peak is the highest. When the initial phase of the harmonic is 90°, the rise rate of the waveform before the second wave peak is the largest, and the peak of the second wave peak is the highest. Therefore, compared with 30° and 60°, the breakdown field strength is lower when the initial phase of harmonics is 0° and 90°.

As shown in Fig. 47.3b and Table 47.2, polypropylene film is applied with power frequency plus 7th harmonic plus 9th harmonic combination voltages. When the initial phase of harmonic is 0°, the breakdown field strength is minimum, and increases with the increase of the initial phase. at different temperature, the breakdown field strength is the largest when the harmonic initial phase is 90°. When the initial phase of harmonic is 0° or 90°, the breakdown field strength is the lowest. As

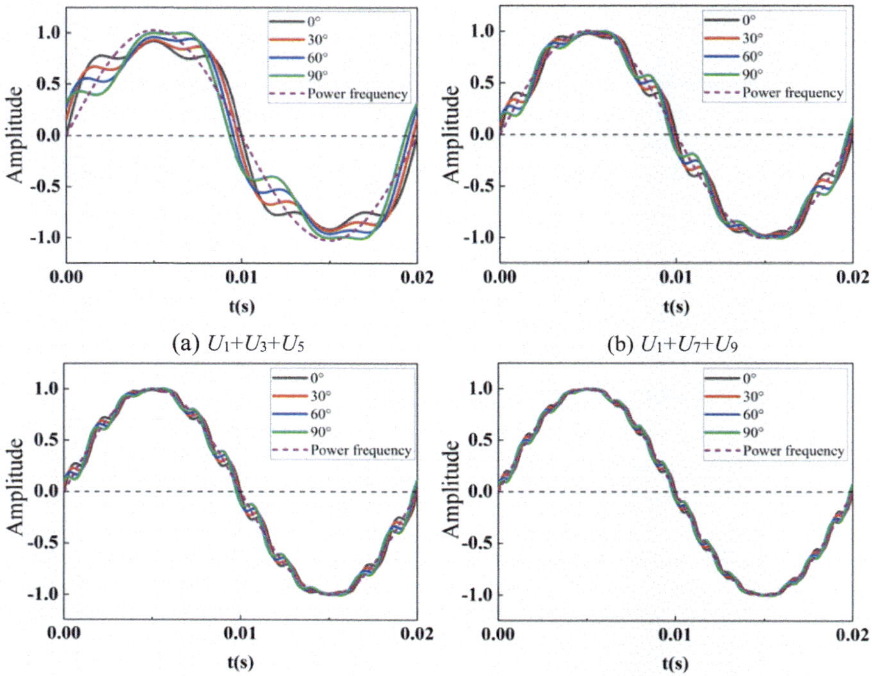

(a) $U_1+U_3+U_5$ (b) $U_1+U_7+U_9$

(c) $U_1+U_{11}+U_{13}$ (d) $U_1+U_{15}+U_{17}$

Fig. 47.2 Waveform of power frequency plus harmonics

shown in Fig. 47.2b, when the power frequency is superimposed with 7th and 9th harmonics, the positive half-cycle contains three peaks, and the positive half-cycle peaks are almost equal at different initial phase. When the initial phase of harmonics is 0°, the rise rate of voltage before the first two wave peaks is the largest, and the peak value of wave peaks is the largest. With the increase of the initial phase, the rise rate of voltage and the peak of the first two peaks decrease. Therefore, when the initial harmonic phase is 0°, the breakdown field strength of the polypropylene film is minimum, and the breakdown field strength increases with the increase of the initial phase.

As shown in Fig. 47.3c, d, under different temperature conditions, when the power frequency is superimposed with 11th and 13th harmonic and power frequency is superimposed with 15th and 17th harmonic voltages are applied to the polypropylene film, when the initial phase of the harmonic is 0°, the breakdown field strength is the smallest, and it is increased under 30°, 60° and 90°. As shown in Fig. 47.2c, d, when the power frequency is superimposed for 11th, 13th, 15th and 17th harmonic voltages, the voltage distortion is significantly smaller and closer to the power frequency waveform than the waveform when the low frequency is superimposed. Therefore, the difference in breakdown field strength under the condition of different initial

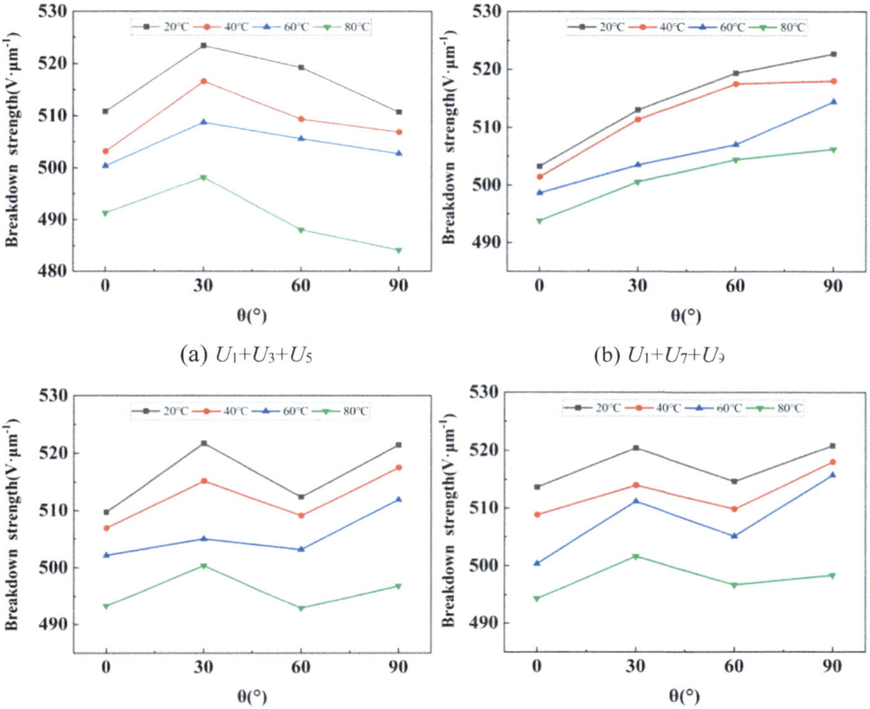

(a) $U_1+U_3+U_5$ (b) $U_1+U_7+U_9$

(c) $U_1+U_{11}+U_{13}$ (d) $U_1+U_{15}+U_{17}$

Fig. 47.3 63.2% probability of breakdown strength of polypropylene film at different temperature, different initial phase of harmonics and different harmonic combinations

phase of harmonics is also relatively small. When the initial phase of harmonic is 0°, the steepness of the voltage and the value of the wave peak are higher, so the breakdown field strength is relatively lower.

47.4 Conclusion

In this paper, the relationship between the breakdown characteristics of polypropylene film for high voltage shunt capacitors and the initial phase of harmonic under multi-frequency harmonic conditions is studied. The voltage distortion rate of each harmonic was determined, and on this basis, the breakdown experiment of polypropylene film under different temperature and voltage conditions was carried out. The data were processed based on Weibull distribution, and the following conclusions were obtained:

(1) Under different harmonic voltage combinations, the breakdown field strength of polypropylene film decreases with the increase of temperature; Compared with power frequency voltage alone, the breakdown field strength under power frequency superimposed harmonics is lower, and the decrease of breakdown field strength decreases with the increase of harmonic frequency.
(2) When the film is subjected to the power frequency superimposed 3rd and 5th harmonic voltage, the initial phase of the harmonic is 0°, 90°, the breakdown field strength of the film is lower; When the voltage condition is power frequency superimposed 7th and 9th harmonic voltage, the breakdown field strength is the lowest when the initial phase is 0°, and the breakdown field strength increases with the increase of the initial phase of harmonics. When the film is subjected to power frequency superimposed 11th, 13th and power frequency superimposed 15th, 17th harmonic voltage, because the distortion rate of higher harmonics is smaller, the impact of the initial phase of harmonic on the breakdown field strength is also smaller, when the initial phase is 0°, the breakdown field strength of the film is the lowest.

Acknowledgements This work was funded by the State Grid Corporation of China Headquarters Science and Technology Project (No. 5500-202214401A-2-0-ZN).

References

1. W. Ying, L. Daijun, X. Xianyong et al., IEEE ICHQP2018 and development directions of power quality. Electric Pow. Automat. Equip. **39**(4), 197–203 (2019)
2. C. Xu, *Research on the Influence of Harmonic on the Life and Protection of Power Capacitor Bank*. North China Electric Power University (2013)
3. L. Qi, L. Manqian, High-temperature polymer dielectrics for film capacitors: review and prospect. High Vol. Eng. **47**(9), 3105–3123 (2021)
4. Z.Y. Ran, B.X. Du, M. Xiao et al., Dielectric property improvement of polypropylene films doped with POSS for HVDC polymer film capacitors. IEEE Trans. Dielectr. Electr. Insul. **28**(4), 1308–1316 (2021)
5. M. Yuwei, C. Lu, L. Hongbo et al., Correlation between aggregation structure of biaxially oriented polypropylene and its breakdown property. Insulating Mat. **56**(1), 1–7 (2023)
6. L. Hongbo, C. Lu, X. Zhe et al., Influence of thermal ageing on structure and electrical properties of polypropylene film for capacitors. Insulating Mat. **56**(2), 96–103 (2023)
7. S.J. Laihonen, U. Gafvert, T. Schutte et al., DC breakdown strength of polypropylene films: area dependence and statistical behavior. IEEE Trans. Dielectr. Electr. Insul. **14**(2), 275–286 (2007)
8. M. Ritamäki, I. Rytöluoto, K. Lahti, in *High Temperature and Ageing Test Methods to Characterize the Dielectric Properties of BOPP Capacitor Films*. 2017 IEEE Conference on Electrical Insulation and Dielectric Phenomenon, pp. 266–269. IEEE, Fort Worth, TX, USA (2017)
9. J. Kurimský, M. Kosterec, B. Vargová, in *Breakdown Voltage of Polypropylene Film During C and Thermal Ageing*. 2017 18th International Scientific Conference on Electric Power Engineering. pp. 1–4. IEEE, Kouty nad Desnou, Czech Republic (2017)
10. S.S.T. Ram, M. Kamaraju, B.P. Singh, in *Flashover Behaviour of Converter Transformer Insulation Subjected to Superimposed AC and DC Voltages*. IEEE Conference on Electrical Insulation and Dielectric Phenomena, pp. 810–815. IEEE, Arlington, TX, USA (1994)

11. L. Shengtao, Z. Tuo, H. Qifeng, Relation among breakdown voltage, frequency and time in oil-paper insulation of transformer. Transformer **46**(4), 39–43 (2009)
12. L. Gang, S. Wei, L. Feng, et al., Study on effect of harmonic voltage on breakdown voltage of insulation paper in transformer. Transf. **57**(6), 6–8+18 (2020)
13. DL/T 840, Specification of high-voltage shunt capacitors for service (2016)

Chapter 48
A Simulation Study on Early Stage Thermal Runaway of Lithium Iron Phosphate Energy Storage Batteries Due to Overcharging

Fan Yang, Zhuofei Wang, Lei Su, Zhichun Yang, Yu Feng, Cheng Zhang, and Tao Shao

Abstract The thermal effects of lithium-ion batteries have always been a crucial concern in the development of lithium-ion battery energy storage technology. To investigate the temperature changes caused by overcharging of lithium-ion batteries, we constructed a 100 Ah experimental platform using lithium iron phosphate ($LiFePO_4$) batteries. Overcharging tests were conducted at a 0.5C rate at different states of charge (SOC), and the resulting temperature evolution was recorded. The experimental results demonstrate that a higher initial SOC leads to an earlier occurrence of thermal runaway in the battery. Once the battery reaches 100% SOC, the temperature rises significantly until thermal runaway occurs. Based on the experimental results of battery discharging at different SOC stages and the heat generation mechanism of lithium iron phosphate batteries during thermal runaway, a simulation model of overcharging-induced thermal runaway in $LiFePO_4$ battery was established. The overcharging-induced thermal runaway process of lithium-ion batteries at different SOC was subsequently examined through simulation and compared with the experimental results. The simulation results indicate that the lithium battery model can accurately predict the temperature distribution on the battery surface following thermal runaway caused by overcharging.

F. Yang · L. Su · Z. Yang
State Grid Hubei Electric Power Research Institute, Wuhan 430074, China

Z. Wang (✉) · Y. Feng · C. Zhang (✉) · T. Shao
Beijing International S&T Cooperation Base for Plasma Science and Energy Conversion, Institute of Electrical Engineering, Chinese Academy of Sciences, Beijing 100190, China
e-mail: wangzhuofei@mail.iee.ac.cn

C. Zhang
e-mail: zhangcheng@mail.iee.ac.cn

Z. Fang et al. (eds.), *Proceedings of the 5th International Symposium on Plasma and Energy Conversion*, Springer Proceedings in Physics 398,
https://doi.org/10.1007/978-981-97-2245-7_48

48.1 Introduction

In today's increasingly pressing global energy landscape, lithium-ion battery-based electrochemical energy storage systems has emerged as a crucial enabling technology for ensuring the secure and stable operation of power grids. Lithium iron phosphate ($LiFePO_4$) batteries are extensively utilized in power grid energy storage systems due to their high energy density and long cycle life. Under extreme conditions such as overcharging, short circuits, or high temperatures, the heat accumulation can lead to a significant rise in battery temperature and trigger a dangerous occurrence called thermal runaway. Consequently, in-depth research into the thermal runaway mechanisms of lithium-ion battery becomes necessary.

Scholars mainly focus on experimental or simulation analysis in the study of thermal runaway behavior of lithium-ion batteries. In terms of experiments, Reference [1] found that during battery overcharging, excessive lithium at the negative electrode can form lithium dendrites, which can penetrate the separator and cause internal short circuits. In literature [2], it is found that batteries at a high SOC lead a more severe exothermic reaction during thermal runaway, resulting in higher temperatures. In literature [3], it is believed that overcharging lithium-ion batteries beyond 120% SOC can induce irreversible internal short circuits. The experimental study in reference [4] found that $LiFePO_4$ batteries have a lower temperature than ternary NCM batteries for thermal runaway when overcharged. In terms of simulation analysis, literature [5] developed an overcharging model that combines a one-dimensional electrochemical model with a three-dimensional thermal abuse model. In literature [6], a two-dimensional electrochemical-thermal coupling method is presented, and simulations are conducted on three different battery sizes. Literature [7] establishes a three-dimensional electrochemical-thermal coupling model for large lithium-ion batteries under needle puncture overcharge conditions. Literature [8, 9] proposes a modified electrochemical model and conducted experiments and simulations on the short-circuit thermal runaway phenomenon of lithium-ion batteries. Literature [10] proposed a simplified electrochemical-thermal model for simulating the thermal runaway of lithium-ion batteries by utilizing the Arrhenius equation which enables the study of dynamic temperature variations of the battery while reducing computation time and complexity.

The cutoff of power during overcharge of lithium-ion batteries at an earlier stage significantly reduces the probability of thermal failure. Therefore, the study of research on the early warning mechanisms of thermal runaway is necessary. In this paper, a lithium-ion battery electrochemical-thermal coupling model is constructed by fitting the battery polarization parameters. Overcharging experiments and simulations are carried out on a $LiFePO_4$ battery. A comparative analysis between the experiments and simulations is performed, considering the opening of the battery safety valve as the boundary.

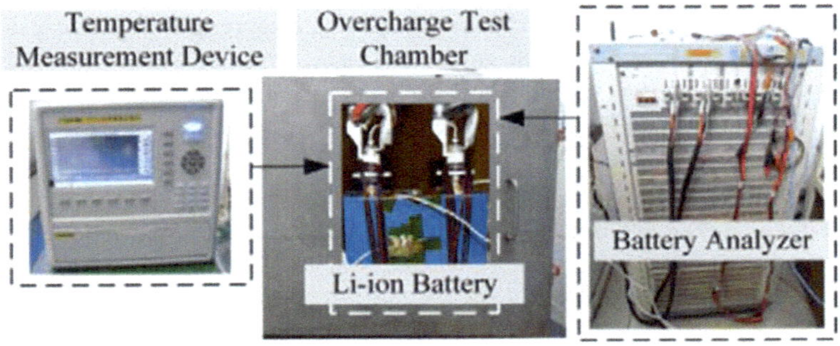

Fig. 48.1 Lithium battery overcharge test platform

48.2 Experimental Platform and Simulation Model in Lithium Iron Phosphate Monomer Battery

48.2.1 Research Object and Experimental Platform

The experiment subject is a hard-shell rectangular $LiFePO_4$ battery with dimensions of 160 mm × 118.5 mm × 50 mm and a rated capacity of 100 Ah. It is equipped with a safety valve on the top. The positive electrode material of the battery is $LiFePO_4$, and the negative electrode material is graphite. The main components of the battery electrolyte are $LiPF_6$ and electrolyte solvents.

The overcharging experimental platform is shown in Fig. 48.1 and mainly consists of a battery analyzer, an overcharge test chamber, and a temperature measurement device.

48.2.2 Thermal Runaway Model of Lithium Iron Phosphate Monomer Battery

In this section, an electrochemical-thermal coupled model was established for $LiFePO_4$ battery. Subsequently, a parameter fitting method for the polarization losses in the model was introduced.

The internal thermal energy of a battery consists of two parts: heat generation and heat dissipation. The heat generation and consumption inside a lithium battery satisfy the energy conservation equation, as shown in Eq. (48.1).

$$\rho c_p \frac{\partial T}{\partial t} = \lambda \nabla^2 T + q_{rea} + q_{act} + q_{ohm} + q_s - q_t \tag{48.1}$$

In which, c_p represents the specific heat capacity, ρ represents the battery density, λ represents the heat rate of each guide, q_{rea}, q_{act}, q_{ohm}, q_s and q_t represents the reaction heat production rate, the polarized heat production rate, the ohmic heat production rate, the total heat production rate of thermal runaway side reaction and the total heat dissipation.

The Eqs. (48.2–48.4) describe the heat generation rates for these mechanisms during typical charge and discharge processes.

$$q_{rea} = S_a j T \frac{\partial U_{eq}}{\partial T} \tag{48.2}$$

$$q_{act} = S_a j \eta \tag{48.3}$$

$$q_{ohm} = -i_s \nabla \Phi_s - i_l \nabla \Phi_l \tag{48.4}$$

In this context, S_a is the molar entropy, j is the electric current density, η is the surface overpotential of the battery, and U_{eq} is the equilibrium potential of the battery. Φ_s is the solid phase potential and i_s is the solid phase current density. In contrast, Φ_l is the liquid phase potential and i_l is the liquid phase current density.

There are three forms of heat dissipation, namely thermal conduction, thermal convection, and thermal radiation. Three forms of heat dissipation equations are presented in the following Eqs. (48.5–48.7):

$$q_r = -\lambda \frac{\partial T}{\partial x} \tag{48.5}$$

$$q_c = hS(T - T_{amb}) \tag{48.6}$$

$$q_{rad} = \delta \sigma A_1 F_{12}(T_a^4 - T_b^4) \tag{48.7}$$

In this context, q_r, q_c and q_{rad} represent thermal flux of three dissipation methods, λ is the thermal conductivity coefficient, h is the convection heat transfer coefficient, S is the surface area of heat transfer, T_{amb} represents the ambient temperature, δ is the emissivity and σ is a Stephen-Boltzmann constant. A_1 is the radiation area of object 1 and F_{12} is the angular coefficient.

Based on the heat generation and dissipation equation mentioned above, the electrochemical-thermal coupling model of $LiFePO_4$ battery was established by COMSOL Multiphysics. Figure 48.2 is the geometry diagram and mesh diagram of the model. The battery module has dimensions of 160 mm in width, 50 mm in thickness, and 115 mm in height (excluding electrodes and connectors).

When a current passes through batteries, various polarization phenomena occur on the electrodes, causing a deviation in the electrode potential, which can be divided into three types, namely ohmic polarization, activation polarization and concentration

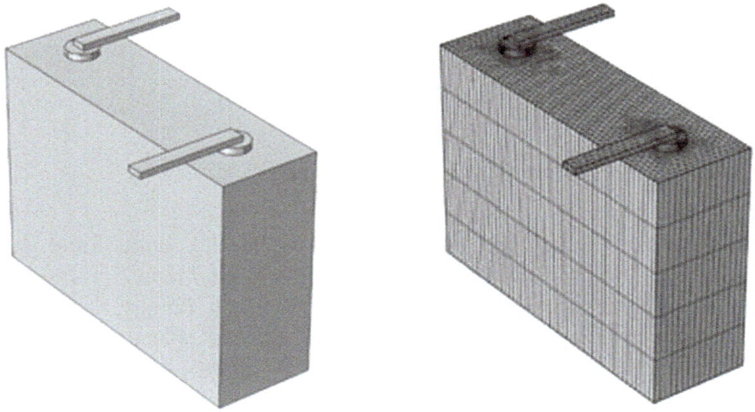

Fig. 48.2 Model geometry and mesh construction

polarization. The corresponding parameters of the losses are ohmic overpotential at 1C (η), dimensionless charge exchange current (J_0) and diffusion time constant (τ).

To determine the over-potential parameters, SOC-OCV measurement experiments and multiple charge–discharge experiments were conducted to obtain the SOC-OCV curve and charge–discharge curve. The optimal value was then fitted using the Levenberg–Marquardt method, and the corresponding parameters for voltage loss were calculated iteratively. The measured SOC-OCV curve and the corresponding fitting voltage curve are depicted in Fig. 48.3. COMSOL Multiphysics calculates the polarization over-potential parameters by fitting curve, as shown in Table 48.1. These parameters serve to enhance the establishment of comprehensive electrochemical models.

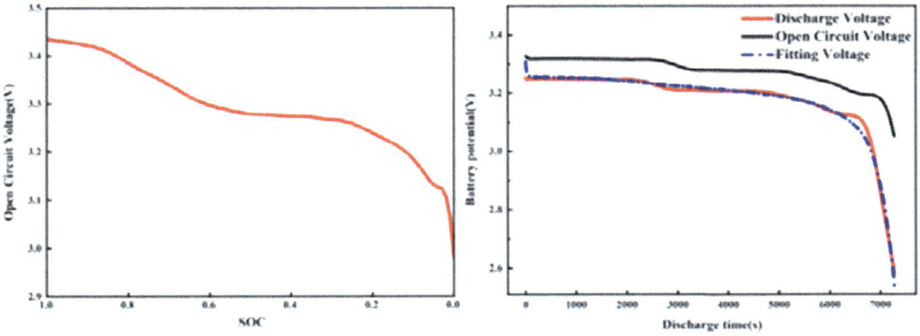

Fig. 48.3 SOC-OCV curve and fitting voltage curve

Table 48.1 Polarization over-potential parameters

Ohmic overpotential	Dimensionless exchange current	Diffusion time constant
0.04952 V	0.0998	443.16 s

48.3 The Experimental and Simulation Analysis

In this section, the experimental and simulation results of a lithium-ion battery is presented firstly. Then, the influence of difference SOC on the overcharge capacity and cumulative temperature rise of the battery is presented. Finally, the comparison between the experimental results and the simulation results is analyzed.

In order to investigate the behavior of lithium-ion batteries at different SOC (0%, 25%, 50%, 75%, and 100%), a charging process was conducted at a constant current (0.5 C) rate. During this process, the temperature of the battery's surface was monitored. As shown in Fig. 48.4, the acquired experimental data was compared with simulation results to assess the accuracy of the model.

In the comparative analysis of the results, it was found that the simulated temperature was slightly higher, with an error margin of less than 10%. This phenomenon can be ascribed to the omission of several crucial factors in the simulation process, namely, the energy dissipation resulting from the activation of the battery safety valve, the energy dissipation arising from internal gas generation within the battery, as well as the heat conduction occurring at the interface between the battery and the chamber.

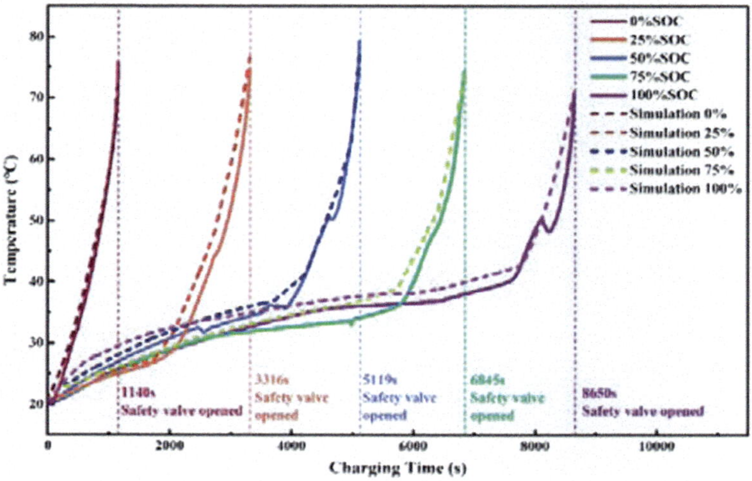

Fig. 48.4 Comparison of experimental and simulated temperatures

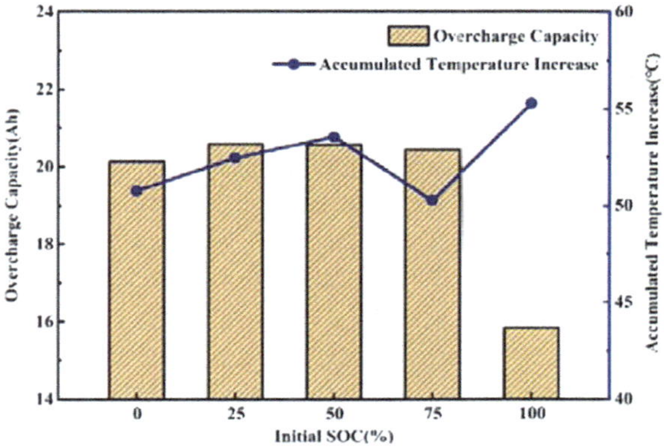

Fig. 48.5 Overcharge capacity and cumulative temperature rise at different SOC

To investigate the impact of overcharging on lithium-ion batteries at different SOC, the overcharge capacity and cumulative temperature rise were recorded during the charging process until the safety valve opened, which are illustrated in Fig. 48.5.

According to Figs. 48.4 and 48.5, it can be observed that the heat generation of the battery exhibits an exponential increase after the onset of overcharging. The overcharging capacity at 100% SOC is slightly lower, and the cumulative temperature elevation is higher compared to the other groups, which can be attributed to the fact that the battery starts overcharging directly at 100% SOC, where the internal resistance of the battery is already high, resulting in more Joule heat generation and faster gas evolution. In addition, the overcharge capacities and cumulative temperature rises of the remaining four groups are closely comparable, indicating that, except for the 100% SOC condition, the batteries at different initial SOC charging to 120% SOC do not exhibit significantly different behaviors under overcharging conditions.

Through research analysis, it has been determined that during overcharging of the lithium-ion battery, the dominant heat generation mechanism is Joule heat. At SOC ≤ 1.2, the total heat generation is primarily determined by Joule heat, without triggering any side reactions.

48.4 Conclusions

By conducting overcharging experiments and electrochemical-thermal coupled simulations on lithium iron phosphate batteries, the early temperature evolution trend of thermal runaway and the influence of different SOC on overcharge capacity and cumulative temperature elevation were obtained. The following findings are concluded:

(1) The constructed lithium-ion battery electro-thermal coupling model, which includes fitting the battery polarization parameters, shows good agreement in terms of the battery's temperature evolution during overcharging, with a slight difference of less than 10%.
(2) The heat generation rate of the battery increases exponentially after the onset of overcharging, primarily due to Joule heat.
(3) Overcharging starting at 100% SOC leads to faster temperature elevation and gas production, causing the safety valve to open earlier, and reducing the overcharge capacity.

References

1. N.D. Rago, J. Bareno, J. Li, et al., Effect of overcharge on Li ($Ni_{0.5}$Mn0.3$Co_{0.2}$)O_2/graphite lithium ion cells with poly (vinylidene fluoride) binder I-Microstructural changes in the anode. J. Pow. Sour. **385**, 148–155 (2018)
2. W. Mingmin, S. Lei, J. Yang et al., Overcharge and thermal runaway characteristics of lithium iron phosphate energy storage battery modules based on gas online monitoring. High Vol. Eng. **47**(1), 279–286 (2021)
3. X. Lai, Y. Zheng, L. Zhou et al., Electrical behavior of overdischarge-induced internal short circuit in lithium-ion cells. Electrochim. Acta **278**, 245–254 (2018)
4. Z. Wang, J. Yuan, X. Zhu et al., Overcharge-to-thermal-runaway behavior and safety assessment of commercial lithium-ion cells with different cathode materials: a comparison study. J. Energy Chem. **55**, 484–498 (2021)
5. C. Qi, Y. Zhu, F. Gao et al., Mathematical model for thermal behavior of lithium ion battery pack under overcharge. Int. J. Heat Mass Transf. **124**, 552–563 (2018)
6. P. Nie, Z. Siwei, R. Aihua, et al., Full-cycle electrochemical-thermal coupling analysis for commercial lithium-ion batteries. Appl. Thermal Eng. **184**, 116258 (2021)
7. Z. Wei, L. Gang, W. Chaoyang, Modeling nail penetration process in large-format li-ion cells. J. Electrochem. Soc. **162**(1), 207–217 (2014)
8. R. Zhao, J. Liu, J. Gu, A comprehensive study on Li-ion battery nail penetrations and the possible solutions. Energy **123**, 392–401 (2017)
9. R. Zhao, J. Liu, J. Gu, Simulation and experimental study on lithium ion battery short circuit. Appl. Energy **173**, 29–39 (2016)
10. P.T. Coman, E.C. Darcy, C.T. Veje et al., Modelling Li-ion cell thermal runaway triggered by an internal short circuit device using an efficiency factor and Arrhenius formulations. J. Electrochem. Soc. **164**(4), A587–A593 (2017)

Chapter 49
Research on Breakdown Characteristics of PP Films for Shunt Capacitors Under Multi-frequency Harmonic Voltages Group

Yingbin Li, Jian Wang, Qiming Sun, Jianjun Zhang, Yong Yu, Xin Chen, Junjie Liu, and Lingyu Zhu

Abstract High voltage AC shunt capacitors bear the electrical stress of power frequency superimposed harmonics when running in the field. With the increase of harmonic content in the power grid, the test of capacitor insulation increases. In order to study the breakdown characteristics of polypropylene film for high voltage shunt capacitors under the condition of power frequency superposition harmonics, this paper carried out the breakdown test of polypropylene (PP) film under different voltage frequency combination and different temperature and other influencing factors. The statistical analysis of the breakdown field strength data of the film was carried out by using Weibull distribution. The results show that the breakdown field strength of polypropylene film decreases with the increase of temperature. The breakdown field strength of power frequency superimposed harmonics is lower than that of power frequency, and with the increase of harmonic frequency, the decline of breakdown field strength decreases.

Y. Li · Q. Sun · J. Liu · L. Zhu
State Key Laboratory of Electrical Insulation and Power Equipment, Xi'an Jiaotong University, Xi'an 710049, Shanxi, China
e-mail: Liybin2021@stu.xjtu.edu.cn

J. Wang (✉)
State Grid Corporation of China, Beijing 100031, China
e-mail: 932987095@qq.com

J. Zhang · Y. Yu · X. Chen
State Grid Tangshan Power Supply Company, Tangshan 063000, China

Z. Fang et al. (eds.), *Proceedings of the 5th International Symposium on Plasma and Energy Conversion*, Springer Proceedings in Physics 398,
https://doi.org/10.1007/978-981-97-2245-7_49

49.1 Introduction

High-voltage AC shunt capacitor is an important reactive power compensation equipment in power system, which plays the role of supporting system voltage, increasing power factor and reducing system loss. As the harmonic ratio in the power grid gradually increases, the harmonic ratio in the operating conditions of high-voltage shunt capacitors also gradually increases, and the defect and failure rates of capacitors simultaneously show a high growth [1].

The existence of harmonics will increase the effective value and peak value of the voltage applied at the capacitor, causing the voltage waveform distortion to be peaked, further improving the internal electric field strength and affecting the insulation characteristics of dielectric materials [2]. Breakdown field strength is a parameter that characterizes the maximum electric field strength that insulating dielectric materials can withstand, and is an important basis for capacitor design and polypropylene film selection [3].

In terms of breakdown characteristics, there are many researches on the breakdown characteristics of capacitor insulating medium under DC and AC at home and abroad. Laihonen from KTH Royal Institute of Technology in Sweden measured the DC breakdown strength of 4–15 μm thick polypropylene film at room temperature using five electrodes with an area of 0.045–9.3 cm^2, and processed the breakdown data using Weibull distribution [4].

Existing studies mainly focus on the production and manufacture of high temperature resistant polypropylene films [5] and their modification [6]. Breakdown tests of thin films and other media only carry out AC and DC voltages, ignoring the influence of power frequency superimposed harmonic voltages tolerated during operation on thin film media.

However, in field operation, high-voltage shunt capacitors usually bear the current and voltage stress of power frequency superimposed harmonics. Therefore, it is necessary to study the effect of harmonics on capacitors from the perspective of electrical breakdown at the film level. In order to study the breakdown characteristics of polypropylene film for high-voltage shunt capacitors under power–frequency superimposed harmonic conditions, this paper combined with the actual operating conditions and disintegration test results of high-voltage shunt capacitors, carried out short-time breakdown tests of polypropylene film under different voltage frequency combinations and different temperature and other influencing factors, and carried out statistical analysis on the breakdown field strength data of the film using Weibull distribution. The mechanism of the effect of harmonics on the breakdown field strength of polypropylene films is revealed.

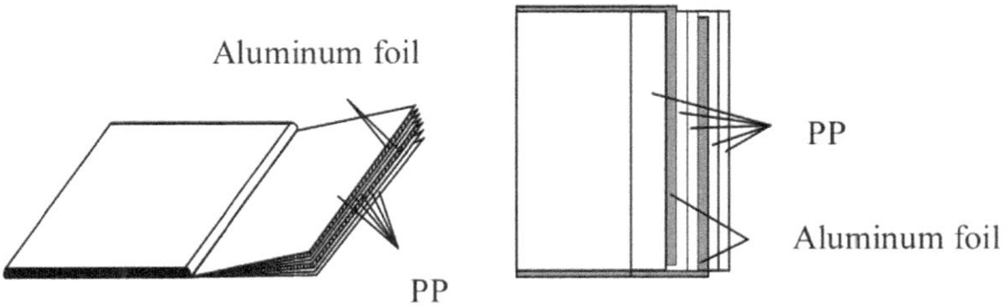

Fig. 49.1 Capacitor element structure

49.2 Operating Condition and Experiment

49.2.1 Operation

The capacitor component is the most core part of the capacitor. The component is usually made of two layers of aluminum foil and multiple layers of polypropylene film after winding and flattening. The basic structure diagram is shown in Fig. 49.1, where the aluminum foil is alternately folded and protruded at the end of the component to realize the isolation and removal of high and low voltage electrodes at the end.

In power system, high voltage shunt capacitor and reactor usually form shunt capacitor devices according to certain configuration principles and are configured in the third winding of the main transformer of substation. The substation is located in the multi-energy collection area, with more nonlinear sources and loads, the harmonic content is large during operation, and the 10 kV capacitor is frequently damaged. Therefore, the bus voltage is measured, and the 95% probability value of the harmonic voltage at both ends of the A, B and C three-phase bus and capacitor at 110 kV voltage level is shown in Fig. 49.2.

As shown in Fig. 49.2, under each voltage level, the odd-harmonic voltage content mainly exceeds the national standard limit. The main harmonic voltage frequency at both ends of the high-voltage shunt capacitor is 3, 5, 7, 11, 13, 15 and 17 times of power frequency, and the content rate exceeds 2%, among which the content rate of the 5th harmonic is the highest, close to 7%. According to the analysis results of the actual operating conditions of the high-voltage shunt capacitor, the loading voltage in the subsequent test can determine the strange harmonics, and the 17th and previous harmonics are mainly.

49.2.2 Disintegration

In order to further understand the internal breakdown characteristics and mechanism of capacitors operating at high frequency harmonics for a long time, this paper disintegrated BAM12/2-500-1W shunt capacitors.

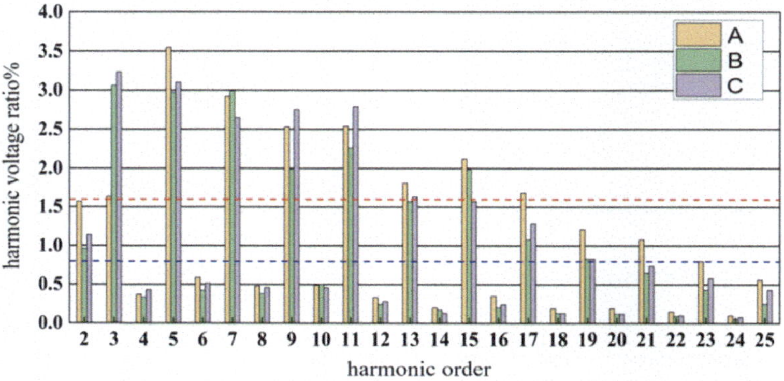

Fig. 49.2 Harmonic voltage content of 110 kV bus and capacitor branches, where the red line is the national standard limit of the strange harmonic content rate, and the blue line is the national standard limit of the even harmonic content rate

As can be seen from the Fig. 49.3, there is component breakdown in the two series of capacitor components, and the breakdown position is respectively on the large face of the component and the edge of the end of the component, which are the places where the component is prone to breakdown. After analysis, the capacitor component due to long-term operation under harmonic voltage, its internal component media aging, transient overvoltage in the case of system voltage fluctuations, breakdown in the weak point of the component media. The capacitance of the capacitor is measured by the capacitance meter. The measured capacitance value drops by 14%, and the capacitor has been seriously damaged.

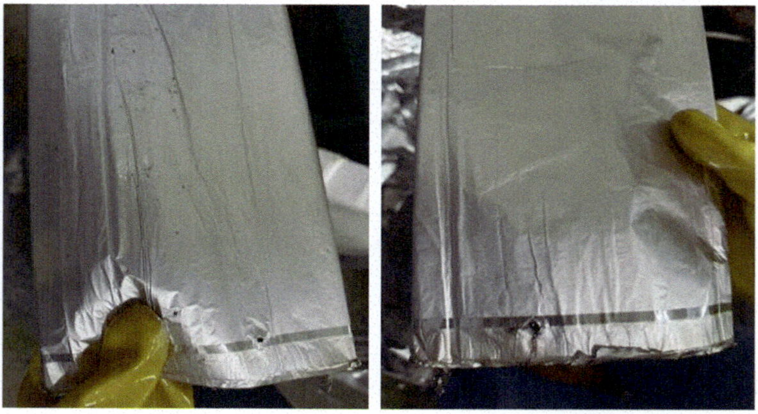

(a) (b)

Fig. 49.3 Shunt capacitor PP film disintegration test picture **a** the first series of capacitor element film breakdown point **b** the second series of capacitor element film breakdown point

According to the disintegration results, it can be seen that for a capacitor with long-term high frequency harmonic operation, the odd-order harmonic current superimposed on the capacitor fundamental current will increase the effective value of the capacitor operating current under high-risk conditions, resulting in breakdown of the capacitor film and obvious abnormal capacitance.

49.2.3 Test Platform

In order to study the effect of the combination of temperature and applied voltage on the breakdown field strength of polypropylene film, the breakdown tests of polypropylene film under different voltage frequency combination and different temperature were carried out. The breakdown test platform of polypropylene film for capacitors mainly includes harmonic loading system, oil cup and measuring system. The multi-frequency synthesis power supply can output the composite current containing 12 kinds of harmonic signal components, and load it to the test film sample in the oil cup through the dry medium frequency transformer. The frequency range of the power supply is 50–2450 Hz, the rated output power is 180 kVA, and the rated output voltage is 340 V.

In order to determine the AC breakdown strength of polypropylene film, the test adopts the continuous voltage boost method with the voltage boost speed of 300 V/s, and the test product is uniformly increased from zero until the test product breaks down. Each set of conditions contained ten samples, and each breakdown occurred between 10 and 20 s. The test conditions are shown in Table 49.1. Un represents the NTH harmonic voltage, where U1 is the power frequency voltage, 50 Hz. The left side of the second row in the table represents the power frequency superimposed 3rd and 5th harmonic voltages.

During the test, 10 times of film sample breakdown were carried out under each group of test conditions, and the peak voltage measured by oscilloscope at the time of breakdown was recorded as the breakdown voltage. The breakdown field strength was calculated by dividing the peak value by the thickness of the film, and the data were analyzed and processed by using the two-parameter Weibull distribution function. The probability distribution function of the two-parameter Weibull distribution is

$$F(E) = 1 - e^{-\left(\frac{E}{\alpha}\right)^{\beta}} \tag{49.1}$$

Table 49.1 Film breakdown test test conditions

Frequency group	Test temperature (°C)
U1 + U3 + U5	20
U1 + U7 + U9	40
U1 + U11 + U13	60
U1 + U15 + U17	80

where, E is the breakdown field strength of the film; F(E) is the breakdown probability of the test product at the field strength E; α is the scale function, representing the characteristic breakdown field strength of the film test product when the breakdown probability is 63.2%. b is a shape parameter, reflecting the dispersion degree of the breakdown field strength, and the larger the b value, the smaller the data dispersion.

49.3 Result and Discussion

The breakdown test was conducted according to the breakdown test conditions set in Table 49.1, and the two-parameter Weibull distribution of the film breakdown field strength at different temperatures and combined voltages of different frequencies was obtained, as shown in Fig. 49.4. Figure 49.4f shows the film breakdown field strength at different temperatures and combined voltages of different frequencies with a breakdown probability of 63.2%. Weibull distribution parameters of the breakdown field strength, including scale parameter α and shape parameter β, are calculated, and the results are shown in Table 49.2.

As shown in Fig. 49.4 and Table 49.2, under the action of power frequency and power frequency superimposed harmonic voltage, the breakdown field strength of the film decreases with the increase of temperature. Under the action of power frequency voltage, the breakdown field strength of the film at 80 °C is 513.85 V/μm, which is decreased by 10.02 V/μm compared with that at 20 °C. Under the U1 + U3 + U5, U1 + U7 + U9, U1 + U11 + U13 and U1 + U15 + U17, the breakdown field strength of the film at 80 °C decreased by 22.82 V/μm, 17.43 V/μm, 19.29 V/μm and 16.34 V/μm respectively. Compared with the power frequency voltage alone, the breakdown field strength under the power frequency superimposed harmonic voltage decreases at all temperature conditions, and the higher the superimposed harmonic frequency, the lower the breakdown field strength decreases.

Both temperature and harmonics accelerate free carrier migration. The increase of temperature will increase the energy of carriers, accelerate the collision of carriers, and accelerate the destruction of molecular chains [7, 8]. The presence of harmonics will increase the steepness of the voltage waveform, compared with the power frequency, in the same time of the voltage rise stage, the free carrier obtains more energy, faster, and more frequent collision with the molecular chain, making the breakdown field strength of the film lower. However, compared with the 3rd and 5th harmonics, under the condition of controlling the harmonic current amplitude, the voltage distortion rate of the higher frequency harmonics is lower, and the harmonic amplitude is smaller, so the impact on the breakdown field strength is smaller.

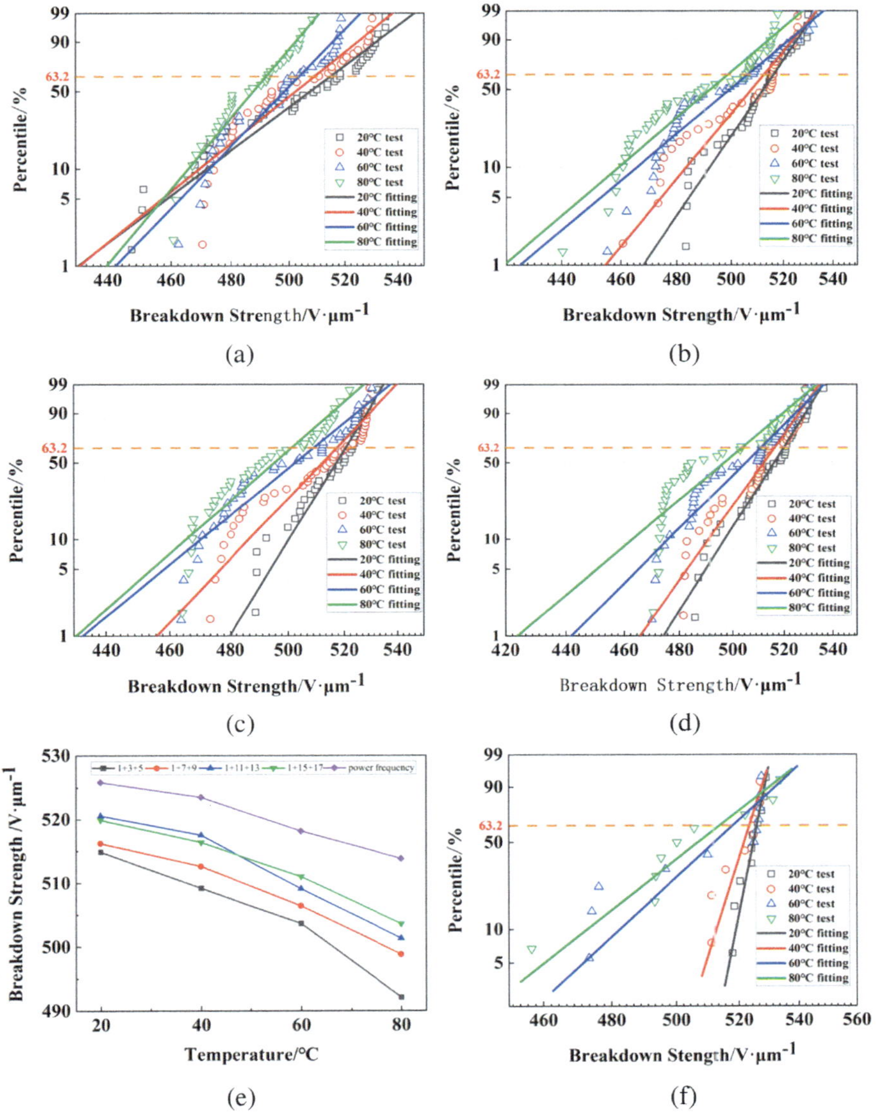

Fig. 49.4 Weibull distribution of film breakdown field strength under different temperature and frequency groups **a** breakdown field strength under U1, U3 and U5 harmonics combination **b** breakdown field strength under U1, U7 and U9 harmonics combination **c** breakdown field strength under U1, U11 and U13 harmonics combination **d** U1, U15 and U17 harmonic combination breakdown field strength **e** power frequency U1 breakdown field strength **f** breakdown probability 63.2% of each harmonic combination breakdown field strength

Table 49.2 Weibull distribution parameter statistics of film breakdown field strength at different temperature and frequency combinations

	Frequency group	U1 + U3 + U5	U1 + U7 + U9	U1 + U11 + U13	U1 + U15 + U17	U1
20 °C	α	514.91	516.28	520.62	519.95	525.87
	β	25.77	46.81	56.41	49.92	172.64
40 °C	α	509.19	512.67	517.54	516.41	523.52
	β	27.64	38.40	36.30	44.46	109.11
60 °C	α	503.64	506.47	509.12	511.04	518.18
	β	35.50	26.77	28.27	31.62	31.55
80 °C	α	492.09	498.85	501.33	503.61	513.85
	β	40.81	27.21	30.24	26.72	26.96

49.4 Conclusion

In this paper, the breakdown characteristics of polypropylene film for high voltage shunt capacitors under harmonic conditions are studied. According to the harmonic frequency distribution and disintegration test results of the withstand voltage of the high-voltage shunt capacitor during actual operation, the voltage distortion rate of each harmonic was determined. On this basis, the breakdown test of polypropylene film under different temperature and voltage conditions was carried out. The data were processed based on Weibull distribution, and the following conclusions were obtained:

(1) Two series of shunt capacitors operating under the action of high frequency harmonic voltage for a long time, the internal component medium of the capacitor aging, the system voltage fluctuation produces instantaneous overvoltage, and the breakdown occurs at the weak point of the component medium, causing the capacitor capacitance value to drop by up to 14%, and the breakdown location mostly occurs at the large surface of the component and the edge of the end of the component.

(2) Under different voltage forms, the breakdown field strength of polypropylene film decreases with the increase of temperature; Compared with power frequency voltage alone, the breakdown field strength under power frequency superimposed harmonics is lower, and the decrease of breakdown field strength decreases with the increase of harmonic frequency.

Acknowledgements This work was funded by the State Grid Corporation of China Headquarters Science and Technology Project (No. 5500-202214401A-2-0-ZN).

References

1. C. Xu, *Research on the Influence of Harmonic on the Life and Protection of Power Capacitor Bank*. North China Electric Power University (2013)
2. Z.Y. Ran, B.X. Du, M. Xiao, et al., in *Breakdown Improvement of PP Films under DC Supercomposed Harmonic Voltages*. 2021 International Conference on Electrical Materials and Power Equipment (ICEMPE), Chongqing, China, 2021, pp. 1–4 (2021). https://doi.org/10.1109/ICEMPE51623.2021.9509158
3. L. Hao, N. Xuefeng, J. Shengbao et al., Study on Insulation Withstand Level of High Voltage Shunt Compensation Capacitor in China. Power Capacitor & Reactive Power Compensation **38**(01), 61–64 (2017)
4. S.J. Laihonen, U. Gafvert, T. Schutte et al., DC breakdown strength of polypropylene films: area dependence and statistical behavior. IEEE Trans. Dielectr. Electr. Insul. **14**(2), 275–286 (2007)
5. P. Cygan, R. Laghari, Models for insulation aging under electrical and thermal multistress. IEEE Trans. Elect. Insul. **25**(5), 923–934 (1990)
6. L. Qi, L. Manqian, High-temperature polymer dielectrics for film capacitors: review and prospect. High Vol. Eng. **47**(9), 3105–3123 (2021)
7. J.H. Zhang, B.X. Du, M. Xiao, et al., in *Evaluation of Breakdown Properties of Polypropylene Films Under Multiple Stresses for DC Capacitors*. 2022 IEEE International Conference on High Voltage Engineering and Applications (ICHVE). IEEE (2022)
8. N. Can, F. Aras, V.A. Alekperov et al., Investigation on aging mechanism of polyester under combined stresses. IEEE Electr. Insul. Mag. **32**(4), 38–42 (2016)

Chapter 50
Thermal Fractionation Technique to Evaluate the Effect of Microstructure Defects on Dielectric Properties of Polypropylene

Yaotian Su, Xinghui Zhang, Zhaoliang Xing, Chong Zhang, Chenchen Zhang, Shaowei Guo, YanGuo Ke, Di Hu, and Chao Pan

Abstract The successive self-nucleation and annealing thermal fractionation (SSA) technique offers a practical way to evaluate the chain crystallization behavior according to the length of polymer crystalline sequence. In this paper, three polypropylene resin (PP) samples with different heat resistance grades were characterized combined SSA and electrical measurement, and the relationship between molecular structure and high temperature dielectric properties of PP was established. The results indicate that the lamellar sequence length distribution coefficient can reflect the molecular interchain defects, which determines the insulation property of the PP resin. The PP-2 sample with high crystallinity and narrow lamellar sequence length distribution shows the most excellent high temperature dielectric property, which can be applied in the field of oil-immersed capacitor film. There is a good correspondence between SSA analysis results and electrical test values, which implies the SSA technique has the advantage in real-time monitoring the molecular structure to effectively evaluate the macroscopic electrical insulation property of PP resin.

Y. Su · Z. Xing (✉) · C. Zhang · S. Guo
State Key Laboratory of Advanced Power Transmission Technology, State Grid Smart Grid Research Institute Co., Ltd., Beijing, China
e-mail: xingzhaoliang007@163.com

X. Zhang
State Grid Corporation of China, Beijing 100031, China

C. Zhang · Y. Ke · D. Hu · C. Pan
State Grid Anhui Electric Power Research Institute, Hefei 230022, Anhui Province, China

Z. Fang et al. (eds.), *Proceedings of the 5th International Symposium on Plasma and Energy Conversion*, Springer Proceedings in Physics 398,
https://doi.org/10.1007/978-981-97-2245-7_50

50.1 Introduction

High voltage oil-immersed capacitor is an important reactive power compensation equipment in power system, which plays the role in supporting system voltage, increasing power factor and reducing system loss [1]. Polypropylene (PP) is a semi-crystalline non-polar material with excellent heat resistance, insulation, chemical stability and mechanical properties, which is widely used in electrical insulation materials and capacitor dielectrics [2]. However, with the rapid development of power electronic systems, higher operating temperatures pose new challenges to the performance of traditional PP materials. For example, When the temperature rises above 85 °C, large dielectric loss can be observed in BOPP film originated from the conductivity loss. In addition, the breakdown field strength of the material is also rapidly reduced, which hinders its further application in the high energy density, low loss, high temperature resistant capacitors [3]. The dielectric properties of PP are largely dependent on the isotactic sequence structure of the molecular chains, i.e. PP with high isotacticity and low interchain defects often possesses better heat resistance characteristics. Therefore, it is significant to acquire more information on molecular chain defects to guide the development and design of PP with excellent high temperature dielectric properties.

At present, the technique to characterize isotactic sequence structures of molecular chain of PP mainly includes X-ray Diffraction (XRD), Infrared Spectroscopy (IR), Nuclear Magnetic Resonance Carbon Spectroscopy (NMR-^{13}C) and Temperature Rising Elution Fractionation (TREF) [4]. However, these techniques are usually complicated and time-consuming. Successive Self-nucleation and Annealing (SSA) is a thermal fractionation technique that is performed by Differential Scanning Calorimetry (DSC). The SSA consists of a thermal protocol that combines isothermal and non-isothermal steps, which induce the molecular segregation of the material without physical separation. The lack of physical separation of the chains during crystallization makes the thermal fractionation techniques sensitive to intra- and inter-molecular heterogeneities, allowing the study of any crystallizable material without damaging the equipment, thereby providing information on chain structures [4].

In this work, three PP samples with different heat resistance grades were characterized combined SSA and electrical measurement. The chain crystallization behavior and dielectric properties were systematically investigated. The results indicate that PP with high crystallinity and narrow crystallization sequence is beneficial for the improvement of high temperature insulation property, and there is a good correspondence between SSA analysis results and high temperature DC- resistivity values.

50.2 Experiment

50.2.1 Materials

Three PP resin samples (PP-1, PP-2, PP-3) with different heat resistance grades (85 °C, 105 °C, < 85 °C,) were provided by foreign and domestic manufacturers. All samples have the same ash content (< 15 ppm) and molecular weight (Mw = 40w, PD = 6).

50.2.2 SSA Methods

The successive self-nucleation and annealing thermal fractionation (SSA) experiment was performed on DSC-Q2000 (TA; USA) by the combination of non-isothermal and isothermal steps as the followings: Firstly, the sample was heated to 200 °C for 5 min to eliminate the thermal history and then the temperature was cooled down to 30 °C at a rate of 20 °C/min for 2 min. Secondly, the sample was heated at the rate of 20 °C/min to 169 °C maintaining for 15 min. Subsequently, the temperature was lowered to 30 °C at the rate of 20 °C/min for 2 min. Repeating the steps illustrated above and the isothermal temperature was set as 169, 165, 161, 157, 153, 149, 145, 141 °C for 15 min, respectively. After the thermal classification was completed, the sample was heated to 200 °C at a rate of 10 °C/min, and the final SSA melting curve was recorded.

Peakfit 4.12 software was employed to fit the peaks of final melting curve of SSA. The lamellar thickness was calculated by thermodynamic method according to Gibbs–Thomson equation (Eq. 50.1) to describe the polydispersity of the lamellar sequence length [4].

$$T = T_{\mathrm{m}}^{0}\left(1 - \frac{2\sigma_e}{\Delta H_m l}\right) \quad (50.1)$$

where, T is the melting point of lamellar with specific thickness, K; l is the lamellar sequence length, nm; $T^0{}_m$ is the equilibrium melting point of polypropylene, and its value is 460.7 K. σ_e is the surface free energy, 0.0496 J/m^2; ΔH is the standard melting heat enthalpy, $1.84 \times 10^8/\mathrm{m}^3$.

Formula (50.2–50.4) is used to calculate the numerical average sequence length L_N, the weight average sequence length L_W and the sequence length distribution coefficient I of lamellar.

$$L_N = \frac{\mathrm{n_1}L_1 + \mathrm{n_2}L_2 + \ldots \mathrm{n_i}L_2}{\mathrm{n_1} + \mathrm{n_2} + \ldots \mathrm{n_i}} = \Sigma f_i L_i. \quad (50.2)$$

$$L_W = \frac{n_1 L_1^2 + n_2 L_2^2 + \ldots n_i L_n^2}{n_1 + n_2 + \ldots n_i} = \Sigma \frac{f_i L_i^2}{f_i L_i}. \tag{50.3}$$

$$I = \frac{L_W}{L_N}. \tag{50.4}$$

50.2.3 Electrical Characterization

DC volume resistivity was measured by Keithley 6517B high resistance meter device with 8009 electrode. The dielectric properties were measured on the impedance analyzer (Concept40; Germany) in a frequency range from 10 Hz to 1 MHz. DC electric breakdown tests were performed on a dielectric strength tester (HT-50, China) and 50 experimental point was selected to calculate the average breakdown strength. The temperatures for the test illustrated above are 25 °C and 85 °C and the thickness of samples are around 200 μm.

50.3 Results and Discussions

50.3.1 Crystallinity

The crystallization behavior of the three different PP was characterized by DSC. Figure 50.1 shows the DSC curves of the sample before SSA procedures. It can be observed that PP-1, which is usually applied on rough film of capacitor exhibits the lowest melting point of 165.8 °C and crystallinity of 37.88%. Meanwhile for PP-2, the highest melting point of 170.8 °C and crystallinity of 47.12% ensures its application at high temperatures. PP-3 also shows a relatively high melting point of 170.2 °C and crystallinity of 45.63%. However, it can only be applied at low temperature less than 85 °C for normal capacitors.

To further investigate the defects in polymer chains of the three PP samples, SSA technique was performed by DSC. The DSC curves of non-isothermal heating process of the last loop are shown in Fig. 50.2 and the crystallization parameters are displayed in Table 50.1. From the Table 50.1, the PP-1 possesses the lowest crystallinity and lamellar thickness, while PP-2 shows the highest crystallinity. As for distribution parameters of lamellar thickness (L_W/L_N), PP-1 and PP-2 have no prominent distinctions. It should be noted that although the crystallinity of PP-3 is slightly less than that of PP-2, its L_W/L_N value is much higher, which implies more defects existing in the intermolecular structures [5]. These intermolecular defects may be the main source of electrical and thermal properties of the three PP materials.

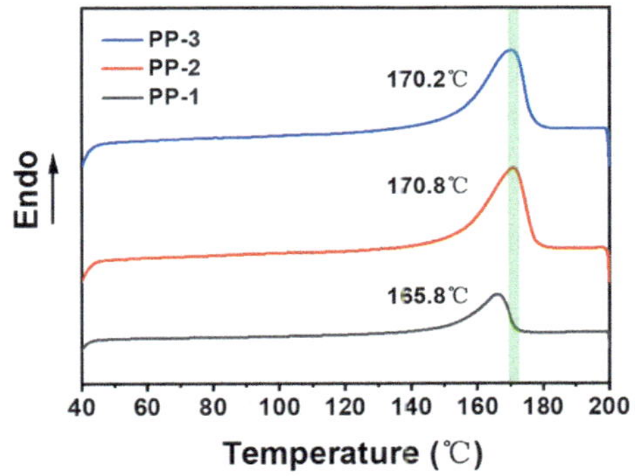

Fig. 50.1 DSC curves of different samples

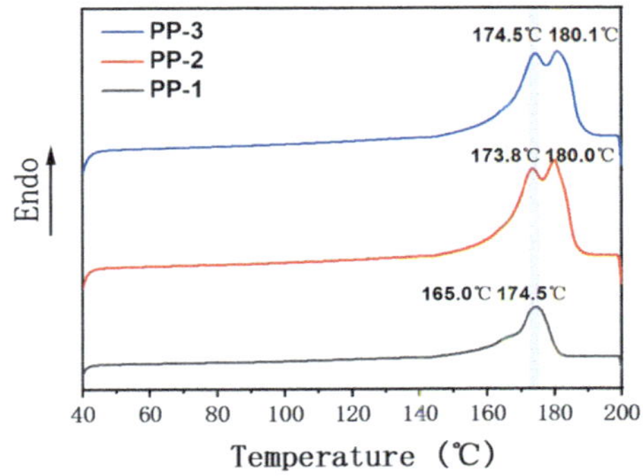

Fig. 50.2 DSC curves of different samples after SSA procedure

Table 50.1 Crystallization parameters calculated from SSA procedures

Sample	Peak 1 (°C)	Peak 2 (°C)	Crystallinity (%)	Area ratio (Peak 1:Peak 2)	L_N (nm)	L_W (nm)	I (L_W/L_N)
PP-1	165.0	174.5	52.2	0.71	15.7	16.7	1.063
PP-2	173.8	180.0	66.0	0.93	26.8	28.8	1.075
PP-3	174.5	180.1	63.4	0.98	28.3	31.3	1.106

50.3.2 Electrical Properties

The DC volume resistivity and broadband dielectric technique was utilized to characterize the difference of electrical properties for the three PP samples. Figure 50.3a shows the DC volume resistivity of the samples in the range of polarization time from 400 to 1000 s. The average volume resistivity of PP-1 and PP-2 are 1.5×10^{15} Ω.m and 1.1×10^{15} Ω.m, respectively, which are higher than that of PP-3 (4×10^{14} Ω.m). This result indicates that high lamellar sequence length distribution is not favorable for the improvement of insulation properties of the PP material, which implies that charge carriers are more inclined to migrate between interchain defects due to the low transmission activation energy [5]. The interchain defects can also be demonstrated by dielectric properties at high temperatures. Figure 50.3b shows the frequency dependent dielectric permittivity and loss tangent of the PP samples. The dielectric permittivity for all samples exhibits the same variation trend and the dielectric constant at 10 Hz is approximately 2.42. However, the dielectric loss for PP-3 is larger than PP-1 and PP-2, and shows frequency sensitivity, which demonstrates defects triggers conductivity loss originated from charge carriers migration.

Figure 50.4 displays the DC breakdown strength of the three PP samples. The highest breakdown strength (588.7 kV/mm) of PP-2 is attributed to the high crystallinity and low sequence length distribution coefficient of lamellar thickness while PP-3 has the worst breakdown strength (441.6 kV/mm) due to its molecular interchain defects.

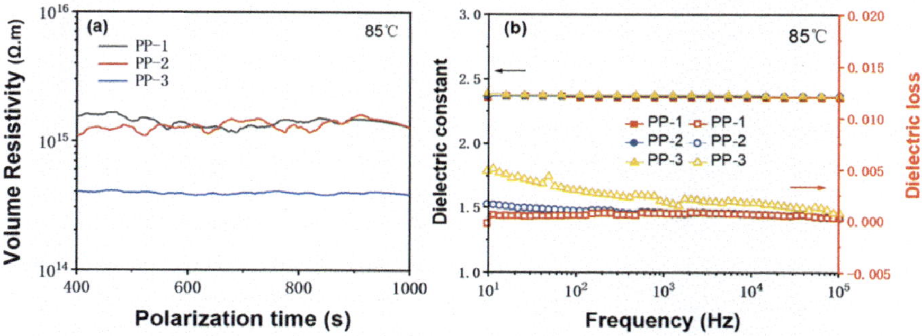

Fig. 50.3 **a** DC volume resistivity and **b** dielectric properties of PP samples at 85 °C

Fig. 50.4 DC breakdown strength of PP sample at 85 °C

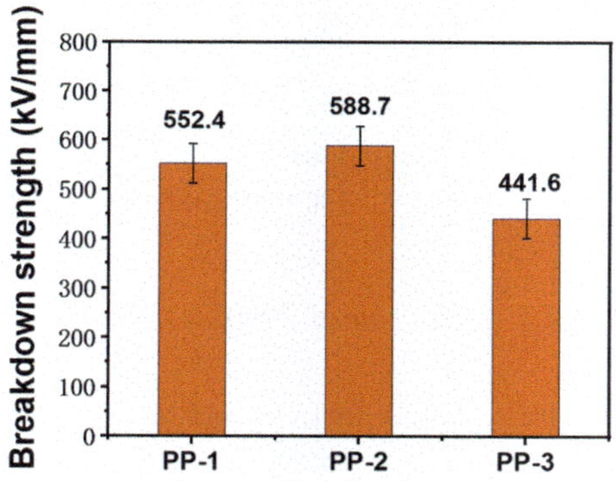

50.4 Conclusion

Three PP resins with different heat resistance grades were characterized combined SSA technique and electrical measurement. The results indicate that the electrical insulation property is closely related to the microstructure defects in polymer chains. The narrow sequence length distribution endows the PP with excellent insulation and dielectric properties at the high temperature (85 °C), which demonstrates that high crystallinity and low interchain defects is beneficial for the improvement of electrical properties. There is a good correspondence between SSA analysis results and electrical test values, which implies the SSA technique has the advantage in real-time monitoring the molecular structure to effectively evaluate the macroscopic electrical insulation property of PP resin.

Acknowledgements This work was funded by the State Grid Corporation of China Headquarters Science and Technology Project (No. 5500-202320114A-1-1-ZN).

References

1. O.G. Gnonhoue, A. Velazquez-Salazar, É. David et al., Review of technologies and materials used in high-voltage film capacitors. Polym. **13**(5), 766 (2021)
2. M.H. El-Husseini, P. Venet, G. Rojat et al., Thermal simulation for geometric optimization of metallized polypropylene film capacitors. IEEE Trans. Ind. Appl. **38**(3), 713–718 (2002)
3. D. Tan, L. Zhang, Q. Chen et al., High-temperature capacitor polymer films. J. Electron. Mater. **43**, 4569–4575 (2014)
4. R.A. Pérez-Camargo, D. Cavallo, A.J. Müller, Recent applications of the successive self-nucleation and annealing thermal fractionation technique. Front. Soft Matt. **2**, 1003500 (2022)
5. S. Hu, H. Yuan, Q. Zhang et al., Deep trap origins, characteristics, and related mechanisms in chemically grafted polypropylene with enhanced direct current volume resistivity. J. Phys. Chem. C **126**(38), 16280–16288 (2022)

Chapter 51
A Compact High Voltage DC Power Supply Design by High-Rate Lithium Batteries

Taizhuang Hu, Huibo Zhang, Zicheng Zhang, and Jingming Gao

Abstract The demand for mobile and portable applications is driving the development of compact and miniaturized pulsed power devices. To detach the pulsed power system from the ground power supply, a compact and lightweight modular high-voltage DC charging power supply technology solution is proposed in this paper. To meet the high voltage and high power input requirements of pulse power devices such as Marx generators, this power supply uses a group of single 3.2 V high-rate lithium batteries for DC power supply, and adopts a two-stage boost solution. Firstly, a 32 V low-voltage lithium battery pack is boosted to 400 V by a non-isolated DC converter, and then the voltage is increased to 50 kV through a phase-shifted full-bridge isolated DC-DC converter. This paper tests the discharge characteristics of single lithium batteries under different discharge rate conditions and explores the output capability of lithium batteries. The design scheme of high-voltage DC power supply is experimentally verified, and the results show that the single-module output voltage is 50 kV, and the output power is about 800W, which can meet the high-power input requirements by stacking multiple modules. This provides technical references for designing high-performance, compact, lightweight and environmentally adaptable high-power pulsed driving sources.

51.1 Introduction

Pulsed power system is an electrophysical technology that converts low-power stored electromagnetic field energy into temporal compression to produce high-power high-density electromagnetic field energy, distinguishing itself from general high-voltage technology by a transient and rapid pulse compression process [1]. With the gradual maturation of pulsed power technology and research systems, in addition to applications in defense and industrial fields such as intense lasers, high-power microwaves,

T. Hu · H. Zhang · Z. Zhang (✉) · J. Gao
College of Advanced Interdisciplinary Studies, National University of Defense Technology, Changsha 410073, China
e-mail: tzhuqqmail@qq.com

Z. Fang et al. (eds.), *Proceedings of the 5th International Symposium on Plasma and Energy Conversion*, Springer Proceedings in Physics 398,
https://doi.org/10.1007/978-981-97-2245-7_51

and flash X-rays [2, 3], they are also widely used in civilian fields such as food sterilization, ozone generation, water treatment, and medical cellular reactions (e.g., electroporation of cell membranes) [4, 5]. Compactness, miniaturization and re-frequency are currently the main research directions in pulsed power technology, and the utility of pulsed power devices often requires operation in mobile, portable environments [6, 7]. Therefore, as the energy supply part of the pulsed power system, the design of the high-voltage DC power supply should be as compact and portable as possible under the requirement of meeting the power supply index of the system. In addition, in working scenarios not equipped with basic power facilities, the use of lithium batteries, which have high energy storage density and are easy to integrate, instead of lithium batteries, has a great advantage as the energy storage part of the system.

At present, most of the high-voltage DC power supplies adopt the technical route of "AC inverter boosting, rectifying and high-voltage output", and its core solution is to use various types of transformer-based DC-DC converters. For example, the use of full-bridge inverters [8], resonant converters [9] and voltage multipliers [10], etc., most of the hundreds of volts within the range of AC or DC voltage, through a single-stage DC-DC converter to achieve the required tens of kilovolts of output voltage. However, limited by the power supply system, the portability of the entire system device is greatly reduced. With the rapid development of new energy technologies and power devices, high energy storage density, high discharge multiplier lithium battery technology has been commercialized. In addition, lithium batteries also have superior cycling performance, fast charging and discharging, and long service life, etc., which have laid the foundation for pulse power technology based on lithium battery energy storage. However, considering the compactness and portability of the entire device, the weight of the battery pack in the energy storage section should not be too large.

Aiming at the energy supply needs of pulse-driven sources in mobile working environments, this paper designs a compact portable high-voltage DC power supply based on the power supply from a high-magnification Li-ion battery pack. The power supply is powered by a 32 V lithium battery pack with high energy storage density, boosted to about 400 V through the intermediate stage of a non-isolated DC-DC boost converter, and then connected to an isolated phase-shifted full-bridge DC-DC converter, outputting a high voltage of 50 kV. The two-stage boosting scheme achieves good performance while taking into account portability.

51.2 Single Battery Test

In the pulsed power system, the primary energy source plays the role of initial energy storage and first-stage compression, which is the foundation of the pulsed power device, and its core parameters such as energy storage density, energy storage efficiency and energy release efficiency largely determine the overall performance of the pulsed power device. In this paper, according to the energy and weight requirements of the pulsed power system, the ANR26650M1B lithium iron phosphate (LFP) power battery produced by A123 Company of the United States was selected as the energy storage battery, and the capacity of a single battery as well as the discharge characteristics of the test.

51.2.1 Capacity Test

The ANR26650M1B LFP power battery is a prime example of a high-rate battery cell. Each individual battery boasts a nominal voltage of 3.2 V and a nominal capacity of 2.5 Ah. With a maximum continuous discharge current of 50 A (20 C) and a pulse discharge current of up to 120 A (48 C) for a duration of 10 s, it delivers impressive performance. Additionally, it exhibits an internal resistance of approximately 6 mΩ and weighs around 76 g. This battery caters to the needs of compact and portable systems, offering a favorable combination of discharge rate, volume, and weight. However, it's worth considering that the purchased battery may experience a decrease in capacity due to extended storage time or usage. Therefore, it is essential to conduct capacity testing to ensure optimal performance.

The battery underwent capacity testing using a charge–discharge integrated machine. The testing procedure consisted of several steps. Initially, the battery was charged at a constant current of 0.83 A until it reached its maximum capacity. Subsequently, a constant voltage charge at 3.6 V was applied to ensure complete charging. Afterwards, the battery was discharged at a constant current of 0.83 A until the voltage dropped below 2 V, indicating complete discharge. This entire process was repeated three times to complete the capacity testing.

By recording the voltage, current, and duration during these cycles, the charging capacity, discharging capacity, and capacity retention rate were calculated. The test results are presented in Table 51.1. Results of capacity test. It can be observed that both the charging and discharging capacities of the battery are approximately 2.45 Ah, which is around 2% lower than the nominal capacity of 2.5 Ah. This suggests that the battery may have experienced a decrease in capacity due to storage or usage. However, the capacity retention rate of the battery remained consistently close to 100% throughout the three cycles, indicating a high-quality battery with a relatively long lifespan.

Table 51.1 Results of capacity test

Cycle	Charging capacity (Ah)	Discharging capacity (Ah)	Capacity retention (%)
1	2.44960	2.44725	99.90407
2	2.45137	2.45140	100.00122
3	2.45223	2.45121	99.95840

51.2.2 Discharge Characteristic Test

In pulsed power systems, the charging process of a modulator (such as a Marx generator) using a high-voltage DC power supply based on battery energy storage requires the power supply to operate at a high-power supply state. Additionally, the battery needs to be in a high-rate discharge state throughout the charging time to meet the power supply requirements of the system. Research indicates that the charging time of a Marx generator is primarily influenced by the power level of the DC power supply and can vary significantly based on the desired charging voltage and the energy storage capacitance, ranging from microseconds to seconds. Therefore, by configuring the battery with different output currents and durations, it can emulate various output power levels, simulating the operational conditions of the power supply during the charging process of the Marx generator.

During the testing process, the batteries were subjected to sequential discharge tests with currents of 50 A for 200 ms, 40 A for 500 ms, and 30 A for 1 s. The recorded voltage variations, shown in Fig. 51.1, were obtained from the battery's terminal voltage as measured by the testing equipment. It is important to note that the recorded voltage reflects the terminal voltage of the battery. Additionally, considering the equivalent circuit model of a lithium battery with a second-order RC network and the presence of polarization capacitance, the battery's terminal voltage during discharge is lower than its rated voltage. Furthermore, as the discharge current increases, the voltage drop becomes more significant. Consequently, the recorded battery voltage represents the discharge envelope in the case of long-duration pulse discharges.

In Fig. 51.1a, the battery's voltage drop reaches 0.6 V when subjected to the maximum continuous discharge current of 50 A. However, in each test cycle, the battery maintains a relatively long discharge time at a voltage plateau of 3.2 V. In Fig. 51.1b, the first discharge test stands out from the remaining three test cycles. This difference may be attributed to the battery's prolonged operation at the maximum continuous discharge current, followed by a relatively short resting time of 30 min. This shorter rest period prevents the battery's internal electrochemical characteristics from fully recovering to their initial state, affecting the first test cycle of the 40 A discharge. Nonetheless, the subsequent three cycles demonstrate consistent discharge performance, indicating the battery's ability to recover effectively. The discharging parameters of the battery, as depicted in Table 51.2. Discharging parameters, showcase impressive characteristics. The power output during a 50A pulse discharge reaches a notable 160W, while maintaining an average duration of 368 s. Additionally, the 30A pulse discharge exhibits an impressive endurance, capable of lasting for up to 10 min.

Taking a comprehensive look at Fig. 51.1a–c and Table 51.2. Discharging parameters, it is evident that the battery exhibits consistent discharge behavior under different discharge conditions. This finding serves as evidence for the discharge stability of lithium batteries during prolonged operation.

Based on capacity testing and discharge characteristic testing, the ANR26650M1B LFP single-cell battery demonstrates commendable stability

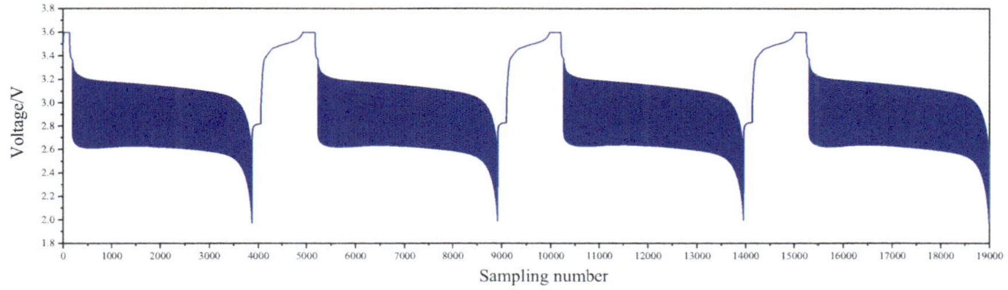

(a) 50 A for 200 ms

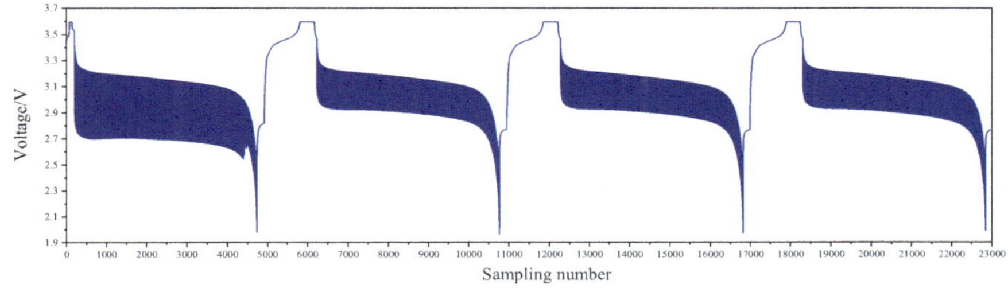

(b) 40 A for 500 ms

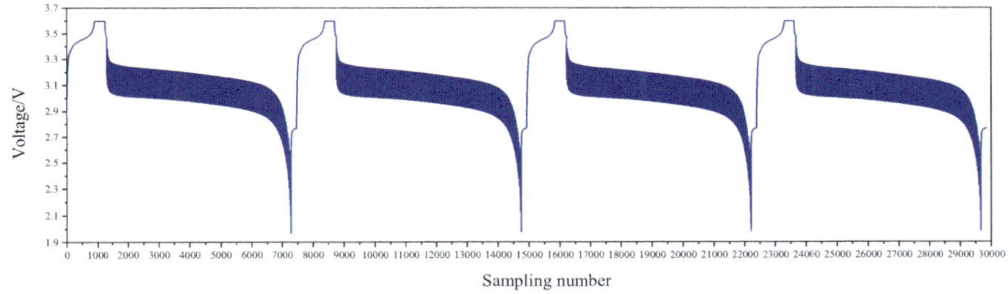

(c) 30 A for 1 s

Fig. 51.1 Recorded voltage variations under different outputs

Table 51.2 Discharging parameters

Discharge	Power (W)	Average duration (s)
50 A for 200 ms	160	368
40 A for 500 ms	128	453
30 A for 1 s	96	598

in terms of capacity and discharge. Notably, it exhibits the ability to sustain voltage for extended durations at higher power levels, especially during high-current pulse discharges that simulate real-world operating conditions. By employing a series and parallel configuration in the battery pack design, it becomes possible to achieve greater power output, fulfilling the pulsed power system's demand for primary energy output and resulting in a more compact and portable system as a whole.

51.3 Boost Circuit Design

51.3.1 Non-isolated DC-DC Converter

This article employs a series connection of 10 high-energy density, high-discharge rate lithium batteries as a low-voltage DC input, with a rated input voltage of 32V. The expected output voltage of this high-voltage DC power supply system is 50 kV, resulting in a gain exceeding 1500. Additionally, the size and weight of the device should be designed reasonably to achieve portability. Therefore, a two-stage boost conversion solution is proposed, as illustrated in Fig. 51.2. The scheme of the high-voltage DC power supply. The second stage consists of an isolated DC-DC converter operating with a fixed voltage gain of 125, capable of achieving an output voltage of 50 kV, and connected to a non-isolated DC-DC converter at the input. Since the gain of the isolated DC-DC converter varies with changes in the load, and the voltage Vbat of the lithium battery pack gradually decreases during long-term continuous discharge, in order to stabilize the output voltage as much as possible, the first stage boost non-isolated DC-DC converter is used as a voltage source regulator, allowing its gain to be dynamically adjusted. The controller analyzes and processes data collected from the battery pack, output voltage Vo, load, etc., and adjusts the gain of the first stage DC-DC converter to achieve stable voltage output. This precise adjustment for the lower gain (around 12.5) and lower output voltage (around 400 V) of the non-isolated DC-DC converter is more convenient and easier to manage than controlling a complex high-gain, high-voltage series resonant isolated DC-DC converter. Moreover, the DC power supply will also control the output voltage according to the needs of the Marx generator, in order to improve the system's environmental adaptability.

Based on the proposed solution, the primary task of the input stage is to serve as a regulating circuit for the input voltage of the battery pack (V_{bat}) with a rated voltage of 32 V and the intermediate voltage ($V_{dc} \approx 320$ V). After researching non-isolated DC-DC converters, a Boost circuit based on coupled inductors was selected for implementation in this system. Figure 51.3 illustrates the circuit diagram of the input stage, which utilizes a coupled inductor Boost-type DC-DC converter with a clamping circuit (enclosed within the dashed box). The addition of the coupled inductor enhances the voltage gain of the converter, thereby avoiding the rectifying

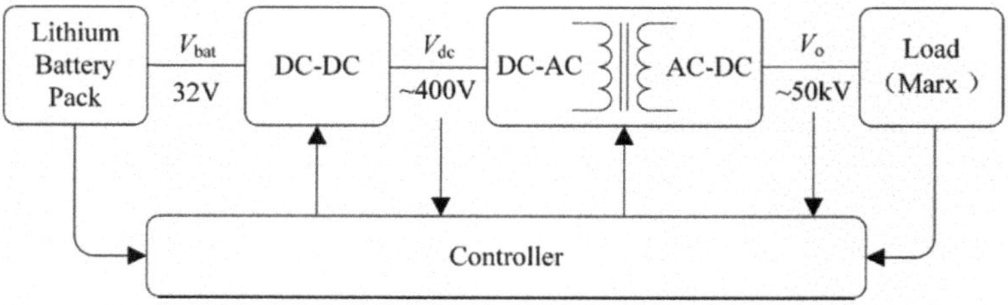

Fig. 51.2 The scheme of the high-voltage DC power supply

diode's reverse recovery problem associated with increasing the duty cycle in traditional Boost converters to raise the output voltage. However, there are still challenges such as high voltage stress on the main switch and low circuit efficiency. To address these issues, a passive clamping circuit consisting of a diode D_c and a small capacitor C_c is introduced. During the operation of the converter, the Diode D_c, together with the coupled windings, reduces the voltage stress on the semiconductor switching devices, thus providing a functionality similar to an active clamp. This significant reduction in circuit complexity and cost is achieved. Furthermore, the capacitor C_c recovers the leakage energy caused by the leakage inductance in the winding, which results in increased overall efficiency.

For the purpose of analysis, we will simplify the circuit by disregarding parasitic parameters and treating the power source, MOSFET switch, diode, and capacitor as ideal components. The coupling coefficient of the coupling inductor in the converter will be assumed to be 1, and an ideal transformer model with a turns ratio of $N = N_2 : N_1$ will be used. The duty cycle (D), which represents the switch's conduction time, will be set. By formulating the volt-second balance equation for the inductor, we can determine the circuit's voltage gain, G_{dc}.

$$G_{dc} = \frac{V_o}{V_{in} = \frac{1+ND}{1-D}} \tag{51.1}$$

In the equation, V_{in} represents the input voltage, while V_o represents the voltage across the output resistor R_L. The traditional Boost circuit has a gain of $1/(1-D)$, which is solely dependent on the duty cycle. However, by using a Boost circuit based on coupled inductors, it becomes possible to achieve higher voltage gains under the same duty cycle, thereby improving the efficiency of voltage boosting. The design parameters are selected as shown in Table 51.3.

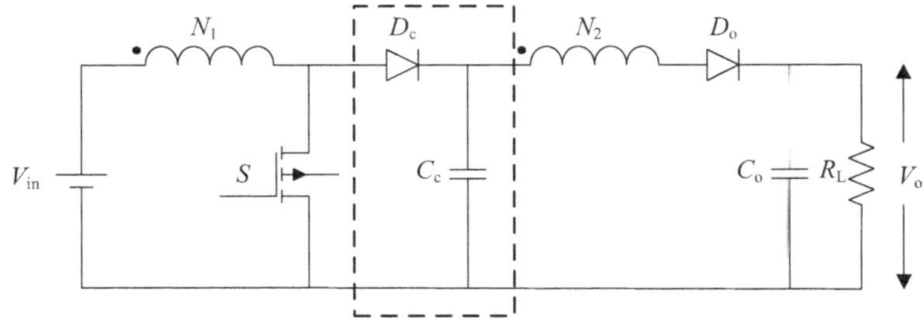

Fig. 51.3 Schematic diagram of the input stage in converter

Table 51.3 Selected parameters of clamp-mode coupled-inductor Boost DC-DC converter

Parameter	Symbol	Value
Input voltage	V_{in}	32 V
Switching frequency	f_s	100 kHz
Duty cycle	D	65–95%
Turns ratio	N	4: 1
Capacitors	C_c, C_o	10 μH, 470 μH
Resistance	R_L	

51.3.2 Isolated Resonant DC-DC Converter

The main purpose of the proposed solution in this article is to boost the voltage from the approximately 400 V output of the non-isolated DC-DC converter in the previous stage to 50 kV. This high-voltage circuit provides electrical isolation between the input and output terminals, thus protecting the front-end electronic components. Through research and analysis of the literature, it has been found that phase-shifted full-bridge technology is a mature solution widely used in the field of switch-mode power supplies. Compared to traditional full-bridge topologies, phase-shifted full-bridge topology can achieve quasi-resonant operation by designing parameters such as inductors and capacitors. This allows for zero-voltage switching (ZVS) during turn-on and zero-current switching (ZCS) during turn-off, thereby reducing power device heat dissipation and improving power supply efficiency. Consequently, it becomes possible to eliminate the need for heat dissipation devices, reducing the size and weight of the power supply and achieving portability for the system.

Therefore, the output stage utilizes advanced phase-shifted full-bridge and series resonant soft-switching technology. The main circuit comprises a phase-shifted full-bridge soft-switching topology and an output rectifier circuit, as illustrated in Fig. 51.4. The basic circuit of the phase-shifted full-bridge involves four IGBT semiconductor main switches (Q_1, Q_2 referred to as leading bridge arm switches, Q_3, Q_4 referred to as lagging bridge arm switches), the main transformer T_1, series resonant inductor L_1, and isolation capacitor C_1. This circuit achieves zero-voltage switching of the switches by leveraging the parasitic capacitance of the primary series resonant inductor L_1 and power devices, effectively minimizing power losses. Adhering to the principles of a conventional full-bridge circuit, the switches alternate conduction to convert the direct current into alternating current, drive the main transformer, and generate high voltages based on the turns ratio.

Furthermore, in contrast to the traditional full-bridge topology where all IGBT switches remain off during periods of no energy transfer, the phase-shifted full-bridge topology ensures one group of IGBTs is always conducting in such instances. This effectively short-circuits the primary side of the main transformer, providing a resonant pathway for current. As a result, the amplitude of the resonant voltage produced on the transformer's secondary side is significantly reduced. This design choice mitigates excessive heating of the rectifier diodes in the output rectifier circuit caused

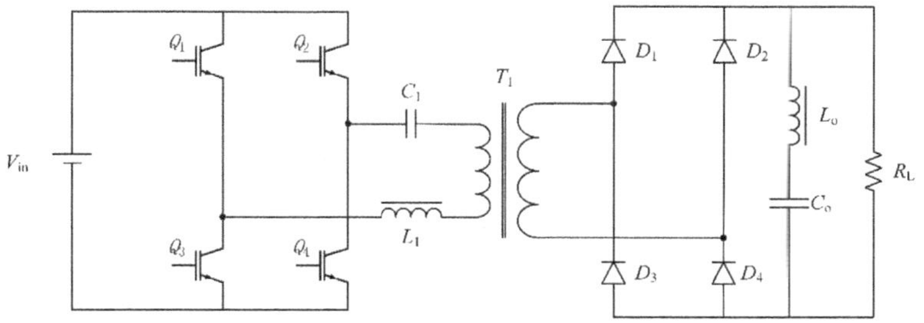

Fig. 51.4 Schematic diagram of the output stage in converter

by high-frequency, high-voltage waveforms. The rectifier circuit itself consists of diodes (D_1-D_4), output capacitor C_o, output commutation inductor L_o, and output load R_L.

The selection of the resonant frequency (f_r) in the circuit is crucial as it influences the choice of switch devices and the overall losses in the circuit. The circuit comprises inductive reactance (X_L) from the presence of inductors and capacitive reactance (X_c) from capacitors.

$$X_L = 2\pi f L \tag{51.2}$$

$$X_C = \frac{1}{2\pi f C} \tag{51.3}$$

In this circuit, the frequency (f) plays a significant role. At lower frequencies, capacitive impedance dominates the circuit, while at higher frequencies, inductive impedance prevails. When the inductive reactance equals the capacitive reactance, the circuit undergoes series resonance, resulting in maximum output power and efficiency. This resonant frequency point is denoted as f_r.

$$f_r = \frac{1}{2\pi\sqrt{LC}} \tag{51.4}$$

Simulation analysis was conducted on the isolated resonant DC-DC converter, and the obtained output parameters are presented in Table 51.4.

Table 51.4 Isolated DC-DC converter simulation output

Input voltage (V)	Output voltage (kV)	Output power (kJ/s)	Operating frequency (kHz)	Capacitor (nF)
400	50	3.5	50	50

51.4 Conclusion

This paper presents a refined design proposal for a lightweight and high-voltage DC power supply, powered by high-capacity lithium batteries. The design is suitable for charging energy storage capacitors in compact pulsed power systems. Capacity and discharging characteristic tests were conducted on individual batteries, indicating excellent performance of the battery model, meeting the output requirements of the pulsed power system for primary energy. The boosting section adopts a two-stage boosting scheme, where the batteries are connected in series to output 50 kV high voltage through a DC-DC converter with a 32 V low voltage input. The input stage is a non-isolated buck-boost DC-DC converter operating in voltage source regulator mode with a clamp-mode coupling inductor, while the output stage is a high-gain isolated DC-DC converter using phase-shifted full-bridge soft-switching topology, improving circuit immunity to interference and output stability. Subsequent installation tests will be carried out on the overall system to achieve compact and lightweight design, meeting the energy supply requirements of the pulse power system.

References

1. Y. Wang, *Pulsed Power Science and Technology* (Beihang University Press, Beijing, 2010)
2. X.S. Liu, *High-Power Pulsed Technology* (National Defense Industry Press, Beijing, 2005)
3. S.H. Jayaram, A.H. El-Hag, F.P. Espino-Cortes et al., Effects of process and product parameters on the shape of nanosecond pulses used in high-field liquid food treatment. IEEE Trans. Industry Appl. **41**(2), 520–526 (2005)
4. J. Rao, Y. Lei, S. Jiang et al., All solid-state rectangular sub-microsecond pulse generator for water treatment application. IEEE Trans. Plasma Sci. **46**(10), 3359–3363 (2018)
5. H.A. Ryan, S. Hirakawa, E. Yang et al., High-voltage, multiphasic, nanosecond pulses to modulate cellular responses. IEEE Trans. Biomed. Circ. S. **12**(2), 338–350 (2018)
6. W. Frey, J.A. White, R.O. Price et al., Plasma membrane voltage changes during nanosecond pulsed electric field exposure. Biophys. J. **90**(10), 3608–3615 (2006)
7. W.H. Jiang, Repetition rate pulsed power technology and its applications: (vii) Major challenges and future trends. High Pow. Laser Part. Beams **27**(1), 8–12 (2015)
8. M. Giesselmann, B. McHale, A. Neuber, in *Rapid Capacitor Chargers for Rep-Rated Operation of Low-Inductance Compact Marx Generators*. Conference Record of the 2006 Twenty-Seventh International Power Modulator Symposium, pp. 588–591. Arlington, USA (2006)
9. A. Pokryvailo, C. Carp, C. Scapellati, A 100 kW high voltage power supply for dual energy computer tomography applications. IEEE Trans. Dielect. El. In. **22**(4), 1945–1953 (2015)
10. S. Mao, in *A High Frequency High Voltage Power Supply*. Proceedings of the 2011 14th European Conference on Power Electronics and Applications (EPE 2011), pp. 12268041 Birmingham, UK (2011)

Chapter 52
Characteristics of Arc Discharge in Transformer Oil

Tianyu Han, Guangjin Zhang, Bo Ma, Tao Wen, Yi Zhao, Ke Wang, and Jiantao Sun

Abstract Internal short-circuit fault of the transformer may cause high-energy arc discharge in the oil, resulting in deformation rupture of the oil tank. As the energy source of transformer explosion accident, arc discharge will decompose the surrounding insulating oil and form bubbles around the arc, resulting in increased pressure in the tank. In this paper, an experimental platform of high energy arc discharge in transformer oil was set up, and electrical measurement, pressure measurement and optical imaging of discharge arc were carried out. The basic characteristics of arc discharge in transformer oil were revealed, including arc resistance model, arc gas production characteristics and pressure characteristics. It laid a foundation for subsequent research on oil tank deformation and explosion pressure relief measures.

T. Han · B. Ma · T. Wen · Y. Zhao
School of Electrical Engineering and Automation, Hefei University of Technology, Hefei 230009, Anhui, China
e-mail: 2022170451@mail.hfut.edu.cn

T. Wen
e-mail: tao-wen@hfut.edu.cn

Y. Zhao
e-mail: zhaoyi@hfut.edu.cn

G. Zhang (✉)
State Key Laboratory of Electric Insulation and Power Equipment, Xi'an Jiaotong University, Xi'an 710049, Shaanxi, China
e-mail: zhangguangjin@stu.xjtu.edu.cn

K. Wang · J. Sun
China Electrical Power Research Institute, Beijing 100192, China
e-mail: wangke1@epri.sgcc.com.cn

J. Sun
e-mail: sunjt@epri.sgcc.com.cn

Z. Fang et al. (eds.), *Proceedings of the 5th International Symposium on Plasma and Energy Conversion*, Springer Proceedings in Physics 398,
https://doi.org/10.1007/978-981-97-2245-7_52

52.1 Introduction

As an important electrical equipment in the power system, transformers play a decisive role in the reliable transmission, flexible distribution, and safe use of electrical energy [1]. When a severe short-circuit fault occurs within a transformer, it results in the vaporisation and decomposition of the surrounding insulating oil by the fault arc. This process generates a significant amount of gas, leading to the formation of bubbles around the arc which compresses the surrounding insulating oil and, in turn, causing a sudden increase in pressure inside the transformer's oil tank. In severe cases, this can lead to bulging, rupture, or even explosion of the oil tank, posing a serious threat to the safety of operator, surrounding equipment, and the environment [2].

Kawamura [3], a Japanese scholar, proposed that due to limited short-circuit energy from power sources and a lack of sufficient testing space, actual arcs were not used in the experiments. Instead, explosions of gunpowder in water-filled casings were used as a substitute for electric arcs. Different masses of gunpowder were used to simulate different energy levels of arc discharges, and this method has been proven effective. Ringle, an American scholar, designed a high-pressure nitrogen release device [4], which releases a certain amount of high-pressure nitrogen at a frequency of 60 Hz. Compared to the explosive energy source of gunpowder, this method is safer and more repeatable. However, since no electrical simulation method was used, the accuracy of this device is questionable.

To investigate the fundamental properties of arc discharge in transformer oil, a high-energy arc discharge experimental platform was established, and arc discharge experiments were conducted under power frequency conditions. The experiments verified calculation model for arc resistance, revealed the identification of dynamic bubble behaviour and growth process, and the analysis of gas composition inside the bubbles. Finally, this study elucidates the essential properties of arc discharge pressure in transformer oil, which provides a basis for further inquiry into measures for relieving oil tank deformation and explosion pressure.

52.2 Experimental Parameter

To simulate the conditions of internal short circuit faults in transformers, an *RLC* discharge circuit was constructed with a capacitance of 0.98 mF and an inductance of 10.2 mH, as shown in Fig. 52.1. This circuit produced an oscillating decaying current waveform with a period of 20 ms. A fine copper-nickel alloy wire with a diameter of 0.12 mm was serially connected in the electrode gap. When a high current passed through the wire, it would fuse, triggering the arc discharge in the oil. The arc voltage was measured using a high-voltage differential probe, the arc current was measured using a Rogowski Coil and the arc discharge process and bubble movement

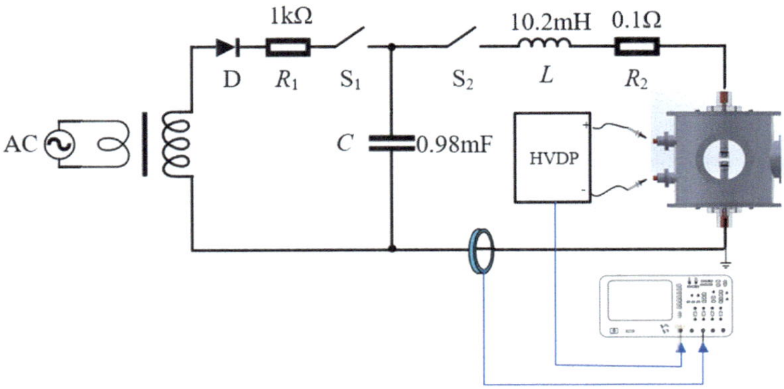

Fig. 52.1 Schematic diagram of the test circuit

were captured using a high-speed camera. Experimental data were collected using an oscilloscope.

52.3 Experimental Results and Analysis

52.3.1 Arc Resistance Model

Based on the study of long air gap arc, Terzija scholars summarized the arc voltage variation rule of air arc [5], as shown in Eq. (52.1).

$$u_{a0}(t) = \left(U_a + U_b\frac{I_0}{i_b(t)} + R_\delta|i_b(t)|\right)\text{sgn}(i_a) + \xi \tag{52.1}$$

where, $\text{i}_\text{b}(t) = \begin{cases} I_0|i_a(t)| < I_0 \\ |i_a(t)||i_a(t)| > I_0 \end{cases}$, $i_a(t) = C + I_1e^{-\alpha t}\sin(w(t - t_1))$ the current is the oscillation attenuation curve.

Where, $U_{a0}(t)$ and $i_a(t)$ are voltage and current signals of fixed length arc. U_a, U_b, I_0 ($I_0 \neq 0$), R_δ are the parameters. Sgn is a sign function and ξ is Gaussian noise with a mean of 0. U_a is obtained by multiplying the arc potential gradient and arc length. $U_b\frac{I_0}{i_b(t)}$ Indicates the starting voltage of an arc. $R_\delta|i_b(t)|$ is the additional quasi-linear part determined by the arc current. R_δ can be called quasi-static portion of arc resistance, but it is only a small part of the actual arc resistance, which is mainly determined by the value of U_a.

According to the piecewise conditions of Terzija's formula, when $|i_a(t)| > I_0$, it can be considered to be in the stable arc burning stage, and the expression of arc voltage at this time can be simplified into the following Eq. (52.2). By dividing the left and right sides of Eq. (52.2) by $i_a(t)$ at the same time, the time-varying formula

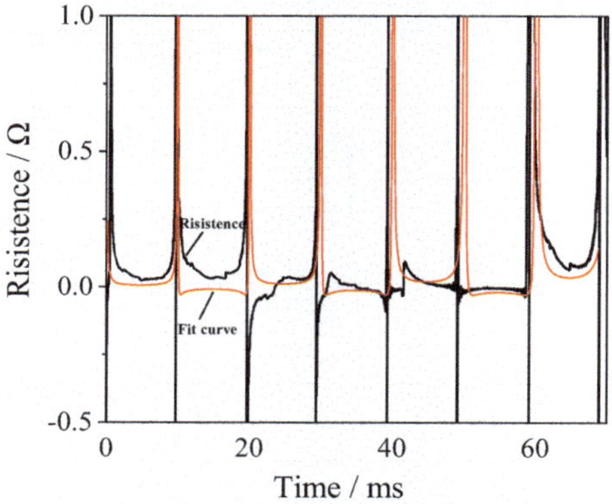

Fig. 52.2 Arc resistance fitting curve

of arc resistance can be obtained, as shown in Eq. (52.3).

$$u_{a0}(t) = U_a + U_b \frac{I_0}{i_a(t)} + R_\delta i_a(t) \tag{52.2}$$

$$R_a(t) = \frac{U_a}{i_a(t)} + U_b \frac{I_0}{i_a^2(t)} + R_\delta \tag{52.3}$$

The resistance curve of this discharge test was simulated according to Eq. (52.3), as shown in Fig. 52.2. The arc resistance curve is a U-shaped curve, and the arc resistance rises sharply near the current crossing point, which has a great impact on the fitting results. Moreover, the power near the crossing point is small and the energy occupied is small, so the arc resistance waveform in this part can be ignored. In the stable arc burning stage, the arc resistance value is stable between 0.02 and 0.1 Ω. With the increase of discharge time, the arc current oscillation decreases, resulting in the arc resistance rising. The arc resistance curve has a good fitting effect during the stable arc phase. The establishment of arc resistance model provides a modeling method for transformer tank fault pressure source model.

52.3.2 *Bubble Behavior and Dynamic Change*

A high-speed camera was utilised to capture images of the behaviour and dynamic changes of the bubble during the discharge process. The morphological features of the bubble are depicted in Fig. 52.3. By combining these high-speed images, it is possible to analyse the subsequent conclusions.

When the initial arc channel forms, numerous shock waves arise around the arc, as depicted in Figs. 52.3(1), 52.3(2), and 52.3(3). Following that, arc energy enters, and the arc decomposes and vaporises the surrounding transformer oil, producing bubbles. Figures 52.3(4) to 52.3(9) illustrates the expansion phase of the bubbles and

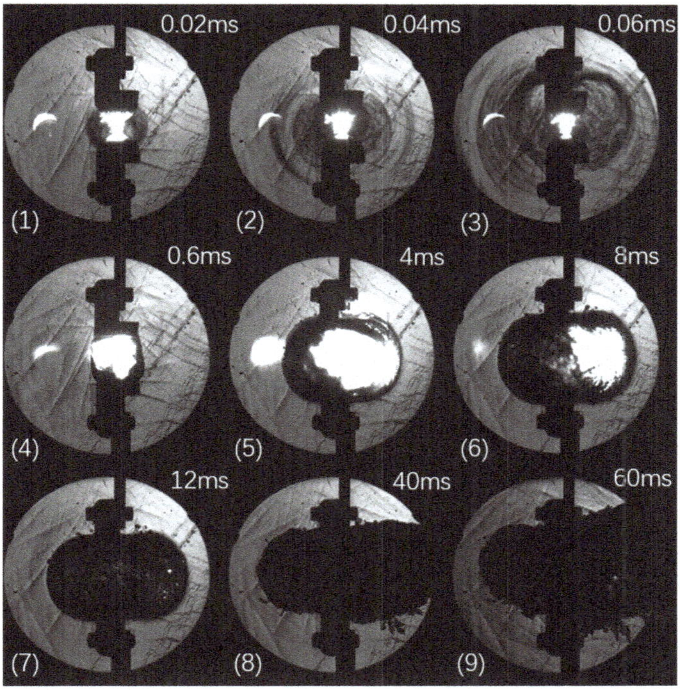

Fig. 52.3 Images of a gas bubble filmed by a high-speed video camera

arc combustion mechanism, as well as the oil gasification process in black, which occurs as a result of the arc's action. At this phase, the bubbles extrude the transformer oil outwards, creating a pressure wave. The arc's maximum size is depicted in Fig. 52.3(9).

Fig. 52.4 Gas composition and proportion

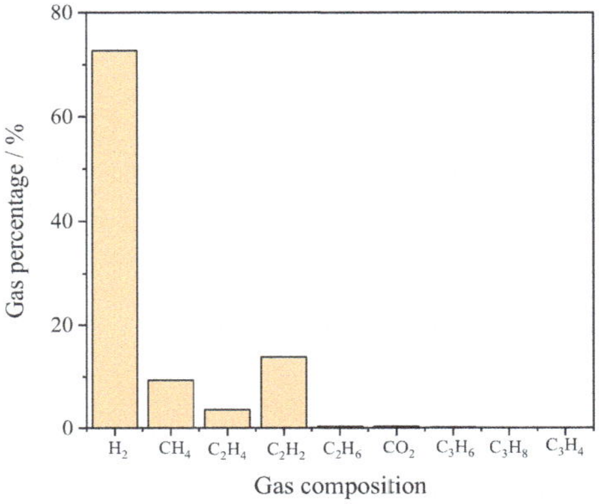

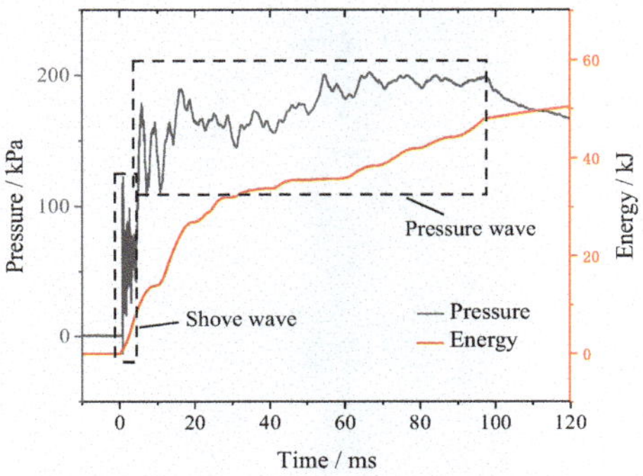

Fig. 52.5 Arc energy and wall pressure curve of experimental tank

52.3.3 Gas Production Characteristics of Arc

After discharge, the gas generated by arc decomposition gasification transformer oil was collected and analyzed by gas chromatography, and the composition and content of the gas were obtained, as shown in Fig. 52.4.

The gas produced by arc decomposition transformer oil is H_2, CH_4, C_2H_2, C_2H_4, C_2H_6, CO_2, C_3H_6, C_3H_8, C_3H_4, of which H_2 takes up the largest proportion of 72.64%. It was followed by C_2H_2, CH_4 and C_2H_4, accounting for 13.82%, 9.35% and 3.53% respectively. The produced hydrogen, acetylene and various alkane gases are flammable and explosive gases, which are prone to explosion accidents after contact with air at high temperatures. Therefore, the detection of gas components in transformer oil can be used as a diagnostic method for transformer faults.

52.3.4 Arc Pressure Characteristic

In this experiment, a pressure sensor was installed on the wall of the experimental tank, and the pressure curve and discharge energy curve of the experimental tank wall were obtained. As shown in Fig. 52.5, high-frequency shock waves were detected on the wall of the tank at the initial moment when the arc channel was formed, and the peak value of the shock wave was 120 kPa. With the continuous injection of arc energy, the arc decomposes and vaporizes the transformer oil to form bubbles, extrudes the liquid to produce pressure waves, and the pressure waves collide with the tank wall and produce refraction and reflection, causing the pressure curve to shake up and down. The peak value of the pressure wave is 200 kPa, about 1.67 times of the shock wave, and the action time of the shock wave is 5 ms, while the action time of the pressure wave is 95 ms, which is much longer than the impact time of the shock wave on the tank wall. Therefore, it can be concluded that the main cause of the deformation and damage of the tank is the impact of the pressure wave on the tank wall.

52.4 Conclusion

(1) Arc resistance model

The arc resistance curve is U-shaped in the stable arc burning stage, the fitting effect is better, and the error is about 10%. The establishment of arc resistance calculation model provides a modeling method for transformer tank fault pressure source.

(2) Bubble shape change

At the initial moment, a large number of shock waves are generated outwards when the arc channel is formed. Later, gas is generated when the transformer oil around the arc is decomposed, and a spherical bubble is formed around the arc. With the injection of deposited energy, the diameter of the bubble continues to increase, and the liquid is extruded outwards to produce pressure waves.

(3) Gas production characteristics

Hydrogen and different kinds of alkane gases were detected in the arc gas production, of which the highest hydrogen content was followed by flammable and explosive gases such as acetylene, methane and ethane. The detection of gas components in transformer oil can be used as a diagnostic method for transformer faults.

(4) Basic characteristics of pressure

The pressure curve is divided into two parts: shock wave and pressure wave. The shock wave is in high frequency jitter state and appears at the initial formation time of the arc channel. The peak value of the pressure wave is 1.67 times that of the shock wave, and the action time is much longer than the shock wave. It is concluded that the main cause of deformation damage of the tank is the impact of pressure wave on the wall of the tank.

Acknowledgements This study was financially supported by the Science and Technology Project of State Grid Corporation of China (SGCC) (5108-202218280A-2-365-XG).

References

1. L. Zehong, Findings in development of±1100kV UHVDC transmission. Prcceed. CSEE **40**(23) (2020)
2. N. Abi-Samra et al., Power transformer tank rupture and mitigation—a summary of current state of practice and knowledge by the task force of IEEE power transformer subcommittee. IEEE Trans. Power Delivery **24**(2), 1959–1967 (2016)
3. T. Kawamura, in *Prevention of Tank Rupture Due to Internal Fault of Oil-Filled Transformer*. In CIGRE International Conference on Large High Voltage Electric Systems, pp. 12.02. 11–12.02.17 (1988)

4. R. Ringlee, Tank pressures resulting from internal explosions. Trans. Am. Inst. Elect. Eng. Part III: Power App. Syst. **78**(4), 1705–1710 (1959)
5. V.V. Terzija, On the modeling of long arc in still air and arc resistance calculation. IEEE Trans. Power Deliv. **19**(3), 1012–1017 (2004)

Part IV
High Voltage Discharge and Applications

Chapter 53
Particles from Electrodes with a Small Radius of Curvature During Pulsed Discharges in Air

Victor Tarasenko, Dmitry Beloplotov, Mikhail Lomaev, Alexei Panchenko, and Dmitry Sorokin

Abstract This work presents a continuation of the study on the scattering of particles in a nanosecond spark discharge between two needle electrodes. Additional data were obtained on the scattering of electrode particles during voltage pulses of positive polarity with a half-maximum duration of 10 ns and a rise time of 2.2 ns, at atmospheric air pressure. Images of a discharge with particles flying away from a high-voltage electrode, obtained using a four-channel ICCD camera, are also presented. The study confirms that the formation of thin luminous tracks during diffuse and spark discharges is associated with the emission of metal particles from the surface of the electrodes. It is hypothesized that this process, similar to discharges in a vacuum, may involve mechanical phenomena that arise under the influence of a high electric field near needle electrodes.

53.1 Introduction

It is known that during the breakdown of vacuum gaps, bright spots form on the cathode and scatter metal particles [1]. The scattering of electrode material particles is also observed during pulsed discharges in air and other gases see, for example, [2]. Studies of the primary breakdown of vacuum gaps between polished electrodes under a uniform electric field, the results of which were analyzed in [3], showed that a decrease in the breakdown voltage cannot be associated only with explosive electron emission [4]. To understand the reasons limiting the electrical strength of vacuum insulation, it is also necessary to take into account the influence of a strong electric field on the mechanical properties of the electrode surface. Calculations carried out in [3] showed that in an unheated region, the size of which significantly exceeds the

V. Tarasenko (✉) · D. Beloplotov · M. Lomaev · A. Panchenko · D. Sorokin
Institute of High Current Electronics, Siberian Branch of the Russian Academy of Sciences, Tomsk 634055, Russia
e-mail: VFT@loi.hcei.tsc.ru

Z. Fang et al. (eds.), *Proceedings of the 5th International Symposium on Plasma and Energy Conversion*, Springer Proceedings in Physics 398,
https://doi.org/10.1007/978-981-97-2245-7_53

size of the center of explosive electron emission, the cathode material is also exposed to destructive mechanical stresses.

When voltage pulses with a steep front are applied to a discharge gap filled with gases, the threshold for explosive electron emission in a non-uniform electric field is reached in a short time [4]. As a result of a significant increase in the electric field due to the use of electrodes having a small radius of curvature, this can occur in less than 1 ns. In [5], it was found that during a nanosecond discharge using electrodes made from needles, thin luminous tracks (TLTs) grow from the area of bright spots on the electrodes in air and argon. These tracks are associated with the scattering of particles. At a pressure of 100 Torr, TLTs were recorded in diffuse and spark nanosecond discharges using an ordinary camera and an ICCD camera. The particle tracks had different shapes, from straight lines to winding ones, and the direction of their movement in some cases changed to the opposite. Some of the TLTs began at the edges of bright spots on the electrodes. However, the data obtained in [5] are preliminary, and further research is required into the conditions for the appearance of particles during diffuse and spark discharges in a non-uniform electric field, as well as studying the images of their expansion.

This work continues the study of the scattering of particles under the conditions of a nanosecond spark discharge between two needle electrodes. At atmospheric air pressure, new data were obtained on the scattering of particles during voltage pulses of positive polarity with a half-maximum duration of 10 ns and a rise time of 2.2 ns.

53.2 Experimental Equipment and Methods

The experimental setup was similar to that used in the work [5]. A diagram of the experimental setup and a time-integrated image of the discharge in air at atmospheric pressure are shown in Fig. 53.1.

Voltage pulses of positive polarity with an amplitude in the incident wave of 15 kV produced by a GIN-50-1 generator were fed to the input of the transmission line of the discharge chamber via a coaxial cable. The emission characteristics of a nanosecond discharge were recorded between two electrodes made of pieces of sewing needles with a base diameter of 0.75 mm. Their length after long-term training was 5.5 mm, and the radii of curvature were approximately 100 μm for the high-voltage, and approximately 300 μm for the grounded needle. During preliminary training of the electrodes, the grounded needle changed its shape and became dull. Time-integrated images of discharge plasma emission were taken with a Canon EOS 2000D SLR camera. Time-resolved ones were taken with a HSFC PRO four-frame ICCD camera. The gas discharge chamber was evacuated by a forevacuum pump and filled with air at atmospheric pressure with a relative humidity of 23%.

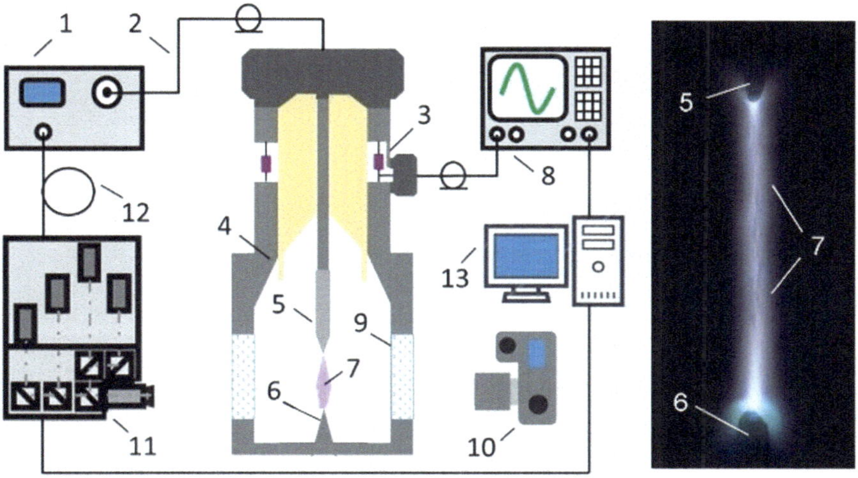

Fig. 53.1 Diagram of the experimental setup and time-integrated image of discharge for one pulse. 1—high-voltage generator, 2—high-voltage cable, 3—shunt, 4—discharge chamber, 5 and 6—electrodes with a small radius of curvature, 7—discharge plasma, 8—oscilloscope, 9—quartz window, 10—camera, 11—ICCD camera, 12—delay line, 13—computer

53.3 Results and Discussion

A typical image of nanosecond spark discharge in air at atmospheric pressure is shown in Fig. 53.1 on the right. Figure 53.2 shows a post-discharge emission and particle tracks.

Particles appeared in a few microseconds at distances from the electrodes sufficient for their observation. The speed of particles with a smaller cross-section, as follows from the length of tracks, was greater ($\approx$25 mm/ms) than that of particles with a larger cross-section. From Fig. 53.2, it can be seen that the particle speed decreased with distance from the electrode. The luminous diffuse areas near the electrodes were forced out to the center of the gap in tens of microseconds due to gas-dynamic processes. It was found that the shape of these diffuse areas changes from pulse to pulse. Analysis of time-resolved images showed that the TLTs were initiated from both the high-voltage and grounded electrodes. However, at atmospheric air pressure, particles were more likely to fly out from the high-voltage electrode (see Figs. 53.2 and 53.3). Figure 53.3 shows the post-discharge emission and particle tracks after their appearance at various times for the second voltage pulse with the same amplitude and duration.

In total, about 50 images of the dynamics of the appearance of metal particles from the electrodes were recorded using the ICCD camera. In addition, more than 100 time-integrated images of discharge were taken with the Canon EOS 2000D SLR camera. From the analysis of the data obtained, it can be assumed that the formation

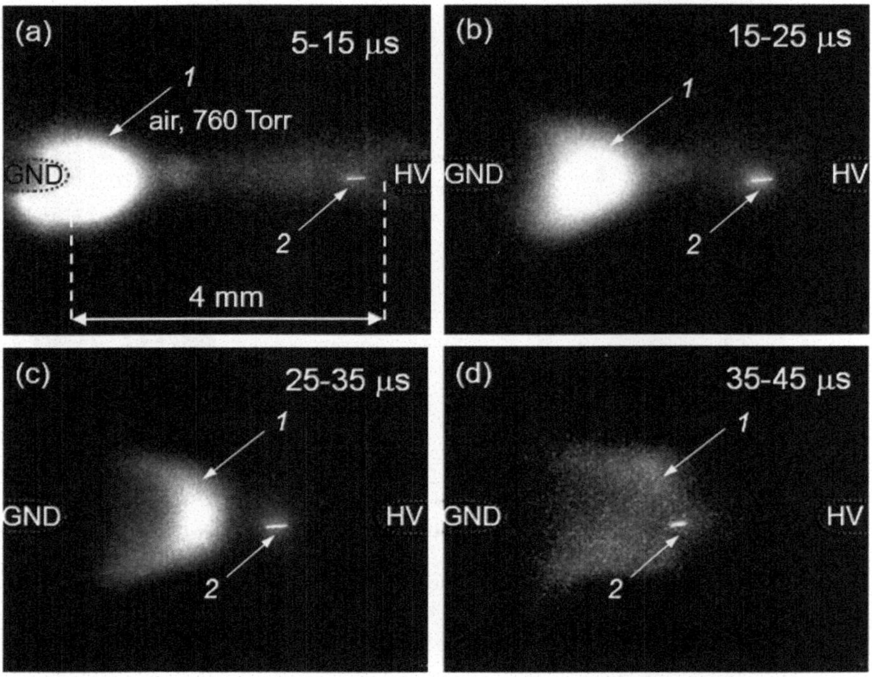

Fig. 53.2 Time-resolved images of post-discharge emission in the gap filled with air at atmospheric pressure at various time intervals (5–15, 15–25, 25–35 and 35–45 μs) after applying a voltage pulse to the gap. GND—grounded electrode, HV—high-voltage electrode, 1—emission of metal vapor during the decay of the spark channel, 2—particle tracks. Implementation 1

of TLTs is associated, among other things, with the emission of metal particles from the surface of the electrodes, including the influence of a high electric field due to mechanical phenomena during the formation of primary plasma as a result of a thermal explosion.

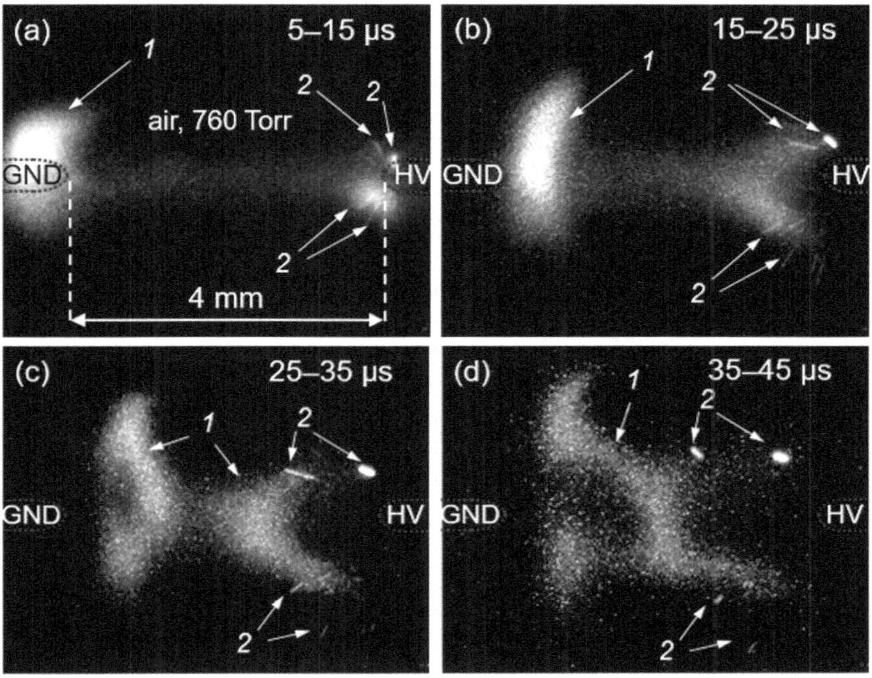

Fig. 53.3 Time-resolved images of post-discharge emission in the gap filled with air at atmospheric pressure at various time intervals (5–15, 15–25, 25–35 and 35–45 μs) after applying a voltage pulse to the gap. GND—grounded electrode, HV—high-voltage electrode, 1—emission of metal vapor during the decay of the spark channel, 2—particle tracks. Implementation 2

53.4 Conclusion

It has been confirmed that the formation of thin luminous tracks during diffuse and spark discharges is associated, with the emission of metal particles from the surface of the electrodes. Using the four-channel ICCD camera, it was established that particle tracks begin to be observed with a delay of ≈ 5 μs or more after the breakdown of the discharge gap. The emission of metal particles from the electrode surface can be caused by a high electric field and explosion of microprotrusions.

Acknowledgements The work was carried out within the framework of the State assignment of the IHCE SB RAS, project No. FWRM-2021-0014.

References

1. I. Beilis, Plasma and Spot Phenomena in Electrical Arcs; Springer Series on Atomic, Optical and Plasma Physiscs, vol. 113 (Springer, Cham Switzerland, 2020), pp. 255–283. https://doi.org/10.1007/978-3-030-44747-2.
2. V. Tarasenko, N. Vinogradov, D. Beloplotov, A. Burachenko, M. Lomaev, D. Sorokin, Influence of nanoparticles and metal vapors on the color of laboratory and atmospheric discharges. Nanomaterials **12**, 652 (2022). https://doi.org/10.3390/nano12040652
3. E.V. Nefedtsev, S.A. Onischenko, Modification of the cathode material around the explosive electron emission centers in the spark stage of vacuum breakdown. Technol. Phys. Lett. **48**, 69–71 (2022). https://doi.org/10.21883/PJTF.2022.22.53804.19307
4. Y.D. Korolev, G.A. Mesyats, Field-Emission and Explosive Processes in Gas Discharges (Novosibirsk, Russia), 255 p. (In Russian) (1982)
5. V.F. Tarasenko, D.V. Beloplotov, A.N. Panchenko, D.A. Sorokin, Thin luminous tracks of particles released from electrodes with a small radius of curvature in pulsed nanosecond discharges in air and argon. Surfaces **6**, 214–226 (2023). https://doi.org/10.3390/surfaces6020014

Chapter 54
The Design Method of an Electron Beam Expansion Device Based on Double Solenoid Lenses

Xiaocheng Hu, Weihao Tie, Yongchun Shao, Jinyong Fang, and Jiangping Wang

Abstract A design method for double solenoid electron beam focusing lenses is presented, based on the beam optics theory. Starting from the exit state of the accelerated electron beam, the method adopts two solenoid magnetic lenses to achieve the goal of increasing the beam's transverse cross-sectional size and reducing its divergence angle. Compared to traditional beam-focusing lenses with quadrupole magnets, the device owns a simpler structure, smaller volume, and clear principles, and holds potential for applications in electron beam irradiation, electron beam processing, electron beam testing, etc. This paper theoretically investigates the transmission laws of relativistic electrons (1 meV) in the different magnetic lenses and subsequently models numerically for verification. More specifically, initial energy dispersion (up to ± 2%), manufacturing errors and assembly errors (± 0.1 mm/1mrad) affect beam transmission are discussed. This paper provides guidance for the engineering design and assembly of the device for further usage.

54.1 Introduction

Humans use electron rays accelerated in electromagnetic fields to irradiate various materials [1] and excite various material molecules through the interaction between high-energy electrons and materials, thereby exploring the structure of materials or developing other applications such as electron beam melting [2]. The electron beam is emitted by an electron-generating device such as an electron gun. After the electrons pass through the accelerating tube, they are accelerated by the electromagnetic field

The original version of the chapter has been revised. A correction to this chapter can be found at https://doi.org/10.1007/978-981-97-2245-7_64

X. Hu (✉) · W. Tie · Y. Shao · J. Fang · J. Wang
Xi'an Branch of China Academy of Space Technology, Xi'an 710100, China
e-mail: 11724015@zju.edu.cn

Z. Fang et al. (eds.), *Proceedings of the 5th International Symposium on Plasma and Energy Conversion*, Springer Proceedings in Physics 398,
https://doi.org/10.1007/978-981-97-2245-7_54

to obtain higher energy. The electron beam envelope at the exit of the accelerating tube is often several millimeters and the divergence angle is about several milliradians [1]. However, applications such as irradiation technology require the electron beam spot size to be in the centimeter range or larger. If the electrons at the exit of the accelerating tube are allowed to diverge freely, a natural drift space of at least 10 m is required, which greatly increases the room occupied by the equipment and limits the application scenarios.

In order to solve this problem, an electron beam expander is usually added behind the accelerator exit to increase the beam envelope scale, and at the same time compress the divergence angle of the electron beam to make it parallel therefore improving the beam quality [3]. The current method is to use quadrupole magnet sets [4, 5] to achieve the above functions. The quadrupole magnet sets are usually optimized and designed based on previous experience or using finite element software simulation methods. There might be a huge waste of computing resources and time consumption, such as trial costs. At the design level, the beam expander magnet composed of multiple sets of quadrupole magnets contains a number of optimization parameters and requires a long optimization time; at the processing and assembly level, the quadrupole magnet set is difficult to process and assemble.

Generally speaking, the focusing ability of quadrupole magnets on high-energy electrons is much stronger than that of solenoid magnetic lenses with the same magnetic strength and size [6, 7]. However, due to the above difficulties, for electron beams with lower energy (below 1 meV), using a solenoid group to expand the electron beam is engineeringly feasible. This solution greatly reduces the difficulty of design, manufacturing, and assembly, and has strong practical application value.

54.2 Designing Method

54.2.1 Magnetic Lenses Layout

The solenoid magnetic lens can generate a nearly uniform magnetic field covering the entire space from the object to the image [8, 9]. Most of the electrons passing through the object point are rotating and moving in a spiral form from the beginning. Regardless of the direction of the axial magnetic field, the solenoid always has a converging effect on the electrons.

As shown in Fig. 54.1, the entrance electron beam is generally in a divergent state, within its phase space P_1, Most electrons are in the first and third quadrants, that is, the electron position and velocity have the same direction. After passing the first solenoid magnetic lens, the long axis OA of the phase diagram rotates clockwise to OB, and the electron phase space after the first solenoid magnetic lens changes to P_2. At this time, the electron beam shrinks to the minimum due to the focusing effect of the magnetic field, and the absolute value of the divergence angle is the maximum. In the second drift section, the focused electron beam becomes divergent after passing

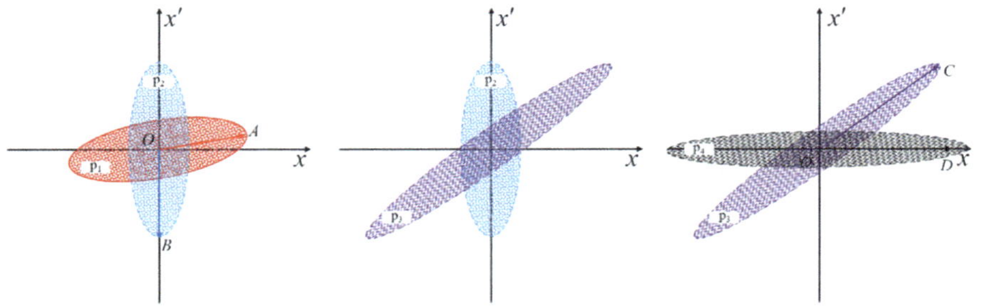

Fig. 54.1 Schematic diagram of electron phase space changes during transmission

through the focus point and drifts in free space at the maximum divergence angle until it reaches the entrance of the second solenoid magnetic lens. At this time, the electron phase diagram is P_3. When the electron beam enters the solenoid, it is focused by the magnetic force. At this time, if the focusing intensity and the effective magnetic path length are reasonably adjusted, the divergent electrons can be emitted in parallel, and the long axis OC of the phase space rotates clockwise to OD. The electron phase diagram at this time is P_4.

In this paper, the space charge effect is ignored since the beam owns a low charge and transmission within a short distance. A physical understanding of the beam dynamics in a solenoid is performed: The electron is regarded as an ideal particle, which means that only axial velocity at the beginning is considered. The axial magnetic field of the solenoid is peaked at its center, while the radial magnetic field is peaked near its physical end. When electrons are about to enter the solenoid, the electron is applied to a force $-eV_zB_r(z)$ in the azimuthal direction. Subsequently, the electron enters the solenoid, the trajectory will be helical. After the electron exits the solenoid, the particle moves in a straight manner with a constant angle θ_r.

In Fig. 54.2 is a double-solenoid beam expander device, in which the inner diameter of solenoid S_1 should be larger than the electron beam spot at the exit of the accelerating tube, and the inner diameter of solenoid S_2 should be larger than the optimized diameter of the target beam spot.

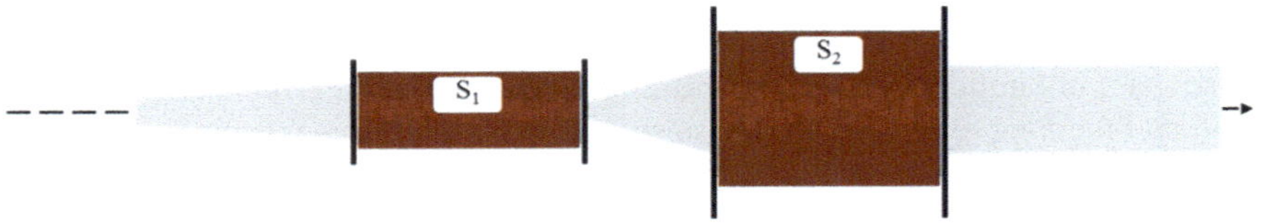

Fig. 54.2 Schematic diagram of double solenoid beam expansion device

54.2.2 Magnetic Field Simulation

For a long solenoid magnetic lens, the magnetic flux density around it can be expressed as [10]:

$$B_\rho = B_0\left[\alpha_+ C(k_+, 1, 1, -1) - \alpha_- C(k_-, 1, 1, -1)\right] \tag{54.1}$$

$$B_z = \frac{B_0 a}{a+\rho}\left[\beta_+ C\left(k_+, \gamma^2, 1, \gamma\right) - \beta_- C\left(k_-, \gamma^2, 1, \gamma\right)\right] \tag{54.2}$$

where B_ρ and B_z are the radial and axial magnetic flux densities respectively. $C(k_c, p, c, s)$ is an integral formula, which can be expressed as $C(k_c, p, c, s) = \int_0^{\pi/2} \frac{(c\cos^2\varphi + s\sin^2\varphi)d\varphi}{(\cos^2\varphi + p\sin^2\varphi)\sqrt{\cos^2\varphi + k_c^2\sin^2\varphi}}$, where k_c, p, c, s are parameters related to the shape of the magnet.$\alpha_+ = \frac{a}{\sqrt{(z+l)^2+(\rho+a)^2}}$, $\alpha_- = \frac{a}{\sqrt{z^2+(\rho+a)^2}}$, $\beta_+ = \frac{z+l}{\sqrt{(z+l)^2+(\rho+a)^2}}$, $\beta_- = \frac{z}{\sqrt{z^2+(\rho+a)^2}}$, $k_+ = \sqrt{\frac{(z+l)^2+(a-\rho)^2}{(z+l)^2+(\rho+a)^2}}$, $k_- = \sqrt{\frac{z^2+(a-\rho)^2}{z^2+(\rho+a)^2}}$, $\gamma = \frac{a-\rho}{a+\rho}$.

B_z is the maximum magnetic induction intensity of the magnet, a is the average radius of the solenoid, and l is the physical axial length of the solenoid. When caring about the distribution on the axis, Eq. (54.2) can be simplified to:

$$B_z = B_0\left[\frac{z+l}{\sqrt{(z+l)^2+a^2}} - \frac{z}{\sqrt{z^2+a^2}}\right] \tag{54.3}$$

Considering that the electron beam spot radius at the exit of the accelerating tube is 2mm and the beam spot radius after expansion is more than 20mm, this paper sets the inner radii of solenoid S_1 and S_2 to be 25 mm and 50 mm respectively, and the length of both solenoids is 200 mm. The lengths of the three drift sections are 20 cm, 100 cm, and 100 cm respectively. In order to ensure the beam expansion effect, the length of the second drift section must be greater than or equal to 100cm.

As mentioned above, according to the solenoid geometric parameters, its effective magnetic circuit lengths L_1 and L_2 determined to be 210.3 mm and 230.6 mm respectively, which are not much different from the physical lengths. Use the dichotomy method to change the excitation current of solenoid S_1 so that the long axis of the P_2 phase space is parallel to the Y axis, and change the excitation current of the solenoid S_2 so that the long axis of the P_4 phase space is parallel to the X axis. As shown in Fig. 54.3, it can be obtained that the physical intervals of solenoid S1 and S2 are 77 mT and 19 mT respectively. The use of permanent magnet solenoid can greatly reduce the axial length of the device and this article will not continue to optimize.

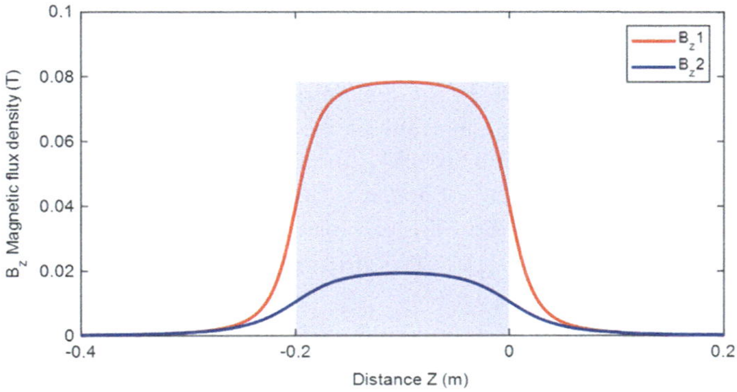

Fig. 54.3 Magnetic flux density distribution inside the double solenoid

54.3 Beam Transport Analysis

54.3.1 Intet Beam

The entrance settings are the common particle beam emission method, and the number of electrons in an envelope is set to 10,000. As shown in Fig. 54.4, the central energy is 1 meV and the energy dispersion at half maximum width is $\pm$ 5‰. The center of the electron beam is the zero point, and the electrons satisfy Gaussian distribution from the beam center to the edge, and there is no coupling in the two directions X and Y, and they are completely symmetrical. Taking the 3σ boundary of the Gaussian distribution for fitting, the radius of the initial beam spot is 1.9 mm, the divergence angle is within 2.8 mrad, and the emissivity is 3.47 mm $\cdot$ mrad.

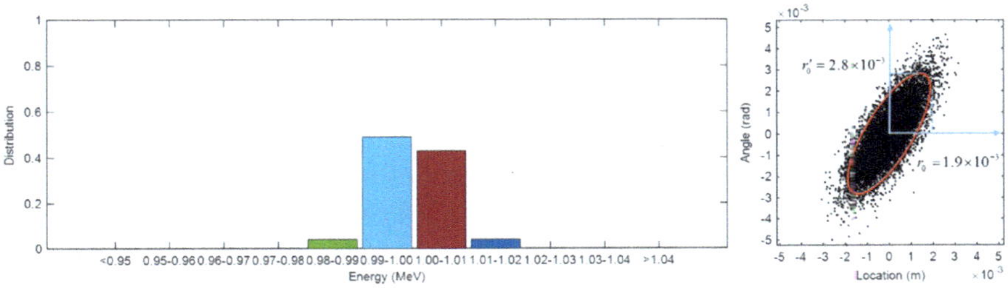

Fig. 54.4 Schematic diagram of the energy dispersion and emittance phase space at the entrance of the device

54.3.2 Beam Simulation

According to the phase space P_1 of the entrance electron $U_0(x_i, xd_i, y_i, yd_i)$, the drift Sect. 54.1 matrix M_{d1} in front of the device, the solenoid S_1 focusing matrix M_{s1}, the drift Sect. 54.2 matrix M_{d2} in the device, the solenoid S_2 focusing matrix M_{s2}, and the drift Sect. 54.3 matrix M_{d3} behind the device, the export electron phase space $U_1(X_i, Xd_i, Y_i, Yd_i)$ can be determined, which is $U_1 = \left(M_{d3}\left(M_{s2}\left(M_{d2}\left(M_{s1}\left(M_{d1}U_0^{\mathrm{T}}\right)\right)\right)\right)\right)^{\mathrm{T}}$. The drift segment matrix and the solenoid focusing matrix can be expressed as:

$$M_d = \begin{bmatrix} 1 & d & 0 & 0 \\ 0 & 1 & 0 & 0 \\ 0 & 0 & 1 & d \\ 0 & 0 & 0 & 1 \end{bmatrix} M_s = \begin{bmatrix} C^2 & \frac{SC}{K} & SC & \frac{S^2}{K} \\ -KSC & C^2 & -KS^2 & SC \\ -SC & -\frac{S^2}{K} & C^2 & \frac{SC}{K} \\ KS^2 & -SC & -KSC & C^2 \end{bmatrix} \tag{54.4}$$

Among them, d is the length of the drift section, in m. Parameters $C = \cos(KL)$, $S = \sin(KL)$, K is the focusing intensity of the solenoid magnetic lens, L is the effective length, and parameter KL depends on $\int B_z \mathrm{d}z/2(B\rho)_0$. where $(B\rho)_0 = P_0/|e|$ is the magnetic stiffness of the ideal electron, $P_0 = \beta\gamma m_0 e$ is the momentum of the ideal electron, m_0 and e are the mass and charge of the stationary electron. B_{z0} is the value where B_z is a constant within the physical range of the solenoid. The focusing intensity is related to the excitation current I (A), the number of unit winding turns n, and the shape of the solenoid (physical length l, average radius a, in m). These specific parameters can be allocated according to engineering requirements.

The electron trajectory calculated based on the theoretical transmission matrix is shown in Fig. 54.5. Within the beam line length of 1.5 m, the device can expand the entrance electron beam radius to more than 20 mm, and emit approximately parallel, meeting the design requirements. Since beam optics adopts common assumptions such as small quantity assumption and ideal field assumption, this solution uses the finite element software COMSOL for simulation modeling verification. It can be seen that the two are basically consistent, proving that the theoretical method is used to carry out twin solenoid beam expansion devices. Rationality of design.

In Fig. 54.6, through the statistics of the electron divergence angle at the device outlet, it can be found that the divergence angle of nearly 90% of the electrons is less than 0.2 mrad. The size of the electron beam envelope after beam expansion is 20 mm, and the beam expansion effect (black and white beam spot) is very obvious.

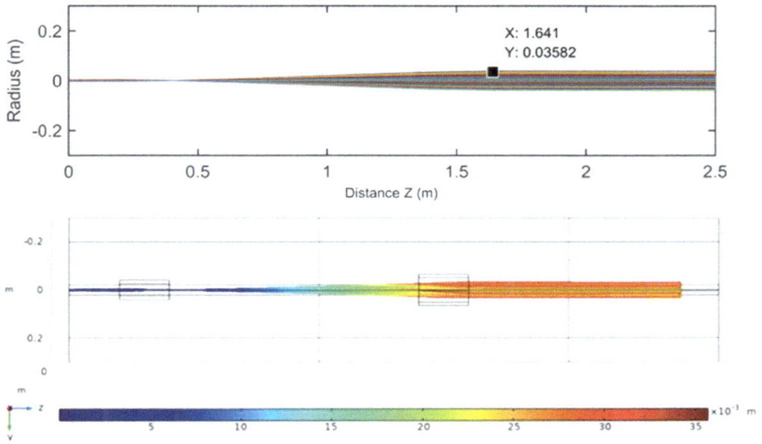

Fig. 54.5 Comparison chart between theoretical electron trajectory and finite element simulation verification electron trajectory

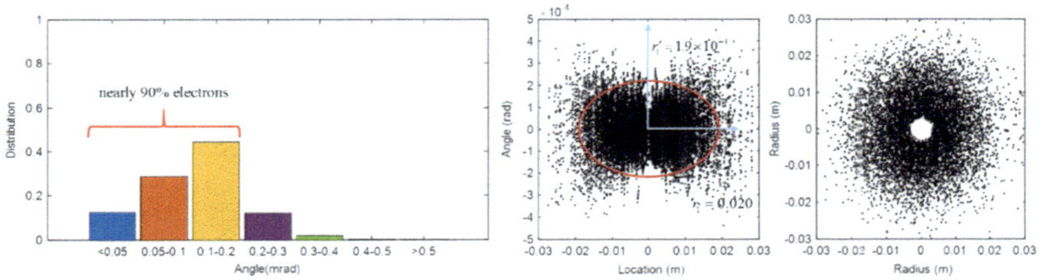

Fig. 54.6 Schematic diagram of electron divergence angle distribution, phase space and beam spot at the device outlet

54.4 Error Analysis

54.4.1 Energy Divergence Error

Since the incident electrons are not always ideal electrons, that is, the energy dispersion exists and the magnetic stiffness is different. High-energy electrons have greater magnetic stiffness, so their deflection radius is smaller under the action of the same magnetic field, making it easier for electrons to maintain their previous state of motion. As shown in Table 54.1, the electron energy dispersion of this scheme changes from ± 5‰ to ± 2%. It can be found that every time the energy dispersion doubles, the proportion of electrons with a divergence angle less than 0.05 mrad decreases by about 5%, and the proportion of electrons with a divergence angle less than 0.2 mrad decreases. The proportion dropped by about 20%. That is to say, in a smaller length scale, the entrance electron energy dispersion has little effect on

Table 54.1 Statistical results of the electron divergence angle to the target and the energy dispersion change of the entrance electrons

Energy divergency	± 5‰	± 1%	± 2%
< 0.05 mrad	14.38	9.81	5.42
0.05–0.1 mrad	32.02	23.29	13.54
0.1–0.2 mrad	42.68	39.61	30.16
Total	89.08	72.71	49.12

the shape of the target electron beam spot, but it will affect the parallelism of the outgoing beam (this indicator is particularly important in long-distance transmission applications of electron beams).

54.4.2 Assembly Error

There are two main types of assembly errors in twin-solenoid electron beam expansion devices, namely concentricity alignment error and coaxiality alignment error. The former is the degree of imperfection between the solenoid inlet center and the ideal geometric axis in the transverse plane, which is ± 0.1 mm; the latter is the degree of imperfection between the solenoid axis and the ideal geometric axis in the longitudinal plane, which is ± 1°. It can be concluded from Table 54.2 that the assembly error of the solenoid magnetic lens will basically not affect the size and divergence angle of the target beam spot, but it will deflect the overall beam direction, similar to the effect of a dipolar magnet on the electron beam.

Table 54.2 The impact of assembly errors on the state of the electron beam reaching the target

Error Count times	The upper limit of concentricity error is ± 0.1 mm, and the upper limit of coaxiality error is ± 1°. Randomly selected, the statistical result is within 3σ
Without error	Beam spot size 18 mm (X deviation 0 mm, Y deviation 0 mm), divergence angle within 0.4 mrad (X deviation 0 mrad, Y deviation 0 mrad)
1	Beam spot size 18 mm (X deviation – 18 mm, Y deviation 2 mm), divergence angle within 0.6 mrad (X deviation – 5.2 mrad, Y deviation – 3.6 mrad)
2	The beam spot size is 18 mm (X deviation – 42 mm, Y deviation 9 mm), the divergence angle is within 0.6 mrad (X deviation – 19.8 mrad, Y deviation – 8.0 mrad)
3	The beam spot size is 18 mm (X deviation is 25 mm, Y deviation is – 6 mm), and the divergence angle is within 0.6 mrad (X deviation is 12.2 mrad, Y deviation is 4.8 mrad)

54.5 Conclusion

If the electron beam drifts freely without being processed by the beam expander, the radius of the electron beam from 1 m to the target will only increase by 3 mm, and it will still be transmitted in a divergent form, which is far from the design requirements. After processing by the double solenoid beam expander designed in this work, the beam spot is expanded by nearly 10 times and the divergence angle is compressed by nearly 10 times, which is a remarkable effect. Compared with the traditional beam expansion device with quadrupole magnets, this device has a higher tolerance for processing and assembly errors, lower energy requirements for the entrance electron beam, and is simpler and more flexible to adjust.

References

1. Z.J. Zhou, T.Y. Chen, K.Y. Sheng, J. Huang, Magnetic field measurements and experimental verification of 0.5 MeV permanent-magnet electron-beam-spread system. AIP Adv. **12**, 095008 (2022)
2. M. Galati, L. Iuliano, A literature review of powder-based electron beam melting focusing on numerical simulations. Addit. Manuf. Manuf. **19**, 1–20 (2018)
3. A. Setty, Electrons beam dynamics of the 100 Mev Preinjector Helios for the Soleil synchrotron, *Proceedings of EPAC* (2004)
4. J. Huang, Y.Q. Xiong, D.Z. Chen, A permanent magnet electron beam spread system used for a low energy electron irradiation accelerator. Chin. Phys. C **38**(10), 107008 (2014)
5. J.K. Lim, P. Frigola, G. Travish, J.B. Rosenzweig, Adjustable, short focal length permanent-magnet quadrupole based electron beam final focus system. Phys. Rev. Special Topics Accel. Beams **8**, 072401 (2005)
6. M. Reiser, *Theory and Design of Charged Particle Beams*, 2nd edn. (WILEY-VCH Verlag GmbH & Co. KGaA, Weinheim, 2008)
7. V. Kumar, Understanding the focusing of charged particle beams in a solenoid magnetic field. Am. J. Phys. **77**, 737 (2009)
8. C. Baumgärtel, R.T. Smith, S. Maher, Accurately predicting electron beam deflections in fringing fields of a solenoid. Sci. Rep. **10**, 10903 (2020)
9. Z.P. Liu, G.Y. Feng, *Beam optics*, 2nd edn. (University of Science and Technology of China Press, Hefei, 2014)
10. N. Derby, S. Olbert, Cylindrical magnets and ideal solenoids. Am. J. Phys. **78**(3), 229 (2010)

Chapter 55
Contour Detection Method for Electron Beam Spot Based on Random Walk, Active Contour, and Improved B-Spline Curve Fitting

Yi Xiong, Baoxi Zhao, Junmin Zhang, Guoliang Li, Wei Tian, and Shunfan He

Abstract With increasingly wider application of electron beam processing, especially in the field of microfabrication, research on the process, diagnosis, and simulation of electron beam spot has become crucial for further technological advancements. In this paper, to respond to the diagnostic needs of electron beam spots formed during processing, an OpenCV based contour detection algorithm is proposed by means of the analysis and processing of electron beam spot images. The algorithm employs the combination of random walks and the contour extraction algorithm incorporating the Canny operator and targeting weak edge in complex background in particular. Additionally, it introduces an active contour algorithm to handle images with gapped edges. By improving the B-spline curve fitting and optimizing curvature, interference caused by material deformation on the contours is effectively addressed. Ultimately, the algorithm overcame disturbances from gaps and noise and obtains, clear and faithfully restored original contours that closely resemble the actual machining. The algorithm is intended preserve edge information as much as possible and restore the original shape of the target contours. Roundness and homogeneity parameters are collected, and the algorithm's superiority is validated through quality factor analysis of electron beam spot image results with different edge structures.

Y. Xiong · B. Zhao · J. Zhang · G. Li · W. Tian (✉) · S. He (✉)
South-Central Minzu University, Wuhan 430074, People's Republic of China
e-mail: 30152473@qq.com

S. He
e-mail: heshunfan@gmail.com

Z. Fang et al. (eds.), *Proceedings of the 5th International Symposium on Plasma and Energy Conversion*, Springer Proceedings in Physics 398,
https://doi.org/10.1007/978-981-97-2245-7_55

55.1 Introduction

Plasma has been widely used in many fields thanks to its unique physical properties. In the research of basic process of discharge plasma, electron beam and its processing are an important part. Electron beams are very useful types of beams due to their many different applications [1]. However, with the increasingly higher machining accuracy requirements and the more severe challenges brought by the complex processing environment and work procedures, the traditional method in observing the shape, size and distribution of electron beam spots with naked eye can no longer meet the current processing demand. To solve the problems in the result detection of electron beam technology, more attention has been drawn to the research of result detection based on computer vision technology.

Image processing for complex background is a hotspot in the field of automatic image processing. Among the existing edge detection algorithms, Canny operator is widely used in many fields because of its simple and easy-to-use characteristics, and is used as a standard algorithm to evaluate the performance of other edge detection algorithms [2]. The main steps of the traditional edge detection operator are: smoothing the image by using Gaussian filter to suppress noise; calculating the gradient amplitude and direction, and generating the gradient map; using the non maximum suppression method to refine the edge and making the edge location more accurate; using the delayed connection to thresholding the gradient graph to get the binary edge graph. However, Gaussian filtering can not eliminate salt and pepper noise well [3]; and the gradient map can only use the intensity change between adjacent pixels, so the utilization of image information is incomplete. It will blur the weak edges and noise in the process of edge detection, resulting in the loss of weak edges. In view of the low utilization rate of gradient map information, literature [4] uses the fuzzy set theory to transform the gradient graph into a membership graph, which improves the utilization of image information and effectively enhances the ability to detect weak edges. However, the algorithm is too complex and introduces more parameters to be set manually. Literature [5] uses hybrid filter instead of Gaussian filter to pretreatment, and considers the spatial and range information of pixels. Combined with Otsu algorithm, it automatically selects high and low thresholds based on the one-dimensional histogram of the image. On the basis of previous studies, literature [6] fused the top hat and Canny operator to obtain the contour, and performed dual threshold screening on the roundness and width parameters of the obtained contour, successfully obtaining a relatively clear electron beam spot contour.

The algorithms in the above literature are well optimized, but there are still problems in terms of the characteristics and difficulties of electron beam spot image edge detection. Therefore, this article proposes an OpenCV based edge detection algorithm that combines random walk, active contour and Canny operator, focusing on the characteristics of weak edges and complex backgrounds. An improved B-spline fitting is also proposed to solve the problems of material deformation and limited accuracy of observation equipment. In the end, roundness and homogeneity of the electron beam spot are obtained. The relevant parameters so obtained can be fed back to the machine to assist the real-time improvement of the electron beam equipment, to establish the relationship between the actual processing effect and the parameter setting, and to enhance the correlation between the two.

55.2 Feature Analysis Based on Electron Beam Spot

55.2.1 Contour Features

In the study of electron beams, Faraday cups are commonly used as measurement tools. The electrons generated by an electron gun are accelerated and focused, and they hit the material surface at a high speed, forming an electron beam spot. As shown in Fig. 55.1, different materials, or even different detectors, may produce different images of the electron beam spots. In this process, the material surface undergoes deformation under high temperature, so it creates a complex background and also distorts the original contour of the electron beam impact. Therefore, this increases the difficulty of the study. To extract the contour of the electron beam spot from the complex background, the image needs to be segmented by a threshold first. The segmented image can improve the accuracy of recognition and processing. This paper evaluates the segmentation performance by using two metrics: peak signal-to-noise ratio (PSNR) and structural similarity (SSIM) between the original and segmented images. A higher PSNR value indicates less distortion and better performance of the segmented image. The positive values of the SSIM index are in [0,1] [7]. A SSIM value closer to 1 indicates higher structural similarity and better segmentation effect.

By comparing the experimental results of applying various edge detection operators, it was found that the contour image from the Canny operator was relatively better. But the contour image was not clear with a lot of interference contours, which needed to be improved. In order to compare and analyze the effect of image edge detection, this paper uses the edge detection performance quality factor [8] for evaluation, which is defined by the following formula:

$$R = \frac{1}{\max\{I_I, I_A\}} \sum_{i=1}^{I_A} \frac{1}{1+ad^2} \tag{55.1}$$

where, I_I represents the sum of pixels in the ideal edge image, I_A represents the sum of pixels in the actual edge image, a is a proportion coefficient, usually 1/9, and d is the perpendicular distance from the ideal edge to the actual edge point. The closer the value of R is to 1, the better the edge detection effect.

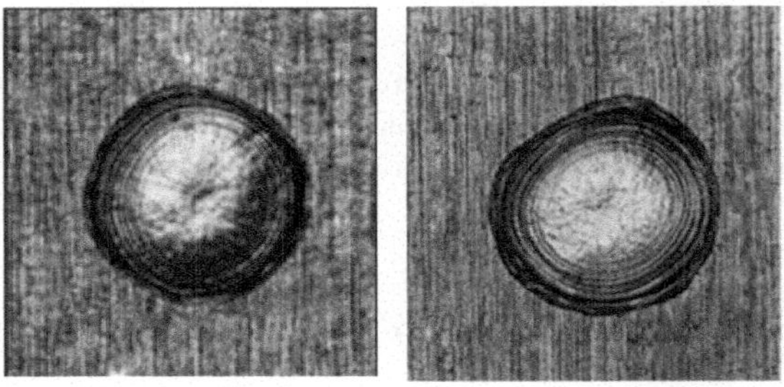

Electron beam spot sample on stainless steel surface (a) Electron beam spot sample on stainless steel surface (b)

Electron beam spot sample on the surface of copper film (c) Electron beam spot sample on the surface of copper film (d) Electron beam spot sample on the surface of copper film (e)

Fig. 55.1 Electron beam spot image

55.2.2 Roundness Features

In mathematics, the definition of roundness D is:

$$D = \frac{4\pi S}{l^2} \tag{55.2}$$

where, S is the area enclosed by the contour, and l is the perimeter of the contour. When the roundness D is equal to 1, the contour is a standard circle; for other shapes of contours, the roundness D is less than 1, and the closer the contour is to a circle, the closer the roundness D is to 1 [9].

55.2.3 Homogeneity Features

Homogeneity is one of the most important basic characteristics of certified reference materials [10], but there is no clear method to analyze the homogeneity of the electron beam spot at present. Therefore, this paper uses a special homogeneity detection standard based on the grid count method. The grid count method is a homogeneity coefficient proposed by Christiansen to describe the sprinkler homogeneity of nozzle water volume [11], which is defined by the following formula:

$$U = 1 - \sqrt{\frac{1}{N}\sum_{i=1}^{N}\left(\frac{p_i}{\overline{p}} - 1\right)^2} \tag{55.3}$$

This method divides the detection region into N small regions uniformly, and calculates the number of homogeneity points n in each small region and the number of points m in the whole region, then $p_i = \frac{n}{m}$, $\overline{p} = \frac{\sum_{i=1}^{N} p_i}{N}$. And by detecting the range X of p_i in the region, which reflects the distribution and dispersion of the electron beam spot, the homogeneity calculation results are verified. The formula is as follows:

$$X = \frac{\max(p_i) - \min(p_i)}{\overline{p}}, \in (1, N) \tag{55.4}$$

55.3 Experiment and Analysis of Electron Beam Spot Image Algorithm

55.3.1 Selection of Threshold Segmentation Method

Otsu method (Otsu) is considered to be a frequently-used algorithm for threshold selection in image segmentation. It is simple for calculation and it is independent of image brightness and contrast. Therefore, it has been widely applied in digital image processing. But the anti-noise ability of one dimensional Otsu algorithm is weak, and it is easy to be polluted by noise [12]. When the target and background have a large size ratio, the inter-class variance function may present double peaks or multiple peaks, and the noise caused by metal texture and processing deformation in the electron beam spot image greatly affects the segmentation effect. Therefore, the random walk segmentation algorithm is introduced. The main idea is to take an input set of voxels marked as foreground and background, referred to as seeds, and expand these sets based on the weighted graph to classify all voxels as either foreground or background [13]. With the weight as the transition probability, the unmarked pixel nodes as the initial points, it calculates the probability of each unmarked node reaching each seed pixel for the first time, and divides the unmarked nodes according to the probability, to obtain the final segmentation result.

In order to objectively compare the segmentation effects of the two segmentation algorithms on the research object of this paper, the results of the two threshold segmentation methods are detected by Canny operator, as shown in Fig. 55.2. Take sample (a) as an example. In Fig. 55.2, the effect using random walk algorithm is compared with the traditional otsu method. It shows that the edge detection contour is relatively clear, and the main subject and background segmentation are also more clear, while the edge features are also preserved more intact.

The final foreground binary image obtained after the random walk threshold segmentation is shown in Fig. 55.3.

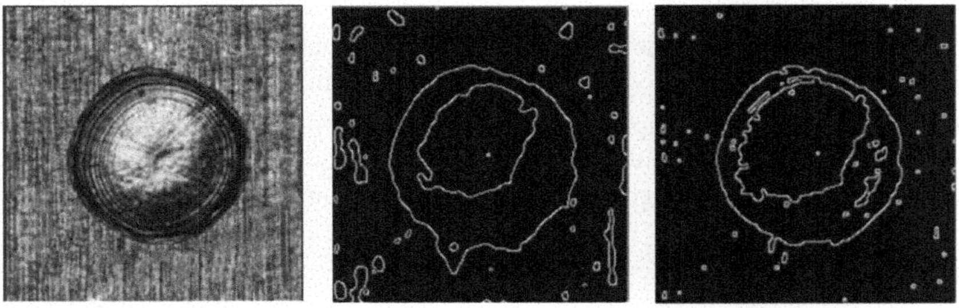

Sample (a) Otsu result of sample (a) Random walk result of sample (a)

Fig. 55.2 Comparison results of threshold segmentation

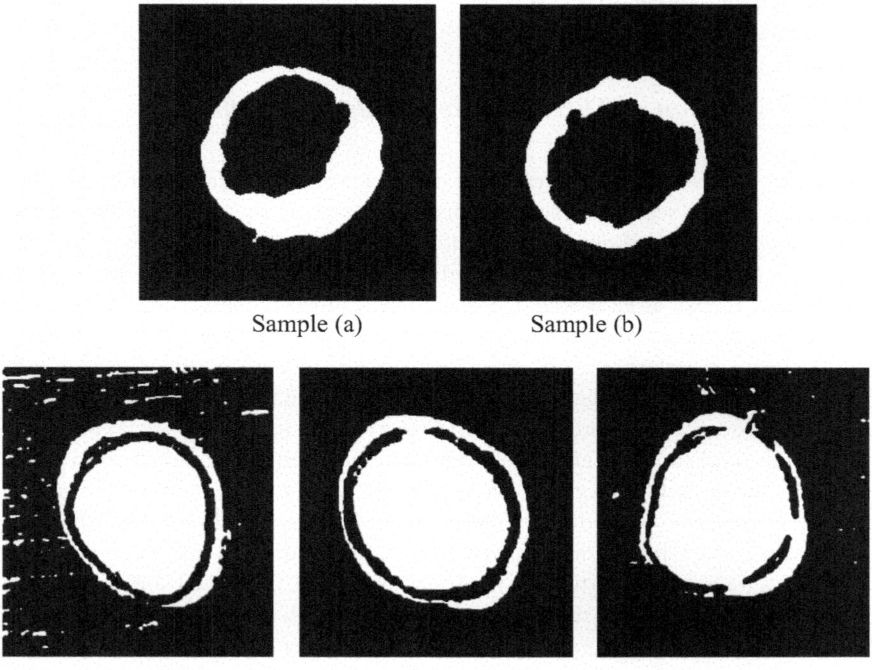

Sample (a) Sample (b)

Sample (c) Sample (d) Sample (e)

Fig. 55.3 Segmentation results of the electron beam spot image using the random walk algorithm

55.3.2 Fusion of Canny Operator and FindContours Function

The foreground binary image obtained through thresholding can be regarded as being composed of multiple contours (sets of points). By using the "findContours" function, contours can be extracted on a relatively smooth basis. However, due to the suboptimal performance of the "findContours" function in regions with unclear edges, it may result in convergence in obvious areas while convergence is yet to occur in less obvious areas. This leads to the appearance of "straight lines," as shown in Fig. 55.4, Sample (a) and (b). Moreover, the detection of unclear and weak edges is adversely affected by the presence of noise near the edges, resulting in poor performance. This can be observed in Fig. 55.4, Sample (e). To address this issue, two approaches can be taken. The first is noise reduction, which can reduce the impact of noise while preserving the important image details and edges. The second is to enhance the weak edges, which can emphasize the edges in the image and make them more distinguishable. Referring back to the previous context, the random walk algorithm has been used for image segmentation to distinguish background noise. Building upon this, the fusion of the Canny operator can be employed to enhance weak edges. Canny operator can not eliminate salt and pepper noise well due to its detection characteristics, which will lead to the failure to effectively distinguish weak edges and noise, resulting in the loss of weak edges. To ensure edge consistency in subsequent contour detection using the "findContours" function, the extended result of Canny operator is superimposed on the original image. Finally, by applying the minimum enclosing circle and minimum convex hull algorithms to the obtained contour information, we can extract the target contours, as shown in Fig. 55.4.

In Table 55.1, the fused approach yielded the optimal PSNR and SSIM values for denoising electron beam spot images. This is because the fusion process with the Canny operator helps fill in the missing regions, while the characteristics of the "findContours" function ensure more complete edge details. As a result, pixels that are far from the edges do not have a significant impact on the edge pixels. Additionally, the pixel values near the edges are well preserved, leading to improved denoising performance.

55.3.3 Active Contours

In addition to the outer contour, the inner contour needs to be extracted for certain electron beam spot images, as shown in samples (c), (d), and (e). In some cases, the inner contour may have discontinuities (as in samples (d) and (e)). The method of finding contours (findContours) can also be used to obtain the inner contour. However, electron beam spot images require the fitting of smooth contours and discontinuities. A commonly used approach is to simultaneously utilize the contour detection method (findContours) and ellipse fitting (fitEllipse). Firstly, the "findContours" function is

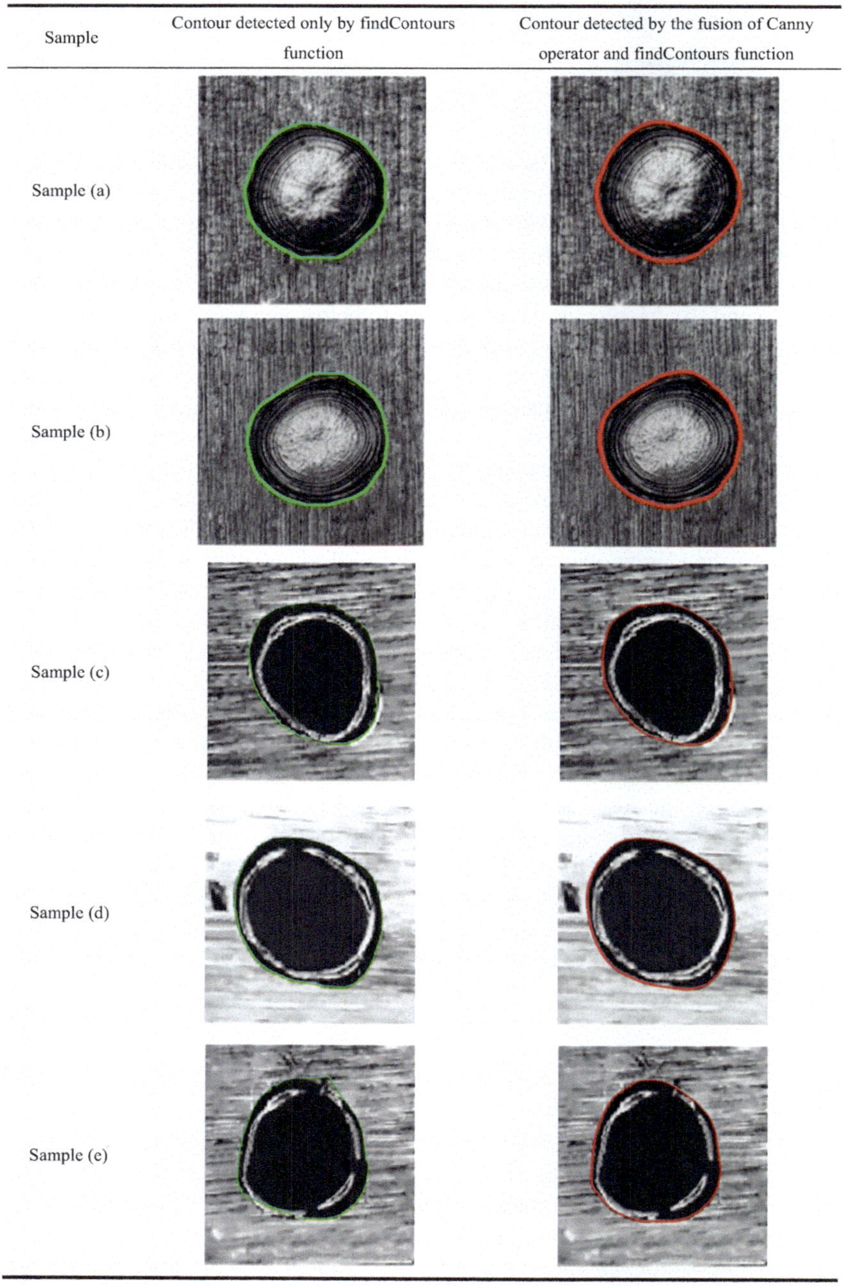

Fig. 55.4 Comparison results of contours detected by two methods

Table 55.1 PSNR and SSIM values of two methods

Method	Sample	PSNR	SSIM
Only by findContours function	a	20.619	0.9136
	b	20.578	0.9088
	c	24.012	0.9706
	d	23.735	0.9659
	e	24.321	0.9707
The fusion of Canny operator and findContours function	a	25.739	0.9356
	b	26.148	0.9380
	c	24.450	0.9778
	d	24.606	0.9692
	e	24.628	0.9756

used to extract the outer contour. This will provide one or more contours, including both inner and outer boundaries. Then, for each contour, the "fitEllipse" function can be employed to fit an ellipse. This function estimates the best-fit ellipse parameters based on the given set of contour points, approximating the shape of the inner contour. By simultaneously utilizing "findContours" and "fitEllipse," a smooth contour and a fitting effect for discontinuities of the inner contour can be achieved, as shown in Fig. 55.5.

Due to the subsequent narrowing of the region caused by the initial detection of the outer contour, the extracted inner contour achieves a more precise fit to the grayscale variations. This, however, contradicts the original intention of excluding material deformation and obtaining smooth contours. Besides, fitellipse fitting has limited fitting effect on the fracture, for the electron beam spot contour itself is not

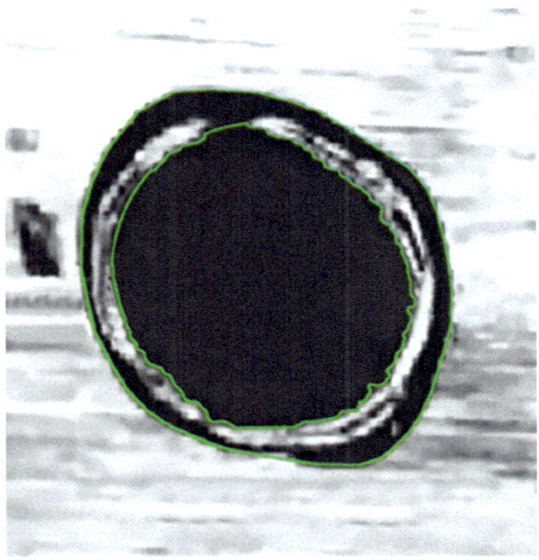

Fig. 55.5 The inner contour obtained by using the findContours function and fitEllipse function

a regular figure. Indeed, adjusting the fitting precision can potentially yield the best fitting results for the inner contour. It is important to note that increasing the fitting precision may also reduce the algorithm's robustness.

To handle these complex scenarios while maintaining a certain level of robustness, this paper introduces the Active Contour algorithm. The main principle of the active contour model is to construct an energy function. Driven by the minimum energy function, the contour curve gradually approaches the edge of the object to be detected, and finally the object is segmented [14]. The Active Contour model, also known as the snake model, derives its name from its ability to locate object boundaries by performing curve evolution. It is designed to obtain continuous and smooth closed segmentation boundaries, even in the presence of high noise.

The model is composed of a set of control points:

$$V(s) = [x(s), y(s)] \quad s \in [0, 1] \tag{55.5}$$

These points are connected by straight lines to create the boundary of the contour. In this context, $x(s)$ and $y(s)$ represent the coordinate positions of each control point in the image, where s is the parameter that describes the boundary in the form of a Fourier transform. The energy function is given by the following equation:

$$E_{snake} = \int_0^1 E_{int} + E_{ext} ds \tag{55.6}$$

In the context of active contours or snakes, the internal energy E_{int} represents the predicted properties of the image objects. It is used to control the elastic deformation of the contour and is composed of the elastic energy (modulus of the first derivative of v) and the bending energy (modulus of the second derivative of v).

The equation for the internal energy is as follows:

$$E_{int} = \alpha\left|\frac{\partial}{\partial s}\overline{v}\right|^2 + \beta\left|\frac{\partial^2}{\partial s^2}\overline{v}\right|^2 \tag{55.7}$$

The external energy term E_{ext} in the context of active contours or snakes represents the energy derived from image features that attract the contour towards the desired object boundaries. It is designed to capture the variations in the target boundaries and guide the contour evolution process.

During the energy function minimization process, the elastic energy term rapidly compresses the contour into a smooth circular shape. This term penalizes large variations in the contour, promoting smoothness and preventing excessive deformations. As a result, the contour tends to converge towards a compact and regular circular form. The bending energy term, on the other hand, drives the contour to become a smooth curve or a straight line. It encourages the contour to conform to the local curvature of the object boundaries, favoring contours that exhibit smooth bending behavior. It promotes the creation of smooth and well-defined curves or straight lines within the contour. The external energy term acts as an attraction force, pulling the

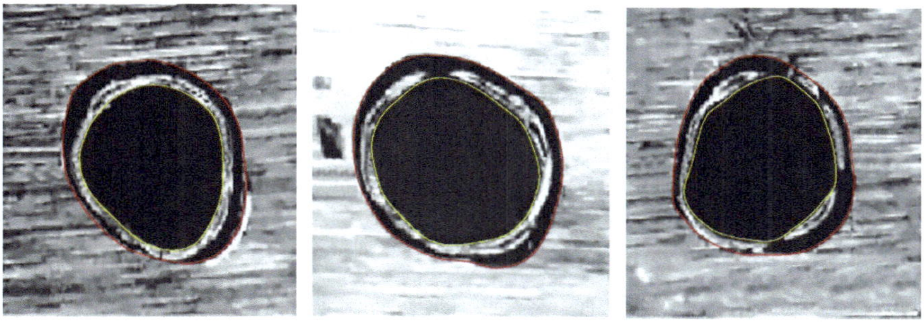

Sample (c) Sample (d) Sample (e)

Fig. 55.6 The inner contour (yellow line) of the detected beam spot obtained using active contour detection

contour towards the boundary positions in the image. It utilizes information from the image features to identify areas of significant changes that likely correspond to object boundaries. By aligning the contour with these high-gradient regions, the external energy term guides the contour to approximate the desired object boundaries more accurately. In contrast to passive models that heavily rely on image information, the active contour model involves the evolution of the contour under the influence of both internal and external energies. The contour curve gradually evolves and eventually settles at the boundaries of the target contour. This process is an iterative one that continues to develop until the energy reaches zero, at which point the movement ceases.

When applied to electron beam spot image processing, the Active Contour Model can be used to locate the outer boundary of the inner contour. Although the outer boundary may have discontinuities, it can be identified by finding the minimum convex hull. The outer boundary is then contracted to half its original size, serving as the initial position of the model. Through the combined influence of internal and external energies, the contour is allowed to expand like a "balloon" until it reaches the boundary points of the inner contour, where the process is halted. While ensuring smoothness of the contour, a good fit to the inner contour can be achieved, as shown in Fig. 55.6.

55.3.4 Spline Fitting

In the process of studying the electron beam spot image, a problem was noticed: both the maximum and minimum convex hull of the outer contour and the active contour of the inner contour are affected by the protruding points. Even if there is a long smooth contour around the protruding point, the contour will start to converge prematurely as soon as it encounters the protruding point. It is possible to eliminate these "spikes" on the contour by morphological transformation, but this will also

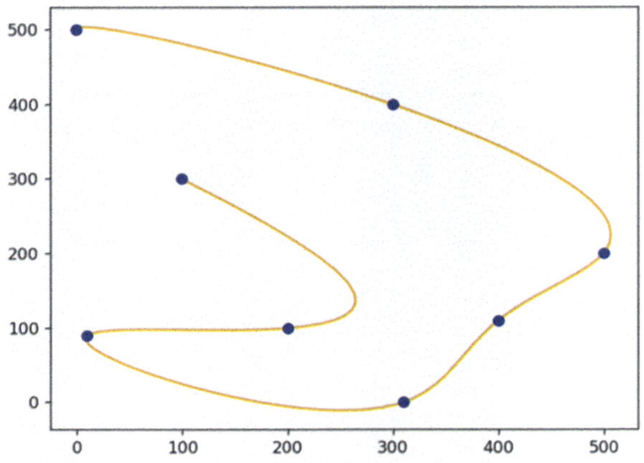

Fig. 55.7 Smoothing effect of ordinary spline interpolation

reduce the robustness of the algorithm. It is better to optimize the curvature of the contour after the detection algorithm gives the result. The curvature optimization of the contour is based on the spline interpolation method, but the ordinary spline interpolation cannot solve the problem of outliers, as shown in Fig. 55.7. Although it is smooth, there will still be concave and convex places.

To solve this problem, this paper proposes a B-spline curve fitting algorithm for the electron beam spot image features. A B-spline curve is a special representation form in the sub-discipline of numerical analysis in mathematics. It is a linear combination of B-spline basis curves, which is a generalization of Bezier curves, and can build accurate models for general geometric objects. It is, in itself, a cubic spline and it has the desirable property of the smallest possible support of any basis for the space of cubic splines [15]. Unlike spline curve interpolation, the fitted curve is as close as possible to the data points, rather than passing through them completely. For each point of the contour (point set) and its adjacent two points, the angle between the two vectors corresponding to the two edges is taken; and it is not smooth where the angle gradient changes greatly. After finding this point, we firstly delete it, and then find the two adjacent points with larger distance. They are the points at both ends of the line, that is, the place where it is not smooth enough. For either the outer contour or the inner contour, the center point of the line connecting these two points must be closer to the center of the contour. Therefore, we take the center point of these two points, and then move outward by a certain distance. Take sample (c) as an example, the effect of spline fitting is shown in Fig. 55.8.

The green contour is the convex hull of the original outer contour, and the blue contour is the result after fitting. It shows that the interference caused by noise and gaps on the contour has been basically corrected and eliminated by the fitting algorithm, and the contour does not tend to pass through the highest point of the protrusion or the lowest point of the depression, but it tries to find a balance between them.

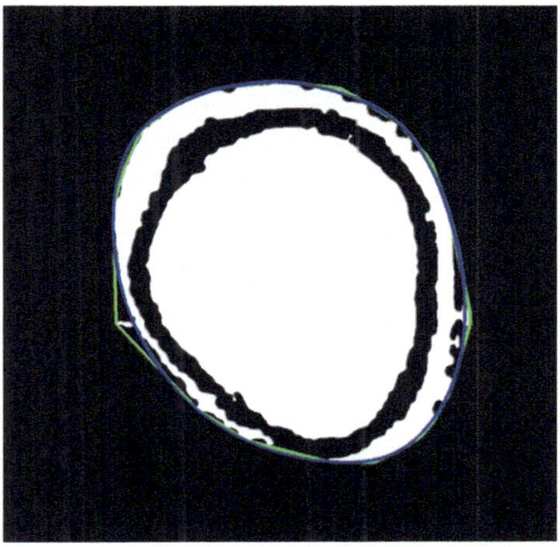

Fig. 55.8 Spline fitting process of the contour of sample (c) (green line before fitting, blue line after fitting)

55.3.5 Experimental Results

55.3.5.1 Detection Results and Related Data

As shown in Fig. 55.9, for different electron beam spot images, the algorithm proposed in this paper can obtain the clear edge of the target contour, fit the gap and smoothen the jaggedness.

This paper compares its detection results with those in literature [5] and literature [6], as shown in Tables 55.2 and 55.3. By using the random walk algorithm for threshold segmentation, the methodology proposed in this paper removes the noise better while preserving the edge information, and makes the edge more complete by fusing the canny operator. It also realizes the inner edge extraction and gap fitting by the active contour algorithm, and proposes the B-spline curve fitting to optimize the curvature of the contour. Therefore, the quality factor of the results obtained by the algorithm in this paper is higher that of the edge detection results in literature [5] and literature [6].

The above experimental comparison can prove that the improved edge detection algorithm has stronger detection ability for weak edges, and can effectively improve the continuity and integrity of edge detection, which has a very important role for electron beam spot image detection and research.

The overall flow chart of this algorithm is as follows (Fig. 55.10):

55.3.5.2 On the Homogeneity of the Electron Beam Spot

For some electron beam spot images, due to their characteristics, homogeneity detection can and should be performed, especially in the study of electron beam properties in laboratory environments. When the sample surface is close to a plane and there is

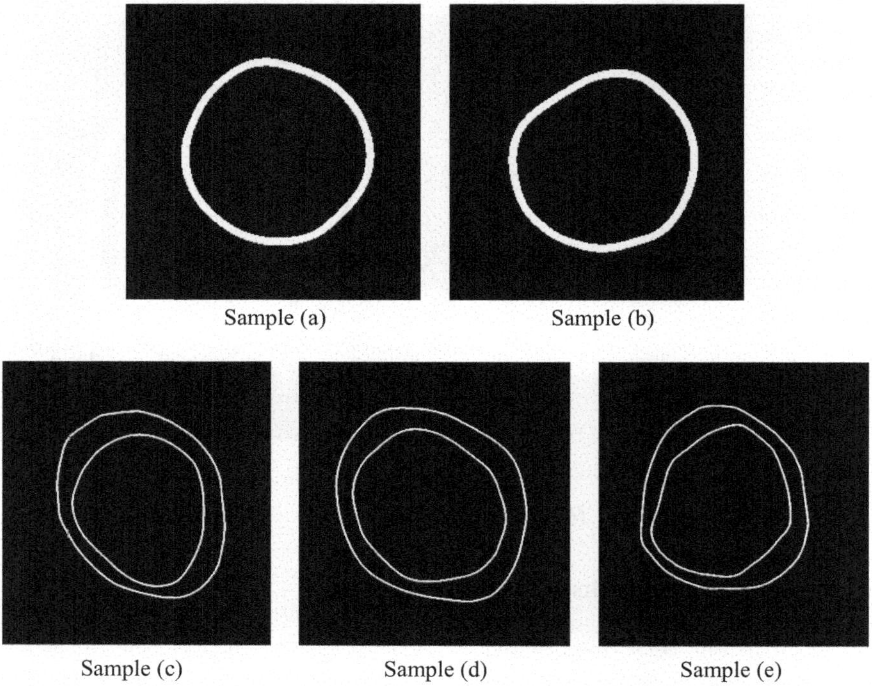

Fig. 55.9 Edge detection images obtained by the algorithm proposed in this paper

Table 55.2 Quality factor R of the edge detection algorithms

Edge detection algorithms	Sample (a)	Sample (b)
Literature [5]	0.8845	0.8358
Literature [6]	0.8985	0.8519
The algorithm proposed in this paper	0.9373	0.9021

Table 55.3 Roundness D of the contours detected by the edge detection algorithms

Edge detection algorithms	Sample (a)	Sample (b)	Sample (c)	Sample (d)	Sample (e)
Literature [5]	0.8387	0.8603	Unable to extract		
Literature [6]	0.8692	0.8872	Unable to extract		
The algorithm proposed in this paper	0.8979	0.8893	External:0.8681 Internal:0.8728	0.8749 0.8842	0.8793 0.8645

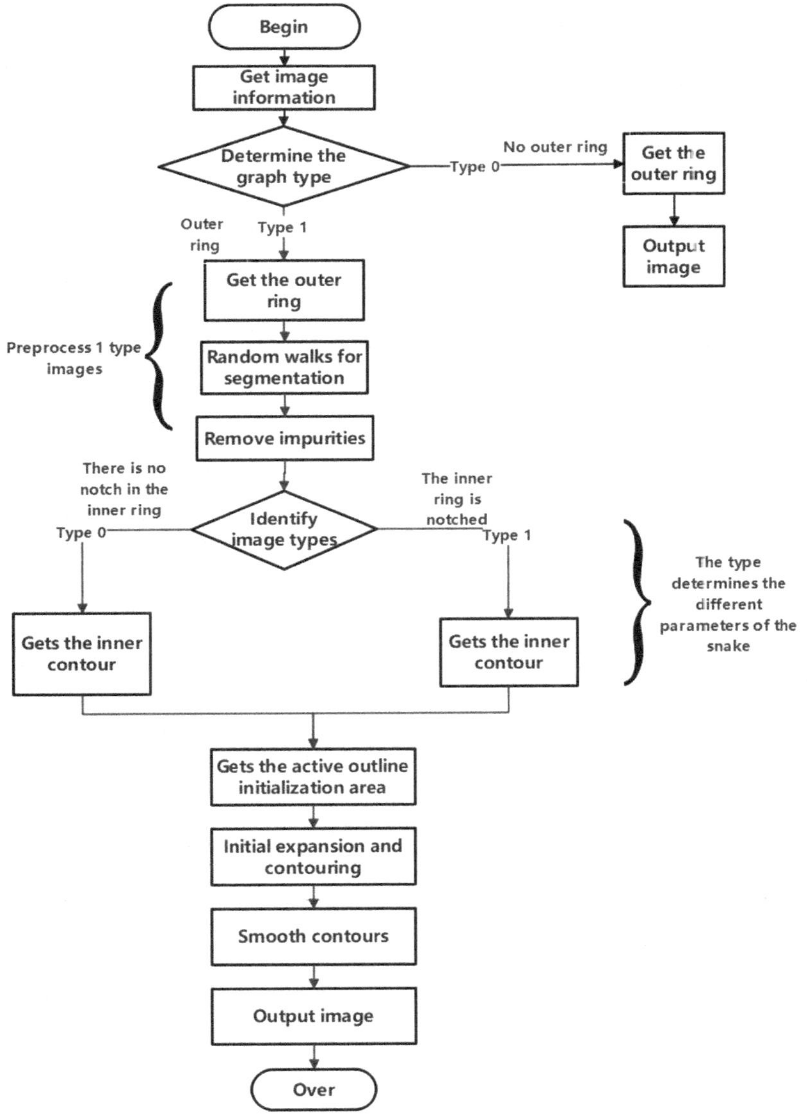

Fig. 55.10 Flowchart of the algorithm proposed in this paper

no deformation or breakdown, the homogeneity can be reflected to some extent by the electron beam spot image, as shown in Fig. 55.11.

For this kind of image, the electron beam spot image homogeneity feature detection based on the grid counting method mentioned earlier is used. Similarly, the image needs to be processed. The process is shown in Fig. 55.12:

Fig. 55.11 Electron beam spot on the surface of aluminum foil

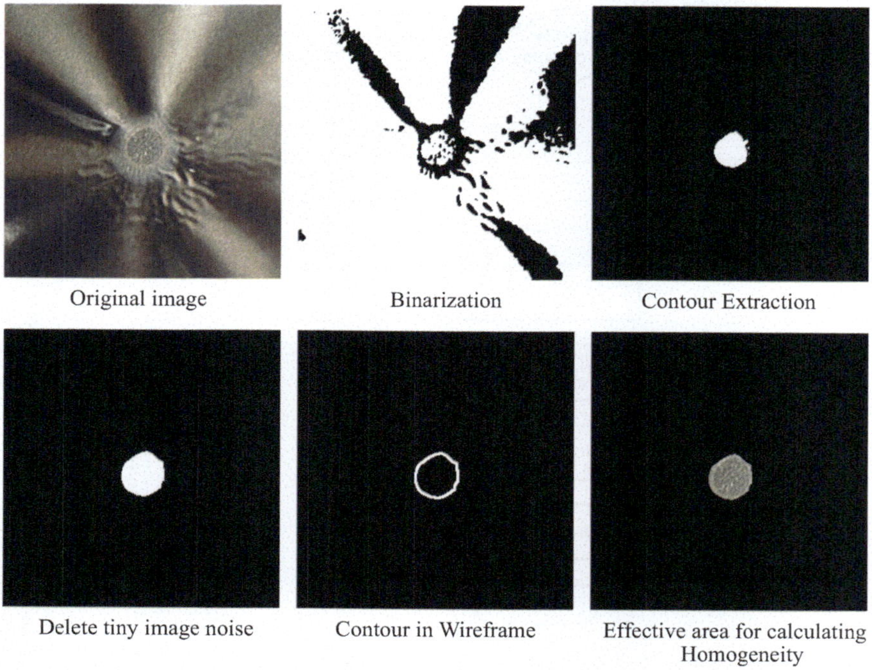

Fig. 55.12 Algorithm processing before calculating homogeneity

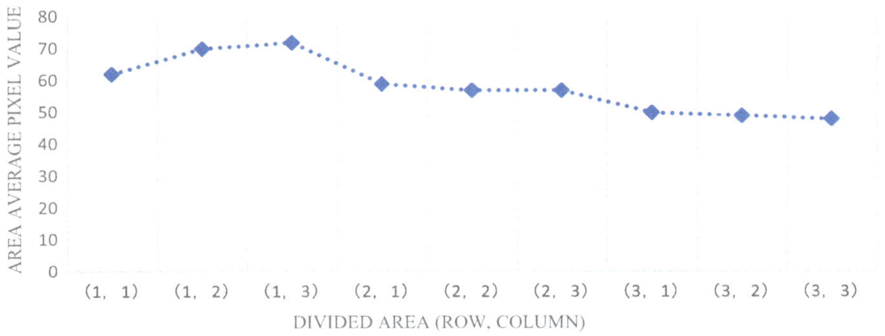

Fig. 55.13 Pixel values of regions

Table 55.4 Homogeneity and range results

Detected image	Homogeneity	Range
Figure 55.11	0.8586	0.4114

With the grid counting method, considering that the electron beam spot is not a normal point set, there will be situations of point overlap and different shapes and sizes of points in the dotting process. The sum of pixel values of each point set in the contour area is taken as n, and the number of point sets contained in each contour area is taken as m. The purpose is to use the average pixel value as the standard, and reduce the error caused by point set overlap or different sizes and shapes of individual points.

The circumscribed rectangle of the obtained contour image is divided into 3×3 equal values. Since some areas at the edge are not inside the contour, these points need to be removed when calculating the number of point sets. Then the area outside the contour within the rectangle is filled with black, so it can be ignored during pixel values adding-up.

With the data in Fig. 55.13 and Table 55.4 combined, the average pixel values of the nine respective regions of the electron beam spot in Fig. 55.11 are close, and the range is small. This indicates that the electron beam spot distributed on the aluminum foil surface is relatively uniform with a small dispersion degree, which, in turn, verifies the good homogeneity calculation result.

55.4 Conclusion

This paper aims at the demand of target edge detection for electron beam spot image. Addressing the problems of complex background noise, weak edge difficult to detect, material deformation interference and so on, it proposes an improved edge detection algorithm for target edge detection. The proposed method proposed comprehensively uses random walk, canny operator, active contour and other image processing methods to replace the traditional algorithm, effectively improving the image information utilization rate and weak edge detection ability. By proposing an adaptive spline fitting method to identify the deformation interference according to the calculated vector angle, it effectively improves the adaptability and practical application ability of edge detection. The electron beam processing effect is reflected in a more intuitive form on the plane. In addition to the roundness and homogeneity data calculated in this paper, a series of electron beam spot related parameters such as deflection angle, maximum and minimum diameter can be easily obtained, providing data for analysis and research. Thus, according to the actual processing effect and the expected effect, the parameters of the electron gun can be set, making it possible to further obtain the appropriate focusing and deflection parameters, thereby improving the processing effect, effectively serving the electron beam or plasma processing industry and research field.

Acknowledgements This work was supported by National Natural Science Foundation of China (No. 52377163 and 51977221).

References

1. F.S. Abdollahi, A. Abdoli-Arani, T. Mohsenpour, Enhancement of gain and modified electron beam trajectories using a relativistic electron beam with an elliptical cross section in free-electron laser with a helical wiggler and ion-channel guiding. IEEE Trans. Plasma Sci. **48**(7), 2396–2407 (2020)
2. Z. Shi, J. Qiao, An automatic loading tagret edge detection method for electric shovels based on improved canny operator. Modern Industrial Econ. Informationization **12**(10), 18–20+48 (2022). https://doi.org/10.16525/j.cnki.14-1362/n.2022.10.007
3. X. Kang et al., A novel adaptive switching median filter for laser image based on local salt and pepper noise density, in *2011 IEEE Power Engineering and Automation Conference*, vol. 3 (IEEE, 2011)
4. C. Lopez-Molina, B. De Baets, H. Bustince, Generating fuzzy edge images from gradient magnitudes. Comput. Vision Image Understanding **115**(11), 1571–1580 (2011)
5. C. Li, G. Chen, Y. Ding, Improved edge detection algorithm for canny operator. J. Chin. Comput. Syst. **41**(8), 1758–1762 (2020)
6. Y. Wang et al., Feature analysis of electron beam spot image by fusion of top-hat and Canny operator. J. South-Central Minzu Univ. (Natural Science Edition) **41**(02), 222–227 (2022)
7. A. Hore, D. Ziou. Image quality metrics: PSNR vs. SSIM, in *2010 20th international conference on pattern recognition* (IEEE, 2010)
8. C. Hongxi, Study on the objective evaluation of the performance of edge detection operator based on the quality factor. Autom. Instrumentation **08**, 8–10 (2015). https://doi.org/10.14016/j.cnki.1001-9227.2015.08.008
9. Z. Zhang, S. Song, Circle locating method based on roundness. J. Shandong Univ. (Eng. Sci.) **01**, 19–22 (2006)
10. X. Wang et al., A review on homogeneity testing techniques for geochemical reference materials in China. Rock Min. Anal. **29**(06), 735–741 (2010). https://doi.org/10.15898/j.cnki.11-2131/td.2010.06.005

11. H. Chang, C. Xie, W. Xu, Analytical methods of quantitative metallography for uniform distribution of particles, in *Proceedings of the 2007 CSM Steel Congress*, ed. by Chinese Society for Metals (Metallurgical Industry Press, 2007), pp. 1191–1192
12. H. Ye, S. Yan, P. Huang, 2D Otsu image segmentation based on cellular genetic algorithm, in *2017 IEEE 9th International Conference on Communication Software and Networks (ICCSN)* (IEEE, 2017)
13. E.J. Lutton, S. Collier, T. Bretschneider, A curvature-enhanced random walker segmentation method for detailed capture of 3D cell surface membranes. IEEE Trans. Med. Imaging **40**(2), 514–526 (2020)
14. D. Hailong, Contour extraction method and implementation of active contour model algorithm based on N-order Bezier curve for cardiac medical images, in *2022 21st International Symposium on Distributed Computing and Applications for Business Engineering and Science (DCABES)* (IEEE, 2022)
15. G. Cipli, F. Sattar, P.F. Driessen, A novel approach to low frequency activity detection in highly sampled hydrophone data based on B-spline approximation, in *2015 IEEE Pacific Rim Conference on Communications, Computers and Signal Processing (PACRIM)* (IEEE, 2015)

Chapter 56
Analysis of Reduced Graphene Oxide for Monitoring the Decomposition of HFO-1336mzz(€)

Nian Tang, Dibo Wang, Dongwei Sun, Yan Luo, Li Li, and Ran Zhuo

Abstract As an alternative SF_6 environmentally friendly insulating gas HFO-1336mzz€, it decomposes under fault conditions such as discharge to produce saturated fluorocarbons such as CF_4, C_2F_6, C_3F_8, and hydrocarbons such as $C_2H_3F_3$ and C_3HF_3. In order to ensure the safe operation of the equipment, it is necessary to monitor its decomposition in real time. Compared with GCMS, which has huge equipment and complicated operation, resistive sensor has more advantages. In this paper, the adsorption characteristics of reduced graphene oxide on CF_4, the typical decomposition products of HFO-1336mzz€, were calculated based on the First-principle, and its potential as an on-line monitoring of the decomposition status of HFO-1336mzz€ was demonstrated by analyzing its adsorption energy, charge transfer amount, and density-of-states function.

56.1 Introduction

SF_6, which is widely used as an insulating gas in the electric power industry, has been listed as one of the six restricted gases in the Kyoto Protocol in 1997 due to its strong greenhouse effect (Global Warming Potentials (GWP) is 23,500 times higher than that of carbon dioxide, and the atmospheric lifespan is up to 3200 years) [1–5]. Moreover, considering China's "dual-carbon" policy [6, 7], it is significant

N. Tang (✉) · D. Sun · L. Li
Electric Power Research Institute of Guangdong Power Grid Co., Ltd., Guangdong 510080, China
e-mail: tangnian_2005@163.com

D. Wang · Y. Luo · R. Zhuo
CSG Electric Power Research Institute Co., Ltd., Guangdong 610663, China
e-mail: wangdb@csg.cn

Y. Luo
e-mail: luoyanl@csg.cn

R. Zhuo
e-mail: zhuoran@csg.cn

Z. Fang et al. (eds.), *Proceedings of the 5th International Symposium on Plasma and Energy Conversion*, Springer Proceedings in Physics 398,
https://doi.org/10.1007/978-981-97-2245-7_56

to find alternative gases to SF_6, in which HFO-1336mzz€ has excellent insulation performance and environmental protection characteristics, its relative SF_6 insulation strength can reach 1.3–1.8 times of SF_6, GWP value of 18, only about seven ten thousandths of SF_6; in addition, HFO-1336mzz€ has good thermal stability and flame retardant characteristics, for equipment failure.

However, in the actual operation of the equipment, due to the presence of overheating and discharge faults, the gas will decompose, and the products mainly include two categories, one is saturated fluorocarbon compounds: CF_4, C_2F_6 and C_3F_8; and one is hydrocarbon-fluorocarbon compounds: CH_3F_3, C_3HF and C_3HF_3 [8]. In order to guarantee the normal operation of gas-insulated equipment, it is necessary to develop suitable HFO-1336mzz€ decomposition condition monitoring equipment. Compared with the use of GCMS and other equipment for detection, resistive sensors have gained widespread attention due to their advantages of fast response speed and small footprint.

Gas-sensitive materials are the core component of resistive gas sensors and the most important factor in determining the performance of gas sensors. Therefore, the selection of suitable gas-sensitive materials is the key to the development of high-performance resistive HFO-1336mzz€ sensors. Among them, graphene's large specific surface area and high electron mobility give it the potential to be used as a gas-sensitive material, but the lack of surface functionality of pristine graphene restricts its applicability, and functionalization of graphene, such as bandgap modulation and functional group modification, can further enhance its sensitivity properties while retaining its original properties [9]. Oxidation of graphene using strong oxidizing agents can yield graphene oxide (GO), which results in similarity to intrinsic graphene, with the difference of having oxygen-containing groups such as epoxy, carboxyl, hydroxyl, and carbonyl groups on the edges and basal surfaces, but its conductivity is lower than that of intrinsic graphene due to the defects introduced in the crystal structure by the use of strong oxidizing agents in the synthesis process, this limitation can be eliminated by reducing graphene oxide to eliminate its surface of organic functional groups to obtain reduced graphene oxide (rGO) with low defects and higher crystallinity to eliminate it [10]. Reduced graphene oxide contains more defects and dangling bonds while having higher electrical conductivity, which enhances the gas-sensing properties and is more suitable for gas-sensing applications [11].

In this paper, the adsorption properties of rGO on CF_4, typical decomposition products of HFO-1336mzz€, were calculated based on the First-principle. By analysing the adsorption energies, charge transfer quantities, and density of states functions, we demonstrated that the rGO gas-sensitive material has the potential to monitor the decomposition status of HFO-1336mzz€ online.

56.2 Calculation Parameters

In this paper, calculations were performed using the $Dmol^3$ module of the software Materials Studio. The van der Waals forces between the surface of the adsorbed material and the gas molecules are corrected using the DFT-D method, and the exchange correlation effect between electrons is handled using the generalized gradient approximation (GGA) method and the Perdew-Burke-Ernzehof (PBE) generalized function. The energy convergence accuracy is 1.0×10^{-5} Ha, and the maximum force and displacement are set to 0.002 Ha/Å and 0.005 Å, respectively. The convergence accuracy of the self-consistent field charge density of 1.0×10^{-6} Ha is used in conjunction with a global orbital cutoff radius of 4.5 Å and a drag-tail value of 0.005 Ha to increase the convergence speed while ensuring accuracy.

The established structural model of graphene monolayer is shown in Fig. 56.1a, while b shows the structural model of reduced graphene oxide containing epoxy groups, and the interaction energy E_{ad} during adsorption is calculated by the following equation:

$$E_{\mathrm{ad}} = E_{m-\mathrm{gas}} - E_m - E_{gas} \tag{56.1}$$

where $E_{m\text{-gas}}$ is the total energy of the system after adsorption of the gas, while E_m and E_{gas} are the energy of the two substances alone before adsorption, respectively. The larger the negative value of the adsorption energy, the stronger the interaction effect.

The electron transfer Qt between the gas molecule and the crystal surface during the adsorption process was calculated by the Hirshfeld method, which was expressed by the charge difference of the gas molecule; if the calculation result was negative, it indicated that the electrons were transferred from the surface of the adsorbed material to the gas molecule during the whole adsorption process, and vice versa indicated that the gas molecule lost the electrons.

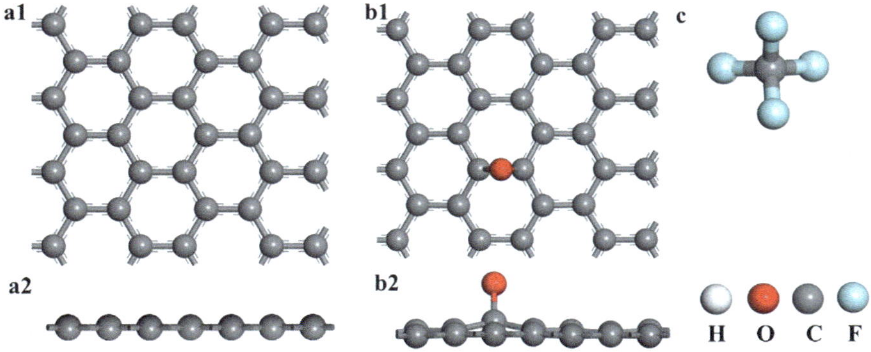

Fig. 56.1 Top view (a1), side view (a2) of monolayer graphene structure, Top view (b1), side view (b2) of rGO, (c) CF_4

Table 56.1 Adsorption energy of different adsorption models

Number	Material	Orientation	Adsorption site	E_{ad}(eV)	Distance(Å)
1	Graphene	–F	–C	– 0.121	2.971
2			Center	– 0.130	3.204
3			–C=C	– 0.122	3.030
4		–CF3	–C	– 0.229	3.286
5			Center	– 0.201	3.248
6			–C=C	– 0.219	3.329
7	rGO	–F	–C	– 0.138	3.012
8			Center	– 0.219	3.075
9			–C=C	– 0.220	2.966
10		–CF3	–C	– 0.215	3.035
11			Center	– 0.153	2.924
12			–C=C	– 0.146	2.956

56.3 Results and Discussion

56.3.1 Optimal Adsorption Structure

For both the intrinsic monolayer graphene model and the rGO adsorption structure containing oxygen-containing group, three adsorption sites were calculated, which were the top of the C atom (–C), the top of the C–C double bond (–C = C), and the center of the C-circle, and for each of the adsorption sites the gas molecules were brought close to each other at two different angles, for a total of twelve adsorption models, controlling that the initial distances of the gas molecules from the surface of the materials were all 2.5 Å. The calculated results are shown in Table 56.1. For the intrinsic graphene, when the gas molecules approach the top of C atoms with $-CF_3$ group, the adsorption energy is the largest, – 0.229 eV, and the distance between the two is 3.286 Å after adsorption, while for the rGO model containing epoxy groups, when the gas molecules approach the C=C group with –F atoms, the adsorption energy is the largest, – 0.220 eV, and the distance between the two is 2.966 Å after adsorption. The two models with the largest adsorption energy are the best adsorption models. models with the maximum adsorption energy are the optimal adsorption structures, as shown in Fig. 56.2.

56.3.2 Adsorption Performance Analysis

The calculated charge transfers of the two best adsorption models during the adsorption process were 0.0148e and 0.0168e, respectively, indicating that the gas molecules

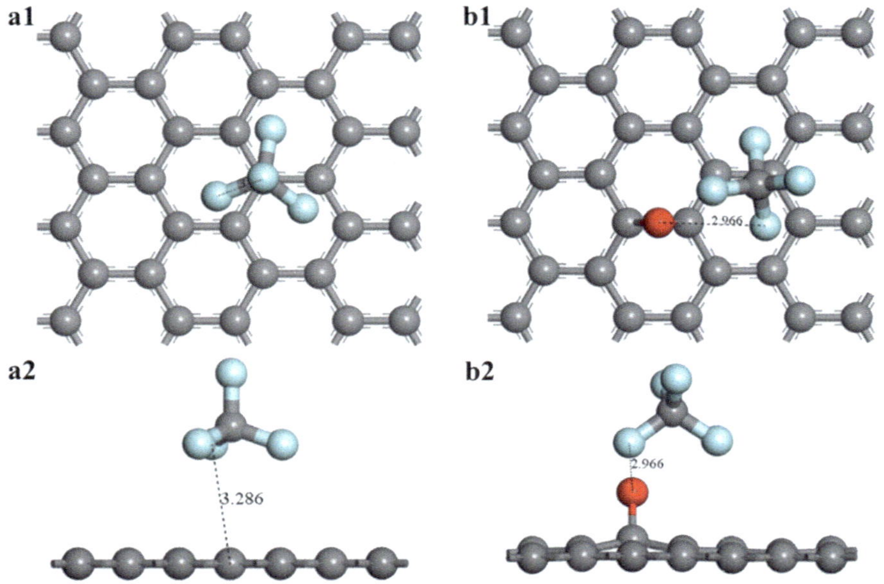

Fig. 56.2 (1) Top view, (2) side view of optimal modeling of CF_4 adsorption on **a** graphene, **b** rGO

lost electrons. To further behavior the adsorption process, the total and partial density of states plots were drawn as shown in Fig. 56.3, and the density of states function as a whole moved slightly to the direction of greater energy after adsorption of the gas molecules, which verified the electron-loss behaviour of the gas molecules during the adsorption process. And it can be seen from the PDOS diagram that the 2p orbitals of the F atoms in the CF_4 molecule overlap with the 2p orbitals of the C atoms and the 2p orbitals of the F atoms of the adsorbent material to a certain extent, which also proves the inter-atomic hybridization.

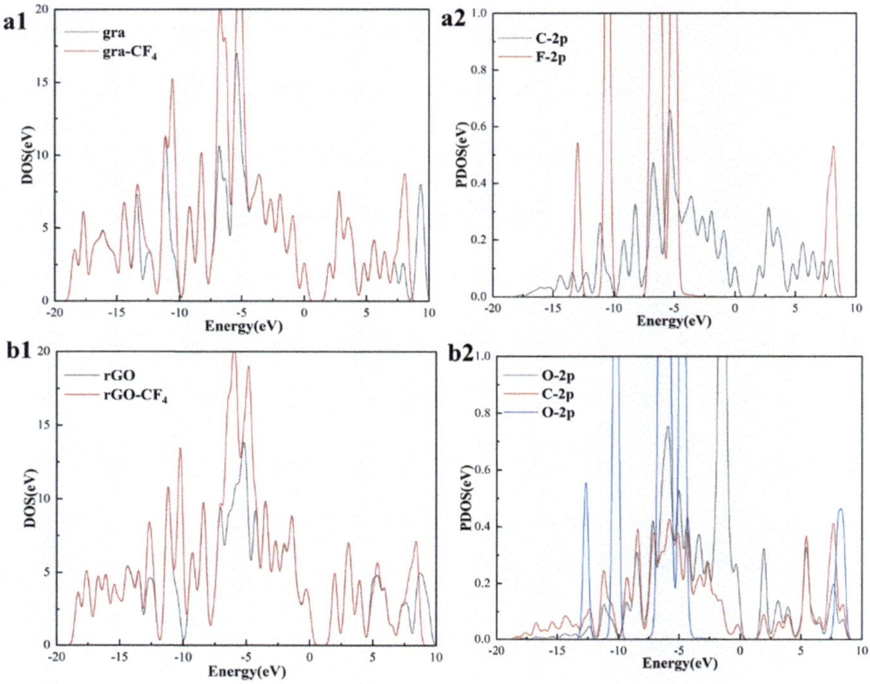

Fig. 56.3 (1) Total density of states (DOS), (2) Partial density of states (PDOS) of CF_4 molecule absorbed on **a** graphene, **b** rGO

56.4 Conclusion

In this paper, the adsorption properties of intrinsic graphene and rGO containing epoxy groups on CF_4, a typical decomposition component of the environmentally friendly insulating gas HFO-1336mzz(E), were calculated based on First-principle. The optimal adsorption structures were found by establishing 12 adsorption models to compare their adsorption energy sizes, and the adsorption energies of intrinsic graphene and rGO on CF_4 were − 0.229 and − 0.220 eV, which belong to physical adsorption. The gas molecules all lost electrons during the adsorption process, 0.0148e and 0.0168e, respectively, and the inter-atomic hybridization also indicates the occurrence of interactions. The synthesis demonstrates the potential of graphene-based gas-sensitive materials for online monitoring of the decomposition status of HFO-1336mzz(E).

Acknowledgements Supported by Southern Power Grid Corporation Technology Project (Project number: GDKJXM20220361)

References

1. G. Xiao, N. He, SF_6 leakage simulation and design optimization of ventilation system in UHV shield tunnel. Guangdong Electric Power **34**(3), 122–129 (2021)
2. P. Wu, Y. Li, S. Xiao*, J. Chen, J. Tang, D. Chen, X. Zhang, SnO_2 nanoparticles based highly sensitive gas sensor for detection of C_4F_7N: A new eco-friendly gas insulating medium. J. Hazardous Mater. **422**(15), 126882 (2022)
3. Y. Li, X. Zhang*, S. Xiao, Q. Chen, J. Tang, D. Chen, D. Wang, Decomposition properties of C_4F_7N/N_2 gas mixture: An environmentally friendly gas to replace SF_6. Indus. Eng. Chemistry Res. **57**(14), 5173–5182 (2018)
4. J. Wang, X. Zhou, Y. Peng et al., Simulation study on interruption process of short-circuit current in SF_6 circuit breaker chamber. Guangdong Electric Power **35**(11), 72–80 (2022)
5. Y. Li, X. Zhang*, J. Zhang, S. Xiao, B. Xie, D. Chen, Y. Gao, J. Tang, Assessment on the toxicity and application risk of C_4F_7N: A new SF_6 alternative gas. J. Hazardous Mater. **368**(15), 653–660 (2019)
6. S. Zhang, Research on deep peak load regulation of coal fired units under the double carbon goals. Guangdong Electric Power **35**(7), 56–63 (2022)
7. M. Zheng, M. Song, J. Wang, Objective and technical analysis of coal fired unit transformation under dual carbon goals. Guangdong Electric Power **35**(7), 14–22 (2022)
8. Z. Gao, Y. Luo*, R. Peng*, X. Wang, P. Yu, W. Zhou, Investigation on insulation properties of HFO-1336mzz(E) and N_2/CO_2 mixtures as SF6 substitutes in gas-insulated electrical applications. High Voltage **8**(1), 48–58 (2023)
9. P. Duan, Q. Duan, Q. Peng, K. Jin*, J. Sun*, Design of ultrasensitive gas sensor based on self-assembled Pd-SnO_2/rGO porous ternary nanocomposites for ppb-level hydrogen. Sens. Actuators: B. Chem. **369**, 132280 (2022)
10. M. Modak, S. Jagtap*, Low temperature operated highly sensitive, selective and stable NO_2 gas sensors using N-doped SnO_2-rGO nanohybrids. Ceramics Int. **48**, 19978–19989 (2022)
11. A. Ahmed, A. Singh, S.-J. Young, V. Gupta, M. Singh, S. Arya*, Synthesis techniques and advances in sensing applications of reduced graphene oxide (rGO) Composites: A review. Composites Part A **165**, 107373 (2023)

Chapter 57
Study on the Metal Tip Partial Discharge in GIS with $C_4F_7N/CO_2/O_2$ Mixed Gas

Peng Xu, Guliang Zhou, Song Zhang, Zhengyong Hu, Zhuoxiao Li, and Yiming Zang

Abstract Gas-Insulated Systems (GIS) have been extensively employed in the electrical industry due to their high reliability and low maintenance costs. However, traditional GIS predominantly uses SF6 as its insulating medium, which has drawn widespread attention due to its greenhouse gas effect on the environment. Consequently, the eco-friendly insulating gas, C_4F_7N, is progressively being regarded as a viable alternative. Nevertheless, research concerning the Partial Discharge (PD) characteristics of a mixture of C_4F_7N with CO_2/O_2 remains somewhat limited. In this context, this study establishes an experimental platform based on a complete GIS setup to investigate the time-domain and spectral features of ultra-high-frequency (UHF) electromagnetic and acoustic signals generated by tip defects within the $C_4F_7N/CO_2/O_2$ mixed gas GIS. Further, the distribution characteristics of PD signal features are explored. Experimental results reveal that compared to SF6-filled GIS, the PD signals from $C_4F_7N/CO_2/O_2$ mixed gas GIS display noticeable differences, particularly with respect to peak locations in the spectrum. From the distribution of feature parameters, a discernible distinction emerges between the PD signals in SF6 gas and those in $C_4F_7N/CO_2/O_2$ mixed gas, thereby establishing an effective means to distinguish discharge signals between the two gases. These findings offer invaluable guidance for optimizing the PD signal pattern recognition algorithms specific to $C_4F_7N/CO_2/O_2$ mixed gas GIS, enhancing the accuracy and monitoring efficiency of pattern recognition.

P. Xu · G. Zhou · S. Zhang · Z. Hu
State Grid Shanghai Municipal Electric Power Company, Shanghai 200122, China

Z. Li (✉) · Y. Zang
Shanghai Jiao Tong University, Shanghai 200240, China
e-mail: zhuoxiaoli@sjtu.edu.cn

Z. Fang et al. (eds.), *Proceedings of the 5th International Symposium on Plasma and Energy Conversion*, Springer Proceedings in Physics 398,
https://doi.org/10.1007/978-981-97-2245-7_57

57.1 Introduction

Gas Insulated Switchgear (GIS) has been extensively utilized in the power industry due to its exceptional reliability, compact design, and low maintenance requirements [1–3]. Historically, sulfur hexafluoride (SF6) has been the primary insulating medium for GIS. However, SF6, being a potent greenhouse gas, poses significant detrimental impacts on global warming and ozone layer depletion. This has urgently underscored the necessity for environmentally benign alternative insulating mediums. In recent times, perfluorobutylnitrile (C_4F_7N) has emerged as a potential alternative to SF6. Notably, the mixture of C_4F_7N with CO_2 and O_2 has demonstrated promising performance potential in electrical insulation and arc quenching. While multinational corporations such as 3M, GE, and ABB have initiated research in this arena, comprehensive studies on the partial discharge characteristics within $C_4F_7N/CO_2/O_2$ mixtures remain in their infancy.

Partial discharge (PD), as an indicator of potential insulation defects within GIS, is pivotal for ensuring its healthy and secure operation [4–6]. Researchers often employ acoustic and ultra-high frequency electromagnetic signals for PD detection and localization. While there exists a thorough understanding of PD signal characteristics in SF6, in different insulating gas environments, the acoustic and electromagnetic signals of PD may manifest varied characteristics. Thus, a profound investigation into the PD signal traits in $C_4F_7N/CO_2/O_2$ mixtures is imperative.

Current research on $C_4F_7N/CO_2/O_2$ mixtures predominantly focuses on the physicochemical and electrical insulating properties, compatibility of the mixture with common materials in electrical equipment, and the characteristics of PD within the mixture. In terms of physicochemical and electrical insulating properties, a study by 3M [7] reported that $C_4F_7N/CO_2/O_2$ mixtures not only possess commendable electrical insulating attributes but also exhibit significantly reduced biotoxicity and greenhouse effects compared to SF6, presenting a trustworthy environmentally-friendly alternative for electrical applications. Scholars from Wuhan University [8] examined the insulating properties of C_4F_7N/CO_2 mixtures at low temperatures and observed a marked decline in breakdown voltage post liquefaction of the mixture. Regarding material compatibility, current findings [9, 10] indicate that most typical materials employed in electrical devices exhibit favorable compatibility with $C_4F_7N/CO_2/O_2$ mixtures, though exceptions like polyester spacers and balsa wood nuts have been noted.

Furthermore, an in-depth understanding of partial discharge characteristics within the $C_4F_7N/CO_2/O_2$ gas mixture is crucial for ensuring the safety and reliability of electrical equipment. Research from Wuhan University [11] discovered that adding an appropriate amount of O_2 to the $C_4F_7N/CO_2/O_2$ mixture significantly suppresses partial discharges and gas decomposition. However, when the oxygen content surpasses a certain threshold, both partial discharge and gas decomposition intensify. A study by Shanghai Jiao Tong University [12] highlighted that, compared to traditional SF6 gas, the surface discharge optical phenomena in the $C_4F_7N/CO_2/$

O_2 mixture are more pronounced, offering a novel perspective for optical detection of partial discharges. Another study from the same university [13] introduced a new multispectral microarray detection technique capable of high-sensitivity detection of partial discharges in the gas mixture. Scholars from Xi'an University of Technology and Xi'an Jiaotong University [14] delved into the corona discharge mechanism in the $C_4F_7N/CO_2/O_2$ mixture and provided comprehensive analysis. Researchers from North China Electric Power University [15] investigated the influence of O_2 volume fraction, gas pressure, and non-uniform electric field coefficient on the partial discharge inception voltage in the mixture. Finally, research conducted by Hefei University of Technology [16] centered on the power frequency breakdown characteristics of the $C_4F_7N/CO_2/O_2$ mixture, yielding results for typical operating pressures, O_2 content, and electric field configurations.

Despite extensive research on the application of the $C_4F_7N/CO_2/O_2$ mixture in the electric power sector, current research on the acoustic and electromagnetic signal characteristics of partial discharges within this gas mixture remains limited and lacks a comprehensive understanding. Therefore, we have established an experimental platform based on a complete GIS system and conducted simulated partial discharge experiments on this platform. Further, by analyzing the obtained experimental data, we have unveiled the ultra-high-frequency electromagnetic wave and acoustic characteristics induced by partial discharges due to metal tip defects inside the $C_4F_7N/CO_2/O_2$ mixture GIS. This research provides a vital theoretical foundation for the precise diagnosis of partial discharges in the $C_4F_7N/CO_2/O_2$ mixture.

57.2 Waveform Analysis Method of Partial Discharge Signal

Due to the inherent complexity and randomness of partial discharge signals, it is challenging to accurately identify and differentiate the differences between partial discharge signals in two types of gases solely by observing their time or frequency domain waveforms. To quantitatively describe the characteristics of partial discharge signals, this study has selected six commonly used feature parameter extraction methods in partial discharge signal research. In the time domain, we employed Skewness and Kurtosis as feature parameters. In the frequency domain, we adopted Gravity Frequency, Average Frequency, Frequency Standard Deviation, and RMS Frequency as feature parameters. The following sections will detail the extraction methods and significance of the aforementioned feature parameters.

Skewness is a statistical metric describing data distribution shape, specifically quantifying data distribution's asymmetry, elucidating the direction of skewness or tilt of the distribution relative to its mean. The formula for Skewness is defined as follows:

$$Skewness(X) = \frac{1}{(n-1)}\sum_{i=1}^{n}\left(\frac{x_i - \overline{x}}{s}\right)^3 \tag{57.1}$$

Herein, n represents the length of the sequence, x_i denotes the i-th item in the sequence, $\overline{x}$ stands for the mean of the sequence, and s is the standard deviation of the sequence. The definitions of these symbols are consistently applied in the subsequent formulas.

Kurtosis similarly describes data distribution shape, representing the sharpness of the peak in the probability density distribution curve. Intuitively, Kurtosis reflects the sharpness of the peak, hence highlighting the impulsiveness of partial discharge signals. The formula for Kurtosis is presented as follows:

$$Kurtosis(X) = \frac{1}{(n-1)}\sum_{i=1}^{n}\left(\frac{x_i - \overline{x}}{s}\right)^4 \tag{57.2}$$

Gravity Frequency offers insight into the weighted center of the signal frequency distribution. In essence, Gravity Frequency signifies the weighted average between the frequency components and their respective amplitudes, illustrating which frequency components predominantly exist in the signal. The formula for Gravity Frequency is defined as follows:

$$F_g = \frac{\sum_{i=1}^{N} f_i \times A_i}{\sum_{i=1}^{N} A_i} \tag{57.3}$$

In this context, f_i indicates the i-th frequency component, A_i corresponds to the amplitude of the i-th frequency component, and N is the total number of frequency components. The definitions of these symbols are also consistently applied in the formulas that follow.

Average Frequency, as an essential parameter in signal frequency characteristics, specifically gauges the average power distribution of various frequency components in the signal. This metric elucidates the average contribution of all frequency components, granting researchers a holistic perspective of the signal's frequency components. For instance, if the signal's average frequency is relatively low, it implies that the majority of its energy is concentrated in low-frequency components. The formula for Average Frequency is defined as follows:

$$F_{Avg} = \frac{\sum_{i=1}^{N} A_i}{N} \tag{57.4}$$

Frequency Standard Deviation is a statistical parameter measuring the dispersion of data points in frequency distribution. It offers researchers a quantitative method to discern the spectrum's shape or its energy distribution. A larger Frequency Standard Deviation indicates a more dispersed signal frequency distribution. The formula for Frequency Standard Deviation is defined as follows:

$$\sigma_f = \sqrt{\frac{\sum_{i=1}^{N}\left(f_i - F_g\right)^2 A_i}{\sum_{i=1}^{N} A_i}} \tag{57.5}$$

RMS Frequency serves as a measure of the uniformity of the signal's energy distribution in the frequency domain. An increase in RMS frequency suggests a more uniform distribution of the signal's energy across the entire frequency range. Conversely, when energy concentrates on a specific frequency, RMS frequency diminishes. Thus, RMS frequency can reflect the signal's distribution uniformity in the frequency domain or the complexity of the signal. The formula for RMS Frequency is defined as follows:

$$F_{RMS} = \sqrt{\frac{\sum_{i=1}^{N} f_i^2 A_i}{\sum_{i=1}^{N} A_i}} \tag{57.6}$$

57.3 Experimental Platform

In light of the nascent and fragmentary research surrounding the local discharge in $C_4F_7N/CO_2/O_2$ mixed gas, this study set forth the establishment of a GIS local discharge experimental platform. This platform is primarily constituted of a non-partial discharge test power source and a GIS model, harmonizing with real-world production environments, which ensures the authenticity and validity of experimental data.

57.3.1 Non-Partial Discharge Test Power Source

The test power system is composed of an isolation transformer (T1), a step-up transformer (T2), and an operational test bench. Specifically, the pivotal role of T1 is to achieve electrical isolation between the test circuit and input power, thereby mitigating harmonics and interference introduced from the power source and enhancing the signal-to-noise ratio of the test circuit. Subsequently, T2, as a non-partial discharge power frequency high-voltage test transformer, is utilized to elevate the voltage. To safeguard the transformer against potential damage in cases of specimen breakdown or short circuits, a $1M\Omega$ protective resistor is connected in series with its high-voltage output. Additionally, a measurement output port is provisioned at the base of the transformer, facilitating the assessment of power frequency phase and output voltage. To curtail discharge levels, both the connection interfaces and the interiors of the power equipment undergo anti-corona treatment, ensuring the overall discharge remains below 2pC. The operational test bench, situated atop an insulating rubber floor, modulates the output voltage by adjusting the tap changer ratio on the transformer, underscoring the imperative of human safety during operations.

57.3.2 GIS Model

The GIS model serves as the apparatus for the local discharge test, wherein various insulating defects are incorporated for research purposes. To bolster the credibility and pertinence of experimental data, tests were conducted on an authentic 110 kV three-phase GIS model provided by a domestic manufacturer, as depicted in Fig. 57.1. This GIS model encompasses an aluminum structural shell, high-voltage internal conductors, basin-type insulators, high-voltage bushings, SF6 gas filling and discharging valves, and gas pressure gauges. Spanning approximately 4 m, the test GIS model consists of three gas chambers, each equipped with windows for UHF sensor installation. We employ embedded UHF sensors alongside external acoustic sensors to capture UHF and acoustic signals generated by local discharges. Signals received by these sensors are amplified by the accompanying conditioning unit and recorded using a digital oscilloscope. A dedicated window in the GIS model facilitates the placement of various simulated insulating defect models, and in this study, a metal-tip discharge defect model was employed. As illustrated in Fig. 57.2, the metal-tip discharge defect model comprises a metal tip welded onto an electrode plate, with a tip diameter of approximately 0.5 mm and a length of about 3 cm.

Fig. 57.1 GIS model

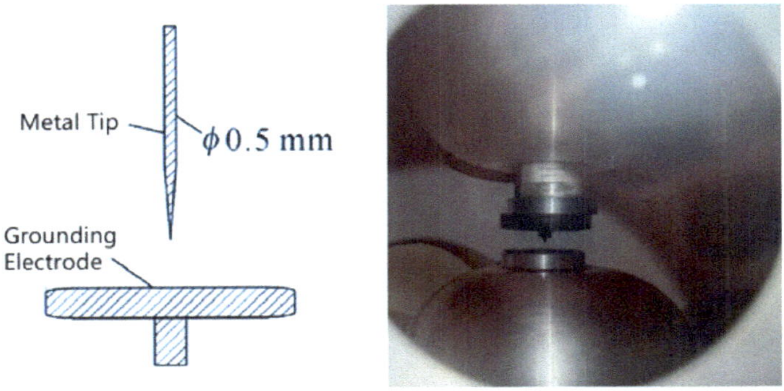

Fig. 57.2 Metal-tip discharge defect model

To discern the differential attributes of acoustic and electromagnetic waves produced by local discharges in diverse insulating mediums, tests were undertaken with both sulfur hexafluoride (SF6) gas and $C_4F_7N/CO_2/O_2$ mixed gas filled in the GIS. The relative pressure of the injected SF6 gas was 0.50 Mpa, while the relative pressure of the $C_4F_7N/CO_2/O_2$ mixed gas was also 0.50 Mpa. with C_4F_7N constituting 8.5%, O_2 at 5.5%, and CO_2 at 86%.

57.4 Analysis of Partial Discharge Signal Characteristics in $C_4F_7N/CO_2/O_2$ Mixed Gas

57.4.1 Time Domain Analysis

Figure 57.3 displays the time domain waveforms of the UHF signals produced by partial discharges in two types of gases. It is evident from this waveform that when a partial discharge occurs in the $C_4F_7N/CO_2/O_2$ mixed gas, the UHF signal duration reaches approximately 1.25 μs. Moreover, as time progresses, this signal demonstrates multiple distinct peaks during its attenuation. In contrast, the UHF signal induced by a partial discharge in sulfur hexafluoride gas has a shorter duration, roughly 0.8 μs, and exhibits a stable attenuation characteristic without prominent peaks.

The time domain waveforms of the acoustic signals produced by partial discharges in both gases are depicted in Fig. 57.4. Observations reveal that the acoustic signal from the $C_4F_7N/CO_2/O_2$ mixed gas exhibits a slow attenuation over time, lasting about 14 ms. On the other hand, the acoustic signal in sulfur hexafluoride gas persists for only about 10 ms and attenuates at a faster rate.

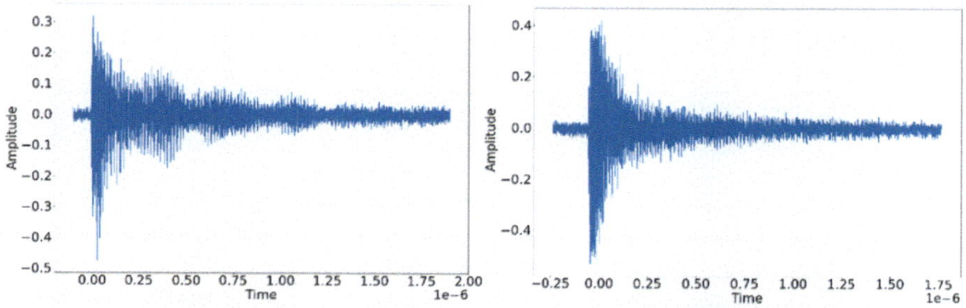

(a) C4F7N/CO2/O2 mixed gas (b) Hexafluoride gas

Fig. 57.3 UHF signal time domain waveform

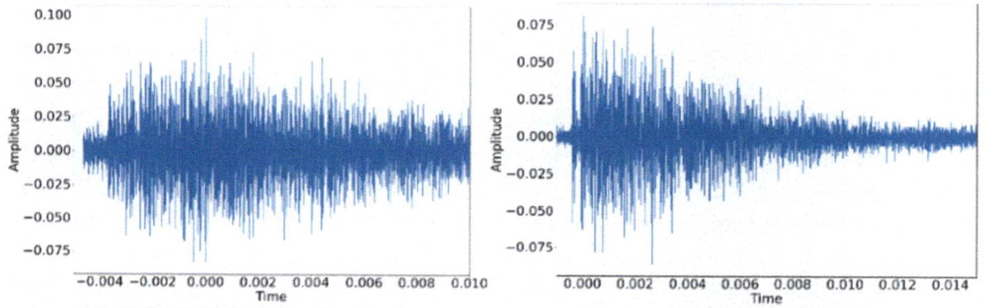

(a) C4F7N/CO2/O2 mixed gas (b) Hexafluoride gas

Fig. 57.4 Acoustic signal time domain waveform

57.4.2 Frequency Domain Analysis

Figure 57.5 compares the spectral content of the UHF signals produced by partial discharges in the two gases. From the figure, it can be discerned that the spectrum below 1.2 GHz shows similarities in both gases, with pronounced characteristic peaks at 0.35, 0.58, and 0.75 GHz. However, significant discrepancies appear in the spectrum above 1.2 GHz, particularly at frequencies like 1.25 and 1.35 GHz, where peaks present in the $C_4F_7N/CO_2/O_2$ mixed gas spectrum are absent in the sulfur hexafluoride spectrum.

Comparing the acoustic signal spectra from the two gases, the characteristic peaks of the $C_4F_7N/CO_2/O_2$ mixed gas spectrum are predominantly found between 10 and 100 kHz, while those in the sulfur hexafluoride spectrum are concentrated between 10 and 80 kHz. Additionally, the most pronounced peaks for sulfur hexafluoride lean more towards the lower frequency range compared to the $C_4F_7N/CO_2/O_2$ mixture (Fig. 57.6).

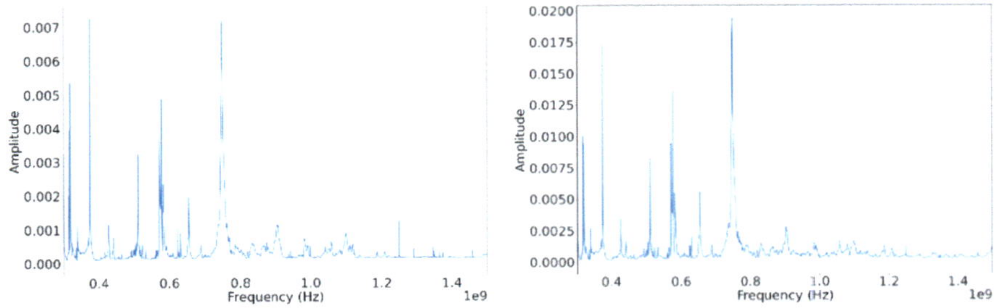

(a) C4F7N/CO2/O2 mixed gas **(b)** Hexafluoride gas

Fig. 57.5 UHF signal frequency domain waveform

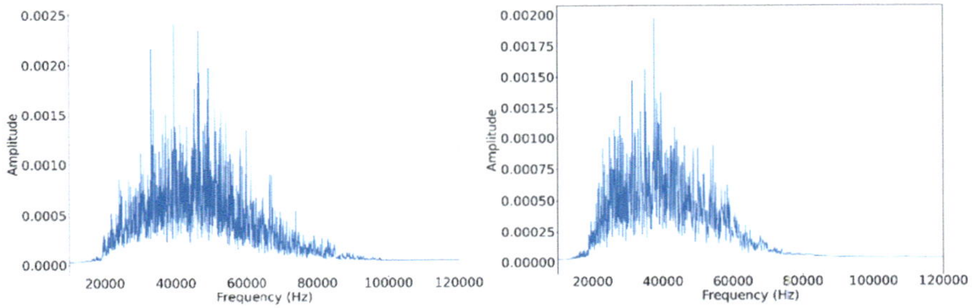

(a) C4F7N/CO2/O2 mixed gas **(b)** Hexafluoride gas

Fig. 57.6 Acoustic signal frequency domain waveform

57.4.3 Statistical Feature Analysis

Multiple sets of UHF signals collected from partial discharge tests in both gases were analyzed using the aforementioned waveform analysis method, extracting six feature parameters: skewness, kurtosis, centroid frequency, average frequency, frequency standard deviation, and root mean square frequency. Distribution charts such as Skewness-Kurtosis, Gravity Frequency-Average Frequency, and Frequency Standard Deviation-RMS Frequency were plotted. From Fig. 57.7, it's evident that all three charts differentiate well between the UHF signals from $C_4F_7N/CO_2/O_2$ mixed gas and sulfur hexafluoride. From a feature perspective, Skewness, Average Frequency, and Frequency Standard Deviation show good differentiation for UHF signals from different gases, while Kurtosis, Gravity Frequency, and RMS Frequency offer weaker distinction. Additionally, UHF signals from sulfur hexafluoride display more consistency across all distribution charts, indicating lesser variation between samples and more stable feature parameters. In contrast, UHF signals from $C_4F_7N/CO_2/O_2$ mixed gas display higher dispersion, especially evident in the Kurtosis and Gravity Frequency, where outliers emerge, suggesting some instability in these signals.

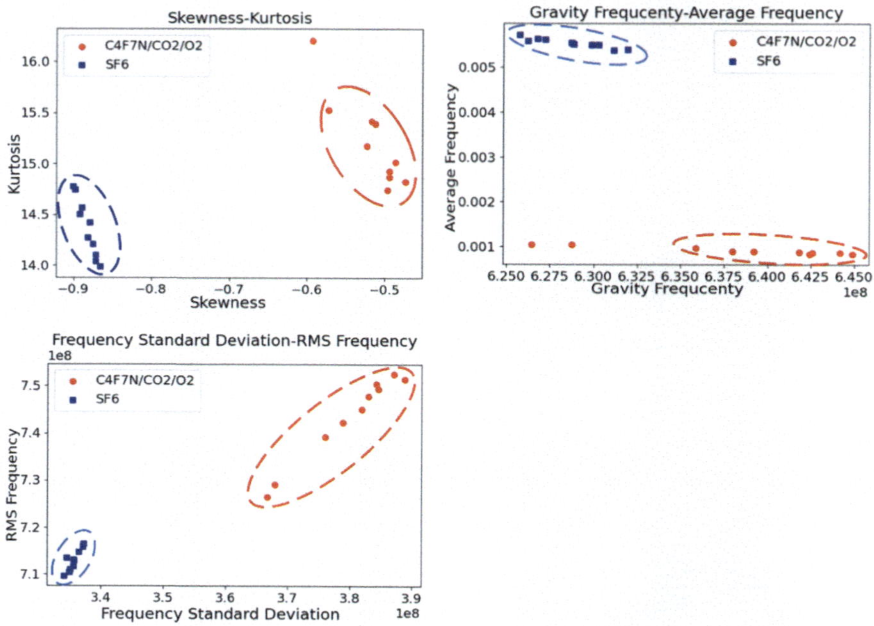

Fig. 57.7 UHF signal feature distribution

Further analysis using the waveform methodology was applied to the acoustic signals produced by partial discharges in both gases, extracting feature parameters and plotting respective distribution charts. As can be seen from Fig. 57.8, the three charts effectively differentiate the acoustic signals from the two gases. Among the various feature parameters, Kurtosis, Gravity Frequency, and RMS Frequency distinguish well between acoustic signals from different gases, while Skewness, Average Frequency, and Frequency Standard Deviation show weaker differentiation. Notably, paralleling the UHF signal findings, the $C_4F_7N/CO_2/O_2$ mixed gas acoustic signals also display outliers in Skewness, Average Frequency, and Frequency Standard Deviation, indicating inherent instabilities.

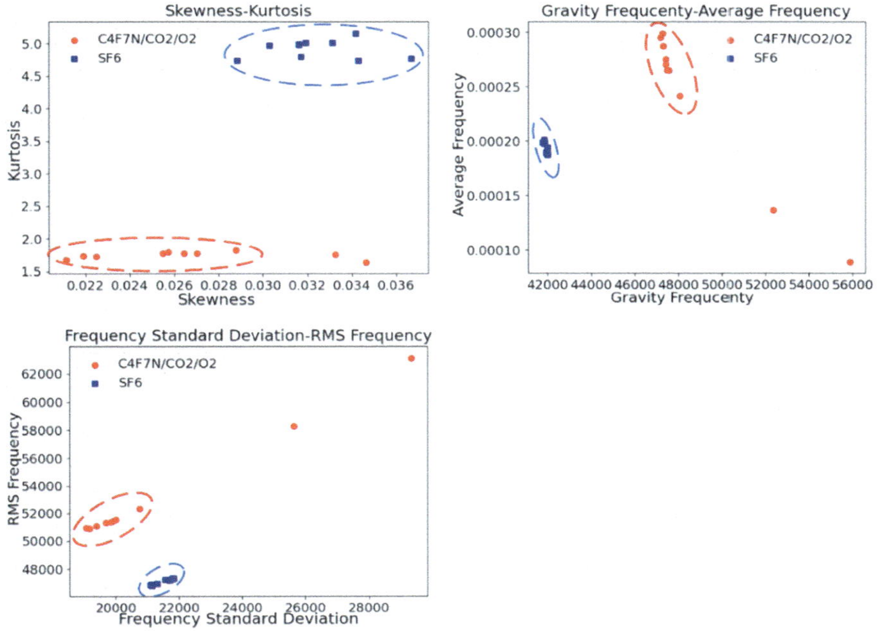

Fig. 57.8 Acoustic signal feature distribution

57.5 Conclusion

(1) In the time-domain waveform, the UHF signals generated by partial discharge in both gases exhibit differences in duration, attenuation speed, and mode. However, these variations are not pronounced. Similarly, acoustic signals generated by partial discharges also display analogous disparities. Notably, due to the randomness of partial discharge signals, these variations in the time domain might not consistently manifest.

(2) In the frequency-domain waveform, the main characteristic peaks in the UHF signal spectrum of partial discharges in both gases are roughly identical, but discrepancies exist in some minor peaks in the higher frequency region. For acoustic signals arising from partial discharges, the frequency bands where the characteristic peaks are concentrated show significant differences, with the spectrum in sulfur hexafluoride gas leaning more towards the low-frequency zone.

(3) From the perspective of the characteristic parameter distribution graph, discharge signals in both gases can be distinctly differentiated, confirming the effectiveness of characteristic parameters in distinguishing discharge signals of these two gases. Moreover, we observed that the UHF and acoustic signals produced by partial discharges in $C_4F_7N/CO_2/O_2$ mixed gas exhibit outliers

in certain characteristic parameters, suggesting potential instability in partial discharges within the $C_4F_7N/CO_2/O_2$ mixed gas.

Acknowledgements Project funded by the State Grid Corporation of China Technology Project (No. 520940220055).

References

1. H.T. Wu, X. Cui, X.F. Liu et al., Characteristics of electromagnetic disturbance for intelligent component due to switching operations via a 1100 kV AC GIS test circuit. IEEE Trans. Power Delivery **32**(5), 2228–2237 (2016)
2. B. Qi, C. Li, Z. Xing et al., Partial discharge initiated by free moving metallic particles on GIS insulator surface: severity diagnosis and assessment. IEEE Trans. Dielectr. Electr. Insul. **21**(2), 766–774 (2014)
3. H. Song, J. Dai, G. Sheng et al., GIS partial discharge pattern recognition via deep convolutional neural network under complex data source. IEEE Trans. Dielectr. Electr. Insul. **25**(2), 678–685 (2018)
4. B.F. Hampton, R.J. Meats, Diagnostic measurements at UHF in gas insulated substations, in *IEE Proceedings C (Generation, Transmission and Distribution)*, vol. 135, no. 2 (IET Digital Library), pp. 137–144
5. J.S. Pearson, O. Farish, B.F. Hampton et al., Partial discharge diagnostics for gas insulated substations. IEEE Trans. Dielectr. Electr. Insul. **2**(5), 893–905 (1995)
6. U. Schichler, W. Koltunowicz, F. Endo et al., Risk assessment on defects in GIS based on PD diagnostics. IEEE Trans. Dielectr. Electr. Insul. **20**(6), 2165–2172 (2013)
7. J. Owens, A. Xiao, J. Bonk et al., Recent development of two alternative gases to SF6 for high voltage electrical power applications. Energies **14**(16), 5051 (2021)
8. Z. Li, W. Ding, K. Gao et al., Insulation characteristics along the surface of epoxy insulator under lightning impulse in C_4F_7N/CO_2 mixed gas. High Voltage Eng. **45**(4), 1071–1077 (2019)
9. L. Li, Q. Yao, P. Wu et al., Study on the compatibility of typical materials in GIT with C_4F_7N/CO_2 gas mixture, in *2022 IEEE 6th Conference on Energy Internet and Energy System Integration (EI2)* (IEEE, 2022), pp. 2337–2340
10. W. Gao, Y. Cao, Y. Wang et al., Materials compatibility study of C_4F_7N/CO_2 gas mixture for medium-voltage switchgear. IEEE Trans. Dielectr. Electr. Insul. **29**(1), 270–278 (2022)
11. F. Ye, X. Zhang, Y. Li et al., Effect of O_2 on AC partial discharge and decomposition behavior of $C_4F_7N/CO_2/O_2$ gas mixture. IEEE Trans. Dielectr. Electr. Insul. **28**(4), 1440–1448 (2021)
12. X. Yan, X. Zhou, Z. Li et al., Surface discharge characteristics and numerical simulation in C_4F_7N/CO_2 mixture. Appl. Sci. **13**(3), 1409 (2023)
13. Y. Zang, Y. Qian, X. Zhou et al., Optical characteristics and fault diagnosis of partial discharge in C_4F_7N/CO_2 gas mixture and SF6 based on novel multispectral microarray detection. IEEE Trans. Dielectr. Electr. Insul.l. **29**(3), 1079–1086 (2022)
14. Q. Gao, X. Wang, K. Adamiak et al. Negative corona discharge mechanism in $C_4F_7N–CO_2$ and $C_4F_7N–N_2$ mixtures. AIP Adv. **12**(9) (2022)
15. D. Tong, Q. Zhao, R. Cao et al., Experimental study on the initiation voltage of partial discharge in $C_4F_7N/CO_2/O_2$ mixed gas. High Voltage Eng. **49**(03), 1007–1014 (2023). https://doi.org/10.13336/j.1003-6520.hve.20220778
16. Y. Yang, K. Gao, S. Yuan et al., Study on the power frequency breakdown characteristics of $C_4F_7N/CO_2/O_2$ mixed gas under typical electric fields. Trans. Elect. Eng. **37**(15), 3913–3922 (2022)

Chapter 58
Dynamic Characteristics of C_4F_7N-CO_2 Arc Based on MHD Simulation

Jiahao Guo, Weiping Zhu, Liguo Liu, Xiaolong Xiao, Fan Wu, and Ning Guo

Abstract C_4F_7N stands out as a promising SF_6 alternative gas with practical applications. This study delves into the arc behavior of a C_4F_7N-CO_2 gas mixture. We initiated a two-dimensional arc model for the C_4F_7N-CO_2 gas mixture employing a magnetohydrodynamic (MHD)-based modeling approach. This model couples equations from multiple physical fields and numerically solves them through the finite volume method. Leveraging this model, we conducted a simulation study on the arc characteristics of the C_4F_7N-CO_2 gas mixture, examining how current magnitude and gas mixture ratios affect these properties. To assess its potential as an SF_6 substitute, we also performed a comparative analysis with SF_6-CO_2 gas mixtures. This research offers valuable insights for the engineering design of high-voltage circuit breakers. This work was supported by the State Grid Jiangsu Electric Power Co., Ltd. Technology Project (J2023026).

58.1 Introduction

SF_6 gas has been widely used in various medium and high voltage electrical equipment since its invention in 1900, due to its excellent insulation and arc extinguishing properties. These include gas circuit breakers (GCB), gas-insulated switchgears (GIS), gas-filled switchgear, and gas-insulated transmission lines (GIL). However, SF_6 gas has a very strong greenhouse effect, with a global warming potential (GWP) of approximately 23,900 times that of CO_2, and it can persist in the atmosphere for

This work was supported by the State Grid Jiangsu Electric Power Co., Ltd. Technology Project (J2023026).

J. Guo (✉) · X. Xiao · F. Wu · N. Guo
Electric Power Research Institute of State Grid Jiangsu Electric Power Co., Ltd, Nanjing 211103, China
e-mail: 2241752161@qq.com

W. Zhu · L. Liu
State Grid Jiangsu Electric Power Co., Ltd, Nanjing 210024, China

Z. Fang et al. (eds.), *Proceedings of the 5th International Symposium on Plasma and Energy Conversion*, Springer Proceedings in Physics 398,
https://doi.org/10.1007/978-981-97-2245-7_58

3200 years [1]. In the Kyoto Protocol established in 1997, SF_6 was listed as one of the six greenhouse gases with restricted use [2]. Although regulations require an annual leakage rate of SF_6 electrical equipment not exceeding 1%, the actual commissioning and maintenance processes can result in SF_6 gas entering the atmosphere. Therefore, it is crucial to find environmentally friendly alternatives to SF_6. In recent years, new insulation gases such as C_4F_7N have attracted widespread attention as potential substitutes for SF_6. Compared to SF_6, C_4F_7N has a lower GWP and superior insulation performance and thermal stability, with an insulation strength exceed twice that of SF_6 [3, 4]. Therefore, C_4F_7N is considered a potential substitute for SF_6.

To better understand the characteristics of C_4F_7N and its mixtures, numerous research teams both domestically and internationally have conducted in-depth studies on its breakdown performance. Owens, an American scholar, found that a mixture of C_4F_7N with a volume fraction of 20% had comparable dielectric strength to SF_6 [5]. General Electric conducted arc extinguishing experiments on C_4F_7N-CO_2 mixtures and SF_6 gas, and found that the arc extinguishing time of the mixture with 4% C_4F_7N at 0.55 MPa was slightly shorter than SF_6, indicating that the arc extinguishing performance of C_4F_7N mixtures is comparable to SF_6 [6]. In order to investigate the dielectric breakdown properties of SF_6 and four carbon-fluoride insulating gases mixed with CO_2, N_2, and CF_4, a study was conducted based on numerical solutions of the Boltzmann equation, The results indicate that c-C4F8 demonstrates the highest dielectric strength, while C_2F_6 exhibits the lowest [7]. A study by A. Hopf and others showed that a mixture with a C_4F_7N content of 7.8% can achieve 70% of the DC breakdown voltage of SF_6 at 0.4 MPa pressure [8]. Kieffel et al. also studied C_4F_7N and found that the thermal decomposition products of C_4F_7N-CO_2 gas mixture include C_2F_6, COF_2, C_2F_3N, C_3F_5N, etc., and the insulation strength of the gas mixture containing 10%C_4F_7N can reach 90% of SF_6 at the same pressure [9]. Researchers investigated the composition, thermodynamic properties, and transport coefficients of high-temperature $C_5F_{10}O$ mixed with CO_2 and O_2. Results show that CO_2 significantly affects the thermodynamic properties and transport coefficients of $C_5F_{10}O$ plasma. However, the addition of O_2 to $C_5F_{10}O$-CO_2 mixture does not have a significant impact on thermodynamic and transport properties [10]. Zhong Linlin and colleagues from Southeast University computed C_4F_7N-CO_2 and C_4F_7N-N_2 mixed gas arcs' physical properties, such as the net radiation coefficient, evaluating their arc extinguishing performance using a one-dimensional model [11]. Although C_4F_7N has become a hot topic in research as a substitute for SF_6, current studies have mainly focused on analyzing the insulation properties of C_4F_7N and its mixtures, while research on its electrical behavior is relatively limited.

Therefore, this paper will use a numerical method based on MHD simulation to simulate the behavior of C_4F_7N arcs. By solving the magnetohydrodynamics equations and considering the effects of magnetic field, current density, and plasma temperature on the behavior of C_4F_7N arcs, the simulation aims to gain in-depth understanding of the formation, propagation, and disappearance of arcs, as well as reveal the underlying physical mechanisms. By comparing and analyzing the simulation results under different parameter conditions, the dynamic characteristics of C_4F_7N arcs will be studied. The study on the electrical behavior of C_4F_7N gas in this paper can provide important theoretical basis for the design of power equipment and the stability of power systems. In addition, it can also provide guidance for the application of new insulation gases, promoting the development and popularization of environmentally friendly power equipment.

58.2 A Two-Dimensional Arc Model of Environmentally Friendly SF_6 Substitute Gas Based on Magnetohydrodynamics (MHD)

58.2.1 Control Equations

The governing equations play a fundamental role in the study of gas-electric behavior, including the conservation equations for mass, energy, and momentum, as well as the electromagnetic field equations. The mass conservation equation describes the conservation property of fluid mass, the energy conservation equation describes the conservation property of fluid energy, and the momentum conservation equation describes the conservation property of fluid momentum. The electromagnetic field equations describe the interaction between the electromagnetic field and the fluid. By applying these equations in a comprehensive manner, it is possible to investigate and analyze the behavior of C_4F_7N gas arcs and provide theoretical support for the promotion and application of environmentally friendly power systems.

(1) Mass conservation equation

$$\frac{\partial \rho}{\partial t} + \frac{\partial \rho v_x}{\partial x} + \frac{\partial \rho v_r}{\partial r} + \frac{\rho v_r}{r} = 0 \tag{58.1}$$

where t represents time, r represents radial distance, x represents axial distance, ρ represents the mass density of environmentally friendly SF_6 substitute gas arc plasma, v_x represents the axial velocity, and v_r represents the radial velocity.

(2) Momentum conservation equation

$$\frac{\partial \rho v_x}{\partial t} + \frac{\partial \left(\rho v_x^2 + p\right)}{\partial x} + \frac{\partial \rho v_x v_r}{\partial r} + \frac{\rho v_x v_r}{r} = \frac{\partial \tau_{xx}}{\partial x} + \frac{\partial \tau_{xr}}{\partial r} + \frac{\tau_{xr}}{r} + j_r B_\theta \tag{58.2}$$

$$\frac{\partial \rho v_r}{\partial t} + \frac{\partial \rho v_x v_r}{\partial x} + \frac{\partial \left(\rho v_x^2 + p\right)}{\partial r} + \frac{\rho v_r^2}{r} = \frac{\partial \tau_{rx}}{\partial x} + \frac{\partial \tau_{rxr}}{\partial r} + \frac{\tau_{rr} - \tau_{\theta\theta}}{r} - j_x B_\theta \tag{58.3}$$

where p represents the pressure of the environmentally friendly gas, τ_{ij} represents the viscosity stress tensor of the environmentally friendly SF_6 substitute gas arc plasma, j_x and j_r represent the axial and radial current densities, and B_θ represents the circumferential magnetic induction intensity. Unlike traditional fluids, an arc is subject to the Lorentz force under the influence of electromagnetic fields. Therefore, this paper incorporates the $j_r B_\theta$ and $j_x B_\theta$ source terms into the two-dimensional momentum equation of the arc to represent the effect of the Lorentz force.

(3) Electric field equations

By solving the Maxwell's equations in the arc plasma, the following potential equation can be obtained:

$$\frac{1}{r}\frac{\partial}{\partial r}\left(r\sigma\frac{\partial\varphi}{\partial r}\right)+\frac{\partial}{\partial x}\left(\sigma\frac{\partial\varphi}{\partial x}\right)=0 \tag{58.4}$$

Whrer φ represents the electric potential. The electric field intensity can be obtained from the potential as follows:

$$E_x=-\frac{\partial\varphi}{\partial x} \tag{58.5}$$

$$E_r=-\frac{\partial\varphi}{\partial r} \tag{58.6}$$

$$E=\sqrt{\frac{\partial^2\varphi}{\partial x^2}+\frac{\partial^2\varphi}{\partial r^2}} \tag{58.7}$$

where E_x represents the axial electric field intensity, E_r represents the radial electric field intensity. The current density can be obtained based on the simplified Ohm's law:

$$j_x=\sigma E_x \tag{58.8}$$

$$j_r=\sigma E_r \tag{58.9}$$

(4) Magnetic Field Equation

The magnetic field equations consist of the axial magnetic vector conservation equation and the radial magnetic vector conservation equation:

$$\frac{1}{r}\frac{\partial}{\partial r}\left(r\frac{\partial A_x}{\partial r}\right)+\frac{\partial}{\partial x}\left(\frac{\partial A_x}{\partial x}\right)=-\mu_0 j_x \tag{58.10}$$

$$\frac{1}{r}\frac{\partial}{\partial r}\left(r\frac{\partial A_r}{\partial r}\right)+\frac{\partial}{\partial x}\left(\frac{\partial A_r}{\partial x}\right)=-\mu_0 j_r+\frac{A_r}{r^2} \tag{58.11}$$

Based on the magnetic vectors, the circumferential magnetic induction intensity B_θ can be obtained as follows:

$$B_\theta = \frac{\partial A_r}{\partial x} - \frac{\partial A_x}{\partial r} \tag{58.12}$$

where A_x represents the axial magnetic vector, A_r represents the radial magnetic vector, μ_0 represents the vacuum magnetic permeability.

(5) Turbulence model

The internal velocity and temperature gradients in electric arc plasma are large and exhibit significant turbulent characteristics. Commonly used turbulence models for describing electric arc plasma include the Prandtl mixing length model, the standard *k*-*ε* model, and the *Chen-Kim* *k*-*ε* model [12]. In this study, the widely applied standard *k*-*ε* turbulence model is adopted.

58.2.2 Two-Dimensional Arc Modeling Process for Environmentally Friendly SF_6 Substitute Gas Based on Magnetohydrodynamics Simulation

The modeling process of a two-dimensional gas arc based on MHD simulation consists of five steps as follows.

Step 1: The geometric model is established by referring to the actual morphology of the arc region and making reasonable simplifications. The experimental chamber designed in this study has a height of 220 mm and a radius of 80 mm. The anode (on the left) and cathode (on the right) have lengths of 100 mm and radii of 2.5 mm. Considering that the arc mainly concentrates in the electrode gap region, the geometric model is reasonably simplified in the simulation model to improve computational efficiency. As shown in Fig. 58.1, the dimensions of the geometric model are AB = 45 mm, AH = BC = 20 mm, GF = DE = 2.5 mm, HG = 29.5 mm, and DC = 14.5 mm. The main arc occurrence region is GDEF, i.e., the intermediate region between the two electrodes.

Step 2: Grid partitioning involves dividing the geometric model into smaller grid elements, and the quality of the grid directly impacts the accuracy and computational speed of the model. In this study, a structured grid approach is utilized, where the blocks are mapped onto the geometry to transform the relatively complex geometric structure into a combination of simpler blocks, simplifying the problem. In the simulation, the grid is denser in the central region of the arc, with a grid edge length of 0.1 mm, while the remaining regions have sparser grids with an edge length of 0.25 mm. The localized grid refinement enhances the simulation accuracy in the arc occurrence region while improving the computational efficiency. Additionally, for the application of dynamic grid techniques, user-defined functions (UDFs) are employed to describe the motion of the electrodes, and a layering method is used for grid updates. Following the experimental setup, the anode is designated as the moving electrode, with a velocity of 8 mm/s, a motion duration of 2 s, and a maximum distance of 17 mm between the electrodes.

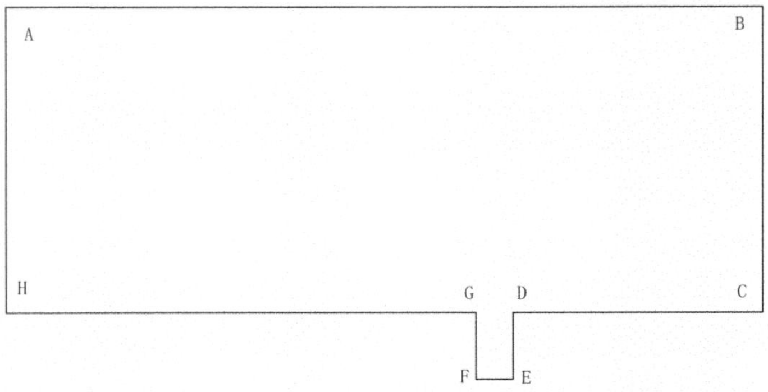

Fig. 58.1 The geometric structures used in the simulation

Step 3: Set simulation parameters, such as defining various physical properties of the environmentally friendly SF_6 replacement gas used for electric arcs. These properties include density, thermal conductivity, viscosity, specific heat, electrical conductivity, etc. Corresponding equations, such as fluid mechanics equations, Lorentz force terms, electromagnetic field equations, and energy equations with Joule heating and radiation terms, are also set up. Assuming a uniform distribution of current density on the anode electrode surface GF, the current density is determined by the arc current per unit area. The corresponding potential boundary condition is set for the anode surface GF. The potential on the cathode surface KJ is set to zero.

Step 4: Initialize and iteratively solve the governing equations. Such as initial temperature and initial pressure. Once the initialization is completed, iterative solution can be performed.

Step 5: After the solution process is completed, post-processing software is used to analyze the computed results and visualize them in a meaningful way.

58.3 Calculation Results and Analysis

This study focuses on the investigation of C_4F_7N pure gas and C_4F_7N-CO_2(1:1) mixed gas as the research objects. A simulation analysis is conducted on the current conditions ranging from 30 to 100 A under four different atmospheric pressure conditions. The aim of this simulation analysis is to explore the behavior characteristics of the arc under different gas compositions and current conditions. Through the simulation analysis, various arc parameters such as arc temperature and arc voltage can be obtained, providing insights into the formation, propagation, and extinction processes of the arc.

58.3.1 *Temperature Characteristics of the Arc*

Temperature is a primary macroscopic characteristic of the radiative transport process in arc plasma. Figure 58.2 illustrates the temporal evolution of the arc temperature distribution for pure C_4F_7N gas under different current conditions (30 A, 50 A, 70 A, 100 A). It can be observed that the arc is extinguished by the movement of the contacts at 30 and 50 A currents, while under other current conditions, the arc remains stable and burns steadily within 2 s. As time progresses, the arc radius increases, and its shape resembles a gourd, with a larger volume closer to the cathode. With increasing current, the arc combustion becomes more intense, resulting in a higher arc temperature and a larger gourd-shaped volume. Figure 58.3 presents the temporal evolution of the arc temperature distribution for the C_4F_7N-CO_2 (1:1) mixed gas under different current conditions. Similar to the behavior of the pure C_4F_7N gas arc, the temperature of the arc plasma increases with higher current, accompanied by a larger arc volume. However, it is worth noting that the C_4F_7N-CO_2 (1:1) gas mixture arc can only be extinguished at a current of 30 A.

The phenomenon described above can be explained as follows: under the same current conditions, pure C_4F_7N gas is more prone to extinguish and undergo contraction compared to C_4F_7N-CO_2 (1:1) gas mixture, and the temperature distribution range of the arc is also smaller. This phenomenon can be attributed to the radiation transfer process in plasma physics and the influence of gas composition. The poor arc

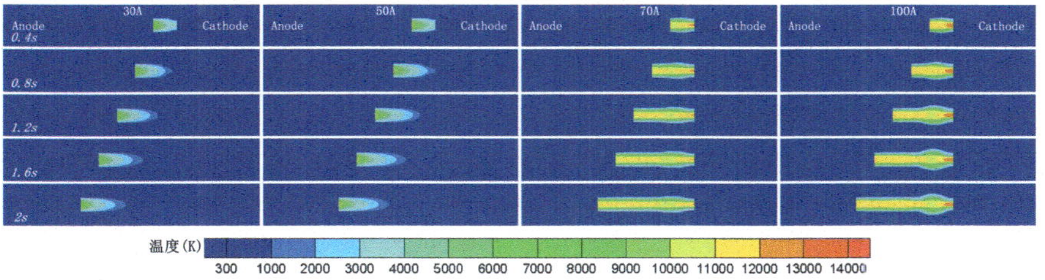

Fig. 58.2 Transient temperature contour of C_4F_7N pure gas two-dimensional electric arc under different current conditions

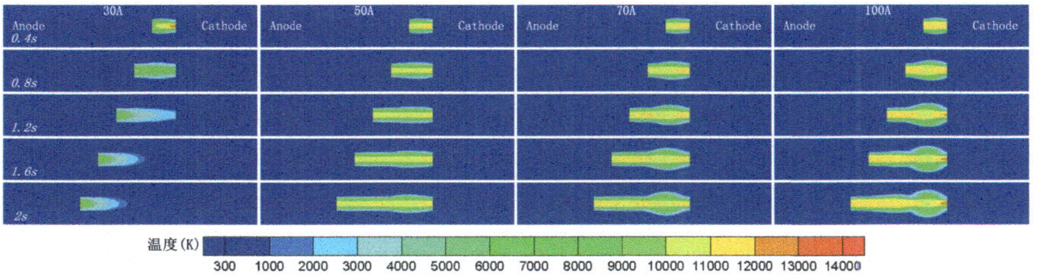

Fig. 58.3 Transient temperature contour of C_4F_7N-CO_2 (1:1) pure gas two-dimensional electric arc under different current conditions

sustainment capability of pure C_4F_7N gas may be due to the chemical composition of the gas, which leads to the instability and extinguishment of the arc. Additionally, pure C_4F_7N gas has a higher molecular collision rate, resulting in a decrease in electron energy, thereby reducing the thermal distribution range of the arc and leading to the occurrence of contraction. In contrast, the C_4F_7N-CO_2 (1:1) gas mixture exhibits better arc sustainment capability. The addition of CO_2 alters the gas transport characteristics, reducing the collision probability and enhancing the stability of the arc. Consequently, the arc remains stable and less prone to extinguishment under the same current conditions, and the temperature distribution range of the arc is wider.

58.3.2 The Voltage-Current Characteristic of the Arc

Considering the voltage distribution of an arc under different currents and different media conditions tends to converge, this study takes the example of a potential cloud map of a C_4F_7N gas arc under 50 A conditions to illustrate. As shown in Fig. 58.4, the potential decreases from the anode region to the cathode region, ultimately approaching zero. The potential difference between the two electrodes is the arc voltage.

Figure 58.5a illustrates the voltage curves of C_4F_7N gas arcs under different current conditions. It can be observed that at a current of 30 A, the voltage of the C_4F_7N arc significantly exceeds the arc voltage under other current conditions, indicating that the arc is in an open-circuit state, or in other words, the arc has been extinguished. This finding is consistent with the conclusion drawn from Fig. 58.2. At lower currents (30 A, 50 A), the arc voltage decreases with increasing current. This can be attributed to the fact that as the current increases, the arc temperature also increases, leading to enhanced conductivity and reduced arc resistance. Conversely, at higher currents (70 A, 100 A), the arc voltage exhibits an increasing trend with increasing current. This can be explained by the increased ionization in the arc due to higher current levels. The presence of ions increases the arc resistance, resulting

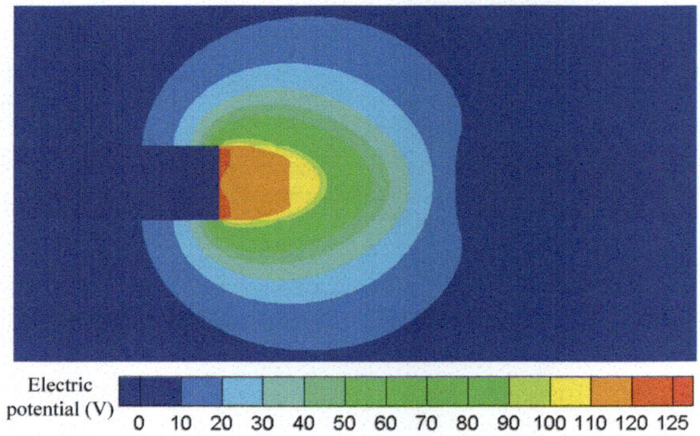

Fig. 58.4 Arc potential distribution cloud at 2 s under 50 A current condition

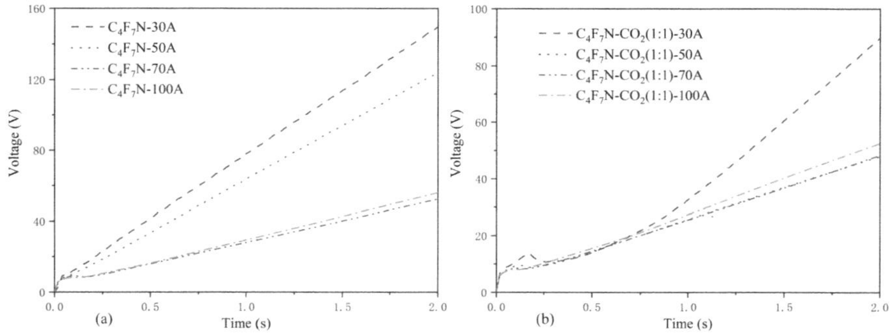

Fig. 58.5 Variation of arc voltage over time for C_4F_7N and C_4F_7N-CO_2 (1:1) under different current conditions

in an increase in arc voltage. In summary, the observed variations in arc voltage with different current conditions can be attributed to the changes in arc temperature and ionization levels. As the current increases, the arc temperature rises, leading to decreased arc resistance and lower arc voltage at lower currents. Conversely, at higher currents, the increased ionization level increases the arc resistance, resulting in a higher arc voltage.

The voltage curve of the C_4F_7N-CO_2 (1:1) gas arc in Fig. 58.5b indicates that at a current of 30 A, the voltage of the C_4F_7N-CO_2 (1:1) arc is significantly higher than the arc voltage under other current conditions, suggesting arc extinction at this point, consistent with the findings in Fig. 58.3. As the current increases (50 A, 70 A, 100 A), although the arc voltage decreases compared to 30 A (indicating sustained arc), there is an overall increasing trend. This is attributed to the increasing concentration of ions within the arc as the current intensifies. The presence of ions results in higher arc resistance, leading to an elevated arc voltage. Translated and refined into academic English: The voltage curve of the C_4F_7N-CO_2 (1:1) gas arc in Fig. 58.5b demonstrates a significant increase in arc voltage at a current of 30 A compared to other current conditions, indicating arc extinguishment. These findings align with the results depicted in Fig. 58.3. As the current rises (50 A, 70 A, 100 A), the arc voltage shows a decreasing trend from the 30 A level, suggesting sustained arc activity. However, the overall trend reveals an increase in arc voltage. This can be attributed to the simultaneous increase in ion concentration within the arc as the current intensifies. The presence of ions elevates the arc's resistance, thus resulting in an elevated arc voltage.

As shown in Fig. 58.6, this study presents the voltage-current characteristics of different gas arcs, specifically C_4F_7N and C_4F_7N-CO_2 (1:1), at a time of 1.5 s. The data reveals that at lower currents (30 A, 50 A, and 70 A), the arc voltage decreases with increasing current. This behavior can be attributed to ionization within the gas, leading to an increase in charged particles and exhibiting a negative voltage-current relationship. However, at higher currents (70 A–100 A), the arc voltage increases with the current. This is likely due to the gas being highly ionized at these levels,

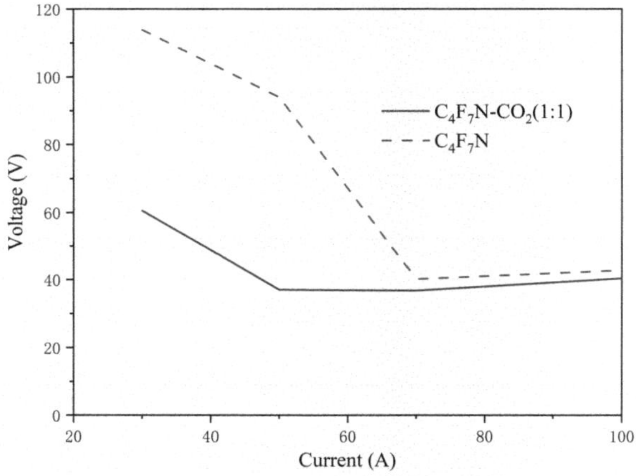

Fig. 58.6 Voltage-current characteristic curve of C_4F_7N and C_4F_7N-CO_2 (1:1) arc at 1.5 s

resulting in minimal changes in the concentration of charged particles and showing a positive voltage-current characteristic.

58.4 Conclusion

Based on the simulation analysis results, we have investigated the behavior characteristics of the arc for pure C_4F_7N gas and C_4F_7N-CO_2 (1:1) gas mixture under different current conditions. Our findings are summarized as follows:

(a) Arc behavior: The arc of pure C_4F_7N gas is more prone to extinguish and undergo contraction compared to the C_4F_7N-CO_2 (1:1) gas mixture. The temperature distribution range of the arc is smaller for pure C_4F_7N gas, indicating a narrower thermal plasma region. In contrast, the C_4F_7N-CO_2 (1:1) gas mixture exhibits better arc sustainment capability and a wider temperature distribution range.

(b) Voltage-current characteristic: The arc voltage of pure C_4F_7N gas is higher than that of the C_4F_7N-CO_2 (1:1) gas mixture under the same current conditions. At lower currents (30 A, 50 A), the arc voltage decreases with increasing current due to enhanced conductivity and reduced arc resistance. However, at higher currents (70 A, 100 A), the arc voltage increases due to increased ionization and higher arc resistance. Similarly, the arc voltage of the C_4F_7N-CO_2 (1:1) gas mixture decreases compared to the 30 A level but shows an overall increasing trend as the current intensifies. In conclusion, both pure C_4F_7N and C_4F_7N-CO_2 (1:1) exhibit different behavior characteristics under different current conditions. However, the C_4F_7N-CO_2 (1:1) gas mixture demonstrates better arc sustainment capability and a wider temperature distribution range compared to pure C_4F_7N gas. Therefore, the C_4F_7N-CO_2 (1:1) gas mixture may be a more suitable substitute for SF_6 gas in electrical equipment applications. These findings contribute to the understanding and development of environmentally friendly alternatives for SF_6 gas.

References

1. Y. Li, S. Tian, L. Zhong et al., Eco-friendly gas insulating medium for next-generation SF_6-free equipment. Energy **2**(1), 14–42 (2023)
2. X. Lin, Y. Xia, G. Sun et al., Study on breakdown characteristic of hot SF_6. High Voltage Apparatus **54**(07), 111–118 (2018)
3. W. Meng, Y. Chang, F. Yin et al., Research of C_4F_7N/CO_2, $C_5F_{10}O/C_6F_{12}O$/air environmental insulation gases performance and its application. High Voltage Apparatus **56**(7), 15–22 (2020)
4. Y. Kieffel, T. Irwin, P. Ponchon et al., Green gas to replace SF_6 in electrical grids. IEEE Power Energ. Mag.Energ. Mag. **14**(02), 32–39 (2016)
5. J.G. Owens, Greenhouse gas emission reductions through use of a sustainable alternative to SF_6, in *IEEE Electrical Insulation Conference (EIC) 2016* (Montreal, QC, Canada), pp. 535–538
6. Y. Kieffel, F. Biquez, SF_6 alternative development for high voltage switchgears. IEEE Electrical Insulation Conference (EIC) 379–383 (2015)
7. L. Zhong, J. Wang, X. Wang et al., Comparison of dielectric breakdown properties for different carbon-fluoride insulating gases as SF_6 alternatives. AIP Adv. **8**(8) (2018)
8. A. Hopf, J.A. Britton, M. Rossner et al., Dielectric strength of SF_6 substitutes, alternative insulation gases and PFC-gas-mixtures, in *IEEE Electrical Insulation Conference (EIC)*, Baltimore, MD, USA (2017), pp. 209–212
9. Y. Kieffel, Characteristics of g3-an alternative to SF_6, in *IEEE International Conference on Dielectrics (ICD)* (2016), pp. 880–884
10. L. Zhong, M. Rong, X. Wang et al., Compositions, thermodynamic properties, and transport coefficients of high-temperature $C_5F_{10}O$ mixed with CO_2 and O_2 as substitutes for SF_6 to reduce global warming potential. AIP Adv. **7**(7) (2017)
11. L. Zhong, J. Wang, J. Xu et al., Effects of buffer gases on plasma properties and arc decaying characteristics of C_4F_7N-N_2 and C_4F_7N-CO_2 arc plasmas. Plasma Chem. Plasma Process. **39**(06), 1379–1396 (2019)
12. S. Guo, G. Wu, G. Chen, Comparison of three high-reynolds-number models in recirculation zones for simulation of computational fluid dynamics. J. Refrigeration **26**(3), 26–29 (2005)

Chapter 59
Comparative Study on Post-arc Dielectric Breakdown Characteristics of C_4F_7N and $C_5F_{10}O$ Gases Mixed with CO_2

Lin Lin, Jiahao Guo, Jianan Wang, Weiwei Su, Wangjie Jiang, and Xiaolong Xiao

Abstract C_4F_7N and $C_5F_{10}O$ are considered promising SF_6 substitute gases with potential applications. However, current research has predominantly focused on analyzing their insulation performance, with less emphasis on their arc extinguishing capabilities. Therefore, this study aims to investigate the post-arc dielectric breakdown properties of C_4F_7N-CO_2 and $C_5F_{10}O$-CO_2 mixed gases and perform a comparative analysis. To begin, we employed the Gibbs free energy minimization method to calculate the particle compositions of heated C_4F_7N-CO_2 and $C_5F_{10}O$-CO_2 gases. Subsequently, we used the two-term approximation Boltzmann equations to calculate key parameters such as the electron energy distribution function, reduced ionization coefficient, reduced attachment coefficient, and reduced critical breakdown field strength of post-arcs. Furthermore, we conducted an analysis to examine the influence of temperature, pressure, and CO_2 content on the post-arc hot gas breakdown characteristics. Our study revealed that the inclusion of CO_2 substantially decreases the concentration of high-energy electrons, and generally, higher pressure levels correspond to a reduction in the probability density of high-energy electrons. This work was supported by the State Grid Jiangsu Electric Power Co., Ltd. Technology Project (J2023026).

59.1 Introduction

SF_6 is a stable, non-toxic, odorless, and noncombustible gas with excellent insulation and arc extinguishing properties. It has been extensively applied in power equipment. Nevertheless, SF_6 possesses a Global Warming Potential (GWP) approximately 23,900 times that of CO_2, and can persist in the atmosphere for up to 3200 years

L. Lin · J. Guo (✉) · J. Wang · W. Su · W. Jiang · X. Xiao
Electric Power Research Institute of State Grid Jiangsu Electric Power Co., Ltd., Nanjing 211103, China
e-mail: 2241752161@qq.com

Z. Fang et al. (eds.), *Proceedings of the 5th International Symposium on Plasma and Energy Conversion*, Springer Proceedings in Physics 398,
https://doi.org/10.1007/978-981-97-2245-7_59

[1]. As a result, It is listed as a restricted greenhouse gas in the Kyoto Protocol [2]. Furthermore, SF_6 can decompose during discharge and produce toxic byproducts, which are harmful to human health. Therefore, it is crucial to find environmentally friendly SF_6 replacement gases for sustainable power industry development. In recent years, Fluorinated ketone gases are regarded as potential substitutes for SF_6 due to their exceptional insulating properties and environmental performance. Among them C_4F_7N has an insulating strength more than twice that of SF_6 and has a GWP value of only 2090 [3], making it a promising replacement option. Meanwhile, $C_5F_{10}O$ possesses an exceptionally high dielectric strength and an extremely low GWP value, which makes it a highly promising substitute for SF_6.

Extensive investigations have been conducted on the breakdown characteristics of C_4F_7N, $C_5F_{10}O$. In 2016, Owens et al. reported the industrial frequency breakdown characteristics of C_4F_7N gas mixtures [4]. Their study revealed that the breakdown voltages of C_4F_7N-CO_2, C_4F_7N-N_2, and C_4F_7N-air gas mixtures increased gradually with increasing air pressure and C_4F_7N content. Some Researchers investigated the breakdown characteristics and potential application of C_4F_7N and $C_5F_{10}O$-CO_2 mixtures [5]. Their study demonstrated that C_4F_7N-CO_2 mixtures meet current requirements for electric power equipment applications and offer superior insulation properties with lower global warming potential. Frequency breakdown tests were conducted by researchers at Hubei University of Technology on two gas mixtures of $C_5F_{10}O$ at various air pressures under a quasi-uniform electric field. The experimental results revealed that the insulation strength of the $C_5F_{10}O$ gas mixture increases with higher gas pressure [6]. Although the substitution of SF_6 with C_4F_7N and $C_5F_{10}O$ as gas replacements has gained attention in recent research, the existing studies predominantly focus on analyzing the insulation properties. In contrast, there is still few research on their post-arc insulation breakdown characteristics.

Therefore, this work utilizes the Gibbs free energy minimization method to calculate gas compositions of C_4F_7N-CO_2 and $C_5F_{10}O$-CO_2 gas mixtures at different pressures and ratios within the temperature range of 300–3500 K. The electron energy distributions are obtained by solving the Boltzmann equation using collision cross-section data and gas compositions. Ionization and attachment coefficients are determined for various temperatures, pressures, and gas ratios. Approximate critical breakdown field strengths $(E/N)_{cr}$ at different temperatures are calculated. These parameters are used to evaluate the insulation breakdown characteristics of C_4F_7N-CO_2 and $C_5F_{10}O$-CO_2 gas mixtures behind the arc under varying pressures and ratios. The study specifically investigates and compares the post-arc insulation breakdown characteristics of C_4F_7N-CO_2 and $C_5F_{10}O$-CO_2 gas mixtures. The results could provide a theoretical reference for the practical application of these gas mixtures in engineering.

59.2 Calculation of Boltzmann Equation

59.2.1 Principles of Calculation

In order to calculate the reduced ionization coefficient, reduced attachment coefficient, and other parameters for C_4F_7N-CO_2, $C_5F_{10}O$-CO_2 thermal mixtures based on component calculations, it is necessary to solve the electron energy distribution function (EEDF) for the mixed gas at different pressures and ratios. The EEDF can be used to determine the reduced ionization coefficient α/N and reduced attachment coefficient η/N of the mixed gas at different reduced electric field strengths. Subsequently, the reduced effective ionization coefficient $(\alpha\text{-}\eta)/N$ can be obtained. The EEDF is obtained by solving the following form of the Boltzmann equation[7]:

$$\frac{\partial f}{\partial t} + v \cdot \nabla f - \frac{e}{m} E \cdot \nabla_v f = C[f] \tag{59.1}$$

where f represents the EEDF, v represents the electron velocity, e is the charge of the electron, m is the mass of the electron, E represents the strength of the electric field, ∇_v is the velocity gradient operator, and C is the rate of change of the EEDF.

Through the derivation of the formula, the equation between the reduced ionization coefficient and the reduced adsorption coefficient and the ionization reaction rate G_{ion} and adsorption reaction rate G_{att} can be obtained as follows:

$$\frac{\alpha}{N} = \frac{G_{ion}}{\mu E} \tag{59.2}$$

$$\frac{\eta}{N} = \frac{G_{att}}{\mu E} \tag{59.3}$$

where μ is the electron mobility; N is the number density of neutral particles.

59.2.2 Collision Cross-Section

The collision cross-section parameters are crucial in solving the Boltzmann equation for the arcs. This paper assumes the plasma is in thermodynamic equilibrium, where collisions between electrons and heavy particles have a significant impact on the generation and loss of electrons in the medium breakdown process. In the process of calculation, the ionization and adsorption of 17 kinds of particles such as C, C_2, C_2F_5N, C_3, C_3F_3N, C_3F_4N, C_3F_5N, C_4F_3N, CF, CF_2, CF_3, CF_4, CO, CO_2, COF_2, F and N_2 are mainly considered, while the photoionization and collision between heavy particles are ignored. Most of the cross-section data are obtained from the

LXCat database, while the ionization cross-section data for C_2F_5N, C_3F_3N, C_3F_4N, C_3F_5N and C_4F_3N are obtained from theoretical calculations [8].

59.2.3 Compositional Calculations Based on Gibbs Free Energy Minimization

In this paper, the Gibbs free energy minimization method is chosen to determine the composition of the post-arc gas. The Gibbs free energy minimization method is a computational method to determine the equilibrium composition of a system by minimizing the Gibbs free energy. The objective is to find the minimum value of Gibbs free energy by taking the derivative of the Gibbs free energy with respect to system composition and setting it equal to zero. As the Gibbs free energy depends on the system composition, numerical optimization algorithms like the Newton–Raphson method can be used to iteratively calculate the Gibbs free energy and adjust the plasma composition until the minimum value is reached. The Gibbs free energy G of a plasma system composed of N particles can be expressed as [9]:

$$G = \sum_{i=1}^{N} n_i \mu_i \tag{59.4}$$

where n_i is the number density of particles, μ_i is the chemical potential of particles.

59.2.4 Calculation Procedure of Post-Arc Insulation Breakdown Characteristics

In order to analyze the post-arc insulation breakdown characteristics of the replacement gases, the following steps are adopted using a Boltzmann equation-based computational approach: Firstly, two-term approximation is made to simplify the Boltzmann equation. The plasma composition parameters and particle collision cross-section parameters were used as inputs to solve the Boltzmann equation. The EEDF of the post-arc gas was obtained. The reduced ionization coefficient and reduced electron attachment coefficient of the post-arc gas were calculated based on the EEDF. Finally, the reduced critical breakdown field strength $(E/N)_{cr}$ was determined by finding the electric field intensity at which the α/N equals the η/N.

59.3 Calculation Results and Comparative Analysis

Firstly, the Gibbs free energy method determines C_4F_7N-CO_2 and $C_5F_{10}O$-CO_2 mixture composition at various ratios, pressures, and temperatures. Then the dominant particles are identified, which are applied to the analysis of Boltzmann equation.

Taking C_4F_7N and $C_5F_{10}O$ gas mixtures as examples, Figs. 59.1 and 59.2 show the composition calculation results at temperature 300–3500 K and pressure 0.1 MPa, respectively. Based on the composition calculations, it is observed that after the temperature drops in the post-arc stage, C_4F_7N gas and C_4F_7N-CO_2(1:1) gas mixture cannot recombine into their initial compositions. Instead, they decompose into C_2F_5N, CF_4,CO_2 and solid carbon particles. Similarly, $C_5F_{10}O$ and $C_5F_{10}O$-CO_2 (1:1) gas mixture undergoes decomposition into CF_4, CO_2, and solid carbon rather than reforming into its original composition.

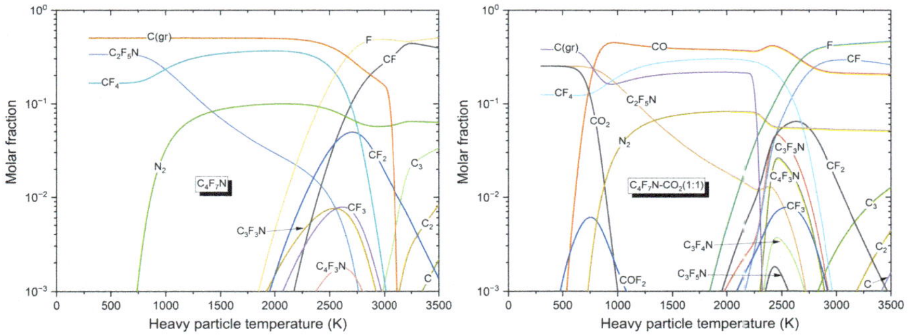

Fig. 59.1 Particle composition of pure C_4F_7N (left) and C_4F_7N-CO_2 (1:1) (right) gases

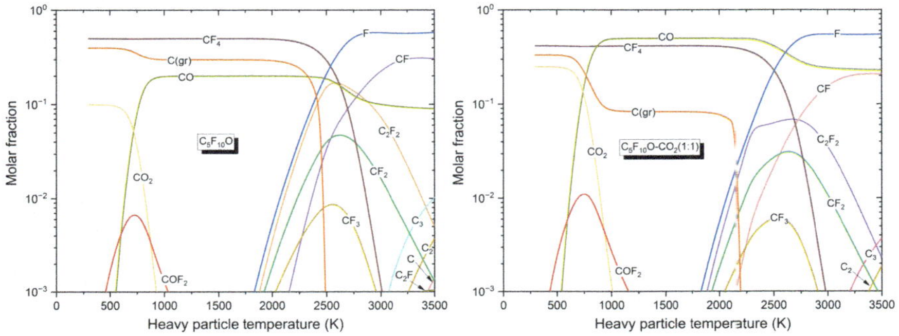

Fig. 59.2 Particle composition of pure $C_5F_{10}O$ (left) and $C_5F_{10}O$-CO_2 (1:1) (right) gases

59.3.1 Electron Energy Distribution Function (EEDF)

Figure 59.3 illustrates the EEDF of pure C_4F_7N and $C_5F_{10}O$ gases at various reduced electric fields (E/N). The graph indicates that as the reduced electric field increases, the EEDF gradually decreases for electron energies below a critical threshold. However, beyond this threshold, the EEDF starts to increase with higher E/N values. This suggests that at higher electric field strengths, electrons can attain higher kinetic energies. The proportion of high-energy electrons increases in the high-energy region, while the proportion of low-energy electrons decreases due to electron scattering or energy loss caused by the stronger electric field.

Figure 59.4 shows the effect of temperature on the EEDF of pure C_4F_7N and $C_5F_{10}O$ gases. It is observed that at temperatures below 2000 K, the corresponding energy values for a given EEDF are relatively small. However, at temperatures above 2000 K, the energy values associated with the same EEDF are significantly larger. This can be attributed to the temperature-dependent energy exchange processes. At lower temperatures, the rate of energy exchange between electrons and gas species is low, resulting in lower electron energies and a narrower energy range within the EEDF. At higher temperatures, the energy exchange rate increases, allowing electrons to gain more energy through collisions with gas species. This leads to a broader energy range within the EEDF, with higher energy values corresponding to the same EEDF.

Figure 59.5 demonstrates the impact of CO_2 ratio on the EEDF of C_4F_7N-CO_2 and $C_5F_{10}O$-CO_2 gas mixtures. CO_2 ratios below 50% have no significant effect on the EEDF. However, CO_2 ratios above 50% notably increase the concentration of high-energy electrons. This is attributed to CO_2's lower ionization energy and its ability, in large quantities, to absorb energy from high-energy electrons. This energy absorption promotes the excitation of electrons in CO_2 to higher energy levels, resulting in the generation of high-energy electrons.

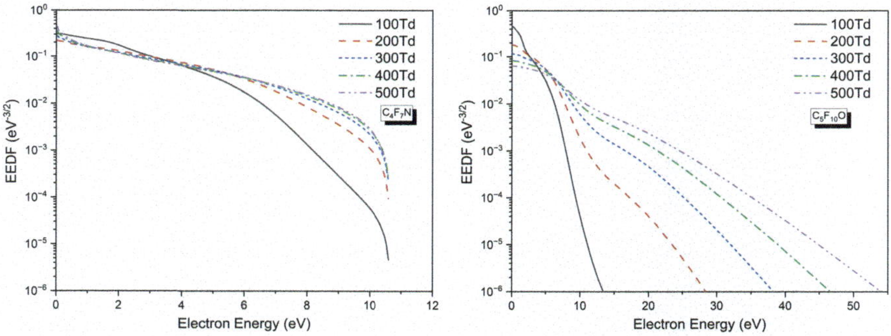

Fig. 59.3 Electron energy distribution functions of pure C_4F_7N and pure $C_5F_{10}O$ gases under different approximate electric fields (temperature 2000 K, pressure 0.1 MPa)

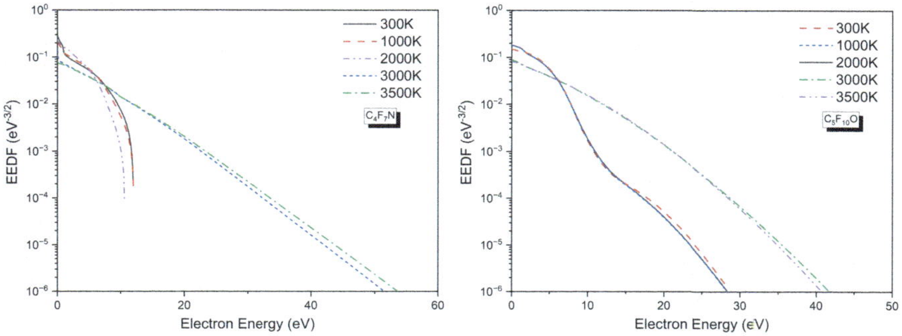

Fig. 59.4 Effect of temperature on the electron energy distribution function of pure C_4F_7N and pure $C_5F_{10}O$ gases (pressure of 0.1 MPa, E/N of 200 Td)

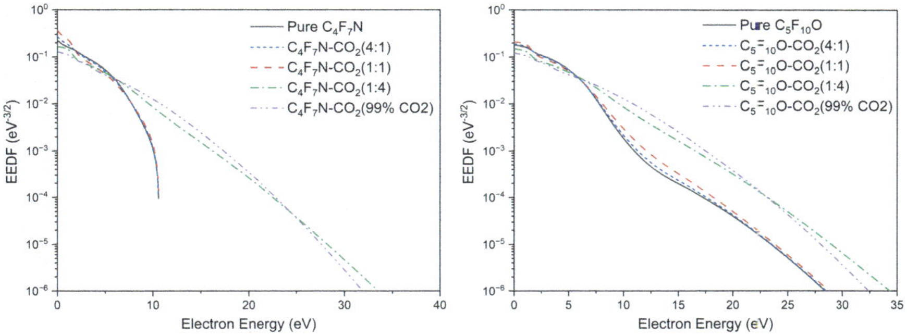

Fig. 59.5 Effect of different CO_2 ratios on the electron energy distribution function of C_4F_7N-CO_2 and $C_5F_{10}O$-CO_2 mixtures (at 2000 K, E/N 200 Td, pressure 0.1 MPa)

59.3.2 Reduced Ionization Coefficient and Reduced Attachment Coefficient

Figure 59.6 illustrates the impact of temperature on the reduced ionization coefficient α/N and reduced attachment coefficient η/N of pure C_4F_7N and $C_5F_{10}O$ gases. At temperatures below 1000 K, there are minimal changes in α/N and η/N for pure C_4F_7N gas at the same *E/N*. However, when the temperature exceeds 1000 K, α/N increases while η/N decreases. Similarly, for pure $C_5F_{10}O$ gas, minimal changes in α/N and η/N are observed at temperatures below 2500 K, but when the temperature exceeds 2500 K, α/N increases and η/N decreases. This is likely due to the increased collision frequency and energy transfer rate between gas molecules at higher temperatures, enhancing the ionization process and leading to an increase in α/N. Additionally, under high temperature and high electric field conditions, as the temperature increases, gas molecules enter excited states with higher electron

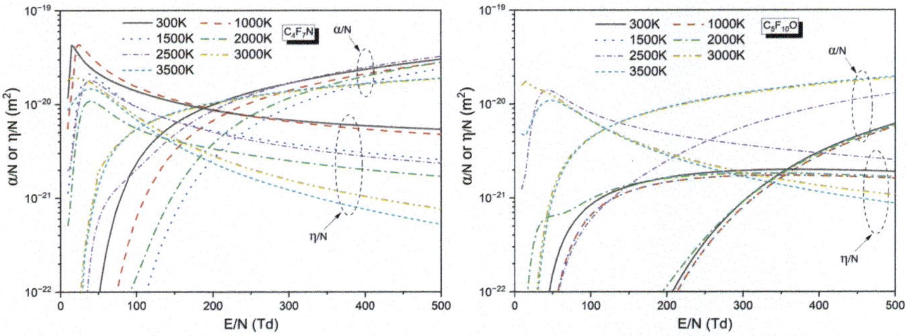

Fig. 59.6 Calculation results of ionization coefficient and adsorption coefficient of pure C_4F_7N and pure $C_5F_{10}O$ gas at different temperature

affinity, making them more likely to capture electrons and form ions, resulting in a decrease in η/N.

Figure 59.7 illustrates the impact of CO_2 content on the α/N and η/N of C_4F_7N-CO_2 and $C_5F_{10}O$-CO_2. The η/N decreases with increasing CO_2 content for both gases, attributed to the higher electron affinity of CO_2 molecules, which facilitates electron attachment reactions in the gas mixture. Conversely, the α/N of C_4F_7N increases with increasing CO_2 content, but when CO_2 content exceeds 50%, α/N is significantly lower compared to when it is below 50%. This is due to the complex interactions between C_4F_7N and CO_2 molecules in the gas mixture. When CO_2 content is below 50%, increased collision frequency between C_4F_7N and CO_2 enhances electron excitation and ionization processes, leading to an increase in α/N. However, when CO_2 content exceeds 50%, increased collision frequency inhibits these processes, resulting in a slower growth rate of α/N. In contrast, the ionization coefficient of $C_5F_{10}O$ increases with increasing CO_2 content due to the increased collision frequency between $C_5F_{10}O$ and CO_2 molecules, promoting electron excitation and ionization processes.

Figure 59.8 shows the impact of pressure on the α/N and η/N values of C_4F_7N-CO_2(1:1) and $C_5F_{10}O$-CO_2(1:1) gas mixtures at 2000 K. At a constant E/N value, the α/N and η/N values of C_4F_7N-CO_2(1:1) increase with higher pressure due to increased gas molecule density. This leads to a higher collision frequency, enhancing energy transfer efficiency, ionization, and excitation processes. Conversely, the α/N and η/N values of $C_5F_{10}O$-CO_2(1:1) decrease with increasing pressure until reaching 0.4 MPa, after which they stabilize. This can be attributed to enhanced molecular interactions as pressure increases. Initially, these interactions cause increased electrostatic repulsion, reducing collision frequency and energy transfer efficiency. However, once pressure reaches a certain level, molecular interactions reach a balanced state, resulting in a stable collision frequency and energy transfer efficiency. Within this pressure range, molecular interactions remain stable, leading to nearly unchanged α/N and η/N values.

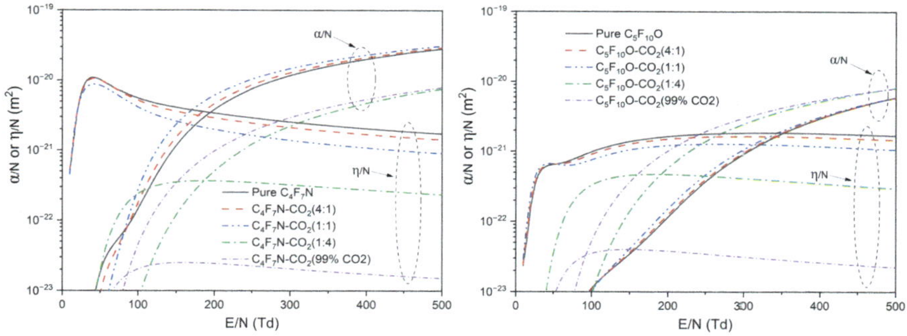

Fig. 59.7 Ionization and adsorption coefficients of C_4F_7N-CO_2 and $C_5F_{10}O$-CO_2 gases with different mixing ratios at 2000 K temperature

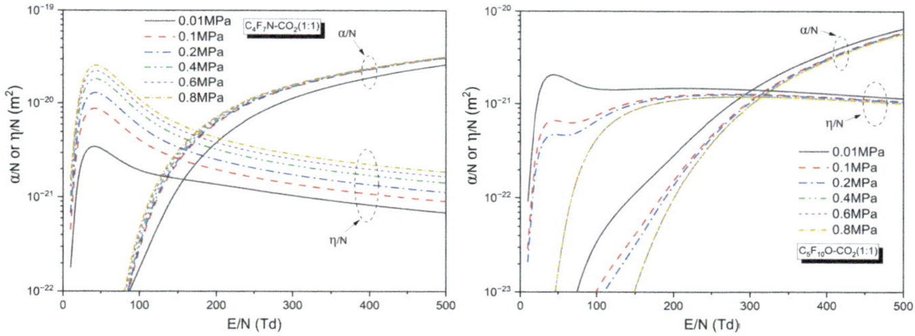

Fig. 59.8 Ionization and adsorption coefficients of C_4F_7N-CO_2 (1:1) and $C_5F_{10}O$-CO_2 (1:1) gas mixtures at different pressures

59.3.3 Critical Breakdown Electric Field (E/N)$_{cr}$

Figure 59.9 shows the impact of pressure on the critical reduced electric field $(E/N)_{cr}$ in C_4F_7N-CO_2(1:1) and $C_5F_{10}O$-CO_2(1:1) gas mixtures.

The C_4F_7N-CO_2(1:1) gas mixture shows minimal pressure dependence, as the mole fraction of main components remains stable at low temperatures. However, at temperatures ranging from 1000 to 2700 K, increasing pressure enhances $(E/N)_{cr}$ and improves the breakdown performance of the mixture. This is due to the increased collision frequency between gas molecules, promoting ionization and dissociation processes. In the high temperature range of 2700–3500 K, the relationship between pressure and $(E/N)_{cr}$ is not apparent, as other factors such as chemical reactions and energy transfer play a dominant role in breakdown performance. This results in a non-linear relationship between $(E/N)_{cr}$ and pressure, leading to an irregular pattern. For the $C_5F_{10}O$-CO_2(1:1), at lower temperatures (300–1700 K), $(E/N)_{cr}$ shows minimal dependence on pressure due to the high ionization energy levels of

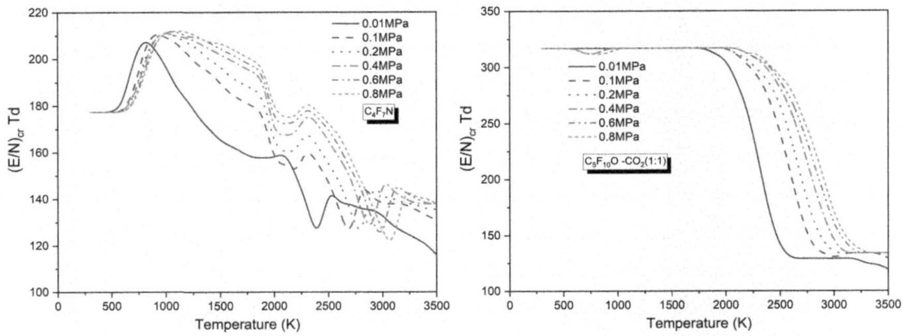

Fig. 59.9 Calculation results of critical breakdown electric field of C_4F_7N-CO_2 (1:1) and $C_5F_{10}O$-CO_2 (1:1) gases at different pressures

$C_5F_{10}O$ molecules, which require more energy for ionization. However, at temperatures ranging from 1700 to 3500 K, increasing pressure can increase the gas molecule density, leading to a higher collision frequency and improved efficiency of ionization and excitation processes. This results in an increase in $(E/N)_{cr}$ and improved breakdown performance.

Figure 59.10 illustrates the impact of CO_2 concentration on $(E/N)_{cr}$ of C_4F_7N-CO_2 and $C_5F_{10}O$-CO_2 mixtures. At temperatures below 700 K, increasing CO_2 concentration in C_4F_7N-CO_2 mixture decreases $(E/N)_{cr}$ due to low collision efficiency hindering effective interaction with charged particles and limiting electron energy dissipation. However, in temperature ranges of 700–1100 K and 2600 K–3500 K, increasing CO_2 concentration enhances $(E/N)_{cr}$ by increasing the number of excited states in CO_2 molecules, providing more channels for electron energy dissipation. Higher electron attachment coefficient in molecular gases also contributes to improved $(E/N)_{cr}$. Therefore, moderate addition of CO_2 enhances electrical insulation performance in these temperature ranges. Conversely, in the range of 1100–2600 K, increasing CO_2 concentration decreases $(E/N)_{cr}$ of C_4F_7N-CO_2 mixture as reduced number of excited states in CO_2 limits electron energy dissipation channels. For $C_5F_{10}O$-CO_2 mixture, increasing CO_2 concentration decreases $(E/N)_{cr}$ in the temperature range of 300–2700 K, possibly due to enhanced synergistic effect among chemical species in the mixture. Higher CO_2 concentration reduces electron energy loss and ionization processes, lowering the mixture's tolerance to electric field. However, in the temperature range of 2700–3500 K, increasing CO_2 concentration boosts the number of excited states in CO_2 molecules and improves collision efficiency with charged particles, resulting in increased electron energy loss and ionization, thus enhancing the mixture's tolerance to electric field.

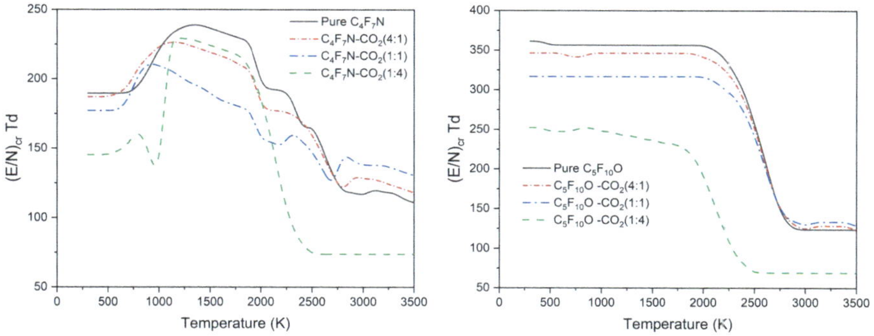

Fig. 59.10 Calculation results of critical breakdown electric field of gases with different ratios of C_4F_7N-CO_2 and $C_5F_{10}O$-CO_2 under 0.1Mpa pressure

59.4 Conclusion

The effects of temperature, pressure, and CO_2 concentration on dielectric breakdown performance of C_4F_7N-CO_2 and $C_5F_{10}O$-CO_2 mixtures was investigated in this paper. Based on the current research, the following conclusions can be drawn:

(a) The EEDF of pure C_4F_7N and $C_5F_{10}O$ gases is influenced by a decreasing electric field intensity. Different ranges of electron energy exhibit different effects. a small amount of CO_2 has no significant effect on the EEDF, while a large amount of CO_2 significantly reduces the concentration of high-energy electrons.

(b) In pure C_4F_7N gas α/N and η/N increase with increasing temperature due to dissociation of C_2F_5N and CF_4. CO_2 addition reduces η/N and increases α/N, but the growth rate is small. Pure $C_5F_{10}O$ gas has small variations below 2500 K, but when it exceeds 2500 K, α/N increases and η/N decreases.

(c) $(E/N)_{cr}$ of C_4F_7N-CO_2 mixtures has minimal dependence on pressure in the 300–500 K temperature range. However, increasing pressure can increase $(E/N)_{cr}$, while no relationship exists between pressure and $(E/N)_{cr}$ in the high temperature range (2700–3500 K). The addition of CO_2 can lower $(E/N)_{cr}$, but the effect varies across temperature ranges.

References

1. Y. Li, S. Tian, L. Zhong et al., Eco-friendly gas insulating medium for next-generation SF_6-free equipment. Energy **2**(1), 14–42 (2023)
2. X. Lin, Y. Xia, G. Sun et al., Study on breakdown characteristic of hot SF_6. High Voltage Apparatus **54**(07), 111–118 (2018)
3. F. Ye, X. Zhang, S. Tian et al., Biosafety and application feasibility assessment of eco-friendly C_4F_7N gas mixture. High Voltage Apparatus **59**(03), 1–7 (2023)

4. J.G. Owens, Greenhouse gas emission reductions through use of a sustainable alternative to SF_6, in *IEEE Electrical Insulation Conference (EIC)*. Montreal, QC, Canada (2016), pp. 535–538
5. X. Li, Y. Deng, X. Jiang et al., Insulation performance and application of enviroment-friendly gases mixtures of C_4F_7N and $C_5F_{10}O$ with CO_2. High Voltage Eng. **43**(03), 708–714 (2017)
6. R. Zhuo, K. Ke, Y. Zhang et al., Insulation characteristics of $C_5F_{10}O$/Air and $C_5F_{10}O/N_2$ under quasi-uniform electric field. Electric Power Eng, Technol. **40**(3), 159–165 (2021)
7. Y. Liu, S. Li, S. Liu et al., Solving the Boltzmann equation to obtain insulation properties of CF_3I/N_2 gas mixtures based on two-term approximation. Electr. Eng. **20**(4), 32–36 (2019)
8. L. Zhong, B. Wu, S. Zheng et al., A database of electron-impact ionization cross sections of molecules composed of H, C, N, O, and F. Phys. Plasmas **28**(8), 083505 (2021)
9. L. Zhong, Q. Gu, B. Wu, Graphite production in two-temperature non-LTE plasmas of C_4F_7N and $C_5F_{10}O$ mixed with CO_2, N_2, and O_2 as eco-friendly SF_6 replacements: a numerical study. Plasma Processes Polym. **18**, e2100036 (2021)

Chapter 60
GCMC Simulation of ZIF-68 Adsorption for C_4F_7N/CO_2 and Its Decomposition Products

Zhengyong Hu, Peng Xu, Guliang Zhou, Song Zhang, and Menglei Jin

Abstract In the realm of gas insulation, the C_4F_7N/CO_2 gas mixture emerges as a promising and environmentally sustainable choice. Ensuring the stability of gas-insulated equipment and the safety of operational and maintenance personnel relies on the judicious selection of materials capable of effectively managing the deleterious decomposition byproducts produced by the C_4F_7N mixture. Among the potential candidates for adsorption, ZIF-68, a type of metal–organic framework, emerges as a viable contender. Therefore, the investigation of intricate interaction mechanisms between ZIF-68 and the C_4F_7N mixture, including its decomposition products, assumes importance in facilitating the informed selection of suitable adsorbents for C_4F_7N/CO_2 gas insulated equipment. In the context of this research endeavor, we employed Grand Canonical Monte Carlo (GCMC) simulations to delve into the adsorption dynamics of the C_4F_7N/CO_2 gas mixture and its nine decomposition products within the confines of ZIF-68. These simulations yielded invaluable insights into the adsorption density distribution, adsorption isotherm, and isosteric heats, collectively enhancing our comprehension of the adsorption process. The outcomes of this comprehensive study underscore ZIF-68's good adsorption performance when confronted with decomposition products. Consequently, ZIF-68 emerges as a potential candidate for adoption as an adsorbent in the realm of C_4F_7N/CO_2 gas-insulated equipment.

Z. Hu · P. Xu · G. Zhou · S. Zhang
State Grid Shanghai Municipal Electric Power Company, Shanghai 200122, China

M. Jin (✉)
School of Electrical Engineering and Automation, Wuhan University, Hubei 430072, China
e-mail: 1277407676@qq.com

Z. Fang et al. (eds.), *Proceedings of the 5th International Symposium on Plasma and Energy Conversion*, Springer Proceedings in Physics 398,
https://doi.org/10.1007/978-981-97-2245-7_60

60.1 Introduction

Perfluoroisobutyronitrile (C_4F_7N) serves as a promising gas insulation medium, renowned for its remarkable insulating characteristics and minimal contribution to global warming. While it exhibits a relatively high liquefaction temperature (4.7 °C), it holds promise for application in medium and low-pressure gas insulation systems when combined with buffer gases like CO_2 [1]. Nevertheless, over prolonged periods of operation in gas-insulated switchgear, the insulating medium may undergo decomposition as a consequence of localized overheating and discharge phenomena. In the case of C_4F_7N mixtures, this decomposition process gives rise to the generation of noxious and corrosive gases, some of which can potentially exert an adverse influence on the insulating properties of the mixtures. As a countermeasure, adsorption materials are typically introduced into the switchgear chamber to selectively capture these decomposition by-products. This proactive measure ensures the prolonged stability and functionality of the gas insulation equipment [2, 3].

In contemporary gas-insulated equipment, porous materials such as molecular sieves and activated alumina have become prevalent choices for use as adsorbents. Nonetheless, extant research has highlighted that the adsorption efficacy of decomposition products within C_4F_7N mixtures is less than optimal [4, 5]. Among the array of metal–organic frameworks (MOFs), ZIF-68 stands out as a porous material characterized by a substantial internal surface area, making it well-suited for diverse applications, including gas storage, separation, and adsorption. This paper undertakes the task of performance prediction from the vantage point of Grand Canonical Monte Carlo (GCMC) simulations. It systematically elucidates the interaction characteristics between ZIF-68 and C_4F_7N/CO_2, along with the decomposition products. This investigation serves to furnish a sound theoretical foundation for the management and detection of components resulting from the decomposition of C_4F_7N/CO_2 gas.

60.2 Molecular Dynamics Simulation

60.2.1 Adsorption Model

Before conducting the GCMC simulations, several key preparatory steps were carried out. Initially, the C_4F_7N/CO_2 gas mixture and its associated nine decomposition products (CO, CF_4, C_2F_6, C_3F_8, C_3F_6, CF_3CN, C_2F_5CN, C_2N_2 and COF_2) underwent geometric optimization through the application of the Forcite code. Electrostatic potential (EPS) charges were subsequently assigned to the guest molecules. Rigorous measures were implemented to ensure the precision of the simulations, including the requirement that the spatial dimensions exceed twice the cutoff radius. To accurately depict the ZIF-68 framework, it was represented as a 2*2*2 supercell. The Universal forcefield was employed for the characterization of MOF parameters,

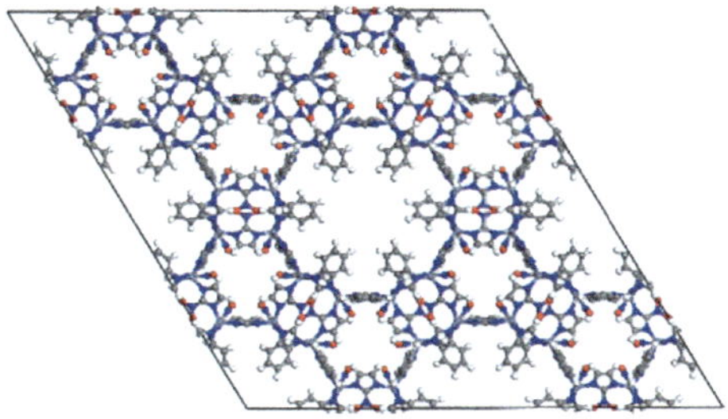

Fig. 60.1 Optimized ZIF-68 supercell structure

and geometry optimization was performed using the Forcite code to achieve stable molecular structures. Figure 60.1 displays the resulting optimized structure of the ZIF-68 supercell.

60.2.2 GCMC Simulation

In the simulation, both the adsorbate molecules and the ZIF-68 framework were considered as rigid, solid structures. To account for electrostatic interactions, the Ewald method was employed, while the atom-based approach was utilized to handle van der Waals interactions. To facilitate the execution of the sorption module, it was imperative to modify the symmetry of ZIF-68 to the P1 symmetry group. The simulation process involved an equilibration phase spanning $1*10^5$ steps to achieve a state of equilibrium, followed by $1*10^6$ production steps for the purpose of sampling thermodynamic properties. These properties encompassed Monte Carlo steps for insertion, deletion, molecular rotation, and translation. Within the sorption module, the Universal forcefield and the Metropolis method were employed. The simulation tasks of Fixed Pressure and Adsorption Isotherm were applied to replicate a scenario of equal proportion competitive adsorption. The temperature was held constant at 298 K, and the fugacity of both the C_4F_7N mixture and its decomposition products remained at consistent levels, spanning the range from 10^{-15} to 100 kPa.

60.3 Results and Discussion

The equal proportion competitive adsorption simulation offers valuable insights into various critical aspects, including the adsorption density distribution, minimum energy configuration, adsorption isotherm, and isosteric heats of adsorption.

The visualization of the distribution of adsorption sites and the density of adsorbate molecules at these sites is readily apparent in both the adsorption density distribution and the minimum energy configuration, as illustrated in Fig. 60.2. A clear examination of Fig. 60.2 reveals that adsorbate molecules are not only present but also well-dispersed within both the larger and smaller micropores of the ZIF-68 structure. This result substantiates the effective adsorption and containment of the adsorbate molecules by both the prominent types of pores present in the ZIF-68 framework.

The adsorption isotherm resulting from the equal proportion competitive adsorption simulation is presented in Fig. 60.3. The figure provides a clear depiction of the capacity of ZIF-68 for adsorbing the decomposition products of the C_4F_7N mixture in the context of equal proportion competitive adsorption. Of note, the competitive adsorption capacity for CO stands out as the highest, reaching 1.4 mmol/g. Additionally, the three highly toxic gases, CF_3CN, COF_2 and C_2N_2, also exhibit substantial competitive adsorption capacities, measuring at 0.73 mmol/g, 0.67 mmol/g, and 0.44 mmol/g, respectively. In contrast, the adsorption capacity for C_4F_7N itself is relatively modest at 0.16 mmol/g. This observation underscores that ZIF-68 tends to preferentially adsorb the decomposition products of the C_4F_7N mixture while exerting minimal impact on the primary gas insulation medium within the equipment.

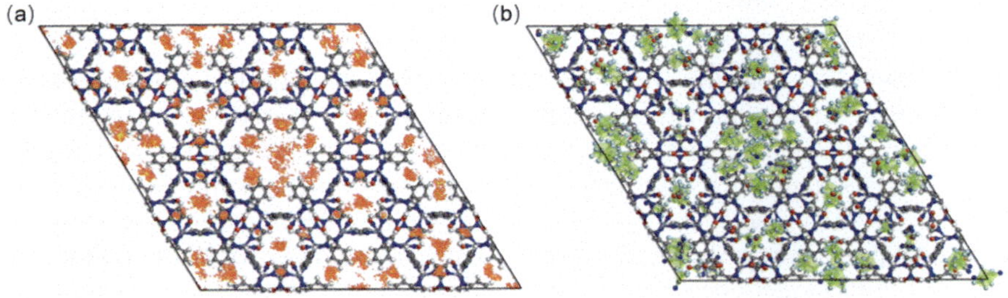

Fig. 60.2 **a** Adsorption density distribution and **b** minimum energy configuration of ZIF-68

Fig. 60.3 Adsorption isotherm of ZIF-68

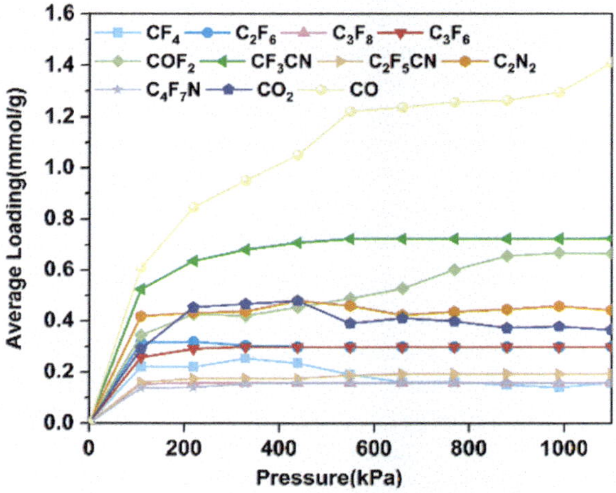

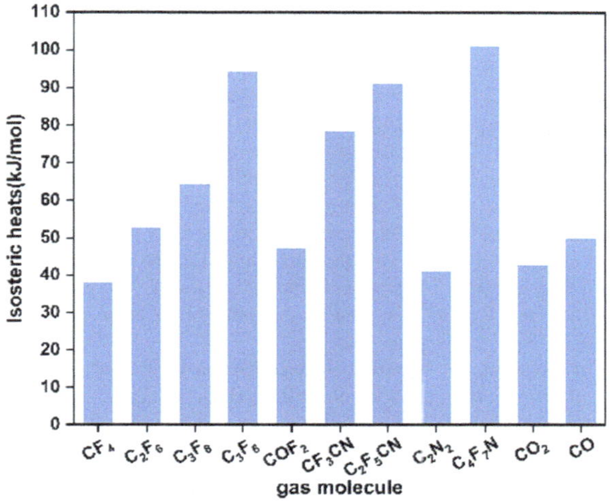

Fig. 60.4 Isosteric heats of ZIF-68 (298 K, 101.3 kPa)

The isosteric heats resulting from the equal proportion competitive adsorption simulation are presented in Fig. 60.4. The adsorption heat levels in the range of 30–100 kJ/mol suggest that the predominant mechanism at play is physical adsorption. Comparing these results with the adsorption capacity data reveals notable distinctions. Remarkably, C_4F_7N exhibits the highest adsorption heat, yet it possesses the lowest adsorption capacity. This contrast highlights that adsorption capacity hinges not only on adsorption heat but also on molecular size. C_4F_7N has a larger molecular dynamics diameter, rendering it more prone to being excluded by the pore cages. In contrast, CO boasts a smaller molecular dynamics diameter, making it more amenable to adsorption at the active sites located within the MOF pore cages.

60.4 Conclusion

In this study, we employed equal proportion competitive adsorption simulations to evaluate the adsorption properties of ZIF-68. The results indicate that ZIF-68 has a minimal impact on the primary gas insulation medium used in the equipment, and it exhibits a notable capacity for adsorbing the decomposition products of the C_4F_7N mixture. To gain a more comprehensive understanding of the adsorption capabilities and practical suitability of ZIF-68, further investigations involving experimental studies are warranted. Additionally, it would be worthwhile to explore alternative adsorbents and conduct comparative experiments to identify the most effective materials for this application.

Acknowledgements This work is supported by the technology project of State Grid Corporation of China (Project ID: 520940220055).

References

1. Y. Kieffel, T. Irwin et al., Green gas to replace SF_6 in electrical grids. IEEE Power Energ. Mag.Energ. Mag. **14**(2), 32–39 (2016)
2. P. Simka, C.B. Doiron et al., Decomposition of alternative gaseous insulation under partial discharge, in *Proceedings of the 20th International Symposium on High Voltage Engineering* 28. Buenos Aires, Argentina (2017)
3. Y. Kieffel, Characteristics of g3-an alternative to SF_6, in *IEEE International Conference on Dielectrics (ICD)*. Montpellier, France (2016), pp 880–884
4. X. Song et al., Adsorption characteristics of γ-Al_2O_3 for the environment-friendly insulating medium C_3F_7CN/N_2 and its decomposition products. High Voltage Eng. **44**(10), 3135–3140 (2018)
5. H. Hua et al., Theoretical investigation on the adsorption of C_4F_7N/CO_2 dielectric gas and decomposition products in zeolite. High Voltage Eng. **45**(04), 1040–1047 (2019)

Chapter 61
Formation of Diffuse Ionization Wave with Streamers

Victor Tarasenko, Nikita Vinogradov, Bangdou Huang, Evgeny Baksht, Cheng Zhang, and Tao Shao

Abstract This article presents the results of studies of diffuse ionization wave (DIW) with streamers during a discharge in low-pressure atmospheric air. The work continues our study of the conditions for laboratory modeling of high-altitude atmospheric discharges, in this case columnar red sprites. Preliminary results obtained using only small-diameter tubes in two setups. In this work, data about the size of DIWs were obtained in a dielectric tube with an inner diameter of 14.2 cm and on the generation of X-rays in pin-to-plate discharge with a gap of 1 cm during a positive polarity discharge in low-pressure air.

61.1 Introduction

The phenomenon of high-altitude atmospheric discharges, which are observed at an altitude of more than 20 km above sea level has been studied in many scientific groups over the past decades. The largest number of publications is devoted to red sprites, the radiation of which is recorded in the mesosphere and stratosphere at altitudes from 40 to 100 km. Sprites have a variety of shapes and, forming at an altitude above 50 km from sea level, emit a red color. Detailed information about the properties of red sprites, as well as the history of their research, can be found in a large number of publications, see [1, 2] and references in [1, 2]. Using high-speed photography, it was shown in [2] that sprites consist of three parts: streamers, glow, and beads. The diameter of the streamer decreases with increasing pressure in proportion, which is described in detail in the review [3].

V. Tarasenko (✉) · N. Vinogradov · E. Baksht
Institute of High Current Electronics, Siberian Branch of the Russian Academy of Sciences, Tomsk 634055, Russia
e-mail: VFT@loi.hcei.tsc.ru

B. Huang (✉) · C. Zhang · T. Shao
Beijing International S&T Cooperation Base for Plasma Science and Energy Conversion, Institute of Electrical Engineering, Chinese Academy of Sciences, Beijing 100190, China
e-mail: huangbangdou@mail.iee.ac.cn

Z. Fang et al. (eds.), *Proceedings of the 5th International Symposium on Plasma and Energy Conversion*, Springer Proceedings in Physics 398,
https://doi.org/10.1007/978-981-97-2245-7_61

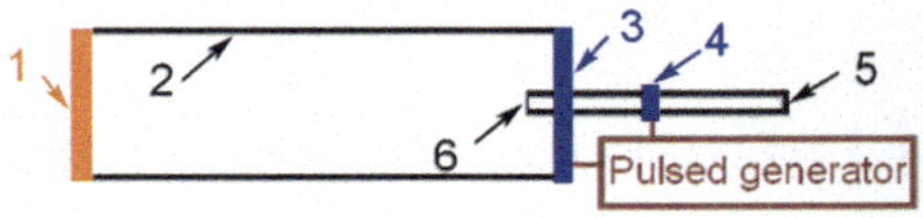

Fig. 61.1 Scheme of the first experimental setup. 1—dielectric flange, with a closed left (relative to the reader) end; 2—dielectric tube with internal diameter 5 or 14.2 cm; 3—grounded aluminum flange electrode; 4—annular high-voltage electrode; 5—quartz tube with internal diameter 1.8 cm and wall thickness of 1.5 mm; 6—the open end of a quartz tube with internal diameter 1.8 cm. The distance between flanges 1 and 3 was 48.5 cm

In this work, the main attention will be paid to experimental modeling of some properties of red sprites, in particular, the influence of the diameter of a dielectric tube on the diameter of analogs of columnar red sprites will be determined. The work continues our research, some of which was published in [4–6].

61.2 Experimental Equipment and Methods

The results were obtained using setups are published in [6]. In new setup #1, created at the Institute of High Current Electronics, the plasma was formed using the dielectric tube, the outer diameter of which was increased from 5.5 to 15 cm. The voltage pulse amplitude of 7 kV and a pulse duration of 2 μs was from generator. This made it possible, at an air pressures $p = 0.4$–3 Torr, to increase the diameter of diffuse ionization wave (DIW) which consisted of a streamer. The design of the first setup is shown in Fig. 61.1.

On setup #2, created at the Institute of Electrical Engineering, the diffuse discharge was generated with a pin-to-plate electrode geometry with a gap distance of ~ 1 cm and air pressure of ~ 3 Torr. Pulsed high voltage (25 kV) with a pulse repetition rate of 2 Hz was used to drive the diffuse discharge, with the pin positively biased. X-ray in the diffuse discharge were investigated with a LaBr3 scintillator.

61.3 Results and Discussion

The influence of the inner diameter of the tube and air pressure on the diameter and length of the formed DIWs was studied using the setup #1. In Fig. 61.2 shows photos of discharge radiation obtained in tubes with an internal diameter of 14.2 and 5 cm.

In the photo (a) brightness of the image was increased, which allows you to see the end flanges of the installation at tube and diameter it side wall. Increasing the diameter of the tube did not affect the formation of DIW, but its length at the same pressure decreased and its diameter increased. From the increase in the diameter of the DIW, it follows that in [4], and in this work, the mode of formation of streamers was implemented at a high pulse repetition rate, in which the charge on their inner

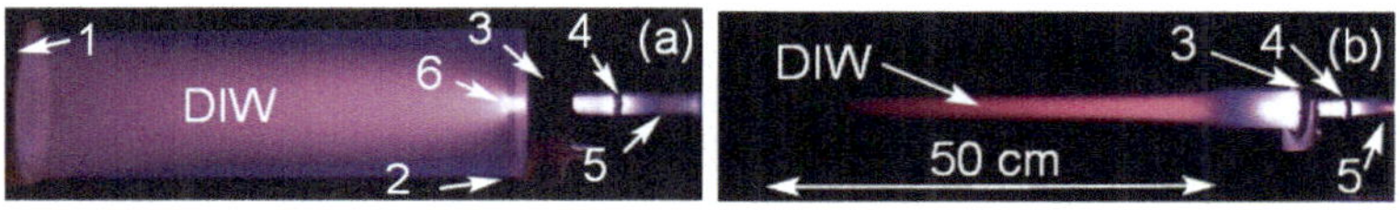

Fig. 61.2 Photos of discharge radiation obtained in tubes with an internal diameter of 14.2 (**a**) and 5 cm (**b**). Polarity of electrode 4 was positive. Air pressure $p = 0.4$ Torr. Pulse repetition frequency $f = 21$ kHz. 1—dielectric flange, with a closed left (relative to the reader) end; 2—dielectric tube with internal diameter 14.2 cm; 3—grounded aluminum flange electrode; 4—annular high-voltage electrode; 5—quartz tube with internal diameter 1.8 cm and wall thickness of 1.5 mm

surface affects their diameter. However, DIWs are formed according to the streamer mechanism and propagate with maximum radiation intensity on the axis of the quartz tube. If remove the side walls, the diameter of the streamer increases at low air pressures. This is consistent with the results of [2, 3]. The effect of air pressure and generator polarity is demonstrated in Fig. 61.3.

The above figures confirm that with an internal diameter of the dielectric tube of 14.5 cm, as well as with a diameter of 5 cm [6], the length of the DIW increases with decreasing air pressure. The diameter of the DIW gradually expands with distance from the end of the thin tube, in which a single-barrier capacitive discharge is formed. The polarity of the high-voltage electrode affects the length of the DIW, and at an air pressure of 3 Torr, the color of the discharge in the DIW. Blue color begins to predominate in the observed and photographed DIW radiation with the positive polarity of the high-voltage electrode (e,f). As was shown in [4], a change in color to blue may be associated with an increase in the reduced electric field strength. An increase in the length of the DIW with a positive polarity of the high-voltage electrode is also observed in a tube with a diameter of 5 cm [6].

At the second installation, additional data were obtained on the generation of X-rays, indicating the formation of high-energy electrons, in low-pressure air. The

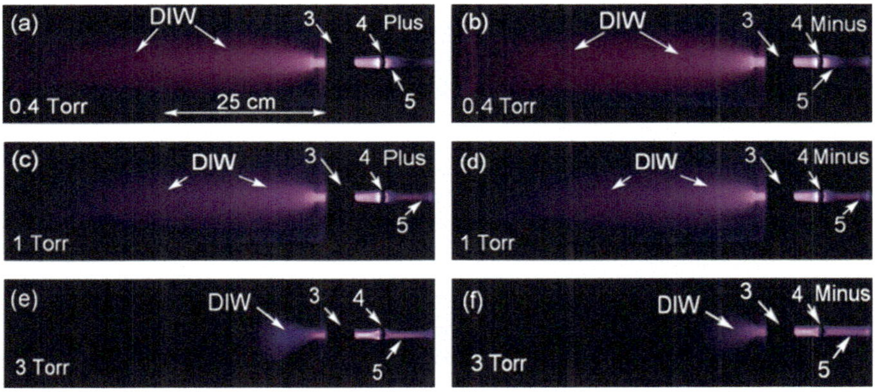

Fig. 61.3 Photos of discharge radiation obtained in a tube with an internal diameter of 14.2 cm. $f = 21$ kHz. Generator voltage 7 kV

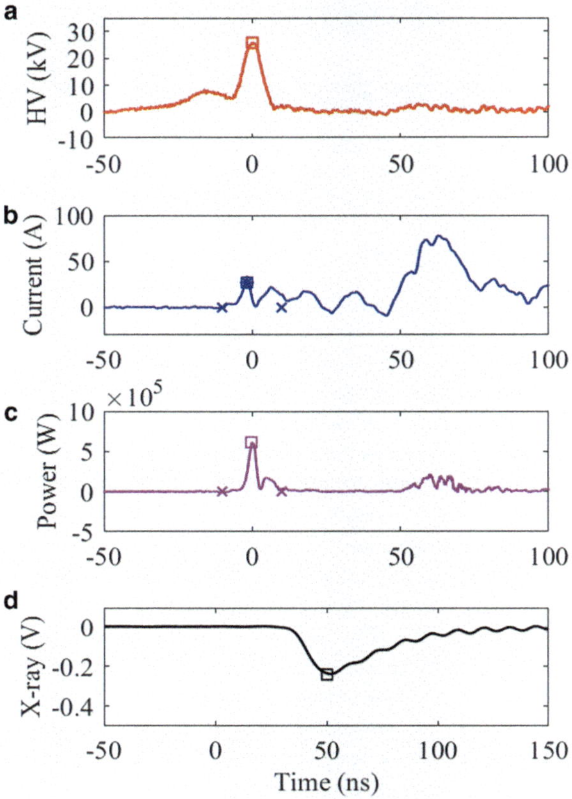

Fig. 61.4 **a** high voltage, **b** discharge current, **c** discharge power, and **d** X-ray signal from diffuse discharge in low-pressure air

oscilloscope traces of high voltage, discharge current, discharge power, and X-ray signal are shown in Fig. 61.4.

It was also confirmed that, the intensity of the X-ray is in proportion to the charge deposition (integration of discharge current) and energy coupling (integration of discharge power) during the breakdown period, but not related to the peak current or peak power. This means that for positive pulsed discharge, the formation of high-energy electrons is related to the collective behavior of charges during breakdown.

61.4 Conclusion

Thus, the studies carried out at the first setup showed that the diameter of the DIW increases with a decrease in air pressure from 3 to 0.4 Torr. At the same time, the diameter of the brightly luminous area does not exceed 10 cm. Research at the second setup made it possible to obtain data on the parameters of the runaway electron beam current and the intensity of X-ray radiation from diffuse discharge at low air pressures.

Acknowledgements The study on setup #1 is funded by the Ministry of Science and Higher Education of the Russian Federation within Agreement no. 075-15-2021-1026, 2021, and on setup #2 is funded by the National Natural Science Foundation of China (Grant No. 51877086).

References

1. V. Pasko, Y. Yair, C. Kuo, Lightning related transient luminous events at high altitude in the Earth's atmosphere: Phenomenology, mechanisms and effects. Space Sci. Rev **168**(1), 475–516 (2012)
2. H. Stenbaek-Nielsen, M. McHarg, R. Haaland, A. Luque, Optical spectra of small-scale sprite features observed at 10,000 fps. J. Geophys. Res. Atmos. **125**, e2020JD033170 (2020)
3. N. Babaeva, G. Naidis, Universal nature and specific features of streamers in various dielectric media. J. Phys. D Appl. Phys. **54**, 223002 (2021)
4. D. Sorokin, V. Tarasenko, E. Baksht, N. Vinogradov, Analogs of columnar sprites initiated in low-pressure air and nitrogen. Phys. Plasmas **30**(8), 083515 (2023)
5. B. Huang, C. Zhang, J. Qiu, X. Zhang, Y. Ding, T. Shao, The dynamics of discharge propagation and x-ray generation in nanosecond pulsed fast ionization wave in 5 mbar nitrogen. Plasma Sourc. Sci. Technnol. **28**(9), 095001 (2019)
6. B. Huang, V. Tarasenko, C. Zhang, E. Baksht, T. Shao, D. Sorokin, Optical and X-ray radiation from pulsed discharge at low-pressure air, in *Springer Proceedings in Physics (ISPEC 2022)* (Springer Nature Singapore, Singapore), pp. 372–376 (2023)

Chapter 62
Analysis of Discharge Characteristics and Gap Factors of Large Electrode in Converter Station

Yujian Ding, Jie Liu, Xiuyuan Yao, Yu Su, and Hongsheng Liang

Abstract With the increase of voltage level, the size of shielding fittings used in the converter station is getting larger and larger. For example, the diameter of the sphere used in the ± 1100 kV DC project reaches 2 m. Its discharge characteristics are greatly different from the typical rod-plane gap and rod-rod gap. It is of great significance to master the discharge characteristics of large-size electrode gap and the change rule of gap factor for the selection of air gap in converter station and the manufacture of high-voltage DC equipment. In this paper, the discharge curves of the sphere-ground air gap of different diameters such as 0.5–2.0 m are obtained by using the full-scale size sphere-ground air gap of the converter station. Then, according to the definition of the gap factors, the variation rules of the diameter of the sphere-ground, the gap distance and the gap factors are analyzed, and the calculation formula of the gap factor is proposed. The research results can provide an important theoretical basis for the choice of air gap between valve halls and DC field in UHV converter station.

62.1 Introduction

Converter station is an important facility in power transmission and transformation projects, which plays a crucial role in the safe and stable operation of UHV projects [1]. In the converter station, as the voltage level increases, the insulation strength needs to be improved accordingly, which requires longer air gap distance and larger fittings (voltage-sharing sphere, etc.) size. The discharge characteristics of air gap have changed, and the previous research on the discharge characteristics of short gap and small size fittings is no longer applicable to the converter station of UHV

Y. Ding · X. Yao · Y. Su
China Electric Power Research Institute, Haidian District, Beijing 100192, China

J. Liu (✉) · H. Liang
State Grid Corporation of China, Xicheng District, Beijing 100032, China
e-mail: hv_cepri@126.com

Z. Fang et al. (eds.), *Proceedings of the 5th International Symposium on Plasma and Energy Conversion*, Springer Proceedings in Physics 398,
https://doi.org/10.1007/978-981-97-2245-7_62

transmission and transformation project [2]. Because of complex and diverse gap structure in the converter station, and long period and great difficulty of full-scale test, it is impossible to exhaust all the research on the insulation strength of air gap through the full-scale test. As one of the typical gaps in the converter station, the sphere-ground gap is a common gap structure in the converter stations. Its research has guiding significance for the study of the discharge characteristics of complex gap structures [3, 4].

International scholars have conducted extensive research on voltage-sharing sphere-ground gaps. In the 1970s, LesRenardières Group carried out a lot of experimental studies [5–7], obtained the critical radius for various sphere gaps. This initial research established a relationship between sphere size and breakdown voltage. Hepworth et al. carried out positive and negative polarity impact tests for sphere-ground gap and rod-plane gap, determining the breakdown voltage for sphere-ground gaps with radius ranging from 0.125 to 0.375 m. They also investigated the influence of electrode size on pre-breakdown currents within the gaps [8]. In 1975, American scholars conducted extensive discharge tests on various-sized sphere-ground gaps and qualitatively analyzed the effects of discharge voltage under dry and wet weather conditions. However, they did not consider the complexity of gaps within converter stations [9]. In 1976, Rizk studied the average positive polarity discharge voltage gradient of large-sized sphere [10]. With the development of technology, there are more and more means to observe air gap discharge, and its discharge mechanism has been studied more deeply. Chen et al. observed the entire discharge process and current waveform of sphere-ground gap and rod-plane gap with sphere radius of 0.55–0.75 m and gap distance of 3–4m, and analyzed its discharge mechanisms [11].

The above-mentioned studies have provided a strong foundation for understanding the discharge characteristics of large-sized electrode-ground gaps within ± 1100kV converter station valve halls and indoor DC fields. However, these studies often dealt with gaps and electrode sizes on the smaller side and may not fully meet the gap distance requirements for the valve halls of ± 1100kV UHVDC converter stations. As a result, providing practical engineering advice on gap distances in converter stations has been challenging.

In this paper, the characteristic curves of the sphere-ground air gap of different diameters such as 0.5–2.0m are obtained by using the full-scale size sphere-ground air gap of the converter station. Then, according to the definition of the gap factors, the variation rules of the diameter of the sphere-ground, the gap distance and the gap factors are analyzed, and the calculation formula of the gap factor is proposed. The experimental study of the critical radius of small-size voltage-sharing sphere was supplemented with the experimental results in the published projects and papers, and the exponential function with the error of fitting the critical radius less than 5% was obtained by the mathematical fitting method. The research results can provide an important theoretical basis for the choice of air gap between valve halls and DC field in UHV converter station.

62.2 Test Layout and Results

62.2.1 Test Layout

In the valve hall of the converter station, there are gap varieties (sphere-net, sphere-ground or sphere-net/ground gaps) at the end of tubular busbar or post insulator. When the voltage level is different, the size of the sphere varies. To research the discharge characteristics of air gaps between spheres with different diameter and the plane, the positive switching impulse discharge test data of 4 kinds of typical voltage-sharing sphere-plane gaps (the diameters of voltage-sharing spheres are 0.5 m, 1.1 m, 1.5 m and 2.0 m respectively; the gap distance is 3.0–10.0 m) are obtained in the UHV DC test base of the State Power Grid.

62.2.2 Discharge Characteristic Curve

We obtained the switching impulse discharge characteristic curves of voltage-sharing spheres with diameters of 0.5, 1.1, 1.5, and 2.0 m at different gap distances. These curves were compared with the typical rod-plane gap discharge characteristic curve, and the results are shown in Fig. 62.1.

Within the range of gap distances tested, the 50% switching impulse discharge voltage at the same gap distance increases significantly as the sphere diameter increases. The switching impulse discharge voltages for the larger diameter sphere-ground gap were significantly higher than those at the rod-plane gap.

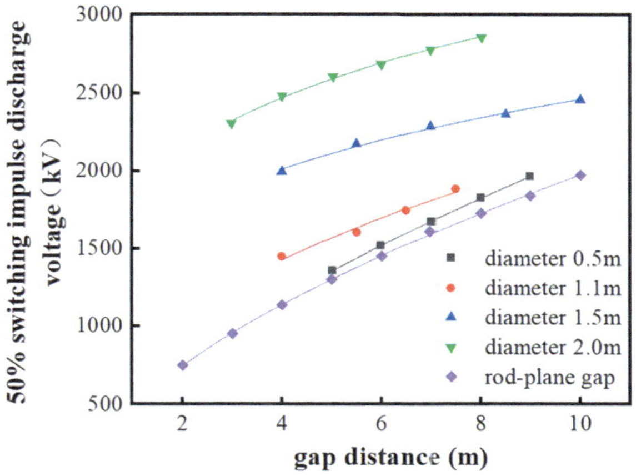

Fig. 62.1 Comparison of discharge characteristic curves between sphere-ground gap and rod-plane gap

62.3 Analysis of Test Results

62.3.1 Critical Radius

To study the impact of spheres with different diameters on sphere-plane gaps, experiments were conducted to analyze the critical radius. Since the sphere sizes of this project is 0.5–2.0 m, in order to ensure the accuracy and comprehensiveness of the critical radius analysis, the test results from the published projects and papers are cited to supplement some of the results of the small-size voltage-sharing spheres.

The test layout is shown in Fig. 62.2. Use an overhead crane to suspend composite insulators with a length of approximately 10 m. Suspend a stainless steel pipe with a length of 10 m and a diameter of 220 mm at the end of the insulator. The upper end of the stainless steel pipe is equipped with voltage-sharing rings that are connected to the high-voltage lead. Suspend stainless steel spheres with radius of 125, 225, 375, and 475 mm at the end of the pipe. Ensure a smooth connection between the spheres and the pipe, so that the discharge path is along the bottom of the sphere towards the ground. In the experiment, stainless steel pipes with a diameter of 38 mm and hemispherical end were used as rod electrodes, and their connection method was the same. Galvanized iron planes are laid on the ground and grounded to serve as plane electrodes. Discharge tests were performed on rod-plane and sphere-ground gaps with gap distances of 2, 3, 4, and 5 m. The applied voltage waveform was a 250 μs/2500 μs positive polarity standard switching impulse voltage.

The characteristic curves between discharge voltage and gap distance for different sphere radius is obtained from the test, as shown in Fig. 62.3.

As can be seen from Fig. 62.3, when the sphere radius in rod tip are 19 mm and 125 mm respectively, the discharge characteristic curves largely overlap. Therefore, when the sphere radius in rod tip is below 125 mm, for rod-plane gap with distances ranging from 2 to 5 m, the positive polarity switching impulse discharge voltage remains essentially the same and is not affected by the radius of the sphere in rod tip. However, when the sphere radius is 225 mm, the three points at gap distances of 2, 3, and 4 m deviate from the previous two curves. While the point at 5 m still lies on the previous two overlapping curves. When the sphere radius is 375 mm, the discharge voltage at the point of 5 m also begins to increase. It can be inferred that when the gap distance is less than 4 m and the sphere radius exceeds 125 mm, the radius of sphere in rod tip has an impact on the discharge voltage. The critical radius of sphere in rod tip within the range of 125–225 mm. When the gap distance is 5 m and the sphere radius exceeds 225 mm, the radius of the sphere in rod tip has an impact on the discharge voltage. The critical radius lies between 225 and 375 mm.

When the sphere radius is small, increasing the gap distance has a significant impact on the discharge voltage. However, when the sphere radius reaches 475 mm, increasing the gap distance has a diminishing effect on the discharge voltage. In Fig. 62.3, the discharge voltage at a sphere radius of 475 mm and a gap distance of 3 m is higher than the discharge voltage at a sphere radius of 375 mm and a gap distance of 5 m. This indicates that when choosing a shielding structure, increasing the

Fig. 62.2 Test layout with sphere radius of 475 mm in rod tip in Beijing

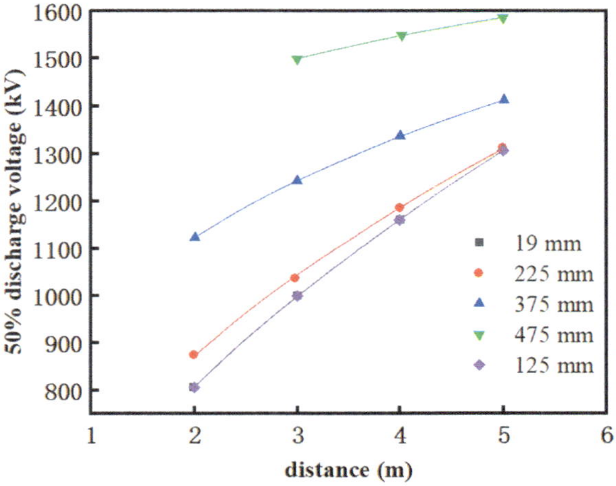

Fig. 62.3 Gap discharge characteristic curves for sphere with different radius

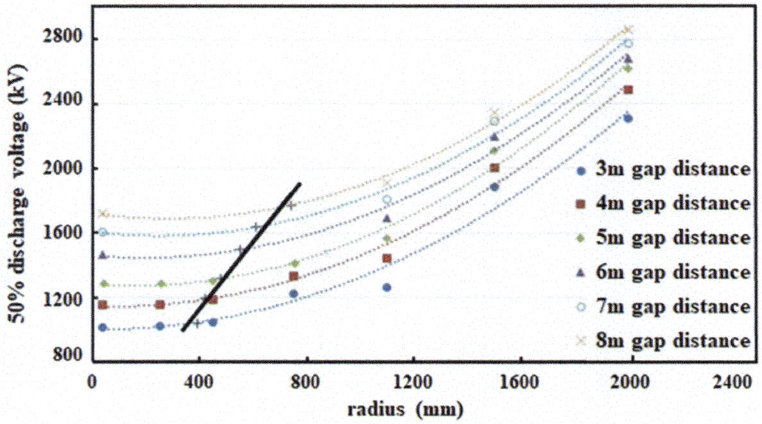

Fig. 62.4 Critical radius at different gap distances

curvature radius of the voltage-sharing sphere can significantly improve the switching impulse discharge voltage and reduce the gap distance between the equipment and the grounding equipment.

At the same gap distance, the discharge voltage increases gradually with the increase of sphere radius. By polynomial fitting the curve in Fig. 62.3, it is found that below the critical radius, the discharge voltage across the sphere-ground gap is almost a horizontal straight line. Define U_{R0} as the discharge voltage across the sphere-ground gap at different gap distances. Take the average discharge voltage values at R = 19 mm and R = 125 mm on the fitting curve as the value of U_{R0}. When the discharge voltage is higher than U_{R0} by 3%, it is considered to be beyond the range of error. Define the radius of sphere in rod tip corresponding to this discharge voltage as the critical radius R_{cr} for this gap distance, and define this discharge voltage as U_{rc}. Calculate the critical radius for different gap distances, as shown in Fig. 62.4.

Within the range of 3–8 m test gaps, when the sphere radius is greater than the critical radius, increasing the sphere radius can effectively reduce the average electric field within the gap, and the discharge voltage of the gap also increases accordingly. It should be noted that when the voltage-sharing sphere radius remains constant, the increase in discharge voltage ΔV caused by increasing the gap distance gradually decreases. The phenomenon was observed in equipotential spheres with diameters of 1.1–2 m when the gap distance was greater than 4 m, which confirms the saturation characteristic of long air gap discharge.

The critical radius can be fitted as an exponential function using mathematical fitting, as shown in Eq. (62.1). D represents the critical diameter of the voltage-sharing sphere (mm), and d represents the gap distance (m).

$$D = 257.5e^{0.13d} \tag{62.1}$$

62.3.2 Gap Factors

When selecting the air gap distance for transmission and transformation engineering, if there is no test with a full-scale equipment model, the gap factor is usually used for calculation. The gap factor for different electrode-shaped air gaps is defined as the ratio of the discharge voltage in that particular air gap to the discharge voltage in a typical rod-plane air gap at the same gap distance. Therefore, the ratio of discharge voltage varies with changes in the gap distance. The gap factor is obtained through impulse discharge tests conducted on gaps with different electrode shapes, and it is applicable to a certain range of distances. The relationship between the discharge voltage of a typical rod-plane air gap and the gap distance d can be approximately represented by Eq. (62.2). The discharge voltage of other electrode gaps can be represented by Eq. (62.3), where K represents the gap factor.

$$U_{50\%} = 500d^{0.6} \tag{62.2}$$

$$U_{50\%} = K500d^{0.6} \tag{62.3}$$

The variation in gap factors for different sizes of voltage-sharing sphere with respect to gap distances is summarized in Table 62.1. It's evident from the table that, at the same gap distance, smaller-sized voltage-sharing sphere have smaller gap factors. Moreover, for similar hardware, the gap factor decreases as the gap distance increases. For long gaps exceeding 2 m, the critical radius is approximately 0.5 m. Consequently, the gap factors for the voltage-sharing spheres in rod tip with radius of 0.5 m are in a critical state, and their discharge voltage is essentially equivalent to that of a rod-plane gap. In such cases, the gap factor is approximately equal to 1.

Based on the above analysis and considering the gap factor of the voltage-sharing sphere-ground, a mathematical expression for the relationship between the gap factor, gap distance, and voltage-sharing sphere size has been obtained through mathematical fitting methods. Within the range of voltage-sharing sphere diameters from 0.1 to 2.0 m and gap distances from 2 to 10 m, the error of the gap factor is within 5%. In the equation, d represents the gap distance (m), and r represents the sphere radius (m).

Table 62.1 The gap factors of large size sphere-ground gap

Gap distance (m)	Diameter of sphere (m)			
	2.0	1.5	1.1	0.5
2	2.82	2.27	1.40	1.02
4	2.15	1.75	1.24	1.03
6	1.84	1.50	1.15	1.04
8	1.65	1.35	1.09	1.05
10	1.51	1.24	1.05	1.05

$$K = 3 \times d^{-0.49} + 0.05 \times r^{3.43} \quad (62.4)$$

62.4 Conclusion

In this paper, tests were conducted on voltage-sharing spheres with diameters of 0.5, 1.1, 1.5 and 2 m, and the characteristic curves of 50% switching impulse discharge voltage and gap distance were obtained respectively, and the conclusions were drawn by comparing with typical rod-plane gap discharge characteristic curves. In the range of gap distances tested, as the sphere diameter increases, the 50% switching impulse discharge voltage at the same gap distance increases significantly. The switching impulse discharge voltage of the larger diameter sphere-ground is significantly higher than that of the rod-plane gap. Combined with the test results in this paper, the critical radius under different gap distances is calculated by $D = 257.5e^{0.13d}$. The gap factors of large size sphere-ground gap under different sphere diameters and gap distances is calculated by $K = 3 \times d^{-0.49} + 0.05 \times r^{3.43}$. Within the range of sphere diameters 0.1–2.0 m and gap distance 2–10 m, the error of gap factors is within 5%. The research results can provide an important theoretical basis for the choice of air gap between valve halls and DC field in UHV converter station.

References

1. F. Lü, B. Zhang, J. Geng, P. Wang, Y. Ding, S. Gao, Streamer discharge characteristics of sphere plate gap under influence of burr defects. Insulating Mater. **55**(02), 67–72 (2022). https://doi.org/10.16790/j.cnki.1009-9239.im.2022.02.009
2. Q. Jin, Z. Qiu, J. Ruan, X. Wang, Y. Quan, Z. Du, Y. Deng, Study on electric field features for the shortest path of sphere gap and the feature selection methods. J. Electr. Power Sci. Technol. **35**(06), 12–20 (2020). https://doi.org/10.19781/j.issn.1673-9140.2020.06.002
3. Les Renardieres Group, Positive discharges in long air gap discharges at Les Renardieres-1975 results and conclusions. Electra **53**, 31–153 (1977)
4. Les Renardieres Group, Negative discharges in long air gap discharges at Les Renardieres-1978 results. Electra **74**, 67–216 (1981)
5. Les Renardi è res Group, Research on long air gap discharges at Les Renardières. Electra **39**(23), 53–157 (1972)
6. Les Renardi è res Group, Research on long air gap discharges at Les Renardières—1973 results. Electra **40**(35), 49–156 (1974)
7. Les Renardieres Group. Positive discharges in long air gaps at Les Renardières—1975 results and conclusions. Electra **43**(53), 31–153 (1977)
8. J.K. Hepworth, R.C. Klewe, B.A. Tozer, Impulse breakdown of large sphere-plane gaps. Proc. Inst. Electr. Eng. **119**(12), 1751–1753 (1972)
9. H.M. Schneider, F.J. Turner, Switching-surge flashover characteristics of long sphere-plane gaps for UHV station design. IEEE Trans. Power Appl. Syst. **94**(2), 551–560 (1975)

10. F.A.M. Rizk, Influence of rain on switching impulse sparkover voltage of large-electrode air gaps. IEEE Trans. Power Appar. Syst. Appar. Syst. **95**(4), 1394–1402 (1976)
11. C. Shen, Z. Rong, Z. Chijie et al., Switching impulse breakdown characteristics of large sphere-plane air gaps compared with rod-plane air gap. IEEE Trans. Dielectr. Electr. Insul. Dielectr. Electr. Insul. **20**(3), 839–844 (2013)

Chapter 63
Study on Discharge Characteristics Between Damping Capacitors in Converter Valve at Different Altitudes

Jie Liu, Yujian Ding, Yu Su, Xiuyuan Yao, and Hongsheng Liang

Abstract Damping capacitor is an important component of DC converter valve, which plays a role in suppressing thyristor switching overvoltage and providing energy for thyristor control unit. Its operational reliability determines the safe and stable operation of the converter valve. Therefore, it is urgent to study the gap insulation characteristics of damping capacitors in converter valves at different altitudes. This article focuses on the damping capacitor gap inside the converter valve and conducts experiments using a true model to verify its margin. The switching impulse, lightning impulse, DC and AC discharge tests of damping capacitors gap in converter valve were carried out in Beijing (altitude of 55 m), Qinghai (altitude of 2500 m) and Tibet (altitude of 4300 m). Comparative analysis was conducted on the discharge voltage of small gaps in damping capacitors at different altitudes and gap distances. The results show that with the increase of altitude, the air density decreases, and the discharge voltage between damping capacitors decreases. At the same altitude, the 50% lightning impulse discharge voltage is the largest. Compared with the lightning impulse discharge voltage, the 50% switching impulse discharge voltage decreases by about 9%, the DC discharge voltage decreases by about 16%, and the AC discharge voltage decreases by about 42%. For the gap distance between damping capacitors, the equipment parameters provided by manufacturers in China have a large margin, and the margin is greater than 227%, which can meet the actual needs of engineering at different altitudes. This article provides an important theoretical basis for improving the insulation reliability of ultra-high voltage DC converter valve towers and optimizing the design of converter valve insulation under minimum safety margin.

J. Liu (✉) · H. Liang
STATE GRID Corporation of China, Beijing 100032, China
e-mail: hv_cepri@126.com

Y. Ding · Y. Su · X. Yao
China Electric Power Research Institute, Beijing 100192, China

Z. Fang et al. (eds.), *Proceedings of the 5th International Symposium on Plasma and Energy Conversion*, Springer Proceedings in Physics 398,
https://doi.org/10.1007/978-981-97-2245-7_63

63.1 Introduction

The ultra-high voltage direct current transmission technology has the advantages of large transmission capacity, long transmission distance, and low line losses [1–4], which can achieve effective consumption of clean energy and help achieve low-carbon energy transformation. The converter valve, as the "heart" of the DC transmission system, can achieve AC/DC conversion and is the core equipment of the entire DC project [5]. Damping capacitors are important components for achieving dynamic voltage sharing and high potential energy extraction of thyristor control units (TCUs) [6, 7]. The safe and reliable operation of the power grid largely depends on the insulation status of transmission and distribution equipment. However, in the production and long-term operation of electrical equipment, hidden dangers such as discharge and even breakdown caused by insulation defects have always been a problem that electrical equipment manufacturers and grid operation and maintenance personnel strive to solve [8–10]. The concentration of electric field stress caused by insulation defects in electrical equipment leads to short gap discharge, posing a safety hazard. Therefore, the study of basic issues related to internal short gap discharge in electrical equipment provides theoretical support for the insulation design and maintenance of high-voltage electrical equipment, and has important practical significance for achieving safe, reliable, and economic operation of the power grid [8].

As the core equipment of ultra-high voltage direct current transmission engineering, it is of great significance to determine the minimum air gap distance required for the outer insulation between layers of the converter valve at different altitudes. Reference [11] conducted operational impulse discharge tests on the external insulation of ± 1100 kV/5000 A ultra-high voltage DC converter valves on virtual walls and virtual ground in the power system electronic laboratory of the Beijing ultra-high voltage DC test base of State Grid Corporation of China (at an altitude of 50 m), obtained corresponding operational impulse discharge characteristic curves, and summarized previous research results on altitude correction factors. Relevant data was extracted through interpolation method, The altitude correction method applicable to ± 1100 kV ultra-high voltage DC converter valves has been studied, but the interlayer altitude correction of converter valves has not been discussed yet. Reference [12] used the Galerkin boundary element method to analyze the surface electric field distribution of the impact body shielding cover, studied the variation of the surface electric field strength of the shielding cover with its radius, and proposed control measures for the field strength to suppress corona generation and prevent breakdown of the air gap in the valve test hall. Reference [13] takes ± 1100 kV DC converter valve as the research object, and uses Solidworks and ANSYS hybrid modeling technology to establish a three-dimensional model of the converter valve tower. The model is subjected to single valve insulation type tests and multiple valve DC withstand voltage tests to obtain the electric field distribution characteristics. The simulation results show that adding beam corners. The chamfer value on the side of the shielding cover and the addition of a grading ring at the interlayer insulation fittings can make the electric field strength at the weak point lower than the control value. Reference [14] takes the interlayer insulator of ± 1100 kV DC converter valve as the research object, and uses SolidWorks and Ansys hybrid modeling technology to establish a three-dimensional model. The simulation results show that the surface field strength distribution of interlayer insulators and insulator hardware is far lower than the field strength control value, and reducing the interlayer spacing can still ensure that the field strength distribution of interlayer insulators and their metal surfaces is within a safe range. The above literature mainly establishes a simulation model for the converter valve tower and verifies it through low altitude test data to obtain the electric field distribution characteristics of the converter valve, and optimizes the external insulation design of the converter valve. However, there is a lack of high altitude test data and a simulation model for the converter valve tower that considers the influence of high altitude factors.

In summary, the construction of converter stations at different altitudes will have a significant impact on the required safe insulation distance for converter valves. Among them, the main atmospheric parameters that affect the external insulation discharge of electrical equipment are air pressure, temperature, and humidity. The level of atmospheric pressure mainly depends on altitude and decreases with increasing altitude, showing a linear relationship; The temperature is related to both altitude and longitude and latitude, but within a fixed regional range, it still decreases with increasing altitude, and the relationship is basically linear; The factors that affect absolute humidity are very complex, related to altitude, longitude and latitude, condensation, evaporation, precipitation, and other factors in the atmosphere[15]. Due to the fact that the DC converter valve is an indoor device located in the valve hall of the converter station, with minimal temperature and humidity changes, the insulation discharge characteristics of the converter valve play a crucial role in affecting the altitude of the air pressure. The existing literature mainly establishes a simulation model for the converter valve module and verifies it through low altitude test data to obtain the electric field distribution characteristics of the converter valve, and optimizes the insulation design of the internal module of the converter valve. However, the distribution of the electric field inside the ultra-high voltage converter valve under different altitude factors has not been considered, and there is a lack of high altitude test data and a simulation model for the converter valve tower that considers the influence of high altitude factors, It is difficult to conduct altitude correction research on typical clearances inside and between layers of the converter valve. Therefore, it is urgent to carry out research on the discharge characteristics of typical gaps inside and between converter valves at different altitudes, and to correct the elevation of typical gaps inside and between converter valves. This provides an important theoretical basis for improving the insulation reliability of ultra-high voltage DC converter valve towers and achieving optimal design of converter valve insulation with minimum safety margin.

63.2 Test Installation and Methods

In order to study the insulation properties of the damping capacitor gap, switching impulse, lightning impulse, DC and AC discharge tests were carried out in Beijing (55 m above sea level), Qinghai (2500 m above sea level) and Tibet (4300 m above sea level), respectively. The test installation is shown in Fig. 63.1.

As shown in Fig. 63.1, two damping capacitors are used. The upper terminal and the shell of each capacitor are short-circuited, one capacitor is connected to the high voltage pole, and one is grounded. The capacitor is fixed on a movable epoxy pad to prevent surface flashover along the insulating rubber at the bottom of the capacitor.

The tests were conducted in accordance with the test and measurement methods specified in GB/T 16927.1-2011 High-voltage test techniques—Part 1: General definitions and test requirements and GB/T 16927.2-2013 High voltage test techniques Part 2: Measurement systems.

63.2.1 *Impulse Voltage Test Method*

The impulse voltage test method is detailed as follows:

(1) The test voltage waveform is the positive polarity 250 μs/2500 μs standard switching impulse wave and the positive polarity 1.2 μs/50 μs standard lightning impulse voltage.

Fig. 63.1 Test layout

(2) The test uses the electric lifting method to obtain the $U_{50\%}$, i.e., 50% of the discharge voltage:

$$U_{50\%} = \frac{\sum (n_i \times U_i)}{n} \tag{63.1}$$

where, U_i is the applied voltage, kV, n_i is the number of tests under the same applied voltage U_i, and n is the total number of effective tests. 40 switching impulse tests and 30 lightning impulse tests are carried out in these tests.

(3) σ, the standard deviation in the tests, is calculated according to Formula (63.2):

$$\sigma = \sqrt{\frac{n\sum_{i=1}^{n} U_i^2 - (\sum_{i=1}^{n} U_i)^2}{n(n-1)}} \tag{63.2}$$

63.2.2 DC and AC Discharge Test Method

The DC and AC discharge test is detailed as follows:

(1) Applying DC or AC voltage to the sample, continuously boosting the voltage until the gap discharge occurs, and recording the voltage value at the moment of discharge. The test should be repeated 20 times.
(2) The average discharge voltage U_a are obtained according to the calculation formula (63.3):

$$U_a = \frac{1}{n}\sum_{i}^{n} U_i \tag{63.3}$$

where, U_i is the discharge voltage of the ith test, and n is the total number of tests.

(3) σ, the standard deviation in the tests, is calculated according to Formula (63.2).

63.3 Analysis of Gap Discharge Characteristics of Damping Capacitors at Different Altitudes

See Figs. 63.2, 63.3, 63.4 and 63.5.

In Beijing (55 m above sea level), Qinghai (2500 m above sea level) and Tibet (4300 m above sea level), switching impulse voltage tests, lightning impulse discharge tests, DC discharge test and AC discharge tests of the damping capacitors with three gap distances, respectively. The analysis of the test results showed that,

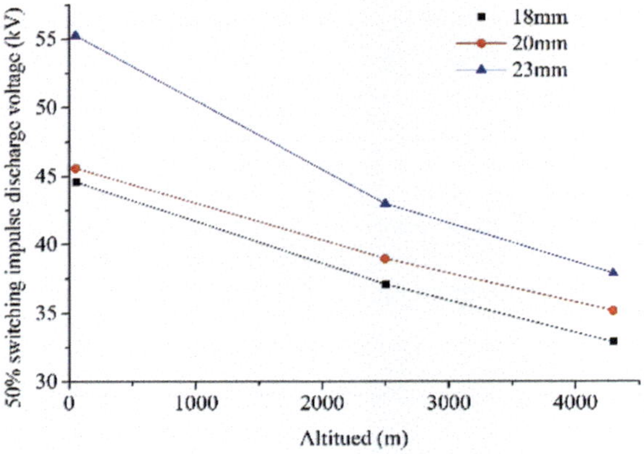

Fig. 63.2 50% switching impulse discharge voltage of damping capacitor gap at different altitudes

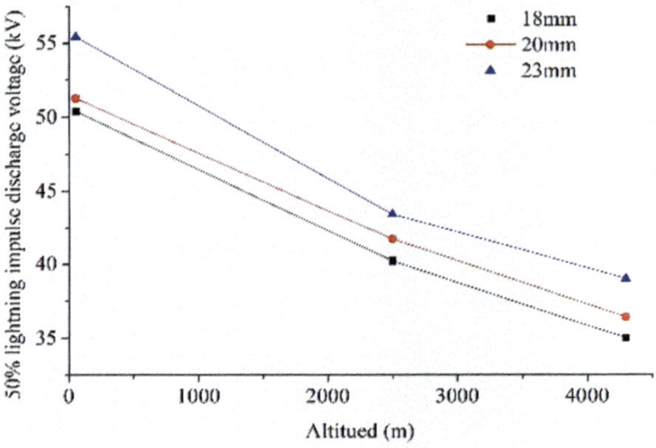

Fig. 63.3 50% lightning impulse discharge voltage of damping capacitor gap at different altitudes

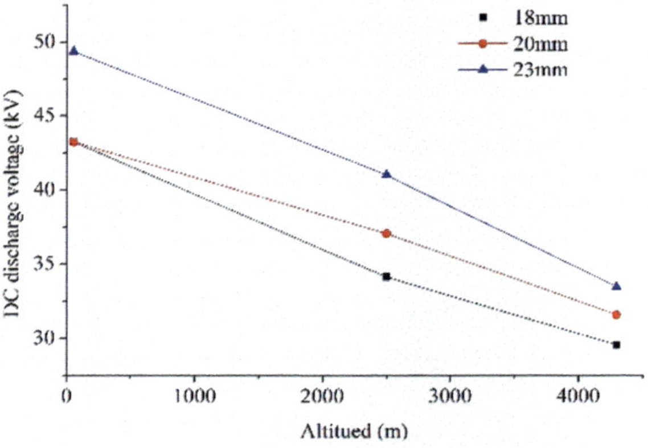

Fig. 63.4 DC discharge voltage of damping capacitor gap at different altitudes

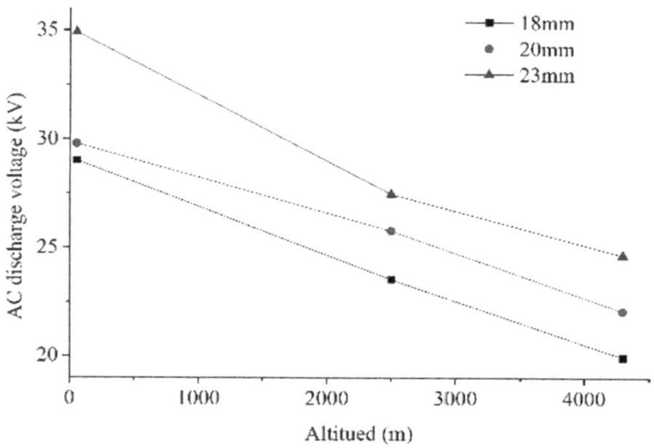

Fig. 63.5 AC discharge voltage of damping capacitor gap at different altitudes

(a) under the switching impulse voltage tests, the discharge voltages of these three damping capacitors at 2500 m altitude reduced by 16.83%, 14.58% and 22.24%, and at 4300 m altitude reduced by 26.27%, 22.93% and 31.48%, respectively, compared with at 55 m altitude; (b) under the lightning impulse discharge tests, the discharge voltages of these three damping capacitors at 2500 m altitude reduced by 20.26%, 18.68% and 21.70%, and at 4300 m altitude reduced by 30.58%, 28.97% and 29.64% respectively; (c) under the DC discharge tests, the discharge voltages of these three damping capacitors at 2500 m altitude reduced by 21.10%, 14.28% and 16.96%, and at 4300 m altitude reduced by 31.63%, 26.94% and 32.25%, respectively; (d) under the AC discharge tests, the discharge voltages of these three damping capacitors at the 2500 m altitude reduced by 18.86%, 13.43% and 21.33%, and at 4300 m altitude reduced by 31.31%, 25.86% and 29.37% respectively.

63.4 Analysis on the Margin of Damping Capacitor Gap Withstand Voltage

The corresponding withstand voltages $U_{\text{withstand}}$ are obtained through $U_{50\%}$ and U_a, the test data of the switching impulse, lightning impulse, DC and AC tests in the small gap between the radiators at different altitudes, according to Formula 63.4:

$$\mathrm{U}_{withstand} = \mathrm{U}_{50\%} \times (1 - 3 \times \theta) \tag{63.4}$$

In the formula, the switching impulse withstand voltage corresponds to $\theta = 0.05$; the lightning impulse withstand voltage to $\theta = 0.03$; the DC withstand voltage to $\theta = 0.01$; the AC withstand voltage corresponds to $\theta = 0.02$.

Based on the four actual withstand voltage values provided by a main manufacturer with a gap distance of 20.6 mm of a damping capacitor in China, the corresponding voltage margins were calculated and shown in Table 63.1:

Table 63.1 Different types of withstand voltages and margins of the small gap between damping capacitors

	Voltage type			
	Switching impulse/kV	Lightning impulse/kV	DC/kV	AC/kV
Beijing (55 m)	41.64	48.23	44.33	29.80
Qinghai (2500 m)	34.22	37.95	36.75	24.35
Tibet (4300 m)	30.20	33.76	30.75	21.12
Actual withstand voltage	8.00	8.35	3.51	4.32
Margins (%)	277.5	304.3	776.1	338.9

It can be seen from Table 63.1 that, at the same altitude, the lightning impulse withstand voltage is the largest, and the AC withstand voltage the smallest with a difference of 18.43 kV at Beijing for example; (b) for the same voltage type, the withstand voltage decreases with the increase of altitude; and (c) for all the four voltage types, the designed gap distance between damping capacitors have a large margin greater than 277%, which means this gap distance meets high altitude application requirement and has greater space for optimization.

63.5 Conclusions

In this paper, the tests of the switching voltage impulse, lightning impulse, DC and AC discharge in the gap between the damping capacitors inside the converter valve were conducted in Beijing (55 m above sea level), Qinghai (2500 m above sea level) and Tibet (4300 m altitude). After comparisons and analysis, the conclusions are as follows:

(1) With the increase of altitude, the air density decreases, and the switching impulse, lightning impulse, DC and AC discharge voltages between damping capacitors all show a downward trend.
(2) At the same altitude, the lightning impulse withstand voltage is the largest, while the AC withstand voltage is the smallest.
(3) For all the four voltage types, the designed gap distance between damping capacitors have a large margin greater than 277% from the standpoint of withstand voltage, which means the existing gap distance meets the application requirement at high altitude areas below 4300 m and has greater space for optimization.

Acknowledgements This paper was supported by the technology project of State Grid Corporation of China (5500-202155470A-0-5-ZN).

References

1. D. Xu, F. Wu, L. Shi et al., Modeling and simulation analysis of ultra-high voltage direct current transmission systems based on PSCAD. Power Eng. Technol. **39**(3), 71–77, 98 (2020)
2. Z. Shen, X. Hui, J. Zhu et al., Modeling and analysis of factors affecting the safe and efficient operation evaluation of ultra-high voltage direct current transmission projects. J. Power Gener. Technol. **42**(1), 48–59 (2021)
3. Y. Liu, J. Wu, J. Xiao et al., Research on the application of active flexible DC transmission technology in the context of global energy internet. Global Energy Internet **3**(2), 107–116 (2020)
4. Y. Zhou, J. Chen, L. Zhang et al., The development opportunities of ultra high voltage transmission technology under the background of "dual carbon and" new infrastructure. High Voltage Technol. **47**(7), 2396–2408 (2021)
5. Z. Wang, M. Ren, J. Guo et al., Research on overvoltage protection of DC transmission converter valve thyristor. Power Syst. Prot. Control **48**(10), 182–187 (2020)
6. C. Zhou, X. Zhang, L. Liu et al. Design and verification of integrated damping capacitors for high voltage DC converter valves. J. Power Capacitors Reactive Comp. **41**(4), 69–75 (2020)
7. Q. Hu, Liu, Y. Tao et al., factors influencing the switching process of thyristor valves in high voltage direct current transmission. J. Southern Power Grid Technol. **14**(1), 6–13 (2020)
8. X. Yu, W. Li, Y. Chen, DC discharge performance and voltage correction of air gaps under high altitude. Trans. China Electrotech. Soc. **33**(9), 2115–2120 (2018). ((in Chinese))
9. H. Chang, J. Fan, System design and its localization of UHVDC transmission project. Power Syst. Technol. **30**(16), 1–5 (2006). ((in Chinese))
10. W. Zhang, Y. Yu, G. Li et al., Researches on UHVDC technology. Proc. CSEE **27**(22), 1–7 (2007). ((in Chinese))
11. X. Wei, W. Zong, G. Wang et al., Research on impulse discharge test and altitude correction of ± 1100 kV UHV DC converter valve external insulation operation. Chin. J. Electr. Eng. **34**(12), 1996–2003 (2014)
12. S. Liu Shili, X. Wei, J. Cao et al,. Research on the impulse voltage test capability of ± 1100 kV ultra high voltage DC converter valve. Chin. J. Electr. Eng. **33**(28), 161–167+24 (2013)
13. L. Qi, X. Wang, C. Li et al., Electric field simulation and optimization under insulation type test of ± 1100 kV ultra-high voltage DC converter valve. High Voltage Technol. **41**(04), 1262–1271 (2015)
14. B. Xie, N. Mei, Z. Zhang et al., Electric field simulation and insulation optimization of internal metal components and interlayer insulators of ± 1100 kV DC converter valve module. High Voltage Electr. Appl. **57**(02), 105–115 (2021)
15. Y. Lou, Y. Ma, Q. Li et al., Study on the reliability life model of thyristors in high altitude DC transmission converter valve systems. High Voltage Electr. Appl. **58**(10), 33–39 (2022). https://doi.org/10.13296/j.1001-1609.hva.2022.10.05

Correction to: The Design Method of an Electron Beam Expansion Device Based on Double Solenoid Lenses

Xiaocheng Hu, Weihao Tie, Yongchun Shao, Jinyong Fang, and Jiangping Wang

Correction to:
Chapter 54 in: Z. Fang et al. (eds.), *Proceedings of the 5th International Symposium on Plasma and Energy Conversion*, Springer Proceedings in Physics 398, https://doi.org/10.1007/978-981-97-2245-7_54

In the original version of the book, the co-author name "Sanhai Ren" has been removed from the author list in Chapter 54. The book and the chapter have been updated with the change.

The updated version of this chapter can be found at https://doi.org/10.1007/978-981-97-2245-7_54

Z. Fang et al. (eds.), *Proceedings of the 5th International Symposium on Plasma and Energy Conversion*, Springer Proceedings in Physics 398,
https://doi.org/10.1007/978-981-97-2245-7_64

The manufacturer's authorised representative in the EU is Springer Nature Customer Service Centre GmbH, Europaplatz 3, 69115 Heidelberg, Germany. If you have any concerns regarding our products, please contact ProductSafety@springernature.com

Printed and bound by CPI Group (UK) Ltd, Croydon, CR0 4YY
01/10/2026
02225937-0019